Current Thoughts on Dementia

Ghulam Md Ashraf • Md. Sahab Uddin
Editors

Current Thoughts on Dementia

From Risk Factors to Therapeutic Interventions

Springer

Editors
Ghulam Md Ashraf
King Fahd Medical Research Center
King Abdulaziz University
Jeddah, Saudi Arabia

Md. Sahab Uddin
Department of Pharmacy
Southeast University
Dhaka, Bangladesh

ISBN 978-981-16-7605-5 ISBN 978-981-16-7606-2 (eBook)
https://doi.org/10.1007/978-981-16-7606-2

This Springer imprint is published by the registered company Springer Nature Singapore Pte Ltd.
The registered company address is: 152 Beach Road, #21-01/04 Gateway East, Singapore 189721, Singapore

Ghulam Md Ashraf would like to dedicate this book to his family members.

Md. Sahab Uddin would like to dedicate this book to his beloved parents, Sabikun Naher and Md. Helal Uddin.

Preface

Dementia is one of the leading causes of disability among older people associated with a decline in memory or other thinking skills to reduce the ability of a person to perform daily activities. Dementia frequently occurs among older adults, but it is not necessarily part of the typical aging process. More than one type of dementia may exist in the same person. Multiple risk factors can affect the risk of developing dementia including aging, gender, genetic factors, and comorbidities. Furthermore, low social engagement, depression, high stress, and altered sleep architecture are also reported to be linked to dementia. Currently, there is no effective cure for dementia but existing therapeutic strategies can provide symptomatic relief and improve the quality of life of the patients. Nowadays, there is a global effort to combat the development and progression of dementia. *Current Thoughts on Dementia Research: From Risk Factors to Therapeutic Interventions* explores new evidence of different risk factors for dementia and how they contribute to disease onset and pathology progression as well as novel therapeutic approaches.

This book represents the copious set of specific research updates. All over the world numerous erudite, experienced, and eminent academicians, researchers, and scientists had participated to write the texts of this book to give a succinct and thorough understanding of the risk factors and their linkage to dementia in terms of molecular and cellular mechanisms as well as therapeutic approaches to combat dementia burden.

This book is suitable for professionals, academicians, students, researchers, scientists, and industrialists around the world. Biomedical, health, and life science departments can use this book as a crucial textbook. Researchers and scientists from research institutes can use this book as efficient research info. Neurologists, pharmacists, physicians, and other healthcare professionals can use this book as a complete reference book. Furthermore, for interested readers, this book is a storehouse of knowledge to comprehend the complexity of dementia. The organization of this book provided profound knowledge and also maintained the reader's interest.

This book contains 22 chapters divided into two parts. The contents of the book cover the molecular and cellular mechanistic aspects of various risk factors affecting dementia and potential therapeutic approaches to reduce the dementia burden.

It is expected that readers shall find this book very informative and enormously useful. Since science is continuously changing, readers are strongly recommended to check the recent updates. The editors are ebulliently ready to accept any comment, suggestion, advice, or critique.

Jeddah, Saudi Arabia Ghulam Md Ashraf
Dhaka, Bangladesh Md. Sahab Uddin

Acknowledgments

Editing is a complex process; the editors would like to thank the people involved in this project for their quality time, expertise, and countless efforts. The editors are extremely indebted to the following:

- The authors for the countless time and expertise that they have put into their texts.
- Those authors who submitted works to contribute to this project but unfortunately rejected.
- The reviewers for their constructive reviews in improving the quality and presentation of the contents.
- The teachers, colleagues, friends, and supporters for their endless inspiration and assists.
- Especially Springer Nature for giving us a great opportunity to edit this book and their aid in the preparation of this edition.
- Last but not least, the editors would like to express the heartiest gratitude to almighty Allah and their parents.

Ghulam Md Ashraf
Md. Sahab Uddin

Contents

About the Editors

Ghulam Md Ashraf is an associate professor at King Fahd Medical Research Center, King Abdulaziz University, Saudi Arabia. His received his PhD in biochemistry from Aligarh Muslim University in India. His current research focuses on understanding antidiabetic drugs' molecular and behavioral mechanisms in various dementia conditions. He has more than 300 publications (citations: 6324, H-index: 41, i10-index: 155) and is involved in many research projects at the King Abdulaziz University. He is also involved in the editorial capacity for many reputed journals.

He is professionally associated with the Royal Society of Medicine (Fellow), Royal Society of Biology (Member), American Society for Biochemistry and Molecular Biology (Member), and Canadian Association of Neuroscience (Member). He has also been approved as a subject of biographical record in Marquis Who's Who in the World (2020). He has been recognized as Expertscape World Expert (top 0.1% scientist) in Alzheimer's disease and Nervous System Diseases. Recently, he has been listed in Top 2% Scientists Worldwide announced by Elsevier BV.

Md. Sahab Uddin is a Registered Pharmacist and a Research Scholar in the Department of Pharmacy at Southeast University, Dhaka, Bangladesh. Md. Uddin's research interest is how neuronal operation can be restored to abate Alzheimer's dementia. He has published copious articles in peer-reviewed international scientific journals. Md. Uddin has been enlisted in the Stanford University Lists of World's Top 2% Influencing Researchers based on the citation impact from Scopus of the single calendar year 2019 and 2020. He has also authored and edited several books related to neuropharmacology and neurodegenerative disorders. Md. Uddin also serves as a guest editor, editorial and reviewer board member of numerous scholarly journals. He is an active ad hoc reviewer for more than 50 journals and reviewed more than 250 papers. Moreover, he is the Founder and Executive Director of the Pharmakon Neuroscience Research Network, an open innovation hub bringing together neuroscientists to advance brain health. He received his BPharm in 2014 securing a first position from the Department of Pharmacy, Southeast University, Bangladesh.

Part I

Risk Factors for Dementia

ApoE: A Risk Factor for Dementia

1

Humira Jeelani, Jahangir Nabi, Yasmeena Akhter, Nahida Tabasum, Dil Afroze, Faheem Hyder Pottoo, and Fasil Ali

Abstract

Dementia, a neurocognitive disorder, is associated with memory impairment and cognitive dysfunction. Recent estimates propose that around 36 million people worldwide are affected by dementia, and by 2050, these figures are estimated to reach up to 150 million. Dementia occurs in several forms, most typical being Alzheimer's disease (AD) and vascular subtypes. Many factors tend to raise the risk of dementia with apolipoprotein E (*ApoE*) playing a prominent role. Mounting evidence suggests the involvement of *ApoE* gene in neurological disorders, including AD, a significant culprit in dementia. Increased levels of *ApoE4* have been observed in AD, resulting in tau protein phosphorylation in the brain. Evidence also suggests the role of ApoE4 in increased predisposition to vascular dementia, cerebrovascular disorders, Parkinson's disease, epilepsy, schizophrenia, migraine, anxiety, and depression. This chapter emphasizes on oversight of *ApoE* and its role as an essential determinant in predisposing to AD and associated dementia.

H. Jeelani · D. Afroze
Department of Advanced Centre for Human Genetics, Sher-I-Kashmir Institute of Medical Sciences, Srinagar, Jammu and Kashmir, India

J. Nabi · Y. Akhter · N. Tabasum (✉)
Department of Pharmaceutical Sciences, Pharmacology Division, University of Kashmir, Hazratbal, Jammu and Kashmir, India

F. H. Pottoo
Department of Pharmacology, College of Clinical Pharmacy, Imam Abdulrahman Bin Faisal University, Dammam, Saudi Arabia

F. Ali
Department of Studies and Research in Biochemistry, Mangalore University, Mangalore, Karnataka, India

G. M. Ashraf, M. S. Uddin (eds.), *Current Thoughts on Dementia*,
https://doi.org/10.1007/978-981-16-7606-2_1

Keywords

Alzheimer's disease · Dementia · Apolipoprotein E · *ApoE* · Lipid metabolism · Neuroinflammation

1.1 Introduction

Dementia is a collective term that includes several diseases associated with the behavior, memory, and cognitive abilities that impair the capacity of an individual to perform daily activities. The US National Institute of Neurological Disorders and Stroke (NINDS) provides a detailed definition stating that dementia is a collection of symptoms that occurs by several disorders, collectively affecting the normal functioning of the brain. It is frequently associated with memory loss and difficulty in communication. Dementia is more commonly seen in older people, and its incidence significantly increases with advanced age. The overall prevalence of dementia was 36 million in 2010, with an increase of about 46 million in 2015, and over 50 million in 2017. These figures are estimated to increase to about 82 million and 150 million in 2030 and 2050, respectively (Fymat 2018).

Dementia is generally categorized into progressive types, including Alzheimer's disease (AD), Lewy body dementia (LBD), frontotemporal lobar degeneration (FTLD), vascular dementia (VD), and mixed dementia (MXD). Dementia is also associated with other conditions like Huntington's disease (HD), traumatic brain injury (TBI), Jacob disease, and Parkinson's disease (PD). Dementia in older people occurs due to neurodegeneration. The aggregation of amyloid-beta (Aβ), increased tau protein phosphorylation, formation of neurofibrillary tangles, oxidative stress, and inflammation are the other factors linked to the pathogenesis of AD and associated dementia (Schneider et al. 2007; Akhter et al. 2020). Factors leading to dementia include advanced age, systemic vascular disease, smoking, drugs, alcohol consumption, comorbidities, and environmental and genetic factors, including apolipoprotein E (ApoE).

ApoE is a prominent and potent genetic factor leading to dementia apart from its indispensable role in lipid metabolism (Prince et al. 2013). To comprehend the role of genetic and environmental influences on the onset of dementia, it is necessary to consider the studies associated with the ApoE and its genetic variants, which will help develop new clinical findings and early diagnostic and therapeutic interventions in treating dementia and associated diseases. Therefore, this chapter describes the understanding of ApoE protein and its role in the brain. Further, various Aβ-related mechanisms linking ApoE4 with AD pathogenesis and effective AD therapy approaches by targeting ApoE have been discussed.

1.2 Types of Dementia

1.2.1 Progressive Dementias

These types of dementias progress and cannot be reversed. These include the following:

1.2.1.1 Alzheimer's Disease

AD is a common type of neurodegenerative dementia. Its prevalence has been reported to be 56% and 30% in individuals aged above 65 and 85 years, respectively (Hebert et al. 2013; Uddin et al. 2020). AD begins with a slow decline in progressive memory, and the mean survival rate after the early onset of AD is only 10–12 years. An enhanced deposition and aggregation of Aβ plaques and neurofibrillary tangles are the significant attributes of AD (Akhter et al. 2020; Mamun et al. 2020; Pandey et al. 2020). As the disease progresses, magnetic resonance imaging (MRI) scans of the AD brain depict temporal lobe atrophy in the hippocampus and other surrounding structures (Mamun et al. 2020; Mendez 2017; Dubois et al. 2016). To date, no satisfactory pharmacological treatment is available for AD. Clinical research primarily focuses on its early detection and various therapeutic targets underlying its histopathology (Aisen et al. 2017).

1.2.1.2 Frontotemporal Lobar Degeneration

FTLD is ranked as third frequent neurodegenerative dementia in patients aged below 65 years. It constitutes nearly 26% of neurodegenerative cases (Perry et al. 2017). The characteristic clinical features of FTLD include corticobasal syndrome, primary progressive aphasia, and progressive supranuclear palsy syndrome (Finger 2016). Sometimes patients present with the signs and symptoms of FTLD along with and amyotrophic lateral sclerosis (ALS), which together constitute ALS spectrum syndrome or frontotemporal dementia (FTD) (Devenney et al. 2015). Another dementia associated with the early personality changes most commonly with features like disinhibited or obsessive behavior, decrease in social/communal compartment, empathy, and problems with the decision-making constitutes behavioral variant frontal temporal dementia (Rascovsky and Grossman 2013). Patients with corticobasal syndrome initially develop asymmetric parkinsonism, executive dysfunction, limb apraxia, and behavioral changes, and with advanced years, patients develop features of aphasia, frequent fall, and gait decline also (Armstrong et al. 2013). Difficulty to maintain posture, axial stiffness, motor, and cognitive deterioration are the characteristic features of the progressive supranuclear palsy syndrome (Lopez et al. 2016). Medical treatment for FTD is focused on alleviating the neuropsychiatric symptoms and motor symptoms of the disease (Karageorgiou and Miller 2014).

1.2.1.3 Alpha-Synucleinopathies

A buildup of the aggregated alpha-synuclein in the neurons is the key hallmark of many neurological disorders, including LBD and PD (McCann et al. 2014). LBD represents the second frequent type of neurodegenerative dementia after AD (Zaccai et al. 2005). The clinical symptoms of LBD include fluctuating cognition, variation in alertness and attention, visual hallucination, and quick eye movement sleep and are also associated with features of PD including tremor at rest, bradykinesia, or rigidity (McKeith et al. 2017). A rare form of alpha-synucleinopathy with manifestations such as PD, cerebral signs, dysautonomia, and pyramidal signs constitutes multiple system atrophy (Gilman et al. 2008). Symptoms such as inspiratory stridor, dysautonomia, including orthostatic hypotension and sexual dysfunction, and sleep behavior disorders, all of which are included in the non-motor category, develop in nearly half of the patients with multiple systemic atrophy. Minimal and cognitive difficulties may also be associated with multiple system atrophy (Fanciulli and Wenning 2015).

1.2.1.4 Vascular Dementia

VD results from factors that increase cerebrovascular dementia chances, e.g., stroke, atrial fibrillation problems associated with the heartbeat rhythm, and raised blood pressures, increased levels of cholesterol, and diabetes mellitus. VD can occur suddenly and then progress or subside throughout the lifetime. It can also co-occur at the same time as AD. Patients with vascular cognitive impairment and dementia display some abnormalities in the brain, seen on MRI scans. These mainly include loss of components of the brain's white matter and thickening of walls of the brain's blood vessels. Problems such as retarded thinking, attention, and problem-solving are dominant in vascular cognitive impairment and dementia, while memory loss is more dominant in AD (Iadecola 2013; Vijayan and Reddy 2016).

Mixed Dementia

People with dementia can eventually develop MXD. Typical combinations of MXD include VD and AD. MXD is the frequent cause of dementia in older people, triggered by changes in the brain resulting from vascular disease-related processes, AD, and neurodegenerative diseases. Evidence shows that about 75% of the elderly brain has two or more pathologies related to vascular damage or neurodegenerative disease (Custodio et al. 2017).

1.2.2 Dementias Associated with Other Conditions

1.2.2.1 Huntington's Dementia

HD is a neurological disorder arising due to aggregation and accumulation of mutant huntingtin (mHtt) protein (Akhter et al. 2020; Pandey et al. 2020). Symptoms include impairment of memory and motor activity that typically appear between 30 and 40 years (Cleret de Langavant et al. 2013).

1.2.2.2 Traumatic Brain Injury

TBI is typically caused by recurrent head trauma. This disorder can cause dementia depending upon the type of brain area affected and manifests as depression, memory impairment, and impaired speech (Nguyen et al. 2018).

1.2.2.3 Creutzfeldt–Jakob Disease

Creutzfeldt–Jakob disease is triggered by the accumulation of prion proteins. This fatal condition's primary symptoms include memory loss, loss of coordination, dysphagia, and ataxia (Knight 2006).

1.2.2.4 Parkinson's Disease

PD is a progressive brain disorder that affects movement. PD affects the nerve cells in the brain that produce dopamine. People suffering from PD inevitably experience dementia termed as Parkinson's dementia (Dementia in parkinsonism 2011).

1.3 Risk Factors for Dementia

1.3.1 Age

The aging process is linked to the occurrence of dementia as has been reported by various studies conducted on the general population (López-Pousa et al. 2004). A study conducted in China found that the occurrence of AD and VD tends to increase with advanced age (Liu et al. 2003). More reports have shown a direct relationship between parental age and dementia risk. The higher age of parents at the time of childbirth has a higher propensity of developing AD, most likely due to the occurrence of chromosomal abnormalities with an increase in reproductive age (Bertram et al. 1998).

1.3.2 Sex

The role of sex in contributing to dementia is prominent. A study found that the AD risk among women increases over 85 years of age relative to men (Andersen et al. 1999). Another study reported a greater frequency of developing AD but not the VD among women than men over or equal to 60 years of age (Liu et al. 2003). However, in another study, this relationship between sex and dementia was found to be insignificant (Ravaglia et al. 2005). Several factors, such as lifestyle, hormones, including steroids, lifestyle factors, and single nucleotide polymorphisms of genes related to the sex, may affect this association. Therefore, to determine the proper association between sex and the occurrence of dementia, it is important to take into account other risk factors as well.

1.3.3 Physical Activity

The correlation between physical activity and dementia was exclusively studied. Studies show that regular physical exercise is directly related to a reduction in cognitive disability of around 30–50% (Karp et al. 2006). A meta-analysis found that exercise training is positively linked to cognitive function (Heyn et al. 2004). One more study reported improved cognitive function with a physical activity training of about 24 weeks (Lautenschlager et al. 2008). A study including men and women aged 70–79 years reported that increased cognitive function is significantly associated with lower pro-inflammatory mediators such as C-reactive protein, tumor necrotic factor-alpha (TNF-α), and interleukins (Reuben et al. 2003). Physical exercise may positively affect vascular disease, endothelial dysfunction, and obesity, further enhancing fitness, neurological health, physical function, and behavior traits (Barnes et al. 2007). Another study found that individuals who took part in four physical activities presented with a decreased risk of dementia than those who participated in only one or did not participate in any physical activity. It was observed to be significant only in non-carriers of *ApoE4* allele (Podewils et al. 2005).

1.3.4 Smoking

There is a controversial relationship associated between smoking and the incidence of dementia. A study observed that smoking is significantly related to the incidence of AD. However, another study found no correlation between smoking, cognitive decline, and dementia (Peters et al. 2009).

Smoking contributes to cerebrovascular disease in dementia; however, the possible role of dementia in cerebrovascular disease has not yet been explored (Aggarwal et al. 2006).

1.3.5 Comorbidity

Comorbid conditions such as hypertension and insulin resistance are potential risk factors for VD (Forette and Boller 1991; Posner et al. 2002). Insulin resistance is a pro-oxidant and a pro-inflammatory state. It is associated with the generation of inflammatory mediators. It is a promoter of endothelial dysfunction, also involved in Aβ formation and aggregation, tau phosphorylation, and appearance of neurofibrillary tangles in the brain. It consequently enhances dementia risk, particularly the risk of VD and AD (Craft 2007; Sun and Alkon 2006). Individuals with infectious diseases are more prone to experience dementia than those who do not have infectious diseases (Dunn et al. 2005). The human immunodeficiency virus (HIV) and hepatitis C virus (HCV) were identified as risk factors for dementia (Corder et al. 1998; Forton et al. 2002). In general, it can be said that inflammatory and vascular dysfunction factors, psychological factors, infectious disease, and head injuries all belong to the same inflammatory pathway that leads to dementia.

1.3.6 Environmental Factors and Nutrition

The contribution of environmental factors in causing dementia is more complicated. Aluminum is among the probable dementia risk factors (Solfrizzi et al. 2006). Other metals, including copper, iron, and zinc, were also linked to increased dementia (Solfrizzi et al. 2006). Various nutrients were also reported to be strongly correlated with the incidence of dementia. Women with dementia have been shown to have low levels of vitamin D (Kipen et al. 1995). One study did not report any impact of vitamin E in cognitive decline at the very early stage of dementia (Petersen et al. 2005). The cognitive function in AD and VD has been linked to macronutrients, glucose, proteins, tyrosine, tryptophan, and polyunsaturated fatty acids. However, reduced antioxidant intake increases the risk of dementia (Solfrizzi et al. 2006).

1.3.7 Drugs

The use of several drugs has been linked with dementia (Table 1.1). A meta-analysis showed that the use of benzodiazepines is linked to dementia risk (Islam et al. 2016). Likewise, previous studies also suggest a strong correlation between the risk of dementia and antidepressant drug therapy (Wang et al. 2018). Furthermore, bladder antimuscarinics have been identified to be linked with an increase in cognitive decline (Moga et al. 2017). Another research has identified a direct correlation between anticholinergic drug intake and dementia risk (Yang et al. 2017). Moreover, many antiepileptic drugs were also found to raise dementia risk (Taipale et al. 2018). It has also been found that opioid users are more prone to dementia risk than non-opioid users (Dublin et al. 2015).

Table 1.1 Drugs affecting dementia

Drug category	Example	Possible mechanism	References
Benzodiazepine	Alprazolam, diazepam,	–	Lagnaoui et al. (2002)
	Chlordiazepoxide	–	Islam et al. (2016)
Antidepressant	Selective serotonin reuptake inhibitor, tricyclic antidepressant, monoamine oxidases	Elevate oxidative and nitrosamine stress, inflammation, mitochondrial dysfunction, and apoptosis	Wang et al. (2018)
Bladder antimuscarinics	Oxybutynin, tolterodine, darifenacin	Inhibit acetylcholine-mediated neurotransmission	Moga et al. (2017)
Antiepileptic drugs	Phenobarbitone, primidone, carbamazepine	Increase GABA-ergic inhibition and decrease neuronal excitability	Taipale et al. (2018)
Opioids	Prescription opioids	Neurodegeneration due to apoptosis-mediated neuronal cell death	Dublin et al. (2015), Hu et al. (2002)

1.3.8 Genetic Effects

The potential associations between several genetic variants to the incidence of dementia were investigated using candidate genetic approaches. Among several genes, the *ApoE* gene was found to have greater significance in predisposing to the risk of developing dementia. Evidence indicates that about 60–80% of cases of AD have a genetic origin (Gatz et al. 2005). Genetic variants of *ApoE* were reported to have a significant contribution to late-onset AD. It was documented that an increased occurrence of *ApoE4* alleles elevates the risk of late-onset AD from 20 to 90% (Corder et al. 1993). In a study including all the ethnic groups belonging to the age group of between 40 and 90 years, *ApoE4* was shown a potent risk factor for AD. This association was more predominant among Japanese than Caucasians but was less significant in African Americans and Hispanics (Pastor and Goate 2004; Farrer et al. 1997).

Inflammation is another key factor involved in AD pathogenesis. Evidence reports that several single nucleotide polymorphisms, e.g., IL-1α-889 and IL-1β-395, present in the promoter and coding regions of inflammatory genes predispose to AD (Du et al. 2000; Griffin and Mrak 2002).

These studies propose that both the *ApoE* and inflammatory genetic polymorphism's joint occurrence may prove AD risk's better predictors.

1.4 Structure of ApoE

ApoE is a 299 amino acid lipoprotein. Its highest level of expression is observed in the liver and brain (Poirier et al. 1993). Within the cerebrospinal fluid (CSF) and plasma, ApoE is generally present as a part of various lipoprotein complexes such as VLDL (very low-density lipoprotein), LDL (low-density lipoprotein), HDL (high-density lipoprotein), and chylomicrons (Phillips 2014). Crystallography investigations displayed that the protein's hinge region links the N-terminal (residues 1–67) and C-terminal domains (residues 206–299) as shown in Fig. 1.1. The LDL receptor-binding site is located at residues 136–150 of the N-terminal domain, while the lipid-binding site is located at residues 244–277 of the C-terminal domain. The three significant isoforms of ApoE (ApoE2, ApoE3, and ApoE4) vary in cysteine/arginine residues at positions 112 and 158. ApoE2 contains cysteine at both places. ApoE3 contains cysteine and arginine at positions 112 and 158, respectively, while ApoE4 contains arginine at both positions (Zannis et al. 1982). These differences in the amino acid sequence are vital for the structural and functional integrity of ApoE (Poirier et al. 1993).

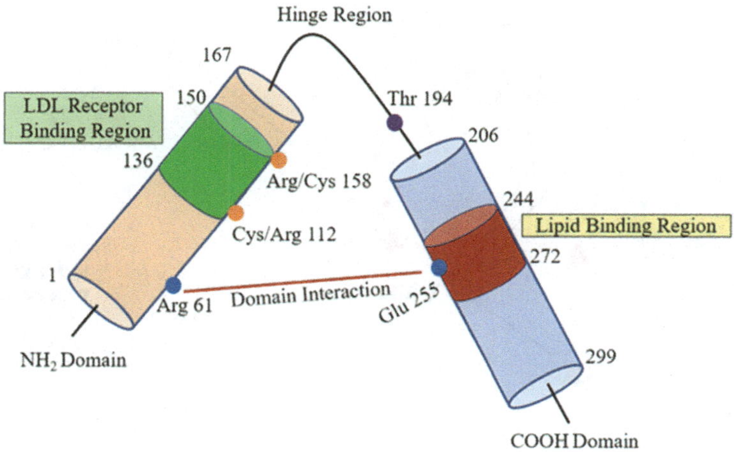

Fig. 1.1 Human ApoE structure and main functional regions. LDL receptor binding site is located at residues 136–150 of the NH2 domain. The lipid-binding site is positioned at residues 244–277 of the COOH domain. ApoE2, ApoE3, and ApoE4 vary in cysteine and arginine residues between positions 112 and 158. ApoE2 contains cysteine at both positions. ApoE3 contains cysteine and arginine at positions 112 and 158 respectively. ApoE4 contains arginine at both positions

1.5 Physiological Functions of ApoE

ApoE is primarily synthesized in the liver; however, its synthesis also occurs in other organs like kidney, spleen, and brain (Czaplińska et al. 2019). Notable ApoE producers are the microglial cells and astrocytes within the brain, while neurons only synthesize it under some pathological conditions (Zhang et al. 2013). The central role of ApoE is to control the metabolism of lipids by transportation and redistribution across different tissues of the body, which is accomplished via its binding to ApoE receptors and other lipid transfer proteins involved in lipolysis (Mahley and Rall 2000).

In plasma, specific binding of ApoE isoforms with other lipoprotein components via LDL receptors plays an essential part in peripheral lipid metabolism, which significantly affects conditions like atherosclerosis hyperlipoproteinemia (Rall et al. 1989; Zende et al. 2013). Within the brain, ApoE is mainly linked with cholesterol and HDL-like particles, where it directs the distribution of phospholipids and cholesterol between the cells via ATP-binding cassette A1 transporter (ABCA1). Once an adequate amount of phospholipids and cholesterol are bound to ABCA1, the dimerization of ABCA1 takes place, which in turn leads to the diffusion of lipids to ApoE. ABCA1 also allows Aβ removal in the brain, possibly via lipidation of ApoE (Koldamova et al. 2014). In support of this, experimental evidence indicates that increased ApoE lipidation prevents Aβ aggregation in amyloid mouse models. Removal of the *ABCA1* gene enhances Aβ deposition, while its overexpression inhibits Aβ accumulation in amyloid mice (Koldamova et al. 2014) (Fig. 1.2).

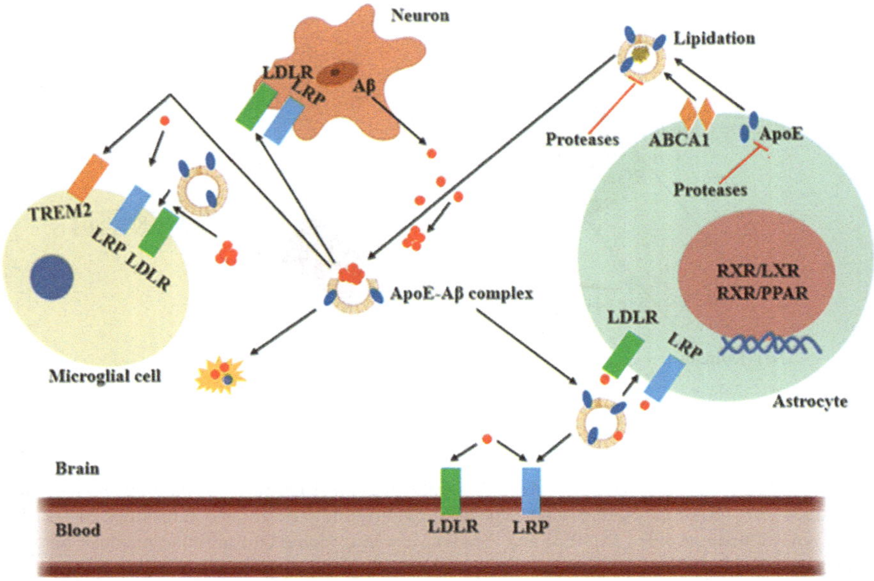

Fig. 1.2 Metabolism of ApoE and Aβ. ApoE transcription is regulated by heterodimers generated by RXR with PPA or LXR. Following lipidation by ABCA1, ApoE is dispersed in the interstitial fluid after lipidation by ABCA1, contributing to lipid diffusion through different cells via receptor-mediated uptake. ApoE also binds to the TREM2 microglial receptor. Aβ is produced in neurons and secreted in the interstitial fluid. ApoE binds Aβ aggregates, which affects Aβ fibrillation. Aβ clearance by ApoE is facilitated by enzymatic degradation and microglia endocytosis

1.6 ApoE in Dementia and Associated Diseases

The essential functions of ApoE in lipid metabolism have been thoroughly explored and studied (LaDu et al. 2012). Besides its function in lipid metabolism, there is evidence for its association with the risk of neurodegenerative and other chronic diseases. Some of the studies summarizing the connection of ApoE and the risk of neurodegenerative disease are given below:

1.6.1 ApoE and Alzheimer's Disease

Alois Alzheimer was the first who described AD in 1907. It is the most frequent age-related dementia known to affect about 36 million individuals globally. ApoE has a function in the transportation of cholesterol in the CNS and is a potential risk of developing late-onset AD. The occurrence of the *ApoE4* allele has been recorded to be greater in AD patients (Liu et al. 2013). Also, a strong correlation of the *ApoE4* allele was observed with Aβ in the brain (Näslund et al. 1995). Studies reported that *ApoE4* binding to Aβ leads to pathological mechanisms that promote β-sheet conformational changes (Wisniewski and Frangione 1992). The ApoE4 has a role in initiating, formation, progression, aggregation, accumulation, and deposition of Aβ and tremendously increasing AD's risk. The presence of low head circumference and the *ApoE4* allele are strong predictors of early-onset AD (EOAD) (Graves et al. 2001). The ApoE4 quantity was reported to be present in great content in the amyloid plaques and neurofibrillary tangles in the AD brains. ApoE4 enhanced phosphorylation of tau proteins, thereby contributing significantly to AD-related neuronal deficits (Brecht et al. 2004). Also, a more substantial increase in the generation of nitric oxide (NO) from the macrophages was reported in AD patients carrying *ApoE4* allele. The enhanced synthesis of NO in *ApoE4* carriers may increase predisposition to nitration and nitrosation and the occurrence of the neuroinflammatory state in AD patients (Colton et al. 2004). Also, *ApoE4* allele carriers have shown decreased glucose metabolism in the brain and an enhanced risk of developing cerebral amyloid angiopathy (CAA) than ApoE3 homozygous carriers (Lambert et al. 2001; Reiman et al. 1996).

Genetic differences in ApoE also lead to differences in ApoE levels in serum. ApoE4 homozygous individuals were found to have reduced ApoE levels in serum (Smit et al. 1990). Besides, *ApoE4* carriers also presented with more inefficient brain metabolism (Reiman et al. 2004).

The decreased metabolic activity was reported in the nucleus of Meynert neurons in AD *ApoE4* allele carrier patients compared to the healthy control individuals (Dubelaar et al. 2004). It has been seen that as AD progresses to later stages of the disease, it promotes neurons' formation with an exclusively small size of the Golgi apparatus irrespective of *ApoE* genotypes. Genome-wide association studies have found the *ApoE4* allele to be linked to an increased tendency to develop dementia (Carrasquillo et al. 2009). *ApoE4* allele is a potential mediator of both familial and sporadic ADs (Corder et al. 1993). In up to 50% of all AD patients, *ApoE4* allele was reported as a potential genetic factor involved in pathogenesis, predisposition, and progression of AD (Selkoe and Schenk 2003). Evidence also indicates that heterozygous *ApoE4* allele carriers are at two- to threefold risk of developing AD, whereas homozygous *ApoE4* allele carriers have 12-fold greater risk of developing AD (Bertram et al. 2007). *ApoE4* allele was also reported to be linked with the vascular system's damage resulting in AD progression and pathogenesis in the brain (Farrer et al. 1997). Although ApoE4 is considered a genetic contributor to AD, the molecular mechanisms leading to dementia still need to be explored.

1.6.1.1 Mechanism of ApoE4 in AD

Aβ Aggregation and Clearance

Various experimental studies have shown that Aβ formation and aggregation are related to ApoE isoforms. Several *in vivo* reports have shown that ApoE4 is less efficient in removing Aβ in comparison with ApoE3 isomeric form (Beffert et al. 1999; Colonna and Wang 2016). It has been found that ApoE isomeric forms regulate cholesterol levels differently and, in turn, modify the function of γ-secretase enzyme and formation of Aβ (Osenkowski et al. 2008). It is known that lower levels of ApoE produced by *ApoE4* carriers are the main contributor to the disease progression in AD patients (Wahrle et al. 2007). Studies reported that activation of retinoid X receptor (RXR) or liver X receptors (LXR) enhances the levels of ApoE protein in circulation and expedites the clearance of Aβ deposits (Cudaback et al. 2011). Recently, a study confirmed that the absence of ABCA1 mediates the transport of lipids to ApoE, impedes the removal of Aβ, and promotes amyloid deposition, which exacerbates memory deficits in *ApoE4* carriers than *ApoE3* carriers (Fitz et al. 2012). This difference in the lipidation status of ApoE isoforms influences the clearance of Aβ plaques in an isoform-dependent manner of *ApoE* alleles. It was found that ApoE can sequester Aβ and enhance their cellular uptake, and finally promote the degradation of ApoE, complexed with Aβ plaques (Kim et al. 2009).

ApoE4 has a lesser affinity to bind with Aβ plaques than the wild ApoE3 isoform, indicating ApoE4 has a low capacity to facilitate the removal of Aβ plaques (LaDu et al. 1994). ApoE also has a function in transporting Aβ from the brain to systemic circulation. ApoE4 has been reported to be more likely to hinder the diffusion of Aβ across the BBB, while ApoE3 is effective in transporting it across the barrier (Du et al. 2015). Studies done in microglial cells showed that ApoE3 efficiently mediates the enzymatic degradation of Aβ more effectively than the ApoE4 isoform of ApoE. These studies also reported that ApoE4 retards the clearance of Aβ plaques less effectively than ApoE3 and ApoE2 isoforms (Jiang et al. 2008). One more study reported decreased ApoE levels in CSF and plasma in *ApoE4* carriers, indicating that decreased ApoE levels may favor the deposition of Aβ in the brain (Shinohara et al. 2013). Another study reported that ApoE4 isoform retards the proteolytic degradation of Aβ by insulin-degrading enzymes and neprilysin, thereby favoring Aβ oligomer formation and Aβ aggregation to a greater degree than ApoE3, which generally impedes the formation of Aβ fibrils from oligomers by sequestering and formation of ApoE3/Aβ complexes (Zekonyte et al. 2016). In the pathogenesis of AD, ApoE4 accords toxic functioning, the failure of neuroprotective activities, or both, as displayed in Fig. 1.3.

Tau Phosphorylation

Hyperphosphorylated tau is considered the main element of neurofibrillary tangles (Damoiseaux et al. 2012). Hyperphosphorylated tau itself can cause neurodegeneration, thereby predisposing to AD neuropathology. Like stress or injury, some factors enhance tau phosphorylation (Hampel et al. 2010). There have been certain mechanisms that correlate with ApoE4 and tau phosphorylation. It is proposed that ApoE4 more effectively binds to phosphorylated tau in comparison with ApoE3, which has a strong affinity to bind to non-phosphorylated tau and, in turn, prevent the deposition of tau and generation of Aβ deposits. Another mechanism suggests that ApoE4 can escape some secretory pathways in neurons, and because of its unique structure, it directly interacts in the cytoplasm with tau to promote its phosphorylation. This mechanism is suggested to be mediated by the specific ApoE4 receptor-driven signaling cascade, which changes the functioning of tau phosphatases and kinases and ApoE4's ability to escape secretory pathways to

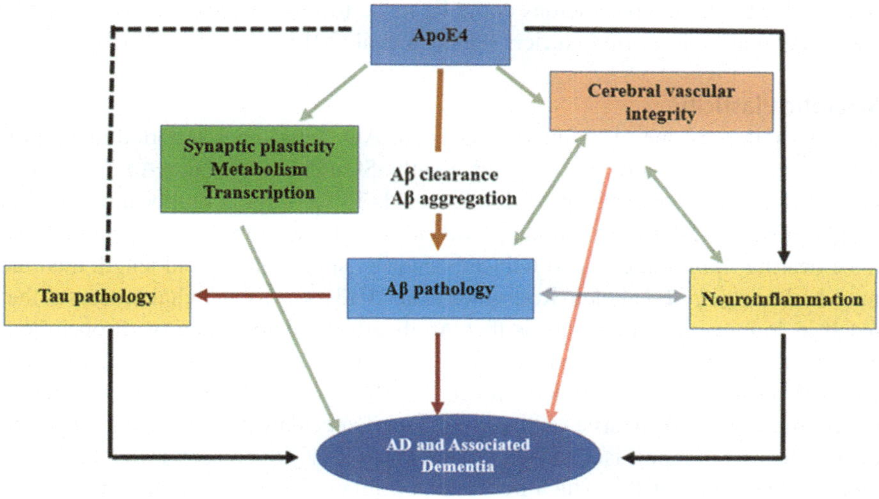

Fig. 1.3 Role of ApoE4 in AD pathogenesis and associated dementia. ApoE affects AD pathogenesis by influencing Aβ pathology. In particular, ApoE modulates neuroinflammation, tau pathology, metabolism, synaptic plasticity, cerebral vascular integrity, and transcription via Aβ independent processes

interact with zinc ions intracellularly to activate ERK kinases to promote tau phosphorylation (Harris et al. 2004; Dubnikov and Cohen 2017).

Neuroinflammation

Neuroinflammation is also the main contributor to neuronal cell death as has been reported by various genome-wide association studies, which show a clear connection between AD and the immunity-associated genes like *CLU* and the *CTREM2* (Zhang et al. 2016; Colonna and Wang 2016). It is observed from studies that neuroinflammation is more predominantly found in *ApoE4* carriers than the other two isoforms ApoE2 and ApoE3 (Cash et al. 2012; Du et al. 2015). An enhanced and prompt neuroinflammatory response is observed in *ApoE4* carriers. It is mainly mediated by ApoE4-mediated activation of microglial cells and increasing the pro-inflammatory cytokine levels (Rodriguez et al. 2014; Fan et al. 2017). It is proposed that the ApoE4 pro-inflammatory functions may be associated with miRNA146a, which is the principal miRNA found in the brain. Studies reported that miRNA146a levels are highly raised in AD brains than the normal mouse models. It is suggested that raised levels of miRNA146a may lead to disturbed feedback patterns of inflammation regulation. It results in increased chronic inflammation (Teter et al. 2016). A study suggested nonsteroidal anti-inflammatory drug (NSAID) beneficial effects on cognition of AD patients and their overall survival rate. Recently, epidemiological studies suggest that carriers of *ApoE4* are better responders to treatment with NSAIDs. The exact underlying mechanisms are not fully understood, but speculations could be made toward increased oxidative stress and inflammation in *ApoE4* carriers (Heneka et al. 2015; Liao et al. 2017).

Synaptic Plasticity

Synaptic failure is one of the major aspects of AD. It has been reported that ApoE isoforms mediate synaptic plasticity differently. Studies reported that *ApoE4* carriers present decreased dendritic spine density levels in the brain's hippocampal region (Androuin et al. 2018). This finding coincides with the study, which showed that mice bearing *ApoE4* alleles had reduced dendritic spine density and length than the mice harboring *ApoE3* alleles (Rodriguez et al. 2013). One of the vital processes that result in low synaptic plasticity is that ApoE affects neurite outgrowth operation. Several studies have shown that ApoE3 is a good promoter of neurite outgrowth in comparison with ApoE4 isomeric form. ApoE4 is reported by several studies to inhibit the process of neurite outgrowth. Several speculations have been suggested behind the role of ApoE isoforms in synaptic plasticity. One reason is that the ApoE receptor known as LRP1, which performs a primary function in regulating neurite outgrowth, is less effectively activated by ApoE4 compared to ApoE3. ApoE starts LRP1 and this activation was improved by binding of ApoE with proteoglycan, namely heparin sulfate, which is reported to occur at a greater speed in *ApoE3* carriers in comparison with *ApoE4* carriers (Sen et al. 2012; Shinohara et al. 2017).

Also, actin polymerization, which forms the base for dendritic spine morphogenesis and neurite outgrowth, is mediated by ApoE receptor 2 (ApoER2), which is activated promptly by ApoE3 compared with ApoE4 isoform of ApoE (Martinelli et al. 2009; Ulrich et al. 2014). Besides, these two isoforms ApoE2 and ApoE3 also differ in intracellular trafficking properties. After endocytosis, the process of retro-endocytosis is readily observed in *ApoE3* carriers, whereas ApoE4 isoform gets trapped in endosomes and clogs the intracellular trafficking (Li et al. 2012; Nuriel et al. 2017). ApoE4 is known to downregulate several receptors such as ApoER, neurotransmitter receptors, receptors for growth factors including insulin, vascular endothelial growth factor (VEGF), and NMDA receptors. The downregulation of these receptors may lead to impaired neuronal plasticity (Lane-Donovan and Herz 2017). In addition to these factors, the process of neurite outgrowth is hindered in *ApoE4* carriers by activation of microglia, phagocytosis, and complement C1q protein in the brain (Rodriguez et al. 2014; Bonham et al. 2016).

Lipid Metabolism

ApoE is the commonly found apolipoprotein in the brain and circulation where it is linked to cholesterol- and phospholipid-rich lipoproteins such as HDL-like particles. It has a vital function in the transport and metabolism of lipids in the brain (Mahley 2016). Many studies have assessed the levels of docosahexaenoic acid (DHA) in the brain and CSF. DHA is an essential fatty acid and contributes to the proper functioning of neurons and the brain. Studies reported decreased levels of DHA in both *ApoE4* carriers and AD patients. ApoE4 isoform affects uptake, distribution, and incorporation of DHA in different regions of the brain. Similar findings were seen in *ApoE4* mice carriers where the pathological effects produced by ApoE4 were counterbalanced by feeding mice with docosahexaenoic acid (DHA) containing a high fish oil diet (Belkouch et al. 2016; Nock et al. 2017; Yassine et al. 2017).

Further, evidence indicates that ApoE4 is associated with dysregulated cholesterol and phospholipid metabolism. People with high cholesterol levels and *ApoE4* genotype are at higher risk of cognitive loss than people having only one among the two (Bangen et al. 2013). Studies done in mouse models reported high cholesterol tends to accentuate the pathological functions of ApoE4 isoform (Kariv-Inbal et al. 2012). Although studies suggested there exists a link between lipids and ApoE4, a precise mechanism is still unknown, nor has a therapeutic target been developed.

1.6.2 Vascular Dementia and Cerebrovascular Disorders

ApoE4 allele frequency was reported higher in patients with VD (Engelborghs et al. 2003). Earlier findings have shown that raised cholesterol levels in plasma and atherosclerosis contribute to VD in *ApoE4* carriers (Noguchi et al. 1993). Studies also reported the *ApoE4* allele as a contributor to multi-infarct dementia. However, some studies observed no significant difference in the occurrence of *ApoE4* allele in patients with AD and VD (Slooter et al. 1996). Increased *ApoE2* allele frequency was associated with CAA-related hemorrhages, indicating that *ApoE2* genotype carriers may have beneficial effects against the development of parenchymal AD. Still, these patients are prone to the breakage of amyloid-laden vessels (Love et al. 2003; Lin et al. 2004). A connection was reported between ApoE2 and vasculopathy in CAA where ApoE2 was found to be a risk factor for the reappearance of CAA. In contrast, the *ApoE4* allele was related to an enhancement in amyloid deposition (Greenberg et al. 1998). Other reports have observed that ApoE4 by itself does not promote an enhanced tendency of CAA but is involved in decreasing longevity and in promoting senescence even without AD or other cerebral hemorrhages (Love et al. 2003).

In some cases, the *ApoE2* allele may have a role in therapeutic responses. It is reported that the intervention for acute ischemic stroke has a better response in *ApoE2* allele carriers than non-*ApoE2* allele carriers (Broderick et al. 2001). Some more studies reported a positive relationship between *ApoE4* allele carriers and an enhanced risk of AD (Yin et al. 2012). Recently, one study found that *ApoE4* carriers were more likely to develop AD than *ApoE3* carriers (Yin et al. 2012). *ApoE4* allele alone is considered to be a significant risk factor for vascular cognitive impairment, even when exempting the influence of other factors like hypertension dyslipidemia, obesity, inflammation, and atherogenesis (Prince et al. 2000).

1.6.3 Parkinson's Disease

The positive association of ApoE4 with the risk of PD was reported in several studies (Pulkes et al. 2011). However, other reports did not find the *ApoE4* allele's relationship with increasing predisposition toward PD and associated dementias. The involvement of ApoE2 was reported to be weakly linked to PD risk (Federoff et al. 2012). However, another study observed a significant connection between *ApoE2* allele PD (Huang et al. 2004). Other researchers also concluded *ApoE4* allele as a predictor of PD development and progression (Li et al. 2004). In the Caucasian population, the *ApoE4* allele was not correlated with the risk of PD (Marder et al. 1994). However, it was found that *ApoE3/E4* and *ApoE4/E4* carriers develop PD at a much earlier age than carriers of *ApoE3/E3* genotype (Zareparsi et al. 2002).

1.6.4 Schizophrenia

ApoE4 allele has been reported as a pathological hallmark of schizophrenia (Harringtona et al. 1995). Also, the early-onset of schizophrenia along with its worst prognosis was reported to be linked with *ApoE4* alleles (Martorell et al. 2001). In contrast, these observations were not reported in other studies (Igata-Yi et al. 1997). Another study reported that ApoE2 is associated with early-onset schizophrenia and a more lousy reaction to neuroleptics (Durany et al. 2000). A comparable frequency of *ApoE4* allele was reported in Centenarians (Howard et al. 1995). It has been stated that the occurrence of psychiatric symptoms is associated with *ApoE* genotype (Scarmeas et al. 2002). *ApoE4* allele was connected with the risk of delusions, but no relationship was reported in behavioral disturbances or depressive symptoms (Durany et al. 2000). However, others have reported the *ApoE4* allele's involvement is connected with increased abnormal psychiatric symptoms in patients with AD (Cacabelos et al. 1996). Some reported elevated amounts of ApoE4 in the frontal cortex of schizophrenia patients. Also, reduced levels of ApoE4 have been seen as a therapeutic modality in response to anti-psychiatric drugs (Dean et al. 2003).

1.6.5 Multiple Sclerosis

Significantly axonal damage and disability progression were found to be associated with multiple sclerosis (MS) (Enzinger et al. 2004). Some reported no association of *ApoE4* allele with the course and progression of multiple sclerosis (Savettieri et al. 2003). *ApoE4* allele was linked to the early onset of the disease but not a risk factor for multiple sclerosis (Santos et al. 2004). Another study, including the Japanese population, did not observe any relationship of ApoE4 with the risk of MS (Niino et al. 2003). The presence of *N*-acetyl aspartate (NAA—exclusively present in mature neurons) was found to be significantly decreased in patients with MS and therefore suggesting axonal loss, neuronal loss, and in general neuronal dysfunction. *ApoE4* allele carriers with multiple sclerosis present with a more significant disability and decreased NAA-to-creatine ratio than non-*ApoE4* MS carriers. *ApoE4* carriers show more relapses and a fivefold increased rate of loss of annual brain volume than non-*ApoE4* allele carriers. *ApoE2* carriers were reported to show loss of brain volume. These findings demonstrate that the *ApoE4* allele in homozygous conditions exerts a more substantial adverse influence on brain volume, predisposing increased brain neurodegenerative tendencies in MS subjects (Enzinger et al. 2004).

1.7 Recent Developments and Future Perspectives

The diverse cellular expression pattern suggests multiple roles of ApoE. The relevance of ApoE4 in AD pathogenesis has been indicated by various cell biological, biochemical, and transgenic animal studies. Depending on the evidence discussed in the above sections of this chapter, it is probable that ApoE4 influences the pathogenesis of AD by engaging with different factors via multiple pathways. Therefore, understanding how the expression of ApoE is governed in the brain may generate necessary insights into numerous roles of ApoE in neurology and neurodegenerative disease, like AD. Therapeutic agents that suppress ApoE4 expression in neurons can eradicate downstream adverse effects.

Drugs that impede ApoE4-stimulated Aβ accumulation might mitigate the A-β-mediated effects of ApoE4 on the neuropathogenesis of AD. Drugs may also be developed to block the enzyme responsible for the fragmentation of ApoE or inhibit ApoE4 fragment's interaction with elements of the cytoskeleton and mitochondria from defending against neurotoxicity caused by fragmented ApoE4. Furthermore, the drug moieties that can improve mitochondrial activity or mitochondrial numbers may be useful in AD. Additionally, besides, ApoE4 domain interaction, which shares several evil effects of ApoE4, is another potential drug target. Small molecules are being designed to inhibit this domain interaction by making ApoE4 quite similar to ApoE3, both structurally and functionally. There is a possibility for combination drug therapy with symptomatic medications and others that may alter the disease onset and progression. Furthermore *ApoE* genomes may be included in clinical studies, as some interventions could only work in particular *ApoE* genotypes.

1.8 Conclusion

ApoE plays a crucial role in Aβ aggregation and clearance as well as facilitates pro-inflammatory responses that can aggravate AD pathogenesis. Studies have reported that the *ApoE4* allele mostly enhances Aβ deposition, formation of neurofibrillary tangles, and hyperphosphorylation of tau proteins. Further, it enhances the process of disease conversion and its progression to a debilitating one. The active participation of *ApoE* offers new insight into AD. Therefore, a deeper understanding of the mechanism of ApoE metabolism's effect will help to explore the disease better and contribute to the development of new therapeutic avenues. Therefore, studying the genetic variants and functions of ApoE may aid in the early identification of high-risk individuals and in developing effective therapeutic strategies.

References

Aggarwal NT, Bienias JL, Bennett DA, Wilson RS, Morris MC, Schneider JA, Shah RC, Evans DA (2006) The relation of cigarette smoking to incident Alzheimer's disease in a biracial urban community population. Neuroepidemiology 26(3):140–146. https://doi.org/10.1159/000091654

Aisen PS, Cummings J, Jack CR Jr, Morris JC, Sperling R, Frölich L, Jones RW, Dowsett SA, Matthews BR, Raskin J, Scheltens P, Dubois B (2017) On the path to 2025: understanding the Alzheimer's disease continuum. Alzheimers Res Ther 9(1):60. https://doi.org/10.1186/s13195-017-0283-5

Akhter Y, Nabi J, Hamid H, Tabassum N, Pottoo F, Sharma A (2020) Protein quality control in neurodegeneration and neuroprotection. In: Quality control of cellular protein in neurodegenerative disorders. IGI-Global, Hershey, pp 1–24. https://doi.org/10.4018/978-1-7998-13170.ch001

Andersen K, Launer LJ, Dewey ME, Letenneur L, Ott A, Copeland JR, Dartigues JF, KraghSorensen P, Baldereschi M, Brayne C, Lobo A, Martinez-Lage JM, Stijnen T, Hofman A (1999) Gender differences in the incidence of AD and vascular dementia: the EURODEM studies. EURODEM Incidence Research Group. Neurology 53(9):1992–1997. https://doi.org/10.1212/wnl.53.9.1992

Androuin A, Potier B, Nägerl UV, Cattaert D, Danglot L, Thierry M, Youssef I, Triller A, Duyckaerts C, El Hachimi KH, Dutar P, Delatour B, Marty S (2018) Evidence for altered dendritic spine compartmentalization in Alzheimer's disease and functional effects in a mouse model. Acta Neuropathol 135(6):839–854. https://doi.org/10.1007/s00401-018-1847-6

Armstrong MJ, Litvan I, Lang AE, Bak TH, Bhatia KP, Borroni B, Boxer AL, Dickson DW, Grossman M, Hallett M, Josephs KA, Kertesz A, Lee SE, Miller BL, Reich SG, Riley DE, Tolosa E, Tröster AI, Vidailhet M, Weiner WJ (2013) Criteria for the diagnosis of corticobasal degeneration. Neurology 80(5):496–503. https://doi.org/10.1212/WNL.0b013e31827f0fd1

Bangen KJ, Beiser A, Delano-Wood L, Nation DA, Lamar M, Libon DJ, Bondi MW, Seshadri S, Wolf PA, Au R (2013) APOE genotype modifies the relationship between midlife vascular risk factors and later cognitive decline. J Stroke Cerebrovasc Dis 22(8):1361–1369. https://doi.org/10.1016/j.jstrokecerebrovasdis.2013.03.013

Barnes DE, Whitmer RA, Yaffe K (2007) Physical activity and dementia: the need for prevention trials. Exerc Sport Sci Rev 35(1):24–29. https://doi.org/10.1097/JES.0b013e31802d6bc2

Beffert U, Cohn JS, Petit-Turcotte C, Tremblay M, Aumont N, Ramassamy C, Davignon J, Poirier J (1999) Apolipoprotein E and beta-amyloid levels in the hippocampus and frontal cortex of Alzheimer's disease subjects are disease-related and apolipoprotein E genotype dependent. Brain Res 843(1–2):87–94. https://doi.org/10.1016/s0006-8993(99)01894-6

Belkouch M, Hachem M, Elgot A, Lo Van A, Picq M, Guichardant M, Lagarde M, BernoudHubac N (2016) The pleiotropic effects of omega-3 docosahexaenoic acid on the hallmarks of Alzheimer's disease. J Nutr Biochem 38:1–11. https://doi.org/10.1016/j.jnutbio.2016.03.002

Bertram L, Busch R, Spiegl M, Lautenschlager N, Müller U, Kurz A (1998) Paternal age is a risk factor for Alzheimer disease in the absence of a major gene. Neurogenetics 1(4):277280. https://doi.org/10.1007/s100480050041

Bertram L, McQueen MB, Mullin K, Blacker D, Tanzi RE (2007) Systematic meta-analyses of Alzheimer disease genetic association studies: the AlzGene database. Nat Genet 39(1):17–23

Bonham LW, Desikan RS, Yokoyama JS (2016) The relationship between complement factor C3, APOE ε4, amyloid and tau in Alzheimer's disease. Acta Neuropathol Commun 4(1):65. https://doi.org/10.1186/s40478-016-0339-y

Brecht WJ, Harris FM, Chang S, Tesseur I, Yu G-Q, Xu Q, Fish JD, Wyss-Coray T, Buttini M, Mucke L (2004) Neuron-specific apolipoprotein e4 proteolysis is associated with increased tau phosphorylation in brains of transgenic mice. J Neurosci 24(10):2527–2534

Broderick J, Lu M, Jackson C, Pancioli A, Tilley BC, Fagan SC, Kothari R, Levine SR, Marler JR, Lyden PD (2001) Apolipoprotein E phenotype and the efficacy of intravenous tissue plasminogen activator in acute ischemic stroke. Ann Neurol 49(6):736–744
Cacabelos R, Rodriguez B, Carrera C, Caamaño J, Beyer K, Lao J, Sellers M (1996) APOErelated frequency of cognitive and noncognitive symptoms in dementia. Methods Find Exp Clin Pharmacol 18(10):693–706
Carrasquillo MM, Zou F, Pankratz VS, Wilcox SL, Ma L, Walker LP, Younkin SG, Younkin CS, Younkin LH, Bisceglio GD (2009) Genetic variation in PCDH11X is associated with susceptibility to late-onset Alzheimer's disease. Nat Genet 41(2):192–198
Cash JG, Kuhel DG, Basford JE, Jaeschke A, Chatterjee TK, Weintraub NL, Hui DY (2012) Apolipoprotein E4 impairs macrophage efferocytosis and potentiates apoptosis by accelerating endoplasmic reticulum stress. J Biol Chem 287(33):27876–27884. https://doi.org/10.1074/jbc.M112.377549
Cleret de Langavant L, Fénelon G, Benisty S, Boissé M-F, Jacquemot C, Bachoud-Lévi A-C (2013) Awareness of memory deficits in early stage Huntington's disease. PLoS One 8(4):e61676–e61676. https://doi.org/10.1371/journal.pone.0061676
Colonna M, Wang Y (2016) TREM2 variants: new keys to decipher Alzheimer disease pathogenesis. Nat Rev Neurosci 17(4):201–207. https://doi.org/10.1038/nrn.2016.7
Colton CA, Needham LK, Brown C, Cook D, Rasheed K, Burke JR, Strittmatter WJ, Schmechel DE, Vitek MP (2004) APOE genotype-specific differences in human and mouse macrophage nitric oxide production. J Neuroimmunol 147(1–2):62–67
Corder EH, Saunders AM, Strittmatter WJ, Schmechel DE, Gaskell PC, Small G, Roses A, Haines J, Pericak-Vance MA (1993) Gene dose of apolipoprotein E type 4 allele and the risk of Alzheimer's disease in late onset families. Science 261(5123):921–923
Corder EH, Robertson K, Lannfelt L, Bogdanovic N, Eggertsen G, Wilkins J, Hall C (1998) HIV-infected subjects with the E4 allele for APOE have excess dementia and peripheral neuropathy. Nat Med 4(10):1182–1184
Craft S (2007) Insulin resistance and Alzheimer's disease pathogenesis: potential mechanisms and implications for treatment. Curr Alzheimer Res 4(2):147–152. https://doi.org/10.2174/156720507780362137
Cudaback E, Li X, Montine KS, Montine TJ, Keene CD (2011) Apolipoprotein E isoformdependent microglia migration. FASEB J 25(6):2082–2091. https://doi.org/10.1096/fj.10176891
Custodio N, Montesinos R, Lira D, Herrera-Perez E, Bardales Y, Valeriano-Lorenzo L (2017) Mixed dementia: a review of the evidence. Dement Neuropsychol 11(4):364–370
Czaplińska M, Ćwiklińska A, Sakowicz-Burkiewicz M, Wieczorek E, Kuchta A, Kowalski R, Kortas-Stempak B, Dębska-Ślizień A, Jankowski M, Król E (2019) Apolipoprotein E gene polymorphism and renal function are associated with apolipoprotein E concentration in patients with chronic kidney disease. Lipids Health Dis 18(1):60–60. https://doi.org/10.1186/s12944-019-1003-x
Damoiseaux JS, Seeley WW, Zhou J, Shirer WR, Coppola G, Karydas A, Rosen HJ, Miller BL, Kramer JH, Greicius MD (2012) Gender modulates the APOE ε4 effect in healthy older adults: convergent evidence from functional brain connectivity and spinal fluid tau levels. J Neurosci 32(24):8254–8262. https://doi.org/10.1523/jneurosci.0305-12.2012
Dean B, Laws SM, Hone E, Taddei K, Scarr E, Thomas EA, Harper C, McClean C, Masters C, Lautenschlager N (2003) Increased levels of apolipoprotein E in the frontal cortex of subjects with schizophrenia. Biol Psychiatry 54(6):616–622
(2011) Dementia in parkinsonism. Ann Indian Acad Neurol 14(suppl 1):S21–S24
Devenney E, Vucic S, Hodges JR, Kiernan MC (2015) Motor neuron disease-frontotemporal dementia: a clinical continuum. Expert Rev Neurother 15(5):509–522. https://doi.org/10.1586/14737175.2015.1034108
Du Y, Dodel R, Eastwood B, Bales K, Gao F, Lohmüller F, Müller U, Kurz A, Zimmer R, Evans R (2000) Association of an interleukin 1α polymorphism with Alzheimer's disease. Neurology 55(4):480–484

Du Z, Jia H, Liu J, Zhao X, Xu W (2015) Effects of three hydrogen-rich liquids on hemorrhagic shock in rats. J Surg Res 193(1):377–382. https://doi.org/10.1016/j.jss.2014.06.051

Dubelaar E, Verwer R, Hofman M, Van Heerikhuize J, Ravid R, Swaab D (2004) ApoE ε4 genotype is accompanied by lower metabolic activity in nucleus basalis of Meynert neurons in Alzheimer patients and controls as indicated by the size of the Golgi apparatus. J Neuropathol Exp Neurol 63(2):159–169

Dublin S, Walker RL, Gray SL, Hubbard RA, Anderson ML, Yu O, Crane PK, Larson EB (2015) Prescription opioids and risk of dementia or cognitive decline: a prospective cohort study. J Am Geriatr Soc 63(8):1519–1526

Dubnikov T, Cohen E (2017) The emerging roles of early protein folding events in the secretory pathway in the development of neurodegenerative maladies. Front Neurosci 11:48. https://doi.org/10.3389/fnins.2017.00048

Dubois B, Hampel H, Feldman HH, Scheltens P, Aisen P, Andrieu S, Bakardjian H, Benali H, Bertram L, Blennow K, Broich K, Cavedo E, Crutch S, Dartigues J-F, Duyckaerts C, Epelbaum S, Frisoni GB, Gauthier S, Genthon R, Gouw AA, Habert M-O, Holtzman DM, Kivipelto M, Lista S, Molinuevo J-L, O'Bryant SE, Rabinovici GD, Rowe C, Salloway S, Schneider LS, Sperling R, Teichmann M, Carrillo MC, Cummings J, Jack CR Jr, Proceedings of the Meeting of the International Working G, the American Alzheimer's Association on "The Preclinical State of AD, July, Washington DC USA (2016) Preclinical Alzheimer's disease: definition, natural history, and diagnostic criteria. Alzheimers Dement 12(3):292–323. https://doi.org/10.1016/j.jalz.2016.02.002

Dunn N, Mullee M, Perry VH, Holmes C (2005) Association between dementia and infectious disease: evidence from a case-control study. Alzheimer Dis Assoc Disord 19(2):91–94

Durany N, Riederer P, Cruz-Sánchez FF (2000) Apolipoprotein E genotype in Spanish schizophrenic patients. Psychiatr Genet 10(2):73–77

Engelborghs S, Dermaut B, Goeman J, Saerens J, Marien P, Pickut B, Van den Broeck M, Serneels S, Cruts M, Van Broeckhoven C (2003) Prospective Belgian study of neurodegenerative and vascular dementia: APOE genotype effects. J Med Genet 40(11):814–814

Enzinger C, Ropele S, Smith S, Strasser Fuchs S, Poltrum B, Schmidt H, Matthews PM, Fazekas F (2004) Accelerated evolution of brain atrophy and "black holes" in MS patients with APOE ε4. Ann Neurol 55(4):563–569

Fan YY, Cai QL, Gao ZY, Lin X, Huang Q, Tang W, Liu JH (2017) APOE ε4 allele elevates the expressions of inflammatory factors and promotes Alzheimer's disease progression: a comparative study based on Han and She populations in the Wenzhou area. Brain Res Bull 132:39–43. https://doi.org/10.1016/j.brainresbull.2017.04.017

Fanciulli A, Wenning GK (2015) Multiple-system atrophy. N Engl J Med 372(3):249–263. https://doi.org/10.1056/NEJMra1311488

Farrer LA, Cupples LA, Haines JL, Hyman B, Kukull WA, Mayeux R, Myers RH, PericakVance MA, Risch N, Van Duijn CM (1997) Effects of age, sex, and ethnicity on the association between apolipoprotein E genotype and Alzheimer disease: a meta-analysis. JAMA 278(16): 1349–1356

Federoff M, Jimenez-Rolando B, Nalls MA, Singleton AB (2012) A large study reveals no association between APOE and Parkinson's disease. Neurobiol Dis 46(2):389392

Finger EC (2016) Frontotemporal dementias. Continuum (Minneap Minn) 22-(2 Dementia):464–489. https://doi.org/10.1212/CON.0000000000000300

Fitz NF, Cronican AA, Saleem M, Fauq AH, Chapman R, Lefterov I, Koldamova R (2012) Abca1 deficiency affects Alzheimer's disease-like phenotype in human ApoE4 but not in ApoE3-targeted replacement mice. J Neurosci 32(38):13125–13136. https://doi.org/10.1523/jneurosci.1937-12.2012

Forette F, Boller F (1991) Hypertension and the risk of dementia in the elderly. Am J Med 90(3a):14s–19s. https://doi.org/10.1016/0002-9343(91)90430-6

Forton DM, Thomas HC, Murphy CA, Allsop JM, Foster GR, Main J, Wesnes KA, TaylorRobinson SD (2002) Hepatitis C and cognitive impairment in a cohort of patients with mild liver disease. Hepatology 35(2):433–439

Fymat A (2018) Dementia: a review. J Clin Psychiatry Neurosci 1(3):2734

Gatz M, Fratiglioni L, Johansson B, Berg S, Mortimer JA, Reynolds CA, Fiske A, Pedersen NL (2005) Complete ascertainment of dementia in the Swedish Twin Registry: the HARMONY study. Neurobiol Aging 26(4):439–447

Gilman S, Wenning GK, Low PA, Brooks DJ, Mathias CJ, Trojanowski JQ, Wood NW, Colosimo C, Dürr A, Fowler CJ, Kaufmann H, Klockgether T, Lees A, Poewe W, Quinn N, Revesz T, Robertson D, Sandroni P, Seppi K, Vidailhet M (2008) Second consensus statement on the diagnosis of multiple system atrophy. Neurology 71(9):670–676. https://doi.org/10.1212/01.wnl.0000324625.00404.15

Graves AB, Mortimer JA, Bowen J, McCormick WC, McCurry SM, Schellenberg GD, Larson E (2001) Head circumference and incident Alzheimer's disease: modification by apolipoprotein E. Neurology 57(8):1453–1460

Greenberg S, Vonsattel J-P, Segal A, Chiu R, Clatworthy A, Liao A, Hyman B, Rebeck G (1998) Association of apolipoprotein E ϵ2 and vasculopathy in cerebral amyloid angiopathy. Neurology 50(4):961–965

Griffin WST, Mrak RE (2002) Interleukin 1 in the genesis and progression of and risk for development of neuronal degeneration in Alzheimer's disease. J Leukoc Biol 72(2):233–238

Hampel H, Blennow K, Shaw LM, Hoessler YC, Zetterberg H, Trojanowski JQ (2010) Total and phosphorylated tau protein as biological markers of Alzheimer's disease. Exp Gerontol 45(1):30–40. https://doi.org/10.1016/j.exger.2009.10.010

Harringtona CR, Roth M, Xuereb JH, McKenna PJ, Wischik CM (1995) Apolipoprotein E type ϵ4 allele frequency is increased in patients with schizophrenia. Neurosci Lett 202(1–2):101–104

Harris FM, Brecht WJ, Xu Q, Mahley RW, Huang Y (2004) Increased tau phosphorylation in apolipoprotein E4 transgenic mice is associated with activation of extracellular signalregulated kinase: modulation by zinc. J Biol Chem 279(43):44795–44801. https://doi.org/10.1074/jbc.M408127200

Hebert LE, Weuve J, Scherr PA, Evans DA (2013) Alzheimer disease in the United States (20102050) estimated using the 2010 census. Neurology 80(19):1778–1783. https://doi.org/10.1212/WNL.0b013e31828726f5

Heneka MT, Carson MJ, El Khoury J, Landreth GE, Brosseron F, Feinstein DL, Jacobs AH, Wyss-Coray T, Vitorica J, Ransohoff RM, Herrup K, Frautschy SA, Finsen B, Brown GC, Verkhratsky A, Yamanaka K, Koistinaho J, Latz E, Halle A, Petzold GC, Town T, Morgan D, Shinohara ML, Perry VH, Holmes C, Bazan NG, Brooks DJ, Hunot S, Joseph B, Deigendesch N, Garaschuk O, Boddeke E, Dinarello CA, Breitner JC, Cole GM, Golenbock DT, Kummer MP (2015) Neuroinflammation in Alzheimer's disease. Lancet Neurol 14(4):388–405. https://doi.org/10.1016/s1474-4422(15)70016-5

Heyn P, Abreu B, Ottenbacher K (2004) The effects of exercise training on elderly persons with cognitive impairment and dementia: a meta-analysis1. Arch Phys Med Rehabil 85(10):1694–1704. https://doi.org/10.1016/j.apmr.2004.03.019

Howard R, Dennehey J, Lovestone S, Birkett J, Sham P, Powell J, Castle D, Murray R, Levy R (1995) Apolipoprotein E genotype and late paraphrenia. Int J Geriatr Psychiatry 10(2):147–150

Hu S, Sheng WS, Lokensgard JR, Peterson PK (2002) Morphine induces apoptosis of human microglia and neurons. Neuropharmacology 42(6):829–836. https://doi.org/10.1016/s00283908(02)00030-8

Huang X, Chen PC, Poole C (2004) APOE-ε2 allele associated with higher prevalence of sporadic Parkinson disease. Neurology 62(12):2198–2202

Iadecola C (2013) The pathobiology of vascular dementia. Neuron 80(4):844–866. https://doi.org/10.1016/j.neuron.2013.10.008

Igata-Yi R, Igata T, Ishizuka K, Kimura T, Sakamoto S, Katsuragi S, Takamatsu J, Miyakawa T (1997) Apolipoprotein E genotype and psychosis. Biol Psychiatry 41(8):906–908

Islam MM, Iqbal U, Walther B, Atique S, Dubey NK, Nguyen P-A, Poly TN, Masud JHB, Li YCJ, Shabbir S-A (2016) Benzodiazepine use and risk of dementia in the elderly population: a systematic review and meta-analysis. Neuroepidemiology 47(3-4):181–191

Jiang Q, Lee CY, Mandrekar S, Wilkinson B, Cramer P, Zelcer N, Mann K, Lamb B, Willson TM, Collins JL, Richardson JC, Smith JD, Comery TA, Riddell D, Holtzman DM, Tontonoz P, Landreth GE (2008) ApoE promotes the proteolytic degradation of Abeta. Neuron 58(5): 681–693. https://doi.org/10.1016/j.neuron.2008.04.010

Karageorgiou E, Miller BL (2014) Frontotemporal lobar degeneration: a clinical approach. Semin Neurol 34(2):189–201. https://doi.org/10.1055/s-0034-1381735

Kariv-Inbal Z, Yacobson S, Berkecz R, Peter M, Janaky T, Lütjohann D, Broersen LM, Hartmann T, Michaelson DM (2012) The isoform-specific pathological effects of apoE4 in vivo are prevented by a fish oil (DHA) diet and are modified by cholesterol. J Alzheimers Dis 28(3):667–683. https://doi.org/10.3233/jad-2011-111265

Karp A, Paillard-Borg S, Wang H-X, Silverstein M, Winblad B, Fratiglioni L (2006) Mental, physical and social components in leisure activities equally contribute to decrease dementia risk. Dement Geriatr Cogn Disord 21(2):65–73. https://doi.org/10.1159/000089919

Kim J, Basak JM, Holtzman DM (2009) The role of apolipoprotein E in Alzheimer's disease. Neuron 63(3):287–303. https://doi.org/10.1016/j.neuron.2009.06.026

Kipen E, Helme RD, Wark JD, Flicker L (1995) Bone density, vitamin D nutrition, and parathyroid hormone levels in women with dementia. J Am Geriatr Soc 43(10):1088–1091

Knight R (2006) Creutzfeldt-Jakob disease: a rare cause of dementia in elderly persons. Clin Infect Dis 43(3):340–346. https://doi.org/10.1086/505215

Koldamova R, Fitz NF, Lefterov I (2014) ATP-binding cassette transporter A1: from metabolism to neurodegeneration. Neurobiol Dis 72:13–21

LaDu MJ, Falduto MT, Manelli AM, Reardon CA, Getz GS, Frail DE (1994) Isoform-specific binding of apolipoprotein E to beta-amyloid. J Biol Chem 269(38):23403–23406

LaDu MJ, Munson GW, Jungbauer L, Getz GS, Reardon CA, Tai LM, Yu C (2012) Preferential interactions between ApoE-containing lipoproteins and Aβ revealed by a detection method that combines size exclusion chromatography with non-reducing gel-shift. Biochim Biophys Acta 1821(2):295–302

Lambert JC, Mann D, Goumidi L, Harris J, Amouyel P, Iwatsubo T, Lendon C, Chartier Harlin MC (2001) Effect of APOE promoter polymorphisms on cerebral amyloid peptide deposition in Alzheimer's disease. Lancet 357(9256):608–609

Lagnaoui R, Bégaud B, Moore N, Chaslerie A, Fourrier A, Letenneur L, Dartigues J-F, Moride Y (2002) Benzodiazepine use and risk of dementia: a nested case–control study. J Clin Epidemiol 55(3):314–318

Lane-Donovan C, Herz J (2017) The ApoE receptors Vldlr and Apoer2 in central nervous system function and disease. J Lipid Res 58(6):1036–1043. https://doi.org/10.1194/jlr.R075507

Lautenschlager N, Cox K, Flicker L, Foster J, Bockxmeer F, Xiao A, Greenop K, Almeida O (2008) Effect of physical activity on cognitive function in older adults at risk for Alzheimer disease: a randomized trial. JAMA 300(9):1027–1037. https://doi.org/10.1001/jama.300.9.1027

Li YJ, Hauser MA, Scott WK, Martin ER, Booze MW, Qin XJ, Walter JW, Nance MA, Hubble JP, Koller WC, Pahwa R, Stern MB, Hiner BC, Jankovic J, Goetz CG, Small GW, Mastaglia F, Haines JL, Pericak-Vance MA, Vance JM (2004) Apolipoprotein E controls the risk and age at onset of Parkinson disease. Neurology 62(11):2005–2009. https://doi.org/10.1212/01.wnl.0000128089.53030.ac

Li J, Kanekiyo T, Shinohara M, Zhang Y, LaDu MJ, Xu H, Bu G (2012) Differential regulation of amyloid-β endocytic trafficking and lysosomal degradation by apolipoprotein E isoforms. J Biol Chem 287(53):44593–44601. https://doi.org/10.1074/jbc.M112.420224

Liao F, Yoon H, Kim J (2017) Apolipoprotein E metabolism and functions in brain and its role in Alzheimer's disease. Curr Opin Lipidol 28(1):60–67. https://doi.org/10.1097/mol.0000000000000383

Lin H-F, Lai C-L, Tai C-T, Lin R-T, Liu C-K (2004) Apolipoprotein E polymorphism in ischemic cerebrovascular diseases and vascular dementia patients in Taiwan. Neuroepidemiology 23(3): 129–134

Liu L, Guo XE, Zhou YQ, Xia JL (2003) Prevalence of dementia in China. Dement Geriatr Cogn Disord 15(4):226–230. https://doi.org/10.1159/000068784

Liu C-C, Kanekiyo T, Xu H, Bu G (2013) Apolipoprotein E and Alzheimer disease: risk, mechanisms and therapy. Nat Rev Neurol 9(2):106–118

Lopez G, Bayulkem K, Hallett M (2016) Progressive supranuclear palsy (PSP): Richardson syndrome and other PSP variants. Acta Neurol Scand 134(4):242–249. https://doi.org/10.1111/ane.12546

López-Pousa S, Vilalta-Franch J, Llinàs-Regla J, Garre-Olmo J, Roman G (2004) Incidence of dementia in a rural community in Spain: the Girona cohort study. Neuroepidemiology 23(4): 170–177. https://doi.org/10.1159/000078502

Love S, Nicoll JA, Hughes A, Wilcock GK (2003) APOE and cerebral amyloid angiopathy in the elderly. Neuroreport 14(11):1535–1536. https://doi.org/10.1097/00001756-200308060-00027

Mahley RW (2016) Central nervous system lipoproteins: ApoE and regulation of cholesterol metabolism. Arterioscler Thromb Vasc Biol 36(7):1305–1315. https://doi.org/10.1161/atvbaha.116.307023

Mahley RW, Rall SC Jr (2000) Apolipoprotein E: far more than a lipid transport protein. Annu Rev Genomics Hum Genet 1:507–537. https://doi.org/10.1146/annurev.genom.1.1.507

Mamun AA, Uddin MS, Bin Bashar MF, Zaman S, Begum Y, Bulbul IJ et al (2020) Molecular insight into the therapeutic promise of targeting APOE4 for Alzheimer's disease. Oxid Med Cell Longev 2020:5086250–5086250. https://doi.org/10.1155/2020/5086250

Marder K, Maestre G, Cote L, Mejia H, Alfaro B, Halim A, Tang M, Tycko B, Mayeux R (1994) The apolipoprotein ϵ4 allele in Parkinson's disease with and without dementia. Neurology 44(7):1330–1330

Martinelli N, Olivieri O, Shen GQ, Trabetti E, Pizzolo F, Busti F, Friso S, Bassi A, Li L, Hu Y, Pignatti PF, Corrocher R, Wang QK, Girelli D (2009) Additive effect of LRP8/APOER2 R952Q variant to APOE epsilon2/epsilon3/epsilon4 genotype in modulating apolipoprotein E concentration and the risk of myocardial infarction: a case-control study. BMC Med Genet 10:41. https://doi.org/10.1186/1471-2350-10-41

Martorell L, Virgos C, Valero J, Coll G, Figuera L, Joven J, Pocovi M, Labad A, Vilella E (2001) Schizophrenic women with the APOE ε4 allele have a worse prognosis than those without it. Mol Psychiatry 6(3):307–310

McCann H, Stevens CH, Cartwright H, Halliday GM (2014) α-Synucleinopathy phenotypes. Parkinsonism Relat Disord 20(suppl 1):S62–S67. https://doi.org/10.1016/s1353-8020(13)70017-8

McKeith IG, Boeve BF, Dickson DW, Halliday G, Taylor JP, Weintraub D, Aarsland D, Galvin J, Attems J, Ballard CG, Bayston A, Beach TG, Blanc F, Bohnen N, Bonanni L, Bras J, Brundin P, Burn D, Chen-Plotkin A, Duda JE, El-Agnaf O, Feldman H, Ferman TJ, Ffytche D, Fujishiro H, Galasko D, Goldman JG, Gomperts SN, Graff-Radford NR, Honig LS, Iranzo A, Kantarci K, Kaufer D, Kukull W, Lee VMY, Leverenz JB, Lewis S, Lippa C, Lunde A, Masellis M, Masliah E, McLean P, Mollenhauer B, Montine TJ, Moreno E, Mori E, Murray M, O'Brien JT, Orimo S, Postuma RB, Ramaswamy S, Ross OA, Salmon DP, Singleton A, Taylor A, Thomas A, Tiraboschi P, Toledo JB, Trojanowski JQ, Tsuang D, Walker Z, Yamada M, Kosaka K (2017) Diagnosis and management of dementia with Lewy bodies: fourth consensus report of the DLB Consortium. Neurology 89(1):88–100. https://doi.org/10.1212/wnl.0000000000004058

Mendez MF (2017) Early-onset Alzheimer disease. Neurol Clin 35(2):263–281. https://doi.org/10.1016/j.ncl.2017.01.005

Moga DC, Abner EL, Wu Q, Jicha GA (2017) Bladder antimuscarinics and cognitive decline in elderly patients. Alzheimers Dement 3(1):139–148

Näslund J, Thyberg J, Tjernberg LO, Wernstedt C, Karlström AR, Bogdanovic N, Gandy SE, Lannfelt L, Terenius L, Nordstedt C (1995) Characterization of stable complexes involving apolipoprotein E and the amyloid β peptide in Alzheimer's disease brain. Neuron 15(1): 219–228

Nguyen TP, Schaffert J, LoBue C, Womack KB, Hart J, Cullum CM (2018) Traumatic brain injury and age of onset of dementia with Lewy bodies. J Alzheimers Dis 66(2):717–723. https://doi.org/10.3233/JAD-180586

Niino M, Kikuchi S, Fukazawa T, Yabe I, Tashiro K (2003) Polymorphisms of apolipoprotein E and Japanese patients with multiple sclerosis. Mult Scler J 9(4):382–386

Nock TG, Chouinard-Watkins R, Plourde M (2017) Carriers of an apolipoprotein E epsilon 4 allele are more vulnerable to a dietary deficiency in omega-3 fatty acids and cognitive decline. Biochim Biophys Acta 1862(10, part A):1068–1078. https://doi.org/10.1016/j.bbalip.2017.07.004

Noguchi S, Murakami K, Yamada N, Payami H, Kaye J, Heston L, Bird T, Schellenberg G (1993) Apolipoprotein E genotype and Alzheimer's disease. Lancet 342(8873):737738

Nuriel T, Peng KY, Ashok A, Dillman AA, Figueroa HY, Apuzzo J, Ambat J, Levy E, Cookson MR, Mathews PM, Duff KE (2017) The endosomal-lysosomal pathway is dysregulated by APOE4 expression in vivo. Front Neurosci 11:702. https://doi.org/10.3389/fnins.2017.00702

Osenkowski P, Ye W, Wang R, Wolfe MS, Selkoe DJ (2008) Direct and potent regulation of gamma-secretase by its lipid microenvironment. J Biol Chem 283(33):22529–22540. https://doi.org/10.1074/jbc.M801925200

Pandey M, Nabi J, Tabassum N, Pottoo F, Khatik R, Ahmad N (2020) Molecular chaperones in neurodegeneration: mechanisms of regulation in cellular proteostasis. In: Quality control of cellular protein in neurodegenerative disorders. IGI-Global, Hershey. https://doi.org/10.4018/978-1-7998-1317-0.ch014

Pastor P, Goate AM (2004) Molecular genetics of Alzheimer's disease. Curr Psychiatry Rep 6(2): 125–133

Perry DC, Brown JA, Possin KL, Datta S, Trujillo A, Radke A, Karydas A, Kornak J, Sias AC, Rabinovici GD, Gorno-Tempini ML, Boxer AL, De May M, Rankin KP, Sturm VE, Lee SE, Matthews BR, Kao AW, Vossel KA, Tartaglia MC, Miller ZA, Seo SW, Sidhu M, Gaus SE, Nana AL, Vargas JNS, Hwang JL, Ossenkoppele R, Brown AB, Huang EJ, Coppola G, Rosen HJ, Geschwind D, Trojanowski JQ, Grinberg LT, Kramer JH, Miller BL, Seeley WW (2017) Clinicopathological correlations in behavioural variant frontotemporal dementia. Brain 140(12): 3329–3345. https://doi.org/10.1093/brain/awx254

Peters R, Poulter R, Warner J, Beckett N, Burch L, Bulpitt C (2009) Smoking, dementia and cognitive decline in the elderly, a systematic review. BMC Geriatr 8(1):8–36. https://doi.org/10.1186/1471-2318-8-36

Petersen RC, Thomas RG, Grundman M, Bennett D, Doody R, Ferris S, Galasko D, Jin S, Kaye J, Levey A (2005) Vitamin E and donepezil for the treatment of mild cognitive impairment. N Engl J Med 352(23):2379–2388

Phillips MC (2014) Apolipoprotein E isoforms and lipoprotein metabolism. IUBMB Life 66(9): 616–623. https://doi.org/10.1002/iub.1314

Podewils L, Guallar E, Kuller L, Fried L, Lopez O, Carlson M, Lyketsos C (2005) Physical activity, APOE genotype, and dementia risk: findings from the cardiovascular health cognition study. Am J Epidemiol 161(7):639–651. https://doi.org/10.1093/aje/kwi092

Poirier J, Davignon J, Bouthillier D, Kogan S, Bertrand P, Gauthier S (1993) Apolipoprotein E polymorphism and Alzheimer's disease. Lancet (London, England) 342(8873):697–699. https://doi.org/10.1016/0140-6736(93)91705-q

Posner HB, Tang MX, Luchsinger J, Lantigua R, Stern Y, Mayeux R (2002) The relationship of hypertension in the elderly to AD, vascular dementia, and cognitive function. Neurology 58(8): 1175–1181. https://doi.org/10.1212/wnl.58.8.1175

Prince M, Lovestone S, Cervilla J, Joels S, Powell J, Russ C, Mann A (2000) The association between APOE and dementia does not seem to be mediated by vascular factors. Neurology 54(2):397–397

Prince M, Bryce R, Albanese E, Wimo A, Ribeiro W, Ferri CP (2013) The global prevalence of dementia: a systematic review and metaanalysis. Alzheimers Dement 9(1):63–75. https://doi.org/10.1016/j.jalz.2012.11.007

Pulkes T, Papsing C, Mahasirimongkol S, Busabaratana M, Kulkantrakorn K, Tiamkao S (2011) Association between apolipoprotein E genotypes and Parkinson's disease. J Clin Neurosci 18(10):1333–1335

Rall SC Jr, Newhouse YM, Clarke HR, Weisgraber KH, McCarthy BJ, Mahley RW, Bersot TP (1989) Type III hyperlipoproteinemia associated with apolipoprotein E phenotype E3/3. Structure and genetics of an apolipoprotein E3 variant. J Clin Invest 83(4):1095–1101. https://doi.org/10.1172/JCI113988

Rascovsky K, Grossman M (2013) Clinical diagnostic criteria and classification controversies in frontotemporal lobar degeneration. Int Rev Psychiatry 25(2):145–158. https://doi.org/10.3109/09540261.2013.763341

Ravaglia G, Forti P, Maioli F, Martelli M, Servadei L, Brunetti N, Dalmonte E, Bianchin M, Mariani E (2005) Incidence and etiology of dementia in a large elderly Italian population. Neurology 64(9):1525–1530. https://doi.org/10.1212/01.wnl.0000160107.02316.bf

Reiman EM, Caselli RJ, Yun LS, Chen K, Bandy D, Minoshima S, Thibodeau SN, Osborne D (1996) Preclinical evidence of Alzheimer's disease in persons homozygous for the ε4 allele for apolipoprotein E. N Engl J Med 334(12):752–758

Reiman EM, Chen K, Alexander GE, Caselli RJ, Bandy D, Osborne D, Saunders AM, Hardy J (2004) Functional brain abnormalities in young adults at genetic risk for late-onset Alzheimer's dementia. Proc Natl Acad Sci U S A 101(1):284289

Reuben DB, Judd-Hamilton L, Harris TB, Seeman TE (2003) The associations between physical activity and inflammatory markers in high-functioning older persons: MacArthur studies of successful aging. J Am Geriatr Soc 51(8):1125–1130. https://doi.org/10.1046/j.1532-5415.2003.51380.x

Rodriguez GA, Burns MP, Weeber EJ, Rebeck GW (2013) Young APOE4 targeted replacement mice exhibit poor spatial learning and memory, with reduced dendritic spine density in the medial entorhinal cortex. Learn Mem 20(5):256–266. https://doi.org/10.1101/lm.030031.112

Rodriguez GA, Tai LM, LaDu MJ, Rebeck GW (2014) Human APOE4 increases microglia reactivity at Aβ plaques in a mouse model of Aβ deposition. J Neuroinflammation 11:111. https://doi.org/10.1186/1742-2094-11-111

Santos M, Costa MC, Rio ME, Jose Sa M, Monteiro M, Valenca A, Sa A, Dinis J, Figueiredo J, Bigotte de Almeida L, Valongueiro A, Coelho I, Matama MT, Pinto-Basto J, Sequeiros J, Maciel P (2004) Genotypes at the APOE and SCA2 loci do not predict the course of multiple sclerosis in patients of Portuguese origin. Mult Scler 10(2):153–157

Savettieri G, Andreoli V, Bonavita S, Cittadella R, Caltagirone C, Fazio MC, Girlanda P, Le Pira F, Liguori M, Logroscino G (2003) Apolipoprotein E genotype does not influence the progression of multiple sclerosis. J Neurol 250(9):1094–1098

Scarmeas N, Brandt J, Albert M, Devanand D, Marder K, Bell K, Ciappa A, Tycko B, Stern Y (2002) Association between the APOE genotype and psychopathologic symptoms in Alzheimer's disease. Neurology 58(8):1182–1188

Schneider JA, Arvanitakis Z, Bang W, Bennett DA (2007) Mixed brain pathologies account for most dementia cases in community-dwelling older persons. Neurology 69(24):21972204. https://doi.org/10.1212/01.wnl.0000271090.28148.24

Selkoe DJ, Schenk D (2003) Alzheimer's disease: molecular understanding predicts amyloidbased therapeutics. Annu Rev Pharmacol Toxicol 43(1):545–584

Sen A, Alkon DL, Nelson TJ (2012) Apolipoprotein E3 (ApoE3) but not ApoE4 protects against synaptic loss through increased expression of protein kinase C epsilon. J Biol Chem 287(19): 15947–15958. https://doi.org/10.1074/jbc.M111.312710

Shinohara M, Petersen RC, Dickson DW, Bu G (2013) Brain regional correlation of amyloid-β with synapses and apolipoprotein E in non-demented individuals: potential mechanisms underlying regional vulnerability to amyloid-β accumulation. Acta Neuropathol 125(4):535–547. https://doi.org/10.1007/s00401-013-1086-9

Shinohara M, Tachibana M, Kanekiyo T, Bu G (2017) Role of LRP1 in the pathogenesis of Alzheimer's disease: evidence from clinical and preclinical studies. J Lipid Res 58(7): 1267–1281. https://doi.org/10.1194/jlr.R075796

Slooter AJ, Breteler MM, Ott A, Van Broeckhoven C, van Duijn CM (1996) APOE genotyping in differential diagnosis of Alzheimer's disease. Lancet 348(9023):334

Smit M, De Knijff P, Van der Kooij-Meijs E, Groenendijk C, van den Maagdenberg A, Leuven JG, Stalenhoef A, Stuyt P, Frants R, Havekes L (1990) Genetic heterogeneity in familial dysbetalipoproteinemia. The E2 (lys146----gln) variant results in a dominant mode of inheritance. J Lipid Res 31(1):45–53

Solfrizzi V, Colacicco AM, D'Introno A, Capurso C, Parigi AD, Capurso SA, Torres F, Capurso A, Panza F (2006) Macronutrients, aluminium from drinking water and foods, and other metals in cognitive decline and dementia. J Alzheimers Dis 10(2–3):303–330

Sun MK, Alkon DL (2006) Links between Alzheimer's disease and diabetes. Drugs Today 42(7): 481–489. https://doi.org/10.1358/dot.2006.42.7.973588

Taipale H, Gomm W, Broich K, Maier W, Tolppanen AM, Tanskanen A, Tiihonen J, Hartikainen S, Haenisch B (2018) Use of antiepileptic drugs and dementia risk—an analysis of Finnish Health Register and German health insurance data. J Am Geriatr Soc 66(6):1123–1129

Teter B, LaDu MJ, Sullivan PM, Frautschy SA, Cole GM (2016) Apolipoprotein E isotypedependent modulation of microRNA-146a in plasma and brain. Neuroreport 27(11): 791–795. https://doi.org/10.1097/wnr.0000000000000608

Uddin MS, Kabir MT, Jeandet P, Mathew B, Ashraf GM, Perveen A et al (2020) Novel anti-Alzheimer's therapeutic molecules targeting amyloid precursor protein processing. Oxid Med Cell Longev 2020:7039138. https://doi.org/10.1155/2020/7039138

Ulrich V, Konaniah ES, Herz J, Gerard RD, Jung E, Yuhanna IS, Ahmed M, Hui DY, Mineo C, Shaul PW (2014) Genetic variants of ApoE and ApoER2 differentially modulate endothelial function. Proc Natl Acad Sci U S A 111(37):13493–13498. https://doi.org/10.1073/pnas.1402106111

Vijayan M, Reddy PH (2016) Stroke, vascular dementia, and Alzheimer's disease: molecular links. J Alzheimers Dis 54(2):427–443. https://doi.org/10.3233/JAD-160527

Wahrle SE, Shah AR, Fagan AM, Smemo S, Kauwe JS, Grupe A, Hinrichs A, Mayo K, Jiang H, Thal LJ, Goate AM, Holtzman DM (2007) Apolipoprotein E levels in cerebrospinal fluid and the effects of ABCA1 polymorphisms. Mol Neurodegener 2:7. https://doi.org/10.1186/1750-1326-2-7

Wang Y-C, Tai P-A, Poly TN, Islam MM, Yang H-C, Wu C-C, Li Y-CJ (2018) Increased risk of dementia in patients with antidepressants: a meta-analysis of observational studies. Behav Neurol 2018:5315098

Wisniewski T, Frangione B (1992) Apolipoprotein E: a pathological chaperone protein in patients with cerebral and systemic amyloid. Neurosci Lett 135(2):235–238

Yang Y-W, Liu H-H, Lin T-H, Chuang H-Y, Hsieh T (2017) Association between different anticholinergic drugs and subsequent dementia risk in patients with diabetes mellitus. PLoS One 12(4):e0175335

Yassine HN, Croteau E, Rawat V, Hibbeln JR, Rapoport SI, Cunnane SC, Umhau JC (2017) DHA brain uptake and APOE4 status: a PET study with [1-(11)C]-DHA. Alzheimers Res Ther 9(1): 23. https://doi.org/10.1186/s13195-017-0250-1

Yin Y-W, Li J-C, Wang J-Z, Li B-H, Pi Y, Yang Q-W, Fang C-Q, Gao C-Y, Zhang L-L (2012) Association between apolipoprotein E gene polymorphism and the risk of vascular dementia: a meta-analysis. Neurosci Lett 514(1):6–11

Zaccai J, McCracken C, Brayne C (2005) A systematic review of prevalence and incidence studies of dementia with Lewy bodies. Age Ageing 34(6):561–566. https://doi.org/10.1093/ageing/afi190

Zannis VI, Breslow JL, Utermann G, Mahley RW, Weisgraber KH, Havel RJ, Goldstein JL, Brown MS, Schonfeld G, Hazzard WR, Blum C (1982) Proposed nomenclature of apoE isoproteins, apoE genotypes, and phenotypes. J Lipid Res 23(6):911–914

Zareparsi S, Camicioli R, Sexton G, Bird T, Swanson P, Kaye J, Nutt J, Payami H (2002) Age at onset of Parkinson disease and apolipoprotein E genotypes. Am J Med Genet 107(2):156–161

Zekonyte J, Sakai K, Nicoll JAR, Weller RO, Carare RO (2016) Quantification of molecular interactions between ApoE, amyloid-beta (Aβ) and laminin: relevance to accumulation of Aβ in Alzheimer's disease. Biochim Biophys Acta 1862(5):1047–1053. https://doi.org/10.1016/j.bbadis.2015.08.025

Zende PD, Bankar MP, Kamble PS, Momin AA (2013) Apolipoprotein e gene polymorphism and its effect on plasma lipids in arteriosclerosis. J Clin Diagn Res 7(10):2149–2152. https://doi.org/10.7860/jcdr/2013/6195.3455

Zhang Z, Mu J, Li J, Li W, Song J (2013) Aberrant apolipoprotein E expression and cognitive dysfunction in patients with poststroke depression. Genet Test Mol Biomarkers 17(1):47–51. https://doi.org/10.1089/gtmb.2012.0253

Zhang S, Li X, Ma G, Jiang Y, Liao M, Feng R, Zhang L, Liu J, Wang G, Zhao B, Jiang Q, Li K, Liu G (2016) CLU rs9331888 polymorphism contributes to Alzheimer's disease susceptibility in Caucasian but not East Asian populations. Mol Neurobiol 53(3):1446–1451. https://doi.org/10.1007/s12035-015-9098-1

2 Infection-Induced Systemic Inflammation and Dementia

Newman Osafo, Silvio Owusu Dei, Nathanael Adote Akwei, and Oduro Kofi Yeboah

Abstract

It has long been recognized that there exists a unique interaction between the peripheral immune system and the brain, an interaction that is increasingly proving to be critical in the normal physiology of the brain and the pathogenesis of neurodegenerative disorders. One such peripheral factor that alters brain function is systemic inflammation. Occurring mostly in response to infections, systemic inflammation induces behavioral deficits, alters cognitive function, and has been hypothesized to contribute significantly to the pathogenesis of dementia. With increasing evidence of the crucial role of inflammation in Alzheimer's disease (AD), coupled with the numerous failed therapeutic attempts in targeting beta-amyloid (Aβ) in AD, the infection-induced systemic inflammation hypothesis is rapidly gaining popularity as a key player in the development of dementia and other neurodegenerative disorders. This chapter summarizes other postulates, and examines them against the infection-induced systemic inflammation hypothesis, discussing their roles in the pathogenesis of neurodegeneration. We describe the effect of infection on both the chronic and acute cognitive changes in dementia, and the implications of various systemic inflammatory markers in the development of dementia. There is a deterioration of both cognitive ability and behavioral functions in demented patients who suffer from infections with a systemic inflammatory component. This chapter delves into the effects of systemic inflammation on the central nervous system (CNS) including activation of microglial and astrocytic cells, as well as the effects of infection on integrity of the blood–brain barrier. We also discuss conventional and novel pharmacological

N. Osafo (✉) · S. O. Dei · N. A. Akwei · O. K. Yeboah
Department of Pharmacology, Kwame Nkrumah University of Science and Technology, Kumasi, Ghana
e-mail: nosafo.pharm@knust.edu.gh

G. M. Ashraf, M. S. Uddin (eds.), *Current Thoughts on Dementia*,
https://doi.org/10.1007/978-981-16-7606-2_2

agents and phytochemicals that are used in targeting infection-induced systemic inflammatory mediation in the management of dementia.

Keywords

Alzheimer's disease · Amyloidosis · Dementia · Excitotoxicity · Infection · Oxidative stress · Systemic inflammation

2.1 Introduction

Dementia, a chronic neurodegenerative disorder, is a clinical syndrome characterized by a series of memory deficits, language disorders, psychological and mental deficits, and impaired functional capabilities (Burns and Iliffe 2009). It is therefore one of the major causes of morbidity and institutionalization among the elderly, thus increases the risk of disability and dependence, posing a severe global public health risk to this population (Alzheimer's Association 2013; Roberts et al. 2018). With the aging worldwide population, over 43.8 million people suffer from dementia and this number is projected to triple by the year 2050 according to recent epidemiological statistics and estimates (Roberts et al. 2018). This relative surge in the aging population has made dementia an alarming issue globally, as it poses a severe disease burden on individuals, their families, and society at large (Chou et al. 2017).

It is widely accepted that the development and progression of dementia involve a complex interaction between several specific molecular pathways that provoke destruction of synaptic connections, neural cell death, and gliosis. The dysfunction of these molecular pathways can cause inflammation and interference of physiological processes involved in cognition, personality, behavioral activities, and sensory and motor functions (Cain et al. 2017).

Alzheimer's disease (AD) accounts for approximately 70% of all reported cases of dementias (Qiu et al. 2009). Aging remains the most important and robust risk factor that influences the development of dementia as over 90% of all dementia cases occur after the age of 65 years. In patients less than 65 years (the cutoff year for early versus late-onset dementia) with dementia, frontotemporal dementia shares the same prevalence rate with AD (Ratnavalli et al. 2002; Harvey et al. 2003). At ages above 75 years, sex is a significant risk factor that affects the development of dementia in AD patients, as a disproportionally higher prevalence is observed in females than in males (Engelhart et al. 2004).

In this chapter, we briefly discuss the effects of systemic inflammation caused by infectious agents in the pathogenesis of dementia. Recent studies are providing increasing evidence that suggests a pivotal role of systemic inflammatory response in the pathogenesis of dementia, and other neurodegenerative disorders such as AD. In subsequent sections, we discuss the various hypothesis linked to the etiology of dementia, especially the systemic infection and inflammation hypothesis.

2.2 Pathogenesis of Dementia: Prevailing Hypotheses and Their Limitations

As a syndrome, dementia comprises loss of memory, and deficits in problem-solving and other cognitive abilities that significantly interfere with an individual's ability to perform daily life activities. Hence, the term dementia applies to a range of symptoms describing a loss of functional capacity in multiple cognitive domains (Cummings and Benson 1992). The symptoms and neurodegeneration that occur in dementia are usually progressive and irreversible. There are cortical or subcortical types, depending on which region of the brain they affect, and maybe differentiated clinically by the spectrum of symptoms exhibited, and with the help of brain scans. A number of pathological features are consistent in several types of dementia. Neuropathological changes and neurotransmitter deficits remain among the most common (Cummings and Benson 1992).

Multiple neuronal pathways are damaged in dementia. Plaques or other misfolded proteins may occur in major neurotransmitter pathways, with damage occurring to any nerve cell population located in or traveling through such plaque-laden areas (George-Hyslop 2000). Widespread cell destruction results in a variety of neurotransmitter deficits, with cholinergic abnormalities being the most prominent in AD (Desai and Grossberg 2005). There is loss of cholinergic neurons in the nucleus basalis of meynert with accompanying loss of nicotinic receptors in the hippocampus and cortex. There is also a decreased concentration of choline acetyltransferase in the hippocampus and neocortex (Wallace et al. 1993). Acetylcholine is important for memory, learning, and other cognitive functions, explaining the neurocognitive decline associated with loss of cholinergic neurons (Garand et al. 2000).

There is also a loss of serotonergic neurons of the raphe nuclei in AD. This decrease in serotonergic neurotransmission is believed to be responsible for some of the non-cognitive behavioral symptoms like depression or aggression that may be present in AD patients (Raskind and Peskind 1994). Effects on noradrenergic systems are less straightforward, with normal, increased, or decreased noradrenergic neurotransmission occurring in different parts of the CNS at different instances in time. In Alzheimer's disease, for instance, it is known that noradrenergic neurons of the locus coeruleus are lost (Garand et al. 2000).

Alterations in CNS dopaminergic pathways are believed to be, at least, partly responsible for the depressive and motor symptoms associated with dementia (Garand et al. 2000). Pathological changes in the substantia nigra are observed in patients with AD with major depression. Dopamine levels in such patients, however, are not significantly different than levels in AD patients without depression. Also, in AD, there is increased expression of monoamine oxidase type B, an enzyme responsible for dopamine metabolism. In other types of dementia like dementia with Lewy bodies (DLBs), the reduced dopaminergic activity is responsible for the parkinsonian symptoms observed (Zubenko et al. 1990).

Whether AD represents primary amyloidosis or is a consequence of an upstream primary defect is still an open debate. Amyloid precursor protein (*APP*) gene mutation, first demonstrated in 1991, was shown to cause early-onset autosomal dominant AD (Goate et al. 1991; Capozzo et al. 2017). Subsequently, mutations in the presenilin genes were also shown to give rise to an autosomal dominant variant (Levy-Lahad et al. 1995). However, there are equally good reasons to believe that AD represents amyloidosis secondary to a primary inciting event. For example, there is an increased formation of an amyloidogenic derivative, representing alterations in the processing of APP when the electron transport chain (ETC) enzyme cytochrome oxidase is inhibited (Swerdlow 2007). Here, amyloidogenesis is secondary to enzyme inhibition. Indeed, Alzheimer himself believed that the histopathologic features of AD were not the primary cause of the disease, but rather were the result of an upstream process (Davis and Chisholm 1999). A number of hypotheses, each with its inherent assumptions as to whether AD is primary or secondary amyloidosis, are discussed below.

2.2.1 Amyloid Cascade Hypothesis

This hypothesis assumes primary amyloidosis as the pathology in AD. Altered processing of APP is recognized as the key etiologic factor. A class of enzymes known as the secretases is responsible for APP processing (Davis and Chisholm 1999). The physiologic pathway for APP processing involves cleavage by an enzyme known as α-secretase. This results mainly in the formation of sAPP (soluble APP), which is thought to have neurotrophic functions. Cleavage by α-secretase precludes subsequent β/γ-secretase proteolytic activity (Swerdlow 2007). A small proportion of APP is, however, physiologically cleaved sequentially by β/γ-secretase enzymes to give rise to Aβ protein. The activity of γ-secretase is highly variable and may cleave at different points, hence generating two main types of Aβ oligomers, $A\beta_{40}$ and $A\beta_{42}$. Physiologically, the predominant form is usually $A\beta_{40,}$ which is weakly amyloidogenic. Mutations in the *APP* gene increase the flux of APP through amyloidogenic pathway, and tamper with the preferred cleavage point, favoring the formation of the more amyloidogenic $A\beta_{42}$ oligomer (O'Brien and Wong 2011).

Presenilin proteins, the translational products of the presenilin genes, are components of the catalytic core of the γ-secretase enzyme. Mutations in these genes increase activity of the enzyme and consequently increase the formation of the $A\beta_{42}$ oligomer (Li et al. 2016). Increased formation of the $A\beta_{42}$ oligomer eventually results in amyloid plaque formation.

It is not known precisely how *ApoE4* gene contributes to late-onset AD. One hypothesis is that the cysteine to arginine substitution found in ApoE4 may compromise its ability to neutralize the effects of oxidative stress (Miyata and Smith 1996). Also, pathologically, the *ApoE4* allele is strongly associated with increased Aβ deposition into the brain (Tiraboschi et al. 2004). Some authors believe the association may be explained entirely by this effect on Aβ deposition (O'Brien and Wong 2011). Finally, some believe the effects of ApoE variations on cholesterol transport (Poirier 2000), and direct effects on Aβ cerebrovascular transport might explain the relationship (DeMattos et al. 2002).

It is believed that the formation of amyloid plaques causes neurotoxicity by inducing oxidative stress and subsequent apoptosis (Deshpande et al. 2006). Another possible mechanism is mediated via neurofibrillary tangles (NFTs), discussed under the tau hypothesis of AD pathogenesis. With regard to the amyloid cascade hypothesis, controversy exists as to which form of the Aβ protein, single oligomer or multimeric complexes of AB, is mainly responsible for the neurotoxicity observed (Shankar et al. 2008). Also, recent versions of the hypothesis advocate that extracellular Aβ, rather than what is sequestered in amyloid plaques, actually drives the disease (Lesné and Kotilinek 2005).

2.2.1.1 Limitations of the Amyloid Cascade Hypothesis

The main limitation of the amyloid cascade hypothesis is that clinical trials using immunotherapeutic approaches have not been promising. In a study by Schenk and colleagues, immunization of AD transgenic mice with Aβ protein did not only prevent plaque formation (Schenk et al. 1999). However, trials in humans had to be terminated because of neuroinflammatory complications. Later on, autopsy findings in the brain of the eight patients who received the vaccine revealed that Aβ pathology was virtually absent in three of the patients (Holmes et al. 2008). However, throughout their life after the trial participation, they still suffered from symptoms of the disease, which may suggest that amyloidosis is probably merely secondary to a more fundamental underlying etiology (Holmes et al. 2008). In addition, results from a phase III clinical trial involving the use of semagacestat, a γ-secretase inhibitor, failed to show effect in the management of AD, ergo raising questions regarding the relevance of γ-secretase in AD dementia.

Lastly, amyloid imaging has shown disparities in the deposition of amyloid plaques in the brains of healthy and AD subjects. These images have provided evidence of the presence of significant amyloid deposits in many healthy brains, and also few deposits in some AD brains. Moreover, unlike other markers of AD progression, such as loss of synapses and tau pathology, the deposition of β-amyloid plateaus, despite declining cognitive function (Engler et al. 2006). Thus, Aβ deposition and neurodegeneration may be unrelated events, and the former may likely be a phenomenon of normal aging (Chételat et al. 2013).

2.2.2 Tau Hypothesis

The tau protein is one of the microtubule-associated proteins that regulate the stability of tubulin assemblies (Kametani and Hasegawa 2018). These microtubules serve as channels for the transport of materials along nerve axons and are considered the cell's transportation and skeletal support system. The tau hypothesis assumes a tauopathy as the etiology of AD. In this hypothesis, there is abnormal phosphorylation of tau by a variety of kinases including glycogen synthase kinase-3β (GSK-3β) and cyclin-dependent kinase 5 (CDK5). This results in tau dissociation from microtubules and intracellular deposition, as paired helical fragments. Lack of functional microtubules accelerates cell death, which leads to the release and aggregation of filaments as extracellular neurofibrillary tangles (Zempel and Mandelkow 2014). Tau phosphorylation is enhanced by the presence of Aβ, possibly by activation of kinases. Conversely, hyperphosphorylated tau favors the formation of Aβ deposits demonstrating an intricate connection between the tau and amyloid cascade hypotheses. The tau hypothesis purports that the principal causative substance of AD is tau (Kametani and Hasegawa 2018). The hyperphosphorylated tau induces destabilization of actin (Fulga et al. 2007), synaptic impairment, and defects in mitochondrial integrity (DuBoff et al. 2012). Hence, tau pathology results in defects in mitochondrial integrity, transport, cytoskeletal, and signaling systems of the cell (Kametani and Hasegawa 2018).

2.2.2.1 Limitations of the Tau Hypothesis

Mutations in the tau gene are associated with another form of dementia, frontotemporal dementia, with parkinsonism (Wider and Wszolek 2008). Although this proves that primary tauopathy can drive neurodegeneration, it does not prove to be a significant factor that drives AD pathogenesis. Thus, tau has not been recognized as a cause of specific AD phenotype (Swerdlow 2007).

2.2.3 Mitochondrial Cascade Hypothesis

The mitochondrial cascade hypothesis assumes that similar physiologic mechanisms underlie AD and brain aging (Swerdlow 2007). It states that, because mitochondrial dysfunction observed in AD is systemic, it cannot simply be a manifestation of neurodegeneration. This hypothesis takes the stance that brain mitochondrial dysfunction is what drives amyloidosis and tau phosphorylation (Swerdlow 2007). According to the free radical theory of aging, cells accumulate damaged products over time from oxidative by-products, and the mitochondria are known sites of free radical production. In multiple AD tissues, defects of three mitochondrial enzymes are reported: reduced activities of pyruvate dehydrogenase complex, α-ketoglutarate dehydrogenase complex, and cytochrome oxidase (Wellings et al. 2013). These mitochondrial defects result in oxidative stress, which drives neurodegeneration. It has been shown that sodium azide, a cytochrome oxidase inhibitor, alters APP processing toward amyloidogenesis under in vitro conditions (Gabuzda et al. 1994). Hence, it is believed that the mitochondrial defects, apart from causing oxidative damage, may also cause proteasomal dysfunction, thus facilitates the increased Aβ formation and disease progression.

Another thing to note if we are to comprehend the mitochondrial cascade is to note that neuronal cell cycle re-entry is a notable feature of AD tissue. Cell cycle re-entry is a phenomenon where differentiated, non-dividing cells manifest molecular changes associated with cell division. There is increased expression of cyclin-dependent kinase (CDK) and DNA content. The AD neurons may get as far as the G_2 state, which immediately precedes mitosis. However, they are unable to undergo mitosis, leaving them in a state called G_2-M arrest (Swerdlow 2007).

In summary, this theory states that mitochondrial durability is determined by inherited mitochondrial ETC differences. More durable mitochondria remain functional much longer than less durable ones, which determine how well an individual ages. Eventually, mitochondrial function deteriorates and the balance between aerobic and anaerobic metabolism tips in favor of anaerobic metabolism. The increased dependence on anaerobic metabolism triggers cell responses, which cause cell cycle re-entry (Swerdlow et al. 2010). This increases CDK and GSK activities, enzymes associated with cell division, which cause hyper-phosphorylation of tau. Furthermore, proteasomal dysfunction as a result of mitochondrial defects results in amyloidosis, which further alters aerobic metabolism and facilitates the process. Ultimately, the neuronal de-differentiation response fails and neurodegeneration results (Swerdlow 2007).

2.2.3.1 Limitations of the Mitochondrial Cascade Hypothesis

This hypothesis states that inherited mitochondrial DNA (mtDNA) differences account for the differences in cytochrome oxidase activity between AD and normal tissues. However distinct homoplasmic or high abundance heteroplasmic mutations of mtDNA cytochrome oxidase account for a very small percentage of AD (Hamblet et al. 2006). Ergo, it is possible that mtDNA polymorphisms may account for this, but such a specific association with regard to cytochrome oxidase has not yet been established.

Also, according to the mitochondrial cascade hypothesis, AD must exhibit clear systemic manifestations, independent and not driven by the CNS effects. Although many studies and reports have been made to show this, such studies were observational or cross-sectional and hence cannot firmly establish causality. Also, it is questionable that such systemic manifestations do not manifest before the onset of neurocognitive symptoms. They may most likely be driven, at least in part, by progressive functional and behavioral decline associated with neurodegeneration (Morris et al. 2014).

2.2.4 NMDA Receptor-Mediated Glutamate Excitotoxicity

Glutamate is the most abundant excitatory neurotransmitter in the mammalian CNS. It elicits its actions by binding to glutamate receptors (Wang and Reddy 2017). There are both metabotropic and ionotropic subtypes. The vast majority of its excitatory neurotransmission is mediated by its ligand-gated ionotropic receptors, which play a role in learning and memory (Riedel et al. 2003). Excessive stimulation of glutamatergic signaling causes excitotoxicity, which damages and kills neurons. The toxicity of excessive glutamatergic stimulation is mediated by excessive calcium entry through NMDA receptor channels. NMDA receptors have a much higher permeability for calcium ions compared to other ionotropic glutamate receptors. The major factors that affect NMDA receptor signaling in AD have to do with glutamate availability and NMDA receptor modulation (Wang and Reddy 2017).

In AD, there is a decrease in the glutamate transporter capacity and protein expression as well as a selective loss of vesicular glutamate transporter (Masliah et al. 1996). Studies that employed a variety of AB peptides in neuronal cell culture demonstrated that AB amyloid protein may impair the glutamate uptake/recycling mechanisms (Fernández-Tomé et al. 2004). These processes increase extracellular glutamate concentration, which leads to excitotoxic actions and neurodegeneration. AB proteins have also been shown to amplify NMDA-mediated synaptic currents and collateral toxicity. AD also has influence on NMDA co-agonists. It is known that binding of NMDA receptors by glutamate solely cannot activate the receptor. Aside from prior activation of AMPA receptors to alleviate the physiologic voltage-dependent Mg^{2+} blockade, a co-agonist (either glycine or D-serine) must bind simultaneously with glutamate to activate NMDA receptors. In the hippocampus of AD patients, there are increased levels of D-serine, as well as increased expression of serine racemase, which converts the S-enantiomer into the D-enantiomer (Wu et al. 2004).

The integrity of presynaptic neurotransmitter release machinery is compromised in AD. Apparently, AB amyloid reduces expression of presynaptic proteins like synaptophysin, synaptotagmin, syntaxin, and proteins, which are important in synaptic vesicle release (Wang and Reddy 2017). This mechanism cannot contribute to excitotoxicity since it would presumably result in reduced instead of increased synaptic concentrations of glutamate. Its essence is merely to support other observations about loss of synaptic transmissions made in AD, and is likely a characterizing rather than an etiologic feature.

2.2.4.1 Limitations of the NMDA Receptor-Mediated Glutamate Excitotoxicity

When first formulated, it was hoped that therapy targeting/antagonizing NMDA receptors would slow disease progression. This hope has not been realized. It may be because most of the effects mediated by NMDA receptor activation depend to a large extent on AB amyloid protein effects. In contrast, no formulation exists, which shows how most of the histopathologic features of AD (Aβ deposits, neurofibrillary tangles) arise from excitotoxic effects, suggesting that it is unlikely glutamate excitotoxicity is a sole or major driver of the disease process, and it likely plays a minor role.

2.2.5 Cholinergic Hypothesis

Basal forebrain cholinergic cell loss is a consistent feature of AD. In AD, there is also a decreased concentration of choline acetyltransferase in the hippocampus and neocortex (Wallace et al. 1993). Acetylcholine is important for memory, learning, and other cognitive functions, explaining the neurocognitive decline associated with loss of cholinergic neurons (Garand et al. 2000). Acetylcholine content and acetylcholinesterase and choline transport in the cortex and hippocampus are all reduced considerably in AD but not in other disorders, such as depression or schizophrenia.

Muscarinic receptor density, determined by binding studies, is not affected, but nicotinic receptors, particularly in the cortex, are reduced. Impaired cortical cholinergic neurotransmission also contributes to AB pathology and increases phosphorylation of tau protein (Mohandas et al. 2009).

2.2.5.1 Limitations of the Cholinergic Hypothesis

There are obvious flaws in the cholinergic hypothesis. Neurotransmitter deficits are a result of neurodegeneration and loss of cells in dominant neurotransmitter pathways. There is a loss of synapses in AD as well. Levels of other neurotransmitters, though not as profoundly as acetylcholine, are altered also in AD. Hence, it is unlikely that merely increasing the neurotransmitter levels would slow down or reverse disease process (Garand et al. 2000).

2.2.6 Oxidative Stress Hypothesis

Reactive oxygen species (ROS) are mainly generated from by-products of the ETC of the mitochondria and cause oxidative injury and cell death (Mohandas et al. 2009). In AD, suppression of the ETC causes accumulation of electrons in complex I and coenzyme Q (CoQ). The electrons may be passed on to molecular oxygen to form the superoxide radical. This radical may react with NO to form peroxynitrate radical ($ONOO^-$). H_2O_2 in the presence of transition metals is converted to toxic $^{\bullet}OH$ radical. The oxidative damage caused by these reactive species damage lipids, proteins, and nucleic acids, all of which are essential for the structure and function of neurons (Smith et al. 2000).

Oxidative stress is intricately linked to many of the dominant hypothesis of AD pathogenesis. Discussions above show how the mitochondrial cascade hypothesis is related to oxidative stress. Many associations with the amyloid cascade hypothesis have been established as well. In various AD transgenic mouse models carrying mutants of APP and PS-1, increased H_2O_2 and NO production along with oxidative modifications of proteins and lipids were observed (Matsuoka et al. 2001). Oxidative stress also decreases α-secretase activity, while the activity of β/γ-secretases is enhanced, increasing the flux of APP through the amyloidogenic pathway (Zhao and Zhao 2013).

Oxidative stress is linked to the tau hypothesis as well. Cells overexpressing the tau protein have an increased susceptibility to oxidative effects. In a *Drosophila* model of human tauopathies expressing a disease-related mutant form of human tau, reduction in gene dosage of thioredoxin reductase or mitochondrial SOD2 enhanced tau-induced neurodegeneration and apoptosis (Zhao and Zhao 2013).

Finally, oxidative stress is associated with metallobiology and the metal homeostasis theory, which highlights the role of metal homeostasis in the pathogenesis of AD. Abnormal levels of metals like Cu, Zn, and Fe have been observed in AD hippocampus and amygdala, as well as within amyloid plaques as well. These metals have a role in Aβ aggregation. The precise cause and nature of these metal imbalances, however, are unknown. However, it is postulated that the interaction between amyloid plaques and these metals may be a source of ROS (Zhao and Zhao 2013). Many of the other theories revolve around the ROS theory because many of them inevitably assume ROS-mediated neuronal dysfunction, either directly or indirectly (Mohandas et al. 2009).

2.2.6.1 Limitations of the Oxidative Stress Hypothesis

Oxidative effects seem to be the final common pathway by which most of the other hypotheses realize neurodegeneration and cell damage (Mohandas et al. 2009). This makes the effect of oxidative stress on neurodegeneration likely secondary to an upstream primary cause. Indeed, oxidative stress as a primary cause of AD has much more scope and grounds when it is incorporated into other theories. The most influential of such theories currently is the mitochondrial cascade hypothesis.

2.3 Infection Theory

AD is the major cause of clinical dementia. In recent years, infectious pathogens such as bacteria, viruses, fungi, and protozoa have been implicated in the pathogenesis of AD (Sochocka et al. 2017). Three different studies conducted in the 1990s gave birth to the infection theory and confirmed the association between microorganisms and AD progression. There were toxic accumulation of Aβ and phosphorylated (*p*)-tau protein in the brain parenchyma of elderly patients with herpes simplex virus (HSV)-1 infection (Jamieson et al. 1991). Postmortem histopathological studies of established AD patients revealed spirochetes in the blood, cerebrospinal fluid, and brain tissue (Uldry et al. 1993). The final study by Balin and colleagues also found *Chlamydia pneumonia* in the autopsy brain samples of AD patients (Balin et al. 1998). In 2008, systemic infection with *Chlamydia pneumonia* was found to be associated with a fivefold increase in AD pathologies (Balin et al. 2008).

A recent AD study involving 128 AD patients with 135 healthy controls provided evidence to support the causal relationship between infectious agents, such as *Helicobacter pylori* and *Herpes simplex virus*, and AD (Bu et al. 2015). A strong link has also been established between increased burden of infectious agents and elevated levels of serum inflammatory cytokines and serum Aβ markers in AD patients. Chronic exposure to these pathogens may result in infections, which can lead to both cardiovascular and cerebrovascular disorders that may ultimately result in the development of AD (Strandberg et al. 2003; Li et al. 2011).

In a retrospective cohort study conducted by Francis Mawanda and his team, the researchers concluded that extra-CNS bacterial infections including septicemia, pneumonia, bacteremia, osteomyelitis, cellulitis, and urinary tract infections (UTIs) were significantly associated with the risk of developing dementia in a large representative sample of US veterans aged 56 years and older (Mawanda et al. 2016). The authors also suggested that the association between systemic CNS bacterial infections and the risk of dementia was higher for severe or systemic infections compared with less severe or localized extra-CNS bacterial infection.

Adjustments were made to take care of confounding parameters such as demographic characteristics and medical and psychiatric comorbidities (Mawanda et al. 2016).

Systemic bacterial infections like periodontitis and viral infections are predisposing factors in the development of AD. Chronic and persistent infections result in low-grade chronic inflammation, which produces exaggerated activation of microglial cells and dysfunction of pro-inflammatory immune response (McGeer et al. 1988). This is clinically relevant especially in the elderly population who are more predisposed to systemic infections and have a weakened immune system that is prone to develop a pro-inflammatory rather than an anti-inflammatory cell phenotype (Liu et al. 2012).

Humans are constantly exposed to a variety of exogenous infectious agents and are affected by a myriad of chronic ailments as we age. Many infectious agents have adverse effects on the physiology of the CNS. Poor treatment of these infectious agents in a human host may produce dysfunction in the central nervous system. Some HIV patients may develop dementia, which shares many similar characteristics as that observed in AD patients. West Nile virus infection may cause prolonged cognitive decline, whereas periodontitis is known to aggravate cognitive decline in AD patients (Holmes et al. 2008; Schroeder and Bäckhed 2016; Vasek et al. 2016).

With the advent of combination antiretroviral therapy (cART), the incidence and prevalence of opportunistic infections associated with human immunodeficiency virus 1 (HIV-1) infection have decreased considerably. However, the occurrence of HIV-associated neurocognitive disorder (HAND) remains unchanged with the use of cART even though the concomitant severe dementia manifestations associated with HAND have reduced significantly. As a result, the cerebral manifestation of HIV infection with respect to disturbances of cognition, behavioral, and autonomic functions is of clinical relevance in the management of HIV-1 patients. The most characteristic clinical feature of HAND is subcortical dementia coupled with a decline in concentration, attention, and memory. Mild neurocognitive disorder (MND) and asymptomatic neurocognitive impairment (ANI) are other presentations of HAND. In treatment-naïve patients, it is imperative to start combination antiretroviral therapy cART as infiltration of the HIV-1 is assumed to cause HAND and cART lowers viral load in the brain parenchyma and the cerebrospinal fluid. HAND patients who are already on cART should have their cART medications modified to include a neuroactive agent like indinavir capable of penetrating the CNS. A lumbar puncture can also be carried out if there is ongoing viral replication while on cART to detect drug-resistant virus (Eggers et al. 2017).

Behavioral deficits, an outcome of some systemic infections and inflammatory responses, are known to precede cognitive decline in AD patients (Dantzer et al. 2008). AD patients who present with delirium, an acute and transient impairment of cognitive function that negatively affects consciousness, attention, spatial orientation, and perception, suffer from exacerbated cognitive deficits (Fong et al. 2009). It is highly prevalent in the aged and demented population, and it is widely accepted that episodes of delirium hasten cognitive and functional decline and is predictive of increased morbidity and mortality (Irving et al. 2006; MacLullich et al. 2009). Recent evidence suggests that systemic inflammation caused by infections, surgery, or injurious insult can provoke episodes of delirium in the elderly or demented patients, which in turn worsens cognitive deficits and neurodegeneration (Beloosesky et al. 2007).

Additionally, emerging evidence from animal and epidemiological studies indicates a causal link between extra-CNS inflammation caused by systemic bacterial infections, and the pathogenesis of dementia. Most notably, the induction of systemic inflammation by extra-CNS injection of bacteria lipopolysaccharide was found to trigger and exacerbate the development and propagation of dementia neuropathology in transgenic mice (Sheng et al. 2003; Kitazawa et al. 2005; Sy et al. 2011). Several other studies indicate that systemic inflammation can provoke various cognitive deficits such as impaired memory (Valero et al. 2014), attention (Holden et al. 2008), and executive function (Culley et al. 2014). These findings and observations support the proposition that extra-CNS infection can induce and exacerbate the development of dementia neuropathology.

The accumulation of misfolded Aβ is pivotal in the complex pathophysiological processes that precede AD pathology. The role of the Aβ as a physiological antimicrobial agent was attributed to Aβ oligomerization, which is a known process that leads to neurodegeneration. The presence of infection may therefore enhance the oligomerization process in an attempt to potentiate the antimicrobial activity of the peptide (Kumar et al. 2016). In order to exhibit its antimicrobial properties, Aβ binds to a microbe and entraps it by forming amyloid fibrils. The presence of microbes therefore provides a suitable surface for nucleation of amyloid aggregates and subsequent amyloid deposition (Golde 2016).

Bacterial and viral infection may also compromise the integrity of the blood–brain barrier (BBB) and lead to diffuse cerebral dysfunction following a systemic inflammatory response with or without direct CNS infection (Cain et al. 2017; Al-Obaidi and Desa 2018). Impaired BBB permeability causes substantial alteration in consciousness by promoting the entry of pro-inflammatory cytokines like tumor necrosis factor α (TNF-α), IL-1β, and chemokines from the periphery into the CNS to cause brain dysfunction (Fig. 2.1) (Semmler et al. 2008).

2.3.1 Infection-Induced Systemic Inflammation and Neuroinflammation

Studies have shown that dysfunction of the central nervous system immune response is paramount in the development and progression of neurodegenerative diseases such as dementia. Current evidence emphasizes neuroinflammation as an integral and prominent contributory factor in the etiology of AD (Fig. 2.1) (Heppner et al. 2015). These studies have provided evidence, which suggests that both acute and chronic systemic inflammations are associated with CNS events that increase cognitive decline, including reduction in hippocampal volume. Evidence from preclinical studies using animal models indicates an association between the degree of

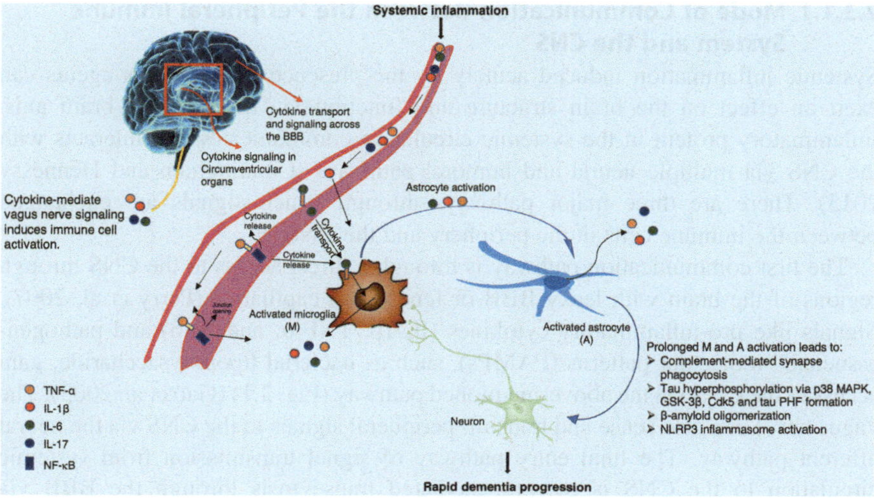

Fig. 2.1 The inflammatory milieu of systemic inflammation, neuroinflammation, and AD-specific pathways. Events such as infection cause systemic inflammation, that is characterized by increased circulating levels of soluble inflammatory mediators (e.g., IL-1β, IL-6, TNF-α, and IL-17). These pro-inflammatory cytokines may cross the BBB, signal across the glial barrier in circumventricular organs such as the area postrema, and also stimulate the vagus nerve to induce signaling within the CNS. Moreover, these pro-inflammatory cytokines can bind to and activate receptors found on endothelial cells, leading to the release of cytokine into the bloodstream, opening of tight junctions between endothelial cells, and the release of cytokines into the brain. Vagal stimulation by peripheral pro-inflammatory cytokines may regulate glutamatergic activation of neural immune cells (microglia and astrocytes), inducing the release of pro-inflammatory cytokines and chemokines in the central nervous system. It is believed that, via Aβ oligomerization, increased tau hyperphosphorylation due to upregulation of kinases, complement-mediated synaptic phagocytosis by microglia, and NLRP3 inflammasome activation, persistent activation of these cells promote neurodegeneration, Alzheimer's disease-related pathologies, and ultimately dementia. *BBB* blood–brain barrier, *CNS* central nervous system, *IL* interleukin, *GSK-3β* glycogen synthase kinase 3β, *NF-κB* nuclear factor kappa light chain enhancer of activated B cells, *NLRP3* NOD-, LRR-, and pyrin domain-containing protein 3, *p38 MAPK* p38 mitogen-activated protein kinase, *PHF* paired helical filaments, *BBB* blood–brain barrier. (Adapted from Walker et al. (2019a, b))

extra-CNS inflammation and the activation of central inflammatory processes that result in hippocampal neurodegeneration and subsequent memory impairment. An inverse association, however, has also been observed between peripheral levels of interleukin-6 (IL-6), a relatively stable biomarker of systemic inflammation and memory function in midlife adults (Yaffe et al. 2004; Marsland et al. 2008).

Systemic production of an inflammatory marker C-reactive protein (CRP) by the liver cells and the pro-inflammatory cytokine TNF-α by macrophages is characteristic of both acute and chronic systemic inflammations. TNF-α modulates the central innate immune response by activating microglial cells, which promotes communication between the immune system and the brain (Perry et al. 2007).

2.3.1.1 Mode of Communication Between the Peripheral Immune System and the CNS

Systemic inflammation induced acutely by the presence of infectious agents can exert an effect on the brain structure and function via the immune–brain axis. Inflammatory protein in the systemic circulation communicates and interacts with the CNS via multiple neural and humoral pathways (Cunningham and Hennessy 2015). There are three major pathways through which signals are exchanged between the immune cells in the periphery and the CNS.

The first communication pathway is through a direct access to the CNS through regions of the brain with leaky BBB or fenestrated capillaries (Perry et al. 2007). Signals like pro-inflammatory cytokines (IL-1β, TNF-α, and IL-6) and pathogen-associated molecular patterns (PAMPs), such as bacterial lipopolysaccharide, gain access to the CNS via the above-mentioned pathway (Fig. 2.1) (Gao et al. 2003). The vagus nerve can also sense and transmit peripheral signals to the CNS via the neural afferent pathway. The final entry pathway of signal transmission from systemic circulation to the CNS is receptor-mediated transcytosis through the BBB via numerous components like endothelial cells (Rivest 2009). Irrespective of the specific mechanism involved, inflammation outside the CNS is believed to amplify inflammatory signals within the CNS, promoting the activation of microglia and astrocytes and their transformation into a neurotoxic pro-inflammatory phenotype (M_1 and A_1, respectively) (Fig. 2.1) (Hennessy et al. 2015).

2.3.2 Infection-Induced Systemic Inflammation and Blood-Brain Barrier Integrity

The BBB is the structural and functional barrier that maintains the highly specialized microenvironment of the central nervous system and modulates communication signals between the brain and the systemic compartment. The BBB structure comprises specialized brain capillaries that possess polarized microvascular endothelial cells with tight junctions, which enables the BBB to maintain a homeostatically balanced environment in the brain. This barrier serves as a regulated interface between the peripheral systemic circulation and the central nervous system (Kniesel and Wolburg 2000). Tight junction proteins are responsible for controlling paracellular transport of molecular substances across the BBB, thus selective permeability of substances to the brain is the main function of the BBB. Additionally, the BBB inhibits the entry of pathogens and strictly controls the influx and efflux of molecules in and out of the brain parenchyma with the help of specialized transporters such as *p*-glycoprotein (Brito et al. 2014). This physiological function is, however, distorted by several CNS pathologies that involve bacterial infection (Hawkins and Davis 2005).

Neuropathological diseases including bacterial meningitis and sepsis are caused by bacterial pathogens. The ability of pathogens to induce meningitis in humans is related to the virulence factor that enables them to escape the host innate immunity, replicate within serum and closely interact with endothelial cells of the blood–

cerebrospinal fluid barrier (Join-Lambert et al. 2010; Iovino et al. 2013). The pathogenesis of bacterial meningitis is characterized by the infiltration into the BBB by bacteria from systemic circulation. In bacterial meningitis, there are increased blood levels of cytokines and chemokines in the patient (Moreillon and Majcherczyk 2003), destruction of endothelium barrier integrity caused by some meningeal pathogens via adherens junction (AJ)/TJ deformation, and subsequently, destruction of BBB (Van Sorge and Doran 2012). In pneumococcal meningitis, the levels of inflammatory mediators such as TNF-α, IL-1β, IL-6, and IL-10 rise in order to augment the immune response and aid in the elimination of the evading pathogen (Ostergaard et al. 2004). An earlier report indicated that the BBB integrity can be distorted due to the release and expression of cytokines, chemokines, and cell-adhesion molecules by invading microorganisms leading to the destruction of the BBB (Kim 2003).

However, contrary reports show some microorganisms like *Streptococcus pneumonia* can invade the endothelial cells without any destruction to the vascular endothelium or the BBB structure. It has been postulated that such pathogens translocate through the BBB through the intracellular or paracellular pathway (Iovino et al. 2013). *Listeria monocytogenes*, on the other hand, produces a toxin called listeriolysin that aids and promotes the expression of cell surface adhesion molecules P- and E-selectin, vascular cell-adhesion protein 1 (VCAM-1), and pro-inflammatory cytokines IL-6, IL-8, and monocyte chemoattractant protein-1 (MCP-1), facilitating neutrophil and monocyte adhesion to the endothelial cells, which ultimately promote BBB disruption (Kayal et al. 2002). It is believed that the complex interaction between the bacteria pathogen and the host is the main cause of neuronal injury and death. The proliferation of bacteria in the CSF causes enhanced BBB permeability that promotes the infiltration of inflammatory cells and the release of a vast range of inflammatory mediators. This in turn destroys neurons, cerebral edema, and secondary neuronal damage (Koedel et al. 2002). Current findings suggest that BBB dysfunction leads to various CNS diseases due to its propensity to cause neuroinflammation (Weiss et al. 2009).

There exist various bacteria–host interactions that facilitate the invasion of the brain by bacterial pathogens (Al-Obaidi and Desa 2018). Pathologies in the CNS cause recruitment of mononuclear leukocytes, monocytes, and macrophages that play supportive and protective roles to that provided by the resident microglia (Davoust et al. 2008). During BBB inflammation, mononuclear cells and neutrophils in circulation are attracted to the infection site, which causes penetration of the barrier and the production of cuffs in the perivascular space around small vessels where the immune response is actively coordinated by this perivascular space. Moreover, when the BBB is inflamed, cytokines and mononuclear cells are recruited to open the tight junctions between the endothelial cells and enter via transcellular and paracellular routes (Konsman et al. 2007).

In normal and non-inflamed BBB, mononuclear cells may penetrate transcellularly through the cytoplasm of the endothelial cells and not through the paracellular route, which involves tight junction opening (Kniesel and Wolburg 2000). IL-17 is capable of disrupting brain endothelium tight junctions, due to clusters of differentiation ($CD4^+$) IL-17-producing T lymphocytes (Th_{17}), which have been recognized as an etiological factor in AD progression (Bailey et al. 2007). Th_{17} lymphocytes can adapt and penetrate the BBB by secreting chemokines and acquiring dendritic cells in the infection sites. Th_{17} lymphocytes are implicated in infection and numerous inflammatory conditions. This immune mediator is capable of transmigrating efficiently across the endothelial cells of the BBB where it becomes neurotoxic and promotes CNS inflammation through $CD4^+$. Permeability of the BBB is thus linked to the infiltration of leukocytes into the brain leading to neuroinflammation (Kebir et al. 2007).

2.3.3 Systemic Inflammation and Alzheimer's Disease Progression

In a prospective study by Holmes et al. (2009), 300 subjects with mild-to-severe AD were cognitively assessed and blood samples were taken to assess for systemic inflammatory markers. Acute systemic inflammatory events in the subjects were associated with elevated serum levels of TNF-α and double the rate of cognitive impairment in subjects over 6 months. Acute systemic inflammation caused by systemic infection or tissue damage occurred in approximately half of the elderly subjects with AD over the 6-month follow period. Elevated baseline levels of TNF-α were linked to a fourfold increase in the rate of cognitive decline. Participants who had low serum levels of TNF-α throughout the study exhibited no signs and symptoms of cognitive decline over the 6 months. They concluded that both acute and chronic systemic inflammations cause elevation in the levels of TNF-α, with subsequent increase in cognitive decline in AD (Holmes et al. 2009).

Cunningham and colleagues performed an experiment using animal models of neuroinflammation, where the presence of neurodegenerative changes activated the brain microglia. It was shown in the study that experimentally induced acute systemic inflammation in such animal models causes an amplified central innate immune response leading to the release of cytotoxic inflammatory mediators with resultant worsening of neurodegeneration (Cunningham et al. 2005).

Further evidence by Solito and colleagues supports the proposition that systemic inflammatory stimuli including infection can activate the central immune response by activating microglia, which results in release of pro-inflammatory mediators like cytokines, chemokines, and lipid metabolites (Solito and Sastre 2012). This process subsequently prompts the autonomic and neuroendocrine responses to initiate tissue repair and healing. Under physiological conditions, resolution of the neuroinflammatory response is a trigger for reversion of the activated microglia into their quiescent states to assume their protective role and eliminate future pathogens and cellular debris (Solito and Sastre 2012). Contrary to this, current evidence suggests that previously "primed" microglia from early-life or sustained inflammatory events become hypersensitive upon responding to future inflammatory stimuli (Perry et al. 2007). Consequentially, a dysregulated and exaggerated immune response may occur in the brain parenchyma causing the release of neurotoxic mediator and cell lysis, which aggravates neurodegeneration.

2.3.4 Role of the Innate and Adaptive Immune System in Dementia Pathology

2.3.4.1 The Innate Immune System

Microglia form the primary immune effectors' cells of the CNS and are paramount cellular modulators of both the innate immune and neuroinflammatory responses in AD pathologies (Solito and Sastre 2012). Microglia possess immunoreceptors (IRs) and other cell surface receptors involved in the recognition of foreign molecules and antigens including pathogens and the activation of the innate immune responses (Linnartz and Neumann 2013). Pattern recognition receptors (PRRs) are forms of immunoreceptors that recognize both PAMPs and damage-associated molecular patterns (DAMPs). PAMPs are molecules and components derived solely from an invading pathogen, whereas DAMPs are intracellular entities released from damaged host cells. Aβ moieties secreted outside the cell into the extracellular compartment are known to bind to these immunoreceptors (Solito and Sastre 2012).

Polymorphism in genes encoding for triggering receptor expressed on myeloid cell-2 (TREM-2) has been identified as a risk factor in the etiology of AD. In mouse models for AD where these immunoreceptors including TREM-2 were genetically knocked out in microglial cells, there were alterations in Aβ deposition prompting the notion that microglia play a pivotal role in the pathogenesis of Aβ neuropathology (Jay et al. 2015; Wang et al. 2015). Interaction between Aβ with these immunoreceptors is postulated to activate intracellular signaling mechanisms and trigger inflammation and phagocytosis by microglia to eliminate Aβ. Thus, alteration of the immune state of microglia may significantly change the progression of AD pathology. Quiescent microglia may be activated to a classic neurotoxic M_1 immunological state, believed to worsen AD, or to a more neuroprotective M_2 immunological phenotype, which halts disease progression (Solito and Sastre 2012; Linnartz and Neumann 2013). Activated M_1 and A_1 glial cells can cause synaptic neuronal loss through complement-mediated phagocytosis of synapses and exacerbation of tau pathology through upregulation of kinases that hyperphosphorylate tau. Hyperphosphorylation of tau causes its misfolding and produces pro-inflammatory cytokines that can induce neuronal and glial cell death via inflammasome activation and autolysis (Walker et al. 2019a, b).

In summary, any etiological factor-like infection that can alter the innate immune responses or promote neuroinflammatory response via microglial activation can increase an individuals' risk of developing AD and dementia.

2.3.4.2 The Adaptive Immune System

Adaptive immunity offers long-lasting protection to the host against specific antigens contrary to the short-lasting antigen non-specific protection offered by the innate immune system. The role of the innate immune system at the site of neuroinflammation is well-established, but the role played by the adaptive immunity in AD pathology is not well-characterized, mainly due to the limited number of T or B cells in the brain parenchyma. Moreover, conflicting results were obtained from two different studies regarding the role of adaptive immune cells in neuroinflammation and AD pathology. PSAPP/Rag2$^{-/-}$ transgenic mice lacking B and T cells exhibited reduced Aβ pathology coupled with highly phagocytic microglia (Späni et al. 2015), whereas B, T, and natural killer (NK) cell-deficient −5xFAD/Rag2$^{-/-}$/Il2rg$^{-/-}$ mice showed increased Aβ plaque deposition and increased neuroinflammation (Marsh et al. 2016).

Nevertheless, a current genetic AD study illustrated a strong link between both the innate and adaptive systems in AD pathology (Gagliano et al. 2016). T-cell involvement has been shown to cause neurodegeneration in other diseases such as PD and amyotrophic lateral sclerosis (Sulzer et al. 2017; Sheean et al. 2018). The remarkable study conducted by Schenk et al. (1999), where Aβ immunization halted amyloid plaque formation, neuronal cell death, and gliosis in PDAPP mice, demonstrated the therapeutic potential of B-cell-mediated immune response and the role of the humoral immune response in AD pathologies. The detection of CD4^{+} and CD8^{+} T cells in the postmortem brains of AD patients, occasionally next to the neuritic plagues or microglia, also provides insight into a possible role of T cells in AD pathologies (Togo et al. 2002).

2.4 Infection and Cognitive Changes in Dementia

The risk for infection-associated disease complications worsens with age because of alterations in host immune response. An increase in the rate of cognitive decline is one such complication. Sometimes, the cognitive decline may be reversible once infection or the underlying neuroinflammation resolves. However, for some, this may lead to long-term cognitive impairment, which usually occurs in people who already are predisposed to developing dementia. Hence, infection is considered by some to unmask symptoms of incipient dementia (Shah et al. 2013). There is sufficient evidence, however, suggesting an infectious/immune component in dementia, particularly AD (Table 2.1).

Table 2.1 Evidence suggesting an infectious/immune component in dementia, particularly Alzheimer's disease

Evidence	References
• Viruses and other microbes are present in the brains of most elderly people, and though dormant may be reactivated when such individuals are stressed or immunosuppressed	Jamieson et al. (1991)
• Encephalitis caused by *Herpes simplex* infection damages regions of the limbic system and is associated with memory, cognitive and affective loss, and the personality changes observed in AD	Roos (2014)
• In the brain of AD patients, pathogen signatures (e.g., HSV1 DNA) coexist with AD histopathology	Roos (2014)
• Features of AD pathology may be transmitted by inoculation of AD brain tissue to primates	Baker et al. (1994)
• Features of AD pathology may be transmitted by inoculation of AD brain tissue to mice	Kane et al. (2000)

There is also evidence for a bidirectional relationship between immunologic dysfunction and dementia. People with dementia are at an increased risk of infections, while individuals who are immunocompromised are at an increased risk of dementia. The immune system thus has a role in both the onset and progression of AD (Shah et al. 2013). Infection evokes a systemic response that comprises elevated levels of circulating inflammatory mediators such as IL-6 and IL-1β. The systemic infection may act as a secondary stimulus to brain microglia, which are already primed due to the presence of Aβ proteins and degenerating neurons (Holmes et al. 2008). The increased level of cytokines in brain tissue initiates a positive feedback loop that results in worsening and accumulation of the histopathological features of AD, accompanied by the cognitive decline which is typical in AD (Holmes et al. 2008).

In young adults who are healthy, a strong response is mounted quickly against an invading pathogen or microbe, which leads to a fast recovery. Aging, however, results in a level of immune system dysfunction, leading to inadequate immune response upon exposure to a microbe. Subsequently, the body launches a very strong, albeit, late response that results in systemic inflammation. This inflammatory response extends to the CNS and drives cognitive decline, possibly by the disruption of synapses and neural communication pathways (Perry et al. 2007). The resultant cognitive decline leads to a number of cognitive symptoms like impairment in language, praxis, judgment, visuospatial function, and memory and related mental activities (Garand et al. 2000). A definite relationship has also been established between intellectual decline and impairment of daily living and presence of myoclonus and extrapyramidal signs in persons with AD (Ditter and Mirra 1987). Some non-cognitive symptoms may occur in conjunction with the cognitive symptoms. These include delusions, hallucinations, and disturbances in effect and behavior (Garand et al. 2000).

2.5 System Inflammatory Markers and Risk of Dementia

Numerous neuropathological studies circumstantially support the fact that inflammation plays a pivotal role in the pathogenesis of AD (Akiyama et al. 2000). Some of these studies revealed that in the diseased brain, there is activation of the complement cascade, upregulation of pro-inflammatory proteins, cytokines, chemokines, and their respective receptors, as well as activation of astrocytes and microglia (Akiyama et al. 2000). Pro-inflammatory cytokines have the propensity to modulate amyloid precursor protein (APP) expression, and subsequently, Aβ induces the release of inflammatory cytokines. Genetic polymorphism of several inflammatory mediators has been linked to the development of AD. Epidemiological studies also support the notion that the use of nonsteroidal anti-inflammatory drugs (NSAIDs) may delay the onset or slow the progression of AD (Dziedzic 2006).

Both epidemiological and translational researches support the idea that inflammation outside the CNS may aggravate neurodegeneration and other Alzheimer's disease-specific pathology within the brain parenchyma. These studies have shown that patients with AD and mild cognitive impairment tend to have elevated levels of cytokines, cytokine receptors, and other inflammatory markers such as IL-6, soluble TNF receptor-1, and CRP in their blood (Walker et al. 2019a, b).

Current findings by Walker et al. (2019a, b) suggest that people with elevated plasma levels of inflammatory mediators during midlife (decades before the typical age of dementia symptom onset) were at a greater risk of developing cognitive deficits during the decades leading up to older adulthood. Moreover, high baseline levels of inflammatory protein during midlife have been associated with increased susceptibility to subsequent dementia-induced decrease in brain volume, notably in gray matter regions vulnerable to AD pathology. These findings are consistent with animal studies, which have proven that exposure to endotoxins or simulated sepsis in animal models can lead to molecular brain changes, such as pro-inflammatory microglial and astrocytic activation, tau phosphorylation, and Aβ oligomerization (Walker et al. 2019a, b).

The Honolulu Asia Aging Study (HAAS) assessed the relationship between midlife CRP levels and the risk of late-age dementia. In this study, there was a significant association between raised CRP levels and an increased risk of vascular dementia (VD) and AD, with or without cardiovascular disease (CVD) contributing to dementia. The risk estimate was comparatively lower after adjusting for cardiovascular risk factors, which indicates that these factors partially explain the association between midlife CRP and late-life dementia (Schmidt et al. 2002).

In the Rotterdam study, elevated levels of inflammatory proteins, especially CRP and IL-6, were found in the plasma of AD patients 5 years prior to the clinical onset of dementia as compared to age-matched controls (Engelhart et al. 2004). In this study, α_1-antichymotrpsin, CRP, and IL-6 were associated with an increased risk of AD and VD (Engelhart et al. 2004). Another epidemiological cohort-based study, the MacArthur Studies of Successful Aging, is comprised of men and women 70–79 years who had good scores in screen cognitive and physical function tests. The full cognitive test included evaluations of language, verbal and spatial memory, abstract concept formation, and visuospatial skills assessed at 2.5 and 7 years from baseline. Subjects with the highest tertile of IL-6 had an increased risk of poor baseline cognitive function, as well as cognitive decline (Weaver et al. 2002; Ashraf et al. 2019).

2.6 Infection-Induced Systemic Inflammatory Mediators as Potential Therapeutic Targets

Dementia is a complex disorder. There is no single therapy up to date that can prevent or cure it. In AD treatment, the US Food and Drug Administration has approved a wide range of medications. Several evidences demonstrate that, by modulating the CNS immune response, which involves communication and signaling between microglia, astrocytes, and neurons, there is a chance to halt neuroinflammation and rescue neurodegeneration that occurs in AD pathologies (Prasad et al. 2002; Blasko and Grubeck-Loebenstein 2003).

2.6.1 Potential Novel and Conventional Pharmacological Agents

2.6.1.1 NSAIDs

Epidemiological studies have always indicated an inverse association between NSAIDs and the risk of developing dementia (Peila and Launer 2006). Many other studies postulate that NSAIDs minimize the risk of AD, possibly through the suppression of activated microglia. Patients who have rheumatoid arthritis and are receiving long-term anti-inflammatory treatment are comparatively less predisposed and have a reduced risk of developing AD. Recent evidence suggests that long-term use of NSAIDs reduces the inflammatory processes in the brain parenchyma by attenuating the production of Aβ peptides, which reduces the risk of developing AD (Szekely and Zandi 2010). NSAIDs have been postulated to not affect people with established AD, and treatment with COX-2 inhibitor has now been shown to increase Aβ levels in the brain. Nevertheless, observational studies in humans provide evidence to support that the use of NSAIDs is associated with minimal AD risk (Yip et al. 2005).

2.6.1.2 Lipoxin A4

The systemic administration of lipoxin A_4 (LXA_4) analog (aspirin-triggered LXA_4) to two separate AD mouse models, Tg2576 and 3 × Tg-AD, effectively converted microglia from the classical pro-inflammatory neurotoxic state to the neuroprotective anti-inflammatory phenotype by suppressing NF-κB signaling in astrocytes. Lipoxin A_4 is an endogenous lipid mediator with potent anti-inflammatory properties and is involved in the acceleration of inflammation resolution, maintenance of homeostasis, and modulation of immune response. Subcutaneous administration of aspirin-triggered lipoxin A_4 decreased NF-κB activation and the level of pro-inflammatory cytokines and chemokines, whereas the level of the beneficial anti-inflammatory cytokine, IL-10, was elevated. These consequentially recruit M_2 microglial phenotype, which promotes clearance of Aβ deposits and ultimately facilitates the reduction in synaptotoxicity and an improvement in cognition (Medeiros et al. 2013; Dunn et al. 2015).

2.6.1.3 Other Therapeutic Agents

Colostrinin, a proline-rich polypeptide (PRP), is a cytokine-like factor that induces interferon-gamma. In preclinical animal studies, colostrinin showed remarkable immunomodulatory properties and has also been shown clinically to improve the outcomes of AD patients with mild-to-moderate dementia (Leszek et al. 2002). A strategy to decrease the hydrolysis of synaptic acetylcholine has also been exploited with the use of acetylcholinesterase inhibitors (AChEIs) such as donepezil, galantamine, and rivastigmine. The use of these AChEIs is considered the first-line treatment for AD (Sochocka et al. 2008). Additionally, AChEIs contribute to the prevention and resolution of neuroinflammation via their action against free radicals and their ability to reduce cytokines from activated microglia, predominantly by promoting acetylcholine action on cholinergic receptors on glial cells. This is supported by findings demonstrating significant reduction in peripheral blood monocyte- and lymphocyte-derived cytokines in AD patients treated with donepezil for 1 month (Reale et al. 2005).

Donepezil showed anti-inflammatory properties in animal studies and ameliorated tau pathology, synaptic loss, and neurodegeneration in a tauopathy (Yoshiyama et al. 2010; Jiang et al. 2013). Donepezil has also been reported to directly attenuate inflammatory NF-κB signaling and reduce the levels of TNF-α and TNF-β in purified microglia (Hwang et al. 2010). Astrocytes and microglia bear cholinergic receptors, which upon activation decrease the release of cytokines from glial cells (Minghetti et al. 2007). The neutralization of TNF-α, a prominent cytokine implicated in RA in peripheral tissues, has shown promising results in AD therapy in animal models. Peripheral administration of XPro1595, a soluble TNF inhibitor changes the profile of immune cells, reduces the amount of Aβ load in mice and ultimately improves impaired long-term potentiation in mice models (MacPherson et al. 2017).

Several other attempts have been made in attenuating dysregulated inflammatory response in AD pathogenesis by pharmacologically targeting the actions of the pro-inflammatory cytokine, IL-1β (Kitazawa et al. 2011). Previous reports indicated that IL-1β increases the expression of APP and tau phosphorylation in both in vitro and in vivo experiments (Yang et al. 1998; Li et al. 2003). In a study conducted by Kitazawa et al. (2011), prolonged systemic administration of anti-IL-1 receptor blocking antibody in aged 3 × Tg-AD mice significantly ameliorated neuroinflammation, reduced tau pathology but not Aβ pathology, and halted cognitive impairment. This action was mediated through astrocytes but not microglia, by causing a decrease in S100B and concomitant restoration of β-catenin signaling in neurons that will enhance tau pathology.

2.6.2 Phytochemicals

Several natural products with anti-inflammatory activity have shown promising results in the treatment of pathogen-mediated systemic inflammation associated with AD pathology. Current evidence demonstrated the efficacy of natural immunosuppressive drugs in AD therapy (Newman and Cragg 2004; Meunier et al. 2006). In 2009, Abraham and Johnson proved that resveratrol, a polyphenol given with diet, alleviated infection-related neuroinflammation, cognitive, and memory deficits in aged mice models (Abraham and Johnson 2009). B-asarone obtained from *Acorus* species and genistein isolated from *Glycine max* exert protection against neuroinflammation by suppressing NF-κB signaling and JNK pathway (Lim et al. 2014; Qian et al. 2015). Another flavonoid from *Epimedium brevicornum*, icariin, exhibited enhanced PI3K/Akt activity, decreased tau protein and GSK-3β hyperphosphorylation, and inhibited Aβ 25–35-induced neurotoxicity in PC-12 cells via PI3K/Akt pathway (Zeng et al. 2010; Zhang et al. 2015). Other useful natural agents that are effective in managing dementia preclinically include melatonin, tripchlorolide, ursolic acid, and xanthoceraside (Dey et al. 2016).

2.7 Recent Developments and Future Perspectives

Systemic inflammation is one of the major causes of aging, and it affects brain structures relevant to dementia, predisposing the individual to functional decline as a result. It has been shown that, during midlife, the association between systemic infection and the occurrence of effects of aging such as impaired neurogenesis and cognitive decline is strong. This is evidenced by the protection that NSAIDs confer on patients against the development of AD and related dementia (Zhang et al. 2013). Recently, the safety and tolerability of etanercept in AD trial (STEADI-09 study) provided the evidence that peripheral blocking of the inflammatory mediator, TNF-α, using etanercept, a fusion protein TNF-α inhibitor, stabilizes cognitive function in AD patients (Butchart et al. 2015).

Recently, it has been discovered that inflammatory mediators altered adipokines and/or increased adiposity induce microgliosis, cytokine secretion (Gao et al. 2014), and neuronal dysfunction and degeneration in the hypothalamus (Thaler et al. 2012). These pathological changes contribute to furthering inflammatory signaling in the hypothalamus, most like the NFκB-IKK-β pathway, thus decreasing neurogenesis, accelerating aging, and increasing cognitive and functional deficits (Zhang et al. 2013). Ablation of Nlrp3, a key subunit of the inflammasome complex responsible for the secretion and regulation of IL-1β, has been shown to reduce functional deficits associated with aging. This technique reduces IL-1β secretion, thus improves glucose metabolism, decreases immune response in the brain such as microgliosis, and improves cognitive function (Youm et al. 2013).

2.8 Conclusion

Increasing evidence from epidemiological and animal studies supports the notion that both systemic inflammations secondary to infection and neuroinflammation are closely associated with the onset and progression of dementia and neurodegenerative disease like AD. Various studies concluded that persistent infection subjects an individual to chronic systemic inflammation in which ultimately predisposes the person to a higher risk of developing dementia in later life. This is mainly attributed to how systemic inflammation impacts the CNS immune system by favoring the activation of microglial and astrocytic cells into a pro-inflammatory state, which causes neural cell death, changes in neurochemical signaling, and synaptotoxicity. The activated microglia become dysregulated, which disrupts the fine balance between pro-inflammatory (IL-1β, IL-6, TNF-α) and anti-inflammatory cytokines (IL-10) produced in the CNS leading to dementia. Several serum peripheral inflammatory markers have been associated with dementia and cognitive decline. Various pharmacological and phytochemical agents targeting these pro-inflammatory cytokines and chemokines have shown promising effects in rescuing cognitive impairment in AD mice models. However, further studies are needed to identify novel therapeutic options. One such option is natural agents, which have proven preclinically to be effective in preventing and rescuing the spectrum of events associated with dementia. Nonetheless, other studies have reported the antimicrobial and anti-inflammatory activities of probiotics, which may serve as alternative therapeutic agents in AD treatment. In this regard, more clinical investigations are needed to develop more potent and safe therapies in dementia treatment.

References

Abraham J, Johnson RW (2009) Consuming a diet supplemented with resveratrol reduced infection-related neuroinflammation and deficits in working memory in aged mice. Rejuvenation Res 12(6):445–453

Akiyama H, Barger S, Barnum S et al (2000) Inflammation and Alzheimer's disease. Neurobiol Aging 21(3):383–421

Al-Obaidi MMJ, Desa MNM (2018) Mechanisms of blood brain barrier disruption by different types of bacteria, and bacterial–host interactions facilitate the bacterial pathogen invading the brain. Cell Mol Neurobiol 38(7):1349–1368

Ashraf GM, Tarasov VV, Makhmutova A et al (2019) The possibility of an infectious etiology of Alzheimer disease. Mol Neurobiol 56(6):4479–4491

Alzheimer's Association (2013) Alzheimer's disease facts and figures. Alzheimers Dement 9(2): 208–245

Bailey SL, Schreiner B, McMahon EJ, Miller SD (2007) CNS myeloid DCs presenting endogenous myelin peptides' preferentially' polarize CD4+ TH-17 cells in relapsing EAE. Nat Immunol 8(2):172–180

Baker HF, Ridley RM, Duchen LW, Crow TJ, Bruton CJ (1994) Induction of β(A4)-amyloid in primates by injection of Alzheimer's disease brain homogenate. Mol Neurobiol 8(1):25–39

Balin BJ, Gérard HC, Arking EJ et al (1998) Identification and localization of Chlamydia pneumoniae in the Alzheimer's brain. Med Microbiol Immunol 187(1):23–42

Balin BJ, Little CS, Hammond CJ et al (2008) Chlamydophila pneumoniae and the etiology of late-onset Alzheimer's disease. J Alzheimers Dis 13(4):371–380

Beloosesky Y, Hendel D, Weiss A et al (2007) Cytokines and C-reactive protein production in hip-fracture-operated elderly patients. J Gerontol Ser A Biol Sci Med Sci 62(4):420–426

Blasko I, Grubeck-Loebenstein B (2003) Role of the immune system in the pathogenesis, prevention and treatment of Alzheimer's disease. Drugs Aging 20(2):101–113

Brito MA, Palmela I, Cardoso FL, Sá-Pereira I, Brites D (2014) Blood–brain barrier and bilirubin: clinical aspects and experimental data. Arch Med Res 45(8):660–676

Bu X, Yao X, Jiao S, Zeng F et al (2015) A study on the association between infectious burden and Alzheimer's disease. Eur J Neurol 22(12):1519–1525

Burns A, Iliffe S (2009) Dementia. BMJ 338:b75

Butchart J, Brook L, Hopkins V et al (2015) Etanercept in Alzheimer's disease: a randomised, placebo controlled, double-blind phase 2 trial. Neurology 84(21):216–218

Cain MD, Salimi H, Gong Y et al (2017) Virus entry and replication in the brain precedes blood-brain barrier disruption during intranasal alphavirus infection. J Neuroimmunol 308:118–130

Capozzo R, Sassi C, Hammer MB et al (2017) Clinical and genetic analyses of familial and sporadic frontotemporal dementia patients in Southern Italy. Alzheimers Dement 13(8):858–869

Chételat G, La Joie R, Villain N et al (2013) Amyloid imaging in cognitively normal individuals, at-risk populations and preclinical Alzheimer's disease. NeuroImage Clin 2:356–365

Chou C-H, Lee J-T, Lin C-C et al (2017) Septicemia is associated with increased risk for dementia: a population-based longitudinal study. Oncotarget 8(48):84300–84308

Culley DJ, Snayd M, Baxter MG et al (2014) Systemic inflammation impairs attention and cognitive flexibility but not associative learning in aged rats: possible implications for delirium. Front Aging Neurosci 6:107

Cummings JL, Benson DF (1992) Dementia: a clinical approach, 2nd edn. Butterworth-Heinemann Medical, Boston
Cunningham C, Hennessy E (2015) Co-morbidity and systemic inflammation as drivers of cognitive decline: new experimental models adopting a broader paradigm in dementia research. Alzheimers Res Ther 7(1):1–13
Cunningham C, Wilcockson DC, Campion S, Lunnon K, Perry VH (2005) Central and systemic endotoxin challenges exacerbate the local inflammatory response and increase neuronal death during chronic neurodegeneration. J Neurosci 25(40):9275–9284
Dantzer R, O'Connor JC, Freund GG, Johnson RW, Kelley KW (2008) From inflammation to sickness and depression: when the immune system subjugates the brain. Nat Rev Neurosci 9(1): 46–56
Davis JN, Chisholm JC (1999) Alois Alzheimer and the amyloid debate. Nature 400(6747):810
Davoust N, Vuaillat C, Androdias G, Nataf S (2008) From bone marrow to microglia: barriers and avenues. Trends Immunol 29(5):227–234
DeMattos RB, Bales KR, Cummins DJ, Paul SM, Holtzman DM (2002) Brain to plasma amyloid-β efflux: a measure of brain amyloid burden in a mouse model of Alzheimer's disease. Science 295(5563):2264–2267
Desai AK, Grossberg GT (2005) Diagnosis and treatment of Alzheimer's disease. Neurology 64(12 suppl 3):S34–S39
Deshpande A, Mina E, Glabe C, Busciglio J (2006) Different conformations of amyloid β induce neurotoxicity by distinct mechanisms in human cortical neurons. J Neurosci 26(22):6011–6018
Dey A, Bhattacharya R, Mukherjee A, Pandey DK (2016) Natural products against Alzheimer's disease: pharmacotherapeutics and biotechnological interventions. Biotechnol Adv 35(2): 178–216
Ditter SM, Mirra SS (1987) Neuropathologic and clinical features of Parkinson's disease in Alzheimer's disease patients. Neurology 37(5):754
DuBoff B, Götz J, Feany MB (2012) Tau promotes neurodegeneration via DRP1 mislocalization in vivo. Neuron 75(4):618–632
Dunn HC, Ager RR, Baglietto-Vargas D et al (2015) Restoration of lipoxin A 4 signaling reduces Alzheimer's disease-like pathology in the 3xTg-AD mouse model. J Alzheimers Dis 43(3): 893–903
Dziedzic T (2006) Systemic inflammatory markers and risk of dementia. Am J Alzheimers Dis Other Dement 21(4):258–262
Eggers C, Arendt G, Hahn K et al (2017) HIV-1-associated neurocognitive disorder: epidemiology, pathogenesis, diagnosis, and treatment. J Neurol 264(8):1715–1727
Engelhart MJ, Geerlings MI, Meijer J et al (2004) Inflammatory proteins in plasma and the risk of dementia: the rotterdam study. Arch Neurol 61(5):668–672
Engler H, Forsberg A, Almkvist O et al (2006) Two-year follow-up of amyloid deposition in patients with Alzheimer's disease. Brain 129(11):2856–2866
Fernández-Tomé P, Brera B, Arévalo M-A, de Ceballos ML (2004) β-Amyloid25-35 inhibits glutamate uptake in cultured neurons and astrocytes: modulation of uptake as a survival mechanism. Neurobiol Dis 15(3):580–589
Fong TG, Jones RN, Shi P et al (2009) Delirium accelerates cognitive decline in Alzheimer disease. Neurology 72(18):1570–1575
Fulga TA, Elson-Schwab I, Khurana V et al (2007) Abnormal bundling and accumulation of F-actin mediates tau-induced neuronal degeneration in vivo. Nat Cell Biol 9(2):139–148
Gabuzda D, Busciglio J, Chen LB, Matsudaira P, Yankner BA (1994) Inhibition of energy metabolism alters the processing of amyloid precursor protein and induces a potentially amyloidogenic derivative. J Biol Chem 269(18):13623–13628
Gagliano SA, Pouget JG, Hardy J et al (2016) Genomics implicates adaptive and innate immunity in Alzheimer's and Parkinson's diseases. Ann Clin Transl Neurol 3(12):924–933
Gao H-M, Liu B, Hong J-S (2003) Critical role for microglial NADPH oxidase in rotenone-induced degeneration of dopaminergic neurons. J Neurosci 23(15):6181–6187

Gao Y, Ottaway N, Schriever SC et al (2014) Hormones and diet, but not body weight, control hypothalamic microglial activity. Glia 62:17–25

Garand L, Buckwalter KC, Hall GR (2000) The biological basis of behavioral symptoms in dementia. Issues Ment Health Nurs 21(1):91–107

George-Hyslop PHS (2000) Piecing together Alzheimer's. Sci Am 283(6):76–83

Goate A, Chartier-Harlin M-C, Mullan M et al (1991) Segregation of a missense mutation in the amyloid precursor protein gene with familial Alzheimer's disease. Nature 349(6311):704–706

Golde TE (2016) Host immune defence, amyloid-β peptide and Alzheimer disease. Nat Rev Neurol 12(8):433–434

Hamblet NS, Ragland B, Ali M, Conyers B, Castora FJ (2006) Mutations in mitochondrial-encoded cytochrome c oxidase subunits I, II, and III genes detected in Alzheimer's disease using single-strand conformation polymorphism. Electrophoresis 27(2):398–408

Harvey RJ, Skelton-Robinson M, Rossor MN (2003) The prevalence and causes of dementia in people under the age of 65 years. J Neurol Neurosurg Psychiatry 74(9):1206–1209

Hawkins BT, Davis TP (2005) The blood-brain barrier/neurovascular unit in health and disease. Pharmacol Rev 57(2):173–185

Hennessy E, Griffin ÉW, Cunningham C (2015) Astrocytes are primed by chronic neurodegeneration to produce exaggerated chemokine and cell infiltration responses to acute stimulation with the cytokines IL-1β and TNF-α. J Neurosci 35(22):8411–8422

Heppner FL, Ransohoff RM, Becher B (2015) Immune attack: the role of inflammation in Alzheimer disease. Nat Rev Neurosci 16(6):358–372

Holden JM, Meyers-Manor JE, Overmier JB, Gahtan E, Sweeney W, Miller H (2008) Lipopolysaccharide-induced immune activation impairs attention but has little effect on short-term working memory. Behav Brain Res 194(2):138–145

Holmes C, Boche D, Wilkinson D et al (2008) Long-term effects of Aβ42 immunisation in Alzheimer's disease: follow-up of a randomised, placebo-controlled phase I trial. Lancet 372(9634):216–223

Holmes C, Cunningham C, Zotova E et al (2009) Systemic inflammation and disease progression in Alzheimer disease. Neurology 73(10):768–774

Hwang J, Hwang H, Lee H-W, Suk K (2010) Microglia signaling as a target of donepezil. Neuropharmacology 58(7):1122–1129

Iovino F, Brouwer MC, van de Beek D, Molema G, Bijlsma JJE (2013) Signalling or binding: the role of the platelet-activating factor receptor in invasive pneumococcal disease. Cell Microbiol 15(6):870–881

Irving K, Fick D, Foreman M (2006) Delirium: a new appraisal of an old problem. Int J Older People Nurs 1(2):106–112

Jamieson GA, Maitland NJ, Wilcock GK, Craske J, Itzhaki RF (1991) Latent herpes simplex virus type 1 in normal and Alzheimer's disease brains. J Med Virol 33(4):224–227

Jay TR, Miller CM, Cheng PJ et al (2015) TREM2 deficiency eliminates TREM2+ inflammatory macrophages and ameliorates pathology in Alzheimer's disease mouse models. J Exp Med 212(3):287–295

Jiang Y, Zou Y, Chen S et al (2013) The anti-inflammatory effect of donepezil on experimental autoimmune encephalomyelitis in C57 BL/6 mice. Neuropharmacology 73:415–424

Join-Lambert O, Morand PC, Carbonnelle E et al (2010) Mechanisms of meningeal invasion by a bacterial extracellular pathogen, the example of Neisseria meningitidis. Prog Neurobiol 91(2): 130–139

Kametani F, Hasegawa M (2018) Reconsideration of amyloid hypothesis and Tau hypothesis in Alzheimer's disease. Front Neurosci 12:25

Kane MD, Lipinski WJ, Callahan MJ et al (2000) Evidence for seeding of beta -amyloid by intracerebral infusion of Alzheimer brain extracts in beta -amyloid precursor protein-transgenic mice. J Neurosci 20(10):3606–3611

Kayal S, Lilienbaum A, Join-Lambert O, Li X, Israël A, Berche P (2002) Listeriolysin O secreted by Listeria monocytogenes induces NF-κB signalling by activating the IκB kinase complex. Mol Microbiol 44(5):1407–1419

Kebir H, Kreymborg K, Ifergan I et al (2007) Human TH 17 lymphocytes promote blood-brain barrier disruption and central nervous system inflammation. Nat Med 13(10):1173–1175

Kim KS (2003) Pathogenesis of bacterial meningitis: from bacteraemia to neuronal injury. Nat Rev Neurosci 4(5):376–385

Kitazawa M, Oddo S, Yamasaki TR, Green KN, LaFerla FM (2005) Lipopolysaccharide-induced inflammation exacerbates tau pathology by a cyclin-dependent kinase 5-mediated pathway in a transgenic model of Alzheimer's disease. J Neurosci 25(39):8843–8853

Kitazawa M, Cheng D, Tsukamoto MR et al (2011) Blocking IL-1 signaling rescues cognition, attenuates tau pathology, and restores neuronal β-catenin pathway function in an Alzheimer's disease model. J Immunol 187(12):6539–6549

Kniesel U, Wolburg H (2000) Tight junctions of the blood–brain barrier. Cell Mol Neurobiol 20(1): 57–76

Koedel U, Scheld WM, Pfister H-W (2002) Pathogenesis and pathophysiology of pneumococcal meningitis. Lancet Infect Dis 2(12):721–736

Konsman JP, Drukarch B, Van Dam A-M (2007) (Peri) vascular production and action of pro-inflammatory cytokines in brain pathology. Clin Sci 112(1):1–25

Kumar DKV, Choi SH, Washicosky KJ et al (2016) Amyloid-β peptide protects against microbial infection in mouse and worm models of Alzheimer's disease. Sci Transl Med 8(340): 340ra72–340ra72

Lesné S, Kotilinek L (2005) Amyloid plaques and amyloid-beta oligomers: an ongoing debate. J Neurosci 25(41):9319–9320

Leszek J, Inglot AD, Janusz M et al (2002) Colostrinin proline-rich polypeptide complex from ovine colostrum--a long-term study of its efficacy in Alzheimer's disease. Med Sci Monit 8(10): PI93–PI96

Levy-Lahad E, Wijsman EM, Nemens E et al (1995) A familial Alzheimer's disease locus on chromosome 1. Science 269(5226):970–973

Li Y, Liu L, Barger SW, Griffin WST (2003) Interleukin-1 mediates pathological effects of microglia on tau phosphorylation and on synaptophysin synthesis in cortical neurons through a p38-MAPK pathway. J Neurosci 23(5):1605–1611

Li J, Wang YJ, Zhang M, Xu ZQ et al (2011) Vascular risk factors promote conversion from mild cognitive impairment to Alzheimer disease. Neurology 76(17):1485–1491

Li N, Liu K, Qiu Y et al (2016) Effect of presenilin mutations on APP cleavage; insights into the pathogenesis of FAD. Front Aging Neurosci 8:51

Lim HW, Kumar H, Kim BW et al (2014) β-Asarone (cis-2,4,5-trimethoxy-1-allylphenyl), attenuates pro-inflammatory mediators by inhibiting NF-κB signaling and the JNK pathway in LPS activated BV-2 microglia cells. Food Chem Toxicol 72:265–272

Linnartz B, Neumann H (2013) Microglial activatory (immunoreceptor tyrosine-based activation motif)-and inhibitory (immunoreceptor tyrosine-based inhibition motif)-signaling receptors for recognition of the neuronal glycocalyx. Glia 61(1):37–46

Liu X, Wu Z, Hayashi Y, Nakanishi H (2012) Age-dependent neuroinflammatory responses and deficits in long-term potentiation in the hippocampus during systemic inflammation. Neuroscience 216:133–142

MacLullich AMJ, Beaglehole A, Hall RJ, Meagher DJ (2009) Delirium and long-term cognitive impairment. Int Rev Psychiatry 21(1):30–42

MacPherson KP, Sompol P, Kannarkat GT et al (2017) Peripheral administration of the soluble TNF inhibitor XPro1595 modifies brain immune cell profiles, decreases beta-amyloid plaque load, and rescues impaired long-term potentiation in 5xFAD mice. Neurobiol Dis 102:81–95

Marsh SE, Abud EM, Lakatos A et al (2016) The adaptive immune system restrains Alzheimer's disease pathogenesis by modulating microglial function. Proc Natl Acad Sci 113(9):E1316–E1325

Marsland AL, Gianaros PJ, Abramowitch SM, Manuck SB, Hariri AR (2008) Interleukin-6 covaries inversely with hippocampal grey matter volume in middle-aged adults. Biol Psychiatry 64(6):484–490

Masliah E, Hansen L, Alford M, Deteresa R, Mallory M (1996) Deficient glutamate tranport is associated with neurodegeneration in Alzheimer's disease. Ann Neurol 40(5):759–766

Matsuoka Y, Picciano M, La Francois J, Duff K (2001) Fibrillar β-amyloid evokes oxidative damage in a transgenic mouse model of Alzheimer's disease. Neuroscience 104(3):609–613

Mawanda F, Wallace RB, McCoy K, Abrams TE (2016) Systemic and localized extra-central nervous system bacterial infections and the risk of dementia among US veterans: a retrospective cohort study. Alzheimers Dement 4:109–117

McGeer PL, Itagaki S, Boyes BE, McGeer EG (1988) Reactive microglia are positive for HLA-DR in the substantia nigra of Parkinson's and Alzheimer's disease brains. Neurology 38(8):1285

Medeiros R, Kitazawa M, Passos GF et al (2013) Aspirin-triggered lipoxin A4 stimulates alternative activation of microglia and reduces Alzheimer disease–like pathology in mice. Am J Pathol 182(5):1780–1789

Meunier J, Ieni J, Maurice T (2006) The anti-amnesic and neuroprotective effects of donepezil against amyloid β25-35 peptide-induced toxicity in mice involve an interaction with the σ1 receptor. Br J Pharmacol 149(8):998–1012

Minghetti L, Carnevale D, Simone RD (2007) Microglia-neuron interaction in inflammatory and degenerative diseases: role of cholinergic and noradrenergic systems. CNS Neurol Disord Targets 6(6):388–397

Miyata M, Smith JD (1996) Apolipoprotein E allele–specific antioxidant activity and effects on cytotoxicity by oxidative insults and β–amyloid peptides. Nat Genet 14(1):55–61

Mohandas E, Rajmohan V, Raghunath B (2009) Neurobiology of Alzheimer's disease. Indian J Psychiatry 51(1):55

Moreillon P, Majcherczyk PA (2003) Proinflammatory activity of cell-wall constituents from gram-positive bacteria. Scand J Infect Dis 35(9):632–641

Morris JK, Honea RA, Vidoni ED, Swerdlow RH, Burns JM (2014) Is Alzheimer's disease a systemic disease? Biochim Biophys Acta 1842(9):1340–1349

Newman DJ, Cragg GM (2004) Marine natural products and related compounds in clinical and advanced preclinical trials. J Nat Prod 67(8):1216–1238

O'Brien RJ, Wong PC (2011) Amyloid precursor protein processing and Alzheimer's disease. Annu Rev Neurosci 34:185–204

Ostergaard C, Brandt C, Konradsen HB, Samuelsson S (2004) Major articles and brief reports-bacteria-differences in survival, brain damage, and cerebrospinal fluid cytokine kinetics due to meningitis caused by 3 different Streptococcus pneumoniae. J Infect Dis 190(7):1212–1220

Peila R, Launer LJ (2006) Inflammation and dementia: epidemiologic evidence. Acta Neurol Scand 114:102–106

Perry VH, Cunningham C, Holmes C (2007) Systemic infections and inflammation affect chronic neurodegeneration. Nat Rev Immunol 7(2):161–167

Poirier J (2000) Apolipoprotein E and Alzheimer's disease a role in amyloid catabolism. Ann N Y Acad Sci 924(1):81–90

Prasad KN, Cole WC, Prasad KC (2002) Risk factors for Alzheimer's disease: role of multiple antioxidants, non-steroidal anti-inflammatory and cholinergic agents alone or in combination in prevention and treatment. J Am Coll Nutr 21(6):506–522

Qian Y, Cao L, Guan T et al (2015) Protection by genistein on cortical neurons against oxidative stress injury via inhibition of NF-kappaB, JNK and ERK signaling pathway. Pharm Biol 26:1–9

Qiu C, Kivipelto M, von Strauss E (2009) Epidemiology of Alzheimer's disease: occurrence, determinants, and strategies toward intervention. Dialogues Clin Neurosci 11(2):111

Raskind MA, Peskind ER (1994) Neurobiologic bases of noncognitive behavioral problems in Alzheimer disease. Alzheimer Dis Assoc Disord 8:54–60

Ratnavalli E, Brayne C, Dawson K, Hodges JR (2002) The prevalence of frontotemporal dementia. Neurology 58(11):1615–1621

Reale M, Iarlori C, Gambi F, Lucci I, Salvatore M, Gambi D (2005) Acetylcholinesterase inhibitors effects on oncostatin-M, interleukin-1β and interleukin-6 release from lymphocytes of Alzheimer's disease patients. Exp Gerontol 40(3):165–171

Riedel G, Platt B, Micheau J, Riedel G, Platt B, Micheau J (2003) Glutamate receptor function in learning and memory. Behav Brain Res 140:1–47

Rivest S (2009) Regulation of innate immune responses in the brain. Nat Rev Immunol 9(6): 429–439

Roberts NLS, Mountjoy-Venning WC, Anjomshoa M, Banoub JAM, Yasin YJ, GBD 2017 Disease and Injury Incidence and Prevalence Collaborators (2018) Global, regional, and national incidence, prevalence, and years lived with disability for 354 diseases and injuries for 195 countries and territories, 1990-2017: a systematic analysis for the Global Burden of Disease Study. Lancet 392:1789. Lancet 393(10190):E44–E44

Roos KL (2014) Encephalitis. In: Handbook of clinical neurology. Elsevier, Amsterdam, pp 1377–1381

Schenk D, Barbour R, Dunn W et al (1999) Immunization with amyloid-β attenuates Alzheimer-disease-like pathology in the PDAPP mouse. Nature 400(6740):173–177

Schmidt R, Schmidt H, Curb JD, Masaki K, White LR, Launer LJ (2002) Early inflammation and dementia: a 25-year follow-up of the Honolulu-Asia Aging Study. Ann Neurol 52(2):168–174

Schroeder BO, Bäckhed F (2016) Signals from the gut microbiota to distant organs in physiology and disease. Nat Med 22(10):1079

Semmler A, Hermann S, Mormann F et al (2008) Sepsis causes neuroinflammation and concomitant decrease of cerebral metabolism. J Neuroinflammation 5(1):1–10

Shah FA, Pike F, Alvarez K et al (2013) Bidirectional relationship between cognitive function and pneumonia. Am J Respir Crit Care Med 188(5):586–592

Shankar GM, Li S, Mehta TH et al (2008) Amyloid-β protein dimers isolated directly from Alzheimer's brains impair synaptic plasticity and memory. Nat Med 14(8):837–842

Sheean RK, McKay FC, Cretney E et al (2018) Association of regulatory T-cell expansion with progression of amyotrophic lateral sclerosis: a study of humans and a transgenic mouse model. JAMA Neurol 75(6):681–689

Sheng JG, Bora SH, Xu G, Borchelt DR, Price DL, Koliatsos VE (2003) Lipopolysaccharide-induced-neuroinflammation increases intracellular accumulation of amyloid precursor protein and amyloid β peptide in APPswe transgenic mice. Neurobiol Dis 14(1):133–145

Smith MA, Rottkamp CA, Nunomura A, Raina AK, Perry G (2000) Oxidative stress in Alzheimer's disease. Biochim Biophys Acta 1502(1):139–144

Sochocka M, Zaczyńska E, Leszek J, Siemieniec I, Błach-Olszewska Z (2008) Effect of donepezil on innate antiviral immunity of human leukocytes. J Neurol Sci 273(1–2):75–80

Sochocka M, Zwolinska K, Leszek J (2017) The infectious etiology of Alzheimer's disease. Curr Neuropharmacol 15(7):996–1009

Solito E, Sastre M (2012) Microglia function in Alzheimer's disease. Front Pharmacol 3:14

Späni C, Suter T, Derungs R et al (2015) Reduced β-amyloid pathology in an APP transgenic mouse model of Alzheimer's disease lacking functional B and T cells. Acta Neuropathol Commun 3(1):1–14

Strandberg TE, Pitkala KH, Linnavuori KH, Tilvis RS (2003) Impact of viral and bacterial burden on cognitive impairment in elderly persons with cardiovascular diseases. Stroke 34(9): 2126–2131

Sulzer D, Alcalay RN, Garretti F et al (2017) Erratum: T cells from patients with Parkinson's disease recognize α-synuclein peptides. Nature 549(7671):292

Swerdlow RH (2007) Pathogenesis of Alzheimer's disease. Clin Interv Aging 2(3):347

Swerdlow RH, Burns JM, Khan SM (2010) The Alzheimer's disease mitochondrial cascade hypothesis. J Alzheimers Dis 20(s2):S265–S279

Sy M, Kitazawa M, Medeiros R et al (2011) Inflammation induced by infection potentiates tau pathological features in transgenic mice. Am J Pathol 178(6):2811–2822

Szekely CA, Zandi PP (2010) Non-steroidal anti-inflammatory drugs and Alzheimer's disease: the epidemiological evidence. CNS Neurol Disord Targets 9(2):132–139

Thaler JP, Yi CX, Schur EA et al (2012) Obesity is associated with hypothalamic injury in rodents and humans. J Clin Invest 122:153–162

Tiraboschi P, Hansen LA, Masliah E, Alford M, Thal LJ, Corey-Bloom J (2004) Impact of APOE genotype on neuropathologic and neurochemical markers of Alzheimer disease. Neurology 62(11):1977–1983

Togo T, Akiyama H, Iseki E et al (2002) Occurrence of T cells in the brain of Alzheimer's disease and other neurological diseases. J Neuroimmunol 124(1–2):83–92

Uldry P-A, Kuntzer T, Bogousslavsky J et al (1993) Early symptoms and outcome ofListeria monocytogenes rhombencephalitis: 14 adult cases. J Neurol 240(4):235–242

Valero J, Mastrella G, Neiva I, Sánchez S, Malva JO (2014) Long-term effects of an acute and systemic administration of LPS on adult neurogenesis and spatial memory. Front Neurosci 8:83

Van Sorge NM, Doran KS (2012) Defense at the border: the blood–brain barrier versus bacterial foreigners. Future Microbiol 7(3):383–394

Vasek MJ, Garber C, Dorsey D et al (2016) A complement–microglial axis drives synapse loss during virus-induced memory impairment. Nature 534(7608):538–543

Walker KA, Ficek BN, Westbrook R (2019a) Understanding the role of systemic inflammation in Alzheimer's disease. ACS Chem Nerosci 10(8):3340–3342

Walker KA, Gottesman RF, Wu A et al (2019b) Systemic inflammation during midlife and cognitive change over 20 years: the ARIC study. Neurology 92(11):e1256–e1267

Wallace W, Ahlers ST, Gotlib J et al (1993) Amyloid precursor protein in the cerebral cortex is rapidly and persistently induced by loss of subcortical innervation. Proc Natl Acad Sci 90(18): 8712–8716

Wang R, Reddy PH (2017) Role of glutamate and NMDA receptors in Alzheimer's disease. J Alzheimers Dis 57(4):1041–1048

Wang Y, Cella M, Mallinson K et al (2015) TREM2 lipid sensing sustains the microglial response in an Alzheimer's disease model. Cell 160(6):1061–1071

Weaver JD, Huang MH, Albert M, Harris T, Rowe JW, Seeman TE (2002) Interleukin-6 and risk of cognitive decline: MacArthur studies of successful aging. Neurology 59(3):371–378

Weiss N, Miller F, Cazaubon S, Couraud P-O (2009) The blood-brain barrier in brain homeostasis and neurological diseases. Biochim Biophys Acta 1788(4):842–857

Wellings K, Jones KG, Mercer CH et al (2013) The prevalence of unplanned pregnancy and associated factors in Britain: findings from the third National Survey of Sexual Attitudes and Lifestyles (Natsal-3). Lancet 382(9907):1807–1816

Wider C, Wszolek ZK (2008) Etiology and pathophysiology of frontotemporal dementia, Parkinson disease and Alzheimer disease: lessons from genetic studies. Neurodegener Dis 5(3–4):122–125

Wu S-Z, Bodles AM, Porter MM, Griffin WST, Basile AS, Barger SW (2004) Induction of serine racemase expression and D-serine release from microglia by amyloid β-peptide. J Neuroinflammation 1(1):2

Yaffe K, Kanaya A, Lindquist K et al (2004) The metabolic syndrome, inflammation, and risk of cognitive decline. JAMA 292(18):2237–2242

Yang Y, Quitschke WW, Brewer GJ (1998) Upregulation of amyloid precursor protein gene promoter in rat primary hippocampal neurons by phorbol ester, IL-1 and retinoic acid, but not by reactive oxygen species. Brain Res Mol Brain Res 60(1):40–49

Yip AG, Green RC, Huyck M, Cupples LA, Farrer LA (2005) Nonsteroidal anti-inflammatory drug use and Alzheimer's disease risk: the MIRAGE Study. BMC Geriatr 5(1):1–6

Yoshiyama Y, Kojima A, Ishikawa C, Arai K (2010) Anti-inflammatory action of donepezil ameliorates tau pathology, synaptic loss, and neurodegeneration in a tauopathy mouse model. J Alzheimers Dis 22(1):295–306

Youm YH, Grant RW, McCabe LR et al (2013) Canonical Nlrp3 inflammasome links systemic low-grade inflammation to functional decline in aging. Cell Metab 18:519–532

Zempel H, Mandelkow E (2014) Lost after translation: missorting of Tau protein and consequences for Alzheimer disease. Trends Neurosci 37(12):721–732

Zeng KW, Ko H, Yang HO, Wang XM (2010) Icariin attenuates β-amyloid-induced neurotoxicity by inhibition of tau protein hyperphosphorylation in PC12 cells. Neuropharmacology 59:542–550

Zhang G, Li J, Purkayastha S et al (2013) Hypothalamic programming of systemic ageing involving IKK-beta, NF-kappaB and GnRH. Nature 497:211–216

Zhang D, Wang Z, Sheng C et al (2015) Icariin prevents amyloid beta-induced apoptosis via the PI3K/Akt pathway in PC-12 cells. Evid Based Complement Alternat Med 2015:235265

Zhao Y, Zhao B (2013) Oxidative stress and the pathogenesis of Alzheimer's disease. Oxid Med Cell Longev 2013:316523

Zubenko GS, Moossy J, Kopp U (1990) Neurochemical correlates of major depression in primary dementia. Arch Neurol 47(2):209–214

Role of Impaired Insulin Signaling in the Pathogenesis of Dementia

3

Abhilasha Ahlawat, Vaibhav Walia, and Munish Garg

Abstract

Insulin in the brain acts as a neuromodulator and neuroprotective agent, which is involved in glucose regulation, cognition, and memory and also regulates neuronal growth and repair. Previously, brain was identified as an insulin-independent organ, but recent research shows that insulin also performs various important functions in the brain and the insulin receptors are distributed in the various regions of the brain including the cortex, hippocampus, olfactory bulb, and hypothalamus. Thus, the impairment of insulin signaling and its downstream pathway leads to cognitive deficits, synaptic dysfunction, hallucinations, and ultimately leads to dementia. Administration of insulin through the nasal route produces improvement in the memory and cognitive impairment in the afflicted patients, suggesting that insulin therapy may ameliorate dementia and other consequences. In the present chapter, the authors describe the function of brain insulin, insulin signaling transduction mechanism, and impaired insulin signaling in the pathogenesis of dementia.

Keywords

Alzheimer's disease · Dementia · GSK3β · Insulin · Parkinson's disease · Phosphatidylinositol 3-kinase · Protein kinase B · Ras-mitogen-activated protein kinase

A. Ahlawat · M. Garg (✉)
Department of Pharmaceutical Sciences, Maharshi Dayanand University, Rohtak, India

V. Walia
Faculty of Pharmacy, SGT University, Gurugram, India

G. M. Ashraf, M. S. Uddin (eds.), *Current Thoughts on Dementia*,
https://doi.org/10.1007/978-981-16-7606-2_3

3.1 Introduction

Dementia is a disorder that causes the impaired cognition, memory, agnosia, apraxia, and impaired executive functions (Chertkow et al. 2013; Hildreth and Church 2015). Recently, according to DSM-V dementia is characterized by amnesia, delirium, and cognitive deficits (American Psychiatric Association 2000). There are mainly four different types of dementia such as Alzheimer's disease (AD), vascular dementia (VaD), Lewy body dementia (LBD), and frontotemporal dementia (FTD). AD is the most common form of dementia, responsible for 60–80% of neurodegenerative disorders, and is characterized by the deposition of Aβ (amyloid-beta) plaques and neurofibrillary tangles in the brain responsible for memory impairment, aphasia, apraxia, agnosia, disorientation, impaired visuospatial and executive functions, depression, aggression, hyperactivity, uncooperativeness, repetitiveness in activities, feeling lost, hallucinations, mutism, inability to sit up, holding the head, or tracking objects with eyes occurs (Duong et al. 2017; Reitz 2015). VD accounts for nearly 20% of cases of dementia (Alzheimer Society of Canada 2016). VD arises due to the deprivation of oxygen of neurons either due to impaired blood supply, and the afflicted individuals often suffer from confusion, loss of vision, executive dysfunctions, disorientation, apraxia, and agnosia (Duong et al. 2017).

LBD accounts for 5–15% of cases of dementia characterized by the presence of Lewy bodies (LBs) (Alzheimer Society of Canada 2016). LBD is region-specific and is responsible for the emergence of the pathology according to the region involved (Braak et al. 2003; Kosaka et al. 1984, 1996). Besides these forms, there is another form of LBD known as incidental LBD (iLBD) found in approximately 10% of individuals of age above 60 years, but no actual neuronal degeneration has been observed in the brain (Adler et al. 2010; Dickson et al. 2008; Frigerio et al. 2011). FTD generally involved the affliction of the frontal and temporal regions of the brain. FTD is also known as Pick's disease and is characterized by symptoms including behavioral deficits, personality changes, and visuospatial dysfunction (Alzheimer Society of Canada 2016; Duong et al. 2017).

Insulin is known to regulate various neurological functions, and insulin signaling impairment is responsible for the emergence of various types of dementia (Fig. 3.1). In the present chapter, the authors represent the role of insulin signaling in the pathogenesis of various dementias.

3.2 Brain Insulin

Brain insulin is known to regulate homeostasis, cognition, and memory and confers neurotrophic and neuroprotective effects (Blázquez et al. 2014). Havrankova and Roth (1979) at first introduced brain insulin and suggested that it is independent of peripheral insulin. Insulin signaling in the brain is known to regulate glucose regulation, feeding behavior, neuronal development, and cognitive functions (Derakhshan and Toth 2013). Insulin mRNA and its receptors have been found in the various regions of the brain such as the hippocampus (Blázquez et al. 2014;

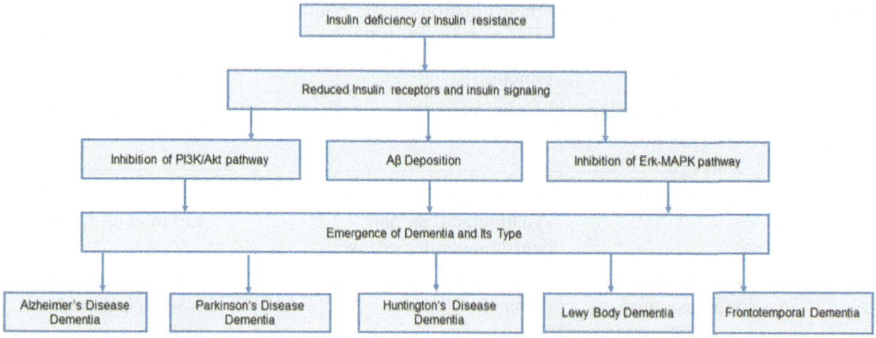

Fig. 3.1 Insulin signaling pathway in the emergence of various types of dementia

Devaskar et al. 1994). It was observed that the peripheral administration of the insulin increases the brain insulin level, suggesting that it may cross the blood–brain barrier (BBB) (Margolis and Altszuler 1967). These findings were confirmed by Woods and colleagues (Woods and Porte Jr 1977).

Duffy and Pardridge (1987) proposed that insulin utilizes the saturable transport system for entry in the brain and follows the nonlinear dependence between blood and cerebrospinal fluid levels. However, the difference in the insulin levels of the different brain regions might be due to the difference in the activity of the transport system of BBB affecting regional insulin permeability (Banks and Kastin 1998). Further, it has been reported that insulin transportation is affected by several conditions like fasting, obesity, aging, diabetes, and AD (Begg 2015). However, one of the findings suggested that the brain itself may produce some amount of insulin (Gerozissis 2008). Insulin is stored inside the synaptic vesicles, released from these vesicles, and regulates neuronal activity (Wei et al. 1990). Both neurons and the beta cells utilize the same mechanism for the release of insulin. The similarity was observed with ATP-sensitive K^+ channel depolarization (Gerozissis 2003). The hormone and the glucose stimulus evoke the release of insulin in the pancreatic cells and neurons due to the depolarization caused by the K^+ ions. However, in presence of calcium ions, the insulin release increases further in neuronal cell cultures and this phenomenon is specific to neuronal cells only (Clarke et al. 1986). Glucose is responsible for the release of insulin in the brain (Santos et al. 1999). Table 3.1 shows the location and functions of various glucose transporters (GLUTs) in the brain.

3.3 Insulin Receptors and Signaling

Insulin receptors (IRs) are present in the various regions of the brain and are encoded by *INSR* gene located on chromosome 19. IRs are of two types such as IR-A and IR-B (Belfiore et al. 2009, 2017). IR-A is found to be present in insulin-independent regions such as the brain, whereas IR-B is present in the insulin-dependent region

Table 3.1 Glucose transporters and their functions in the brain

Glucose transporter	Location	Function	References
GLUT 1	Brain endothelial cells and astrocytes	Insulin regulation	Levin et al. (2004), Simpson et al. (1999)
GLUT 2	Hypothalamus	Acts as a glucose sensor, regulates synaptic activity and neurotransmitter release	Jurcovicova (2014), Levin et al. (2004)
GLUT 3	Cortex and hippocampus	Regulate local cerebral glucose demand	Jurcovicova (2014), Nagamatsu et al. (1992)
GLUT 4	Hippocampus, cortex, and hypothalamus	Regulate glucose and insulin	Talbot et al. (2012), Vannucci et al. (1998)
GLUT 5	Microglia	Involved in brain hexose transport and fructose	Devraj (2010)
GLUT 8	Hippocampus and hypothalamus	Glucose homeostasis	Ibberson et al. (2002), McEwen and Reagan (2004)

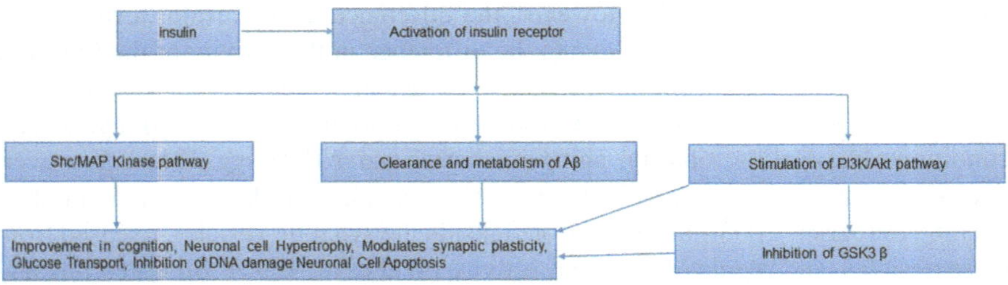

Fig. 3.2 Insulin signaling pathway and cellular functions

including muscles and livers (Vienberg et al. 2011). The exact functions of IR in the brain are completely not known, but it is suggested that the brain insulin and its receptors are known to influence and regulate cholinergic signaling, olfaction, appetite, autonomic functions, etc. (Hill et al. 1986; Landau et al. 1983; Werther et al. 1987, 1989). Peripheral IRs are downregulated in the presence of excess insulin, whereas such downregulation is not seen in brain IR (Heidenreich et al. 1983; Zahniser et al. 1984). IRs consist of α- and β-subunit (Schulingkamp et al. 2000). Insulin binds on the α-subunit and leads to the activation of tyrosine kinase activity (Pomytkin et al. 2018; Wilden et al. 1992), responsible for the phosphorylation of tyrosine residues and the subsequent phosphorylation and activation of IRS (Heffetz and Zick 1986; Pilch et al. 1983; White 2003). Insulin receptor substrate (IRS) proteins consist of various tyrosine and Ser/Thr phosphorylation sites (Myers Jr et al. 1994), which further interact with the other downstream signaling pathways including MAPK, PI3K, and JNK (Fig. 3.2) (Aguirre et al. 2002; White 2003).

In the hippocampus region, the IR is co-expressed with a substrate for tyrosine kinase, i.e., p58/53 (IRSp53) responsible for synaptic growth and maintenance (Abbott et al. 1999; Frölich et al. 1998; Lee et al. 2011). Insulin by interaction with IR phosphorylates IRSp53 that causes the growth of neurites (Choi et al. 2005; Miyahara et al. 2003). Therefore, IRSp53-deficient animal shows neuronal and cognitive deficits (Mackie and Aitken 2005; Sawallisch et al. 2009). Insulin also performs its downstream function through PI3K/AKT pathway (Lemmon and Schlessinger 2010). Insulin leads to the activation of the enzyme PI3K, which then activates the enzyme Akt/PKB, expressed widely in hippocampus and astrocytes and inhibit Bax, Bad, caspase-9, and glycogen synthase kinase-3 (GSK-3) to promote cell survival, proliferation, differentiation, growth, and inhibit cellular apoptosis (Boucher et al. 2014; De Meyts 2000; Kim and Feldman 2012; Levenga et al. 2017; Manna and Jain 2013; Noguchi and Suizu 2012).

Insulin further activates another signaling pathway, known as Ras-MAPK pathway (Kim and Feldman 2012; Lemmon and Schlessinger 2010). Binding of the insulin to the insulin receptors results in the phosphorylation of Tyr, which results in the binding of SH2 and Grb2, which then interact with SOS (son of sevenless) and activate Ras by the replacement of GDP with GTP. Ras-GTP then stimulates Raf, which then stimulates MEK and ultimately ERK, important for cellular proliferation, differentiation, cytoskeletal reorganization, etc. (Boucher et al. 2014; Cargnello and Roux 2011). It has been reported that the SOCS, JAK-STAT, Grb10/Grb14, and PC-1 inhibit insulin signaling (De Meyts 2000). Further, insulin resistance involves the upregulation of pathways including ERK, JNK, and p38 MAPK pathways (Nandipati et al. 2017).

AMP-activated protein kinase (AMPK) regulates the sensitivity of insulin, and AMPK pathway dysfunction is responsible for the emergence of insulin resistance. AMPK activation is responsible for the expression and translocation of GLUT4-mediated glucose uptake and downregulation of gluconeogenesis, inhibits JNK/NFκB pathway, reduces mTOR phosphorylation and upregulation of PI3K pathway, and thus counteracts the insulin resistance (Nandipati et al. 2017).

3.4 Functions of Insulin in Brain

3.4.1 Maintain Blood–Brain Barrier Function

Brain endothelial cells (BECs) act as the binding site and transporter of insulin (Burgos-Ramos et al. 2012; Miller et al. 1994). Insulin affects the transportation of various molecules from the blood to the brain by altering these transporters (Kastin and Akerstrom 2001; Liu et al. 2009; Tagliamonte et al. 1975).

3.4.2 Glucose Homeostasis

Glucose plays various important roles in the brain, and it acts as fuel for the brain and regulates feeding, energy utilization, insulin release, sugar regulation, etc. (Schulingkamp et al. 2000; White and Venkatesh 2011). Glucokinase enzyme functions as a glucose sensor in the neurons (Marty et al. 2007) and thus mediates insulin release, which then causes the opening of ATP-sensitive K^+ channels resulting in glycemic control in the brain (Cotero and Routh 2009; Pocai et al. 2005). Insulin resistance thus affects brain glucose homeostasis (Obici et al. 2002a, b).

3.4.3 Cognition and Memory

Hippocampus is involved in the formation of spatial memory and damage to the hippocampus may lead to severe learning and memory impairments (Martin and Clark 2007). Insulin regulates neurotransmitters (including acetylcholine, noradrenaline), cognitive functions, brain potential, spatial and verbal memory, etc. (Benedict et al. 2004; Kern et al. 1999; Kopf and Baratti 1999; Park et al. 2000). Insulin clears Aβ from BBB, prevents tau phosphorylation, and improves memory by promoting synaptic function and long-term potentiation (Ribe and Lovestone 2016).

3.4.4 Inhibit Excitotoxicity and Neuronal Protection

Glutamate results in the activation of the ionotropic glutamate receptors responsible for the excitotoxicity. It has been reported that the administration of insulin for a shorter period protects against glutamate-induced neurotoxicity (Pinelis et al. 2019). Insulin improves synaptic communication by decreasing glutamate and increasing GABA (Yu and Pei 2015). Insulin provides neuroprotection by regulating neurotransmitters, enhancing glycogen synthesis, inhibiting neuronal cell death, and preventing hypoglycemia and hyperglycemia (Yu and Pei 2015). Insulin also improves neuronal glucose metabolism and antioxidant defense and restores the IR/IGF-1R signaling (Duarte et al. 2005; Duarte et al. 2008; Fulop et al. 2003). Insulin prevents the formation of Aβ fibril and protects the cells against Aβ toxicity (Rensink et al. 2004). Insulin also antagonizes the effects of oxidative stress and diminishes oxidative stress by promoting glucose uptake and pyruvate formation (Duarte et al. 2006).

Insulin promotes lactate deposition, acidosis, and mitochondrial impairment and restores intracellular ATP formation (Duarte et al. 2006; Garg et al. 2006). Insulin also provides antioxidant effect by increasing uric acid, glutathione, and vitamins C and E (Sevanian et al. 1991). Insulin also protects from ischemia either by increasing GABA during transient ischemia or by reducing blood glucose concentration (Auer 1998; Shuaib et al. 1995). Insulin alters glucose metabolism and decreases lactic acidosis (Grunstein et al. 1985). Insulin stimulates the Na^+/K^+ATP pump and alters neuronal firing rate by decreasing K^+ and Na^+ ions and thus prevents water accumulation and edema (Voll and Auer 1991). Insulin-mediated antiapoptotic effects may be due to inhibition of SAPK, GSK-3β, and caspase inactivation (Chin et al. 2005; Kim and Han 2005). Insulin-mediated antiapoptotic effects are based on PI3K/Akt/mTOR pathway (Ryu et al. 1999).

3.5 Insulin Signaling Impairment Led to Dementia and Related Pathologies

Insulin signaling dysfunction is responsible for the emergence of dementia and related pathologies as shown in Fig. 3.3. Brain insulin resistance is responsible for cognitive dysfunction and neuroinflammation in afflicted patients (Correia et al. 2012; Gaspar et al. 2016). Insulin also regulates the functioning of brain microglial cells, and the microglial cells play an important role in neuroinflammation (Joers et al. 2017) and confer neuronal protection (Sekiyama et al. 2012). Various evidences have shown impaired transport of insulin in the brain (de la Monte 2009) due to a decrease in insulin levels in plasma or CSF or IR expression (Moloney et al. 2010). Defects in insulin or IGF 1 signaling alter PI3K/Akt pathway and lead to decreased expression and activation of GLUTs, which may lead to a decrease in brain glucose and mitochondrial metabolism and ATP production (Bosco et al. 2011).

Dementia in PD is generally linked with IR, and it may be an independent risk factor for cognitive decline (Plastino et al. 2010). IR may exert an effect on cognitive function either through glucose metabolism (as exogenous insulin may improve brain glucose metabolism and brain functions like memory (Bingham et al. 2002) or through neurotransmitter modulation (as insulin can oppose anamnestic effects of inhibition of cholinergic impulses (Blanchard and Duncan 1997) and raise norepinephrine levels in the brain (Watson et al. 2006)). More impairment in IGF 1 and

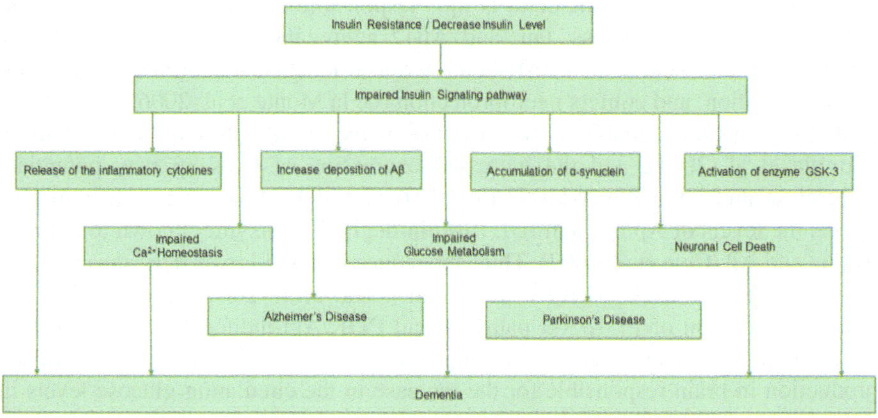

Fig. 3.3 Impaired insulin signaling in the emergence of dementia

IGF 2 and neurotrophin signaling is also observed in LBD than in PD (Tong et al. 2009). Further, inflammation leads to insulin desensitization in the brain in AD, and PD as cytokines (TNF) will block insulin/IGF 1 signaling (Clark and Vissel 2014) causing neuronal loss as neurogenesis will not occur and neurons will accumulate and undergo neurodegeneration (Ziabreva et al. 2006). Insulin resistance also results in the Aβ deposition and hyperphosphorylation of tau protein (Barroso et al. 2013). Insulin resistance leads to the sustained activation of enzyme GSK3β (Lyman et al. 2014) and the release of TNF-α and IL-6 (Beurel and Jope 2009; Wang et al. 2010). Impaired insulin signaling is responsible for oxidative stress (Duarte et al. 2012), metabolic disturbance (de la Monte 2017), neuroinflammation (Duarte et al. 2012), and subsequent neuronal damage (de la Monte and Wands 2005).

3.5.1 Insulin Signaling in Dementia Associated with Alzheimer's Disease

AD is mainly responsible for the emergence of dementia in later stages of life and is characterized by the deposition of Aβ plaques and NFTs in various regions of the brain. Neuronal cell death in AD is characterized by increased neuroinflammation, Aβ aggregation, tau protein hyperphosphorylation, etc. (Bosco et al. 2011). Recent studies suggested the involvement of the impaired insulin signaling pathway in the pathogenesis of AD and related dementias. Insulin signaling impairment results in neuroinflammation and oxidative stress responsible for the neuronal damage in the brain (Akhtar and Sah 2020). Prolonged neuronal inflammation has been observed in type 2 diabetes (T2D), and AD and is considered as one of the main features of these pathologies (Calle and Fernandez 2012; Heneka et al. 2015). Insulin resistance-mediated neuroinflammation involves the downregulation of peroxisome proliferator activator receptor (PPAR) delta (maintains the insulin sensitivity) (Collino et al. 2008; de la Monte and Wands 2006).

Downregulation of PPAR-delta is thus responsible for neuronal inflammation, astrogliosis, oxidative stress, tau, and Aβ42 aggregation (Barroso et al. 2013). Further, the treatment with PPAR-delta agonists reduces neuroinflammation and Aβ42 deposition, and confers neuroprotection (de la Monte et al. 2006; de la Monte et al. 2017). Insulin signaling is also responsible for the inhibition of GSK-3 activity (Correia et al. 2011), and the increased expression of GSK-3β may potentiate c-secretase and intracellular levels of Aβ (Kim and Feldman 2012). Insulin also affects the levels of Aβ through IDE (insulin-degrading enzyme) known to degrade insulin and Aβ (Cole et al. 2007). Thus, the reduced expression of IDE is responsible for the increased levels of Aβ in the brain (Zhao et al. 2007). Besides this, the reduced activation of PI3K/Akt pathway and PI3K/Akt-induced GLUT activation lead to reduced glucose metabolism, impaired mitochondrial metabolism, and ATP production in brain responsible for the increase in the circulating glucose levels in the brain responsible for the increased formation of AGEs that confers neurotoxic effects (Bosco et al. 2011).

Akt inhibits the expression of GSK3β, which mainly mediates neuronal cell apoptosis (Wu et al. 2007). Therefore, Akt inhibition is often responsible for the increased expression of the enzyme GSK-3 due to the decreased phosphorylation at serine 9 (Malagelada et al. 2008; Wang et al. 2007). AD and related dementia are often characterized by the impaired PI3K-Akt signaling mechanism (Liu et al. 2011a, b). Further in AD, the increased formation and deposition of Aβ result in the activation of TNF-α and JNK pathway responsible for the attenuation of insulin signaling, emergence of insulin resistance and subsequent neuronal inflammation, oxidative damage, and neurodegeneration (Bomfim et al. 2012; de la Monte et al. 2017; Gregor and Hotamisligil 2011).

3.5.2 Insulin Signaling in Dementia Associated with Parkinson's Disease

Parkinson's disease (PD) is another major cause of dementia and is characterized by the degeneration of neurons in the midbrain region. The afflicted individuals often suffer from symptoms including motor dysfunction, cognitive defect, and dementia. PD dementia (PDD) involves impairments in execution, recognition, vision, and other functions. However, it is largely seen that executive dysfunction occurs prior to motor dysfunction. PDD involves damage to the dopaminergic and cholinergic neurons and structural damage to the hippocampus and posterior occipital cortices (Bohnen et al. 2003; Emre 2003). PDD frequently affects the patients in the advanced stages and has been observed in approximately 80–90% of patients of the age above 90 and contributes to high morbidity and mortality (Gratwicke et al. 2015).

Recent studies show a possible link between the defect in the insulin signaling pathways and the development of dementia in PD (de la Monte 2017). Insulin is known to regulate glucose metabolism, food intake, memory, etc. (Lee et al. 2005). Further, insulin treatment alleviates neuronal inflammation and repairs the inflammation-induced damages. Insulin treatment improves motor function in pre-clinical models of PD (Hölscher 2014; Pang et al. 2016), and the reduced insulin signaling increases the risk of PD and related dementia (Ma et al. 2015). Insulin and IGF-1-mediated PI3K/Akt pathway has been shown to exert a positive effect on dopaminergic neurons (Kao 2009). Attenuation of insulin signaling is responsible for α-synuclein aggregation, neurotoxicity, and neuronal apoptosis (Kao 2009). GSK-3 is the major target of Akt, and Akt inhibits GSK-3 to inhibit the GSK-3 mediated neuronal degeneration and spatial learning defects (Lucas et al. 2001). Therefore, the counteraction of insulin resistance contributes to the alleviation of the neuronal degeneration in PD and related disorders (Athauda and Foltynie 2016).

3.5.3 Insulin Signaling in Dementia Associated with Huntington's Disease

Mutant *HTT/mHTT* is an important inducer of Huntington's disease (HD) (Warby et al. 2009) responsible for impaired mitochondrial functions, production of reactive oxygen species (ROS), and neuronal cell death (Jodeiri Farshbaf and Ghaedi 2017). *mHTT* affects the insulin secretions and mediates insulin resistance and is responsible for the formation of aggregates in the pancreas and pathological changes in the hypothalamus (Martin et al. 2009). HD is characterized by the reduced level and signaling of IGF-1 in the striatum (Pouladi et al. 2010). Defective insulin signaling has been found to be involved in the emergence of HD (Lalić et al. 2008).

Insulin is known to activate the enzyme Akt, which promotes mitochondrial function (Li et al. 2013), protects the neurons against oxidative stress (Duarte et al. 2008), and promotes neuronal cell survival (Pouladi et al. 2010). It has been reported that the mutation in the *HTT* gene and the mutant CAG repeats reduces the sensitivity to the insulin (Aziz et al. 2010) leading to defective insulin signaling (Montojo et al. 2017) and further leading to cognitive deficits and neuronal cell death (Pomytkin et al. 2018). Further, it has been reported that potentiating the brain insulin and its downstream signaling alleviates metabolic and mitochondrial dysfunction and exerts a beneficial effect in the treatment of HD (Naia et al. 2016; Ribeiro et al. 2014; Yamamoto et al. 2006)

3.5.4 Insulin Signaling in Dementia Associated with Lewy Body Disease

Lewy body dementia (LBD) accounts for 5–15% of cases of dementia characterized by the presence of Lewy bodies (LBs) (Alzheimer Society of Canada 2016). LBDs are region-specific and are responsible for the emergence of the pathology according to the region involved. For example, in brain stem-type LBD the individual often experiences PD; diffuse-type LBs are found in the cerebral cortex and amygdale and dementia often precedes PD in this type of LBD, whereas the transitional LBD refers to the pathology brainstem type and diffuse LBD (Braak et al. 2003; Kosaka et al. 1984, 1996). Besides these forms, there is another form of LBD known as incidental LBD (iLBD) found in approximately 10% of individuals of age above 60 years, but no actual neuronal degeneration has been observed in the brain (Adler et al. 2010; Dickson et al. 2008; Frigerio et al. 2011). α-Synuclein is known to play an important role in LBD and PD (Kim et al. 2014). Aβ is known to promote α-synuclein accumulation and aggravate cognitive dysfunction (Marsh and Blurton-Jones 2012). Both Aβ and α-Syn suppress AMPK signaling, which is responsible for the impairment of brain insulin signaling responsible for oxidative stress and neurotoxicity (Chang et al. 2018).

Further, in dementia associated with LBD, phosphorylated tau is found adjoining with phosphorylated α-synuclein at the position Ser 129 in the cerebral cortex region of the brain (Fujiwara et al. 2002) that promotes fibril formation (Fujiwara et al.

2002) phosphorylation of GSK3 (Credle et al. 2015) and aggravates neurodegeneration (Sato et al. 2011). Thus, α-synuclein is known to activate GSK3 and it mediates neurodegeneration (Yuan et al. 2010) by causing impairment in mitochondrial function, cholinergic neurotransmission, BDNF/TrkB activity, and insulin/IGF-1 signaling (Albeely et al. 2018). Further, in LBD impaired insulin signaling and reduced IR expression are localized to the amygdala and frontal white matter, whereas in LBD associated with AD, there are reduced expressions of IGF-1 and IGF-2 receptor in the frontal cortex (Tong et al. 2009). Thus, the treatment with the insulin and potentiation of the downstream signaling improve neuronal and cognitive functions in the patients afflicted with LBD (Tong et al. 2009).

3.5.5 Insulin Signaling in Frontotemporal Dementia

FTD having unknown etiology, responsible for 80% of dementia cases (Liou et al. 2019), results in increased mortality and reduced survival (Bang et al. 2015), characterized by cortical atrophy (Rascovsky et al. 2011), glial cells hypertrophy, and accumulation of mutated and misfolded protein in the cytoplasm of the neuronal cell (Kumar-Singh 2011). Impaired insulin signaling has been implicated in FTD and is responsible for oxidative stress (Duarte et al. 2012), metabolic disturbance (de la Monte 2017), neuroinflammation (Duarte et al. 2012), and subsequent neuronal damage in FTD (de la Monte and Wands 2005).

3.6 Neuroinflammation and Impaired Insulin Signaling in Dementia

Neuroinflammation is an important feature of various neurodegenerative disorders. Similar to systemic disease, here also inflammation exacerbates insulin resistance (Vykoukal and Davies 2011). Insulin resistance-mediated neuronal inflammation has been seen in neurodegenerative disorders such as AD and PD (Talbot et al. 2012; Tong et al. 2009). Brain insulin resistance is responsible for cognitive dysfunction and neuroinflammation in afflicted patients (Gaspar et al. 2016). PPAR is known to regulate insulin sensitivity and exerts anti-inflammatory effects (Collino et al. 2008). Therefore, the decrease in the PPAR-δ signaling pathway is responsible for insulin resistance and neuroinflammation (Cimini et al. 2005; de la Monte and Wands 2006). Further, the downregulation of insulin signaling and insulin resistance has been observed in AD patients (Cimini et al. 2005; de la Monte and Wands 2006). Insulin resistance in AD is also characterized by neuroinflammation, glial cell atrophy, oxidative stress, Aβ deposition, and hyperphosphorylation of tau protein (Barroso et al. 2013) responsible for the neuronal dysfunctions.

Further, the treatment with PPAR-δ agonists counteracts insulin resistance-mediated neuronal inflammation, Aβ aggregation, and neurotoxicity (de la Monte et al. 2006; de la Monte et al. 2017). Aβ is involved in the neuronal inflammation mediated by insulin resistance (Yao et al. 2004). It has been reported that the insulin resistance resulted in the decreased clearance of Aβ, and promotes its aggregation responsible for neuroinflammation and further neuronal damage (van Himbergen et al. 2012). Aβ deposition, itself also, results in the development of insulin resistance (Moloney et al. 2010; Najem 2014) leading to the sustained activation of enzyme GSK3β (Lyman et al. 2014) and release of TNF-α and IL-6 (Beurel and Jope 2009; Wang et al. 2010). These cytokines further contribute to neuronal inflammation, alteration of the BBB integrity, and neuronal apoptosis (Ramirez et al. 2010). Also, the increased production of the inflammatory cytokines evokes insulin resistance (Zúñiga et al. 2010) and causes tau phosphorylation (Li et al. 2003; Quintanilla et al. 2004) responsible for synaptic dysfunction and neuronal cell death. Insulin also regulates the functioning of brain microglial cells, and the microglial cells play an important role in neuroinflammation (Joers et al. 2017) and confer neuronal protection (Sekiyama et al. 2012). However, the prolonged activation of microglial cells is responsible for the activation and aggregation of α synuclein protein responsible for chronic inflammation and neuronal apoptosis. Thus, the brain insulin signaling impairment may contribute to neuronal inflammation and damage in PD (Jha et al. 2015). Thus, insulin resistance and impaired insulin signaling contribute to neuronal inflammation, oxidative stress, and apoptosis in the pathologies such as AD and PD (Dineley et al. 2014).

3.7 Insulin in the Treatment of Dementia

Insulin regulates the various function of the brain, and insulin treatment has been shown to counteract oxidative stress, apoptosis, autophagy, and neuroinflammation in various regions of the brain (Gerozissis 2003). Also, the administration of insulin through the nasal route improves memory deficits, cognition, and abolishes neuronal cell death in AD patients (Benedict et al. 2004; Reger et al. 2008a, b).

Insulin thus improves learning, memory, cognition, markers of neuronal degeneration, and Aβ40/42 ratio in AD patients (de la Monte 2012; Reger et al. 2008a, b). Daily administration of insulin for 2 months also produces improvement in verbal memory and mood enhancement (Benedict et al. 2008). Insulin treatment also evokes beneficial effects in PD, improves amnesic defects (Vanhanen et al. 1999), and protects the dopaminergic neurons against the 6-OHDA toxicity (Pang et al. 2016).

3.8 Recent Developments and Future Perspectives

Despite the evidence obtained from the various studies, the exact role of insulin in the pathogenesis of dementia is not known. However, none of the studies reported the level below which the reduction in insulin led to the emergence and the development of dementia. Further, the majority of the pathways presented by the studies are often linked with the classical and the conventional pathways of the disease thus complicated the pathways and the exact role of the insulin. Further limited studies were conducted to determine the therapeutic role of insulin in the treatment of dementia. The result is that despite the beneficial role of insulin in demented patients, acetylcholinesterase inhibitors are the choice of drugs for the treatment of cognitive dysfunction in demented patients. It was observed that the treatment with the anti-TNF agents promotes insulin signaling and Aβ clearance, improves cognitive functions, and counteracts the neuronal cell damage in these pathologies (Łuc et al. 2019).

Further, the treatment with the antidiabetic drugs also shows improvement in the clinical symptoms of AD (Frozza et al. 2018). Also, the anti-glycemic agents reduce the risk of PD development in clinical settings (Svenningsson et al. 2016). Treatment of the PD patients with the antidiabetic agents reduces neuronal inflammation and insulin resistance and improves insulin signaling pathway in the brain of PD patients (Athauda and Foltynie 2016). Further, the treatment with protein tyrosine phosphatase-1B (PTP1B) inhibitors improves peripheral insulin resistance (Liu et al. 2010; Qin et al. 2016) and has been used in the early stages of diabetes and PD (Athauda and Foltynie 2016). Also, there is a great and urgent need for specific preclinical models of dementia for the evaluation of potential and antidementia agents. Further, the targeting of dementia by stages and the treatment in earlier stages confers beneficial effects in the patients (Łuc et al. 2019).

3.9 Conclusion

It is concluded that brain insulin is responsible for the regulation of cognition and memory. Insulin signaling has been found to be de-sensitized in the brains of dementia patients. Therefore, impaired insulin signaling is responsible for the emergence of dementia and related pathologies. Insulin resistance in the brain finally leads to apoptosis and neurodegeneration. Deregulated glucose metabolism mediated through insulin receptors leads to dementia-like symptoms of AD. Further, the administration of insulin produces improvement in the memory and cognitive outcomes in dementia patients, suggesting that insulin might be the milestone therapeutics for dementia or related conditions in near future.

References

Abbott MA, Wells DG, Fallon JR (1999) The insulin receptor tyrosine kinase substrate p58/53 and the insulin receptor are components of CNS synapses. J Neurosci 19(17):7300–7308

Adler CH, Connor DJ, Hentz JG, Sabbagh MN, Caviness JN, Shill HA, Noble B, Beach TG (2010) Incidental Lewy body disease: clinical comparison to a control cohort. Mov Disord 25(5):642–646

Aguirre V, Werner ED, Giraud J, Lee YH, Shoelson SE, White MF (2002) Phosphorylation of Ser307 in insulin receptor substrate-1 blocks interactions with the insulin receptor and inhibits insulin action. J Biol Chem 277(2):1531–1537

Akhtar A, Sah SP (2020 Feb) Insulin signaling pathway and related molecules: role in neurodegeneration and Alzheimer's disease. Neurochem Int 21:104707

Albeely AM, Ryan SD, Perreault ML (2018) Pathogenic feed-forward mechanisms in Alzheimer's and Parkinson's disease converge on GSK-3. Brain Plast 4(2):151–167

Alzheimer Society of Canada (2016) Vascular dementia. Available http://www.alzheimer.ca/en/Aboutdementia/Dementias/Vascular-Dementia. Accessed 19 Sep 2016

American Psychiatric Association (2000) Diagnostic and statistical manual of mental disorders, 4th edn. American Psychiatric Publishing, Washington, DC

Athauda D, Foltynie T (2016) Insulin resistance and Parkinson's disease: a new target for disease modification? Prog Neurobiol 145:98–120

Auer RN (1998) Insulin, blood glucose levels, and ischemic brain damage. Neurology 51(3 suppl 3):S39–S43

Aziz NA, Pijl H, Frölich M, Snel M, Streefland TC, Roelfsema F, Roos RA (2010) Systemic energy homeostasis in Huntington's disease patients. J Neurol Neurosurg Psychiatry 81(11):1233–1237

Bang J, Spina S, Miller BL (2015) Frontotemporal dementia. Lancet 386(10004):1672–1682

Banks WA, Kastin AJ (1998) Differential permeability of the blood–brain barrier to two pancreatic peptides: insulin and amylin. Peptides 19(5):883–889

Barroso E, del Valle J, Porquet D, Santos AM, Salvadó L, Rodríguez-Rodríguez R, Gutiérrez P, Anglada-Huguet M, Alberch J, Camins A, Palomer X (2013) Tau hyperphosphorylation and increased BACE1 and RAGE levels in the cortex of PPARβ/δ-null mice. Biochim Biophys Acta 1832(8):1241–1248

Begg DP (2015) Insulin transport into the brain and cerebrospinal fluid. In: Vitamins & hormones, vol 98. Academic Press, Boston, pp 229–248

Belfiore A, Frasca F, Pandini G, Sciacca L, Vigneri R (2009) Insulin receptor isoforms and insulin receptor/insulin-like growth factor receptor hybrids in physiology and disease. Endocr Rev 30(6):586–623

Belfiore A, Malaguarnera R, Vella V, Lawrence MC, Sciacca L, Frasca F, Morrione A, Vigneri R (2017) Insulin receptor isoforms in physiology and disease: an updated view. Endocr Rev 38(5):379–431

Benedict C, Hallschmid M, Hatke A, Schultes B, Fehm HL, Born J et al (2004) Intranasal insulin improves memory in humans. Psychoneuroendocrinology 29(10):1326–1334

Benedict C, Kern W, Schultes B, Born J, Hallschmid M (2008) Differential sensitivity of men and women to anorexigenic and memory-improving effects of intranasal insulin. J Clin Endocrinol Metabol 93(4):1339–1344

Beurel E, Jope RS (2009) Lipopolysaccharide-induced interleukin-6 production is controlled by glycogen synthase kinase-3 and STAT3 in the brain. J Neuroinflammation 6(1):1–1

Bingham EM, Hopkins D, Smith D, Pernet A, Hallett W, Reed L, Marsden PK, Amiel SA (2002) The role of insulin in human brain glucose metabolism: an 18fluoro-deoxyglucose positron emission tomography study. Diabetes 51(12):3384–3390

Blanchard JG, Duncan PM (1997) Effect of combinations of insulin, glucose and scopolamine on radial arm maze performance. Pharmacol Biochem Behav 58(1):209–214

Blázquez E, Velázquez E, Hurtado-Carneiro V, Ruiz-Albusac JM (2014) Insulin in the brain: its pathophysiological implications for States related with central insulin resistance, type 2 diabetes and Alzheimer's disease. Front Endocrinol 5:161

Bohnen NI, Kaufer DI, Ivanco LS, Lopresti B, Koeppe RA, Davis JG, Mathis CA, Moore RY, DeKosky ST (2003) Cortical cholinergic function is more severely affected in parkinsonian dementia than in Alzheimer disease: an in vivo positron emission tomographic study. Arch Neurol 60(12):1745–1748

Bomfim TR, Forny-Germano L, Sathler LB, Brito-Moreira J, Houzel JC, Decker H, Silverman MA, Kazi H, Melo HM, McClean PL, Holscher C (2012) An anti-diabetes agent protects the mouse brain from defective insulin signaling caused by Alzheimer's disease–associated Aβ oligomers. J Clin Invest 122(4):1339–1353

Bosco D, Fava A, Plastino M, Montalcini T, Pujia A (2011) Possible implications of insulin resistance and glucose metabolism in Alzheimer's disease pathogenesis. J Cell Mol Med 15(9):1807–1821

Boucher J, Kleinridders A, Kahn CR (2014) Insulin receptor signaling in normal and insulin-resistant states. Cold Spring Harb Perspect Biol 6(1):a009191

Braak H, Del Tredici K, Rüb U, De Vos RA, Steur EN, Braak E (2003) Staging of brain pathology related to sporadic Parkinson's disease. Neurobiol Aging 24(2):197–211

Burgos-Ramos E, González-Rodríguez Á, Canelles S, Baquedano E, Frago LM, Revuelta-Cervantes J, Gomez-Ambrosi J, Frühbeck G, Chowen JA, Argente J, Valverde AM (2012) Differential insulin receptor substrate-1 (IRS1)-related modulation of neuropeptide Y and proopiomelanocortin expression in nondiabetic and diabetic IRS2−/− mice. Endocrinology 153(3):1129–1140

Calle MC, Fernandez ML (2012) Inflammation and type 2 diabetes. Diabete Metab 38(3):183–191

Cargnello M, Roux PP (2011) Activation and function of the MAPKs and their substrates, the MAPK-activated protein kinases. Microbiol Mol Biol Rev 75(1):50–83

Chang CC, Li HH, Chang YT, Ho YJ, Hsieh LJ, Chiu PY, Cheng YS, Lin CL, Lai TJ (2018) Aβ exacerbates α-synuclein-induced neurotoxicity through impaired insulin signaling in α-synuclein-overexpressed human SK-N-MC neuronal cells. CNS Neurosci Ther 24(1):47–57

Chertkow H, Feldman HH, Jacova C, Massoud F (2013) Definitions of dementia and predementia states in Alzheimer's disease and vascular cognitive impairment: consensus from the Canadian conference on diagnosis of dementia. Alzheimers Res Ther 5(1):1–8

Chin PC, Majdzadeh N, D'Mello SR (2005) Inhibition of GSK3β is a common event in neuroprotection by different survival factors. Mol Brain Res 137(1–2):193–201

Choi J, Ko J, Racz B, Burette A, Lee JR, Kim S, Na M, Lee HW, Kim K, Weinberg RJ, Kim E (2005) Regulation of dendritic spine morphogenesis by insulin receptor substrate 53, a downstream effector of Rac1 and Cdc42 small GTPases. J Neurosci 25(4):869–879

Cimini A, Benedetti E, Cristiano L, Sebastiani P, D'Amico MA, D'Angelo B et al (2005) Expression of peroxisome proliferator-activated receptors (PPARs) and retinoic acid receptors (RXRs) in rat cortical neurons. Neuroscience 130(2):325–337

Clark IA, Vissel B (2014) Inflammation-sleep interface in brain disease: TNF, insulin, orexin. J Neuroinflammation 11(1):51

Clarke DW, Mudd L, Boyd FT Jr, Fields M, Raizada MK (1986) Insulin is released from rat brain neuronal cells in culture. J Neurochem 47(3):831–836

Cole AR, Astell A, Green C, Sutherland C (2007) Molecular connexions between dementia and diabetes. Neurosci Biobehav Rev 31(7):1046–1063

Collino M, Patel NS, Thiemermann C (2008) PPARs as new therapeutic targets for the treatment of cerebral ischemia/reperfusion injury. Ther Adv Cardiovasc Dis 2(3):179–197

Correia SC, Santos RX, Perry G, Zhu X, Moreira PI, Smith MA (2011) Insulin-resistant brain state: the culprit in sporadic Alzheimer's disease? Ageing Res Rev 10(2):264–273

Correia SC, Santos RX, Carvalho C, Cardoso S, Candeias E, Santos MS, Oliveira CR, Moreira PI (2012) Insulin signaling, glucose metabolism and mitochondria: major players in Alzheimer's disease and diabetes interrelation. Brain Res 1441:64–78

Cotero VE, Routh VH (2009) Insulin blunts the response of glucose-excited neurons in the ventrolateral-ventromedial hypothalamic nucleus to decreased glucose. Am J Physiol Endocrinol Metab 296(5):E1101–E1109

Credle JJ, George JL, Wills J, Duka V, Shah K, Lee YC et al (2015) GSK-3β dysregulation contributes to parkinson's-like pathophysiology with associated region-specific phosphorylation and accumulation of tau and α-synuclein. Cell Death Differ 22(5):838–851

De Meyts P (2000) The insulin receptor and its signal transduction network. [Updated 2016 Apr 27]. In: Feingold KR, Anawalt B, Boyce A et al. (eds) Endotext [Internet]. MDText.com, Inc., South Dartmouth

Derakhshan F, Toth C (2013) Insulin and the brain. Curr Diabetes Rev 9(2):102–116

Devaskar SU, Giddings SJ, Rajakumar PA, Carnaghi LR, Menon RK, Zahm DS (1994) Insulin gene expression and insulin synthesis in mammalian neuronal cells. J Biol Chem 269(11):8445–8454

Devraj K (2010) The glucose transporters that facilitate the passage of glucose across the blood-brain barrier. The Pennsylvania State University, State College

Dickson DW, Fujishiro H, DelleDonne A, Menke J, Ahmed Z, Klos KJ, Josephs KA, Frigerio R, Burnett M, Parisi JE, Ahlskog JE (2008) Evidence that incidental Lewy body disease is pre-symptomatic Parkinson's disease. Acta Neuropathol 115(4):437–444

Dineley KT, Jahrling JB, Denner L (2014) Insulin resistance in Alzheimer's disease. Neurobiol Dis 72:92–103

Duarte AI, Santos MS, Oliveira CR, Rego AC (2005) Insulin neuroprotection against oxidative stress in cortical neurons—involvement of uric acid and glutathione antioxidant defenses. Free Radic Biol Med 39(7):876–889

Duarte AI, Proença T, Oliveira CR, Santos MS, Rego AC (2006) Insulin restores metabolic function in cultured cortical neurons subjected to oxidative stress. Diabetes 55(10):2863–2870

Duarte AI, Santos P, Oliveira CR, Santos MS, Rego AC (2008) Insulin neuroprotection against oxidative stress is mediated by Akt and GSK-3β signaling pathways and changes in protein expression. Biochim Biophys Acta 1783(6):994–1002

Duarte AI, Moreira PI, Oliveira CR (2012) Insulin in central nervous system: more than just a peripheral hormone. J Aging Res 2012:384017

Duffy KR, Pardridge WM (1987) Blood-brain barrier transcytosis of insulin in developing rabbits. Brain Res 420(1):32–38

Duong S, Patel T, Chang F (2017) Dementia: what pharmacists need to know. Can Pharm J 150(2):118–129

Emre M (2003) Dementia associated with Parkinson's disease. Lancet Neurol 2(4):229–237

Frigerio R, Fujishiro H, Ahn TB, Josephs KA, Maraganore DM, DelleDonne A, Parisi JE, Klos KJ, Boeve BF, Dickson DW, Ahlskog JE (2011) Incidental Lewy body disease: do some cases represent a preclinical stage of dementia with Lewy bodies? Neurobiol Aging 32(5):857–863

Frölich L, Blum-Degen D, Bernstein HG, Engelsberger S, Humrich J, Laufer S, Muschner D, Thalheimer A, Türk A, Hoyer S, Zöchling R (1998) Brain insulin and insulin receptors in aging and sporadic Alzheimer's disease. J Neural Transm 105(4–5):423–438

Frozza RL, Lourenco MV, De Felice FG (2018) Challenges for Alzheimer's disease therapy: insights from novel mechanisms beyond memory defects. Front Neurosci 12:37

Fujiwara H, Hasegawa M, Dohmae N, Kawashima A, Masliah E, Goldberg MS et al (2002) A-Synuclein is phosphorylated in synucleinopathy lesions. Nat Cell Biol 4(2):160–164

Fulop T, Larbi A, Douziech N (2003) Insulin receptor and ageing. Pathol Biol 51(10):574–580

Garg R, Chaudhuri A, Munschauer F, Dandona P (2006) Hyperglycemia, insulin, and acute ischemic stroke: a mechanistic justification for a trial of insulin infusion therapy. Stroke 37(1):267–273

Gaspar JM, Baptista FI, Macedo MP, Ambrosio AF (2016) Inside the diabetic brain: role of different players involved in cognitive decline. ACS Chem Nerosci 7(2):131–142

Gerozissis K (2003) Brain insulin: regulation, mechanisms of action and functions. Cell Mol Neurobiol 23(1):1–25

Gerozissis K (2008) Brain insulin, energy and glucose homeostasis; genes, environment and metabolic pathologies. Eur J Pharmacol 585(1):38–49

Gratwicke J, Jahanshahi M, Foltynie T (2015) Parkinson's disease dementia: a neural networks perspective. Brain 138(pt 6):1454–1476

Gregor MF, Hotamisligil GS (2011) Inflammatory mechanisms in obesity. Annu Rev Immunol 29: 415–445

Grunstein HS, James DE, Storlien LH, Smythe GA, Kraegen EW (1985) Hyperinsulinemia suppresses glucose utilization in specific brain regions: in vivo studies using the euglycemic clamp in the rat. Endocrinology 116(2):604–610

Havrankova J, Roth J (1979) Concentrations of insulin and of insulin receptors in the brain are independent of peripheral insulin levels: studies of obese and streptozotocin-treated rodents. J Clin Invest 64(2):636–642

Heffetz D, Zick Y (1986) Receptor aggregation is necessary for activation of the soluble insulin receptor kinase. J Biol Chem 261(2):889–894

Heidenreich KA, Zahniser NR, Berhanu P, Brandenburg D, Olefsky JM (1983) Structural differences between insulin receptors in the brain and peripheral target tissues. J Biol Chem 258(14):8527–8530

Heneka MT, Golenbock DT, Latz E (2015) Innate immunity in Alzheimer's disease. Nat Immunol 16(3):229–236

Hildreth KL, Church S (2015) Evaluation and management of the elderly patient presenting with cognitive complaints. Med Clin North Am 99(2):311–335

Hill JM, Lesniak MA, Pert CB, Roth J (1986) Autoradiographic localization of insulin receptors in rat brain: prominence in olfactory and limbic areas. Neuroscience 17(4):1127–1138

Hölscher C (2014) Insulin, incretins and other growth factors as potential novel treatments for Alzheimer's and Parkinson's diseases. Biochem Soc Trans 42:593–599

Ibberson M, Riederer BM, Uldry M, Guhl B, Roth J, Thorens B (2002) Immunolocalization of GLUTX1 in the testis and to specific brain areas and vasopressin-containing neurons. Endocrinology 143(1):276–284

Jha SK, Jha NK, Kar R, Ambasta RK, Kumar P (2015) p38 MAPK and PI3K/AKT signalling cascades in Parkinson's disease. Int J Mol Cell Med 4(2):67

Jodeiri Farshbaf M, Ghaedi K (2017) Huntington's disease and mitochondria. Neurotox Res 32(3):518–529

Joers V, Tansey MG, Mulas G, Carta AR (2017) Microglial phenotypes in Parkinson's disease and animal models of the disease. Prog Neurobiol 155:57–75

Jurcovicova J (2014) Glucose transport in brain - effect of inflammation. Endocr Regul 48(1):35–48

Kao SY (2009) Rescue of α-synuclein cytotoxicity by insulin-like growth factors. Biochem Biophys Res Commun 385(3):434–438

Kastin AJ, Akerstrom V (2001) Glucose and insulin increase the transport of leptin through the blood-brain barrier in normal mice but not in streptozotocin-diabetic mice. Neuroendocrinology 73(4):237–242

Kern W, Born J, Schreiber H, Fehm HL (1999) Central nervous system effects of intranasally administered insulin during euglycemia in men. Diabetes 48(3):557–563

Kim B, Feldman EL (2012) Insulin resistance in the nervous system. Trends Endocrinol Metab 23(3):133–141

Kim SJ, Han Y (2005) Insulin inhibits AMPA-induced neuronal damage via stimulation of protein kinase B (Akt). J Neural Transm 112(2):179–191

Kim WS, Kågedal K, Halliday GM (2014) Alpha-synuclein biology in Lewy body diseases. Alzheimers Res Ther 6(5–8):73

Kopf SR, Baratti CM (1999) Effects of posttraining administration of insulin on retention of a habituation response in mice: participation of a central cholinergic mechanism. Neurobiol Learn Mem 71(1):50–61

Kosaka K, Iseki E, Odawara T, Yamamoto T (1996) Cerebral type of Lewy body disease. Neuropathology 16(1):32–35

Kosaka K, Yoshimura M, Ikeda K, Budka H (1984) Diffuse type of Lewy body disease: progressive dementia with abundant cortical Lewy bodies and senile changes of varying degree–a new disease? Clin Neuropathol 3(5):185–192

Kumar-Singh S (2011) Progranulin and TDP-43: mechanistic links and future directions. J Mol Neurosci 45(3):561

de La Monte SM (2012) Early intranasal insulin therapy halts progression of neurodegeneration: progress in Alzheimer's disease therapeutics. Aging Health 8(1):61–64

Lalić NM, Marić J, Svetel M, Jotić A, Stefanova E, Lalić K, Dragašević N, Miličić T, Lukić L, Kostić VS (2008) Glucose homeostasis in Huntington disease: abnormalities in insulin sensitivity and early-phase insulin secretion. Arch Neurol 65(4):476–480
Landau BR, Takaoka Y, Abrams MA, Genuth SM, van Houten M, Posner BI, White RJ, Ohgaku S, Horvat A, Hemmelgarn E (1983) Binding of insulin by monkey and pig hypothalamus. Diabetes 32(3):284–292
Lee CC, Huang CC, Wu MY, Hsu KS (2005) Insulin stimulates postsynaptic density-95 protein translation via the phosphoinositide 3-kinase-Akt-mammalian target of rapamycin signaling pathway. J Biol Chem 280(18):18543–18550
Lee CC, Huang CC, Hsu KS (2011) Insulin promotes dendritic spine and synapse formation by the PI3K/Akt/mTOR and Rac1 signaling pathways. Neuropharmacology 61(4):867–879
Lemmon MA, Schlessinger J (2010) Cell signaling by receptor tyrosine kinases. Cell 141(7):1117–1134
Levenga J, Wong H, Milstead RA, Keller BN, LaPlante LE, Hoeffer CA (2017) AKT isoforms have distinct hippocampal expression and roles in synaptic plasticity. Elife 6:e30640
Levin BE, Routh VH, Kang L, Sanders NM, Dunn-Meynell AA (2004) Neuronal glucosensing: what do we know after 50 years? Diabetes 53(10):2521–2528
Li Y, Liu L, Barger SW, Griffin WS (2003) Interleukin-1 mediates pathological effects of microglia on tau phosphorylation and on synaptophysin synthesis in cortical neurons through a p38-MAPK pathway. J Neurosci 23(5):1605–1611
Li C, Li Y, He L, Agarwal AR, Zeng N, Cadenas E, Stiles BL (2013) PI3K/AKT signaling regulates bioenergetics in immortalized hepatocytes. Free Radic Biol Med 60:29–40
Liou CJ, Tong M, Vonsattel JP, de la Monte SM (2019) Altered brain expression of insulin and insulin-like growth factors in frontotemporal lobar degeneration: another degenerative disease linked to dysregulation of insulin metabolic pathways. ASN Neuro 11:1759091419839515
Liu H, Yang H, Wang D, Liu Y, Liu X, Li Y, Xie L, Wang G (2009) Insulin regulates P-glycoprotein in rat brain microvessel endothelial cells via an insulin receptor-mediated PKC/NF-κB pathway but not a PI3K/Akt pathway. Eur J Pharmacol 602(2–3):277–282
Liu Z, Chai Q, Li YY, Shen Q, Ma LP, Zhang LN, Wang X, Sheng L, Li JY, Li J, Shen JK (2010) Discovery of novel PTP1B inhibitors with antihyperglycemic activity. Acta Pharmacol Sin 31(8):1005–1012
Liu J, Stevens PD, Gao T (2011a) mTOR-dependent regulation of PHLPP expression controls the rapamycin sensitivity in cancer cells. J Biol Chem 286(8):6510–6520
Liu J, Stevens PD, Li X, Schmidt MD, Gao T (2011b) PHLPP-mediated dephosphorylation of S6K1 inhibits protein translation and cell growth. Mol Cell Biol 31(24):4917–4927
Łuc M, Woźniak M, Helemejko M, Rymaszewska J (2019) Tackling Alzheimer's disease: Hypothetical synergism between anti-inflammatory and anti-diabetic agents. Life Sci 231:116483
Lucas JJ, Hernández F, Gómez-Ramos P, Morán MA, Hen R, Avila J (2001) Decreased nuclear β-catenin, tau hyperphosphorylation and neurodegeneration in GSK-3β conditional transgenic mice. EMBO J 20(1–2):27–39
Lyman M, Lloyd DG, Ji X, Vizcaychipi MP, Ma D (2014) Neuroinflammation: the role and consequences. Neurosci Res 79:1–2
Ma J, Jiang T, Tan L, Yu JT (2015) TYROBP in Alzheimer's disease. Mol Neurobiol 51(2):820–826
Mackie S, Aitken A (2005) Novel brain 14-3-3 interacting proteins involved in neurodegenerative disease. FEBS J 272(16):4202–4210
Malagelada C, Jin ZH, Greene LA (2008) RTP801 is induced in Parkinson's disease and mediates neuron death by inhibiting Akt phosphorylation/activation. J Neurosci 28(53):14363–14371
Manna P, Jain SK (2013) PIP3 but not PIP2 increases GLUT4 surface expression and glucose metabolism mediated by AKT/PKCzeta/lambda phosphorylation in 3T3L1 adipocytes. Mol Cell Biochem 381(1-2):291–299
Margolis RU, Altszuler N (1967) Insulin in the cerebrospinal fluid. Nature 215(5108):1375–1376

Marsh SE, Blurton-Jones M (2012) Examining the mechanisms that link β-amyloid and α-synuclein pathologies. Alzheimers Res Ther 4(2):1–8
Martin SJ, Clark RE (2007) The rodent hippocampus and spatial memory: from synapses to systems. Cell Mol Life Sci 64(4):401–431
Martin B, Golden E, Carlson OD, Pistell P, Zhou J, Kim W et al (2009) Exendin-4 improves glycemic control, ameliorates brain and pancreatic pathologies, and extends survival in a mouse model of Huntington's disease. Diabetes 58:318–328
Marty N, Dallaporta M, Thorens B (2007) Brain glucose sensing, counterregulation, and energy homeostasis. Physiology (Bethesda) 22:241–251
McEwen BS, Reagan LP (2004) Glucose transporter expression in the central nervous system: relationship to synaptic function. Eur J Pharmacol 490(1–3):13–24
Miller DW, Keller BT, Borchardt RT (1994) Identification and distribution of insulin receptors on cultured bovine brain microvessel endothelial cells: possible function in insulin processing in the blood-brain barrier. J Cell Physiol 161(2):333–341
Miyahara A, Okamura-Oho Y, Miyashita T, Hoshika A, Yamada M (2003) Genomic structure and alternative splicing of the insulin receptor tyrosine kinase substrate of 53-kDa protein. J Hum Genet 48(8):410–414
Moloney AM, Griffin RJ, Timmons S, O'Connor R, Ravid R, O'Neill C (2010) Defects in IGF-1 receptor, insulin receptor and IRS-1/2 in Alzheimer's disease indicate possible resistance to IGF-1 and insulin signalling. Neurobiol Aging 31(2):224–243
de la Monte SM (2009) Insulin resistance and Alzheimer's disease. BMB Rep 42(8):475
de la Monte SM (2017) Insulin resistance and neurodegeneration: progress towards the development of new therapeutics for Alzheimer's disease. Drugs 77(1):47–65
de la Monte SM, Wands JR (2005) Review of insulin and insulin-like growth factor expression, signaling, and malfunction in the central nervous system: relevance to Alzheimer's disease. J Alzheimers Dis 7(1):45–61
de la Monte SM, Wands JR (2006) Molecular indices of oxidative stress and mitochondrial dysfunction occur early and often progress with severity of Alzheimer's disease. J Alzheimers Dis 9(2):167–181
de la Monte SM, Tong M, Lester-Coll N, Plater M Jr, Wands JR (2006) Therapeutic rescue of neurodegeneration in experimental type 3 diabetes: relevance to Alzheimer's disease. J Alzheimers Dis 10(1):89–109
de la Monte SM, Tong M, Schiano I, Didsbury J (2017) Improved brain insulin/IGF signaling and reduced neuroinflammation with T3D-959 in an experimental model of sporadic Alzheimer's disease. J Alzheimers Dis 55(2):849–864
Montojo MT, Aganzo M, González N (2017) Huntington's disease and diabetes: chronological sequence of its association. J Huntingtons Dis 6(3):179–188
Myers MG Jr, Sun XJ, White MF (1994) The IRS-1 signaling system. Trends Biochem Sci 19(7):289–293
Nagamatsu S, Kornhauser JM, Burant CF, Seino S, Mayo KE, Bell GI (1992) Glucose transporter expression in brain. cDNA sequence of mouse GLUT3, the brain facilitative glucose transporter isoform, and identification of sites of expression by in situ hybridization. J Biol Chem 267(1):467–472
Naia L, Ribeiro M, Rodrigues J, Duarte AI, Lopes C, Rosenstock TR, Hayden MR, Rego AC (2016) Insulin and IGF-1 regularize energy metabolites in neural cells expressing full-length mutant huntingtin. Neuropeptides 58:73–81
Najem D (2014) Insulin, cholesterol and A-beta: roles and mechanisms in Alzheimer's disease (Doctoral dissertation, Université d'Ottawa/University of Ottawa)
Nandipati KC, Subramanian S, Agrawal DK (2017) Protein kinases: mechanisms and downstream targets in inflammation-mediated obesity and insulin resistance. Mol Cell Biochem 426 (1–2):27–45
Noguchi M, Suizu F (2012) Regulation of AKT by phosphorylation of distinct threonine and serine residues. Adv Med Biol 47:139–162

Obici S, Feng Z, Karkanias G, Baskin DG, Rossetti L (2002a) Decreasing hypothalamic insulin receptors causes hyperphagia and insulin resistance in rats. Nat Neurosci 5(6):566–572
Obici S, Zhang BB, Karkanias G, Rossetti L (2002b) Hypothalamic insulin signaling is required for inhibition of glucose production. Nat Med 8(12):1376–1382
Pang Y, Lin S, Wright C, Shen J, Carter K, Bhatt A, Fan LW (2016) Intranasal insulin protects against substantia nigra dopaminergic neuronal loss and alleviates motor deficits induced by 6-OHDA in rats. Neuroscience 318:157–165
Park CR, Seeley RJ, Craft S, Woods SC (2000) Intracerebroventricular insulin enhances memory in a passive-avoidance task. Physiol Behav 68(4):509–514
Pilch PF, Shia MA, Benson RJ, Fine RE (1983) Coated vesicles participate in the receptor-mediated endocytosis of insulin. J Cell Biol 96(1):133–138
Pinelis VG, Surin AM, Fisenko AP, Boyarkin D, Balyasin M, Demchenko A, Asanov A, Pomytkin I, Krasilnikova I (2019) Insulin protects cortical neurons against glutamate excitotoxicity. Front Neurosci 13:1027
Plastino M, Fava A, Pirritano D, Cotronei P, Sacco N, Sperlì T, Spanò A, Gallo D, Mungari P, Consoli D, Bosco D (2010) Effects of insulinic therapy on cognitive impairment in patients with Alzheimer disease and diabetes mellitus type-2. J Neurol Sci 288(1–2):112–116
Pocai A, Lam TK, Gutierrez-Juarez R, Obici S, Schwartz GJ, Bryan J et al (2005) Hypothalamic K (ATP) channels control hepatic glucose production. Nature 434(7036):1026–1031
Pomytkin I, Costa-Nunes JP, Kasatkin V, Veniaminova E, Demchenko A, Lyundup A, Lesch KP, Ponomarev ED, Strekalova T (2018) Insulin receptor in the brain: mechanisms of activation and the role in the CNS pathology and treatment. CNS Neurosci Ther 24(9):763–774
Pouladi MA, Xie Y, Skotte NH, Ehrnhoefer DE, Graham RK, Kim JE, Bissada N, Yang XW, Paganetti P, Friedlander RM, Leavitt BR (2010) Full-length huntingtin levels modulate body weight by influencing insulin-like growth factor 1 expression. Hum Mol Genet 19(8):1528–1538
Qin XY, Zhang SP, Cao C, Loh YP, Cheng Y (2016) Aberrations in peripheral inflammatory cytokine levels in Parkinson disease: a systematic review and meta-analysis. JAMA Neurol 73(11):1316–1324
Quintanilla RA, Orellana DI, González-Billault C, Maccioni RB (2004) Interleukin-6 induces Alzheimer's-type phosphorylation of tau protein by deregulating the cdk5/p35 pathway. Exp Cell Res 295(1):245–257
Ramirez SH, Fan S, Zhang M, Papugani A, Reichenbach N, Dykstra H, Mercer AJ, Tuma RF, Persidsky Y (2010) Inhibition of glycogen synthase kinase 3β (GSK3β) decreases inflammatory responses in brain endothelial cells. Am J Pathol 176(2):881–892
Rascovsky K, Hodges JR, Knopman D, Mendez MF, Kramer JH, Neuhaus J, Van Swieten JC, Seelaar H, Dopper EG, Onyike CU, Hillis AE (2011) Sensitivity of revised diagnostic criteria for the behavioural variant of frontotemporal dementia. Brain 134(9):2456–2477
Reger MA, Watson GS, Green PS, Baker LD, Cholerton B, Fishel MA, Plymate SR, Cherrier MM, Schellenberg GD, Frey WH 2nd, Craft S (2008a) Intranasal insulin administration dose-dependently modulates verbal memory and plasma amyloid-beta in memory-impaired older adults. J Alzheimers Dis 13(3):323–331
Reger MA, Watson GS, Green PS, Wilkinson CW, Baker LD, Cholerton B, Fishel MA, Plymate SR, Breitner JC, DeGroodt W, Mehta P, Craft S (2008b) Intranasal insulin improves cognition and modulates beta-amyloid in early AD. Neurology 70(6):440–448
Reitz C (2015) Genetic diagnosis and prognosis of Alzheimer's disease: challenges and opportunities. Expert Rev Mol Diagn 15(3):339–348
Rensink AA, Otte-Höller I, de Boer R, Bosch RR, Hans J, de Waal RM, Verbeek MM, Kremer B (2004) Insulin inhibits amyloid β-induced cell death in cultured human brain pericytes. Neurobiol Aging 25(1):93–103
Ribe EM, Lovestone S (2016) Insulin signalling in Alzheimer′ s disease and diabetes: from epidemiology to molecular links. J Intern Med 280(5):430–442

Ribeiro M, Rosenstock TR, Oliveira AM, Oliveira CR, Rego AC (2014) Insulin and IGF-1 improve mitochondrial function in a PI-3K/Akt-dependent manner and reduce mitochondrial generation of reactive oxygen species in Huntington's disease knock-in striatal cells. Free Radic Biol Med 74:129–144

Ryu BR, Ko HW, Jou I, Noh JS, Gwag BJ (1999) Phosphatidylinositol 3-kinase-mediated regulation of neuronal apoptosis and necrosis by insulin and IGF-I. J Neurobiol 39(4):536–546

Santos MS, Pereira EM, Carvaho AP (1999) Stimulation of immunoreactive insulin release by glucose in rat brain synaptosomes. Neurochem Res 24(1):33–36

Sato H, Arawaka S, Hara S, Fukushima S, Koga K, Koyama S et al (2011) Authentically phosphorylated-synuclein at Ser129 accelerates neurodegeneration in a rat model of familial Parkinson's disease. J Neurosci 31(46):16884–16894

Sawallisch C, Berhörster K, Disanza A, Mantoani S, Kintscher M, Stoenica L, Dityatev A, Sieber S, Kindler S, Morellini F, Schweizer M (2009) The insulin receptor substrate of 53 kDa (IRSp53) limits hippocampal synaptic plasticity. J Biol Chem 284(14):9225–9236

Schulingkamp RJ, Pagano TC, Hung D, Raffa RB (2000) Insulin receptors and insulin action in the brain: review and clinical implications. Neurosci Biobehav Rev 24(8):855–872

Sekiyama K, Sugama S, Fujita M, Sekigawa A, Takamatsu Y, Waragai M, Takenouchi T, Hashimoto M (2012) Neuroinflammation in Parkinson's disease and related disorders: a lesson from genetically manipulated mouse models of α-synucleinopathies. Parkinsons Dis 2012: 271732

Sevanian A, Davies KJ, Hochstein P (1991) Serum urate as an antioxidant for ascorbic acid. Am J Clin Nutr 54(6):1129S–1134S

Shuaib A, Ijaz MS, Waqar T, Voll C, Kanthan R, Miyashita H et al (1995) Insulin elevates hippocampal GABA levels during ischemia. This is independent of its hypoglycemic effect. Neuroscience 67(4):809–814

Simpson IA, Appel NM, Hokari M, Oki J, Holman GD, Maher F, Koehler-Stec EM, Vannucci SJ, Smith QR (1999) Blood—brain barrier glucose transporter: effects of hypo-and hyperglycemia revisited. J Neurochem 72(1):238–247

Svenningsson P, Wirdefeldt K, Yin L, Fang F, Markaki I, Efendic S, Ludvigsson JF (2016) Reduced incidence of Parkinson's disease after dipeptidyl peptidase-4 inhibitors—a nationwide case-control study. Mov Disord 31(9):1422–1423

Tagliamonte A, DeMontis MG, Olianas M, Onali PL, Gessa GL (1975) Possible role of insulin in the transport of tyrosine and tryptophan from blood to brain. Pharmacol Res Commun 7(6):493–499

Talbot K, Wang HY, Kazi H, Han LY, Bakshi KP, Stucky A, Fuino RL, Kawaguchi KR, Samoyedny AJ, Wilson RS, Arvanitakis Z (2012) Demonstrated brain insulin resistance in Alzheimer's disease patients is associated with IGF-1 resistance, IRS-1 dysregulation, and cognitive decline. J Clin Invest 122(4):1316–1338

Tong M, Dong M, de la Monte SM (2009) Brain insulin-like growth factor and neurotrophin resistance in Parkinson's disease and dementia with Lewy bodies: potential role of manganese neurotoxicity. J Alzheimers Dis 16(3):585–599

Van Himbergen TM, Beiser AS, Ai M, Seshadri S, Otokozawa S, Au R, Thongtang N, Wolf PA, Schaefer EJ (2012) Biomarkers for insulin resistance and inflammation and the risk for all-cause dementia and Alzheimer disease: results from the Framingham Heart Study. Arch Neurol 69(5):594–600

Vanhanen M, Kuusisto J, Koivisto K, Mykkänen L, Helkala EL, Hänninen T, Riekkinen P, Soininen H, Laakso M (1999) Type-2 diabetes and cognitive function in a non-demented population. Acta Neurol Scand 100(2):97–101

Vannucci SJ, Koehler-Stec EM, Li K, Reynolds TH, Clark R, Simpson IA (1998) GLUT4 glucose transporter expression in rodent brain: effect of diabetes. Brain Res 797(1):1–11

Vienberg SG, Bouman SD, Sørensen H, Stidsen CE, Kjeldsen T, Glendorf T, Sørensen AR, Olsen GS, Andersen B, Nishimura E (2011) Receptor-isoform-selective insulin analogues give tissue-preferential effects. Biochem J 440(3):301–308

Voll CL, Auer RN (1991) Insulin attenuates ischemic brain damage independent of its hypoglycemic effect. J Cereb Blood Flow Metab 11(6):1006–1014
Vykoukal D, Davies MG (2011) Vascular biology of metabolic syndrome. J Vasc Surg 54(3):819–831
Wang W, Yang Y, Ying C, Li W, Ruan H, Zhu X, You Y, Han Y, Chen R, Wang Y, Li M (2007) Inhibition of glycogen synthase kinase-3β protects dopaminergic neurons from MPTP toxicity. Neuropharmacology 52(8):1678–1684
Wang MJ, Huang HY, Chen WF, Chang HF, Kuo JS (2010) Glycogen synthase kinase-3β inactivation inhibits tumor necrosis factor-α production in microglia by modulating nuclear factor κB and MLK3/JNK signaling cascades. J Neuroinflammation 7(1):99
Warby SC, Montpetit A, Hayden AR, Carroll JB, Butland SL, Visscher H, Collins JA, Semaka A, Hudson TJ, Hayden MR (2009) CAG expansion in the Huntington disease gene is associated with a specific and targetable predisposing haplogroup. Am J Hum Genet 84(3):351–366
Watson GS, Bernhardt T, Reger MA, Cholerton BA, Baker LD, Peskind ER, Asthana S, Plymate SR, Frölich L, Craft S (2006) Insulin effects on CSF norepinephrine and cognition in Alzheimer's disease. Neurobiol Aging 27(1):38–41
Wei L, Matsumoto H, Rhoads DE (1990) Release of immunoreactive insulin from rat brain synaptosomes under depolarizing conditions. J Neurochem 54(5):1661–1662
Werther GA, Hogg A, Oldfield BJ, McKinley MJ, Figdor R, Allen AM et al (1987) Localization and characterization of insulin receptors in rat brain and pituitary gland using in vitro autoradiography and computerized densitometry. Endocrinology 121(4):1562–1570
Werther GA, Hogg A, Oldfield BJ, McKinley MJ, Figdor R, Mendelsohn FA (1989) Localization and characterization of insulin-like growth factor-I receptors in rat brain and pituitary gland using in vitro autoradiography and computerized densitometry* a distinct distribution from insulin receptors. J Neuroendocrinol 1(5):369–377
White MF (2003) Insulin signaling in health and disease. Science 302(5651):1710–1711
White H, Venkatesh B (2011) Clinical review: ketones and brain injury. Crit Care 15(2):219
Wilden PA, Kahn CR, Siddle K, White MF (1992) Insulin receptor kinase domain autophosphorylation regulates receptor enzymatic function. J Biol Chem 267:16660–16668
Woods SC, Porte DA Jr (1977) Relationship between plasma and cerebrospinal fluid insulin levels of dogs. Am J Physiol Endocrinol Metab 233(4):E331
Wu Y, Shang Y, Sun S, Liang H, Liu R (2007) Erythropoietin prevents PC12 cells from 1-methyl-4-phenylpyridinium ion-induced apoptosis via the Akt/GSK-3β/caspase-3 mediated signaling pathway. Apoptosis 12(8):1365–1375
Yamamoto A, Cremona ML, Rothman JE (2006) Autophagy-mediated clearance of huntingtin aggregates triggered by the insulin-signaling pathway. J Cell Biol 172(5):719–731
Yao Y, Chinnici C, Tang H, Trojanowski JQ, Lee VM, Praticò D (2004) Brain inflammation and oxidative stress in a transgenic mouse model of Alzheimer-like brain amyloidosis. J Neuroinflammation 1(1):1–9
Yu LY, Pei Y (2015) Insulin neuroprotection and the mechanisms. Chin Med J (Engl) 128(7):976
Yuan Y, Sun J, Zhao M, Hu J, Wang X, Du G et al (2010) Overexpression of alpha-synuclein down-regulates BDNF expression. Cell Mol Neurobiol 30(6):939–946
Zahniser NR, Goens MB, Hanaway PJ, Vinych JV (1984) Characterization and regulation of insulin receptors in rat brain. J Neurochem 42(5):1354–1362
Zhao Z, Xiang Z, Haroutunian V, Buxbaum JD, Stetka B, Pasinetti GM (2007) Insulin degrading enzyme activity selectively decreases in the hippocampal formation of cases at high risk to develop Alzheimer's disease. Neurobiol Aging 28(6):824–830
Ziabreva I, Perry E, Perry R, Minger SL, Ekonomou A, Przyborski S, Ballard C (2006) Altered neurogenesis in Alzheimer's disease. J Psychosom Res 61(3):311–316
Zúñiga LA, Shen WJ, Joyce-Shaikh B, Pyatnova EA, Richards AG, Thom C, Andrade SM, Cua DJ, Kraemer FB, Butcher EC (2010) IL-17 regulates adipogenesis, glucose homeostasis, and obesity. J Immunol 185(11):6947–6959

Sex Hormones as Risk Factors for Dementia

4

Priscilla Kolibea Mante and Nana Ofori Adomako

Abstract

Generally, estrogens and androgens are the two main classes of sex hormones. These typically exert their actions through interaction with nuclear receptors, mediating reproductive functions in mammalian systems. The brain, a sex hormone-responsive organ, is an important target and source of sex hormones. Sex hormones modulate an essential number of biological processes including neuronal function. In the CNS, these hormones mediate neuroprotection, neuroplasticity, and brain energy metabolism. Normal aging is accompanied by significant depletion of sex hormones such as 17-beta-estradiol, progesterone, and testosterone. Hormone-responsive tissues and organs therefore risk susceptibility to disease as a result of this depletion. Aging is considered the most important risk factor for developing neurodegenerative conditions like Alzheimer's and Parkinson's disease. These conditions constitute the primary causes of dementia. This chapter covers sex hormones and how their influence on synaptic plasticity, neurosteroidogenesis, neuroinflammation, DNA repair, neuroprotection, and interaction with growth factors and mitochondria impacts on dementia.

P. K. Mante (✉)
Department of Pharmacology, Kwame Nkrumah University of Science and Technology, Kumasi, Ghana
e-mail: pkmante.pharm@knust.edu.gh

N. O. Adomako
Department of Pharmacy Practice, Kwame Nkrumah University of Science and Technology, Kumasi, Ghana

G. M. Ashraf, M. S. Uddin (eds.), *Current Thoughts on Dementia*,
https://doi.org/10.1007/978-981-16-7606-2_4

Keywords

Estrogen · Progesterone · Testosterone · Neurosteroids · Neurodegeneration · Synaptic plasticity · Alzheimer's disease · Aging

4.1 Introduction

Aging is a dynamic, complex process and may be viewed as an accumulation of deficits in an individual organism, in a variety of ways and rates for different organ systems, influenced by an interaction of intrinsic and extrinsic factors (Harman, 1991; Kritsilis et al. 2018). Hence, aging is associated with gradual homeostatic dysregulation resulting in an organism becoming less and ultimately non-resilient (Barron and Pike 2012; Zárate et al. 2017). All animal species, both male and female, experience aging. This includes humans. Sex differences, however, exist in the aging process (Austad and Fischer 2016). Among mammals, females rather than males tend to have longer expectancies of life (Barford et al. 2006; Glei and Horiuchi 2007; Seifarth et al. 2012). This is not an exclusive phenomenon in humans. These differences in life expectancies most likely reflect definite biological characteristics of both sexes, rather than socioeconomic factors (Zárate et al. 2017).

Tauopathies such as Alzheimer's disease (AD) and Parkinson's disease (PD) are neurodegenerative conditions that have a strong correlation with age increase (Ishihara et al. 1999; Prvulovic et al. 2005; Spittau 2017). AD has been considered as the primary cause of dementia and even considered to be a major subtype of dementia (Fratiglioni et al. 2000; Walker et al. 2015); other common causes being Lewy body, frontotemporal dementia, vascular dementia, and vascular cognitive impairment (Ebly et al., 1994; Sancesario and Bernardini, 2018). Incidence and prevalence of dementia considerably surge in later life making age an important risk factor. Currently, there are no effective interventions known that modify the advancement of the disease after clinical onset (Matthews et al. 2005; Plassman et al. 2007).

About one-third of dementia cases have been considered preventable by provision of interventions aimed at modifiable risk factors (Livingston et al. 2017; Winblad et al. 2016). Given that sex differences in the genetic risk and prevalence of dementia have been recognized previously (Gatz et al. 2003; Lövheim et al. 2009), potentially modifiable factors that are often overlooked are sex hormones. Dementia, commonly characterized by a decline of cognitive ability, results from damage or loss of once-healthy neurons (Donev et al. 2009). With advancement in age, loss of neurons occurs; patients of dementia and related conditions, however, suffer greater loss (Raz et al. 2016; Selkoe 2001). The influence of sex hormones on neuronal structure and function impacts significantly on cognition (Gurvich et al. 2018), and their exploration may yield valuable therapeutic interventions for dementia management.

This chapter explores how neuroactive steroid hormones produced either in peripheral tissues or directly in neuronal cells, modulate classical nuclear receptors and/or non-classical steroid receptors, and act as key physiological regulators of nervous function, affecting the pathophysiology of dementia and neurodegenerative conditions. The influence of these hormones on synaptic plasticity, neurosteroidogenesis, neuroinflammation, DNA repair, neuroprotection, and interaction with growth factors and mitochondria and how these interactions contribute to dementia are covered.

4.2 Sex Hormones

Sex hormones also referred to as gonadal steroids or gonadocorticoids are fat-soluble compounds with a similar structure to mineralocorticoids and glucocorticoids, which are primary steroidal groups (Baker 2003; Guerriero 2009). Sex steroids have the ability to control sexual maturity and growth. Naturally, they are produced by the gonads (testes produce androgens or ovaries produce estrogens and progestin) (Brook 1999), in the adrenal glands or through enzymatic conversion in the liver or fatty tissues (Comitato et al. 2015; Feldman et al. 2002). Figure 4.1 shows the biosynthesis of sex hormones covered in this chapter.

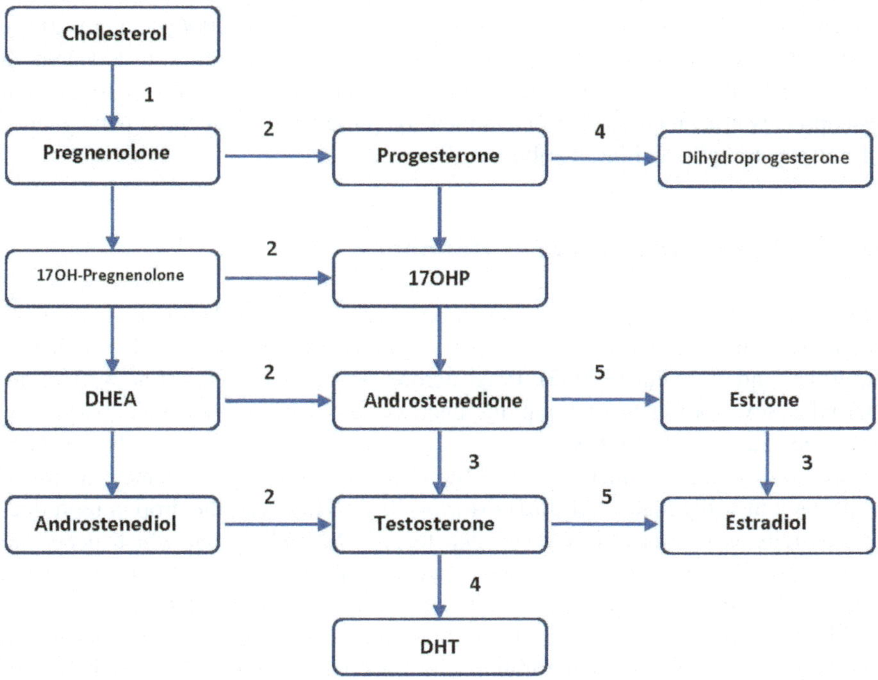

Fig. 4.1 The pathway of sex steroid biosynthesis. Enzymes are numbered in order of appearance as follows: (1) cleavage enzyme of P450 side chain; (2) 3β-hydroxysteroid dehydrogenase; (3) 17-β-hydroxysteroid dehydrogenase; (4) 5α-reductase; and (5) aromatase. Not all pathways, enzymes, and intermediate steroids are included in the diagram. The dashed arrow is indicative of poor flux

Gonadal synthesis of estrogen is modulated by the follicle-stimulating hormone whose function is also controlled by the hypothalamic gonadotropin-releasing hormone, while synthesis of progesterone is controlled by the luteinizing hormone (Catenaccio et al. 2016; Micevych et al. 2007). The pituitary glands and hypothalamus are the main modulators of testosterone production sending signals to the testes as to the quantities of testosterone needed to be produced.

Sex steroids interact with mammalian steroid hormone receptors (Acconcia and Marino, 2018; Amenyogbe et al. 2020). The effects produced in the biological system are affected generally through interaction with nuclear receptors. Generally, estrogens and androgens are the two main classes of sex steroids, but progesterone may be included as the third class. Androgens are typically referred to as the male sex hormones, while progesterone and estrogen are known to be female sex hormones, even though men and women have all the hormones present but in different quantities (ElAttar and Hugoson 1974; Gurvich et al. 2018).

Through genomic and non-genomic receptor binding, they show activity in human brains and are also involved in various molecular processes (Foradori et al. 2008; Wilkenfeld et al. 2018) and have been shown to be essential in neuroprotection (Thakur and Paramanik, 2009). Relatedly, these sex hormones can affect neural and behavioral functions such as cognition, motor coordination, blood pressure regulation, pain opioid sensitivity, and mood. These sex steroids also possess other non-reproductive effects (Aloisi and Bonifazi 2006; Kimura 2002). Estrogen antagonizes the action of parathyroid hormones resulting in a decreased loss of calcium from the bones. Testosterone has also been shown to prevent osteoporosis in men (Jones 2009) and their role in red cell production, maintenance of muscle strength, and fat distribution (Bann et al. 2015).

4.2.1 Aging and Loss of Sex Hormones

Normal aging is accompanied by depletion of sex hormones. Hormone-responsive tissues risk vulnerability to disease as a result of this depletion (Fillit and Luine 1997; Kaufman and Vermeulen 2005). Progesterone (P4) and 17-beta-estradiol (E2) are steroidal sex hormones that rapidly decrease in women after menopause has occurred (Santoro 2005). Likewise, there is a substantial reduction in the levels of plasma testosterone in men, due to age. This process has been named androgen deficiency in aging males (ADAM) (Morley 2001). However, the drop in reproductive activity is not inevitably connected to ADAM, unlike that which occurs in menopause. Bioavailable testosterone concentrations decline at ~3% annually from the 30th year of life (Feldman et al. 2002; Swerdloff and Wang 2004).

Plasma levels and brain levels of hormones differ significantly because of determinants such as sequestration by sex hormone-binding globulin or neurosteroidogenesis (Rosario et al. 2011; Stoffel-Wagner 2001). E2 happens to be an exception to this norm. In comparison with premenopausal women, E2 levels considerably decrease in the menopausal brain, but with very little further, decrease after menopause (Bixo et al. 1995). Age-related decreases in testosterone levels in

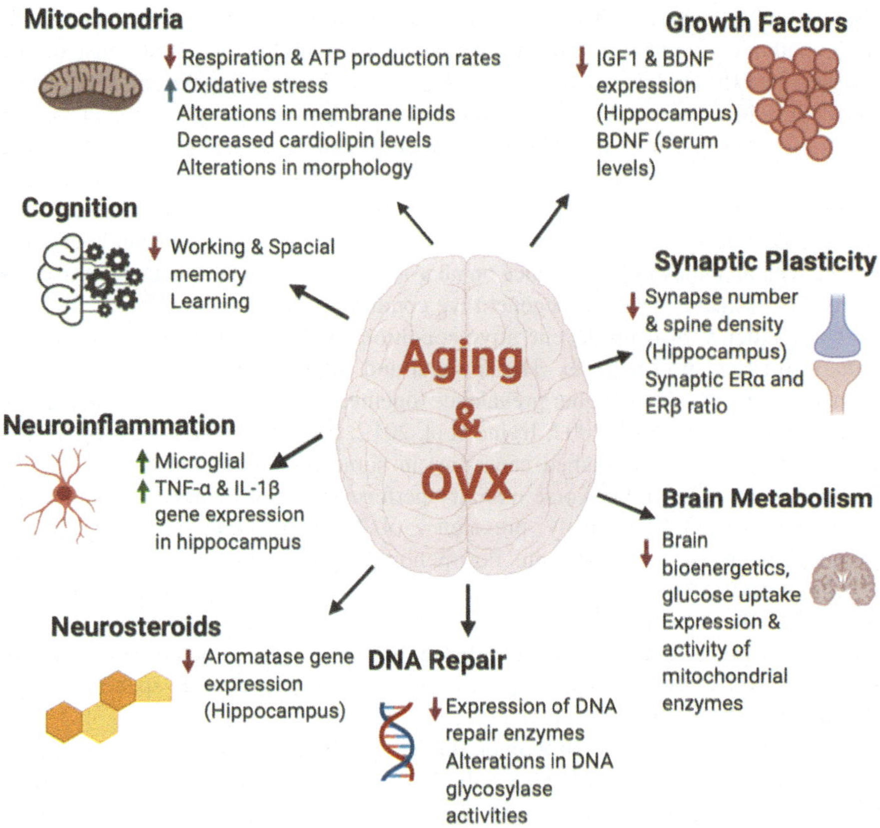

Fig. 4.2 Changes in organismal, cellular, and molecular integrity in the aging brain as observed in ovariectomized (OVX) animal brains. The diagram shows commonly shared features as aging occurs and absence of brain sex hormones (Zárate et al. 2017)

men's brains also occur with depletion to very low levels. Men, in contrast to women, show reduced levels of plasma testosterone, which are, however, functionally significant even at an advanced age (Kaufman and Vermeulen 2005; Muller et al. 2003). Aging results in profound changes in brain morphological and function of mammals, a majority of these modifications happen as a result of loss of sex hormones (see Fig. 4.2).

4.2.2 Age-Related Neurodegeneration and Sex Hormones

A foremost risk factor for developing AD is advanced age (Johnson 2015). Equally, in individuals with genetic risk factors and autosomal dominant mutations linked to AD, the single most prominent risk factor is an increase in age (2010; Launer et al. 1999). AD tends to develop during middle or advanced age in these individuals

(Barron and Pike 2012). The specific factors coupled with regular aging, which leads to the pathogenesis of AD, have not yet been clearly understood. Notwithstanding, research has shown that estrogen depletion together with progesterone loss during menopause in women and the more slow but steady loss of androgens in men are risk factors for AD (Chen et al. 2006; Haaxma et al. 2007; Janowsky 2006). As living organisms grow, a moderately fast decrease in the ovarian sex hormones in females following menopause, and a steady but substantial drop in testosterone levels, occurs in men. Therefore, there is an expectation that reproductive senescence in both sexes comes with an undesirable influence on nervous function and has been recognized as a crucial risk factor for neurodegenerative conditions (Barron and Pike 2012).

The incidence of neurodegenerative conditions with advanced age also shows a pattern of association with sex (Hanamsagar and Bilbo 2016). Parkinson's disease, in males, presents with a higher prevalence together with earlier onset in comparison with females (Elbaz et al. 2002; Irvine et al. 2012; Plassman et al. 2011). Men aged over 40 years have diminished capacities in some regions of the brain, e.g., the hippocampus, and exhibit worse memory performance than women of similar age (Jack et al. 2015). Conversely, prevalence of AD in postmenopausal women is higher than in age-matched men. Postmenopausal women as well exhibit faster cognitive deterioration after disease onset in AD (Zagni et al. 2016). However, the beneficial effects exhibited by estrogen on different molecular processes, like cardiovascular protection, antioxidant defense, immunity, and telomere conservation, have been well shown, at least in part, to result in an extended life span in women (Shadyab et al. 2017).

AD prevalence is higher in women due to the abrupt and substantial loss of E2 and P4 during menopause (Li and Singh 2014). This results in a more severe disease presentation characterized by severe cognitive deficits in women as compared to men. Amyloid-beta (Aβ) neuropathology has been key in this process although tau pathology in men has been increasingly reported (Grimm et al. 2016a; Overk et al. 2012). The female brain exhibits increased vulnerability to AD mainly associated with the depletion of sex steroid hormones, which are believed to be neuroprotective. However, evidence has been provided implicating the developmental effects of the sex hormones as influencing susceptibility to AD (Pike 2017).

A comparison of hormone levels in both women and men shows that loss of sex hormones connected with age is significantly associated with an increase in risk of AD. Lower plasma and brain levels of E2 occur in women with AD, while lower levels of plasma and brain testosterone have been detected in men with AD (Manly et al. 2000; Paoletti et al. 2004). This reduction in hormone levels usually precedes the onset of AD and is more probably a contributory factor instead of a result of the disease process (Barron and Pike 2012).

In AD, brain responsiveness to sex hormones may also be impaired due to changes in sex hormone receptor density and dissemination caused by heightened immunoreactivity of receptors particularly in the hippocampus of AD brains. In men, polymorphisms observed in the ERα and the androgen receptor have also been linked to the incidence of AD (Lehmann et al. 2003; Lu et al. 2003).

4.3 Sex Hormones and the Brain

The brain, in the past, was not regarded as sex hormone-responsive, except for the hypothalamus, which is required for regulation of reproductive function. The entire brain is nonetheless now widely accepted to be equally a target and a source of sex hormones (Barron and Pike 2012; Furigo et al. 2017). Sex hormones exert antioxidants and several protective effects in the brain of an adult (Sanchez-Rodriguez et al. 2018). This enhances neural function, builds elasticity, and promotes neuronal survival (Zárate et al. 2017).

Like other biological steroidal hormones, sex steroid hormones bind to nuclear receptors responsible for regulating the expression of certain target genes (McDevitt et al. 2008). ERα and ERβ are two classical steroid receptors in which estrogen binds in order to elicit a response. ER is associated with membranes and proteins such as membrane G-protein-coupled ER 1 (GPER1), which have also recently evolved in their interactions with estrogens (Dumitriu et al. 2010; Hara et al. 2015; Rettberg et al. 2014; Zárate and Seilicovich 2010). Likewise, progestin and androgen receptors also exist as both non-nuclear and nuclear forms in biological systems (Brinton et al. 2008; Zarif and Miranti 2016). Their actions in higher centers are therefore complex. The brain is highly endowed with sex hormone receptors, in both nuclear and non-nuclear sections, which include the mitochondria and are highly expressed in brain areas that regulate reproductive activity and neuroendocrine functions such as the hypothalamus (Boulware et al. 2007; McEwen and Milner 2017; Rettberg et al. 2014).

Steroid sex hormones modulate an essential number of biological processes including neuronal function. Proliferation and maturation of oligodendrocytes are influenced by estrogen and progesterone (Avila et al. 2018; Marin-Husstege et al. 2004). The two major sex hormones have additionally been described to modulate processes associated with localized inflammation, which are regulated by microglia and astrocytes (Arevalo et al. 2010). During a stroke, estrogens are considered neuroprotective due to their beneficial effects on the cerebral vasculature; these effects include reducing leukocyte adhesion and suppression of inflammatory markers such as COX-2 with enhanced endothelial nitric oxide synthase action (Suzuki et al. 2009). Progesterone produces similar actions in the vasculature (Yu et al. 2017).

4.3.1 Sex Hormones and Synaptic Plasticity

Estrogens and ERs in the hippocampus and prefrontal cortex (PFC) perform key roles in the synaptic basis of cognitive functions (Dumitriu et al. 2010; Hara et al. 2015, 2014). Estrogen carries out its effects on synapses through activation of membrane-associated ERα and ERβ in synaptic terminals (McEwen and Milner 2007). Glial cell processes, dendritic shafts, and dendritic spines also possess these ERs (Milner et al. 2005). A decrease in number of synapses and density of the dendritic spine occurring either naturally or through loss by surgical removal of sex steroid hormones found in the CA1 region of the hippocampus cannot be sustainably restored with estrogen treatment in the aged females (Adams et al. 2001; Sheppard et al. 2019). This is as a result of the unresponsiveness of the hippocampus to estrogen as aging occurs (Adams et al. 2001). A decrease in the number of synapses has been suggested as being the main cause of loss of brain plasticity observed with an increase in age (Adams et al. 2002).

Comparably, the brain of a male is also sensitive to variability in synaptic androgen levels. Androgen depletion and replacement are reported to modulate dendritic spine density in the hippocampus *in vivo* (Jia et al. 2013, 2016; Leranth et al. 2003; Pan et al. 2016). This is suggestive of a direct action of testosterone through activation of its receptors rather than an indirect mechanism via biosynthesis of local estradiol in hippocampal synapses.

4.3.2 Sex Hormones and Mitochondrial Function

An estimate of about a fifth of all ATP produced by the body is used by the brain, emphasizing its high requirement for energy (Rettberg et al. 2014). With this said, properly functioning mitochondria are relevant for brain activity. Cell-type-specific estrogen receptors coordinate the activities of estrogen in the mitochondria. Estrogen exerts advantageous actions in the mitochondria mainly in tissues that demand a higher level of energy to perform cellular activities like the CNS. Notable functions of estrogen in the mitochondria include control of metabolic activities, biogenesis, apoptosis, and morphology (Brinton 2008). How estrogen controls mitochondria function is poorly understood, but research has demonstrated direct and indirect mechanisms that may be useful in understanding these processes. A direct transcription of mitochondrial DNA (mtDNA) due to the presence of ERs in the mitochondria could be a possible way estrogen regulates mitochondrial activity (Rettberg et al. 2014). ERβ found in mitochondria binds to estrogen response element-like (ERE) proteins in mtDNA.

However, the action of estrogen on mitochondria is not entirely dependent on these mechanisms. Estrogens are relevant in the generation of ATP and in brain energy metabolism. Research has shown that there is a reduction in bioenergetics associated with decreased ovarian hormones as seen in women undergoing menopause and in animal models with reproductive senescence induced naturally or surgically (Yao et al. 2010, 2012, 2009). AD and PD that are associated with aging correlate with dysfunctional mitochondria (See Fig. 4.3) (Johri and Beal 2012). As aging occurs, there is a reduction in mitochondrial activities, observed as impaired oxidative phosphorylation, a decrease in expression of respiratory chain complexes, and a decrease in antioxidant activities (Grimm et al. 2016a). The loss of ovarian hormones is directly linked to a reduction in brain mitochondrial activity as seen in reproductive senescent females, and this has been proven by extensive research (Yao et al. 2010, 2012). During steroidogenesis, the first step occurs in the mitochondria. Neurodegeneration can occur as a result of the accumulation of mitochondrial deficits associated with age as a result of its detrimental effect in

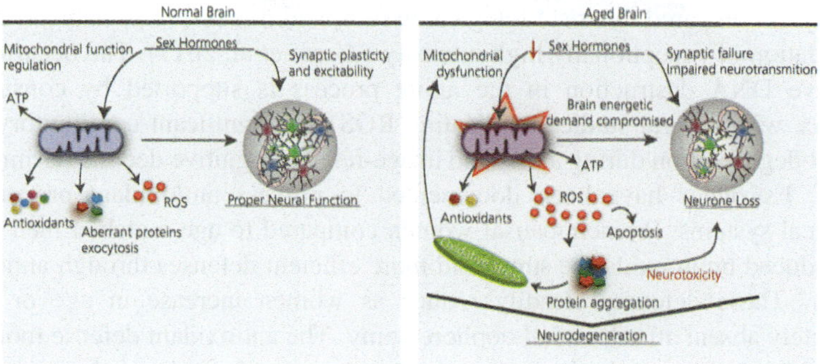

Fig. 4.3 Pathological mechanisms of neurodegeneration accompanying the aging process. Sex steroid hormones control several activities of the brain and mechanisms, such as brain bioenergetics, mitochondrial activities, and synaptogenesis. As a consequence of aging, sex hormone decreases may compromise some of these events, contributing to mitochondrial dysfunction (Duarte et al. 2016)

steroid biosynthesis (Velarde 2014). In summary, premature aging can occur from the loss of ovarian hormones through a reduction in mitochondrial bioenergetics (Yao et al. 2009).

Estrogen controls mitochondrial functions. This has been shown to occur through conventional nuclear mechanisms specifically through nuclear transcription of various key proteins influencing activity of the mitochondrion involving the peroxisome proliferator-activated receptor-gamma coactivator 1 (PCG-1) and nuclear respiratory factor-1 (NRF-1) (Kemper et al. 2013). These mechanisms are important to activate nuclear genes that encode proteins associated with biogenesis of mitochondria and complexes of the electron transport chain of the mitochondrion. Estrogen additionally controls the transcription of mitochondrial transcription factor A (TFAM). TFAM initiates the transcription and replication of mtDNA after translocation into mitochondria (Kang et al. 2007).

The activity of estrogen in the brain mitochondria may be critical since the mtDNA mutation accumulation, as well as the associated mitochondrial abnormality, is important in processes associated with brain aging and in the onset of AD and PD (Cantuti-Castelvetri et al. 2005). A reduction in the capacity to repair mtDNA is a major contributory factor in mtDNA unpredictability during brain aging and also in neurodegenerative disease, and this can be seen in both the mitochondria and nucleus (Kujoth et al. 2007). Oxidation, alkylation, and deamination are the main modifications to mtDNA, which can be repaired through a major pathway by BER occurring in the mitochondria (Jeppesen et al. 2011).

The mitochondria of a cell are a major location for the synthesis of reactive oxygen species (ROS) under normal physiological conditions. As respiration occurs in the mitochondria, electrons leaking from the electron transport chain react with molecular oxygen to generate the superoxide anion, which consequently is converted to hydrogen peroxide (H_2O_2) by the enzyme, superoxide dismutase (SOD) (Grimm et al. 2016b, c). The brain tends to exhibit very low antioxidant capacity as compared to most biological tissues, and it is therefore liable to damage by oxidation of exceptionally high intensity (Zárate et al. 2017). The occurrence of oxidative DNA destruction in the aging process is supported by considerable evidence with several studies implicating ROS as a significant contributory factor to neurodegeneration during aging and in age-related cognitive decline (Grimm et al. 2016c). Estrogens have been documented to possess antioxidant properties in biological systems. Premenopausal women compared to age-matched men present with reduced brain oxidative stress and more efficient defenses through antioxidant activity. These defenses steadily reduce as women increase in age or maybe completely absent after bilateral oophorectomy. The antioxidant defense molecules, glutaredoxin and peroxiredoxin 5, occur in the mitochondria of the brain. Their expression is increased by estrogens (Nilsen et al. 2007). Similarly, circulating testosterone levels in males positively correlate with the action of enzymes with antioxidant properties occurring in both the hippocampus and serum (Cunningham et al. 2014). Generally, the hippocampus appears to be one of the first targets of aging processes, oxidative stress, and loss of sex hormones, as is the PFC (Hara et al. 2015). The PFC is strongly associated with cognitive function in the human brain. Aging induces structural alterations in mitochondria from the PFC producing a mitochondrial phenotype accompanying heightened oxidative stress and ROS production (Hara et al. 2014). Notably, this mitochondrial phenotype is associated with a deterioration of working memory. Studies therefore conclude that estrogen's positive action on PFC-associated memory may be a result of its antioxidant ability, which improves the health of the mitochondria (Hara et al. 2014).

4.3.3 Interaction of Sex Hormones and Growth Factors

Growth factors, for example, insulin-like growth factor-1 (IGF-1) or BDNF, interact (functionally) in the brain with ERs (Cardona-Gómez et al. 2002). The glucose transporter GLUT-4 mediates glucose transport in the brain, which is mainly induced by estrogen (Garcia-Segura et al. 2010; Huffman et al. 2017; Sohrabji 2015). Clinical research has demonstrated that individuals with AD tend to possess reduced densities of insulin receptors with a subsequent deficiency in insulin signaling in areas of the brain prone to AD (Schiöth et al. 2012). Estrogen's hippocampal and memory activities are moderated by the signaling pathway of the IGF-1 receptor. Figitumumab, a IGF-1 receptor inhibitor, can block estrogen-induced improvement in learning and memory (Witty et al. 2013). Research has shown that compromised IGF-1 and estrogen receptor systems in higher centers may explain, at least partly, a well-documented susceptibility of women to the development of AD following menopause (Zárate et al. 2017). In contrast, a functional association between androgens and IGF-1 in the brain has been far less researched on. However, a few studies currently demonstrate that IGF-1/androgen interactions result in advantageous actions such as neuroprotection in men (Huffman et al. 2017). Additionally, ERα and IGF-1 coupling confer protection against stroke and stimulate receptor adult hippocampal neurogenesis (Cardona-Gómez et al. 2002).

Brain-derived neurotrophic factor (BDNF) is an important molecule in hippocampal formation of memory and synaptic plasticity (Bekinschtein et al. 2014). This factor activates several similar signaling pathways as estrogens; pathways result in neural growth induction, neural plasticity, and learning and memory. Gene expression of BDNF may be promoted by estrogens through direct activation of an ERE on the gene of BDNF or through the upregulation of BDNF specifically (Scharfman and MacLusky 2006). Levels of circulating BDNF tend to fall both in women and men when they advance in age. A very significant reduction is seen in women after menopause (Shimada et al. 2014).

4.3.4 Effect of Sex Hormones on Neurosteroidogenesis

Synthesis of neurosteroids is affected by aging. During steroidogenesis, the first step that happens to be a rate-limiting step involves the movement of cholesterol molecules from the mitochondrial outer membrane to its inner membrane (Jefcoate et al. 1992). The two molecules known to be involved in this trafficking are translocator protein of 18 kDa (TSPO) and steroidogenic acute regulatory protein (StAR) (Acaz-Fonseca et al. 2016). TPSO was initially referred to as the peripheral benzodiazepine receptor. Genetic-mediated reduction in TSPO levels in *in vivo* and *in vitro* experimental paradigms shows that TSPO may not be a relevant enzyme for steroidogenesis (Selvaraj et al. 2015). Research has shown the importance of TPSO in the maintenance of androgen levels during aging, but it may, however, be redundant in achieving baseline steroidogenesis, functional-wise (Barron and Pike 2012). After a brain injury and during aging, there is usually an enhancement of the activity and expression of both enzymes. Therefore, producing these brain steroids may serve as a protective barrier to help deal with pathological conditions or a reduction in the equivalent peripheral steroids (Veiga et al. 2004). Enzymes responsible for the catalysis of various stages in neurosteroidogenesis are expressed differentially in glia and neurons per the area of interest or the pathological condition under consideration (Acaz-Fonseca et al. 2016).

Concerning the pathophysiology of the condition, these neurons are mainly the site in the brain for the production of estrogen and they rely on high aromatase expression. After a brain injury in the PFC of humans during the late phases of AD, there is enhanced expression of aromatase in astrocytes and this shows that neuronal impairment can stimulate the production of estrogen to confer protection to the glial cells against neuronal death (Luchetti et al. 2011). Zhao et al. (2017) reported that the hippocampus of aged female mice had reduced aromatase levels in comparison with the levels in adult mice of other ages. Also, an increase in risk of AD has been seen in individuals with genetic variants in human aromatase. As aromatase levels were depleted in AD animal models, an early incidence with high severity of neuropathology was observed as compared to that in control OVX mice (Azcoitia et al. 2003). This shows that a reduction in levels of estrogen derived from the brain as opposed to estrogen from peripheral plasma is a very unequivocal and relevant risk factor in the development of AD and shows how important it is that neurosteroidogenesis be preserved in the healthy brain aging process (Cui et al. 2013). Newly identified pharmacological targets with the aim to prevent a decline of brain activity during aging and to halt development of neurodegenerative disorders utilize key enzymes responsible for the production of estrogen.

4.3.5 Sex Hormones and Neuroinflammation

Neuroinflammation is facilitated by astroglia and microglia; these glial cells are significantly affected by the process of aging (O'Callaghan and Sriram 2005; Refolo and Stefanova 2019). Microglia are the major immunocompetent cells present in the CNS, and they regulate the inflammatory response during a pathological process (Katsumoto et al. 2014). Microglia also have the ability to maintain balance in a healthy brain mediated by immune inspection of the brain parenchyma (Nissen 2017). As aging ensues, there is a resultant impairment of physiological functions of glial cells as a result of the dysregulation of microglia as shown in Fig. 4.4 (Wu et al. 2013). This impairment occurs together with a chronic inflammatory

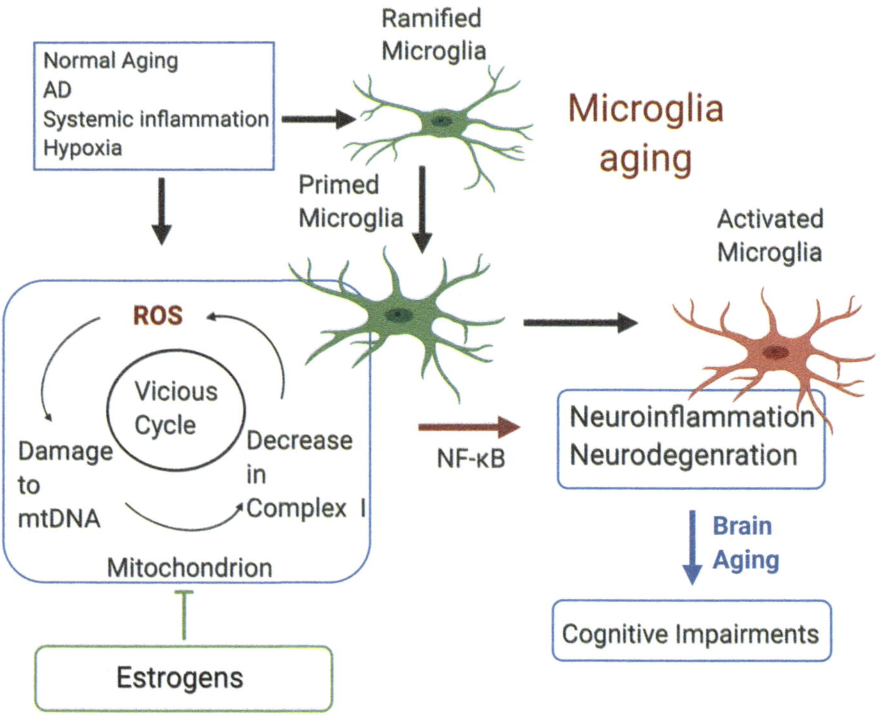

Fig. 4.4 Schematic representation of neurodegeneration and neuroinflammation resulting from increased microglial mitochondria-derived ROS through the activation of the NF-κB signaling pathway, leading to cognitive impairments during aging and neurodegenerative states. Estrogens exert antioxidant action to prevent impairment. (Adapted from Wu et al. 2013)

process of mild intensity distinguished by an inflammatory cytokines storm (Nissen 2017). In the event of injuries or mild stimulatory events, the aged microglia are stimulated and respond aggressively to the signals produced whether they are local or peripheral (Nissen 2017). Studies have shown the frontal cortex of postmenopausal women is well endowed with macrophage-related genes than in women at the premenopausal stage and this is suggestive of a loss of ovarian hormone that is responsible for the change in microglial phenotype from the resting to a reactive state (Christensen and Pike 2015; Njie et al. 2012). Pro-inflammatory/neurodegenerative M1 or anti-inflammatory/neuroprotective M2 phenotypes are by-products of polarized, activated microglia produced to respond to several physiological and pathological stimuli. This polarization can be triggered by estrogen to form an M2 phenotype (Habib and Beyer 2015; Villa et al. 2016). This action of estrogen is particularly important during chronic inflammation processes, where the prolongation of microglial pro-inflammatory state can result in neuronal damage confirming the neuroprotective activity of estrogen during aging and in neurodegenerative states (Dimayuga et al. 2005; Siani et al. 2017). Also, the renin–angiotensin system has been implicated in aging along with other processes believed to be facilitated by activation of microglia and neuroinflammation (Bhat et al. 2016; Torika et al. 2016). Estrogen is particularly beneficial in lessening neuroinflammation, oxidative stress, and/or neurodegeneration occurring in dopaminergic neurons in murine models of PD, and this partly explains the lower risk associated with developing PD in women at the premenopausal stage in comparison with men and postmenopausal women (Labandeira-Garcia et al. 2016; Villar-Cheda et al. 2014).

Studies have shown that there is enhanced inflammation in the female brain modulated by estrogen as aging occurs in comparison with males of comparable age, especially in the hippocampus (Mangold et al. 2017). ERs are targets of estrogen in microglia as estrogen replacement treatment (ERT) halts ovariectomy (OVX)-induced effects (Sárvári et al. 2012, 2014). Both ERα and ERβ agonists imitate most of these effects (Sárvári et al. 2011). Moreover, the augmented release of pro-inflammatory cytokines including IL-1β and TNF-α in the hippocampus of old mice is associated with both OVX and aging (Benedusi et al. 2012). In recent times, the ability of microglia to manage inflammation that occurs as a result of aging has been impaired in animal models with some forms of estrogen resistance (Villa et al. 2016). During the process of aging, astroglial cells also exhibit pro-inflammatory phenotype (Salas et al. 2020). This is mediated through the expression and secretion of augmented magnitude of inflammatory mediators including IL-1β, IL-6, and TNF-α (Salminen et al. 2011). Astroglial cells are also associated with age-related increase in the concentrations of intermediate glial fibrillary acidic protein (GFAP), vimentin filaments, and an increase in the accumulation of proteotoxic aggregates (Hol and Pekny 2015; Zamanian et al. 2012). Preclinical and clinical data have shown that menopause and aging may result in increased neuroinflammation, and this constitutes one of the foremost causative factors of age-associated neurodegenerative diseases including AD and stroke (Zárate and Seilicovich 2010). A typical change seen in astroglial cells is an enhanced expression of GFAP induced by age, and research shows that an increase in the number of dystrophic astrocytes may be the main contributory factor in biological brain aging and the initial phases of neurodegenerative conditions (Kulijewicz-Nawrot et al. 2012; Olabarria et al. 2011). The action of glutamine synthetase is reduced coupled with a decreased capacity for glutamine uptake with a resultant reduction in neuroprotection and homeostatic ability because the cells are relatively smaller with less complexity (Verkhratsky et al. 2014).

4.3.6 DNA Repair, Sex Hormones, and Brain Aging

There is constant damage to brain cell DNA, and removal of the damage is necessary to prevent consequences, which include compromised genomic stability. Sources of DNA damage can either be exogenous—which may result in replication failure and breaking of the strands or—endogenous by ROS resulting in DNA lesions in brain cells, which are non-replicating. Also, various DNA lesions that significantly impact genome stability have been recently recognized. These include the inclusion of ribonucleoside triphosphate (rNTP) into DNA, transcription-associated genome damage from R-loop formation, and abnormal topoisomerase activity. Topoisomerase I may result in DNA destruction through the cleavage complexes formed during transcription. Since transcription rates are higher in neurons, this is a relevant process in these cells. The brain has both non-dividing and dividing cells. The difference in replicative status partly affects the destruction and repair procedures in cells. Depending on their differentiation status, glial cells exist in a developmental or non-developmental condition, while neurons occur in a post-mitotic state. Since the two are important in the performance of several challenging brain functions, it is beneficial that all cell types are in the required configuration and number.

Exposure to ROS may cause oxidation-mediated base modifications. 8-oxo-dG is a principal and well-distinguished ROS in mammalian systems. An estimate of about 180 guanine bases is oxidized to produce 8-oxo-dG per mammalian genome in a day. This may result in G:C to T:A transversion mutation as 8-oxo-dG forms base pairs with adenine and cytosine during replication of DNA. Generation of protein mutated species may occur through an event referred to as transcriptional mutagenesis (TM). 8-oxo-dG may contribute to this occurrence by facilitating a misincorporation of adenine. TM that is responsible for α-synuclein aggregation is implicated in the pathogenesis of PD. ROS may cause DNA single-strand breaks (SSBs) being one of the most familiar DNA lesions, which approximately occurs in tens of thousands per cell per day. Constant DNA SSBs may block transcription and cause the replication fork to collapse during duplication of the chromosome. ROS may form double-strand breaks (DSBs). Despite the fact that these lesions rarely occur, DSBs may cause loss of genomic information and translocations rendering it one of the severest forms of DNA damage. Lastly, lipid peroxidation can occur as a result of the action of ROS and there can be a reaction between the by-products and DNA to produce exocyclic DNA lesions. Research has shown that DNA lesions left unrepaired accumulate in the CNS during progressive neurodegeneration and during both normal aging and accelerated aging. This change is, however, region-specific and also depends on the type of cell and subcellular location of the DNA.

Mitochondrial DNA exhibits vulnerability to oxidative stress due to its proximity to the location where the majority of the ROS is generated cellularly. Also, age-related DNA destruction is mostly shown for mtDNA. Research shows that there is a continuous rise in the quantities of 8-oxo-dG in human brain cells. However, mtDNA has a tenfold higher chance of age-associated destruction as compared to nDNA.

DNA repair is essential since the damage can result in a cascade of events resulting in cell senescence and death as well as the occurrence of genomic instability. These are associated with aging. DNA instability can occur during normal aging as a result of an increase in the levels of mutations in the brain DNA and other tissues. The BER pathway has four definite steps. First of all, DNA glycosylases remove and recognize the bases, which have been modified. The DNA glycosylases, for example, Ogg1 and NTH1, are definite in their substrate specificities. They provide an AP site (apurinic/apyrimidinic site), mainly controlled by the AP endonuclease (APE1). Within the nuclei and mitochondria, BER subsequently advances through two distinct sub-pathways. These pathways can be differentiated based on the differences in the quantity of nucleotides consolidated into the gap by a DNA polymerase enzyme. This step involves several accessory proteins. DNA strand ligation by the enzyme DNA ligase constitutes the final step.

A major contributory factor of brain aging is DNA instability. Research has shown that accumulation of DNA mutations and reduction in DNA repair ability have a role to play in the aging of the brain and age-associated neurodegenerative diseases (Jeppesen et al. 2011; Sanders et al. 2014). These sex hormones are known to cause neuroprotection. This could be as a result of their ability to directly impact DNA repair. Sensors of damage to DNA such as poly-(ADP-ribose) polymerases (PARP-1) enzymes are present in cell nuclei in abundance, and several researchers have shown sex differences in the activity of PARP-1. Experimental paradigms using OVX animal models have shown the neuroprotective action of estrogen and progesterone through the regulation of DNA repair mechanisms. BER enzymes have been the focus of study on the ability of estrogen levels to affect repair of DNA. Through transcription and translocation, estrogen can regulate different cellular compartments of BER. Estrogen exerts neuroprotective effects in the cerebral cortex through oxidative reduction in DNA after hypoxia (Rao et al. 2011), and this effect has a related improvement in transcription of DNA repair enzymes such as APE1 present in that site of the brain (Dietrich et al. 2013). The dorsal raphe of old OVX female rodents experienced an increase in DNA transcription levels of different repair enzymes after estrogen supplementation (Bethea et al. 2017). A substantial increase in DNA transcription of various pathways including BER (NTH1, APE1, etc.) and NER (GTF2H5 and RAD23, a subunit of THIIF) has been shown. However, co-administration of progesterone and estrogen caused a decrease or complete halt of that effect. This confirms research that has already established an antagonism of progesterone on estrogen as seen in cognitive function of OVX rats administered with sex hormones. Progesterone also causes a reduction in estrogen-associated improvement of BDNF in OVX rats (Bimonte-Nelson et al. 2004). The importance of the effect cannot be overemphasized since BDNF is important in enhancing neuronal survival partly by inducing DNA transcription repair enzymes such as APE1. Estrogen causes the expression of BDNF, and as a result, the DNA repair effect of estrogen is associated with BDNF levels. Several studies have attributed the benefits of estrogen administration to upregulation of Nrf-2 through the Pl3K/Akt signaling pathway. Nrf-2 is associated with transcription of antioxidant feedback mechanisms, which includes enzymes associated with DNA repair, rendering the attribution significantly relevant (Geismann et al. 2014; Zhang et al. 2013).

4.3.7 Neuroprotective Effects Exhibited by Sex Hormones

Estrogen is known to play a relevant role in sex-related dissimilarities in neurodegeneration and aging of the brain. The brain is well endowed with estrogen receptors. In addition to estrogen's many brain functions such as cognition, its neuroprotective function has been shown (Kowal et al. 2013). The delayed onset of symptoms of PD in females may be associated with neuroprotection. To buttress the neuroprotective activity of sex steroids, various animal models have shown the benefits of hormone replacement therapy (HRT) (Ding et al. 2013; Lu et al. 2018; Yun et al. 2018). Lower estrogen level is connected to the inhibition of the enzyme aromatase. This enzyme is responsible for the catalysis of the non-reversible conversion of aromatizable androgens into estrogens resulting in an alteration of amyloid-beta deposition and an increase in stroke severity in AD (Overk et al. 2012). Aromatase knockout (KO) mice have exhibited similar effects (Yue et al. 2005). Even though several preclinical, clinical, and epidemiological studies have affirmed the important role of HRT on cognition and memory as well as to a reduction in risk for AD, a preliminary appraisal of findings from the Women's Health Initiative (WHI) trial, the most extensive clinical trial with women at the postmenopausal stage currently, stated otherwise (Rettberg et al. 2014). Subsequent reviews, however, have observed that women in the study population used for the WHI trial were different from women in past observational research in terms of age or years after menopause as well as onset of treatment (Miller and Harman 2017; Rettberg et al. 2014). The WHI trial together with other epidemiological studies after a sub-analysis showed evidence indicating that there is a fixed time or window around onset of menopause where HRT can provide improvements in functions of the brain. The "window of opportunity" hypothesis thus proposes the advantageous actions of estrogens that are only evident on a healthy brain, i.e., starting HRT before menopause or on its onset (Miller and Harman 2017; Scott et al. 2012).

Conversely, initiation of HRT years following the start of menopause brings about deleterious effects on the functions of the brain and may cause an increased risk of developing AD (Miller and Harman 2017; Scott et al. 2012). Research has also shown that the effectiveness of HRT depends on other factors including formulations of treatments taking into consideration the type of molecule (conjugated equine estrogen or lower estrogen levels 17β-estradiol, medroxyprogesterone acetate, or progesterone), regimen (continuous vs. cyclic), or mode of delivery (oral vs. transdermal) (Miller and Harman 2017). The brain is a steroidogenic organ and a target for sex steroids. Studies are being conducted to understand how the brain synthesizes neuroactive steroids and how this affects its function. These neuroactive steroids can still be detected in the brain even after the peripheral steroidogenic organs are removed (Mellon and Griffin 2002; Sorwell et al. 2012). Their levels are, additionally, affected by changes in the quantities of serum neuroactive steroids in a site, time after gonadectomy, and sex type, suggesting a brain-specific compensatory mechanism to counteract the detrimental effects of peripheral hormone loss (Arevalo et al. 2015). Levels of estrogen in the brain have been shown to be a determinant of how estrogen therapy induces neuroprotective effects in the development of AD (Chamniansawat and Chongthammakun 2012). These neurosteroids are synthesized *de novo* from cholesterol and through the conversion of blood peripheral steroid hormones into potent by-products with enhanced neuromodulator activities (Veiga et al. 2004). Androgens and progesterone possess neuroprotective actions on the brain, even though less researched. During periods of glucose deprivation in the hippocampal regions and in astroglial cells, testosterone and its metabolites perform neuroprotective functions in these areas (Ishihara et al. 2016; Toro-Urrego et al. 2016). After surrounding neurons undergo induced death, testosterone can halt dendritic atrophy in motoneurons (Cai et al. 2017). Continuous stress conditions have been described to enhance susceptibility to oxidative insults in several sites of the brain, and this is associated with depletion of testosterone levels. According to Colciago et al. (2015), androgens possess a positive impact on cognition and may be associated with testosterone and its metabolites in the hippocampus since hippocampal CA1 pyramidal cells have a high density of androgen receptors. Testosterone and dihydrotestosterone administration can restore hippocampal spine density. BDNF expression is associated with the activity of testosterone on maturation and spine density (Li et al. 2012).

4.4 Progesterone in Neurodegeneration

Progesterone (P4) is a major regulator of reproduction, and the brain is well endowed with its receptors. P4 receptors can also be found in all neuronal cell types (Baulieu et al. 1996). Progesterone and its metabolites exert various physiological actions in the brain such as the control of neuronal growth of Purkinje cells in cerebellum, differentiation and proliferation of oligodendrocytes, and synaptogenesis and neuronal plasticity (Ghoumari et al. 2005; Rossetti et al. 2016). Besides, progesterone and its metabolites are known to exert advantageous actions in several animal paradigms to show neurodegeneration and damage to the brain, such as in PD, stroke, traumatic brain injury, and demyelination (El-Etr et al. 2015; Singh and Su 2013).

Progesterone according to research has a direct or indirect correlation in regulating risk of AD alongside estrogen and androgens. The estrogen-based HRT for postmenopausal women usually possesses a progesterone moiety to prevent the possible oncogenic effect of estrogen on the uterus (Dai et al. 2002; Persson et al. 1989). Even though more is known regarding the relevance of estrogen and androgens in AD than progesterone alone, it may positively impact neural activities associated with AD (Brinton et al. 2008; Schumacher et al. 2007). Both natural (P4) and synthetic (medroxyprogesterone) progesterones can regulate E2 neuroprotection, and the treatment condition determines whether there will be alternate negation or an improvement in the estrogen effect (Barron and Pike 2012). Important determinants in progesterone and estrogen interaction include time and duration of exposure to the hormones. Interaction between P4 and E2 in behavioral models causes a decrease in the effects of E2. Progesterone countered the valuable tonic and cyclic actions of E2 administration on spatial reference memory in middle-aged OVX rats. P4 halted E2-mediated performance impairment in a conditioned avoidance task (Chesler and Juraska 2000). In all experimental models, no antagonism of E2-mediated cognitive benefits by P4 has been observed; however, a combination of E2 and P4 enhanced spatial memory execution in aged OVX rats, and this was also seen in the group administered with E2 alone (Rice et al. 2000). Researches in human populations have shown the possibility of deleterious effects that come with combined estrogen and progesterone HRT observed as impairment in cognition. However, administration of estrogen alone had no impact on cognition in aged postmenopausal women. P4 can therefore regulate neuroprotection associated with E2 experimental paradigms of neural injury (Gibbs 2000).

P4, however, blocked an E2-mediated increase in neurotrophic factors such as NT3, NGF, and BDNF in the entorhinal cortex of the brain. Brain mitochondrial function was enhanced when E2 and P4 were administered alone. Conversely, co-administration of E2 and P4 resulted in a decrease in brain mitochondrial function (Irwin et al. 2008). P4 enhanced the protective effect of estrogen in some experimental models. This was observed when co-administration of P4 and E2 offered neuroprotective effects against glutamate toxicity in primary cultures of hippocampal neurons as compared to the separate administration of the two hormones (Carroll et al. 2008).

Several mechanisms account for the neuroprotective action of progesterone. These include the activation of the *MAPK/ERK* and *PI3K/Akt* pathways, recognized survival signaling pathways (Baudry et al. 2013). Similarly, progesterone expends neurotrophic activities by controlling the expression of neurotrophins including BDNF (Coughlan et al., 2009). Progesterone also regulates neuroinflammation, and this anti-inflammatory effect has broadly been studied in research in multiple sclerosis in murine models: autoimmune experimental encephalomyelitis (EAE). Sex hormones, allopregnanolone (a reduced metabolite of progesterone), and progesterone are known to decrease inflammatory markers, cause inhibition of activation of microglia, and halt infiltration of circulating macrophages and lymphocytes in the CNS (Irwin et al. 2008; Noorbakhsh et al. 2014). The effect is perceived as a reduction in clinical severity, a decrease in demyelination, and an improvement in neuronal function after administration of progesterone. Conversely, co-administration of progesterone and estrogen does not always yield positive outcomes. Progesterone controls the action of estrogen, especially at the reproductive level, but at the CNS level, several researches have shown that co-administration of these hormones will result in progesterone antagonizing rather than potentiating the action of estrogen. Even though the exact mechanism underlying this antagonism remains poorly understood, regulation of ER expression has been described as the most likely mechanism (Graham and Clarke 1997).

4.5 Recent Developments and Future Perspectives

A combination treatment of estrogen and progestin raises concern about potential adverse effects. This, therefore, led to an interest in the use of selective estrogen receptor modulators (SERMs) as potentially safer alternatives (Kauffman and Bryant 1995; Petersen and Briggs 2005). SERMs function as partial agonists at estrogen receptors found in nervous tissue and bone. SERMs may exert antagonistic effects in breast tissue while producing mixed actions on vaginal and uterine tissue. The breast cancer drug, tamoxifen, a first-generation SERM, is an agonist on ERs in tissues of the bone and the uterus (Denk et al. 2015; Jordan 2003). Raloxifene is a second-generation SERM used in the treatment of osteoporosis (Ensrud et al. 2008). It is a comparatively pure antagonist in reproductive tissue such as the uterus with negligible agonistic effects but possesses arguably full agonist properties in bone tissue (Bryant et al. 1996). The precise effects of raloxifene in the nervous tissue are not wholly known. Raloxifene has, however, been shown to prevent dopaminergic neuron loss in the myenteric plexus partly through its anti-inflammatory effects (Poirier et al. 2016). Additionally, it may affect mood and cognition due to both pre- and postsynaptic modulation of cholinergic and serotonergic neurotransmission (Nickelsen et al. 1999; Sánchez et al. 2010). Raloxifene may also regulate GABAergic neurotransmission (Schroeder et al. 2017).

Bazedoxifene is a third-generation SERM, which possesses similar effects to those produced by raloxifene (Silverman et al. 2012). It acts by binding to both ERα and ERβ intracellular subtypes. Bazedoxifene has an indole base and a 2-phenyl ring system that improves its receptor selectivity compared to other SERMs. It displays estrogen agonist activity in the absence of estradiol, whereas it has an estrogen-blocking action when estradiol is present (Silverman et al. 2008). In response to the positive outcomes of trialing SERMs in schizophrenia, they have been considered for the management of other neurological disorders. Bazedoxifene has been demonstrated to cross the blood–brain barrier after an IP injection (Hill et al. 2020). The clinical relevance of SERMs is further augmented by evidence of beneficial cognitive action in the Y-maze spatial memory task in addition to the ability to increase hippocampal and spine density of the medial prefrontal cortex (González-Burgos et al. 2012; Schroeder et al. 2017; Wu et al. 2015).

4.6 Conclusion

Decreases in levels of sex hormones are directly associated with the loss of neuronal performance, resulting in the onset of neurodegenerative diseases. Experimental models using female rodents have shown that negative effects in brain functionality caused by OVX can be reversed by supplementation with E2, hence confirming the neuroprotective effects of estrogen. The type of estrogen receptor, as well as the time of estrogen administration, is relevant to achieving this effect. ERβ plays an enormous role in preventing age-related detrimental effects of the brain through its continuous action on plasticity on the brain as females' age and its pronounced effect on mitochondrion function. Further research is required to comprehend the processes backing the neuroprotective activities of sex hormones as they could be employed in the management of age-related neuronal decline and neurodegenerative conditions.

References

Acaz-Fonseca E, Avila-Rodriguez M, Garcia-Segura LM, Barreto GE (2016) Regulation of astroglia by gonadal steroid hormones under physiological and pathological conditions. Prog Neurobiol 144:5–26

Acconcia F, Marino M (2018) Steroid hormones: synthesis, secretion, and transport. In: Principles of endocrinology and hormone action. Springer, New York, pp 43–72

Adams MM, Shah RA, Janssen WG, Morrison JH (2001) Different modes of hippocampal plasticity in response to estrogen in young and aged female rats. Proc Natl Acad Sci 98:8071–8076

Adams MM, Fink SE, Shah RA, Janssen WG, Hayashi S, Milner TA, McEwen BS, Morrison JH (2002) Estrogen and aging affect the subcellular distribution of estrogen receptor-α in the hippocampus of female rats. J Neurosci 22:3608–3614

Aloisi AM, Bonifazi M (2006) Sex hormones, central nervous system and pain. Horm Behav 50:1–7

(2010) Alzheimer's disease facts and figures. Alzheimers Dement 6:158–194

Amenyogbe E, Chen G, Wang Z, Lu X, Lin M, Lin AY (2020) A review on sex steroid hormone estrogen receptors in mammals and fish. Int J Endocrinol 2020:5386193

Arevalo M-A, Santos-Galindo M, Bellini M-J, Azcoitia I, Garcia-Segura LM (2010) Actions of estrogens on glial cells: implications for neuroprotection. Biochim Biophys Acta 1800:1106–1112

Arevalo M-A, Azcoitia I, Garcia-Segura LM (2015) The neuroprotective actions of oestradiol and oestrogen receptors. Nat Rev Neurosci 16:17–29

Austad SN, Fischer KE (2016) Sex differences in lifespan. Cell Metab 23:1022–1033

Avila M, Bansal A, Culberson J, Peiris AN (2018) The role of sex hormones in multiple sclerosis. Eur Neurol 80:93–99

Azcoitia I, Sierra A, Veiga S, Garcia-Segura LM (2003) Aromatase expression by reactive astroglia is neuroprotective. Ann N Y Acad Sci 1007:298–305

Baker ME (2003) Evolution of adrenal and sex steroid action in vertebrates: a ligand-based mechanism for complexity. Bioessays 25:396–400

Bann D, Wu FC, Keevil B, Lashen H, Adams J, Hardy R, Muniz G, Kuh D, Ben-Shlomo Y, Ong KK (2015) Changes in testosterone related to body composition in late midlife: findings from the 1946 British birth cohort study. Obesity 23:1486–1492

Barford A, Dorling D, Smith GD, Shaw M (2006) Life expectancy: women now on top everywhere. BMJ 332:808

Barron AM, Pike CJ (2012) Sex hormones, aging, and Alzheimer's disease. Front Biosci (Elite Ed) 4:976–997

Baudry M, Bi X, Aguirre C (2013) Progesterone–estrogen interactions in synaptic plasticity and neuroprotection. Neuroscience 239:280–294

Baulieu EE, Schumacher M, Koenig H, Jung-Testas I, Akwa Y (1996) Progesterone as a neurosteroid: actions within the nervous system. Cell Mol Neurobiol 16:143–154

Bekinschtein P, Cammarota M, Medina JH (2014) BDNF and memory processing. Neuropharmacology 76:677–683

Benedusi V, Meda C, Della Torre S, Monteleone G, Vegeto E, Maggi A (2012) A lack of ovarian function increases neuroinflammation in aged mice. Endocrinology 153:2777–2788

Bethea CL, Mueller K, Reddy AP, Kohama SG, Urbanski HF (2017) Effects of obesogenic diet and estradiol on dorsal raphe gene expression in old female macaques. PLoS ONE 12(6):e0178788. https://doi.org/10.1371/journal.pone.0178788

Bhat SA, Goel R, Shukla R, Hanif K (2016) Angiotensin receptor blockade modulates NFκB and STAT3 signaling and inhibits glial activation and neuroinflammation better than angiotensin-converting enzyme inhibition. Mol Neurobiol 53:6950–6967

Bimonte-Nelson HA, Singleton RS, Williams BJ, Granholm ACE (2004) Ovarian hormones and cognition in the aged female rat: II. Progesterone supplementation reverses the cognitive enhancing effects of ovariectomy. Behav Neurosci 118(4):707–714. https://doi.org/10.1037/0735-7044.118.4.707

Bixo M, Bäckström T, Winblad B, Andersson A (1995) Estradiol and testosterone in specific regions of the human female brain in different endocrine states. J Steroid Biochem Mol Biol 55: 297–303

Boulware MI, Kordasiewicz H, Mermelstein PG (2007) Caveolin proteins are essential for distinct effects of membrane estrogen receptors in neurons. J Neurosci 27:9941–9950

Brinton RD (2008) The healthy cell bias of estrogen action: mitochondrial bioenergetics and neurological implications. Trends Neurosci 31:529–537

Brinton RD, Thompson RF, Foy MR, Baudry M, Wang J, Finch CE, Morgan TE, Pike CJ, Mack WJ, Stanczyk FZ, Nilsen J (2008) Progesterone receptors: form and function in the brain. Front Neuroendocrinol 29:313–339

Brook C (1999) Mechanism of puberty. Horm Res Paediatr 51:52–54

Bryant HU, Glasebrook AL, Yang NN, Sato M (1996) A pharmacological review of raloxifene. J Bone Miner Metab 14:1–9

Cai Y, Chew C, Muñoz F, Sengelaub DR (2017) Neuroprotective effects of testosterone metabolites and dependency on receptor action on the morphology of somatic motor neurons following the death of neighboring motoneurons. Dev Neurobiol 77:691–707

Cantuti-Castelvetri I, Lin MT, Zheng K, Keller-McGandy CE, Betensky RA, Johns DR, Beal MF, Standaert DG, Simon DK (2005) Somatic mitochondrial DNA mutations in single neurons and glia. Neurobiol Aging 26:1343–1355

Cardona-Gómez GP, Mendez P, DonCarlos LL, Azcoitia I, Garcia-Segura LM (2002) Interactions of estrogen and insulin-like growth factor-I in the brain: molecular mechanisms and functional implications. J Steroid Biochem Mol Biol 83:211–217

Carroll JC, Rosario ER, Pike CJ (2008) Progesterone blocks estrogen neuroprotection from kainate in middle-aged female rats. Neurosci Lett 445:229–232

Catenaccio E, Mu W, Lipton ML (2016) Estrogen-and progesterone-mediated structural neuroplasticity in women: evidence from neuroimaging. Brain Struct Funct 221:3845–3867

Chamniansawat S, Chongthammakun S (2012) A priming role of local estrogen on exogenous estrogen-mediated synaptic plasticity and neuroprotection. Exp Mol Med 44:403–411

Chen S, Nilsen J, Brinton RD (2006) Dose and temporal pattern of estrogen exposure determines neuroprotective outcome in hippocampal neurons: therapeutic implications. Endocrinology 147: 5303–5313

Chesler EJ, Juraska JM (2000) Acute administration of estrogen and progesterone impairs the acquisition of the spatial morris water maze in ovariectomized rats. Horm Behav 38:234–242
Christensen A, Pike CJ (2015) Menopause, obesity and inflammation: interactive risk factors for Alzheimer's disease. Front Aging Neurosci 7:130
Colciago A, Casati L, Negri-Cesi P, Celotti F (2015) Learning and memory: steroids and epigenetics. J Steroid Biochem Mol Biol 150:64–85
Comitato R, Saba A, Turrini A, Arganini C, Virgili F (2015) Sex hormones and macronutrient metabolism. Crit Rev Food Sci Nutr 55:227–241
Coughlan T, Gibson C, Murphy S (2009) Progesterone, BDNF and neuroprotection in the injured CNS. Int J Neurosci 119:1718–1740
Cui J, Shen Y, Li R (2013) Estrogen synthesis and signaling pathways during aging: from periphery to brain. Trends Mol Med 19:197–209
Cunningham RL, Singh M, O'Bryant SE, Hall JR, Barber RC (2014) Oxidative stress, testosterone, and cognition among Caucasian and Mexican-American men with and without Alzheimer's disease. J Alzheimers Dis 40:563–573
Dai D, Wolf DM, Litman ES, White MJ, Leslie KK (2002) Progesterone inhibits human endometrial cancer cell growth and invasiveness: down-regulation of cellular adhesion molecules through progesterone B receptors. Cancer Res 62:881–886
Denk F, Ramer LM, Erskine EL, Nassar MA, Bogdanov Y, Signore M, Wood JN, McMahon SB, Ramer MS (2015) Tamoxifen induces cellular stress in the nervous system by inhibiting cholesterol synthesis. Acta Neuropathol Commun 3:74
Dietrich AK, Humphreys GI, Nardulli AM (2013) 17β-Estradiol increases expression of the oxidative stress response and DNA repair protein apurinic endonuclease (Ape1) in the cerebral cortex of female mice following hypoxia. J Steroid Biochem Mol Biol 138:410–420. ISSN 0960-0760. https://doi.org/10.1016/j.jsbmb.2013.07.007
Dimayuga FO, Reed JL, Carnero GA, Wang C, Dimayuga ER, Dimayuga VM, Perger A, Wilson ME, Keller JN, Bruce-Keller AJ (2005) Estrogen and brain inflammation: effects on microglial expression of MHC, costimulatory molecules and cytokines. J Neuroimmunol 161:123–136
Ding F, Yao J, Zhao L, Mao Z, Chen S, Brinton RD (2013) Ovariectomy induces a shift in fuel availability and metabolism in the hippocampus of the female transgenic model of familial Alzheimer's. PLoS One 8:e59825
Donev R, Kolev M, Millet B, Thome J (2009) Neuronal death in Alzheimer's disease and therapeutic opportunities. J Cell Mol Med 13:4329–4348
Duarte A, Hrynchak M, Gonçalves I, Quintela T, Santos C (2016) Sex hormone decline and amyloid β synthesis, transport and clearance in the brain. J Neuroendocrinol 28
Dumitriu D, Rapp P, McEwen B, Morrison J (2010) Estrogen and the aging brain: an elixir for the weary cortical network? Ann N Y Acad Sci 1204:104
Ebly EM, Parhad IM, Hogan DB, Fung TS (1994) Prevalence and types of dementia in the very old. Results from the Canadian Study of Health and Aging. Neurology 44:1593–1593
ElAttar T, Hugoson A (1974) Comparative metabolism of female sex steroids in normal and chronically inflamed gingiva of the dog. J Periodontal Res 9:284–289
Elbaz A, Bower JH, Maraganore DM, McDonnell SK, Peterson BJ, Ahlskog JE, Schaid DJ, Rocca WA (2002) Risk tables for parkinsonism and Parkinson's disease. J Clin Epidemiol 55:25–31
El-Etr M, Rame M, Boucher C, Ghoumari AM, Kumar N, Liere P, Pianos A, Schumacher M, Sitruk-Ware R (2015) Progesterone and nestorone promote myelin regeneration in chronic demyelinating lesions of corpus callosum and cerebral cortex. Glia 63:104–117
Ensrud KE, Stock JL, Barrett-Connor E, Grady D, Mosca L, Khaw K-T, Zhao Q, Agnusdei D, Cauley JA (2008) Effects of raloxifene on fracture risk in postmenopausal women: the raloxifene use for the heart trial. J Bone Miner Res 23:112–120
Feldman HA, Longcope C, Derby CA, Johannes CB, Araujo AB, Coviello AD, Bremner WJ, McKinlay JB (2002) Age trends in the level of serum testosterone and other hormones in middle-aged men: longitudinal results from the Massachusetts male aging study. J Clin Endocrinol Metab 87:589–598

Fillit H, Luine V (1997) The neurobiology of gonadal hormones and cognitive decline in late life. Maturitas 26:159–164

Foradori C, Weiser M, Handa RJ (2008) Non-genomic actions of androgens. Front Neuroendocrinol 29:169–181

Fratiglioni L, Launer L, Andersen K, Breteler M, Copeland J, Dartigues J-F, Lobo A, Martinez-Lage J, Soininen H, Hofman A (2000) Incidence of dementia and major subtypes in Europe: a collaborative study of population-based cohorts. Neurology 54:S4–S9

Furigo IC, Metzger M, Teixeira PD, Soares CR, Donato J (2017) Distribution of growth hormone-responsive cells in the mouse brain. Brain Struct Funct 222:341–363

Garcia-Segura LM, Arévalo M-A, Azcoitia I (2010) Interactions of estradiol and insulin-like growth factor-I signalling in the nervous system: new advances. In: Progress in brain research. Elsevier, Amsterdam, pp 251–272

Gatz M, Fiske A, Reynolds CA, Wetherell JL, Johansson B, Pedersen NL (2003) Sex differences in genetic risk for dementia. Behav Genet 33:95–105

Geismann C, Arlt A, Sebens S, Schäfer H (2014) Cytoprotection "gone astray": Nrf2 and its role in cancer. Onco Targets Ther 7:1497–1518

Ghoumari A, Baulieu E, Schumacher M (2005) Progesterone increases oligodendroglial cell proliferation in rat cerebellar slice cultures. Neuroscience 135:47–58

Gibbs RB (2000) Long-term treatment with estrogen and progesterone enhances acquisition of a spatial memory task by ovariectomized aged rats. Neurobiol Aging 21:107–116

Glei DA, Horiuchi S (2007) The narrowing sex differential in life expectancy in high-income populations: effects of differences in the age pattern of mortality. Popul Stud 61:141–159

González-Burgos I, Rivera-Cervantes MC, Velázquez-Zamora DA, Feria-Velasco A, Garcia-Segura LM (2012) Selective estrogen receptor modulators regulate dendritic spine plasticity in the hippocampus of male rats. Neural Plast 2012:309494

Graham JD, Clarke CL (1997) Physiological action of progesterone in target tissues. Endocr Rev 18:502–519

Grimm A, Biliouris EE, Lang UE, Götz J, Mensah-Nyagan AG, Eckert A (2016a) Sex hormone-related neurosteroids differentially rescue bioenergetic deficits induced by amyloid-β or hyperphosphorylated tau protein. Cell Mol Life Sci 73:201–215

Grimm A, Friedland K, Eckert A (2016b) Mitochondrial dysfunction: the missing link between aging and sporadic Alzheimer's disease. Biogerontology 17:281–296

Grimm A, Mensah-Nyagan AG, Eckert A (2016c) Alzheimer, mitochondria and gender. Neurosci Biobehav Rev 67:89–101

Guerriero G (2009) Vertebrate sex steroid receptors: evolution, ligands, and neurodistribution. Ann N Y Acad Sci 1163:154–168

Gurvich C, Hoy K, Thomas N, Kulkarni J (2018) Sex differences and the influence of sex hormones on cognition through adulthood and the aging process. Brain Sci 8:163

Haaxma CA, Bloem BR, Borm GF, Oyen WJ, Leenders KL, Eshuis S, Booij J, Dluzen DE, Horstink MW (2007) Gender differences in Parkinson's disease. J Neurol Neurosurg Psychiatry 78:819–824

Habib P, Beyer C (2015) Regulation of brain microglia by female gonadal steroids. J Steroid Biochem Mol Biol 146:3–14

Hanamsagar R, Bilbo SD (2016) Sex differences in neurodevelopmental and neurodegenerative disorders: focus on microglial function and neuroinflammation during development. J Steroid Biochem Mol Biol 160:127–133

Hara Y, Yuk F, Puri R, Janssen WG, Rapp PR, Morrison JH (2014) Presynaptic mitochondrial morphology in monkey prefrontal cortex correlates with working memory and is improved with estrogen treatment. Proc Natl Acad Sci 111:486–491

Hara Y, Waters EM, McEwen BS, Morrison JH (2015) Estrogen effects on cognitive and synaptic health over the lifecourse. Physiol Rev 95:785–807

Harman D (1991) The aging process: major risk factor for disease and death. Proc Natl Acad Sci 88: 5360–5363

Hill RA, Kouremenos K, Tull D, Maggi A, Schroeder A, Gibbons A, Kulkarni J, Sundram S, Du X (2020) Bazedoxifene – a promising brain active SERM that crosses the blood brain barrier and enhances spatial memory. Psychoneuroendocrinology 121:104830

Hol EM, Pekny M (2015) Glial fibrillary acidic protein (GFAP) and the astrocyte intermediate filament system in diseases of the central nervous system. Curr Opin Cell Biol 32:121–130

Huffman J, Hoffmann C, Taylor GT (2017) Integrating insulin-like growth factor 1 and sex hormones into neuroprotection: implications for diabetes. World J Diabetes 8:45

Irvine K, Laws KR, Gale TM, Kondel TK (2012) Greater cognitive deterioration in women than men with Alzheimer's disease: a meta analysis. J Clin Exp Neuropsychol 34:989–998

Irwin RW, Yao J, Hamilton RT, Cadenas E, Brinton RD, Nilsen J (2008) Progesterone and estrogen regulate oxidative metabolism in brain mitochondria. Endocrinology 149:3167–3175

Ishihara T, Hong M, Zhang B, Nakagawa Y, Lee MK, Trojanowski JQ, Lee VMY (1999) Age-dependent emergence and progression of a tauopathy in transgenic mice overexpressing the shortest human tau isoform. Neuron 24:751–762

Ishihara Y, Fujitani N, Sakurai H, Takemoto T, Ikeda-Ishihara N, Mori-Yasumoto K, Nehira T, Ishida A, Yamazaki T (2016) Effects of sex steroid hormones and their metabolites on neuronal injury caused by oxygen-glucose deprivation/reoxygenation in organotypic hippocampal slice cultures. Steroids 113:71–77

Jack CR, Wiste HJ, Weigand SD, Knopman DS, Vemuri P, Mielke MM, Lowe V, Senjem ML, Gunter JL, Machulda MM (2015) Age, sex, and APOE ε4 effects on memory, brain structure, and β-amyloid across the adult life span. JAMA Neurol 72:511–519

Janowsky JS (2006) The role of androgens in cognition and brain aging in men. Neuroscience 138: 1015–1020

Jefcoate CR, McNamara BC, Artemenko I, Yamazaki T (1992) Regulation of cholesterol movement to mitochondrial cytochrome P450scc in steroid hormone synthesis. J Steroid Biochem Mol Biol 43:751–767

Jeppesen DK, Bohr VA, Stevnsner T (2011) DNA repair deficiency in neurodegeneration. Prog Neurobiol 94:166–200

Jia J, Kang L, Li S, Geng D, Fan P, Wang L, Cui H (2013) Amelioratory effects of testosterone treatment on cognitive performance deficits induced by soluble Aβ1–42 oligomers injected into the hippocampus. Horm Behav 64:477–486

Jia J-X, Cui C-L, Yan X-S, Zhang B-F, Song W, Huo D-S, Wang H, Yang Z-J (2016) Effects of testosterone on synaptic plasticity mediated by androgen receptors in male SAMP8 mice. J Toxicol Environ Health A 79:849–855

Johnson IP (2015) Age-related neurodegenerative disease research needs aging models. Front Aging Neurosci 7:168

Johri A, Beal MF (2012) Mitochondrial dysfunction in neurodegenerative diseases. J Pharmacol Exp Ther 342:619–630

Jones TH (2009) Advances in the management of testosterone deficiency. Karger Medical and Scientific Publishers, Basel

Jordan VC (2003) Tamoxifen: a most unlikely pioneering medicine. Nat Rev Drug Discov 2:205–213

Kang D, Kim SH, Hamasaki N (2007) Mitochondrial transcription factor A (TFAM): roles in maintenance of mtDNA and cellular functions. Mitochondrion 7:39–44

Katsumoto A, Lu H, Miranda AS, Ransohoff RM (2014) Ontogeny and functions of central nervous system macrophages. J Immunol 193:2615–2621

Kauffman RF, Bryant H (1995) Selective estrogen receptor modulators. Drug News Perspect 8: 531–559

Kaufman JM, Vermeulen A (2005) The decline of androgen levels in elderly men and its clinical and therapeutic implications. Endocr Rev 26:833–876

Kemper MF, Zhao Y, Duckles SP, Krause DN (2013) Endogenous ovarian hormones affect mitochondrial efficiency in cerebral endothelium via distinct regulation of PGC-1 isoforms. J Cereb Blood Flow Metab 33:122–128

Kimura D (2002) Sex hormones influence human cognitive pattern. Neuroendocrinol Lett 23:67–77
Kowal SL, Dall TM, Chakrabarti R, Storm MV, Jain A (2013) The current and projected economic burden of Parkinson's disease in the United States. Mov Disord 28:311–318
Kritsilis M, Rizou SV, Koutsoudaki PN, Evangelou K, Gorgoulis VG, Papadopoulos D (2018) Ageing, cellular senescence and neurodegenerative disease. Int J Mol Sci 19:2937
Kujoth GC, Bradshaw PC, Haroon S, Prolla TA (2007) The role of mitochondrial DNA mutations in mammalian aging. PLoS Genet 3:e24
Kulijewicz-Nawrot M, Verkhratsky A, Chvátal A, Syková E, Rodríguez JJ (2012) Astrocytic cytoskeletal atrophy in the medial prefrontal cortex of a triple transgenic mouse model of Alzheimer's disease. J Anat 221:252–262
Labandeira-Garcia JL, Rodriguez-Perez AI, Valenzuela R, Costa-Besada MA, Guerra MJ (2016) Menopause and Parkinson's disease. Interaction between estrogens and brain renin-angiotensin system in dopaminergic degeneration. Front Neuroendocrinol 43:44–59
Launer LJ, Andersen K, Dewey ME, Letenneur L, Ott A, Amaducci LA, Brayne C, Copeland JR, Dartigues JF, Kragh-Sorensen P, Lobo A, Martinez-Lage JM, Stijnen T, Hofman A (1999) Rates and risk factors for dementia and Alzheimer's disease: results from EURODEM pooled analyses. EURODEM Incidence Research Group and Work Groups. European Studies of Dementia. Neurology 52:78–84
Lehmann D, Butler H, Warden D, Combrinck M, King E, Nicoll J, Budge M, de Jager CA, Hogervorst E, Esiri M (2003) Association of the androgen receptor CAG repeat polymorphism with Alzheimer's disease in men. Neurosci Lett 340:87–90
Leranth C, Petnehazy O, MacLusky NJ (2003) Gonadal hormones affect spine synaptic density in the CA1 hippocampal subfield of male rats. J Neurosci 23:1588–1592
Li R, Singh M (2014) Sex differences in cognitive impairment and Alzheimer's disease. Front Neuroendocrinol 35:385–403
Li M, Masugi-Tokita M, Takanami K, Yamada S, Kawata M (2012) Testosterone has sublayer-specific effects on dendritic spine maturation mediated by BDNF and PSD-95 in pyramidal neurons in the hippocampus CA1 area. Brain Res 1484:76–84
Livingston G, Sommerlad A, Orgeta V, Costafreda SG, Huntley J, Ames D, Ballard C, Banerjee S, Burns A, Cohen-Mansfield J (2017) Dementia prevention, intervention, and care. Lancet 390: 2673–2734
Lövheim H, Sandman P-O, Karlsson S, Gustafson Y (2009) Sex differences in the prevalence of behavioral and psychological symptoms of dementia. Int Psychogeriatr 21:469–475
Lu Y-P, Zeng M, Hu X-Y, Xu H, Swaab DF, Ravid R, Zhou J-N (2003) Estrogen receptor α-immunoreactive astrocytes are increased in the hippocampus in Alzheimer's disease. Exp Neurol 183:482–488
Lu H, Ma K, Jin L, Zhu H, Cao R (2018) 17β-estradiol rescues damages following traumatic brain injury from molecule to behavior in mice. J Cell Physiol 233:1712–1722
Luchetti S, Bossers K, Van de Bilt S, Agrapart V, Morales RR, Frajese GV, Swaab DF (2011) Neurosteroid biosynthetic pathways changes in prefrontal cortex in Alzheimer's disease. Neurobiol Aging 32:1964–1976
Mangold CA, Wronowski B, Du M, Masser DR, Hadad N, Bixler GV, Brucklacher RM, Ford MM, Sonntag WE, Freeman WM (2017) Sexually divergent induction of microglial-associated neuroinflammation with hippocampal aging. J Neuroinflammation 14:141
Manly J, Merchant C, Jacobs D, Small S, Bell K, Ferin M, Mayeux R (2000) Endogenous estrogen levels and Alzheimer's disease among postmenopausal women. Neurology 54:833–837
Marin-Husstege M, Muggironi M, Raban D, Skoff RP, Casaccia-Bonnefil P (2004) Oligodendrocyte progenitor proliferation and maturation is differentially regulated by male and female sex steroid hormones. Dev Neurosci 26:245–254
Matthews F, Brayne C, Medical Research Council Cognitive Function and Ageing Study Investigators (2005) The incidence of dementia in England and Wales: findings from the five identical sites of the MRC CFA study. PLoS Med 2:e193

McDevitt MA, Glidewell-Kenney C, Jimenez MA, Ahearn PC, Weiss J, Jameson JL, Levine JE (2008) New insights into the classical and non-classical actions of estrogen: evidence from estrogen receptor knock-out and knock-in mice. Mol Cell Endocrinol 290:24–30
McEwen BS, Milner TA (2007) Hippocampal formation: shedding light on the influence of sex and stress on the brain. Brain Res Rev 55:343–355
McEwen BS, Milner TA (2017) Understanding the broad influence of sex hormones and sex differences in the brain. J Neurosci Res 95:24–39
Mellon SH, Griffin LD (2002) Neurosteroids: biochemistry and clinical significance. Trends Endocrinol Metab 13:35–43
Micevych PE, Chaban V, Ogi J, Dewing P, Lu JK, Sinchak K (2007) Estradiol stimulates progesterone synthesis in hypothalamic astrocyte cultures. Endocrinology 148:782–789
Miller VM, Harman SM (2017) An update on hormone therapy in postmenopausal women: mini-review for the basic scientist. Am J Physiol Heart Circ Physiol 313:H1013–H1021
Milner TA, Ayoola K, Drake CT, Herrick SP, Tabori NE, McEwen BS, Warrier S, Alves SE (2005) Ultrastructural localization of estrogen receptor β immunoreactivity in the rat hippocampal formation. J Comp Neurol 491:81–95
Morley JE (2001) Androgens and aging. Maturitas 38:61–71; discussion 71–73
Muller M, den Tonkelaar I, Thijssen JH, Grobbee DE, van der Schouw YT (2003) Endogenous sex hormones in men aged 40-80 years. Eur J Endocrinol 149:583–589
Nickelsen T, Lufkin EG, Lawrence Riggs B, Cox DA, Crook TH (1999) Raloxifene hydrochloride, a selective estrogen receptor modulator: safety assessment of effects on cognitive function and mood in postmenopausal women. Psychoneuroendocrinology 24:115–128
Nilsen J, Irwin RW, Gallaher TK, Brinton RD (2007) Estradiol in vivo regulation of brain mitochondrial proteome. J Neurosci 27:14069–14077
Nissen JC (2017) Microglial function across the spectrum of age and gender. Int J Mol Sci 18:561
Njie EG, Boelen E, Stassen FR, Steinbusch HWM, Borchelt DR, Streit WJ (2012) Ex vivo cultures of microglia from young and aged rodent brain reveal age-related changes in microglial function. Neurobiol Aging 33:195.e191–195.e112
Noorbakhsh F, Baker GB, Power C (2014) Allopregnanolone and neuroinflammation: a focus on multiple sclerosis. Front Cell Neurosci 8:134
O'Callaghan JP, Sriram K (2005) Glial fibrillary acidic protein and related glial proteins as biomarkers of neurotoxicity. Expert Opin Drug Saf 4:433–442
Olabarria M, Noristani HN, Verkhratsky A, Rodríguez JJ (2011) Age-dependent decrease in glutamine synthetase expression in the hippocampal astroglia of the triple transgenic Alzheimer's disease mouse model: mechanism for deficient glutamatergic transmission? Mol Neurodegener 6:55
Overk CR, Lu P-Y, Wang Y-T, Choi J, Shaw JW, Thatcher GR, Mufson EJ (2012) Effects of aromatase inhibition versus gonadectomy on hippocampal complex amyloid pathology in triple transgenic mice. Neurobiol Dis 45:479–487
Pan W, Han S, Kang L, Li S, Du J, Cui H (2016) Effects of dihydrotestosterone on synaptic plasticity of the hippocampus in mild cognitive impairment male SAMP8 mice. Exp Ther Med 12:1455–1463
Paoletti A, Congia S, Lello S, Tedde D, Orrù M, Pistis M, Pilloni M, Zedda P, Loddo A, Melis GB (2004) Low androgenization index in elderly women and elderly men with Alzheimer's disease. Neurology 62:301–303
Persson I, Adami HO, Bergkvist L, Lindgren A, Pettersson B, Hoover R, Schairer C (1989) Risk of endometrial cancer after treatment with oestrogens alone or in conjunction with progestogens: results of a prospective study. BMJ 298:147–151
Petersen NM, Briggs AL (2005) Selective estrogen receptor modulators. Clin Rev Bone Miner Metab 3:19–30
Pike CJ (2017) Sex and the development of Alzheimer's disease. J Neurosci Res 95:671–680
Plassman BL, Langa KM, Fisher GG, Heeringa SG, Weir DR, Ofstedal MB, Burke JR, Hurd MD, Potter GG, Rodgers WL, Steffens DC, Willis RJ, Wallace RB (2007) Prevalence of dementia in

the United States: the aging, demographics, and memory study. Neuroepidemiology 29:125–132

Plassman BL, Langa KM, McCammon RJ, Fisher GG, Potter GG, Burke JR, Steffens DC, Foster NL, Giordani B, Unverzagt FW (2011) Incidence of dementia and cognitive impairment, not dementia in the United States. Ann Neurol 70:418–426

Poirier A-A, Côté M, Bourque M, Morissette M, Di Paolo T, Soulet D (2016) Neuroprotective and immunomodulatory effects of raloxifene in the myenteric plexus of a mouse model of Parkinson's disease. Neurobiol Aging 48:61–71

Prvulovic D, Van de Ven V, Sack AT, Maurer K, Linden DEJ (2005) Functional activation imaging in aging and dementia. Psychiatry Res 140:97–113

Rao X, Di Leva G, Li M, Fang F, Devlin C, Hartman-Frey C, Burrow ME, Ivan M, Croce CM, Nephew KP (2011) MicroRNA-221/222 confers breast cancer fulvestrant resistance by regulating multiple signaling pathways. Oncogene 30(9):1082–1097

Raz L, Knoefel J, Bhaskar K (2016) The neuropathology and cerebrovascular mechanisms of dementia. J Cereb Blood Flow Metab 36:172–186

Refolo V, Stefanova N (2019) Neuroinflammation and glial phenotypic changes in alpha-synucleinopathies. Front Cell Neurosci 13:263

Rettberg JR, Yao J, Brinton RD (2014) Estrogen: a master regulator of bioenergetic systems in the brain and body. Front Neuroendocrinol 35:8–30

Rice MM, Graves AB, McCurry SM, Gibbons LE, Bowen JD, McCormick WC, Larson EB (2000) Postmenopausal estrogen and estrogen-progestin use and 2-year rate of cognitive change in a cohort of older Japanese American women: the Kame Project. Arch Intern Med 160:1641–1649

Rosario ER, Chang L, Head EH, Stanczyk FZ, Pike CJ (2011) Brain levels of sex steroid hormones in men and women during normal aging and in Alzheimer's disease. Neurobiol Aging 32:604–613

Rossetti MF, Cambiasso MJ, Holschbach MA, Cabrera R (2016) Oestrogens and progestagens: synthesis and action in the brain. J Neuroendocrinol 28

Salas IH, Burgado J, Allen NJ (2020) Glia: victims or villains of the aging brain? Neurobiol Dis 143:105008

Salminen A, Ojala J, Kaarniranta K, Haapasalo A, Hiltunen M, Soininen H (2011) Astrocytes in the aging brain express characteristics of senescence-associated secretory phenotype. Eur J Neurosci 34:3–11

Sancesario GM, Bernardini S (2018) Diagnosis of neurodegenerative dementia: where do we stand, now? Ann Transl Med 6:340–340

Sánchez MG, Bourque M, Morissette M, Di Paolo T (2010) Steroids-dopamine interactions in the pathophysiology and treatment of CNS disorders. CNS Neurosci Ther 16:e43–e71

Sanchez-Rodriguez M, Zacarias-Flores M, Correa-Muñoz E, Mendoza-Nuñez VM (2018) 304 - Changes in antioxidant components in midlife women. Is it age or menopause? Free Radic Biol Med 128:S127

Sanders DW, Kaufman SK, DeVos SL, Sharma AM, Mirbaha H, Li A, Barker SJ, Foley AC, Thorpe JR, Serpell LC, Miller TM, Grinberg LT, Seeley WW, Diamond MI (2014) Distinct tau prion strains propagate in cells and mice and define different tauopathies. Neuron 82 (6):1271–1288. https://doi.org/10.1016/j.neuron.2014.04.047

Santoro N (2005) The menopausal transition. Am J Med 118:8–13

Sárvári M, Hrabovszky E, Kalló I, Solymosi N, Tóth K, Likó I, Széles J, Mahó S, Molnár B, Liposits Z (2011) Estrogens regulate neuroinflammatory genes via estrogen receptors α and β in the frontal cortex of middle-aged female rats. J Neuroinflammation 8:82

Sárvári M, Hrabovszky E, Kalló I, Solymosi N, Likó I, Berchtold N, Cotman C, Liposits Z (2012) Menopause leads to elevated expression of macrophage-associated genes in the aging frontal cortex: rat and human studies identify strikingly similar changes. J Neuroinflammation 9:264

Sárvári M, Kalló I, Hrabovszky E, Solymosi N, Liposits Z (2014) Ovariectomy and subsequent treatment with estrogen receptor agonists tune the innate immune system of the hippocampus in middle-aged female rats. PLoS One 9:e88540

Scharfman HE, MacLusky NJ (2006) Estrogen and brain-derived neurotrophic factor (BDNF) in hippocampus: complexity of steroid hormone-growth factor interactions in the adult CNS. Front Neuroendocrinol 27:415–435

Schiöth HB, Craft S, Brooks SJ, Frey WH, Benedict C (2012) Brain insulin signaling and Alzheimer's disease: current evidence and future directions. Mol Neurobiol 46:4–10

Schroeder A, Hudson M, Du X, Wu YWC, Nakamura J, van den Buuse M, Jones NC, Hill RA (2017) Estradiol and raloxifene modulate hippocampal gamma oscillations during a spatial memory task. Psychoneuroendocrinology 78:85–92

Schumacher M, Guennoun R, Stein DG, De Nicola AF (2007) Progesterone: therapeutic opportunities for neuroprotection and myelin repair. Pharmacol Ther 116:77–106

Scott E, Zhang Q-g, Wang R, Vadlamudi R, Brann D (2012) Estrogen neuroprotection and the critical period hypothesis. Front Neuroendocrinol 33:85–104

Seifarth JE, McGowan CL, Milne KJ (2012) Sex and life expectancy. Gend Med 9:390–401

Selkoe DJ (2001) Alzheimer's disease results from the cerebral accumulation and cytotoxicity of amyloid beta-protein. J Alzheimers Dis 3:75–80

Selvaraj V, Stocco DM, Tu LN (2015) Minireview: translocator protein (TSPO) and steroidogenesis: a reappraisal. Mol Endocrinol 29:490–501

Shadyab AH, Macera CA, Shaffer RA, Jain S, Gallo LC, Gass ML, Waring ME, Stefanick LM, LaCroix ZA (2017) Ages at menarche and menopause and reproductive lifespan as predictors of exceptional longevity in women: the women's health initiative. Menopause 24:35–44. https://doi.org/10.1097/GME.0000000000000710

Sheppard PAS, Choleris E, Galea LAM (2019) Structural plasticity of the hippocampus in response to estrogens in female rodents. Mol Brain 12:22

Shimada H, Makizako H, Yoshida D, Tsutsumimoto K, Anan Y, Uemura K, Lee S, Park H, Suzuki T (2014) A large, cross-sectional observational study of serum BDNF, cognitive function, and mild cognitive impairment in the elderly. Front Aging Neurosci 6:69

Siani F, Greco R, Levandis G, Ghezzi C, Daviddi F, Demartini C, Vegeto E, Fuzzati-Armentero M-T, Blandini F (2017) Influence of estrogen modulation on glia activation in a murine model of Parkinson's disease. Front Neurosci 11:306

Silverman SL, Christiansen C, Genant HK, Vukicevic S, Zanchetta JR, de Villiers TJ, Constantine GD, Chines AA (2008) Efficacy of bazedoxifene in reducing new vertebral fracture risk in postmenopausal women with osteoporosis: results from a 3-year, randomized, placebo-, and active-controlled clinical trial. J Bone Miner Res 23:1923–1934

Silverman S, Chines A, Kendler D, Kung A, Teglbjaerg C, Felsenberg D, Mairon N, Constantine G, Adachi J, Bazedoxifene Study Group (2012) Sustained efficacy and safety of bazedoxifene in preventing fractures in postmenopausal women with osteoporosis: results of a 5-year, randomized, placebo-controlled study. Osteoporos Int 23:351–363

Singh M, Su C (2013) Progesterone and neuroprotection. Horm Behav 63:284–290

Sohrabji F (2015) Estrogen-IGF-1 interactions in neuroprotection: ischemic stroke as a case study. Front Neuroendocrinol 36:1–14

Sorwell KG, Kohama SG, Urbanski HF (2012) Perimenopausal regulation of steroidogenesis in the nonhuman primate. Neurobiol Aging 33(1487):1487.e1413

Spittau B (2017) Aging microglia—phenotypes, functions and implications for age-related neurodegenerative diseases. Front Aging Neurosci 9:194

Stoffel-Wagner B (2001) Neurosteroid metabolism in the human brain. Eur J Endocrinol 145:669–679

Suzuki S, Brown CM, Wise PM (2009) Neuroprotective effects of estrogens following ischemic stroke. Front Neuroendocrinol 30:201–211

Swerdloff RS, Wang C (2004) Androgens and the ageing male. Best Pract Res Clin Endocrinol Metab 18:349–362

Thakur M, Paramanik V (2009) Role of steroid hormone coregulators in health and disease. Horm Res Paediatr 71:194–200

Torika N, Asraf K, Roasso E, Danon A, Fleisher-Berkovich S (2016) Angiotensin converting enzyme inhibitors ameliorate brain inflammation associated with microglial activation: possible implications for Alzheimer's disease. J Neuroimmune Pharmacol 11:774–785

Toro-Urrego N, Garcia-Segura LM, Echeverria V, Barreto GE (2016) Testosterone protects mitochondrial function and regulates neuroglobin expression in astrocytic cells exposed to glucose deprivation. Front Aging Neurosci 8:152

Veiga S, Melcangi RC, DonCarlos LL, Garcia-Segura LM, Azcoitia I (2004) Sex hormones and brain aging. Exp Gerontol 39:1623–1631

Velarde MC (2014) Mitochondrial and sex steroid hormone crosstalk during aging. Longev Healthspan 3:1–10

Verkhratsky A, Rodríguez JJ, Parpura V (2014) Neuroglia in ageing and disease. Cell Tissue Res 357:493–503

Villa A, Vegeto E, Poletti A, Maggi A (2016) Estrogens, neuroinflammation, and neurodegeneration. Endocr Rev 37:372–402

Villar-Cheda B, Dominguez-Meijide A, Valenzuela R, Granado N, Moratalla R, Labandeira-Garcia JL (2014) Aging-related dysregulation of dopamine and angiotensin receptor interaction. Neurobiol Aging 35:1726–1738

Walker Z, Possin KL, Boeve BF, Aarsland D (2015) Lewy body dementias. Lancet 386:1683–1697

Wilkenfeld SR, Lin C, Frigo DE (2018) Communication between genomic and non-genomic signaling events coordinate steroid hormone actions. Steroids 133:2–7

Winblad B, Amouyel P, Andrieu S, Ballard C, Brayne C, Brodaty H, Cedazo-Minguez A, Dubois B, Edvardsson D, Feldman H (2016) Defeating Alzheimer's disease and other dementias: a priority for European science and society. Lancet Neurol 15:455–532

Witty CF, Gardella LP, Perez MC, Daniel JM (2013) Short-term estradiol administration in aging ovariectomized rats provides lasting benefits for memory and the hippocampus: a role for insulin-like growth factor-I. Endocrinology 154:842–852

Wu Z, Zhu A, Wu S, Nakanishi H (2013) Preventing and reversing "microglia-aging" by nature materials for slow brain-aging. J Neurol Disord 2:143

Wu Y, Du X, Van Den Buuse M, Hill R (2015) Analyzing the influence of BDNF heterozygosity on spatial memory response to 17β-estradiol. Transl Psychiatry 5:e498–e498

Yao J, Irwin RW, Zhao L, Nilsen J, Hamilton RT, Brinton RD (2009) Mitochondrial bioenergetic deficit precedes Alzheimer's pathology in female mouse model of Alzheimer's disease. Proc Natl Acad Sci 106:14670–14675

Yao J, Hamilton RT, Cadenas E, Brinton RD (2010) Decline in mitochondrial bioenergetics and shift to ketogenic profile in brain during reproductive senescence. Biochim Biophys Acta 1800: 1121–1126

Yao J, Irwin R, Chen S, Hamilton R, Cadenas E, Brinton RD (2012) Ovarian hormone loss induces bioenergetic deficits and mitochondrial β-amyloid. Neurobiol Aging 33:1507–1521

Yu P, Li S, Zhang Z, Wen X, Quan W, Tian Q, Gao C, Su W, Zhang J, Jiang R (2017) Progesterone-mediated angiogenic activity of endothelial progenitor cell and angiogenesis in traumatic brain injury rats were antagonized by progesterone receptor antagonist. Cell Prolif 50:e12362

Yue X, Lu M, Lancaster T, Cao P, Honda S-I, Staufenbiel M, Harada N, Zhong Z, Shen Y, Li R (2005) Brain estrogen deficiency accelerates Aβ plaque formation in an Alzheimer's disease animal model. Proc Natl Acad Sci 102:19198–19203

Yun J, Yeo IJ, Hwang CJ, Choi D-Y, Im H-S, Kim JY, Choi WR, Jung MH, Han S-B, Hong JT (2018) Estrogen deficiency exacerbates Aβ-induced memory impairment through enhancement of neuroinflammation, amyloidogenesis and NF-κB activation in ovariectomized mice. Brain Behav Immun 73:282–293

Zagni E, Simoni L, Colombo D (2016) Sex and gender differences in central nervous system-related disorders. Neurosci J 2016:2827090

Zamanian JL, Xu L, Foo LC, Nouri N, Zhou L, Giffard RG, Barres BA (2012) Genomic analysis of reactive astrogliosis. J Neurosci 32:6391–6410

Zárate S, Seilicovich A (2010) Estrogen receptors and signaling pathways in lactotropes and somatotropes. Neuroendocrinology 92:215–223

Zárate S, Stevnsner T, Gredilla R (2017) Role of estrogen and other sex hormones in brain aging. Neuroprotection and DNA repair. Front Aging Neurosci 9:430

Zarif JC, Miranti CK (2016) The importance of non-nuclear AR signaling in prostate cancer progression and therapeutic resistance. Cell Signal 28:348–356

Zhang T, Liang X, Shi L, Wang L, Chen J, Kang C, Zhu J, Mi M (2013) Estrogen receptor and PI3K/Akt signaling pathway involvement in S-(-)equol-induced activation of Nrf2/ARE in endothelial cells. PLoS One 8:e79075

Zhao Y, He L, Zhang Y, Zhao J, Liu Z, Xing F, Liu M, Feng Z, Li W, Zhang J (2017) Estrogen receptor alpha and beta regulate actin polymerization and spatial memory through an SRC-1/mTORC2-dependent pathway in the hippocampus of female mice. J Steroid Biochem Mol Biol 174:96–113

Mitophagy Impairments as Culprit of Alzheimer's Disease

5

Shalini Mani, Geeta Swargiary, Manisha Singh, and Mahima Rawal

Abstract

Dementia is considered as a common term which includes different disease known to affect memory and other cognitive abilities too. Amongst different brain diseases, which can cause dementia, Alzheimer's disease (AD) is reported to be the most common form and nearly 60–70% of dementia cases are AD patients. The major events of AD primarily include protein aggregation, neuronal injury, synaptic failure, and consequently neuronal death in the hippocampal region of the brain. In the recent past, defects in mitochondrial function are also reported as a central aspect of different neurodegenerative diseases, including AD. To maintain the mitochondrial dynamics and neuronal health, the aged and dysfunctional mitochondria are efficiently removed by a specific process called mitophagy. Several studies have revealed that mitochondrial turnover is regulated by the integration of different cellular pathways. In this chapter, the mechanism of different mitophagy pathways and the possible molecular defect leading to mitophagy impairment in AD are discussed. This chapter also provides a glimpse of different therapeutic approaches to enhance mitophagy in neurons of AD brain and combat such diseases. Though there are lots of developments in mitophagy-based therapeutic interventions of AD; however, the associated limitations need to be addressed.

Keywords

Dementia · Alzheimer's disease · Mitochondria · Mitophagy · Parkin · Tau · Amyloid β

S. Mani (✉) · G. Swargiary · M. Singh · M. Rawal
Centre for Emerging Diseases, Department of Biotechnology, Jaypee Institute of Information Technology, Noida, India
e-mail: shalini.mani@jiit.ac.in

G. M. Ashraf, M. S. Uddin (eds.), *Current Thoughts on Dementia*,
https://doi.org/10.1007/978-981-16-7606-2_5

5.1 Introduction

Dementia is a clinical syndrome representing a group of symptoms demonstrated by loss of memory, disturbances in language, psychological and psychiatric changes, etc. Across the world, the number of dementia cases in the year 2017 are reported to be around 50 million, which was expected to rise to 82 million in 2030 and 152 million in 2050 (WHO 2019). In terms of the global burden of diseases, dementia is considered to be the main cause of age-associated disability. As per the description provided by the World Health Organization (WHO 2019), dementia can be considered an umbrella term which includes such disease, known to affect memory and other cognitive abilities too. Patients suffering from dementia generally suffer from disabilities to maintain their activities related to daily life. Generally, dementia is considered a normal part of aging but it is not true as not every aged person develops dementia. Dementia is not a specific disease but different types of brain diseases which can result in causing a long-term and mostly steady decrease in mental abilities. The different disabilities which jointly define dementia include emotional problems, language difficulties, loss of memory, and decreased motivation too. Different brain-related diseases which can commonly cause dementia include Alzheimer's disease (AD), Parkinson's disease (PD), frontotemporal dementia (FTD), Lewy body dementia (LBD), vascular dementia (VD), syphilitic dementia, mixed dementia (MD), senility dementia, and even stroke (Fig. 5.1). As the symptoms are quite variable in case of dementia and thus support the fact that the type of symptom experienced by the dementia patient is completely dependent upon the area of the brain affected. However, some of the primary symptoms, indicating

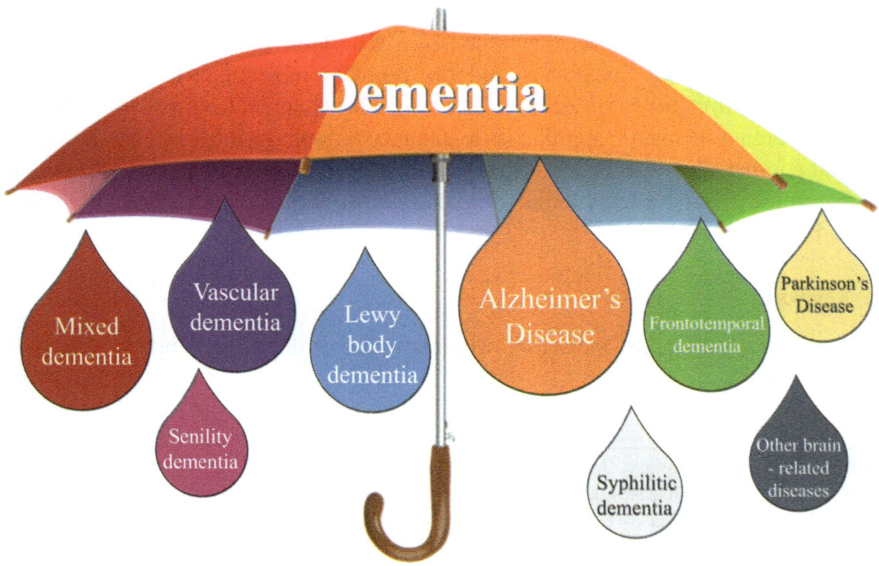

Fig. 5.1 Different types of brain-related diseases commonly cause dementia

the cognitive and functional impairment in dementia may be common in different types of dementia. Amongst different brain diseases, which can cause dementia, AD is reported to be the commonest form of dementia. It has been observed that approximately 60–70% of dementia cases are AD patients (WHO 2019).

AD is known to be caused by different factors but the protein aggregates in neurons of the hippocampal region, and mutations in *APOE* gene are most commonly known to be associated with the pathogenesis of AD (Bekris et al. 2010). The protein aggregates in neurons of AD patients are suggested to trigger mitochondrial dysfunction in them (Weidling and Swerdlow 2020). On the contrary, different studies also revealed that mitochondrial defects in neurons lead to the characteristic aggregate formation of different proteins in AD (Weidling and Swerdlow 2020) patients. In the recent past also, the defects in mitochondrial metabolism and dynamics are reported to be associated with different neurodegenerative disorders including AD (Kim and Mook-Jung 2019). It is thus critically very important to maintain mitochondrial metabolism in neurons as these cells require the highest amount of energy for their proper functioning and any perturbation in mitochondrial activity may severely affect these cells. To maintain the mitochondrial dynamics, three different processes of mitochondrial fission, fusion, and mitophagy (removal of damaged mitochondria) need to be in tight coordination (Kerr et al. 2017). However, as per different studies, the elimination of damaged mitochondria through mitophagy is impaired in neurons of many AD patients (Kerr et al. 2017). This impairment in mitophagy may lead to excessive accumulation of reactive oxygen species (ROS), ATP depletion, consequently leading to neuronal cell death.

This chapter is an attempt to highlight the importance of healthy mitochondrial metabolism and mitophagy for neuronal plasticity. It also aims to describe the consequences of impaired mitophagy on neuronal functions and specifically its association with the pathogenesis of AD. In the end, the authors indicated the need to design the strategies aiming to trigger mitophagy in neuron cells of AD brain, as it may help to combat this most common form of dementia.

5.2 Alzheimer's Disease: A Common Form of Dementia

AD is most commonly characterized by neurodegenerative disorder (NDD) marked by a sequential chain of events, tightly linked with each other. The major events include protein aggregation, neuronal injury, synaptic failure (reduction in the density of synapses), and finally succeeding in neuronal death in the hippocampal region of the brain and thus leading to cognitive decline (Karch and Goate 2015). AD represents two clinical forms, the first one where it is mostly triggered due to aging is called sporadic AD (SAD). However, in other forms, the rare gene mutation has been identified and this is called the familial AD (FAD) (Eckert et al. 2003; Hsieh and Yang 2013). These two forms of AD are proposed to be triggered by different factors; however, they represent similar clinical and neuropathological features. These features commonly include extracellular senile plaques, composed of β-amyloid (Aβ) deposits, intracellular neurofibrillary tangles (NFTs); aggregates

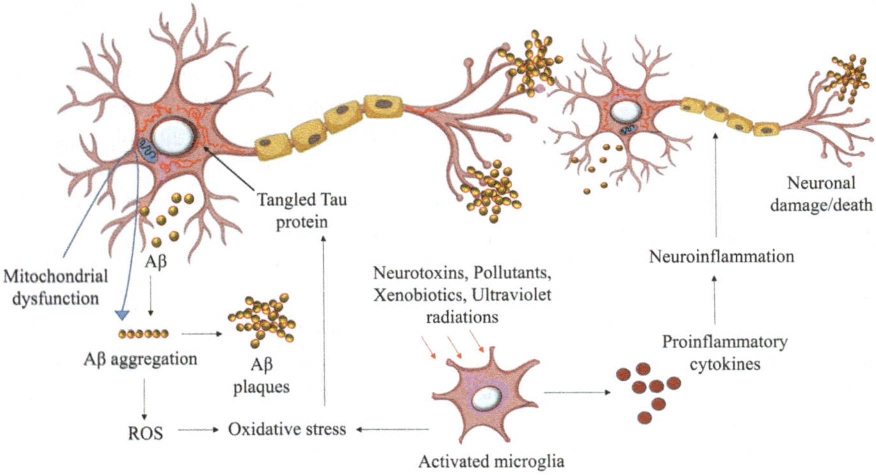

Fig. 5.2 Pathological hallmarks of AD comprising of Aβ aggregation, tau protein tangles, and neuroinflammation. The mitochondrial dysfunctions cause aggregation of Aβ, which may lead to the production of more ROS and further oxidative stress in the neuron that causes tangles of tau protein. Pathogens such as neurotoxins, pollutants, etc. activate the microglia that causes oxidative stress to the neurons. These defects ultimately lead to neuronal damage and/or death

of hyperphosphorylated tau protein including loss of neurons as well (Anandatheerthavarada and Devi 2007) (Fig. 5.2).

The tau proteins are composed of a total of six protein isoforms, which are soluble in nature. These proteins are important to stabilize the microtubules in axons and are predominantly present in the neurons of the central nervous system. The degree of phosphorylation of tau protein decides its involvement in promoting and regulation of microtubule assembly process. In normal conditions, 2–3 mol of phosphate are reported to be present along with each mole of tau protein present in the adult human brain. The hyperphosphorylated form of tau is further known to be less biologically active. Interestingly, in comparison to the normal brain; the tau proteins present in AD brain are three to four times more phosphorylated. Because of its hyperphosphorylated state, the systemic solubility of tau protein decreasing and as a result, the process of agglomeration and sequestration of inclusion bodies gets initiated. This entire process eventually leads to the formation of NFTs in neuronal cells (Serrano-Pozo et al. 2011). The hyperphosphorylated tau protein gets polymerized and form paired helical filaments which are further mixed with straight filaments. This structure altogether forms NFTs. This accumulation of tau proteins is observed to be continued throughout the progression of the disease and the amount of abnormal tau in the AD brain is mostly the representative of severity and stage of the disease.

The second main component is the accretion of Aβ peptide, the main component of senile plaques. These Aβ peptides are produced from the proteolytic cleavage of amyloid-beta precursor protein (APP). APP is an important glycoprotein and being membrane protein in nature, it plays a vital role in different types of biological activities. Some of the important functions of APP include the development of neurons, intracellular transport, and maintaining the homeostasis of neurons. Though APP is expressed in different tissues, it is predominantly present in the synapse region of neurons. Different studies revealed that different cleavage products of APP may be involved in causing different neurological problems. Deposits of Aβ peptides are mainly observed in the cerebrovasculature region, hippocampus, and neocortex region as well (Viswanathan and Greenberg 2011). As per the amyloid hypothesis, Aβ is considered as the main cause of the disease. This hypothesis also emphasizes the fact that the extracellular accumulation of misfolded Aβ proteins in the form of senile plaques and the intracellular deposition of misfolded tau protein as NFTs are the main culprit to cause memory loss, confusion, and cognitive decline over time (Chen et al. 2017; Selkoe and Hardy 2016). According to a body of evidence, Aβ seems to be responsible for raising the vulnerability of neuronal cells toward oxidative stress as well as defects in electron transport chain ETC (Moreira et al. 2007). Several studies also indicate that these protein aggregates may further affect the mitochondrial dynamics and its functions (Bonda et al. 2010; Pagani and Eckert 2011). As brain cells require maximum energy for its function, hence such perturbations in mitochondrial activities may further deteriorate the pathological symptoms in AD.

5.3 Mitochondria and Neuroplasticity in Alzheimer's Disease

The mitochondria are important organelle in eukaryotic cells and mainly responsible to generate energy in these cells. These organelles also play a crucial role in different cellular processes such as regulation of cell cycle, calcium homeostasis, and ROS production (Sas et al. 2007). In addition to these functions, mitochondria are known to be a potential regulator of cell death pathways, which is a characteristic of a neuronal cell in AD and other forms of dementia (Lin and Beal 2006). The defects in mitochondrial activities may cause a reduction in cellular energy levels as well as frequent leakage of electrons from different respiratory chain complexes. These leaked electrons will lead to the formation of excessive and eventually damaging the proteins, membrane lipids, and nucleic acids (Wallace 1999). Increased ROS production might also influence the integral mitochondrial parameters such as calcium uptake, ATP generation, membrane integrity, and activation of permeability transition pore (MPTP) as well.

The term neuroplasticity includes different process such as the making of neurons from neural progenitor cells, the development of axons and dendrites, and construction as well as the restructuring of synapses. Interestingly, different studies support the importance of mitochondrial activity in maintaining the same. These processes in maintaining the neuroplasticity are regulated by different paracrine and endocrine system-based mechanisms that help in converting the environmental inputs to various neuronal cell responses. Mitochondria are highly movable organelles traveling within and between the subcellular compartments playing important role in neuroplasticity. The main compartments areas are synaptic terminals, dendrites, cell body, and the axon. Mitochondria also emit molecular signals such as ROS and proteins as well as lipid mediators which can act either locally or can also travel to distant targets including the nucleus. By generating ATP as well as NAD+, controlling subcellular Ca^{2+} level, and maintaining redox homeostasis, this tiny cellular organelle is considered to play a very important role in monitoring the fundamental processes of neuroplasticity. It can be justified from the fact that mitochondria help in buffering the level of cytosolic calcium and thus promote the polymerization of axonal microtubules (Mattson et al. 2008) during early neuronal differentiation. Thus it can be confirmed that mitochondria may play an important role in regulating the differentiation and growth of the axons. These observations were further supported by a study conducted by Cheng et al. (2012). In their study, they inhibited the mitochondrial biogenesis by knocking down the PGC-1α by RNAi-mediated approach. As a result, they observed that the inhibition of mitochondrial biogenesis further affected the process of neuronal circuit development and synapses formation and the maintenance of synapses in the hippocampus region of the adult brain (Cheng et al. 2012). Additionally, PGC-1α expression and mitochondrial biogenesis are known to be stimulated by a brain-derived neurotrophic factor (BDNF). It is an important protein and plays important role in maintaining the synaptic plasticity in the hippocampal region, thus also helps in regulating learning, memory, and neuronal stress resistance (Cheng et al. 2012). Various studies conducted on experimental models of AD as well as affected patients propose that deficiency in BDNF-mediated signaling may cause synaptic dysfunction and neuronal degeneration in them (Marosi and Mattson 2014). Defects in mitochondrial functions and related signaling pathways may cause impairment in neuroplasticity and may also trigger neuronal degeneration in AD (Cheng et al. 2010). It is important to mention that the cytochrome *c* may be released out of damaged mitochondria and thus it can further lead to Caspase-3-dependent neuronal apoptosis and hence neurodegeneration as well (Lin et al. 2012).

The first indication of the connection of mitochondria in neurodegenerative diseases was described with detection of complex I deficiency in mitochondria isolated from substantia nigra and platelet of PD patients (Parker et al. 1989; Schapira et al. 1990). Additional supporting evidence was also found from complex I and cytochrome *c* oxidase deficiency in AD patients (Morán et al. 2012). The biochemical studies conducted on postmortem AD brains also exhibited the impaired function of various Krebs's cycle-related enzymes such as pyruvate dehydrogenase, isocitrate dehydrogenase, and α-ketoglutarate dehydrogenase. These changes were correlated with the clinical state and diminished brain metabolism of patients (Petrozzi et al. 2007). The relationship between mitochondrial dysfunction and neuronal metabolism was proposed as early features of AD by Blass and Gibson (1991). Since then, hundreds of studies have been reported which indicate the significance of mitochondrial anomalies in AD and also expose the possible molecular mechanisms and cellular significances of mitochondrial deficits. It is important to highlight that in 2004, Swerdlow and Khan proposed the mitochondrial cascade hypothesis, which stated that the mitochondrial function of each individual which is mostly influenced by its genetics and environment, is the key factor inducing late-onset AD pathology (Swerdlow and Khan 2004). Studies conducted in postmortem brain samples as well as living AD patients have also delivered enough evidence to support the impairment in mitochondrial function of neurons present in affected regions of the brain. PET brain scans of such patients also indicated that there was a reduction in radiolabeled glucose uptake into neurons. The biochemical analysis also revealed the reduction of activity of mitochondrial enzymes that are involved in Kreb's cycle and oxidative phosphorylation (Kapogiannis and Mattson 2011).

Recent findings further suggest the impairment in mitochondrial biogenesis in AD and it was confirmed by determining the level of PGC-1α, which was reduced in such AD patients (Katsouri et al. 2016). Studies also support the fact that defective mitochondria accumulate in neurons and cause an elevation in ROS level, which can further amplify the mitochondrial damage and abnormal processing of APP as well as tau (Mattson et al. 2008; Mattson 2004). These processes subsequently lead to the formation of AD-defining Aβ plaques and NFTs. In order to the contrary, different studies also showed that accumulations of NFTs and Aβ may further trigger mitochondrial defects (Lu et al. 2018). In the case of sporadic AD, dysfunction in mitochondria is proposed to be the primary event leading to deposition of Aβ, synaptic degeneration, and NFTs formation (Swerdlow et al. 2010). Low ATP levels are also observed as a characteristic feature of both peripheral cells derived from AD patients and AD brains (Beal 2005; Gibson et al. 1998; Manczak et al. 2004). Besides, the activities of the three key Kreb's cycle enzyme complexes, pyruvate dehydrogenase, isocitrate dehydrogenase, and alpha-ketoglutarate dehydrogenase, are also reported to be compromised in AD brain and fibroblasts samples obtained from AD patients (Bubber et al. 2005). Activities of respiratory chain enzymes such as complex I, III, and IV had also been described in platelets and lymphocytes of AD patients (Bosetti et al. 2002; Kish et al. 1992).

Thus, it seems that the maintenance of a healthy mitochondrial pool is essential for neuronal health. As per various studies, multiple mechanisms called mitochondrial quality control (MQC) pathways are known to be utilized by mitochondria to maintain their homeostasis and regulate their quality (Amadoro et al. 2014; Misgeld and Schwarz 2017). Most importantly, the constant fusion, fission, and removal of damaged mitochondria through mitophagy are crucial to preserving the mitochondrial homeostasis (Cai and Tammineni 2016; Menzies et al. 2015).

5.4 Mitophagy in Maintaining Mitochondrial Homeostasis and Alzheimer's Disease

Cells have developed different types of well-regulated and integrated processes to sustain the homeostasis of mitochondria via balancing the biogenesis and removing degraded mitochondria. Mitophagy is a type of macroautophagy and selectively eliminates the damaged/superfluous mitochondria and helps in achieving steady-state mitochondrial homeostasis (Palikaras et al. 2018). Though mitophagy is a basal phenomenon, however, it can also be triggered by different external factors that cause mitochondrial damages such as hypoxia, starvation, energy stress, cell senescence, ROS, abnormal protein aggregates, and exposure to harmful medications (Angelini et al. 2009; Fischer et al. 2012; Manczak and Reddy 2012a; Tuppen et al. 2010; Vuda and Kamath 2016). Mitophagy serves a crucial role in neurodegeneration resistance, neuroprotection, and most importantly mitochondrial homeostasis and during the last few decades, different studies conducted in various reference models, including mice, flies, yeast, and nematodes, demonstrated the massive impact in age-related disorders due to degraded activity of mitochondria (Fang et al. 2017). Additionally, the compromised mitophagy has further been implicated in different neurodegenerative disorders including AD (Fang et al. 2016; Palikaras et al. 2015; Rubinsztein 2011).

During mitophagy, the defective mitochondria are selectively degraded through the process of engulfment into a double-membraned autophagosome that fuses with the lysosomes. Lysosomes consist of hydrolytic enzymes that are responsible for the complete degradation of the engulfed components. However, the actual mechanism of their selectivity is yet not clear. With the progress of research in this area, the various mechanisms of mitophagy have been suggested. Collectively all the mechanisms proposed to date can be broadly divided into Parkin-dependent and Parkin-independent pathways. Further, the Parkin-independent pathway is grouped based upon the factor through which it is mediated such as receptor-mediated, ubiquitin ligase-mediated, and lipid-mediated mitophagy (Fig. 5.3). The detailed mechanisms of these different pathways are discussed below.

5.4.1 Mitophagy Regulated by Parkin-Dependent Pathway

Parkin-dependent mitophagy is also known as the PINK1/Parkin-dependent mitophagy, where PINK1 stands for PTEN-induced kinase 1 that is a serine/threonine kinase encoded by *PARK6* gene. On the other hand, Parkin is E3 Ub ligase encrypted by the gene *PARK2*. Variations in *PARK6* and *PARK2* genes were first identified in autosomal recessive case of PD. In addition to these mutations, degradation of PINK1 and Parkin functionality is reported to induce malfunction of mitochondria in PD pathology (Clark et al. 2006; Morais et al. 2014). The PINK1/Parkin pathway is activated in response to the dysfunctional and/or damaged mitochondria, thereby initiating the signals for activation of MQC. The assembly of Ub chains present in the mitochondria recognizes the dysfunctional and/or damaged mitochondria and further leads to their clearance. This clearance process includes sensing mitochondrial damages by PINK1, signal amplification by Parkin, and detection of the signal by Ub chains (Harper et al. 2018).

PINK1 mainly consists of mitochondrial targeting sequence (MTS) that helps the mitochondria to initiate mitophagy by sensing the mitochondrial depolarization or ROS accumulation. Normally, PINK1 is translocated to the mitochondrial matrix through TIM23 and TOM complexes. TIM23 is an abbreviated form of the translocase of inner membrane 23 and TOM is translocase of the outer membrane (TOM). Firstly, translocation of MTS through TIM23 is supported by the

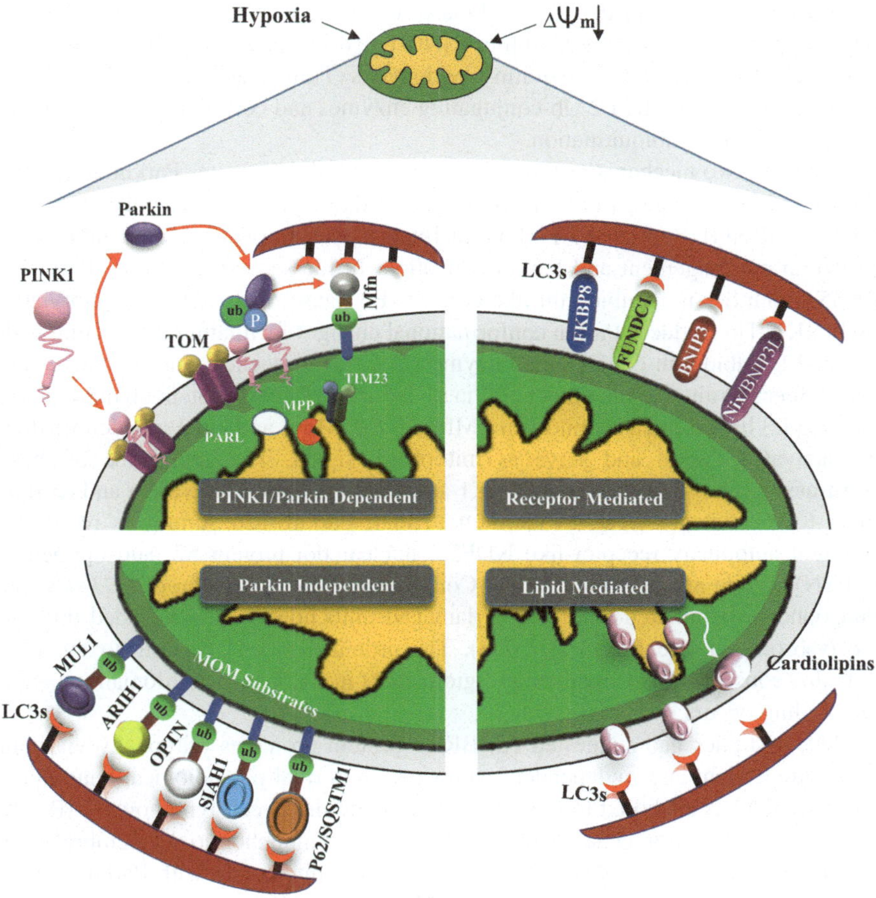

Fig. 5.3 Schematic representation of different pathways that mediate mitophagy. Broadly the mitophagy is categorized as PINK1/Parkin-dependent and Parkin-independent, where the receptor-mediated and lipid-mediated pathways are the subgroups of Parkin-independent mitophagy

mitochondrial membrane potential (Δψm) across the MIM. Once translocation of MTS is onto the matrix of mitochondria, it is sheared through the mitochondrial processing peptidase (MPPα/β) present in the matrix. Secondly, the TM domain of PINK1 is sheared by the protease Presenilin-associated rhomboid-like (PARL) of the MIM. As a result, a truncated 52 kDa PINK1 is retranslocated to the cytoplasm and damaged by protein catabolism pathway (Meissner et al. 2015; Yamano and Youle 2013). These mechanisms are altered in the condition of various mitochondrial stresses such as depolarized mitochondrial membrane, dysfunctional mitochondrial complexes, and proteotoxicity. These stresses do not allow the PINK1 to translocate to the TIM23 complex. As a consequence, PINK1 remains uncleaved and thereby aggregating at the MOM. The uncleaved PINK1 forms a large complex consisting of subunits from the TOM complex and dimerized PINK1 induces auto-phosphorylation in the PINK1 kinase domain (Lazarou et al. 2012). On the other hand, Parkin is composed of four RING-like domains namely RING0, RING1, IBR, and RING2 as well as ubiquitin-like domains (Trempe et al. 2013). The RING-like domain binds to the E2 Ub-conjugating enzymes and contributes to the transfer of Ub required for ubiquitination.

There are two mechanisms through which the PINK1 activates Parkin during the mitochondrial damages. In the first condition, PINK1 phosphorylates Ub at Ser65, binds to RING1 domain of Parkin, facilitates phosphorylation, and induces its structural rearrangement and Parkin activation. In the second condition, Parkin at Ser65 which lies in its ubiquitin-like domain (E3 ligase) is directly phosphorylated by PINK1. This induces Parkin conformational changes facilitating the binding with charged E2 ubiquitin-conjugating enzymes. As a consequence, Parkin is activated and further recruited to the mitochondria to initiate the MOM protein ubiquitination. Various E3 ligases such as mitofusin (Mfn), TOM, Miro, and Drp1 are identified by the activated Parkin and serve as mitophagy signal. The abundance of these substrates to phosphorylation of PINK1 amplifies the Parkin activation and recruitment to the mitochondria. In addition to that, Parkin also stimulates recruiting different autophagy receptor-like NDP52, nuclear dot protein 52, and optineurin (OPTN) (Wong and Holzbaur 2015). Consequently, the formation of LC3-positive phagophores is initiated and thereby damaged mitochondria are degraded through the lysosomes (Narendra et al. 2008). The proteins directly binding to the LC3 domains consist of LC3-interacting region (LIR) in its N-terminal domain wherein the binding occurs.

Recent studies had suggested AMBRA1 protein promotes mitophagy via both Parkin-dependent and -independent pathways. In normal physiological conditions, the AMBRA1 is anchored to the mitochondria and the B-cell lymphoma 2 (Bcl-2) inhibits its proautophagic activity. When the mitochondrial membrane is depolarized, the Bcl-2 releases AMBRA1 allowing interacting with Parkin to promote clearance of damaged mitochondria. Moreover, AMBRA1 can directly interact with the LC3 and lead to Parkin-mediated mitophagy. AMBRA1 recruits the dysfunctional and/or damaged mitochondria to autophagosomes through its LC3 interactions (Strappazzon et al. 2015). This further leads to the unification of lysosomes with autophagosomes and as a result degrades the damaged mitochondria engulfed within it.

5.4.2 Mitophagy Regulated by Parkin-Independent Pathway

Parkin is an important regulator of mitophagy, yet recent studies had suggested that mitophagy can even occur when Parkin is not present. Such mitophagy is mediated by other Ub ligases, receptors, and lipids of the mitochondria and is called the Parkin-independent mitophagy. Earlier studies suggested that Parkin is mandatory for mitophagy. However, further development in the studies of mitophagy revealed that PINK1 and Parkin are indeed important for mitophagy, but other important players can induce mitophagy (Uddin and Ashraf 2020).

5.4.2.1 Ub Ligase-Induced Mitophagy

PINK1 directly induces mitophagy without relying on the Parkin and recruits autophagy receptors including the OPTN and NDP52 to the mitochondrial damage (Lazarou et al. 2015). These autophagy receptors interact with the LC3 and induce mitophagy. Apart from Parkin, other E3 Ubligases have the potential to mediate clearance of dysfunctional and/or damaged mitochondria through the mechanism of Parkin-independent mitophagy. Some of the potential E3 Ub ligases involved in mitophagy are Glycoprotein 78 (Gp78), mitochondrial E3 Ub protein ligase 1 (MUL1), SIAH1, Synphilin-1, and ARIH1 (Fu et al. 2013; Georgakopoulos et al. 2017; Szargel et al. 2016; Villa et al. 2017). MUL1 and Gp78 ubiquitinate Mfn1 and Mfn2 and stimulate the proteasomal degradation that mediated the regulation of mitochondrial dynamics and mitophagy (Fu et al. 2013; Georgakopoulos et al. 2017). MUL1 also interacts with dynamin-related protein 1 (Drp1) and another mitochondrial fission GTPase protein that are the two important substrates of Parkin (Braschi et al. 2009; Yun et al. 2014).

However, MUL1 can only suppress the mutant phenotypes of PINK1/Parkin in neurons without affecting the PINK1/Parkin-mediated mitophagy (Georgakopoulos et al. 2017). The RING domain of MUL1 also consists of LIR motif that engages with GABAA receptor-associated protein (GABARAP) which is a descendant of Atg8 family and is involved in the autophagy and mitophagy regulations (Ambivero et al. 2014). PINK1 recruits synphilin-1 to the mitochondria and further SIAH1 is also recruited in the absence of Parkin. Various mitochondrial proteins are ubiquitinated by this complex and also lead to the recruitment of LC3 for mitophagy induction (Szargel et al. 2016). ARIH1 is another E3 ligase that actively participated in mitophagy in PINK1-dependent manner (Villa et al. 2017). Therefore, E3 Ub ligases localized in MOM flag the mitochondria with Ub chains allowing the recruitment of autophagy adaptors like OPTN, NDP52, and SQSTM1/p62 that recruit other autophagic proteins including ULK1, DFCP1, and WIPI1. As a result, it facilitates phagophore biogenesis, thereby mediating the expansion of the autophagosomal membrane (Lazarou et al. 2015). The autophagy adapters' LIR motifs interact with the autophagosomal LC3 and completely engulfing the Ub flagged mitochondria into the autophagosomes. Moreover, the phosphorylation of OPTN, NDP52, and SQSTM1/p62 involved in Parkin-independent mitophagy is regulated by TBK, a serine/threonine-protein kinase protein. Subsequently, it increases the binding affinity of the autophagy adaptors to the Ub chains, enhancing the clearance of damaged mitochondria (Heo et al. 2015; Richter et al. 2016).

5.4.2.2 Receptor-Induced Mitophagy and Lipid-Induced Mitophagy

Another aspect of Parkin-independent mitophagy is mediated by the different receptors of mitochondria including the FUN14 domain-containing protein 1 (FUNDC1), Bcl-2/adenovirus E1B 19 kDa-interacting protein 3 (BNIP3), NIX/BNIP3L, and FK506-binding protein 8 (FKBP8) (Okamoto et al. 2009; Kanki and Klionsky 2010; Bhujabal et al. 2017). Upon stressed signals and hypoxia, this mitochondrial receptor induces mitophagy by Parkin-independent pathway.

These proteins consist of LIR motif that enables the adhesion to autophagosomal LC3 proteins crucial for the selection of cargo for mitophagy (Liu et al. 2014).

In hypoxic conditions, hypoxia-inducible factor and forkhead box O3 (FOXO3) regulate the mitochondrial receptors NIX/BNIP3L and BNIP3 suggesting their role as a mediator of mitophagy induction by hypoxia (Mammucari et al. 2007; Sowter et al. 2001). Moreover, NIX/BNIP3L and BNIP3 are up-regulated during the neuronal stress and overexpressed NIX/BNIP3L was shown to reinstate mitophagy in PD-derived fibroblasts with PARK2 or PARK6 mutations (Schweers et al. 2007; Yeh et al. 2011). Yet the exact mechanism of NIX/BNIP3L and BNIP3-mediated mitophagy in neurons are still elusive. FUNDC1 is a receptor for mitophagy that activates during anoxia. To induce mitophagy, the interaction of FUNDC1 and LC3 are promoted by the phosphorylation at Ser17 (Chen et al. 2014). Dephosphorylation of FUNDC1 occurs at Ser13 and also can be blocked through the inhibition of PGAM5 phosphatase activity by BCL2L1/Bcl-XL. Under the conditions of no stress, FUNDC1 interacts with the OPA1 and DNM1L/DRP1 for the management of mitochondrial mitophagy, fusion, and fission. However, in hypoxic stress, FUNDC1 detaches from OPA1, binds to DNM1L, and facilitates mitophagy (Chen et al. 2014).

Therefore, FUNDC1 regulates the biogenesis of mitochondria and MQC. Bcl2-like 13 (BCL2L13) is a vital role player in the fragmentation induced by mitochondrial damage and mitophagy. BCL2L13 protein is a functional counterpart of Atg32 and at Ser272 position, it is phosphorylated which stimulates the interaction of BCL2L13 to LC3 (Murakawa et al. 2015). BCL2L13 has shown to promote the fragmentation of mitochondria when DNM1L/Drp1 was not present and even in the utter lack of PARK2/Parkin the mitophagy occurred (Otsu et al. 2015). Similarly, FKBP8 serves a crucial role in mitochondrial fragmentation through its LIR domain and engages LC3 to mediate Parkin-independent mitophagy (Bhujabal et al. 2017; Yoo et al. 2019).

There is considerable evidence that suggests that lipids such as cardiolipin can induce mitophagy via interaction with LC3 that acts as a signal for the elimination of dysfunctional mitochondria. Cardiolipins are mitochondrial phospholipid that is localized in the MIM of healthy mitochondria and mediates the stabilization of mitochondrial cristae to increase the assembly and function of OXPHOS complexes. Cardiolipins are activated in response to mitochondrial depolarization. Although the activation process is similar to that of Parkin, their activation is in response to different degrees of mitochondrial depolarization (Kagan et al. 2016). The LC3-binding domain of cardiolipins interacts with the N-terminal of LC3. Cardiolipins when moved out from MIM to MOM, it functions as a mitophagy transmitter by recruiting LC3, promotes the autophagosome mechanism and facilitating mitophagy (Chu et al. 2014). Moreover, the cardiolipin migration from the MIM to the MOM acts as a signal for mitochondrial dysfunction and enables to eliminate the dysfunctional and/or damaged mitochondria via mitophagy (Chu et al. 2014).

5.5 Impaired Mitophagy in Alzheimer's Diseases

It has been observed that cellular ability to discard impaired organelles potentially declines in numerous cells and tissues during the process of aging (Palikaras et al. 2015). This decline may consequently lead to extreme macroautophagy deficits, deterioration of energy homeostasis, and deposition of the weakened mitochondrial population (Palikaras et al. 2017) in these cells. Mitophagy deficiencies are proven to induce accumulation of degraded mitochondria accompanied with increased susceptibility of cells toward any kind of stress (Fang et al. 2014; Kirienko et al. 2015; Palikaras et al. 2015). The study conducted in different model systems also supported that mitochondrial elimination is reduced with aging in mice and human cells (García-Prat et al. 2016; Sun et al. 2015). Studies also show that hippocampal neurons in old mice undergo about 70% reduction in mitophagy events (Fang et al. 2019). These observations demonstrate the potential function of mitophagy in this specific region of the brain that is interestingly known to be susceptible to early-stage degradation in AD. Some recent studies have shown that mitophagy is inhibited in tau and Aβ-based mouse models, *C. elegans,* and even in postmortem hippocampal tissues of the brain from AD patients. These findings showed that mitophagy deficiencies are a typical feature of the pathogenesis of AD (Cummins et al. 2019; Fang et al. 2019).

The pathological hallmarks of AD include cumulative extracellular Aβ plaques derived from APP, tau tangles, and neuronal inflammation (Davidson et al. 2018). Past decades of studies have also observed a significant involvement of mitochondrial abnormality in AD patients. Therefore, exploring the underlying mechanism of the mitochondrial abnormalities and their possible association with other pathological hallmarks of AD has become an important aspect in the study of AD. Interestingly, the proteins engaged in mitochondrial function, biogenesis, and mitophagy were altered in AD patient's brain and animal models (Manczak et al. 2006). Mutations in presenilin-1 that can induce early-onset AD had also been shown to impair mitophagy by disruption of lysosomal maturation in the neurons derived from *PSEN-1* conditional knockout mice (Lee et al. 2010). Thus, alterations in PSEN-1 may indirectly affect the function of mitochondria by impairing its recycling by mitophagy (García-Escudero et al. 2013). Studies also showed that diffusible ligands derived from Aβ also called as ADDLs or soluble Aβ oligomers induced fragmentation of mitochondria and mitophagy in AD (Nakamura et al. 2010; Ryu et al. 2015; Wang et al. 2009). Damaged mitochondria and synaptic dysfunction are primary features of AD that are raised due to the formations of Aβ plaques or tau tangles (Guo et al. 2017; Tang et al. 2019). Dysfunctional mitochondria may trigger Aβ and tau pathology and therefore, mitochondria are also a major player in AD pathogenesis (Albensi 2019; Guo et al. 2017; Swerdlow 2018). Moreover, mitophagy impairment is observed in the AD pathogenesis and studies have highlighted the amalgamated effect of Aβ, tau, and impaired mitophagy in AD. The study conducted by transducing the human unmodified fibroblasts with lentivectors encoding APP and tau demonstrated a compromised accumulation of PINK1 and Parkin on the mitochondrial membrane leading to mitophagy impairment (Martin-Maestro et al.

2019). Increased accumulation of mitophagosomes was also identified in the neurons of APP transgenic mouse models. Damaged mitochondria were retained in a large, organized, and grouped lysosomal membrane-associated protein 1-positive vesicle. Thus, this evidence indicated that mitophagy was induced in the early stages of AD, but the damaged mitochondrial cargo was not degraded efficiently which may be due to the compromised activity of the lysosome (Kerr et al. 2017; Orr and Oddo 2013).

Besides, the evidence of impaired mitophagy in AD, few other studies showed that along with the progression of AD, the cytosolic Parkin levels were also decreased in the brain samples of AD patients (Martín-Maestro et al. 2016). Therefore, it denotes that the mitophagy is not efficient in the advanced phases of AD and may further enhance the severity of the AD phenotype. Parkin is an effective collaborator in the elimination of compromised mitochondria during AD progression, thus studies suggested that Parkin overexpression can potentially increase the AD symptoms (Ye et al. 2015). Indeed, the aggregation of dysfunctional and/or damaged mitochondria affects the neurons in AD, but mitophagy in microglia cells of AD mouse models was decreased by 60% (Fang et al. 2019). Microglial cells are important for the removal of neurotoxins by phagocytosis and there is a higher energy requirement for their functioning. Hence, compromised rate of mitophagy in microglial cells can affect the level of energy and efficiency. Interestingly, the mitophagy-initiating proteins including the phosphor-TBK1 and phospho-ULK1 were inactive in some of the AD patients (Fang et al. 2019). These findings were reinforced by another research of cortical neuronal cultures derived from iPSC, generated from sporadic and familial cases of AD. The results showed decreased expression of FUNDC1, Bcl2L13, AMBRA1, and MUL1 that induces receptor-mediated mitophagy (Fang et al. 2019). The concentration of BNIP3L/NIX, Bcl2L13, PINK1, and mitophagy-related proteins was also decreased in AD patient's brain samples (Fang et al. 2019). Overall, the above-discussed evidence suggests that impaired mitophagy is a major player in the etiology of AD and its relationship with the other hallmarks of AD is discussed below.

5.5.1 Aβ and Mitophagy

The cognitive impairment in AD is associated with the amount of Aβ aggregation inside the mitochondria and the mitochondrial anomalies (Dragicevic et al. 2010). Past decades of studies had highlighted that Aβ is an important agent in modulating mitochondrial damages and impairs multiple facets of the mitochondrial system such as the function of ETC (Manczak et al. 2010), production of ROS (Manczak et al. 2006) and mitochondrial dynamics (Calkins et al. 2011; Rui et al. 2006). Aβ may enter the mitochondria through the TOM complex of the MOM or mitochondrial-associated ER membrane (Petersen et al. 2008; Schreiner et al. 2015). Aggregating Aβ peptide treatment had led the mitochondrial permeability transition pore (mPTP) to open in the cultured cortical neural progenitor cells. Moreover, the transitory opening of the mPTP lowers cell proliferation while the protracted opening of the mPTP elicits cell death (Hou et al. 2014). Research on the lived AD mouse model disclosed that scattered and dysfunctional mitochondria are restricted only to the extracellular Aβ plaques in closest proximity. The irregular aggregation of Aβ plaques, inside the mitochondria, further leads to exacerbating damage to mitochondria (Xie et al. 2013). Other studies also suggested that the association of Aβ with Aβ-binding alcohol dehydrogenase (ABAD) and cyclophilin D (CypD) mediated Aβ-induced cytotoxicity (Du et al. 2008; Lustbader et al. 2004). ABAD is a protein in the matrix of mitochondria that are reported as up-regulated in the neurons of AD. ABAD overexpression aggravates Aβ-triggered cellular oxidative stress and death of the cells. However, on the other side, CypD is an element of the mPTP that interacts with the Aβ in AD human patients' cortical mitochondria and transgenic mAPP mice. The observations of this study also highlighted that lack of CypD prevents the neurons from cell death induced by Aβ- and oxidative stress (Du et al. 2008). Also, CypD deficiency greatly enhanced the neurotransmitter activity along with memory and learning in the AD mouse model and increased the Aβ-regulated reduction in LTP (Du et al. 2008). Collectively these findings indicate that the interaction of Aβ with CypD modulates AD-linked mitochondrial deficits. Additionally, this evidence also demonstrates that the abnormal aggregation of Aβ inside the mitochondria has a crucial role in the dysfunctioning of mitochondria in AD.

Studies in AD models also revealed that the affected neurons in AD undergo impairment in the mitophagy contributing to Aβ pathologies. Accumulation of Aβ may aggravate the impaired mitophagy and similarly, impaired mitophagy may also aggravate the Aβ accumulation (Du et al. 2017; Kerr et al. 2017). Moreover, impairment in the mitochondrial unfolded protein response (UPR) pathway is also correlated with the Aβ proteotoxicity (Sorrentino et al. 2017). Under mitochondrial stress, the protein known as the activating transcription factor associated with stress (ATFS-1) plays a critical role in the regulation of mitochondrial proteostasis including mitochondrial UPR and mitochondrial functions (Nargund et al. 2012). A study in RNAi knockdown of ATFS-1 showed repressed mitophagy and aggravated Aβ proteotoxicity in *C. elegans* model (GMC101). Also, restoration of the mitochondrial UPR reduced the AD pathology in *C. elegans* as well as AD mouse model (3xTg AD) (Sorrentino et al. 2017). The mitochondrial UPR is regulated by the MTS of the ATFS-1. In physiological conditions, after the ATFS-1 importation into the matrix of mitochondria, its MTS is sheared off and the remaining protein is degraded. This mechanism suggests that the efficiency of the mitochondrial import negatively regulates the activation of mitochondrial UPR (Melber and Haynes 2018). However, the importation of ATFS-1 into the mitochondrial matrix is decreased under the state of mitochondrial dysfunction and hence enables the aggregation of ATFS-1 in the cytosol. Interestingly, ATFS-1 also consists of nuclear localization sequence (NLS) domain, due to which the ATFS-1 is trafficked to the nucleus, whereby activating transcriptional responses that protect the mitochondrial function and AD pathology is eliminated (Melber and Haynes 2018). Additionally, mutations in the ATFS-1 MTS inhibit the importation of ATFS-1 into the mitochondrial matrix and thereby, activate the mitochondrial UPR (Rauthan et al. 2013).

Studies had also shown impaired mitochondrial homeostasis in the mutant APP-HT22 cells. The mitochondrial fission protein levels Drp1 and Fis1 increased, and the fusion protein levels Mfn1, Mfn2, and OPA1 decreased (Manczak et al. 2018; Reddy et al. 2018). In addition to the compromised mitochondrial UPR, deficient mitophagy is also a significant cause of deficiency in proteostasis of mitochondria and proteinopathy of Aβ in AD. Studies had shown impaired mitophagy in the AD patients' postmortem brain tissues, iPSC-derived human AD neurons, and AD transgenic mice model (APP/PS1) (Fang et al. 2019). Moreover, in an attempt to restore mitophagy in the neurons and microglia improved the Aβ proteinopathy and memory loss in the APP/PS1 AD mouse models that reveal the crucial role of impaired mitophagy in AD (Fang et al. 2019). Another study had shown that the LC3-binding mitophagy protein disrupted-in-schizophrenia-1 (DISC1) was reduced in the AD brain samples and APP/PS1 mouse model. Overexpression of the DISC1 in APP/PS1 mice rescued the Aβ-regulated mitochondrial malfunction, spines loss, and suppressed potentiation for the long-term (Wang et al. 2019). Therefore, the overall observations indicate that the degraded mitochondrial proteostasis potentially contributes to Aβ-induced neurotoxicity through the deficit mitochondrial UPR and weakened mitophagy, yet the exact molecular mechanism is unknown.

5.5.2 Tau and Mitophagy

Many studies had associated tau with mitochondrial damages. A pathogenic form of tau is reported to induce mitochondrial damages in the neuronal cells. Moreover, studies had also demonstrated that phosphor-tau impairs complex I of the ETC leading to loss of Δψm, enhanced production of ROS, reduced activities of detoxifying enzymes, and lipid peroxidation (Eckert et al. 2014; Spires-Jones and Hyman 2014). Additionally, mutant human tau protein (htauP301L) overexpression reduced the levels of ATP and in neuroblastoma cells increased the propensity to oxidative stress (SY5Y) (Schulz et al. 2012). Alterations in the compositions of the mitochondrial enzyme and their disrupted activity were observed in the P301S mouse model of tauopathy (Dumont et al. 2011). Studies of overexpression of htauP301L in pR5 mice revealed mitochondrial malfunction depicted by reduced complex I activity, increased ROS levels, disrupted respiration of mitochondria, and synthesis of ATP (David et al. 2005; Götz et al. 2001). Studies had shown that phosphorylated-tau (p-tau) directly interacts with VDAC in AD brains which may inhibit the opening of mitochondrial pores resulting in impaired mitochondrial function (Manczak and Reddy 2012b). Moreover, mitochondrial stress had been shown to raise tau hyperphosphorylation in the SOD knocked-out AD mouse model (Tg2576) (Melov et al. 2007).

Tau is closely associated with the complex I activity of mitochondria in AD. When the complex I function is inhibited, the tau is redistributed from axon to soma and then triggers cell death (Escobar-Khondiker et al. 2007). Therefore, the toxicity of tau on mitochondria is reciprocating and the deficiency of mitochondria might be a crucial contributor in understanding the tau pathology. The impact of pathogenic tau on the function of mitochondria is due to its deleterious effects on mitochondrial dynamics. The research on the tissues of AD postmortem brain from human patients and mouse models has shown that the levels of mitochondrial fission proteins Drp1 and Fis1 increased, and the levels of fusion proteins Mfn1, Mfn2, and OPA1 decreased (Manczak et al. 2011). Moreover, overproduction of Aβ, accumulation of tau, and abnormal interaction of Drp1 with Aβ or tau induce enhanced mitochondrial fragmentation and fission, thereby increasing the progression of AD (Kandimalla et al. 2016; Reddy et al. 2011). Overexpression of mutant tau in neurons is associated with FTD with Parkinsonism that is chromosome 17 linked, revealing a decrease in the amount of fusion and fission and making the neurons more susceptible to oxidative stress (Schulz et al. 2012). Interestingly, reduced Drp1 expression can safeguard the neurons from mutated tau-triggered mitochondrial dysfunction (Kandimalla et al. 2016). Therefore, transmissible forms of tau could degrade the function of mitochondria by directly interacting with VDAC and indirectly via their toxic effect on the dynamics of mitochondria.

Tau also links and maintains microtubules while leading to multiple physiological processes including outgrowth of a neurite, production in neurons, axonal transportations, and synaptogenesis (Dixit et al. 2008; Ballatore et al. 2007). Experimental research conducted on different models demonstrated that Aβ and p-tau can adversely affect the mitochondrial function and integrity that form a vicious cycle reaction known as the vicious cycle hypothesis (Kerr et al. 2017). Recently, research demonstrated that Parkin-dependent mitophagy is compromised in AD pathogenesis mediated by tau (Cummins et al. 2019). The mutant human tau (hP301L) expressed Parkin translocation as reduced in neuroblastoma cells to the disrupted mitochondria, subsequently leading to Parkin accumulation in the cytosol and increased susceptibility to oxidative stress. The same study validated the outcomes in the *C. elegans* nervous system (Cummins et al. 2019). Additionally, the occurrence of tau-regulated impaired mitochondria, the abnormal tau-Parkin interaction can also impact mitophagy subjecting to the pathology of AD. Moreover, the tau with truncated N-terminal appears to cause aberrant recruitment of Parkin, resulting in severe mitophagy leading to the failure synapse (Corsetti et al. 2015). Furthermore, overexpressed Aβ and tau lead to impaired mitophagy in human unmodified fibroblasts and hippocampus tissues from AD transgenic mice (3xTgAD) with both Aβ and tau proteinopathies (Fang et al. 2019; Martin-Maestro et al. 2019). Studies had also shown that restoration of mitophagy by various pharmacological compounds such as urolithin A, nicotinamide mononucleotide, and actinonin decreased tau phosphorylation at Ser262, Thr181, Thr231, and Ser202/Thr205 (Fang et al. 2019). The overall evidence discussed indicates that tau's pathological category inhibits mitophagy which highlights impaired mitophagy as a potential therapeutic option for AD.

5.5.3 Inflammation and Mitophagy

Inflammation is the new addition to the list of AD hallmarks. Various preclinical and clinical studies had demonstrated activation of immune responses in AD through numerous cytokines and microglia that trigger and drive the AD pathophysiology (Heppner et al. 2015). Moreover, mitochondria under stressed conditions release damage-associated molecular patterns (DAMPs) that induce innate immunity through cyclic GMP-AMP synthase-STING pathway. This pathway mainly regulates type I interferon response to the cytosolic DNA (Chen et al. 2016; Ishikawa et al. 2009; Ishikawa and Barber 2008).

Also, mitophagy alleviates inflammatory response by reducing the release of cytokines and by controlling the homeostasis of immune cells associating inflammation with autoimmune disorder pathogenesis at multiple levels (Xu et al. 2019). Additionally, various studies had shown the regulatory effect of Parkin and PINK1 in both adaptive and innate immunity. Firstly, strong inflammatory phenotypes in Parkin and PINK1 mice were suggested to be the central regulators of mitophagy. Moreover, Parkin and alleviated STING regulated inflammation and retrieved dopaminergic deficiency from substantia nigra in Parkin and PINK1 mice following the intensive exercise (Sliter et al. 2018). Secondly, immunity regulatory effect of PINK1 and Parkin is accomplished by suppressing the mitochondrial antigen presentation through the mitochondrial-derived vesicles (Matheoud et al. 2016). The role of mitochondrial antigen presentation and STING in the AD phenotype of inflammation is uncertain, but the impaired PINK1/Parkin pathway in AD indicates a possible overlaying effect between the AD and PD (Fang et al. 2019; Sliter et al. 2018).

Overall observations of the association of mitophagy with Aβ, tau proteinopathies, and inflammation indicate that more research is required to reinforce the studies discussed above. However, it is an immense lead for the therapeutic interventions of AD by targeting the mitophagy in such cases.

5.6 Triggering Mitophagy for Therapeutic Interventions in Alzheimer's Diseases

Lately, studies on different therapeutics aspects for various neurodegenerative disorders and especially AD have been focusing upon the significance of mitophagy as a target. Most importantly, impaired mitophagy observed in AD conditions has been the base for proposing mitophagy as a target for AD. Various studies discussed in the highlight of the above section that the restoration of mitophagy in neurons can potentially reduce AD pathologies. Improved mitochondrial health and mitophagy restoration can be fulfilled through various therapeutic interventions. There are also various mitophagy inducers in AD models that enhance mitochondrial resistance toward oxidative stress, prolong the health span of mitochondria, and improve AD pathology (Madeo et al. 2018; Schondorf et al. 2018). For example, the supplementation of NAD+ precursor, UA, and AC exhibited restored mitophagy in AD mice with APP/PS1 by enhancing the extracellular microglial phagocytosis, Aβ plaques and reduced inflammatory cytokines that are released by the active microglia (Fang et al. 2019).

Studies had shown that astrocytes are the key players in the mitophagy degradation of dysfunctional and/or damaged mitochondria in the neuronal cell (Davis et al. 2018). Hence, mitophagy induction may also improve the functions of astrocytes in AD. Very interestingly, the development of 3D human AD triculture model consisting of neurons, astrocytes, and microglia provides a suitable environment for mimicking the human brain (Park et al. 2018). Such a specific environment may endow the studies of mitophagy which is specific to the cell type in AD. Recent studies had shown that resveratrol, a natural compound isolated from the skin of grapes, blueberries, raspberries, mulberries, and peanuts is a remarkable mitophagy enhancer in AD (Wang et al. 2018). It tends to minimize the development and progression of AD. Mitophagy may also be boosted by mediators that can persuade slight bioenergetics stress or inhibit mTOR pathway. APP-mutant mouse model treated with rapamycin, an mTOR inhibitor-induced mitophagy and enhances cognitive ability by reducing the Aβ pathology (Lee et al. 2017). Collectively, these shreds of evidence may provide a basis for the screening of compounds that may induce and trigger mitophagy in preclinical AD models.

5.7 Recent Developments and Future Perspectives

To date, a large number of small drug molecules have been predicted to enhance mitophagy and some of them seem to be quite promising as well (Fang et al. 2014, 2019; Schondorf et al. 2018; Wang et al. 2018). But the molecular agent which can completely ameliorate the disease condition and improve the survival in AD is still missing. Many of such studied drugs are also known to exhibit cellular toxicity in healthy cells. Thus to get the perfect therapy to improve mitophagy in AD patients, certain issues should be addressed. Firstly the detailed mitophagy pathways and the interconnection and coordination between different mitophagy pathways should be explored in great detail. Secondly, the balance between different pathways to regulate the mitochondrial quality such as mitochondrial biogenesis, mitochondrial fusion-fission processes, and mitophagy should be investigated. The balancing role of mitophagy in determining the survivability of neuronal cells should be nicely explored before designing any therapy to improve mitophagy in AD patients. Additionally, the association of mitophagy with other neurodegenerative disorders in a large cohort should be studied.

Apart from small compound-based mitophagy induction, change in lifestyle, physical activities, and caloric restriction may also improve mitochondrial metabolism and hence neuronal health. Rodents-based studies suggested that fasting and exercise may help in lowering the mitochondrial ROS as well as induce mitophagy and mitochondrial biogenesis in neurons (Longo and Mattson 2014). Current studies also revealed that SIRT3 was overexpressed in cortical and hippocampal and neurons of mice after they ran on a treadmill. It suggested the role of exercise in neuroprotection (Cheng et al. 2016). Thus, important lifestyle changes should also be considered as an important factor while designing drug-based therapies for the treatment of AD.

5.8 Conclusion

In conclusion, the defects in mitochondrial metabolism, downstream oxidative stress, and impaired mitophagy activity are the main factors involved in neurodegeneration. Mitophagy acts as one of the important mechanisms for controlling the quality of mitochondria in neurons. At the molecular level, different types of pathways are responsible for mitophagy and thus play a crucial role in maintaining neuronal health; however, the PINK1/Parkin pathway is the only pathway that has been investigated in greater detail. It has also been studied extensively for its role in the regulation of neurodegeneration AD. As per different studies, the impairment in mitophagy has been recognized to play a pivotal role in AD and interestingly the activation of mitophagy in such disease conditions has further been shown to provide neuroprotection. Thus it seems that regulating the process of mitophagy may behave as an interesting therapeutic target for AD. Hence analyzing the increasing prevalence and severity of this form of dementia, there is an urgent need to explore the mitophagy-triggering ability of different synthetic and natural compounds to combat the disease.

References

Albensi BC (2019) Dysfunction of mitochondria: implications for Alzheimer's disease. Int Rev Neurobiol 145:13–27

Amadoro G, Corsetti V, Florenzano F, Atlante A, Bobba A, Nicolin V et al (2014) Morphological and bioenergetic demands underlying the mitophagy in post-mitotic neurons: the pink–parkin pathway. Front Aging Neurosci 6:18. https://doi.org/10.3389/fnagi.2014.00018

Ambivero CT, Cilenti L, Main S, Zervos AS (2014) Mulan E3 ubiquitin ligase interacts with multiple E2 conjugating enzymes and participates in mitophagy by recruiting GABARAP. Cell Signal 26:2921–2929. https://doi.org/10.1016/j.cellsig.2014.09.004

Anandatheerthavarada HK, Devi L (2007) Amyloid precursor protein and mitochondrial dysfunction in Alzheimer's disease. Neuroscientist 13:626–638. https://doi.org/10.1177/1073858407303536

Angelini C, Bello L, Spinazzi M, Ferrati C (2009) Mitochondrial disorders of the nuclear genome. Acta Myol 28:16–23

Ballatore C, Lee VM, Trojanowski JQ (2007) Tau-mediated neurodegeneration in Alzheimer's disease and related disorders. Nat Rev Neurosci 8:663–672. https://doi.org/10.1038/nrn2194

Beal MF (2005) Mitochondria take center stage in aging and neurodegeneration. Ann Neurol 58: 495–505. https://doi.org/10.1002/ana.20624

Bekris LM, Yu CE, Bird TD, Tsuang DW (2010) Genetics of Alzheimer disease. J Geriatr Psychiatry Neurol 23:213–227. https://doi.org/10.1177/0891988710383571

Bhujabal Z, Birgisdottir ÅB, Sjøttem E, Brenne HB, Øvervatn A, Habisov S et al (2017) FKBP8 recruits LC3A to mediate Parkin-independent mitophagy. EMBO Rep 18:947–961. https://doi.org/10.15252/embr.201643147

Blass JP, Gibson GE (1991) The role of oxidative abnormalities in the pathophysiology of Alzheimer's disease. Rev Neurol (Paris) 147:513–525

Bonda DJ, Wang X, Perry G, Smith MA, Zhu X (2010) Mitochondrial dynamics in Alzheimer's disease. Drugs Aging 27:181–192. https://doi.org/10.2165/11532140-000000000-00000

Bosetti F, Brizzi F, Barogi S, Mancuso M, Siciliano G, Tendi EA et al (2002) Cytochrome c oxidase and mitochondrial F1F0-ATPase (ATP synthase) activities in platelets and brain from patients with Alzheimer's disease. Neurobiol Aging 23:371–376. https://doi.org/10.1016/s0197-4580(01)00314-1

Braschi E, Zunino R, McBride HM (2009) MAPL is a new mitochondrial SUMO E3 ligase that regulates mitochondrial fission. EMBO Rep 10:748–754. https://doi.org/10.1038/embor.2009.86

Bubber P, Haroutunian V, Fisch G, Blass JP, Gibson GE (2005) Mitochondrial abnormalities in Alzheimer brain: mechanistic implications. Ann Neurol 57:695–703. https://doi.org/10.1002/ana.20474

Cai Q, Tammineni P (2016) Alterations in mitochondrial quality control in Alzheimer's disease. Front Cell Neurosci 10:24. https://doi.org/10.3389/fncel.2016.00024

Calkins MJ, Manczak M, Mao P, Shirendeb U, Reddy PH (2011) Impaired mitochondrial biogenesis, defective axonal transport of mitochondria, abnormal mitochondrial dynamics and synaptic degeneration in a mouse model of Alzheimer's disease. Hum Mol Genet 20:4515–4529. https://doi.org/10.1093/hmg/ddr381

Chen G, Han Z, Feng D, Chen Y, Chen L, Wu H et al (2014) A regulatory signaling loop comprising the PGAM5 phosphatase and CK2 controls receptor-mediated mitophagy. Mol Cell 54:362–377. https://doi.org/10.1016/j.molcel.2014.02.034

Chen Q, Sun L, Chen ZJ (2016) Regulation and function of the cGAS STING pathway of cytosolic DNA sensing. Nat Immunol 17:1142–1149. https://doi.org/10.1038/ni.3558

Chen GF, Xu TH, Yan Y, Zhou YR, Jiang Y, Melcher K et al (2017) Amyloid beta: structure, biology and structure-based therapeutic development. Acta Pharmacol Sin 38:1205–1235. https://doi.org/10.1038/aps.2017.28

Cheng A, Hou Y, Mattson MP (2010) Mitochondria and neuroplasticity. ASN neuro 2(5):e00045. https://doi.org/10.1042/AN20100019

Cheng A, Wan R, Yang JL, Kamimura N, Son TG, Ouyang X et al (2012) Involvement of PGC-1a in the formation and maintenance of neuronal dendritic spines. Nat Commun 3:1250. https://doi.org/10.1038/ncomms2238

Cheng A, Yang Y, Zhou Y, Maharana C, Lu D, Peng W et al (2016) Mitochondrial SIRT3 mediates adaptive responses of neurons to exercise and metabolic and excitatory challenges. Cell Metab 23:128–142. https://doi.org/10.1016/j.cmet.2015.10.013

Chu CT, Bayır H, Kagan VE (2014) LC3 binds externalized cardiolipin on injured mitochondria to signal mitophagy in neurons: implications for Parkinson disease. Autophagy 10:376–378. https://doi.org/10.4161/auto.27191

Clark IE, Dodson MW, Jiang C, Cao JH, Huh JR, Seol JH et al (2006) Drosophila pink1 is required for mitochondrial function and interacts genetically with parkin. Nature 441:1162–1166. https://doi.org/10.1038/nature04779

Corsetti V, Florenzano F, Atlante A, Bobba A, Ciotti MT, Natale F et al (2015) NH2-truncated human tau induces deregulated mitophagy in neurons by aberrant recruitment of Parkin and UCHL-1: implications in Alzheimer's disease. Hum Mol Genet 24:3058–3081. https://doi.org/10.1093/hmg/ddv059

Cummins N, Tweedie A, Zuryn S, Bertran-Gonzalez J, Götz J (2019) Disease-associated tau impairs mitophagy by inhibiting Parkin translocation to mitochondria. EMBO J 38:e99360. https://doi.org/10.15252/embj.201899360

David DC, Hauptmann S, Scherping I, Schuessel K, Keil U, Rizzu P et al (2005) Proteomic and functional analyses reveal a mitochondrial dysfunction in P301L tau transgenic mice. J Biol Chem 280:23802–23814. https://doi.org/10.1074/jbc.M500356200

Davidson YS, Robinson A, Prasher VP, Mann DM (2018) The age of onset and evolution of Braak tangle stage and Thal amyloid pathology of Alzheimer's disease in individuals with Down syndrome. Acta Neuropathol Commun 6(1):56. https://doi.org/10.1186/s40478-018-0559-4
Davis SA, Itaman S, Khalid-Janney CM, Sherard JA, Dowell JA, Cairns NJ et al (2018) TDP-43 interacts with mitochondrial proteins critical for mitophagy and mitochondrial dynamics. Neurosci Lett 678:8–15. https://doi.org/10.1016/j.neulet.2018.04.053
Dixit R, Ross JL, Goldman YE, Holzbaur EL (2008) Differential regulation of dynein and kinesin motor proteins by tau. Science 319:1086–1089. https://doi.org/10.1126/science.1152993
Dragicevic N, Mamcarz M, Zhu Y, Buzzeo R, Tan J, Arendash GW et al (2010) Mitochondrial amyloid-beta levels are associated with the extent of mitochondrial dysfunction in different brain regions and the degree of cognitive impairment in Alzheimer's transgenic mice. J Alzheimers Dis 20(Suppl 2):535–550. https://doi.org/10.3233/JAD-2010-100342
Du H, Guo L, Fang F, Chen D, Sosunov AA, McKhann GM et al (2008) Cyclophilin D deficiency attenuates mitochondrial and neuronal perturbation and ameliorates learning and memory in Alzheimer's disease. Nat Med 14:1097–1105. https://doi.org/10.1038/nm.1868
Du F, Yu Q, Yan S, Hu G, Lue LF, Walker DG et al (2017) PINK1 signalling rescues amyloid pathology and mitochondrial dysfunction in Alzheimer's disease. Brain 140:3233–3251. https://doi.org/10.1093/brain/awx258
Dumont M, Stack C, Elipenahli C, Jainuddin S, Gerges M, Starkova NN et al (2011) Behavioral deficit, oxidative stress, and mitochondrial dysfunction precede tau pathology in P301S transgenic mice. FASEB J 25:4063–4072. https://doi.org/10.1096/fj.11-186650
Eckert A, Keil U, Marques CA, Bonert A, Frey C, Schüssel K et al (2003) Mitochondrial dysfunction, apoptotic cell death, and Alzheimer's disease. Biochem Pharmacol 66:1627–1634. https://doi.org/10.1016/j.bbadis.2009.10.006
Eckert A, Nisbet R, Grimm A, Götz J (2014) March separate, strike together—role of phosphorylated TAU in mitochondrial dysfunction in Alzheimer's disease. Biochim Biophys Acta 1842:1258–1266. https://doi.org/10.1016/j.bbadis.2013.08.013
Escobar-Khondiker M, Höllerhage M, Muriel MP, Champy P, Bach A, Depienne C et al (2007) Annonacin, a natural mitochondrial complex I inhibitor, causes tau pathology in cultured neurons. J Neurosci 27:7827–7837. https://doi.org/10.1523/JNEUROSCI.1644-07.2007
Fang EF, Scheibye-Knudsen M, Brace LE, Kassahun H, SenGupta T, Nilsen H et al (2014) Defective mitophagy in XPA via PARP-1 hyperactivation and NAD(+)/SIRT1 reduction. Cell 157:882–896. https://doi.org/10.1016/j.cell.2014.03.026
Fang EF, Scheibye-Knudsen M, Chua KF, Mattson MP, Croteau DL, Bohr VA (2016) Nuclear DNA damage signaling to mitochondria in ageing. Nat Rev Mol Cell Biol 17:308–321. https://doi.org/10.1038/nrm.2016.14
Fang EF, Lautrup S, Hou Y, Demarest TG, Croteau DL, Mattson MP et al (2017) NAD(+) in aging: molecular mechanisms and translational implications. Trends Mol Med 23:899–916. https://doi.org/10.1016/j.molmed.2017.08.001
Fang EF, Hou Y, Palikaras K, Adriaanse BA, Kerr JS, Yang B et al (2019) Mitophagy inhibits amyloid-β and tau pathology and reverses cognitive deficits in models of Alzheimer's disease. Nat Neurosci 22:401–412. https://doi.org/10.1038/s41593-018-0332-9
Fischer F, Hamann A, Osiewacz HD (2012) Mitochondrial quality control: an integrated network of pathways. Trends Biochem Sci 37:284–292. https://doi.org/10.1016/j.tibs.2012.02.004
Fu M, St-Pierre P, Shankar J, Wang PT, Joshi B, Nabi IR (2013) Regulation of mitophagy by the Gp78 E3 ubiquitin ligase. Mol Biol Cell 24:1153–1162. https://doi.org/10.1091/mbc.e12-08-0607
García-Escudero V, Martín-Maestro P, Perry G, Avila J (2013) Deconstructing mitochondrial dysfunction in Alzheimer disease. Oxidative Med Cell Longev 2013:162152. https://doi.org/10.1155/2013/162152
García-Prat L, Martínez-Vicente M, Perdiguero E, Ortet L, Rodríguez-Ubreva J, Rebollo E et al (2016) Autophagy maintains stemness by preventing senescence. Nature 529:37–42. https://doi.org/10.1038/nature16187

Georgakopoulos ND, Wells G, Campanella M (2017) The pharmacological regulation of cellular mitophagy. Nat Chem Biol 13(2):136–146. https://doi.org/10.1038/nchembio.2287

Gibson GE, Sheu KF, Blass JP (1998) Abnormalities of mitochondrial enzymes in Alzheimer disease. J Neural Transm 105:855–870. https://doi.org/10.1007/s007020050099

Götz J, Chen FV, Van Dorpe J, Nitsch RM (2001) Formation of neurofibrillary tangles in P301l tau transgenic mice induced by Abeta 42 fibrils. Science 293:1491–1495. https://doi.org/10.1126/science.1062097

Guo L, Tian J, Du H (2017) Mitochondrial dysfunction and synaptic transmission failure in Alzheimer's disease. J Alzheimers Dis 57:1071–1086. https://doi.org/10.3233/JAD-160702

Harper JW, Ordureau A, Heo JM (2018) Building and decoding ubiquitin chains for mitophagy. Nat Rev Mol Cell Biol 19(2):93–108. https://doi.org/10.1038/nrm.2017.129

Heo JM, Ordureau A, Paulo JA, Rinehart J, Harper JW (2015) The PINK1-PARKIN mitochondrial ubiquitylation pathway drives a program of OPTN/NDP52 recruitment and TBK1 activation to promote mitophagy. Mol Cell 60:7–20. https://doi.org/10.1016/j.molcel.2015.08.016

Heppner FL, Ransohoff RM, Becher B (2015) Immune attack: the role of inflammation in Alzheimer disease. Nat Rev Neurosci 16:358–372. https://doi.org/10.1038/nrn3880

Hou Y, Ghosh P, Wan R, Ouyang X, Cheng H, Mattson MP et al (2014) Permeability transition pore-mediated mitochondrial superoxide flashes mediate an early inhibitory effect of amyloid beta1-42 on neural progenitor cell proliferation. Neurobiol Aging 35:975–989. https://doi.org/10.1016/j.neurobiolaging.2013.11.002

Hsieh HL, Yang CM (2013) Role of redox signaling in neuroinflammation and neurodegenerative diseases. Biomed Res Int 2013:484613. https://doi.org/10.1155/2013/484613

Ishikawa H, Barber GN (2008) STING is an endoplasmic reticulum adaptor that facilitates innate immune signalling. Nature 455:674–678. https://doi.org/10.1038/nature07317

Ishikawa H, Ma Z, Barber GN (2009) STING regulates intracellular DNA-mediated, type I interferon-dependent innate immunity. Nature 461:788–792. https://doi.org/10.1038/nature08476

Kagan VE, Jiang J, Huang Z, Tyurina YY, Desbourdes C, Cottet-Rousselle C et al (2016) NDPK-D (NM23-H4)-mediated externalization of cardiolipin enables elimination of depolarized mitochondria by mitophagy. Cell Death Differ 23:1140–1151. https://doi.org/10.1038/cdd.2015.160

Kandimalla R, Manczak M, Fry D, Suneetha Y, Sesaki H, Reddy PH (2016) Reduced dynamin-related protein 1 protects against phosphorylated Tau-induced mitochondrial dysfunction and synaptic damage in Alzheimer's disease. Hum Mol Genet 25:4881–4897. https://doi.org/10.1093/hmg/ddw312

Kanki T, Klionsky DJ (2010) The molecular mechanism of mitochondria autophagy in yeast. Mol Microbiol 75:795–800. https://doi.org/10.1111/j.1365-2958.2009.07035.x

Kapogiannis D, Mattson MP (2011) Disrupted energy metabolism and neuronal circuit dysfunction in cognitive impairment and Alzheimer's disease. Lancet Neurol 10:187–198. https://doi.org/10.1016/S1474-4422(10)70277-5

Karch CM, Goate AM (2015) Alzheimer's disease risk genes and mechanisms of disease pathogenesis. Biol Psychiatry 77:43–51. https://doi.org/10.1016/j.biopsych.2014.05.006

Katsouri L, Lim YM, Blondrath K, Eleftheriadou I, Lombardero L, Birch AM et al (2016) PPARg-coactivator-1a gene transfer reduces neuronal loss and amyloid-beta generation by reducing b-secretase in an Alzheimer's disease model. Proc Natl Acad Sci U S A 113:12292–12297. https://doi.org/10.1073/pnas.1606171113

Kerr JS, Adriaanse BA, Greig NH, Mattson MP, Cader MZ, Bohr VA et al (2017) Mitophagy and Alzheimer's disease: cellular and molecular mechanisms. Trends Neurosci 40:151–166. https://doi.org/10.1016/j.tins.2017.01.002

Kim DK, Mook-Jung I (2019) The role of cell type-specific mitochondrial dysfunction in the pathogenesis of Alzheimer's disease. BMB Rep 52(12):679–688. https://doi.org/10.5483/BMBRep.2019.52.12.282

Kirienko NV, Ausubel FM, Ruvkun G (2015) Mitophagy confers resistance to siderophore-mediated killing by Pseudomonas aeruginosa. Proc Natl Acad Sci U S A 112:1821–1826. https://doi.org/10.1073/pnas.1424954112

Kish SJ, Bergeron C, Rajput A, Dozic S, Mastrogiacomo F, Chang LJ et al (1992) Brain cytochrome oxidase in Alzheimer's disease. J Neurochem 59:776–779. https://doi.org/10.1111/j.1471-4159.1992.tb09439.x

Lazarou M, Jin SM, Kane LA, Youle RJ (2012) Role of PINK1 binding to the TOM complex and alternate intracellular membranes in recruitment and activation of the E3 ligase Parkin. Dev Cell 22:320–333. https://doi.org/10.1016/j.devcel.2011.12.014

Lazarou M, Sliter DA, Kane LA, Sarraf SA, Wang C, Burman JL et al (2015) The ubiquitin kinase PINK1 recruits autophagy receptors to induce mitophagy. Nature 524:309–314. https://doi.org/10.1038/nature14893

Lee JH, Yu WH, Kumar A, Lee S, Mohan PS, Peterhoff CM et al (2010) Lysosomal proteolysis and autophagy require presenilin 1 and are disrupted by Alzheimer-related PS1 mutations. Cell 141:1146–1158. https://doi.org/10.1016/j.cell.2010.05.008

Lee HK, Kwon B, Lemere CA, de la Monte S, Itamura K, Ha AY et al (2017) mTORC2 (Rictor) in Alzheimer's disease and reversal of amyloid-β expression-induced insulin resistance and toxicity in rat primary cortical neurons. J Alzheimers Dis 56:1015–1036. https://doi.org/10.3233/JAD-161029

Lin MT, Beal MF (2006) Mitochondrial dysfunction and oxidative stress in neurodegenerative diseases. Nature 443:787–795. https://doi.org/10.1038/nature05292

Lin TK, Cheng CH, Chen SD, Liou CW, Huang CR, Chuang YC (2012) Mitochondrial dysfunction and oxidative stress promote apoptotic cell death in the striatum via cytochrome c/caspase-3 signaling cascade following chronic rotenone intoxication in rats. Int J Mol Sci 13:8722–8739. https://doi.org/10.3390/ijms13078722

Liu L, Sakakibara K, Chen Q, Okamoto K (2014) Receptor-mediated mitophagy in yeast and mammalian systems. Cell Res 24:787–795. https://doi.org/10.1038/cr.2014.75

Longo VD, Mattson MP (2014) Fasting: molecular mechanisms and clinical applications. Cell Metab 19:181–192. https://doi.org/10.1016/j.cmet.2013.12.008

Lu MH, Zhao XY, Yao PP, Xu DE, Ma QH (2018) The mitochondrion: a potential therapeutic target for Alzheimer's disease. Neurosci Bull 34:1127–1130. https://doi.org/10.1007/s12264-018-0310-y

Lustbader JW, Cirilli M, Lin C, Xu HW, Takuma K, Wang N et al (2004) ABAD directly links Abeta to mitochondrial toxicity in Alzheimer's disease. Science 304:448–452. https://doi.org/10.1126/science.1091230

Madeo F, Eisenberg T, Pietrocola F, Kroemer G (2018) Spermidine in health and disease. Science 359(6374):eaan2788. https://doi.org/10.1126/science.aan2788

Mammucari C, Milan G, Romanello V, Masiero E, Rudolf R, Del Piccolo P et al (2007) FoxO3 controls autophagy in skeletal muscle in vivo. Cell Metab 6:458–471. https://doi.org/10.1016/j.cmet.2007.11.001

Manczak M, Reddy PH (2012a) Abnormal interaction between the mitochondrial fission protein Drp1 and hyperphosphorylated tau in Alzheimer's disease neurons: implications for mitochondrial dysfunction and neuronal damage. Hum Mol Genet 21:2538–2547. https://doi.org/10.1093/hmg/dds072

Manczak M, Reddy PH (2012b) Abnormal interaction of VDAC1 with amyloid beta and phosphorylated tau causes mitochondrial dysfunction in Alzheimer's disease. Hum Mol Genet 21:5131–5146. https://doi.org/10.1093/hmg/dds360

Manczak M, Park BS, Jung Y, Reddy PH (2004) Differential expression of oxidative phosphorylation genes in patients with Alzheimer's disease: implications for early mitochondrial dysfunction and oxidative damage. NeuroMolecular Med 5:147–162. https://doi.org/10.1385/NMM:5:2:147

Manczak M, Anekonda TS, Henson E, Park BS, Quinn J, Reddy PH (2006) Mitochondria are a direct site of Aβ accumulation in Alzheimer's disease neurons: implications for free radical generation and oxidative damage in disease progression. Hum Mol Genet 15:1437–1449. https://doi.org/10.1093/hmg/ddl066

Manczak M, Mao P, Calkins MJ, Cornea A, Reddy AP, Murphy MP et al (2010) Mitochondria-targeted antioxidants protect against amyloid-beta toxicity in Alzheimer's disease neurons. J Alzheimers Dis 20(Suppl 2):609–631. https://doi.org/10.3233/JAD-2010-100564

Manczak M, Calkins MJ, Reddy PH (2011) Impaired mitochondrial dynamics and abnormal interaction of amyloid beta with mitochondrial protein Drp1 in neurons from patients with Alzheimer's disease: implications for neuronal damage. Hum Mol Genet 20:2495–2509. https://doi.org/10.1093/hmg/ddr139

Manczak M, Kandimalla R, Yin X, Reddy PH (2018) Hippocampal mutant APP and amyloid beta-induced cognitive decline, dendritic spine loss, defective autophagy, mitophagy and mitochondrial abnormalities in a mouse model of Alzheimer's disease. Hum Mol Genet 27:1332–1342. https://doi.org/10.1093/hmg/ddy042

Marosi K, Mattson MP (2014) BDNF mediates adaptive brain and body responses to energetic challenges. Trends Endocrinol Metab 25:89–98. https://doi.org/10.1016/j.tem.2013.10.006

Martín-Maestro P, Gargini R, Perry G, Avila J, García-Escudero V (2016) PARK2 enhancement is able to compensate mitophagy alterations found in sporadic Alzheimer's disease. Hum Mol Genet 25:792–806. https://doi.org/10.1093/hmg/ddv616

Martin-Maestro P, Gargini R, Garcia E, Simon D, Avila J, Garcia-Escudero V (2019) Mitophagy failure in APP and Tau overexpression model of Alzheimer's disease. J Alzheimers Dis 70:525–540. https://doi.org/10.3233/JAD-190086

Matheoud D, Sugiura A, Bellemare-Pelletier A, Laplante A, Rondeau C, Chemali M et al (2016) Parkinson's disease-related proteins PINK1 and parkin repress mitochondrial antigen presentation. Cell 166:314–327. https://doi.org/10.1016/j.cell.2016.05.039

Mattson MP (2004) Pathways towards and away from Alzheimer's disease. Nature 430:631–639. https://doi.org/10.1038/nature02621

Mattson MP, Gleichmann M, Cheng A (2008) Mitochondria in neuroplasticity and neurological disorders. Neuron 60:748–766. https://doi.org/10.1016/j.neuron.2008.10.010

Meissner C, Lorenz H, Hehn B, Lemberg MK (2015) Intramembrane protease PARL defines a negative regulator of PINK1-and PARK2/Parkin-dependent mitophagy. Autophagy 11:1484–1498. https://doi.org/10.1080/15548627.2015.1063763

Melber A, Haynes CM (2018) UPR^{mt} regulation and output: a stress response mediated by mitochondrial-nuclear communication. Cell Res 28:281–295. https://doi.org/10.1038/cr.2018.16

Melov S, Adlard PA, Morten K, Johnson F, Golden TR, Hinerfeld D et al (2007) Mitochondrial oxidative stress causes hyperphosphorylation of tau. PLoS One 2:e536. https://doi.org/10.1371/journal.pone.0000536

Menzies FM, Fleming A, Rubinsztein DC (2015) Compromised autophagy and neurodegenerative diseases. Nat Rev Neurosci 16:345–357. https://doi.org/10.1038/nrn3961

Misgeld T, Schwarz TL (2017) Mitostasis in neurons: maintaining mitochondria in an extended cellular architecture. Neuron 96:651–666. https://doi.org/10.1016/j.neuron.2017.09.055

Morais VA, Haddad D, Craessaerts K, De Bock PJ, Swerts J, Vilain S et al (2014) PINK1 loss-of-function mutations affect mitochondrial complex I activity via NdufA10 ubiquinone uncoupling. Science 344:203–207. https://doi.org/10.1126/science.1249161

Morán M, Moreno-Lastres D, Marín-Buera L, Arenas J, Martín MA, Ugalde C (2012) Mitochondrial respiratory chain dysfunction: implications in neurodegeneration. Free Radic Biol Med 53:595–609. https://doi.org/10.1016/j.freeradbiomed.2012.05.009

Moreira PI, Santos MS, Oliveira CR (2007) Alzheimer's disease: a lesson from mitochondrial dysfunction. Antioxid Redox Signal 9:1621–1630. https://doi.org/10.1089/ars.2007.1703

Murakawa T, Yamaguchi O, Hashimoto A, Hikoso S, Takeda T, Oka T et al (2015) Bcl-2-like protein 13 is a mammalian Atg32 homologue that mediates mitophagy and mitochondrial fragmentation. Nat Commun 6:1–14. https://doi.org/10.1038/ncomms8527

Nakamura T, Cieplak P, Cho DH, Godzik A, Lipton SA (2010) S-nitrosylation of Drp1 links excessive mitochondrial fission to neuronal injury in neurodegeneration. Mitochondrion 10: 573–578. https://doi.org/10.1016/j.mito.2010.04.007

Narendra D, Tanaka A, Suen DF, Youle RJ (2008) Parkin is recruited selectively to impaired mitochondria and promotes their autophagy. J Cell Biol 183:795–803. https://doi.org/10.1083/jcb.200809125

Nargund AM, Pellegrino MW, Fiorese CJ, BakerBM HCM (2012) Mitochondrial import efficiency of ATFS-1 regulates mitochondrial UPR activation. Science 337:587–590. https://doi.org/10.1126/science.1223560

Okamoto K, Kondo-Okamoto N, Ohsumi Y (2009) Mitochondria-anchored receptor Atg32 mediates degradation of mitochondria via selective autophagy. Dev Cell 17:87–97. https://doi.org/10.1016/j.devcel.2009.06.013

Orr ME, Oddo S (2013) Autophagic/lysosomal dysfunction in Alzheimer's disease. Alzheimer Res Therapy 5:1–9. https://doi.org/10.1186/alzrt217

Otsu K, Murakawa T, Yamaguchi O (2015) BCL2L13 is a mammalian homolog of the yeast mitophagy receptor Atg32. Autophagy 11:1932–1933. https://doi.org/10.1080/15548627.2015.1084459

Pagani L, Eckert A (2011) Amyloid-Beta interaction with mitochondria. Int J Alzheimers Dis 2011: 925050. https://doi.org/10.4061/2011/925050

Palikaras K, Lionaki E, Tavernarakis N (2015) Coordination of mitophagy and mitochondrial biogenesis during ageing in C. elegans. Nature 521:525–528. https://doi.org/10.1038/nature14300

Palikaras K, Daskalaki I, Markaki M, Tavernarakis N (2017) Mitophagy and age-related pathologies: development of new therapeutics by targeting mitochondrial turnover. Pharmacol Ther 178:157–174. https://doi.org/10.1016/j.pharmthera.2017.04.005

Palikaras K, Lionaki E, Tavernarakis N (2018) Mechanisms of mitophagy in cellular homeostasis, physiology and pathology. Nat Cell Biol 20:1013–1022. https://doi.org/10.1038/s41556-018-0176-2

Park J, Wetzel I, Marriott I, Dreau D, D'Avanzo C, Kim DY et al (2018) A 3D human triculture system modeling neurodegeneration and neuroinflammation in Alzheimer's disease. Nat Neurosci 21:941–951. https://doi.org/10.1038/s41593-018-0175-4

Parker WD Jr, Boyson SJ, Parks JK (1989) Abnormalities of the electron transport chain in idiopathic Parkinson's disease. Ann Neurol 26:719–723. https://doi.org/10.1002/ana.410260606

Petersen CA, Alikhani N, Behbahani H, Wiehager B, Pavlov PF, Alafuzoff I et al (2008) The amyloid beta-peptide is imported into mitochondria via the TOM import machinery and localized to mitochondrial cristae. Proc Natl Acad Sci U S A 105:13145–13150. https://doi.org/10.1073/pnas.0806192105

Petrozzi L, Ricci G, Giglioli NJ, Siciliano G, Mancuso M (2007) Mitochondria and neurodegeneration. Biosci Rep 27:87–104. https://doi.org/10.1007/s10540-007-9038-z

Rauthan M, Ranji P, Aguilera Pradenas N, Pitot C, Pilon M (2013) The mitochondrial unfolded protein response activator ATFS-1 protects cells from inhibition of the mevalonate pathway. Proc Natl Acad Sci U S A 110:5981–5986. https://doi.org/10.1073/pnas.1218778110

Reddy PH, Reddy TP, Manczak M, Calkins MJ, Shirendeb U, Mao P (2011) Dynamin-related protein 1 and mitochondrial fragmentation in neurodegenerative diseases. Brain Res Rev 67: 103–118. https://doi.org/10.1016/j.brainresrev.2010.11.004

Reddy PH, Yin X, Manczak M, Kumar S, Pradeepkiran JA, Vijayan M et al (2018) Mutant APP and amyloid beta-induced defective autophagy, mitophagy, mitochondrial structural and functional changes and synaptic damage in hippocampal neurons from Alzheimer's disease. Hum Mol Genet 27:2502–2516. https://doi.org/10.1093/hmg/ddy154

Richter B, Sliter DA, Herhaus L, Stolz A, Wang C, Beli P et al (2016) Phosphorylation of OPTN by TBK1 enhances its binding to Ub chains and promotes selective autophagy of damaged mitochondria. Proc Natl Acad Sci U S A 113:4039–4044. https://doi.org/10.1073/pnas.1523926113

Rubinsztein DC (2011) Autophagy and aging. Cell 146:682–695. https://doi.org/10.1016/j.cell.2011.07.030

Rui Y, Tiwari P, Xie Z, Zheng JQ (2006) Acute impairment of mitochondrial tracking by beta-amyloid peptides in hippocampal neurons. J Neurosci 26:10480–10487. https://doi.org/10.1523/JNEUROSCI.3231-06.2006

Ryu SW, Han EC, Yoon J, Choi C (2015) The mitochondrial fusion-related proteins Mfn2 and OPA1 are transcriptionally induced during differentiation of bone marrow progenitors to immature dendritic cells. Mol Cells 38(1):89–94. https://doi.org/10.14348/molcells.2015.2285

Sas K, Robotka H, Toldi J, Vécsei L (2007) Mitochondria, metabolic disturbances, oxidative stress and the kynurenine system, with focus on neurodegenerative disorders. J Neurol Sci 257:221–239. https://doi.org/10.1016/j.jns.2007.01.033

Schapira AHV, Cooper JM, Dexter D, Clark JB, Jenner P, Marsden CD (1990) Mitochondrial complex I deficiency in Parkinson's disease. J Neurochem 54:823–827. https://doi.org/10.1111/j.1471-4159.1990.tb02325.x

Schondorf DC, Ivanyuk D, Baden P, Sanchez-Martinez A, De Cicco S, Yu C et al (2018) The NAD + precursor nicotinamide riboside rescues mitochondrial defects and neuronal loss in iPSC and fly models of Parkinson's disease. Cell Rep 23:2976–2988. https://doi.org/10.1016/j.celrep.2018.05.009

Schreiner B, Hedskog L, Wiehager B, Ankarcrona M (2015) Amyloid-beta peptides are generated in mitochondria-associated endoplasmic reticulum membranes. J Alzheimers Dis 43:369–374. https://doi.org/10.3233/JAD-132543

Schulz KL, Eckert A, Rhein V, Mai S, Haase W, Reichert AS et al (2012) A new link to mitochondrial impairment in tauopathies. Mol Neurobiol 46:205–216. https://doi.org/10.1007/s12035-012-8308-3

Schweers RL, Zhang J, Randall MS, Loyd MR, Li W, Dorsey FC et al (2007) NIX is required for programmed mitochondrial clearance during reticulocyte maturation. Proc Natl Acad Sci U S A 104:19500–19505. https://doi.org/10.1073/pnas.0708818104

Selkoe DJ, Hardy J (2016) The amyloid hypothesis of Alzheimer's disease at 25 years. EMBO Mol Med 8:595–608. https://doi.org/10.15252/emmm.201606210

Serrano-Pozo A, Frosch MP, Masliah E, Hyman BT (2011) Neuropathological alterations in Alzheimer disease. Cold Spring Harb Perspect Med 1(1):a006189. https://doi.org/10.1101/cshperspect.a006189

Sliter DA, Martinez J, Hao L, Chen X, Sun N, Fischer TD et al (2018) Parkin and PINK1 mitigate STING-induced inflammation. Nature 561:258–262. https://doi.org/10.1038/s41586-018-0448-9

Sorrentino V, Romani M, Mouchiroud L, Beck JS, Zhang H, Moullan N et al (2017) Enhancing mitochondrial proteostasis reduces amyloid-b proteotoxicity. Nature 552:187–193. https://doi.org/10.1038/nature25143

Sowter HM, Ratcliffe PJ, Watson P, Greenberg AH, Harris AL (2001) HIF-1-dependent regulation of hypoxic induction of the cell death factors BNIP3 and NIX in human tumors. Cancer Res 61:6669–6673

Spires-Jones TL, Hyman BT (2014) The intersection of amyloid beta and tau at synapses in Alzheimer's disease. Neuron 82:756–771. https://doi.org/10.1016/j.neuron.2014.05.004

Strappazzon F, Nazio F, Corrado M, Cianfanelli V, Romagnoli A, Fimia GM et al (2015) AMBRA1 is able to induce mitophagy via LC3 binding, regardless of PARKIN and p62/SQSTM1. Cell Death Differ 22:419–432. https://doi.org/10.1038/cdd.2014.139

Sun N, Yun J, Liu J, Malide D, Liu C, Rovira II et al (2015) Measuring in vivo mitophagy. Mol Cell 60:685–696. https://doi.org/10.1016/j.molcel.2015.10.009

Swerdlow RH (2018) Mitochondria and mitochondrial cascades in Alzheimer's disease. J Alzheimers Dis 62:1403–1416. https://doi.org/10.3233/JAD-170585

Swerdlow RH, Khan SM (2004) A "mitochondrial cascade hypothesis" for sporadic Alzheimer's disease. Med Hypotheses 63:8–20. https://doi.org/10.1016/j.mehy.2003.12.045

Swerdlow RH, Burns JM, Khan SM (2010) The Alzheimer's disease mitochondrial cascade hypothesis. J Alzheimers Dis 20(Suppl 2):265–279. https://doi.org/10.3233/JAD-2010-100339

Szargel R, Shani V, Abd Elghani F, Mekies LN, Liani E, Rott R et al (2016) The PINK1, synphilin-1 and SIAH-1 complex constitutes a novel mitophagy pathway. Hum Mol Genet 25:3476–3490. https://doi.org/10.1093/hmg/ddw189

Tang J, Oliveros A, Jang MH (2019) Dysfunctional mitochondrial bioenergetics and synaptic degeneration in Alzheimer disease. Int Neurourol J 23(Suppl 1):S5. https://doi.org/10.5213/inj.1938036.018

Trempe JF, Sauvé V, Grenier K, Seirafi M, Tang MY, Ménade M et al (2013) Structure of parkin reveals mechanisms for ubiquitin ligase activation. Science 340:1451–1455. https://doi.org/10.1126/science.1237908

Tuppen HA, Blakely EL, Turnbull DM, Taylor RW (2010) Mitochondrial DNA mutations and human disease. Biochim Biophys Acta 1797:113–128. https://doi.org/10.1016/j.bbabio.2009.09.005

Uddin MS, Ashraf GM (2020) Quality control of cellular protein in neurodegenerative disorders. IGI Global, Pennysylvania. https://doi.org/10.4018/978-1-7998-1317-0

Villa E, Proïcs E, Rubio-Patiño C, Obba S, Zunino B, Bossowski JP et al (2017) Parkin-independent mitophagy controls chemotherapeutic response in cancer cells. Cell Rep 20:2846–2859. https://doi.org/10.1016/j.celrep.2017.08.087

Viswanathan A, Greenberg SM (2011) Cerebral amyloid angiopathy in the elderly. Ann Neurol 70:871–880. https://doi.org/10.1002/ana.22516

Vuda M, Kamath A (2016) Drug-induced mitochondrial dysfunction: mechanisms and adverse clinical consequences. Mitochondrion 31:63–74. https://doi.org/10.1016/j.mito.2016.10.005

Wallace DC (1999) Mitochondrial diseases in man and mouse. Science 283:1482–1488. https://doi.org/10.1126/science.283.5407.1482

Wang X, Su BO, Lee HG, Li X, Perry G, Smith MA et al (2009) Impaired balance of mitochondrial fission and fusion in Alzheimer's disease. J Neurosci 29:9090–9093. https://doi.org/10.1523/JNEUROSCI.1357-09.2009

Wang H, Jiang T, Li W, Gao NA, Zhang T (2018) Resveratrol attenuates oxidative damage through activating mitophagy in an in vitro model of Alzheimer's disease. Toxicol Lett 282:100–108. https://doi.org/10.1016/j.toxlet.2017.10.021

Wang ZT, Lu MH, Zhang Y, Ji WL, Lei L, Wang W et al (2019) Disrupted-in-schizophrenia-1 protects synaptic plasticity in a transgenic mouse model of Alzheimer's disease as a mitophagy receptor. Aging Cell 18:e12860. https://doi.org/10.1111/acel.12860

Weidling IW, Swerdlow RH (2020) Mitochondria in Alzheimer's disease and their potential role in Alzheimer's proteostasis. Exp Neurol 330:113321. https://doi.org/10.1016/j.expneurol.2020.113321

WHO (2019) Dementia: World Health Organization. https://www.who.int/news-room/fact-sheets/detail/dementia. Accessed 30 Aug 2020

Wong YC, Holzbaur E (2015) Temporal dynamics of PARK2/parkin and OPTN/optineurin recruitment during the mitophagy of damaged mitochondria. Autophagy 11:422–424. https://doi.org/10.1080/15548627.2015.1009792

Xie H, Guan J, Borrelli LA, Xu J, Serrano-Pozo A, Bacskai BJ (2013) Mitochondrial alterations near amyloid plaques in an Alzheimer's disease mouse model. J Neurosci 33:17042–17051. https://doi.org/10.1523/JNEUROSCI.1836-13.2013

Xu Y, Shen J, Ran Z (2019) Emerging views of mitophagy in immunity and autoimmune diseases. Autophagy 16:3–17. https://doi.org/10.1080/15548627.2019.1603547

Yamano K, Youle RJ (2013) PINK1 is degraded through the N-end rule pathway. Autophagy 9:1758–1769. https://doi.org/10.4161/auto.24633

Ye X, Sun X, Starovoytov V, Cai Q (2015) Parkin-mediated mitophagy in mutant hAPP neurons and Alzheimer's disease patient brains. Hum Mol Genet 24:2938–2951. https://doi.org/10.1093/hmg/ddv056

Yeh SH, Ou LC, Gean PW, Hung JJ, Chang WC (2011) Selective inhibition of early-but not late-expressed HIF-1α is neuroprotective in rats after focal ischemic brain damage. Brain Pathol 21: 249–262. https://doi.org/10.1111/j.1750-3639.2010.00443.x

Yoo SM, Yamashita SI, Kim H, Na D, Lee H, Kim SJ et al (2019) FKBP8 LIRL-dependent mitochondrial fragmentation facilitates mitophagy under stress conditions. FASEB J 34:2944–2957. https://doi.org/10.1096/fj.201901735R

Yun J, Puri R, Yang H, Lizzio MA, Wu C, Sheng ZH (2014) MUL1 acts in parallel to the PINK1/parkin pathway in regulating mitofusin and compensates for loss of PINK1/parkin. eLife 3: e01958. https://doi.org/10.7554/eLife.01958

Parkinson's Disease: Neurochemistry and Pharmacological Treatment

6

Shradha Bisht, Mamta F. Singh, and Aman Upaganlawar

Abstract

Parkinson's disease (PD) is an age-related neurodegenerative disorder that is characterized by bradykinesia, muscular rigidity, tremors in the hand, and postural imbalance which lead to disturbance of gait and posture. The primary neurological marker of Parkinsonism is dopaminergic neuron degeneration in the substantia nigra pars compacta which is further associated with lack of dopamine and bradykinesia or akinesia in PD patients. Dopaminergic neurons in basal ganglia that are involved in behavioral functions are affected but most importantly neurons involved in movement are affected. This chapter describes neurochemical disturbances of dopamine and other neurotransmitters particularly acetylcholine, noradrenaline, serotonin, GABA, substance P, adenosine, and polypeptides take place in Parkinsonism. Protein misfolding and their aggregation, excitotoxicity, apoptosis, mitochondrial dysfunction and energy disturbances, oxidative stress, calcium dyshomeostasis, and other neurological diseases are the other major role players in PD that are also presented. The chapter further explores currently used drugs along with various surgical approaches like deep brain stimulation, pallidotomy, and thalamotomy, and gene therapy evolved to reduce PD symptoms. However, despite extensive study, p*resent therapy* used for PD is symptomatic; none of the drugs inhibit the development and progression of the disease.

S. Bisht (✉) · M. F. Singh
Sardar Bhagwan Singh University, Dehradun, Uttarakhand, India

A. Upaganlawar
SNJB's Shriman Sureshdada Jain College of Pharmacy, Neminagar, Chandwad, Nashik, Maharashtra, India

G. M. Ashraf, M. S. Uddin (eds.), *Current Thoughts on Dementia*,
https://doi.org/10.1007/978-981-16-7606-2_6

Keywords

Acetylcholine · Dopamine · Lewy body · Mitochondrial dysfunction · Neurotransmitter · Parkinson's disease

6.1 Introduction

Parkinson's disease (PD) is a multifactorial neurodegenerative, age-associated disorder, and second-largest CNS disorder. It is mostly sporadic and in less than 10% cases, biological inheritance is the cause. Vital motor symptoms of PD include rigidity, bradykinesia, difficulty in speaking, abnormal posture, and resting tremor that occur due to degeneration of midbrain dopamine (DA) neurons of substantia nigra (SN) (Francesca et al. 2016). This damage to vulnerable neurons caused by cellular distress results by misfolding of proteins and their aggregation, disturbance of autophagy, endoplasmic reticulum (ER), stress, dysfunctional mitochondria, or imbalance of Ca^{2+} in the brain. Except for dopamine, neurochemical disturbances of other neurotransmitters particularly acetylcholine (ACh), serotonin, noradrenaline, gamma-aminobutyric acid (GABA), and glutamate also take place in PD. Cholecystokinin (CK), dynorphin, neurotensin, and substance P also played a significant role in the pathogenesis of PD. There are many drugs available for the management of the symptoms of PD but the drugs inhibiting the pathogenesis of the PD are not available presently. Further research should be done to explore the different mechanisms involved in the development and progression of the PD, so that, a definite treatment to cure the disease could be developed.

Several factors and conditions may lead to PD. Different types of viral encephalitis, neurological disorders, such as Alzheimer's disease, dementia, supranuclear palsy, CNS disorders associated with disturbances in the arrangement of structure, such as brain tumors, strokes, repeated head injury, Wilson disease mainly in young people play a major role in PD (Noyce et al. 2012; Stacey et al. 2014). Drugs used to treat nausea, especially metoclopramide and prochlorperazine, antipsychotic drugs, some toxic substances like manganese, methanol, and carbon monoxide may cause Parkinsonism. Certain medicines that block dopamine's action are the major cause of the progression of Parkinsonism. For example, antipsychotic drugs, used for the treatment of schizophrenia, block the action of DA neurons. Various genetic mutations may raise the risk of PD (McGee et al. 2018).

Primary motor symptoms of PD include mild tremors, posture abnormality, difficulty in speech, lack of movement in limbs, facial muscle incoordination, lack of concentration in thought and speech, fatigue, depression, and irritability, without any cause (Santens et al. 2003). Sometimes the patient may become stiff, unstable, and as the neurodegeneration progresses, tremor occurs, begins unilaterally, and eventually spreads bilaterally. These symptoms are not easy to detect as they progress slowly (Jankovic 2008). Secondary motor symptoms include a round-backed posture, dystonia, fatigue, motor incoordination, cramping, affected arm swing, akathisia, excess salivation, difficulty in swallowing and mastication, and sexual dysfunction (Davie 2008).

Depression, insomnia, cognitive dysfunction, gesture and emotional changes, difficulty in the urinary system, constipation, sweating, skin problem, increased incidences of cardiovascular diseases, and pain in muscles and joints are frequently observed nonmotor symptoms felt by PD patients (Tsukamoto et al. 2013). Therefore the objective of this chapter is to describe the neurobiology and neurotransmitters involved in PD and to discuss various treatment approaches and advancements in treatment with surgical interventions and gene therapy in PD.

6.2 Pathogenesis of Parkinson's Disease

Various mechanisms explain the pathology of PD. Degeneration of dopaminergic neurons may happen as a result of cellular distress caused by misfolded proteins and their aggregation, autophagic catabolism disruption, ER stress, dysfunctional mitochondria and energy disturbances, and oxidative stress or calcium dyshomeostasis in some parts of the brain. In PD, the symptoms are noticed when there is up to 80% deficiency of DA in substantia nigra pars compacta (SNc) and basal ganglia, i.e., putamen and caudate nucleus occur (Varanese et al. 2011). Neurotransmitters, such as ACh, serotonin, glutamate, and GABA, also have a crucial role in PD. In reality, there is a deficiency of DA and GABA and an excess of ACh and glutamate in the extrapyramidal tract. DA and ACh function as postsynaptic exciting neurotransmitters in the brain.

Furthermore, GABA influences a decreased presynaptic inhibitory function through GABA receptors, and glutamate strives for excitotoxic effect and inhibitory presynaptic effect through *N*-methyl D-aspartate (NMDA) and metabotropic glutaminergic receptors. PD pathophysiology also involves neuropeptides as well (Werner and Coveñas 2011). In this context, the concentration of neurotensin and cholecystokinin in the extrapyramidal tract has risen and the concentration of dynorphin and substance P has fallen.

6.2.1 α-Synuclein and Formation of Lewy Bodies

In 1997, it was found that Lewy bodies (LBs) are the main characteristic of sporadic PD neuropathology. LBs consist of misfolded and aggregated α-synuclein (α-syn), a tiny protein of 140 amino acids usually abundant in the presynaptic compartments especially in the brain, while in the core, muscle, and other tissues, lower quantities are discovered (Spillantini et al. 1997; Iwai et al. 1995). α-Syn is encoded by *SNCA* gene (Werner and Coveñas 2012). α-Syn controls the dynamics of vesicles and the release of neurotransmitters (Burré et al. 2010).

α-Syn plays a crucial role in between DNA repair and PD (Abugable et al. 2019). α-Syn activates ataxia-telangiectasia mutated (ATM), a signaling kinase that repairs major DNA damage. It attaches to breaks that occur in double-stranded DNA and helps the DNA repair mechanism of nonhomologous end-joining (Schaser et al. 2019). It was suggested that cytoplasmic aggregation of LBs reduces nuclear levels of α-syn, resulting in decreased DNA repair, increased double-strand breaks in DNA, and increased apoptosis in neurons (Schaser et al. 2019). Lost functions of mutated α-syn result in impaired dopamine vesicular storage resulting in increased cytosolic dopamine. Cytosolic DA degradation produces species of reactive oxygen and therefore results in neurotoxicity.

Protein kinase C (PKC) is a serine-threonine kinase that phosphorylates various target proteins and thus regulates variable processes for example, apoptosis. PKC is highly susceptible to oxidative stress and begins an apoptotic pathway in DA neurons. It has been observed that α-syn can shut off the proteolytic cascade by down-regulation of PKC delta (PKCd) gene expression by deactivating nuclear factor kappa B that ultimately decreases PKCd transcription (Guzman et al. 2010). Thus, α-syn can be regarded as a neuroprotective in these neurons. In many studies, it was observed that accumulation of aggregated α-syn causes dopamine cell death, and aggregate process reduction has been appeared to decrease neurodegeneration and enhance motor defects in some species, including, rodents, and nonhuman primates (Mónica et al. 2020). Mutated α-syn loses its functionality resulting in impaired dopamine vesicular storage resulting in increased cytosolic DA. Cytosolic DA degradation produces species of reactive oxygen and therefore results in neurotoxicity. Although it is now well recognized that α-syn may occur under physiological circumstances in multiple conformational and oligomeric states that are insoluble and accumulate in LBs (Danzer et al. 2007; Winner et al. 2011). At the beginning of the PD, LBs first appear in the olfactory bulb, pontine tegmentum, and medulla oblongata, and as the disease progresses, it spreads in the SN, and basal forebrain, and the neocortex.

6.2.1.1 Association Between α-Synuclein and Autophagy-Lysosome Pathway

Autophagosomes are double-membrane-bound structures that phagocytose damaged organelles and wastes and transport them to the lysosomal environment for degradation. The degraded products of the process called autophagy are then transported to the cytoplasm for either used to synthesize fresh molecules or to utilize as an energy source (Boya et al. 2013). In PD patients, autophagosomes are increased and lysosomes are reduced in SN. DA neurons indicate interrupted autophagy lysosomes pathway in the patients (Chu et al. 2009). Various genes related to PD have been associated with autophagy dysfunction (Winslow et al. 2010) especially α-syn and leucine-rich repeat kinase 2 (LRRK2). Moreover, recessive PD-associated proteins, such as PINK1 and PARKIN, have a significant role in the mitophagy cycle (Alegre-Abarrategui and Wade-Martins 2009). LRRK2 overexpression reduces autophagosome and lysosomal pH levels, thereby decreasing the activity of lysosomal hydrolytic enzymes (Gómez-Suaga et al. 2012). In vivo gene-targeting studies in mice indicate that the Atg-7 (enzyme needed for autophagosome formation) leads to α-syn accumulation in presynaptic cells.

Chaperone-mediated autophagy (CMA) includes lysosome membrane-mediated direct translocation of unfolded proteins. Some chaperones are heat-shock proteins (e.g., HSP 70). These are generally involved in proper protein folding. It has been observed that proteins α-syn and LKKR2 cannot reach inside the lysosome for CMA (Orenstein et al. 2013). This is because of atypical relationships between these enzymes and the LAMP-2A CMA receptor. In patients with PD, both LAMP2A and HSC70 concentrations are decreased by providing extra proof for decreased CMA activity in PD (Alvarez-Erviti et al. 2010). It is also found that mutations in genes encoded endosomal/lysosomal system proteins, vacuolar protein sorting-35 (VPS35), type 5 P-type ATPase, and glucocerebrosidase (GBA) results in PD.

6.2.1.2 Synucleinopathy and Endoplasmic Reticulum Stress Response

The ER is essential for the folding of proteins in the cells, and any disturbances that alter the ER environment may lead to interruption of the process of protein folding and deposition of misfolded/unfolded proteins. This situation is called ER stress, generally, gives rise to a protective physiological reaction, called unfolded protein response (UPR), to restore homeostasis of ER and normal functioning of cells if it exerts transiently (Hetz 2012). However, the deposition of improperly folded proteins and continuous UPR activity under chronic ER stress triggers the activation of apoptotic pathways that leads to cell death, completely removing damaged cells. Some of the findings revealed that the accumulation of misfolded α-syn and their aggregation is a major factor that truly plays a role in initiating harmful chronic ER stress response in PD (Matus et al. 2011).

6.2.2 Disturbances in Mitochondrial Functions

Increasing proof promotes the association between mitochondrial dysfunction and sporadic PD. α-syn is related to the membrane of various organelles, including mitochondria, and has been shown to affect the structure and functions of mitochondria. α-Syn is present in a specific structure that forms a coherence between ER and mitochondria that is very necessary for controlling Ca^{2+} signaling and apoptosis. A reduction in binding to the mitochondria-associated membrane (MAM) and enhanced mitochondrial fragmentation was discovered in wild type or pathogenic mutations in α-syn, indicating the effect of α-syn in controlling mitochondrial morphology (Guardia-Laguarta et al. 2014), thereby Ca^{2+} exchange impairs and there is reduced mitochondrial energy production (Paillusson et al. 2017).

Recently, a study described that α-syn can affect mitochondrial physiology by regulating peroxisome proliferator-activated receptor gamma coactivator 1-α (PGC1α). Different models over-expressing PD-related mutant LRRK2 showed enhanced destructibility to mitochondrial toxins, mitochondrial dysfunction, and reduced ROS output. In DA neurons (Reinhardt et al. 2013) and knock-in mice (Yue et al. 2015), physiological concentrations of the prevalent LRRK2 G2019S mutant were consistently discovered in combination with mitochondrial defects. By interacting with mitochondrial fission or fusion proteins, vacuolar protein-associated protein 35 (VPS35) in mitochondria appears to control mitochondrial function. Any mutation in VPS35 may result in fragmentation of the mitochondria which may lead to further neurodegeneration. This can be triggered either by a reduction in mitochondrial E3 ubiquitin ligase 1 (MUL1) degradation, which subsequently improves mitofusin degradation or by an increase in DRP1 complex turnover through mitochondrial-derived vesicle-dependent lysosome trafficking (Wang et al. 2016).

Coiled-coil-helix-coiled-coil-helix domain-containing 2 (CHCHD2) is an intermembrane protein in the mitochondria and nucleus with a dual function. Under ordinary circumstances, CHCHD2 primarily associated with mitochondrial complex IV (MC-IV), bounds to mitochondria, and decreased expression of CHCHD2 has been continuously shown to reduce the activity of MC-IV, leading to increased oxidative stress and mitochondrial fragmentation (Aras et al. 2015).

It is now well explained that PINK1 and parkin genes are likely to play a crucial role in the control and functions of mitochondria. PINK1 and parkin importantly control the removal of abnormally functioned mitochondria through mitophagy (Deas et al. 2011; Bertolin et al. 2013). Based on these researches, it was demonstrated that mitophagy impairment due to PINK1 or parkin mutations could become lethal through the collection of dysfunctional mitochondria (Narendra et al. 2010).

6.2.3 Iron Accumulation and Oxidative Stress

Iron is an important component needed for normal brain function. Many enzymes engaged in the synthesis of neurotransmitters have iron as a prosthesis group, for example monoamine oxidases A and B engaged in the dopamine metabolism, tryptophan hydroxylase, needed for serotonin synthesis, tyrosine hydroxylase, essential for DA and norepinephrine synthesis, glutamate decarboxylase catalyzes the synthesis of GABA, and glutamate transaminase needed in the synthesis of L-glutamate, all have iron in their structure. There is pretty clear evidence that iron accumulation is a characteristic of many neurological disorders (Zecca et al. 2008), but the evidence is not available to ensure whether it is the main cause of the disease or a result of a prior dysfunction. An increase in iron level stimulates the formation of hydroxyl radical, enhancing lipid peroxidation, and increased α-syn aggregation, glutathione intake, and alteration of nucleic acid that results in cell death.

Several neurodegenerative diseases were correlated with mitochondrial dysfunction including Alzheimer's disease, Huntington's disease, and PD (Moreira et al. 2010; Grubman et al. 2013). The outcomes of mitochondrial dysfunction are reduced synthesis of ATP and reduced synthesis of iron-sulfur mass and prosthetic groups heme. There are some evidences of accumulation of mitochondrial iron in experimental PD models (Lee et al. 2009; Mena et al. 2011).

6.2.4 Inflammation and Gliosis

In 1988, McGeer and collaborators (McGeer et al. 1988) first defined the dopaminergic degenerative areas in the brain of patients who died due to PD showing clear signs of neuroinflammation, characterized by microglial cell activation. In 1999, the first postmortem analysis was published in three patients of so-called frozen addicts, taking heroin contaminated with 1-methyl-4-phenyl-1,2,3,6-tetrahydropyridine (MPTP) neurotoxin that injected intravenously into primates causing a continuous Parkinsonian syndrome. Postmortem studies disclosed the depletion of SN cells, as was the case in patients with PD (Langston et al. 1999). The dopaminergic degeneration regions showed active microglia, expressing elevated HLA-DR levels, identical to the findings of Mc Geer in PD brains described years earlier (Langston et al. 1999).

This information showed that a still active degeneration of the nerves continued many years after the neurotoxic insult, indicating that microglial cells could initiate a neuroinflammatory cycle in the degeneration areas that add to neuronal death. In the stress condition, these cells produce potentially toxic chemicals, including proinflammatory cytokines, prostaglandins, and reactive oxygen and nitrogen species (Teismann et al. 2003). Association between different pathways of DA neurons degeneration in PD is given in Fig. 6.1 (Maiti et al. 2017).

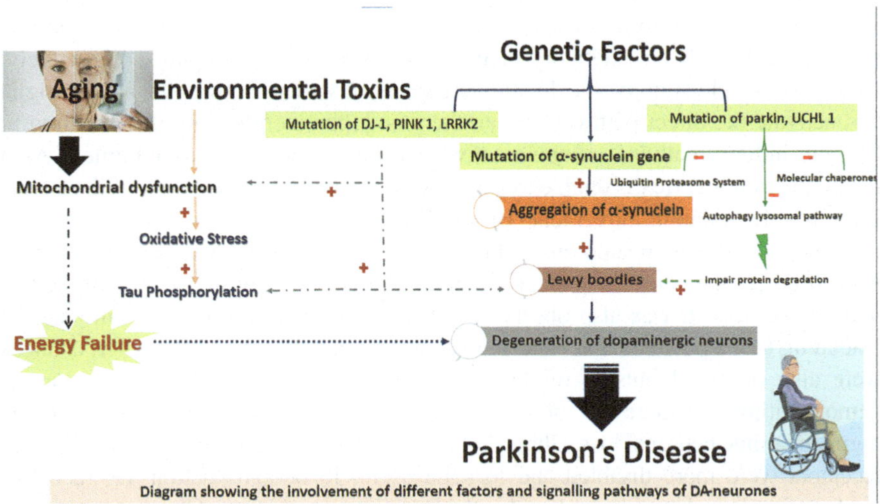

Fig. 6.1 Different factors and signaling pathways are involved in degeneration of dopaminergic neurons in Parkinson's disease (Maiti et al. 2017). *DJ1* protein deglycase, *LRRK2* leucine-rich repeat kinase 2, *PINK1* phosphatase and tensin homolog-induced kinase 1, *UCHL1* ubiquitin C-terminal hydrolase L1

6.2.5 Neurotransmitters and Neuropeptides Associated with Parkinson's Disease

Presently, PD is considered as a multisystem neurodegenerative disease, prominently affecting dopamine neurons. Growing evidence reveals that nondopaminergic neurons like serotoninergic, glutamatergic, cholinergic neurons are affected as well and play an important role in the pathophysiology of the disease. However, the major motor symptoms like bradykinesia, rigidity, and dyskinesia, are linked to dopamine transmission, rationalize the main focus on the nigrostriatal neurons degeneration but tremor and most of the nonmotor symptoms do not respond to dopaminergic treatment. Some studies described that imbalance in the nondopaminergic system has a direct association to nonmotor symptoms of PD, such as depression, fatigue, and hallucinations (Langston et al., 1999). Moreover, some evidences also described the changes in neuropeptides, e.g., cholecystokinin, dynorphin, neurotensin, substance P in PD.

6.2.5.1 Dopamine

The movement of the body is mainly controlled by the particular area of the brain, *basal ganglia*. It receives a signal from the overlying motor cortex and sending the signal to the muscles via the spinal cord. There are two pathways in basal ganglia involved in the processing of movement. The direct pathway initiates movement; while the indirect pathway blocks the movement (Langston et al. 1999). In the direct pathway, the thalamus gets excited and active, while in the indirect pathway thalamus remains inhibited. The thalamus receives signals from the direct or indirect pathway and sends an excitatory or inhibitory message to the cortex respectively. Under normal physiology, the dopaminergic neurons release DA in the basal ganglia and activate the direct pathway. When there is dopaminergic neurodegeneration in PD, the indirect pathway remains active that keeps the thalamus inhibited. As a result, the motor cortex is unable to get activated. This is the reason why PD patients have trouble initiating movement (Brooks 2001).

Loss of the dopamine neurons is directly proportional to the disease progression. It was observed that when the patient is diagnosed with PD, almost 50–60% of the DA neurons are lost. It was also observed that the axon parts of the dopamine neuron extending to the putamen and caudate nucleus moderately vanish. However, changes were also observed outside of the extrapyramidal system in dopamine neurons, demonstrating that there are more other brain defects in addition to the substantia nigra dopamine neuronal loss. Clinically, patients with a reduced amount of striatal D_2 receptors were more disabled and lost the useful levodopa reaction. D_3 receptor-binding sites have been also reduced in the Parkinsonian striatum. Changes in striatum cholinergic-muscarinic receptors appear to be associated with changes in D_2 receptors, and supersensitivity of muscarinic receptors has appeared in cortical areas.

Dopamine release from somatodendritic cells in the SNc and the dorsal striatum, both are required for the motor behaviors and many cognitive functions, those are mediated by the expression of basal ganglia. The striatal dopamine innervations in both rats and monkeys are obtained not only from SNc, but also from the lateral ventral tegmental area (VTA) and retrorubral cells. Because of this, the term mesostriatal dopamine pathway is referred to include all elements of the DA system

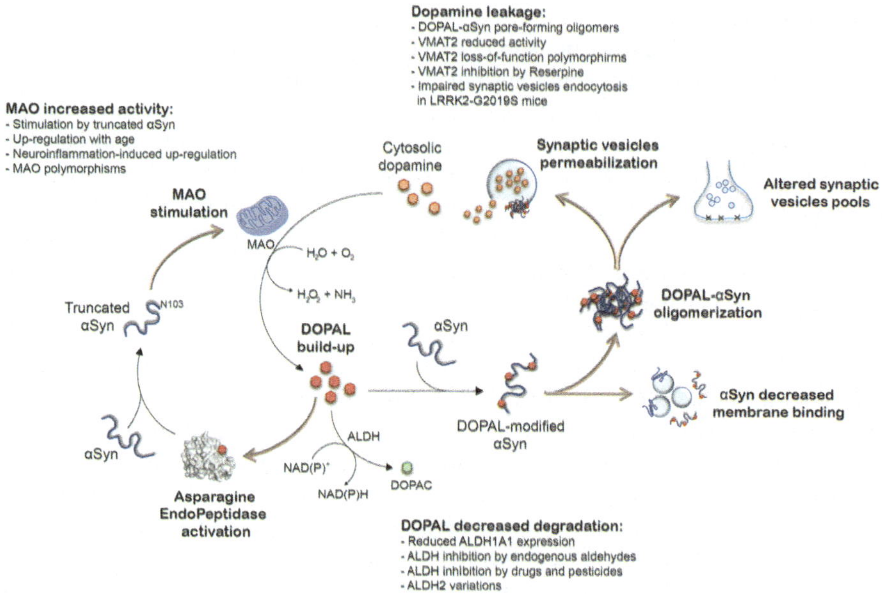

Fig. 6.2 DOPAL at the presynaptic terminals covalently modifies α-synuclein lysines, decreasing α-synuclein affinity for membrane binding and resulting in synaptic vesicle pools redistribution. α-synuclein-DOPAL oligomers permeabilize membrane of synaptic vesicles, resulting in cytosolic DA release, released DA is further metabolized into DOPAL by MAO. DOPAL activates AEP also, which breaks α-synuclein at N103 which is more prone to aggregation and also increases MAO activity. This path leads to α-synuclein aggregation and degeneration of synapses (Masato et al. 2019). *α-syn* α-synuclein, *AEP* asparagine endopeptidase, *ALDH* aldehyde dehydrogenase, *DOPAC* 3,4-dihydroxyphenylacetic acid, *DOPAL* 3,4-dihydroxyphenylacetaldehyde, *MAO* monoamine oxidase, *NAD(P)H* nicotinamide adenine dinucleotide phosphate, *VMAT2* vesicular monoamine transporter 2

of the midbrain extending to the striatum (Björklund and Dunnett 2007). In the last decades, different studies described the association between PD and metabolites of dopamine in various PD models and some autoptic samples (Goldstein et al. 2014). Among the several metabolites, a toxic 3,4-dihydroxyphenylacetaldehyde (DOPAL), was found to be related to PD. In 2003, Burke et al. (2003) provided considerable evidence of DOPAL in in vivo neurotoxicity. The catechol and aldehyde moieties of DOPAL make DOPAL highly reactive that leads to alteration of protein residue, oxidative stress, protein aggregation, and cell death.

Interestingly, α-syn a major element of the PD is a primary target of DOPAL modification. It initiates α-syn oligomerization resulting in impaired synaptic physiology (Dexter et al. 1989). Many factors are involved in the DOPAL aggregation at the presynaptic terminals, such as leakage of DA from synaptic vesicles, increased rate of DA metabolism, and decreased degradation of DOPAL by aldehyde dehydrogenases (Masato et al. 2019). Several researchers demonstrate that the decreased amount of aldehyde dehydrogenases in the brain, as well as genetic variability, increases the risk of development of PD. Thus, the imbalance of these enzymes may be a contributing factor in the pathogenesis of PD (Fig. 6.2).

6.2.5.2 Acetylcholine

Besides the significant reductions in DA level, various studies are explaining initial changes in the cholinergic neurotransmitter pathway in PD. Braak et al. (2003) have been noticed that there is an early deposition of α-syn in cholinergic neurons of the basal forebrain in patients of PD. This may be linked with the LBs and neurons degeneration in the SN (Braak et al. 2003). Significant loss of nucleus basalis of Meynert (nbM) cholinergic neurons is recorded in the brain of PD patients (Rogers et al. 1985; Tagliavini et al. 1984). Arendt et al. (1983) discovered that there was a higher nbM neuron loss in PD than in AD, showing that cholinergic deficiency could be crucial in PD. In the brain, the higher levels of acetylcholine (ACh) and dopamine are present in the striatum. These are involved in cognitive mechanisms, motor responses, and reward and both interact bidirectionally at the presynaptic as well as the postsynaptic level (Threlfell and Cragg 2011; Threlfell et al. 2012). The cholinergic interneurons play a key role in this pathway, and they represent about 1% of the total neurons in striatal (Gonzales and Smith 2015; Cachope and Cheer 2014).

Both nicotinic (nAChRs) and muscarinic (mAChRs) acetylcholine receptors (AChRs) are present in the striatum. Many evidences explain that nAChRs activation may initiate DA release from the nigrostriatal tract, whereas the opposite effect is exerted by activation of the mAChRs, located presynaptically on corticostriatal and nigrostriatal tract. Furthermore, DA inhibits striatal cholinergic interneurons (SCIs), which are hence overactive in PD. There is also subsequent inhibition of DP secretion by mAChRs excitation, thus forming a vicious cycle. There is an indirect association between DAergic and cholinergic neurons, describing the significant role of extracellular DA, released from nigrostriatal terminals (Goto et al. 2007).

Recently, it has shown that PD patients do not vary in the degree of nigrostriatal dopamine neuron distribution but have significantly reduced thalamic cholinergic distribution relative to normal individuals (Bohnen et al. 2009). Thalamic acetylcholinesterase (AChE) activity is derived primarily from brainstem pedunculopontine nucleus (PPN) neuron terminals which play the main role in controlling movement. The PPN is situated in the dorsolateral section of the pontomesencephalic tegmentum and sends efferent cholinergic fibers to various thalamic nuclei; especially those are connected to the limbic system. Dysfunction of PPN is associated with DA-resistant akinesia in PD. Loss of thalamic AChE describes PPN neuron degeneration or dysfunction (Nicolaas and Roger 2011). Degeneration or dysfunction of the cholinergic system of the cortical and subcortical area may also demonstrate the significant postural imbalance and gait in PD with LBs formation and these patients are having a high risk of developing dementia.

6.2.5.3 Serotonin

Tremor and most of the nonmotor symptoms are not managed sufficiently by dopaminergic treatment. Recent evidence indicates that the serotonin (5-HT) system could help toward lessening the nonmotor symptoms including depression, exhaustion, changes in weight, and visual hallucinations of PD. In a postmortem of PD brains, a decreased amount of striatal 5-HT and its 5-hydroxyindoleacetic acid (5-HIAA) metabolite is detected (Stephen et al. 2008). In the latest PD postmortem study (Bédard et al. 2011) and animal models of PD, however striatal hyperinnervation of serotonin was noted. Although in an unregulated manner 5-HT converts levodopa (L-DOPA) to DA, these findings may reflect a secondary mechanism involved in the generation of DP when there is a loss of striatal DA occurs (Zeng et al. 2010; Rylander et al. 2010).

Braak et al. (2003) suggested that the PD initially impacts the dorsal motor nucleus and expands toward structures of the midbrain and forebrain (stage 1). At the next stage (stage 2), LBs accumulation takes place within the raphe nuclei, and in stage 3, the SN, amygdala, and hypothalamus get affected. According to it, caudal 5HT neurons are degenerated followed by degeneration of DAergic neurons of the midbrain (Braak et al. 2003). However, in vivo PET (positron emission tomography) data (Politis et al. 2010) demonstrate that there is no degeneration of caudal and rostral raphe nuclei takes place until the disease progresses at a later stage. So far, a few pieces of evidence are available related to the functional effect of LBs and neuronal deposition in the serotonergic neurons. Possibly α-syn accumulates in serotonergic neurons at the initial stage but the effect is less severe as compared to DA neurons.

It is understood that comparing to bradykinesia and rigidity, tremor is a more challenging symptom to control PD. It is because the nondopaminergic system plays a role in its pathophysiology. One PET imaging study including 26 PD patients and 8 normal individuals was conducted to find out the relation between tremor and 5-HT system (Doder et al. 2003) and observed that midbrain raphe 5-HT1A binding in PD patients is decreased by 27% relative to healthy individuals. The researchers also described a relationship between decreased midbrain raphe 5-HT1A binding and severity of tremor as measured by the unified PD rating scale (UPDRS). According to researcher, the decreased 5-HT1A binding could indicate the loss of 5-HT neuronal cell bodies, probably due to LBs. Further in vivo research to explore the impact of serotonergic dysfunction in tremor in PD is required.

6.2.5.4 Role of GABA/Ca^{2+} System

Depression, gastrointestinal, and some systemic disturbances, and other nonmotor symptoms in PD (Pellicano et al. 2007; Stefanis 2012) are associated with GABA deficiency. There is also a complex association among GABA, Ca^{2+}-dependent neurotransmission, and Ca^{2+}-dependent metabolism in the neurons of the brain. In order to prevent the neurons, the GABAergic pathway monitors the Ca^{2+} influx directly as well as indirectly, via GABA receptor and astrocytes/glial cells, respectively (Yamakage and Namiki 2002; Allaman et al. 2011). Excitation of the presynaptic $GABA_A$ and $GABA_B$ receptors causes hyperpolarization in these cells and thus inhibition of neurotransmitter release (Yamakage and Namiki 2002). During hyperpolarization, the voltage-sensitive Ca^{2+} channels remain inactivated, preventing the toxicity of Ca^{2+} in neurons and giving time to remove the excess Ca^{2+} ions (Hurley et al. 2013). However, removal of Ca^{2+} ions from the cytoplasm and mitochondria needs a considerable quantity of energy. As a result, production of adenosine triphosphate (ATP) in the mitochondria decreases speedily and excess Ca^{2+} cannot be eliminated from the cell. Prolonged intracellular Ca^{2+} retention causes increased oxidative stress in mitochondria that may initiate degeneration in neurons (Surmeier and Schumacker 2013). These networks of pathological processes are monitored by GABA that may regulate the concentration of Ca^{2+} influxes into the cell (Glass et al. 2010; Mosharov et al. 2009; Walker and Semyanov 2007). Decreased Ca^{2+} buffering capacity is associated with neuronal loss in the SN in PD (Hurley et al. 2013; Surmeier and Schumacker 2013).

Another pathway involves the secretion of neurotransmitters within neuronal networks that excites neighboring astrocytes resulting in the inhibition of Ca^{2+} ions influx into the presynaptic neurons (Rial et al. 2016). Hyperactivity of neurons is initially controlled by increased GABA-mediated inhibition and, if unsuccessful, the numbers of GABA synaptic receptors and Ca^{2+} channels are decreased (Surmeier and Schumacker 2013). When all these processes fail, the neurons kill itself by a process of apoptosis when the amount of neurotransmitter around the synapses cannot be reduced and reaches a greater level (i.e., owing to neuronal overexcitation).

A normal level of GABA helps to keep the blood–brain barrier tight and selective. Long-term GABA deficiency may change the permeability of the blood–brain barrier and may induce inflammation in blood vessels and brain tissues therefore significantly increasing neurodegeneration. The damaged or dysfunctional blood–brain barrier is more prone to inflammatory responses that may impart to the development of PD (Glass et al. 2010). The efficiency of GABA is directly proportional to its amount, uptake, and enzymes and cofactors associated with its processing (Richerson and Wu 2003).

6.2.5.5 Glutamate

Glutamate is the main exciting neurotransmitter in the basal ganglia that acts via ionotropic *N*-methyl-D-aspartate (NMDA), kainate, and alpha-amino-3-hydroxy-5-methyl-4-isoxazole- propionic acid (AMPA) receptors and metabotropic receptor subtypes coupled with G-protein. Increased glutamate neurotransmission is associated with the development and progression of the levodopa-induced disease (LID) in both Park's models of 6-hydroxydopamine-injured rodents and 1-methyl-4-phenyl-1,2,3,6-tetrahydropyridine (MPTP)-injured primates (Oh et al. 1998; Calon et al. 2002; Nash and Brotchie 2002).

However, evidences explaining a relation between glutamate imbalance and dyskinesia in PD patients are inconclusive. In the pathophysiology of PD, altered glutamate concentration and neuronal metabolic dysfunction appear to be crucial. The SNc, the region where the main pathological lesion occurs, is especially subjected to oxidative stress and harmful metabolic insults. A decreased ability to deal with metabolic requirements, potentially associated with impaired mitochondrial function, can make substantia nigra extremely susceptible to glutamate impacts, which acts as a neurotoxin in this condition. Glutamate can thus be involved in PD pathogenesis.

During DA depletion in the nigrostriatal tract, the glutamatergic transmission from the subthalamic nucleus to the basal ganglia becomes overexcited, and some regulatory changes take place in glutamate receptors. In some animal models, glutamate receptors inactivation improves the motor symptoms of PD. Therefore, the correlation between the glutamate system and pathogenesis of PD provides a potential treatment approach for this neurodegenerative disorder (Blandini et al. 1996).

6.2.5.6 Cholecystokinin

Cholecystokinin (CK) is secreted by enteroendocrine I-cells from the lining of the duodenum and is also secreted by some brain cells. CK receptors (A and B) are located throughout the gut and CNS nervous system. In the mesolimbic pathway, DA and CK both are present in DA neurons. CK regulates the secretion of dopamine and thus monitors DA-associated behavior. Since CK and its receptors, CK-A and CK-B have a physiological correlation with DA neurotransmission, changes in it may be predisposing to PD.

One hundred and sixty patients of PD and same numbers of healthy subjects were recruited and CK-A receptor and CK-B receptor gene polymorphisms were studied using polymerase chain reaction-restriction fragment length polymorphism analyses. No correlation between these three gene polymorphisms and risk of PD was found; however, the CKCT/TT genotype was associated with a 4.429-fold increased risk of visual hallucinations in PD. CK-A and -B receptor polymorphisms, also showed no correlation with PD associated hallucinations; however, a combined impact was discovered in patients with hallucinations containing both CT/TT CK and TC/CC receptor CK-A genotypes. Patients with PD who harbor this genotype have an enhanced possibility of visual hallucinations 5.922-fold. These findings indicate that visual hallucinations in PD are associated with CK polymorphism in Chinese, and this association has been noted in the presence of the TC/CC genotype CK-A receptor, indicating a possible relation of these genes in the visual hallucination in PD (Wang et al. 2003).

6.2.5.7 Dynorphin

In PD, neurotransmitters other than DA, including neuropeptides, may have significant pathophysiological and clinical roles. Both met-enkephalin, the striatopallidal pathway's primary transmitter, and dynorphin, one of the striatonigral pathway's cotransmitters, show complicated anatomical and biochemical relationships with the basal DA ganglion system. In one research, the concentration of a proenkephalin derivative, Met5enkephalin-Arg6-Gly7-Leu8 (MERGL), in the cerebrospinal fluid (CSF) was observed considerably at small levels in PD patients after overnight withdrawal of treatment compared to healthy individuals and did not alter after at least 16 h of optimum doses of levodopa administration (Langston et al. 1999). With advancing age, MERGL concentrations improved among ordinary people but not among PD patients. On the other hand, the concentration of dynorphin A did not distinguish between the two research groups, and not altered with levodopa treatment. These results, accordant with postmortem research in brains of PD patients, do not support the results in animal models of PD, indicate that enkephalin system abnormality in PD may be caused by the participation of these striatal neurons in the main pathological phase (Baronti et al. 1991).

6.2.5.8 Neurotensin

Neurotensin is a neuropeptide, present in different parts of the brain but in a large amount, it is present in the hypothalamus, amygdala, and nucleus accumbens. There is a strong relation between neurotensin and the dopaminergic system. It is observed that neurotensin regulates dopamine release from striatal, nucleus accumbens, and retinal slices. It may increase or decrease the DA release that may depend on the area where it is injected. By using electrochemical detection and a carbon fiber electrode, it was observed that neurotensin when applied intracerebroventricularly, increases the level of DA in the nucleus accumbens (Blaha and Phillips 1992). Administration of neurotensin in the nucleus accumbens by microdialysis results in a $GABA_A$ receptor-dependent local decrease in DA release. Neurotensin decreases the D_2 receptor-mediated presynaptic inhibition at dopaminergic axon terminals resulting in DA release in the accumbens. Finally, neurotensin injected in the VTA results in an increased release of DA in the nucleus accumbens. This is also described by an increased firing rate of dopaminergic neurons by neurotensin. However, the possible explanation of a decrease in somatodendritic autoreceptor functions is required to be evaluated (Baronti et al. 1991).

Various studies have tried to determine the association between PD and neurotensinergic system. Results of biochemical as well as histological studies of brain tissues of dead PD patients show a decreased number of neurotensin-binding sites in the SN, caudate and putamen, VTA, and globus pallidus compared to healthy individuals (Chinaglia et al. 1990; Fernandez et al. 1994). More research must be carried out to determine the correlation between decreased number of neurotensin receptors expressed by striatal medium spiny neurons and PD (Filomeni et al. 2015).

6.2.5.9 Substance P

Substance P (subs-P) is an undecapeptide that belongs to the tachykinin of the peptide family and primarily uses the NK1 receptor to perform its function. In the SN, there is a positive link between the concentration of substance P and secretion of DA. In the caudate nucleus, dynorphin inhibits the subs-P-containing neurons via kappa receptors. Therefore, there is less excitation of GABA neurons in the globus pallidus (Emma et al. 2010).

6.3 Treatment Approach for Parkinson's Disease

There are different medicines with different mechanisms of action that are used to manage the symptoms of PD. These medicines are either used alone or in combination with each other, depending on whether symptoms are mild or advanced. Some major drugs for PD along with category uses and adverse events are listed in Table 6.1.

Table 6.1 Treatment approach for the management of Parkinson's disease (Goetz et al. 2005; Maiti et al. 2017)

Category	Drug	Indications	Adverse effects
Dopamine agonist	Levodopa	Primary treatment of PD	Hypotension, restlessness, drowsiness
Anticholinergics	Benzotropil, trihexyphenidyl	Symptomatic control of PD	Dry mouth, hypotension, constipation, cognitive impairment, and urinary retention
COMT inhibitors	Entacapone, tolcapone	Motor fluctuations in patients taking levodopa	Diarrhea aggravates the adverse effects of levodopa; monitoring of liver function is required
Dopamine agonists	Apomorphine	Early and advanced disease	Nausea, dizziness, headache
	Bromocriptine, pergolide	Initial treatment and as adjuvant therapy with levodopa	Somnolence, hallucination, edema, fibrosis of lungs and cardiac valves, and lungs
	Pramipexole, ropinirole	Motor fluctuation	Hallucination, edema, sleep attack, and hypotension.
MAO-B inhibitors	Selegiline	Symptomatic and as adjuvant therapy	Nausea, insomnia, interacts with MAO inhibitors/tyramine
	Rasagiline		Dry mouth, weight loss, interaction with MAO inhibitors/tyramine, and low BP
NMDA receptor inhibitor	Amantadine	Akinesia, tremor, rigidity	Nausea, mental confusion, hypotension, hallucination, and edema
Boost mitochondrial function	Creatine	Prevents neuronal damage	Nausea, muscle cramps, diarrhea, insufficient breathing, swelling in the face, tongue, and throat, and weight gain
COX-2 inhibitors	Rofecoxib	Prevents inflammation	Headache, diarrhea, dizziness, heartburn, weakness, running nose, swelled legs and feet, blurred vision, and constipation
Serotonin receptors blocker	ACP-103 (pimavanserin)	Levodopa-associated complications	Motor incoordination, akathisia, hyperprolactinemia, sexual dysfunction, and restlessness
Tyrosine kinase A dimerization and increases neurotrophic factors	GM1, gangliosides	Increases dopamine concentration	–
D_2R and 5-HT_2R blocker	Quetiapine	Psychosis and agitation	Involuntary movements, agitation, anxiety, tremor, hypertonia, dyskinesia, hyperkinesias, confusion,

(continued)

Table 6.1 (continued)

Category	Drug	Indications	Adverse effects
			amnesia, myoclonus, apathy, hemiplegia, and aphasia
Antioxidant	Ubiquinone or coenzyme Q10	Improves mitochondrial function	Hypotension, muscle pain, hemorrhage, difficulty in breathing, backache, bronchitis, constipation, coughing, diarrhea, fatigue, hearing problem, heart attack, insomnia, respiratory tract and urinary tract infection, and sore throat
Methylates phospholipid and increases nerve-cell communication	*S*-adenosyl-methionine (SAM)	Improves dopamine transmission, decreases depression	Gastrointestinal manifestations and anxiety

6.4 Recent Advancement and Future Perspectives

6.4.1 Surgical Treatments

When there is no symptomatic relief after the use of medicines, surgical treatments may be done to reduce motor symptoms, especially when the disease is at an advanced stage (Groiss et al. 2009). Presently, there are two approaches for surgical treatments in PD such as deep brain stimulation (DBS) and pallidotomy and/or a thalamotomy. In DBS, small electrodes are implanted into either the subthalamic nucleus or the globus pallidus, involved in motor function. This implanted device initiates small electrical impulses in all parts of the body. Electrode located on the left side of the brain will monitor the function of the right part of the body and vice versa. DBS may be an effective treatment for all primary motor dysfunctions of PD. In pallidotomy, a wire probe is inserted into the globus pallidus to form the lesions. These lesions help to restore the balance needed for normal movement. Pallidotomy may help to eradicate drugs (used for PD treatment) associated with dyskinesia, muscle rigidity, tremor, and gradual loss of spontaneous movement. In thalamotomy, radiofrequency energy currents are used to a degenerate specific portion of the thalamus. This treatment is used more often in patients with essential tremors, rather than PD (Iacono et al. 1995).

6.4.2 Neural Transplantation Therapy

Over 20 past years, many investigators transplanted DA neurons produced from rodent embryonic stem cells, into the striata of some animal models of PD, for example, adrenal medullary DA neurons (Solari et al. 2013; Dauer and Przedborski 2003). Interestingly, these neurons integrate into the brain system and improve the behavioral abnormalities in animals (Kim et al. 2002; Kim 2011). When human fetus-derived DA tissues are transplanted into the striatum of the PD patients, increased level of DA indicates that these stem cells are capable to survive and can differentiate into DA neurons (Lindvall and Bjorklund 2011). But few patients are cured with neural transplantation, while in others no improvement is seen and more complications are developed in some patients.

6.4.3 Gene Therapy

As mutations in various genes are related to PD, gene modification may be a promising treatment approach for PD (Coune et al. 2012). Presently, various viral vectors containing target genes are inserted into the animal brain either directly or via systemic injections. These viruses are integrated inside the host cells, resulted in the expression of certain genes, and thus prevent degeneration of DA-neurons and consequently, increase the level of DA (Coune et al. 2012), for example, adenovirus, lentivirus, nonlentivirus, and recombinant adeno-associated virus (rAAV).

Aromatic amino acid decarboxylase (AADC), tyrosine hydroxylase (TH), and guanosine triphosphate cyclohydrolase (GCH) are the dopamine-synthesizing enzymes. The behavioral changes are observed when AADC-TH-GCH genes are transferred into the striatal neurons of PD patients (Jarrya 2009). RNA interference-based therapy is used to inhibit some genes like SNCA, PINK, or parkin genes that are associated with PD. In a study, the polyethylene glycol-polyethyleneimine (PEG/PEI) siSNCA complex has been transferred into pheochromocytoma cell-12 (PC12) cells of the rat. There significant decrease in SNCA-mRNA expression, protecting MPTP-induced apoptosis is observed (Liu et al. 2014).

CRISPR/Cas-9 system is a new technique used for adding, breaking, or changing the sequence of specific genes (Kolli 2017a, b). Researchers developed a stable cell line of SNCA by using CRISPR/Cas9-mediated genome editing. Though gene therapy seems a very promising treatment, more research should be carried out to observe its effective role in PD patients. Until a proper cure is discovered, presently available drugs and certain lifestyle modifications may help to manage a better life of PD patients.

6.4.4 Experimental Research

Researchers are currently trying to produce DA neurons from human stem cells in the research laboratories to transplant into the patients of PD. Embryonic stem cells, derived from several day-old embryos, produced through the process of in vitro fertilization are used in these experiments. It is believed that these cells may be manipulated into any cells of the body, to replace the degenerated cells form during the development of the disease (Curt et al. 2011). The successful generation and supply of unlimited DA neurons may lighten up hope for better treatment for PD patients.

Neurotrophic factors, a family of peptides or proteins biomolecules. These biomolecules support the growth, differentiation, and survival of both developing as well as mature cells. Human research of neurotrophic factors is also being carried out. In some studies on animals, these biomolecules have restored quiescent brain cells to synthesize dopamine, and significant improvement in symptoms is observed (Palasz et al. 2020).

6.5 Conclusion

PD is a neurodegenerative disease involving neuronal pathways, various neurotransmitters and neuropeptides, and genetic factors. Advancement in the understanding of PD has led to progress in its medical management. Currently, various classes of drugs are effective in managing the symptoms of PD. Deep brain stimulation by surgical procedure, neuronal cell transplantation, and genetic modifications are other approaches evolving as an effective treatment for PD. However despite extensive study, p*resent therapy* available for *PD* is symptomatic, none of the drugs affect the pathological pathway of the disease. As patients with PD are increasing significantly, it becomes essential to develop new treatments by exploring deep knowledge of the pathways that lead to this disease.

References

Abugable AA, Morris JL, Palminha NM, Zaksauskaite R, Ray S, E-Khamisy SF (2019) DNA repair and neurological disease: from molecular understanding to the development of diagnostics and model organisms. DNA Repair 81:102669

Alegre-Abarrategui J, Wade-Martins R (2009) Parkinson disease, LRRK2 and the endocytic-autophagic pathway. Autophagy 5:1208–1210

Allaman I, Belanger M, Magistretti PJ (2011) Astrocyte–neuron metabolic relationships: for better and for worse. Trends Neurosci 34:76–87

Alvarez-Erviti L, Rodriguez-Oroz MC, Cooper JM, Caballero C, Ferrer I, Obeso JA et al (2010) Chaperone-mediated autophagy markers in Parkinson disease brains. Arch Neurol 67:1464–1472

Aras S, Bai M, Lee I, Springett R, Huttemann M, Grossman LI (2015) MNRR1 (formerly CHCHD2) is a bi-organellar regulator of mitochondrial metabolism. Mitochondrion 20:43–51

Arendt T, Bigl V, Arendt A, Tennstedt A (1983) Loss of neurons in the nucleus basalis of Meynert in Alzheimer's disease, paralysis agitans and Korsakoff's disease. Acta Neuropathol (Berl) 61:101–108

Baronti F, Conant KE, Giuffra M, Davis TL, Brughitta G, Iadarola MJ et al (1991) Opioid peptides in Parkinson's disease: effects of dopamine repletion. Brain Res 27(560):92–96

Bédard C, Wallman MA, Pourcher E, Gould PV, Parent A, Parent M (2011) Serotonin and dopamine striatal innervation in Parkinson's disease and Huntington's chorea. Parkinsonism Relat Disord 17(8):593–598

Bertolin G, Ferrando-Miguel R, Jacoupy M, Traver S, Grenier K, Greene AW et al (2013) The TOMM machinery is a molecular switch in PINK1 and PARK2/PARKIN-dependent mitochondrial clearance. Autophagy 9:1801–1817
Björklund A, Dunnett SB (2007) Dopamine neuron systems in the brain: an update. Trends Neurosci 30:194–202
Blaha CD, Phillips AG (1992) Pharmacological evidence for common mechanisms underlying the effects of neurotensin and neuroleptics on in vivo dopamine efflux in the rat nucleus accumbens. Neuroscience 49:867–877
Blandini F, Porter RHP, Greenamyre JT (1996) Glutamate and Parkinson's disease. Mol Neurobiol 12(1):73–94
Bohnen NI, Muller ML, Koeppe RA, Studenski SA, Kilbourn MA, Frey KA et al (2009) History of falls in Parkinson disease is associated with reduced cholinergic activity. Neurology 73:1670–1676
Boya P, Reggiori F, Codogno P (2013) Emerging regulation and functions of autophagy. Nat Cell Biol 15:713–720
Braak H, Del Tredici K, Rub U, de Vos RA, Jansen Steur EN, Braak E (2003) Staging of brain pathology related to sporadic Parkinson's disease. Neurobiol Aging 24:197–211
Brooks DJ (2001) Functional imaging studies on dopamine and motor control. J Neural Transm 108(11):1283–1298
Burke WJ, Li SW, Williams EA, Nonneman R, Zahm DS (2003) 3,4-Dihydroxyphenylacetaldehyde is the toxic dopamine metabolite in vivo: implications for Parkinson's disease pathogenesis. Brain Res 989:205–213
Burré J, Sharma M, Tsetsenis T, Buchman V, Etherton MR, Südhof TC (2010) Alpha-synuclein promotes SNARE-complex assembly in vivo and in vitro. Science 329:1663–1667
Cachope R, Cheer JF (2014) Local control of striatal dopamine release. Front Behav Neurosci 8:188
Calon F, Morissette M, Ghribi O, Goulet M, Grondin R, Blanchet PJ et al (2002) Alteration of glutamate receptors in the striatum of dyskinetic 1-methyl-4-phenyl-1,2,3,6-tetrahydropyridine-treated monkeys following dopamine agonist treatment. Progr Neuropsychopharmacol Biol Psychiatry 26:127–138
Chinaglia G, Probst A, Palacios JM (1990) Neurotensin receptors in Parkinson's disease and progressive supranuclear palsy: an autoradiographic study in basal ganglia. Neuroscience 39: 351–360
Chu Y, Dodiya H, Aebischer P, Olanow CW, Kordower JH (2009) Alterations in lysosomal and proteasomal markers in Parkinson's disease: relationship to alpha-synuclein inclusions. Neurobiol Dis 35:385–398
Coune PG, Schneider BL, Aebischer P (2012) Parkinson's disease: gene therapies. Cold Spring Harb Med 2(4):a009431
Curt RF, Wenbo Z, Breeze RE (2011) Dopamine cell transplantation for Parkinson's disease: the importance of controlled clinical trials. Neurotherapeutics 8(4):549–561
Danzer KM, Haasen D, Karow AR, Moussaud S, Habeck M, Giese A, Kretzschmar H, Hengerer B, Kostka M (2007) Different species of alpha-synuclein oligomers induce calcium influx and seeding. J Neurosci 27:9220
Dauer W, Przedborski S (2003) Parkinson's disease: mechanisms and models. Neuron 39(6): 889–909
Davie CA (2008) A review of Parkinson's disease. Br Med Bull 86:109–127
Deas E, Wood NW, Plun-Favreau H (2011) Mitophagy and Parkinson's disease: the PINK1-parkin link. Biochim Biophys Acta 1813:623–633
Dexter DT, Wells FR, Lees AJ, Agid F, Agid Y, Jenner P et al (1989) Increased nigraliron content and alterations in other metal ions occurring in brain in Parkinson's disease. J Neurochem 52: 1830–1836
Doder M, Rabiner EA, Turjanski N, Lees AJ, Brooks DJ (2003) Tremor in Parkinson's disease and serotonergic dysfunction: an 11C-WAY 100635 PET study. Neurology 60(4):601–605

Emma T, Tuyet TBT, Robert V (2010) A substance P-mediated pathway contributes to 6-hydroxydopamine induced cell death. Neurosci Lett 481(1):64–67
Fernandez A, de Ceballos ML, Jenner P, Marsden CD (1994) Neurotensin, substance P, delta and mu opioid receptors are decreased in basal ganglia of Parkinson's disease patients. Neuroscience 61:73–79
Filomeni G, De Zio D, Cecconi F (2015) Oxidative stress and autophagy: the clash between damage and metabolic needs. Cell Death Differ 22:377–388
Francesca M, Alessandro P, Pierluigi T, Angela F, Laura R, Nicola S, Giampietro Z, Stefano T (2016) Pathophysiology of motor dysfunction in Parkinson's disease as the rationale for drug treatment and rehabilitation. Parkinsons Dis 2016:9832839
Glass CK, Saijo K, Winner B, Marchetto MC, Gage FH (2010) Mechanisms underlying inflammation in neurodegeneration. Cell 140(6):918–934
Goetz CG, Poewe W, Rascol O, Sampaio C (2005) Evidence-based medical review update: pharmacological and surgical treatments of Parkinson's disease: 2001 to 2004. Mov Disord 20:523–539
Goldstein DS, Kopin IJ, Sharabi Y (2014) Catecholamine autotoxicity. Implications for pharmacology and therapeutics of Parkinson disease and related disorders. Pharmacol Ther 144:268–282
Gómez-Suaga P, Luzón-Toro B, Churamani D, Zhang L, Bloor-Young D, Patel S et al (2012) Leucine-rich repeat kinase 2 regulates autophagy through a calcium-dependent pathway involving NAADP. Hum Mol Genet 21:511–525
Gonzales KK, Smith Y (2015) Cholinergic interneurons in the dorsal and ventralstriatum: anatomical and functional considerations in normal and diseased conditions. Ann N Y Acad Sci 1349:1–45
Goto Y, Otani S, Grace AA (2007) The Yin and Yang of dopamine release: a new perspective. Neuropharmacology 53:583–587
Groiss SJ, Wojtecki L, Sudmeyer M, Schnitzler A (2009) Deep brain stimulation in Parkinson's disease. Ther Adv Neurol Disord 2(6):20–28
Grubman A, White AR, Liddell JR (2013) Mitochondrial metals as a potential therapeutic target in neurodegeneration. Br J Pharmacol 171(8):2159–2173
Guardia-Laguarta C, Area-Gomez E, Rub C, Liu Y, Magrane J, Becker D et al (2014) alpha-Synuclein is localized to mitochondria-associated ER membranes. J Neurosci 34(1):249–259
Guzman JN, Sanchez-Padilla J, Wokosin D, Kondapalli J, Ilijic E, Schumacker PT et al (2010) Oxidant stress evoked by pacemaking in dopaminergic neurons is attenuated by DJ-1. Nature 468:696–700
Hetz C (2012) The unfolded protein response: controlling cell fate decisions under ER stress and beyond. Nat Rev Mol Cell Biol 13:89–102
Hurley MJ, Brandon B, Gentleman SM, Dexter DT (2013) Parkinson's disease is associated with altered expression of CaV1 channels and calcium-binding proteins. Brain 136:2077–2097
Iacono RP, Henderson JM, Lonser RR (1995) Combined stereotactic thalamotomy and posteroventral pallidotomy for Parkinson's disease. J Image Guid Surg 1:133–140
Iwai A, Masliah E, Yoshimoto M, Ge N, Flanagan L, de Silva HA, Kittel A et al (1995) The precursor protein of non-A beta component of Alzheimer's disease amyloid is a presynaptic protein of the central nervous system. Neuron 14:467–475
Jankovic J (2008) Parkinson's disease: clinical features and diagnosis. J Neurol Neurosurg Psychiatry 79(4):368–376
Kim HJ (2011) Stem cell potential in Parkinson's disease and molecular factors for the generation of dopamine neurons. Biochim Biophys Acta 1812(1):1–11
Kim JH, Auerbach JM, Rodriguez-Gomez JA, Velasco I, Gavin D, Lumelsky N, Lee SH, Nguyen J, Sanchez-Pernaute R, Bankiewicz K et al (2002) Dopamine neurons derived from embryonic stem cells function in an animal model of Parkinson's disease. Nature 418(6893):50–56
Kolli N, Lu M, Maiti P, Rossignol J, Dunbar GL (2017a) Application of the gene editing tool, CRISPR-Cas9, for treating neurodegenerative diseases. Neurochem Int 112:187

Kolli N, Lu M, Maiti P, Rossignol J, Dunbar GL (2017b) CRISPR-Cas9 mediated gene-silencing of the mutant huntingtin gene in an in vitro model of Huntington's disease. Int J Mol Sci 18(4):754
Langston JW, Forno LS, Tetrud J, Reeves AG, Kaplan JA, Karluk D (1999) Evidence of active nerve cell degeneration in the substantia nigra of humans years after 1-methyl-4-phenyl-1,2,3,6-tetrahydropyridine exposure. Ann Neurol 46:598–605
Lee DW, Kaur D, Chinta SJ, Rajagopalan S, Andersen JK (2009) A disruption in iron-sulfur center biogenesis via inhibition of mitochondrial dithiol glutaredoxin 2 may contribute to mitochondrial and cellular iron dysregulation in mammalian glutathione-depleted dopaminergic cells: implications for Parkinson's disease. Antioxid Redox Signal 11:2083–2094
Lindvall O, Bjorklund A (2011) Cell therapeutics in Parkinson's disease. Neurotherapeutics 8(4): 539–548
Liu YY, Yang XY, Li Z, Liu ZL, Cheng D, Wang Y et al (2014) Characterization of polyethylene glycol-polyethyleneimine as a vector for alpha-synuclein siRNA delivery to PC12 cells for Parkinson's disease. CNS Neurosci Ther 20(1):76–85
Maiti P, Manna J, Dunbar GL (2017) Current understanding of the molecular mechanisms in Parkinson's disease: targets for potential treatments. Transl Neurodegeneration 6:28
Masato A, Plotegher N, Boassa D, Bubacc L (2019) Impaired dopamine metabolism in Parkinson's disease pathogenesis. Mol Neurodegener 14:35
Matus S, Glimcher LH, Hetz C (2011) Protein folding stress in neurodegenerative diseases: a glimpse into the ER. Curr Opin Cell Biol 23:239–252
McGee DJ, Lu XH, Disbrow EA (2018) Stomaching the possibility of a pathogenic role for Helicobacter pylori in Parkinson's disease. J Parkinsons Dis 8(3):367–374
McGeer PL, Itagaki S, Boyes BE (1988) McGeer, reactive microglia are positive for HLA-DR in the substantia nigra of Parkinson's and Alzheimer's disease brains. Neurology 38(8):1285–1291
Mena NP, Bulteau AL, Salazar J, Hirsch EC, Núñez MT (2011) Effect of mitochondrial complex I inhibition on Fe-S cluster protein activity. Biochem Biophys Res Commun 409:241–246
Mónica GB, Noelia G, Patricia GS, Anne M, Mireille D, Rosario M (2020) Modeling Parkinson's disease with the alpha-synuclein protein. Front Pharmacol 11:356
Moreira PI, Zhu X, Wang X, Lee HG, Nunomura A, Petersen RB et al (2010) Mitochondria: a therapeutic target in neurodegeneration. Biochim Biophys Acta 1802:212–220
Mosharov EV, Larsen KE, Kanter E, Phillips KA, Wilson K, Schmitz Y et al (2009) Interplay between cytosolic dopamine, calcium, and alphasynuclein causes selective death of substantia nigra neurons. Neuron 62:218–229
Narendra DP, Jin SM, Tanaka A, Suen DF, Gautier CA, Shen J et al (2010) PINK1 is selectively stabilized on impaired mitochondria to activate Parkin. PLoS Biol 8(1):e1000298
Nash JE, Brotchie JM (2002) Characterisation of striatal NMDA receptors involved in the generation of parkinsonian symptoms: intrastriatal microinjection studies in the 6-OHDA-lesioned rat. Mov Disord 17:455–466
Nicolaas IB, Roger LA (2011) The cholinergic system and Parkinson disease. Behav Brain Res 221(2):564–573
Noyce AJ, Bestwick JP, Moriyama SL, Moriyama LS, Hawkes CH, Giovannoni G, Lees AJ et al (2012) Meta-analysis of early nonmotor features and risk factors for Parkinson disease. Ann Neurol 72(6):893–901
Oh JD, Russell DS, Vaughan CL, Chase TN, Russell D (1998) Enhanced tyrosine phosphorylation of striatal NMDA receptor subunits: effect of dopaminergic denervation and L-DOPA administration. Brain Res 813:150–159
Orenstein SJ, Kuo SH, Tasset I, Arias E, Koga H, Fernandez-Carasa I et al (2013) Interplay of LRRK2 with chaperone-mediated autophagy. Nat Neurosci 16:394–406
Paillusson S, Gomez-Suaga P, Stoica R, Little D, Gissen P, Devine MJ et al (2017) alpha-Synuclein binds to the ER-mitochondria tethering protein VAPB to disrupt Ca2+ homeostasis and mitochondrial ATP production. Acta Neuropathol 134(1):129–149
Palasz E, Wysocka A, Gasiorowska A, Chalimoniuk M, Niewiadomski W, Niewiadomska G (2020) BDNF as a promising therapeutic agent in Parkinson's disease. Int J Mol Sci 21:1170

Pellicano C, Benincasa D, Pisani V, Buttarelli FR, Giovannelli M, Pontieri FE (2007) Prodromal non-motor symptoms of Parkinson's disease. Neuropsychiatry Dis Treat 3:145–152
Politis M, Wu K, Loane C, Brooks DJ, Kiferle L, Turkheimer F et al (2010) Staging of serotonergic dysfunction in Parkinson's disease: an in vivo 11CDASB PET study. Neurobiol Dis 40(1): 216–221
Reinhardt P, Schmid B, Burbulla LF, Schondorf DC, Wagner L, Glatza M et al (2013) Genetic correction of a LRRK2 mutation in human iPSCs links parkinsonian neurodegeneration to ERK-dependent changes in gene expression. Cell Stem Cell 12(3):354–367
Rial D, Lemos C, Pinheiro H, Duarte JM, Gonçalves FQ, Real JI et al (2016) Depression as aglial-based synaptic dysfunction. Front Cell Neurosci 9:521
Richerson GB, Wu Y (2003) Dynamic equilibrium of neurotransmitter transporters: not just for reuptake anymore. J Neurophysiol 90:1363–1374
Rogers JD, Brogan D, Mirra SS (1985) The nucleus basalis of Meynert in neurological disease: a quantitative morphological study. Ann Neurol 17:163–170
Rylander D, Parent M, O'Sullivan SS, Dovero S, Lees AJ, Bezard E et al (2010) Maladaptive plasticity of serotonin axon terminals in levodopa induced dyskinesia. Ann Neurol 68(5): 619–628
Santens P, Boon P, Van Roost D, Caemaert J (2003) The pathophysiology of motor symptoms in Parkinson's disease. Acta Neurol Belg 103(3):129–134
Schaser AJ, Osterberg VR, Dent SE, Stackhouse TL, Wakeham CM, Boutros SW et al (2019) Alpha-synuclein is a DNA-binding protein that modulates DNA repair with implications for Lewy body disorders. Sci Rep 9(1):10919
Solari N, Bonito-Oliva A, Fisone G, Brambilla R (2013) Understanding cognitive deficits in Parkinson's disease: lessons from preclinical animal models. Learn Mem 20(10):592–600
Spillantini MG, Schmidt ML, Lee VM, Trojanowski JQ, Jakes R, Goedert M (1997) Alpha-synuclein in Lewy bodies. Nature 388:839–840
Stacey ES, Jose AS, Hope B, Judith AP (2014) The emerging role of nutrition in Parkinson's disease. Front Aging Neurosci 6:36
Stefanis L (2012) a-Synuclein in Parkinson's disease. Cold Spring Harb Perspect Med 2:a009399
Stephen JK, Junchao T, Oleh H, Ali R, Li-Jan C, Mark G, Yoshiaki F (2008) Preferential loss of serotonin markers in caudate versus putamen in Parkinson's disease. Brain 131(1):120–131
Surmeier DJ, Schumacker PT (2013) Calcium, bioenergetics, and neuronal vulnerability in Parkinson's disease. J Biol Chem 288:10736–10741
Tagliavini F, Pilleri G, Bouras C, Constantinidis J (1984) The basal nucleus of Meynert in idiopathic Parkinson's disease. Acta Neurol Scand 70:20–28
Teismann P, Tieu K, Choi DK, Wu DC, Naini A, Hunot S et al (2003) Cyclooxygenase-2 is instrumental in Parkinson's disease neurodegeneration. Proc Natl Acad Sci 100:5473–5478
Threlfell S, Cragg SJ (2011) Dopamine signaling in dorsal versus ventral striatum: the dynamic role of cholinergic interneurons. Front Syst Neurosci 5:11
Threlfell S, Tatjana L, Nicola JP, Katie AJ, Karl D, Stephanie JC (2012) Striatal dopamine release is triggered by synchronized activity in cholinergic interneurons. Neuron 75:58–64
Tsukamoto T, Kitano Y, Kuno S (2013) Blood pressure fluctuation and hypertension in patients with Parkinson's disease. Brain Behav 3(6):710–714
Varanese S, Birnbaum Z, Rossi R, Di Rocco A (2011) Treatment of advanced Parkinson's disease. Parkinsons Dis 2010:480260
Walker MC, Semyanov A (2007) Regulation of excitability by extrasynaptic GABAA receptors. In: Darlison MG (ed) Inhibitory regulation of excitatory neurotransmission. Springer-Verlag, Berlin, pp 29–48
Wang J, Si YM, Liu ZL, Yu L (2003) Cholecystokinin, cholecystokinin-A receptor and cholecystokinin-B receptor gene polymorphisms in Parkinson's disease. Pharmacogenetics 13(6):365–369

Wang W, Wang X, Fujioka H, Hoppel C, Whone AL, Caldwell MA et al (2016) Parkinson's disease-associated mutant VPS35 causes mitochondrial dysfunction by recycling DLP1 complexes. Nat Med 22(1):54–63

Werner FM, Coveñas R (2011) Classical neurotransmitters and neuropeptides involved in generalized epilepsy: a focus on antiepileptic drugs. Curr Med Chem 18:4933–4948

Werner FM, Coveñas R (2012) Possible therapeutic options in Parkinson's disease according to a neuronal network. Neurol Rehabil 18:420–421

Winner B, Jappelli R, Maji SK, Desplats PA, Boyer L, Aigner S et al (2011) In vivo demonstration that alpha-synuclein oligomers are toxic. Proc Natl Acad Sci 108:4194–4199

Winslow AR, Chen CW, Corrochano S, Acevedo-Arozena A, Gordon DE, Peden AA et al (2010) α-Synuclein impairs macroautophagy: implications for Parkinson's disease. J Cell Biol 190: 1023–1037

Yamakage M, Namiki A (2002) Calcium channels-basic aspects of their structure, function and gene encoding; anesthetic action on the channels—a review. Can J Anesth 49:151–164

Yue M, Hinkle KM, Davies P, Trushina E, Fiesel FC, Christenson TA et al (2015) Progressive dopaminergic alterations and mitochondrial abnormalities in LRRK2 G2019S knock-in mice. Neurobiol Dis 78:172–195

Zecca L, Wilms H, Geick S, Claasen JH, Brandenburg LO, Holzknecht C et al (2008) Human neuromelanin induces neuroinflammation and neurodegeneration in the rat substantia nigra: implications for Parkinson's disease. Acta Neuropathol 116:47–55

Zeng BY, Iravani MM, Jackson MJ, Rose S, Parent A, Jenner P (2010) Morphological changes in serotoninergic neurites in the striatum and globus pallidus in levodopa primed MPTP treated common marmosets with dyskinesia. Neurobiol Dis 40(3):599–607

Emerging Roles of TREM2 in Neurodegenerative Diseases

7

Md. Tanvir Kabir

Abstract

Variants of triggering receptor expressed on myeloid cells 2 (*TREM2*) have been detected as important risk factors for various neurodegenerative diseases (NDs) including Alzheimer's disease (AD). Variants of *TREM2* are expressed predominantly in myeloid cells that significantly increase AD risk, associating innate immune system, and microglia as key factors in case of AD pathological process. Furthermore, detection of these *TREM2* variants exclusively confirms the active contribution of an immune response in the pathological processes of NDs. Indeed, these *TREM2* variants involve the highest risk for AD development; therefore, better understanding regarding TREM2 activity may give important insights regarding ND pathogenesis and mediate opportunities for the development of novel immune-associated ND therapeutics and biomarkers. TREM2's function, signaling, and expression pattern in case of NDs have been widely studied to find out the function of immune function in the progression and pathogenesis of diseases. It has been demonstrated that TREM2 has a proinflammatory role and its contribution to inflammation is complex. In this chapter, TREM2's structure, expression pattern, regulation, and signaling have been discussed. Moreover, TREM2-dependent cellular processes, actions of disease-linked mutations on TREM2 activity, roles of TREM2 in NDs, and possible therapeutic interventions have also been presented.

Keywords

TREM2 · Neurodegenerative diseases · Frontotemporal dementia · Alzheimer's disease · Parkinson's disease · Amyotrophic lateral sclerosis · Biomarkers

M. T. Kabir (✉)
Department of Pharmacy, Brac University, Dhaka, Bangladesh

G. M. Ashraf, M. S. Uddin (eds.), *Current Thoughts on Dementia*,
https://doi.org/10.1007/978-981-16-7606-2_7

7.1 Introduction

Various human genetic analyses have provided vital insights regarding the pathogenesis of several neurodegenerative diseases (NDs). Indeed, Alzheimer's disease (AD) is an important example of how progress in genetic technologies have improved our knowledge regarding the etiology of various NDs. Former studies used genetic linkage methods to identify familial mutations in proteins that are associated with the amyloid beta (Aβ) generation, presenilin 1 and presenilin 2, amyloid precursor protein (APP), and late-onset AD (LOAD) risk variant apolipoprotein E4 (APOE4) (Guerreiro and Hardy 2014; Kabir et al. 2020c). It has been observed in these studies that Aβ plays a vital role in AD pathogenesis. So far, numerous novel AD-linked genetic variants have been detected via case-control genome-wide association studies (GWAS) (Grupe et al. 2007; Karch and Goate 2015). Even though most of these studies discuss solely regarding the modestly raised risk for AD development, but generally these studies give wide insight regarding the LOAD-associated mechanisms and pathways. Numerous genetic linkages have been detected that are associated with modifying immune activity (Wes et al. 2016), which is confirming a vital contribution of the immune function in case of AD (Uddin et al. 2020c, f). In recent times, next-generation sequencing techniques are making it possible to identify rare variants, even some of them might confer an increased risk of disease and thus can offer significant understanding regarding genes that play significant roles in disease (Guerreiro and Hardy 2014). Use of GWAS with imputation and exome sequencing (Guerreiro et al. 2013b) based on anticipated genetic links (Jonsson et al. 2013) with AD resulted in the detection of comparatively rare variants in triggering receptor expressed on myeloid cells 2 (*TREM2*) gene that are linked with an increased risk for the development of AD.

Interestingly, heterozygous *TREM2* variants offer comparable AD risk as one *APOE4* copy. AD-linked variants of *TREM2* are mainly coding variants, which are contrary to most of the single nucleotide polymorphisms (SNPs) detected in GWAS (Cuyvers and Sleegers 2016). In addition, variants of *TREM2* were found to be associated with other NDs, which are indicating that TREM2 plays a vital role in shared disease processes. Detection of these AD-linked variants of TREM2 also has provided a strong relationship between the pathogenesis of NDs and innate immune system. It was reported that the immune system is impaired in case of AD and other NDs. However, it was not well reported whether this aforesaid impairment played role in the pathogenesis and progression of the diseases or was simply a secondary response to AD-associated pathogenesis. It has been observed that TREM2 is most commonly found in immune cells, which is suggesting that immune system impairment can play a significant role in the pathogenesis of NDs (Neumann and Takahashi 2007; Ransohoff 2016). Therefore, ND-linked variants of TREM2 can provide more opportunity to identify the vital functions that the immune system has in case of neurodegeneration (Hickman and El Khoury 2014). Following identification of AD risk-associated TREM2 variants, numerous researchers have developed various study techniques that primarily focused on understanding the structure, function, signaling, expression, and genetics of TREM2, and its link with pathogenesis of NDs and utilized these outcomes to clinical therapeutics and biomarkers.

Advances in these areas have improved our knowledge of TREM2 receptor biology. Earlier it was regarded that expression of TREM2 was reduced via proinflammatory stimuli and facilitated anti-inflammatory activities, but its activities are more complex. It has been revealed by in vitro studies that inflammatory stimuli reduce the expression of TREM2; however, in vivo studies showed that expression of TREM2 is elevated in inflammatory settings. Many studies have reported that TREM2 exerts proinflammatory action, which is indicating the context- and cell type-dependent activities of the receptor. In recent times, analyses have revealed the new features of TREM2 biology, which require reinterpretation and reassessment of former studies. In a disease progression-dependent manner, a soluble form of TREM2 is generated in case of AD (Suárez-Calvet et al. 2016a, b). Also, this soluble TREM2 (sTREM2) might possess different biological activities (Zheng et al. 2017; Zhong et al. 2017). Studies are also intensely studying the other fundamental features of TREM2 biology, such as ontogenesis of TREM2-expressing cells in the brain, post-translational and epigenetic modification of TREM2 that can influence activity and expression, and how noncanonical signaling mechanisms might play roles in the functions of TREM2 (Gervois and Lambrichts 2019; Zhong and Chen 2019). In this chapter, the roles of TREM2 in several neurodegenerative diseases, as well as therapeutic inventions are discussed.

7.2 Expression Pattern and Structure of TREM2

TREM2 is a member of a family of cell surface receptors known as triggering receptors expressed on myeloid cells (TREM). TREM family members are cell-surface transmembrane glycoproteins containing cytoplasmic tails and extracellular domains of V-immunoglobulin (Bouchon et al. 2000). *TREM2* genes encode a 230-amino acid glycoprotein and are found on human chromosome 6p21 (Allcock et al. 2003). In addition, *TREM2* genes are expressed in a subgroup of myeloid cells such as granulocytes, dendritic cells, and tissue-specific macrophages including osteoclasts, alveolar macrophages, and Kupffer cells (Bouchon et al. 2001; Cella et al. 2003; Paloneva et al. 2003; Humphrey et al. 2006; Koth et al. 2010; Gonçalves et al. 2013; Hu et al. 2014). Microglia induce the expression of TREM2 in the brain (Sessa et al. 2004; Kiialainen et al. 2005; Takahashi et al. 2005; Neumann and Takahashi 2007; Garcia-Alloza et al. 2013; Forabosco et al. 2013), but there is a lack of consistency about the level of its translation and expression (Schmid et al. 2002; Satoh et al. 2011; Jay et al. 2015; Lue et al. 2015) in humans and mice. TREM2 expression differs based on the specific areas of the central nervous system (CNS) (Sessa et al. 2004; Chertoff et al. 2013). Among these areas, an increased expression was observed in white matter, spinal cord, and hippocampus (Forabosco et al. 2013). Furthermore, TREM2 expression is modulated via inflammation, but inflammatory activities were found to be the opposite in case of in vitro and in vivo *studies*. It has been observed that TREM2 expression can be increased by the expression of anti-inflammatory molecules (Turnbull et al. 2006), whereas expression of TREM2 was decreased in vitro by the expression of proinflammatory molecules such as tumor necrosis factor-alpha, lipopolysaccharide (LPS), or interleukin 1 beta (Bouchon et al. 2001; Bhattacharjee et al. 2016; Zheng et al. 2016). Expression of TREM2 is increased in case of various NDs including AD (Lue et al. 2015; Celarain et al. 2016; Ma et al. 2016; Perez et al. 2017), amyotrophic lateral sclerosis (ALS) (Cady et al. 2014), and Parkinson's disease (PD) (Liu et al. 2016). Increased TREM2 expression has been demonstrated in individuals with AD (Lue et al. 2015; Celarain et al. 2016; Ma et al. 2016; Perez et al. 2017) and mouse models containing tau and amyloid pathology (Jiang et al. 2014; Matarin et al. 2015; Jay et al. 2015; Savage et al. 2015; Raha et al. 2016) and appears to be linked with the employment of microglia in Aβ plaques (Perez et al. 2017; Yeh et al. 2017). Aging also elevates the expression of TREM2 in humans and mouse models (Forabosco et al. 2013; Raha et al. 2016).

7.3 Regulation of TREM2 Signaling Pathways

TREM2 contains an extracellular Ig-like ligand-binding domain. On the other hand, a transmembrane region of TREM2 contains a positively charged lysine residue that has been reported to interact with TYRO protein tyrosine kinase-binding protein (TYROBP) (a signaling partner of TREM2). Nevertheless, no activity has been reported regarding the TREM2's short cytoplasmic tail. Ligation of TREM2 stimulates the phosphorylation of the immunoreceptor tyrosine-based activation motif (ITAM) in TYRO protein tyrosine kinase-binding protein (TYROBP) through Src family kinases. In cells of humans and mice, it has been confirmed that the tyrosine-phosphorylated ITAM domain employs spleen tyrosine kinase (SYK) and/or Zeta-chain-associated protein kinase 70 to transduce downstream signals via intracellular calcium signaling pathway, phosphoinositide 3-kinase (PI3K), extracellular signal-regulated kinase, and actin reorganization pathway, which can further result in various cellular functions such as induction of phagocytosis and transcriptional activation by nuclear factor of activated T cells (NFAT) (Bouchon et al. 2001; Takahashi et al. 2005; N'Diaye et al. 2009; Peng et al. 2010; Wang et al. 2015) (Fig. 7.1). It has been suggested that the signaling adaptor DNAX-activating Protein 10 (DAP10) and DNAX-activating protein of 12 kDa (DAP12) have mediated signal transduction in TREM2 (Fig. 7.1). Besides, DAP10 is essential for the PI3K recruitment in the membrane signaling complex (Peng et al. 2010). Interestingly, SH2 domain-containing inositol phosphatase limited TREM2-TYROBP signaling via binding with TYROBP and prevented the recruitment of PI3K. Although it is well understood how calcium signaling mediates the NFAT translocation to the nucleus for downstream transcriptional target activation (Gwack et al. 2007), while more studies are required regarding the dynamic cytoskeletal and membrane activities through which phagocytosis is attained.

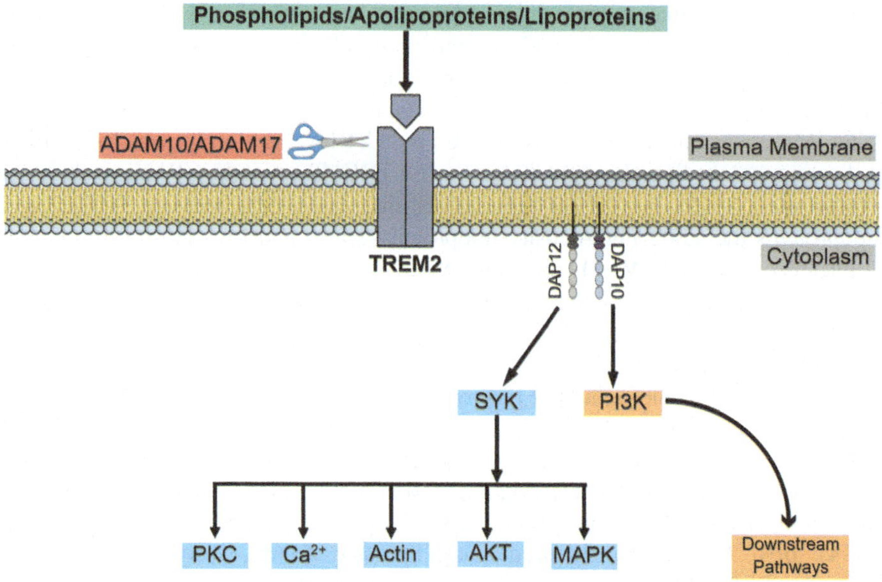

Fig. 7.1 TREM2 Signaling Pathways. TREM2 interacts with lipoproteins, apolipoproteins, and phospholipids, which then leads to signal propagation through DAP10 and DAP12, which further leads to activation of PI3K or SYK, respectively. On the surface, TREM2 cleavage via ADAM10/17 results in sTREM2 and halts TREM2 signaling event. Furthermore, SYK initiates downstream signaling cascades including mobilization of PKC, Ca^{2+}, MAPK, actin, and AKT. ADAM10/17, disintegrin and metalloproteinase domain-containing protein 10/17; AKT, protein kinase B; DAP10, DNAX-activating Protein 10; DAP12, DNAX-activating protein of 12 kDa; MAPK, mitogen-activated protein kinase; PI3K, phosphoinositide 3-kinase; PKC, protein kinase C; sTREM2, soluble TREM2; SYK, spleen tyrosine kinase; TREM2, triggering receptor expressed on myeloid cells 2

7.4 TREM2-Dependent Cellular Responses

Following agonists (such as zymosan, CpG, and LPS)-mediated Toll-like receptor (TLR) stimulation, former experiments with the cellular activities of TREM2 showed that bone marrow-derived macrophages (which were isolated from TREM2-deficient mouse models) secreted more inflammatory cytokines including interleukin-6 and tumor necrosis factor in comparison with the wild-type cells (Hamerman et al. 2006; Turnbull et al. 2006). Furthermore, following TLR stimulation, exaggerated response was observed in TYROBP-deficient bone marrow-derived mouse macrophages, which is indicating that the TREM2-TYROBP signaling pathway may have contribution to anti-inflammatory activity via suppressing the TLR signaling pathway (Hamerman et al. 2005; Turnbull et al. 2006). In human monocyte-derived dendritic cells, TREM2 activation via antibody ligation stimulates intracellular mobilization of calcium (Bouchon et al. 2001) and subsequent in vitro nuclear factor of activated T cells (NFAT) activation (Colonna

2003). Calcium signaling pathway was found to be activated in NFAT reporter cell line expressing TREM2 via various phospholipids (such as phosphatidylserine, phosphatidylglycerol, phosphatidylcholine, phosphatidylinositol, and phosphatidic acid), and glycolipids from cells and myelin (sphingomyelin, sulfatide, and cerebroside). Moreover, this activity was blocked by a TREM2 antibody, which is further indicating that the ligands might directly cause TREM2 activation (Wang et al. 2015; Poliani et al. 2015). In a NFAT reporter cell line, interaction between TREM2 and lipoproteins including high-density lipoprotein (HDL) and low-density lipoprotein (LDL) also activated TREM2 signaling pathway (Song et al. 2017).

The role of TREM2 has also been linked with phagocytosis. TREM2 knockdown in mice decreased phagocytosis of microsphere beads and apoptotic neurons via microglia (Takahashi et al. 2005). It has been confirmed by various studies that TREM2 knockout microglia and macrophages decreased the capacity to phagocytose bacteria, cellular debris, and apoptotic neurons (N'Diaye et al. 2009; Kleinberger et al. 2014; Atagi et al. 2015). Furthermore, gain-of-function experiments confirmed that TREM2 expression in heterologous cells enabled internalization of bacteria and apoptotic cells (N'Diaye et al. 2009; Kleinberger et al. 2014). In a TREM2-dependent manner, lipidated APOE, acetylated LDL, and clusterin (CLU/APOJ) were internalized into cells in HEK-293 cells containing the expression of inducible TREM2. Interestingly, microglia cultured from TREM2 knockout mouse models exerted decreased uptake rates for CLU and LDL (Yeh et al. 2016). Interaction between APOE isoforms and TREM2 caused no noticeable alteration in the extent of internalization of lipidated APOE2, APOE3, or APOE4 via TREM2-expressing HEK-293 cells (Yeh et al. 2016). As CLU, APOE, and lipoprotein particles have been linked with the soluble form of Aβ (Strittmatter et al. 1993; Calero et al. 2000), CLU and APOE transported soluble form of Aβ to microglia and helped in Aβ phagocytosis via TREM2 binding. As compared to the free form of Aβ, Aβ aggregates complexed either with LDL or lipidated CLU were more effectively taken up via microglia (Yeh et al. 2016). Indeed, TREM2 knockout microglia markedly decreased uptake of Aβ-lipoprotein complexes and, particularly, heterozygous knockout microglia were moderately weakened in this process (Yeh et al. 2016). Therefore, it needs to be considered that whether or not presence of CLU and APOE in Aβ plaques (Namba et al. 1991; Calero et al. 2000) may help in TREM2-facilitated phagocytosis of insoluble Aβ through microglia.

7.5 Neurodegenerative Disease-Associated Mutations in TREM2

Interaction between human TREM2 and its various lipoprotein ligands (such as CLU, APOE, and LDL) was weakened due to AD-associated TREM2 mutations (including D87N, R62H, and R47H) and was totally eradicated via NHD/FTD-associated mutations (T66M and Y38C), in direct binding assays with purified recombinant proteins (therefore prevented cell expression and trafficking) (Atagi et al. 2015; Bailey et al. 2015; Yeh et al. 2016). Collectively, these observations suggest that disease-associated mutations are loss-of-function mutations in nature and these findings indicate the notion that weakened binding of lipoproteins might underlie the disease risk of these variants of TREM2 (Table 7.1). In comparison with the macrophages from matched control subjects containing wild-type TREM2, blood monocyte-derived macrophages obtained from human subjects heterozygous for the TREM2 R62H variant exhibited decreased uptake of Aβ–lipoprotein complexes (Yeh et al. 2016). Indeed, these findings also increase the chance that weakened uptake of Aβ–lipoprotein complexes might play role in the elevated AD risk in individuals containing TREM2 mutations. It has been estimated via NFAT reporter assays that LDL, HDL, phosphatidylinositol, phosphatidylserine, phosphatidylglycerol, phosphatidic acid, and sulfatide signaling via TREM2 were decreased with R62H and R47H mutations (Wang et al. 2015; Song et al. 2017). NFAT reporter function was increased by the TREM2 T96K and D87N mutants, which is indicating a gain-of-function activity of these mutations (Table 7.1) (Song et al. 2017). In order to identify the mechanistic relationship between AD and TREM2 variants, more data are required on how these variants influence in vivo TREM2 signaling mechanism, disease pathogenesis in preclinical models, and AD risk in the human population.

7.6 TREM2-Mediated Demyelination and Remyelination

Loss of axons and myelin detected in Nasu-Hakola disease (NHD) indicates that TREM2 might be associated with the white matter maintenance. Levels of TREM2 mRNA and protein were found to be increased in microglia of the spinal cord in the experimental autoimmune encephalomyelitis (EAE) mouse model of multiple sclerosis (MS) (Piccio et al. 2007). In mouse models, intraperitoneal injection of anti-TREM2 blocking antibody exacerbated the severity of the disease, which is suggesting that TREM2 may have a protective function in the disease (Piccio et al. 2007). At the peak of the disease, introduction of myeloid cells overexpressing TREM2 decreased the severity of the disease in comparison with the control myeloid cells (Takahashi et al. 2007). TREM2-overexpressing cells facilitated to restrict axonal damage and demyelination in the spinal cord (Takahashi et al. 2007).

Collectively, these findings indicate that TREM2 may exert protective function in MS (as like AD), which is further suggesting that therapeutic agents targeting TREM2 might be beneficial in MS treatment. TREM2-deficient mouse models showed less activated microglia in the corpus callosum in comparison with the wild-type mouse models in the cuprizone-stimulated model of demyelinating disease, which is indicating the defect in myelin removal and rise in axonal pathology as confirmed in an immunohistochemistry study (Cantoni et al. 2015). After the injury, remyelination in TREM2 knockout mouse models was also affected, which was also observed in microglial transcriptional response for genes associated with lipid metabolism and phagocytosis (Poliani et al. 2015). TREM2 ligands including cerebroside, sulfatide, and sphingomyelin might be secreted and cause activation of signaling mechanisms that mediate removal of damaged myelin and help

Table 7.1 Variants of human TREM2

Amino acid change	Single-nucleotide polymorphism	Surface expression	Signaling	Lipoprotein binding	Lipoprotein uptake	Reference
Q33X	rs104894002	Undetermined	Undetermined	Undetermined	Undetermined	Klünemann et al. (2005), Guerreiro et al. (2013c)
Y38C	NA	Eradicated	Undetermined	Eradicated	Weakened	Kleinberger et al. (2014), Yeh et al. (2016)
R47H	rs75932628	Weakened	Weakened	Weakened	Weakened	Guerreiro et al. (2013b), Jonsson et al. (2013), Cuyvers et al. (2014), Kleinberger et al. (2014), Jin et al. (2014), Wang et al. (2015), Atagi et al. (2015), Bailey et al. (2015), Yeh et al. (2016), Song et al. (2017), Del-Aguila et al. (2019)
R62H	rs143332484	Undetermined	Weakened	Weakened	Weakened	Guerreiro et al. (2013b), Cuyvers et al. (2014), Jin et al. (2014), Yeh et al. (2016), Song et al. (2017)
T66M	rs201258663	Eradicated	Abolished	Abolished	Weakened	Kleinberger et al. (2014), Yeh et al. (2016), Song et al. (2017)
D86V	NA	Undetermined	Undetermined	Undetermined	Weakened	Guerreiro et al. (2013c)
D87N	rs142232675	Undetermined	Increased	Weakened	Weakened	Guerreiro et al. (2013b), Jin et al. (2014), Yeh et al. (2016), Song et al. (2017)
T96K	rs2234253	Undetermined	Increased	Undetermined	Undetermined	Guerreiro et al. (2013b), Jin et al. (2014), Song et al. (2017)
R136Q	rs149622783	Undetermined	Undetermined	Undetermined	Undetermined	Guerreiro et al. (2013b), Song et al. (2017)
W198X	rs2234258	Undetermined	Undetermined	Undetermined	Undetermined	Giraldo et al. (2013)
L211P	rs2234256	Undetermined	Wild-type	Undetermined	Undetermined	Jin et al. (2014), Song et al. (2017)

remyelination of oligodendrocytes via the secretion of growth factors (Miron et al. 2013). Incapacity to detect myelin damage and trigger proper regeneration and clearance programs in the nonappearance of TREM2 might elucidate the demyelinating lesions present in the white matter pathogenesis of NHD (Verloes et al. 1997).

7.7 Role of TREM2 in Neurodegenerative Diseases

7.7.1 Frontotemporal Dementia

Biallelic mutations in the *TREM2* gene have been identified in 10 families with frontotemporal dementia (FTD) (without polycystic lipomembranous osteodysplasia with sclerosing leukoencephalopathy (PLOSL) bone phenotypes) (Chouery et al. 2008; Giraldo et al. 2013; Guerreiro et al. 2013a, c; Chee et al. 2017; Redaelli et al. 2018; Carmona et al. 2018). Previously, some of these variants were found to trigger PLOSL, while other variants were newly detected causes of the disease. Mutation type and linked phenotype were not linked, since missense, splice-site, and nonsense mutations were found to trigger both the diseases (Guerreiro et al. 2013b). Personality change with dementia, resembling behavioral variant FTD, appears to be a very common characteristic of all FTD individuals containing biallelic *TREM2* mutations. Family history of consanguinity along with atypical characteristics, including corpus-callosum atrophy and seizures, in an individual with a behavioral FTD phenotype, ought to prompt the detection of biallelic mutations in the gene. Also, this genetic cause ought to be introduced in the differential diagnosis of individuals with early-onset dementia containing seizures (Guerreiro et al. 2013b). Indeed, novel heterozygous variants of *TREM2* in individuals with clinical presentations that indicate FTD ought to be carefully evaluated and not supposed to be disease causative (Pesaran et al. 2016; Starita et al. 2017).

There is a controversy regarding the relationship between heterozygous variants of TREM2 and FTD risk. As compared to healthy controls, p.Thr96Lys variant was linked with FTD subtypes (Thelen et al. 2014), and an increased burden of heterozygous variants of TREM2 was observed in FTD patients (Borroni et al. 2014; Cuyvers et al. 2014). In a North American cohort, p.Arg47His variant was linked with risk for FTD (Rayaprolu et al. 2013). Moreover, a large meta-analysis of individuals containing p.Arg47His did not find an important link with FTD risk (Lill et al. 2015). Collectively these results indicate a link between FTD risk and *TREM2* at the gene level; however, no reliable relationships were observed at the variant level. In addition, these indecisive findings are perhaps because of the rarity of variants of *TREM2* and small cohort analyses in which cases might not be well characterized.

7.7.2 Alzheimer's Disease

Alzheimer's disease (AD) is a ND and clinical characteristics of AD include progressive loss of memory and cognitive declines that are observed along with neurofibrillary tangles (NFTs) and pathological deposition of senile plaques in the brain (Duyckaerts et al. 2009; Jankowsky and Zheng 2017; Kabir et al. 2020b, d; Uddin et al. 2020d). Indeed, AD is an age-associated disorder that occurs at an increased rate in individuals over 65 years. It has been projected that the occurrence rate of AD doubles every five years. Furthermore, it has been expected that one new AD case will develop every 33 s by 2050, which will lead to about 115 million cases in total and one million new cases every year (Gaugler et al. 2016). Since there are huge social and economic burdens linked with the care of individuals with AD, thus noticeable efforts have been made to develop early diagnostic techniques, recognize AD pathogenesis, and identify effective therapeutic agents (Kabir et al. 2020a; Uddin et al. 2020a, e, f). It has been suggested by genetic studies that the innate immune system has a significant contribution to AD pathogenesis. As TREM2 variants involve increased genetic risk in case of AD, thus numerous experiments have focused on TREM2 to detect the activities of the immune system in AD (Jay et al. 2017; Shi and Holtzman 2018; Wolfe et al. 2019). AD-associated TREM2 variants are located in its extracellular ligand-binding domain, interfere with signaling and binding of newly identified ligands including polyanionic phospholipids and lipoproteins (including CLU and APOE) (Dean et al. 2019; Yang et al. 2020). Receptor engagement can result in uptake of its ligands and mediates signaling that can result in cellular responses toward Aβ, microglia survival, and the generation of a protective barrier of microglia around Aβ plaque (Fig. 7.2).

It is clear from the extensive mouse and genetic studies that TREM2-induced actions are closely associated with NDs, injury repair, and aging (Yeh et al. 2017). Even though variants of TREM2 associated with NHD are totally impaired, but variants that are linked with altered function of TREM2 are linked with regulation of risk for LOAD (Ulrich et al. 2017). In this regard, crossing TREM2-deficient mouse models with neurodegeneration models revealed certain TREM2 activities including mediating survival of microglia (Ulland et al. 2017) and their phenotypic transition to disease-associated microglia (DAM), a subtype limiting growth of plaque via making a physical barrier around aggregates of Aβ, which is characterized through increased phagocytosis and signature of lipid metabolism (Keren-Shaul et al. 2017). PS2APP transgenic mouse models of AD were compared to TREM2-deficient PS2APP mouse models and it was revealed that exacerbated neuritic dystrophy is a more steady consequence of TREM2 lacking as compared to load of Aβ plaque, which is indicating the role of DAM in packing Aβ into dense plaques is a significant neuroprotective function (Meilandt et al. 2020; Uddin et al. 2020g).

DAM was also detected in other neurodegeneration models, such as EAE, MS model, ALS mouse model, and natural aging (Keren-Shaul et al. 2017; Krasemann et al. 2017; Deczkowska et al. 2018). In GWASs, unique transcriptomic characteristic of DAM presented multiple other genes linked with AD risks, such as a potential modifier of TREM2 signaling (CD33) and TREM2 ligands (CLU/APOJ and APOE)

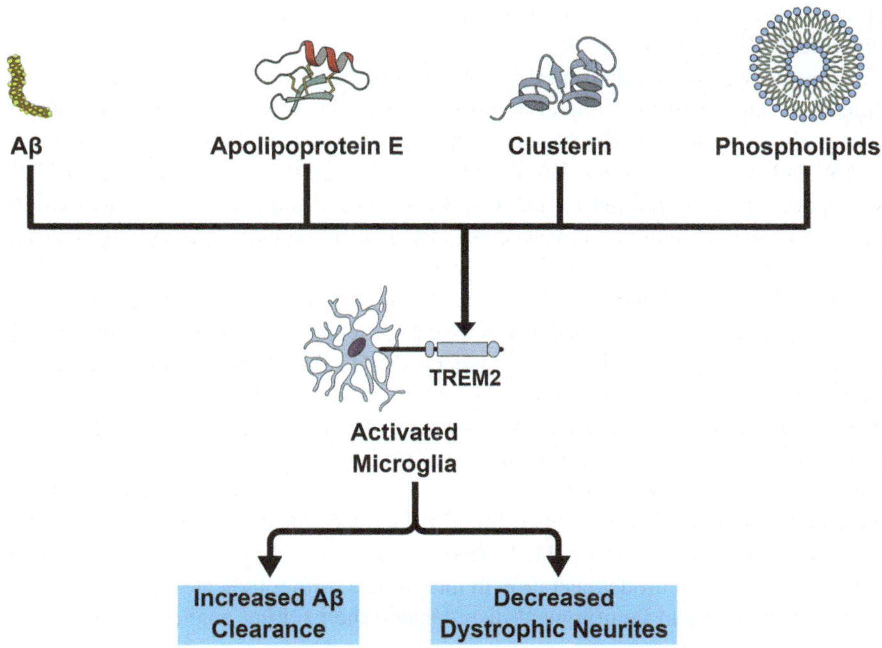

Fig. 7.2 Role of TREM2 in Alzheimer's disease. Lipoproteins, apolipoproteins (apolipoprotein E or clusterin), and phospholipids bind with TREM2, which promote microglial activation, survival, and generation of a microglia barrier around Aβ plaques. Furthermore, the link of Aβ with apolipoproteins might aid the delivery of these complexes to microglia, which further results in increased Aβ clearance. In addition, TREM2 activity can also prevent neuritic dystrophy

(Yeh et al. 2016; Deczkowska et al. 2018; Griciuc et al. 2019; Bhattacherjee et al. 2019). Collectively, these findings indicate a significant function of DAM and TREM2 in AD and possibly in other neurodegeneration cases. In microglia, targeted human TREM2 overexpression in an AD mouse model improved pathological phenotypes, such as cognitive deficit and Aβ plaque load (Lee et al. 2018; Uddin et al. 2020b). R47H is the most common AD-associated TREM2 variant, which is a mutation that confers loss of ligand-binding capacity, and therefore phagocytosis of TREM2-interacting ligands, including CLU/APOJ (Yeh et al. 2016), APOE, and Aβ (Zhong et al. 2018), and plaque-linked lipids (Wang et al. 2015). In diseases of CNS, complex association of R47H and other variants of *TREM2* gene and their precise role in disease outcomes and microglial phenotypes have been explained (Yeh et al. 2017; Hansen et al. 2018; Ulland and Colonna 2018; Hammond et al. 2019). In human brains, staining for the DAM markers CD11c and LPL detects their existence near Aβ plaques (Keren-Shaul et al. 2017), single-nucleus RNA-seq (snRNA-seq) and scRNA-seq experiments are inconsistent about the presence of DAM in case of human AD. No difference was observed between control human microglia and AD in a study (Alsema et al. 2020), while others observed DAM-like signature (Mathys et al. 2019; Grubman et al. 2019) or different signature in the brains of AD individuals (Zhou et al. 2020).

It has been observed that interferon regulatory factor 8 induces the human DAM signature, particularly peripheral nerve injury-related signature (Zhou et al. 2020). Comparison of microglia obtained from AD individuals containing common variants of TREM2 or the risk variants R62H and R47H showed that fully functional TREM2 still appears to be important for DAM development (Zhou et al. 2020). Inconsistent findings of the signature and presence of DAM might be elucidated by the observation that separation of nuclei and single cells from postmortem brain tissue involves severe technical challenges (Bakken et al. 2018; Frigerio 2020). In human brains, comparison of snRNA-seq and scRNA-seq particularly revealed that snRNA-seq could not identify the major genes associated with the DAM signature, including SPP1, CST3, and APOE (Frigerio 2020). Unlike samples of AD brains, scRNA-seq analysis of human adipose tissue that can be easily derived from individuals (which also provides acceptable yield of high-quality cells) showed an increased level of similarity between the human and mouse lipid-associated macrophage signature, which in mouse models have been found to be mechanistically reliant on TREM2 signaling (Jaitin et al. 2019). Furthermore, finding of altered sTREM2 levels in various aspects of neurodegeneration in individuals further indicates the significance of TREM2 in human neurological disorder and the exciting chance of therapeutic TREM2 targeting.

7.7.3 Parkinson's Disease

After AD, PD is the most common ND and around one percent of the population over the age of 60 is affected by this ND. PD is usually regarded as a disorder of motor function. While in diagnosis, 1 in 4 individuals meet criteria for mild cognitive impairment (Muslimović et al. 2005; Aarsland et al. 2009) and around 80% ultimately develop dementia during PD (Aarsland et al. 2003; Hely et al. 2008; Cholerton et al. 2013). It has been observed that dopamine therapy could not ameliorate these nonmotor symptoms and play a significant role in late morbidity, mortality, and loss of quality of life. Even though in initial diagnosis, about 6.5% of PD individuals showed aberrant cerebrospinal fluid (CSF) levels of tau and Aβ-42 (Marek et al. 2018) at autopsy, but about 60–80% of PD patients were likely to develop brain pathology as like AD, along with a significant buildup of Aβ plaques and/or tau-containing NFTs (Tsuang et al. 2013; Robinson et al. 2018; Dickson et al. 2018).

It has been reported that the advancement of cognitive deficit and Parkinsonism is increased in patients with AD and/or cerebrovascular disease (Tsuang et al. 2013; Dickson et al. 2018; Buchman et al. 2019). For clinical trials, identification of a biomarker that will be able to predict the alteration in PD-associated cognitive function may provide an important tool for the clinical management and measurement of outcomes. Since PD is heterogeneous, thus there is an urgent need for the development of novel biomarkers that may analyze PD patient subgroups as per the underlying copathologies to give more details on the individual clinical trajectory (Chen-Plotkin et al. 2018). After primary results of a link between AD and *TREM2* in 2013, a link has also been reported between risk of PD and p.Arg47His (Benitez and Cruchaga 2013). Nonetheless, in this study, occurrence of variant in healthy controls was lesser as compared to what would be anticipated frequencies in a known population. As frequency of p.Arg47His variant differs across population, thus association tests should only be done (for this variant and others), in adequately large-scale and well-matched case-control groups (Reitz et al. 2013). It has been observed that there is an important link between p.Arg47His and risk for PD in a discovery cohort of North American PD individuals and healthy controls. Nonetheless, in that same study (Rayaprolu et al. 2013) or other association studies (Lill et al. 2015; Mengel et al. 2016), these findings could not be replicated. In a group of Chinese individuals, this link was also studied and no noticeable activity on risk was observed, and no variant was detected in these individuals (Feng et al. 2014; Chen et al. 2015; Li et al. 2016; Tan et al. 2016).

7.7.4 Polycystic Lipomembranous Osteodysplasia with Sclerosing Leukoencephalopathy

PLOSL is a rare inherited disorder and the characteristics of this disorder include early-onset dementia and spontaneous bone fractures (Paloneva et al. 2002). It is estimated that the occurrence rate of PLOSL is 1–2 cases per one million individuals in Finland (Pekkarinen et al. 1998). Over 100 PLOSL cases have been detected in Japanese individuals (Pekkarinen et al. 1998), even though PLOSL seems to be less common in other parts of the world (Klünemann et al. 2005). Indeed, PLOSL is a genetically heterogeneous syndrome that takes place due to biallelic (homozygous or compound heterozygous) mutations in two genes including *TYROBP* and *TREM2* (Paloneva et al. 2002). Eleven *TREM2* mutations have been found to be responsible to trigger PLOSL. The natural course of PLOSL is the same in cases triggered via *TREM2* or *TYROBP* mutations, which can be divided into four phases including latent, osseous, early neurological, and late neurological (Paloneva et al. 2002). First report associated mutations of *TREM2* with PLOSL (out of the 39 studied individuals), mutations of *TYROBP* were detected in 31 (79%) individuals, and mutations of *TREM2* in eight (21%) individuals (Paloneva et al. 2002). It was observed in 25 Finnish individuals that all these individuals carried similar 5.3 kb deletion encompassing exons 1–4 of *TYROBP* (Paloneva et al. 2002), which is further indicating that this mutation is an initial mutation in this population, and ought to be particularly studied in Finnish population. In a different genetic study, Finnish family carrying *TYROBP* deletion showed coexistence of other variants in neurologically linked genes, including an expansion of a *C9orf72* gene, which was regarded as a related finding along with no clinical significance (perhaps because of increased frequency of both genetic alterations in this population), and a novel *EPM2A* mutation, this gene is associated with chronic myoclonic epilepsy type 2 (Lafora disease), which is accountable for severe epilepsy in the individual who carried this mutation (Solje et al. 2014). AD pathogenesis has been assessed in PLOSL patients. Amyloid positron emission tomography scan showed severe deposition of Aβ in the gray matter of the occipital lobes and inferior frontal of an Italian individual who had a *TREM2* homozygous p.Gln33X mutation (Ghezzi et al. 2017).

Functional and neuropsychological nuclear imaging studies in heterozygous people in a family who carried this same mutation exhibited impaired visuospatial memory linked with hypoperfusion in the basal ganglia, which is suggesting an overlap of pathogenic processes between PLOSL and AD (Montalbetti et al. 2005). Nonetheless, immunohistochemistry studies of brains if five PLOSL patients could not reveal amyloid angiopathy or Aβ plaques. Furthermore, only a small number of tau tangle-containing neurons (commonly in hippocampus), which are indicating that loss of function of *TREM2* does not induce AD pathogenesis (Satoh et al. 2018). In addition, studies on *TYROBP* genetic variability did not show links with the cognitive deficit in a Finnish cohort (Kaivola et al. 2018). Similarly, in Turkish patients, *TYROBP* variability did not cause dementia (Darwent et al. 2017). Involvement of *TREM2* and *TYROBP* mutations in PLOSL resulted in the detection of signaling mechanisms accountable for this disease (Paloneva et al. 2002). Since these genes are found to be expressed in osteoclasts and microglia, which partially elucidated irregular tissue distribution of symptoms. Nevertheless, still it remains unclear why some individuals develop bone cysts and pathological fractures linked with dementia, while others exhibit FTD. Although it is tempting to establish that genetic modifiers play role in these disparate phenotypes, but it will be challenging to achieve sample sizes that are big enough to detect variants of modifier that possess such a low frequency in the population. In spite of these challenges, it is important to understand which factors affect these differences in presentation, since these factors might be protective and play a role as potential targets for therapies that will provide prevention or symptom management.

7.7.5 Amyotrophic Lateral Sclerosis

ALS is a fatal disorder that takes place ought to the chronic degeneration of lower and upper motor neurons. Pathological characteristics of ALS include activated microglia around the degenerating neurons (Engelhardt and Appel 1990); however, it is still unclear whether this activation is an injurious contributor or beneficial response to the disease process. In this regard, studies with mouse models revealed that microglia simultaneously express both neurotoxic and neuroprotective factors (Chiu et al. 2013) and might change from a neuroprotective phenotype at symptom onset to become more neurotoxic later in the disease process (Liao et al. 2012). Cady et al. (2014) demonstrated that p.R47H (a rare *TREM2* variant) significantly increases the risk for ALS. Furthermore, in order to detect a novel risk factor for ALS, this observation indicates first association between microglial activation and genetic variation in ALS pathogenesis. In a different study, it has been demonstrated that increased levels of microglial activation on the pathological study were associated with both a more rapid progression of disease and extent of upper motor neuron symptoms (Brettschneider et al. 2012). A similar trend was observed in the expression of TREM2 in ALS spinal cord, along with increased TREM2 levels correlated with shorter survival. Moreover, expression of TREM2 was also elevated in spinal cords from $SOD1^{G93A}$ mouse models is consistent with other studies of isolated microglia from this same model (Chiu et al. 2013), which further indicates that experiments of microglial activation in this model might deliver insights pertinent to human ALS.

Indeed, p.R47H increases the risk for various disorders including AD, FTD, and PD (Benitez et al. 2013; Gonzalez Murcia et al. 2013; Pottier et al. 2013; Rayaprolu et al. 2013; Guerreiro et al. 2013b; Jonsson et al. 2013). Still, it is not well-known how p.R47H influences TREM2 activity and inclines to neurodegeneration. As TREM2 signaling induces vital neuroprotective microglial functions including secretion of anti-inflammatory cytokines and phagocytosis of apoptotic cells, thus one model theorized that p.R47H variant is a loss-of-function allele. Counter-productive inflammation and insufficient removal of cellular debris are likely to incline to symptomatic disease. It has been reported that p.R47H variant is found in TREM2's extracellular domain, where it may interfere with binding to unknown ligand(s) or disturb signaling by TYROBP. Since TREM2 signaling dysregulation can confer risk for multiple NDs, thus observations obtained from the experiment of TREM2 in ALS may be appropriate for other disorders and vice versa. Furthermore, this involves a significant chance that manipulation of microglial activation or TREM2 signaling would be a useful therapeutic approach (Cady et al. 2014).

7.7.6 Other Neurodegenerative Diseases

It has been suggested that TREM2 is a member of a functional network involved in various NDs (Benitez and Cruchaga 2013). Therefore, multiple studies have evaluated the link between variants of *TREM2* and risk of developing NDs, other than AD. Various studies have assessed the activities of *TREM2* variants in risk for multiple system atrophy (Chen et al. 2015), ischemic stroke (Rayaprolu et al. 2013), progressive supranuclear palsy (Rayaprolu et al. 2013), Creutzfeldt-Jakob disease (Slattery et al. 2014), posterior cortical atrophy (Carrasquillo et al. 2016), Lewy body dementia (Walton et al. 2016), and essential tremor (Ortega-Cubero et al. 2015). However, no conclusive results have been obtained from these studies due to the rarity of *TREM2* variants and because only small groups of individuals have been studied. Therefore, the link between the risk of NDs (except AD) development and *TREM2* has not been strongly replicated so far. In order to explain these potential links, more studies are required containing larger cohorts of controls and cases.

7.8 Therapeutic Interventions to Tackle TREM2

As a therapeutic target, there are several potent interventions to tackle TREM2 pathway. Among them, the most promising one is to target the active domain of the receptor directly by a particular small molecule or monoclonal antibody that would activate or block the downstream signaling pathway. In case of AD, targeting TREM2 is considered the most advanced TREM2 therapeutic approach. Furthermore, biotech companies and researchers are focusing to activate the receptor to induce microglia to phagocytose and to remove the deposits of Aβ. In this regard, Alector, LLC has used AL002 (a TREM2 agonistic monoclonal antibody) which is now in phase I clinical trial for AD individuals (Deczkowska et al. 2020). Another effective intervention approach is to target condition- and/or tissue-specific TREM2 ligands to use the treatment where it is required. Our understanding is limited regarding such specific ligands of TREM2. Since TREM2 is continuously shed by the ADAM (a disintegrin and metalloproteinase), suppressing or inducing this mechanism is another approach of intervention (Voytyuk et al. 2018).

As there are several substrates for ADAM10/17, suppressing a particular activity of protease is usually linked with unwanted effects. Better understanding regarding the regulatory processes including regulation of expression, localization, and function of these metalloproteases will possibly reveal appropriate targets to permit more particular and fine-tuned regulation to block the shedding of TREM2. In a complementary approach, therapeutics targeted the TREM2's cleavage site to disturb receptor shedding via ADAM, which increased the membrane level of the receptor and increased its signaling function. In a different study, 4D9 (an antibody) was used that binds with the TREM2's stalk region and decreases its proteolytic shedding (Schlepckow et al. 2017). It has also been reported that 4D9 activates TREM2 signaling and elevates Aβ uptake and microglia survival. In an APP knockin mouse model of AD, it was observed that this antibody reduced the load of cortical plaques; however, any alterations in cognitive function were reported (Schlepckow et al. 2020). However, more studies are required to understand the signaling mechanism, its downstream effector genes, and regulators of TREM2 expression are likely to identify novel targets and alternative approaches to control TREM2 activity.

7.9 Recent Developments and Future Perspectives

Even though links between *TREM2* variants p.Arg62His and p.Arg47His and AD risk are well-known, but more studies are essential to establish probable links between other *TREM2* variants and AD, and the variants that are found within the broader *TREM* locus. In addition, it is also important to understand the link between variants of *TREM2* and the development of other NDs. Indeed, large-scale studies are required to fully understand the activity of *TREM2* variability in the genetic design of AD and other NDs. Moreover, more large-scale studies are essential to ameliorate our understanding regarding the activity of variants of *TREM2* in clinical presentation, neuropathology, and progression. So far, not many studies have indicated that AD patients carrying p.Arg47His exhibit fairly common AD symptoms, even though earlier age of onset and shorter duration of the disease as compared to commonly observed cases have been stated (Slattery et al. 2014; Korvatska et al. 2015). Besides, autopsy reports revealed decreased clustering of microglia around plaques in p.Arg47His carrying individuals (Yuan et al. 2016; Krasemann et al. 2017).

As there is a low occurrence rate of *TREM2* variants, thus it is difficult to arrange adequately larger cohorts to provide robust correlations between genotype-phenotype. Even though sTREM2 level is increased in the CSF of AD patients as compared to healthy controls, but levels of sTREM2 do not have the discriminative control essential for diagnostic techniques (Brosseron et al. 2018). Nonetheless, sTREM2 level in CSF seems to exert a dynamic reaction during the progression of the disease (Suárez-Calvet et al. 2016a, b), which is probably indicating microglial function. Indeed, sTREM2 has the potential to be a biomarker of disease progression if validated, mainly for the transition from preclinical AD to dementia. Biomarkers for Aβ and tau may be beneficial for the AD differential diagnosis. In addition, sTREM2 can be utilized to identify the disease characteristics and stage. More studies are also required to find out how variants of *TREM2* influence AD risk and result in FTD and PLOSL. It has been demonstrated that variants of *TREM2* that cause loss of protein function may weaken microglial activity to stressors, including AD-related Aβ accumulation (Ulland et al. 2017). Growing evidence indicates TREM2 as an effective therapeutic target. TREM2 overexpression in vivo improves neuropathological features of AD and AD-associated memory declines (Lee et al. 2018). Therefore, activation of TREM2 signaling mechanism may provide a new therapeutic technique, such as, via the use of small molecules that mimic TREM2 ligands.

During the course of the disease, it is important to estimate when it will be beneficial to increase the activation or expression of TREM2. In this regard, a study of AD's familial forms might be useful due to the likelihood of the age at symptom onset. Both peripherally and centrally, it is crucial to systematically characterize the activities of controlling TREM2 levels. Furthermore, TREM2-targeted therapeutics might demonstrate to be a novel target for NDs. Multiple factors can be considered to develop TREM2 therapeutics. Firstly, TREM2 variants involve a strong risk for AD development as one copy of the ApoE4 allele, whereas minor allele frequency of TREM2 variants is considerably lower (<1% for TREM2 to around 20% for ApoE4) (Gonzalez Murcia et al. 2013; Cuyvers and Sleegers 2016). Although it has been indicated by some studies that therapeutics may want to reinstate wild-type TREM2 activity in these variant carriers as potential therapeutic, thus correcting variants of TREM2 are not likely to be a widely applicable therapeutic technique. Studying variants of TREM2 that involve risk for NDs will reveal components of the immune response that are mainly significant in immune modulation of pathology, and play a role as a prerequisite to develop immune-targeted therapeutics.

Up until now, various potential altering activities for immune cell action during AD progression, and perhaps detected important activity for peripherally derived immune cells in AD pathogenesis, which will significantly help in therapeutically targeting the immune cells associated with AD pathogenesis. Common TREM2 activities have been detected in various NDs, which indicate that therapeutic targets might be relevant to several diseases. As per its disease progression-dependent activities, it is not clear that simply suppressing or activating TREM2 would be helpful even in AD. Interestingly, there might be gender-specific TREM2 activities, since in some (Piccio et al. 2016) but not all (Suárez-Calvet et al. 2016a, b) cases have revealed differences in levels of sTREM2 in CSF between men and women. Similarly, a variant of TREM2 was linked with systemic inflammation markers in women not men, which was also hormone-independent.

7.10 Conclusion

It is essential to systematically integrate available data including biomarkers, proteomics, expression, and sequencing, in order to completely understand the complex function of TREM2 in the risk for NDs. More studies are required to better understand the expression pattern, function, and signaling of TREM2 and their roles in various NDs. In addition, it is also important to identify how a diverse range of TREM2 variants plays role in the risk for NDs. At present, although these findings regarding variants of TREM2 might not directly lead to TREM2-targeted therapeutics or biomarkers; however, understanding regarding how the immune system actively plays roles in ND pathology assures to reveal many approaches for a new class of immune-targeted therapeutic targets for NDs.

References

Aarsland D, Andersen K, Larsen JP et al (2003) Prevalence and characteristics of dementia in Parkinson disease: an 8-year prospective study. Arch Neurol 60:387–392. https://doi.org/10.1001/archneur.60.3.387

Aarsland D, Brønnick K, Larsen JP et al (2009) Cognitive impairment in incident, untreated Parkinson disease: the norwegian parkwest study. Neurology 72:1121–1126. https://doi.org/10.1212/01.wnl.0000338632.00552.cb

Allcock RJN, Barrow AD, Forbes S et al (2003) The human TREM gene cluster at 6p21.1 encodes both activating and inhibitory single lgV domain receptors and includes NKp44. Eur J Immunol 33:567–577. https://doi.org/10.1002/immu.200310033

Alsema A, Jiang Q, Kracht L, et al (2020) Profiling microglia from AD donors and non-demented elderly in acute human post-mortem cortical tissue. bioRxiv. https://doi.org/10.1101/2020.03.18.995332

Atagi Y, Liu CC, Painter MM et al (2015) Apolipoprotein E is a ligand for triggering receptor expressed on myeloid cells 2 (TREM2). J Biol Chem 290:26043–26050. https://doi.org/10.1074/jbc.M115.679043

Bailey CC, Devaux LB, Farzan M (2015) The triggering receptor expressed on myeloid cells 2 binds apolipoprotein E. J Biol Chem 290:26033–26042. https://doi.org/10.1074/jbc.M115.677286

Bakken TE, Hodge RD, Miller JA et al (2018) Single-nucleus and single-cell transcriptomes compared in matched cortical cell types. PLoS One 13:e0209648. https://doi.org/10.1371/journal.pone.0209648

Benitez BA, Cruchaga C (2013) TREM2 and neurodegenerative disease. N Engl J Med 369:1564–1570. https://doi.org/10.1056/nejmc1306509

Benitez BA, Cooper B, Pastor P et al (2013) TREM2 is associated with the risk of Alzheimer's disease in Spanish population. Neurobiol Aging 34:1711.e15–1711.e17. https://doi.org/10.1016/j.neurobiolaging.2012.12.018

Bhattacharjee S, Zhao Y, Dua P et al (2016) MicroRNA-34α-mediated down-regulation of the microglial-enriched triggering receptor and phagocytosis-sensor TREM2 in age-related macular degeneration. PLoS One 11:e0150211. https://doi.org/10.1371/journal.pone.0150211

Bhattacherjee A, Rodrigues E, Jung J et al (2019) Repression of phagocytosis by human CD33 is not conserved with mouse CD33. Commun Biol 2:1–13. https://doi.org/10.1038/s42003-019-0698-6

Borroni B, Ferrari F, Galimberti D et al (2014) Heterozygous TREM2 mutations in frontotemporal dementia. Neurobiol Aging 35:934.e7–934.e10. https://doi.org/10.1016/j.neurobiolaging.2013.09.017

Bouchon A, Dietrich J, Colonna M (2000) Cutting edge: inflammatory responses can be triggered by TREM-1, a novel receptor expressed on neutrophils and monocytes. J Immunol 164:4991–4995. https://doi.org/10.4049/jimmunol.164.10.4991

Bouchon A, Hernández-Munain C, Cella M, Colonna M (2001) A DAP12-mediated pathway regulates expression of CC chemokine receptor 7 and maturation of human dendritic cells. J Exp Med 194:1111–1122. https://doi.org/10.1084/jem.194.8.1111

Brettschneider J, Toledo JB, van Deerlin VM et al (2012) Microglial activation correlates with disease progression and upper motor neuron clinical symptoms in amyotrophic lateral sclerosis. PLoS One 7:e39216. https://doi.org/10.1371/journal.pone.0039216

Brosseron F, Traschütz A, Widmann CN et al (2018) Characterization and clinical use of inflammatory cerebrospinal fluid protein markers in Alzheimer's disease. Alzheimers Res Ther 10:25. https://doi.org/10.1186/s13195-018-0353-3

Buchman AS, Yu L, Wilson RS et al (2019) Progressive Parkinsonism in older adults is related to the burden of mixed brain pathologies. Neurology 92:E1821–E1830. https://doi.org/10.1212/WNL.0000000000007315

Cady J, Koval ED, Benitez BA et al (2014) TREM2 variant p.R47H as a risk factor for sporadic amyotrophic lateral sclerosis. JAMA Neurol 71:449–453. https://doi.org/10.1001/jamaneurol.2013.6237

Calero M, Rostagno A, Matsubara E et al (2000) Apolipoprotein J (clusterin) and Alzheimer's disease. Microsc Res Tech 50:305–315. https://doi.org/10.1002/1097-0029(20000815)50:4<305::AID-JEMT10>3.0.CO;2-L

Cantoni C, Bollman B, Licastro D et al (2015) TREM2 regulates microglial cell activation in response to demyelination in vivo. Acta Neuropathol 129:429–447. https://doi.org/10.1007/s00401-015-1388-1

Carmona S, Zahs K, Wu E et al (2018) The role of TREM2 in Alzheimer's disease and other neurodegenerative disorders. Lancet Neurol 17:721–730. https://doi.org/10.1016/S1474-4422(18)30232-1

Carrasquillo MM, Barber I, Lincoln SJ et al (2016) Evaluating pathogenic dementia variants in posterior cortical atrophy. Neurobiol Aging 37:38–44. https://doi.org/10.1016/j.neurobiolaging.2015.09.023

Celarain N, Sánchez-Ruiz de Gordoa J, Zelaya MV et al (2016) TREM2 upregulation correlates with 5-hydroxymethycytosine enrichment in Alzheimer's disease hippocampus. Clin Epigenetics 8:37. https://doi.org/10.1186/s13148-016-0202-9

Cella M, Buonsanti C, Strader C et al (2003) Impaired differentiation of osteoclasts in TREM-2-deficient individuals. J Exp Med 198:645–651. https://doi.org/10.1084/jem.20022220

Chee KY, Gaillard F, Velakoulis D et al (2017) A case of TREM2 mutation presenting with features of progressive non-fluent aphasia and without bone involvement. Aust N Z J Psychiatry 51: 1157–1158. https://doi.org/10.1177/0004867417707821
Chen Y, Chen X, Guo X et al (2015) Assessment of TREM2 rs75932628 association with Parkinson's disease and multiple system atrophy in a Chinese population. Neurol Sci 36: 1903–1906. https://doi.org/10.1007/s10072-015-2279-x
Chen-Plotkin AS, Albin R, Alcalay R et al (2018) Finding useful biomarkers for Parkinson s disease. Sci Transl Med 10:eaam6003. https://doi.org/10.1126/scitranslmed.aam6003
Chertoff M, Shrivastava K, Gonzalez B et al (2013) Differential modulation of TREM2 protein during postnatal brain development in mice. PLoS One 8:e72083. https://doi.org/10.1371/journal.pone.0072083
Chiu IM, Morimoto ETA, Goodarzi H et al (2013) A neurodegeneration-specific gene-expression signature of acutely isolated microglia from an amyotrophic lateral sclerosis mouse model. Cell Rep 4:385–401. https://doi.org/10.1016/j.celrep.2013.06.018
Cholerton BA, Zabetian CP, Quinn JF et al (2013) Pacific Northwest Udall center of excellence clinical consortium: study design and baseline cohort characteristics. J Parkinsons Dis 3:205–214. https://doi.org/10.3233/JPD-130189
Chouery E, Delague V, Bergougnoux A et al (2008) Mutations in TREM2 lead to pure early-onset dementia without bone cysts. Hum Mutat 29:E194–E204. https://doi.org/10.1002/humu.20836
Colonna M (2003) Trems in the immune system and beyond. Nat Rev Immunol 3:445–453. https://doi.org/10.1038/nri1106
Cuyvers E, Sleegers K (2016) Genetic variations underlying Alzheimer's disease: evidence from genome-wide association studies and beyond. Lancet Neurol 15:857–868. https://doi.org/10.1016/S1474-4422(16)00127-7
Cuyvers E, Bettens K, Philtjens S et al (2014) Investigating the role of rare heterozygous TREM2 variants in Alzheimer's disease and frontotemporal dementia. Neurobiol Aging 35:726.e11–726.e19. https://doi.org/10.1016/j.neurobiolaging.2013.09.009
Darwent L, Carmona S, Lohmann E et al (2017) Mutations in TYROBP are not a common cause of dementia in a Turkish cohort. Neurobiol Aging 58:240.e1–240.e3. https://doi.org/10.1016/j.neurobiolaging.2017.06.019
Dean HB, Roberson ED, Song Y (2019) Neurodegenerative disease-associated variants in TREM2 destabilize the apical ligand-binding region of the immunoglobulin domain. Front Neurol 10: 1252. https://doi.org/10.3389/fneur.2019.01252
Deczkowska A, Keren-Shaul H, Weiner A et al (2018) Disease-associated microglia: a universal immune sensor of neurodegeneration. Cell 173:1073–1081. https://doi.org/10.1016/j.cell.2018.05.003
Deczkowska A, Weiner A, Amit I (2020) The physiology, pathology, and potential therapeutic applications of the TREM2 signaling pathway. Cell 181:1207–1217. https://doi.org/10.1016/j.cell.2020.05.003
Del-Aguila JL, Benitez BA, Li Z et al (2019) TREM2 brain transcript-specific studies in AD and TREM2 mutation carriers. Mol Neurodegener 14:18. https://doi.org/10.1186/s13024-019-0319-3
Dickson DW, Heckman MG, Murray ME et al (2018) APOE 4 is associated with severity of Lewy body pathology independent of Alzheimer pathology. Neurology 91:E1182–E1195. https://doi.org/10.1212/WNL.0000000000006212
Duyckaerts C, Delatour B, Potier MC (2009) Classification and basic pathology of Alzheimer disease. Acta Neuropathol 118:5–36. https://doi.org/10.1007/s00401-009-0532-1
Engelhardt JI, Appel SH (1990) Igg reactivity in the spinal cord and motor cortex in amyotrophic lateral sclerosis. Arch Neurol 47:1210–1216. https://doi.org/10.1001/archneur.1990.00530110068019
Feng S, Nie K, Gan R et al (2014) Triggering receptor expressed on myeloid cells 2 variants are rare in Parkinson's disease in a Han Chinese cohort. Neurobiol Aging 35:1780.e11–1780.e12. https://doi.org/10.1016/j.neurobiolaging.2014.01.142

Forabosco P, Ramasamy A, Trabzuni D et al (2013) Insights into TREM2 biology by network analysis of human brain gene expression data. Neurobiol Aging 34:2699–2714. https://doi.org/10.1016/j.neurobiolaging.2013.05.001

Frigerio CS (2020) Single nucleus sequencing fails to detect microglial activation in human tissue. bioRxiv. https://doi.org/10.1101/2020.04.13.035386

Garcia-Alloza M, Borrelli LA, Thyssen DH et al (2013) Four-dimensional microglia response to anti-Aβ treatment in APP/PS1xCX3CR1/GFP mice. IntraVital 2:e25693. https://doi.org/10.4161/intv.25693

Gaugler J, James B, Johnson T et al (2016) 2016 Alzheimer's disease facts and figures. Alzheimers Dement 12:459–509. https://doi.org/10.1016/j.jalz.2016.03.001

Gervois P, Lambrichts I (2019) The emerging role of triggering receptor expressed on myeloid cells 2 as a target for immunomodulation in ischemic stroke. Front Immunol 10:1668. https://doi.org/10.3389/fimmu.2019.01668

Ghezzi L, Carandini T, Arighi A et al (2017) Evidence of cns b-amyloid deposition in nasu-hakola disease due to the TREM2 Q33X mutation. Neurology 89:2503–2505. https://doi.org/10.1212/WNL.0000000000004747

Giraldo M, Lopera F, Siniard AL et al (2013) Variants in triggering receptor expressed on myeloid cells 2 are associated with both behavioral variant frontotemporal lobar degeneration and Alzheimer's disease. Neurobiol Aging 34:2077.e11–2077.e18. https://doi.org/10.1016/j.neurobiolaging.2013.02.016

Gonçalves LA, Rodrigues-Duarte L, Rodo J et al (2013) TREM2 governs Kupffer cell activation and explains belr1 genetic resistance to malaria liver stage infection. Proc Natl Acad Sci U S A 110:19531–19536. https://doi.org/10.1073/pnas.1306873110

Gonzalez Murcia JD, Schmutz C, Munger C et al (2013) Assessment of Trem2 Rs75932628 association with Alzheimer's disease in a population-based sample: the cache county study. Neurobiol Aging 34:2889.e11–2889.e13. https://doi.org/10.1016/j.neurobiolaging.2013.06.004

Griciuc A, Patel S, Federico AN et al (2019) TREM2 acts downstream of CD33 in modulating microglial pathology in Alzheimer's disease. Neuron 103:820–835.e7. https://doi.org/10.1016/j.neuron.2019.06.010

Grubman A, Chew G, Ouyang JF et al (2019) A single-cell atlas of entorhinal cortex from individuals with Alzheimer's disease reveals cell-type-specific gene expression regulation. Nat Neurosci 22:2087–2097. https://doi.org/10.1038/s41593-019-0539-4

Grupe A, Abraham R, Li Y et al (2007) Evidence for novel susceptibility genes for late-onset Alzheimer's disease from a genome-wide association study of putative functional variants. Hum Mol Genet 16:865–873. https://doi.org/10.1093/hmg/ddm031

Guerreiro R, Hardy J (2014) Genetics of Alzheimer's disease. Neurotherapeutics 11:732–737. https://doi.org/10.1007/s13311-014-0295-9

Guerreiro R, Bilgic B, Guven G et al (2013a) A novel compound heterozygous mutation in TREM2 found in a Turkish frontotemporal dementia-like family. Neurobiol Aging 34:2890.e1–2890.e5. https://doi.org/10.1016/j.neurobiolaging.2013.06.005

Guerreiro R, Wojtas A, Bras J et al (2013b) TREM2 variants in Alzheimer's disease. N Engl J Med 368:117–127. https://doi.org/10.1056/NEJMoa1211851

Guerreiro RJ, Lohmann E, Brás JM et al (2013c) Using exome sequencing to reveal mutations in TREM2 presenting as a frontotemporal dementia-like syndrome without bone involvement. Arch Neurol 70:78–84. https://doi.org/10.1001/jamaneurol.2013.579

Gwack Y, Feske S, Srikanth S et al (2007) Signalling to transcription: store-operated Ca2+ entry and NFAT activation in lymphocytes. Cell Calcium 42:145–156. https://doi.org/10.1016/j.ceca.2007.03.007

Hamerman JA, Tchao NK, Lowell CA, Lanier LL (2005) Enhanced toll-like receptor responses in the absence of signaling adaptor DAP12. Nat Immunol 6:579–586. https://doi.org/10.1038/ni1204

Hamerman JA, Jarjoura JR, Humphrey MB et al (2006) Cutting edge: inhibition of TLR and FcR responses in macrophages by triggering receptor expressed on myeloid cells (TREM)-2 and DAP12. J Immunol 177:2051–2055. https://doi.org/10.4049/jimmunol.177.4.2051

Hammond TR, Marsh SE, Stevens B (2019) Immune signaling in neurodegeneration. Immunity 50: 955–974. https://doi.org/10.1016/j.immuni.2019.03.016

Hansen DV, Hanson JE, Sheng M (2018) Microglia in Alzheimer's disease. J Cell Biol 217:459–472. https://doi.org/10.1083/jcb.201709069

Hely MA, Reid WGJ, Adena MA et al (2008) The Sydney multicenter study of Parkinson's disease: the inevitability of dementia at 20 years. Mov Disord 23:837–844. https://doi.org/10.1002/mds.21956

Hickman SE, El Khoury J (2014) TREM2 and the neuroimmunology of Alzheimer's disease. Biochem Pharmacol 88:495–498. https://doi.org/10.1016/j.bcp.2013.11.021

Hu N, Tan MS, Yu JT et al (2014) Increased expression of TREM2 in peripheral blood of Alzheimer's disease patients. J Alzheimers Dis 38:497–501. https://doi.org/10.3233/JAD-130854

Humphrey MB, Daws MR, Spusta SC et al (2006) TREM2, a DAP12-associated receptor, regulates osteoclast differentiation and function. J Bone Miner Res 21:237–245. https://doi.org/10.1359/JBMR.051016

Jaitin DA, Adlung L, Thaiss CA et al (2019) Lipid-associated macrophages control metabolic homeostasis in a Trem2-dependent manner. Cell 178:686–698.e14. https://doi.org/10.1016/j.cell.2019.05.054

Jankowsky JL, Zheng H (2017) Practical considerations for choosing a mouse model of Alzheimer's disease. Mol Neurodegener 12:89. https://doi.org/10.1186/s13024-017-0231-7

Jay TR, Miller CM, Cheng PJ et al (2015) TREM2 deficiency eliminates TREM2+ inflammatory macrophages and ameliorates pathology in Alzheimer's disease mouse models. J Exp Med 212: 287–295. https://doi.org/10.1084/jem.20142322

Jay TR, Von Saucken VE, Landreth GE (2017) TREM2 in neurodegenerative diseases. Mol Neurodegener 12:1–33. https://doi.org/10.1186/s13024-017-0197-5

Jiang T, Tan L, Zhu XC et al (2014) Upregulation of TREM2 ameliorates neuropathology and rescues spatial cognitive impairment in a transgenic mouse model of Alzheimer's disease. Neuropsychopharmacology 39:2949–2962. https://doi.org/10.1038/npp.2014.164

Jin SC, Benitez BA, Karch CM et al (2014) Coding variants in TREM2 increase risk for Alzheimer's disease. Hum Mol Genet 23:5838–5846. https://doi.org/10.1093/hmg/ddu277

Jonsson T, Stefansson H, Steinberg S et al (2013) Variant of TREM2 associated with the risk of Alzheimer's disease. N Engl J Med 368:107–116. https://doi.org/10.1056/NEJMoa1211103

Kabir MT, Uddin MS, Abdeen A et al (2020a) Evidence linking protein misfolding to quality control in progressive neurodegenerative diseases. Curr Top Med Chem 20:2025–2043. https://doi.org/10.2174/1568026620666200618114924

Kabir MT, Uddin MS, Al Mamun A et al (2020b) Combination drug therapy for the management of Alzheimer's disease. Int J Mol Sci 21:3272. https://doi.org/10.3390/ijms21093272

Kabir MT, Uddin MS, Setu JR et al (2020c) Exploring the role of PSEN mutations in the pathogenesis of Alzheimer's disease. Neurotox Res 38(4):833–849

Kabir MT, Uddin MS, Zaman S et al (2020d) Molecular mechanisms of metal toxicity in the pathogenesis of Alzheimer's disease. Mol Neurobiol 58:1–20

Kaivola K, Jansson L, Saarentaus E et al (2018) Heterozygous TYROBP deletion (PLOSLFIN) is not a strong risk factor for cognitive impairment. Neurobiol Aging 64:159.e1–159.e4. https://doi.org/10.1016/j.neurobiolaging.2017.12.008

Karch CM, Goate AM (2015) Alzheimer's disease risk genes and mechanisms of disease pathogenesis. Biol Psychiatry 77:43–51

Keren-Shaul H, Spinrad A, Weiner A et al (2017) A unique microglia type associated with restricting development of Alzheimer's disease. Cell 169:1276–1290.e17. https://doi.org/10.1016/j.cell.2017.05.018

Kiialainen A, Hovanes K, Paloneva J et al (2005) Dap12 and Trem2, molecules involved in innate immunity and neurodegeneration, are co-expressed in the CNS. Neurobiol Dis 18:314–322. https://doi.org/10.1016/j.nbd.2004.09.007

Kleinberger G, Yamanishi Y, Suárez-Calvet M et al (2014) TREM2 mutations implicated in neurodegeneration impair cell surface transport and phagocytosis. Sci Transl Med 6:243ra86. https://doi.org/10.1126/scitranslmed.3009093

Klünemann HH, Ridha BH, Magy L et al (2005) The genetic causes of basal ganglia calcification, dementia, and bone cysts: DAP12 and TREM2. Neurology 64:1502–1507. https://doi.org/10.1212/01.WNL.0000160304.00003.CA

Korvatska O, Leverenz JB, Jayadev S et al (2015) R47H variant of TREM2 associated with Alzheimer disease in a large late-onset family clinical, genetic, and neuropathological study. JAMA Neurol 72:920–927. https://doi.org/10.1001/jamaneurol.2015.0979

Koth LL, Cambier CJ, Ellwanger A et al (2010) DAP12 is required for macrophage recruitment to the lung in response to cigarette smoke and chemotaxis toward CCL2. J Immunol 184:6522–6528. https://doi.org/10.4049/jimmunol.0901171

Krasemann S, Madore C, Cialic R et al (2017) The TREM2-APOE pathway drives the transcriptional phenotype of dysfunctional microglia in neurodegenerative diseases. Immunity 47:566–581.e9. https://doi.org/10.1016/j.immuni.2017.08.008

Lee CYD, Daggett A, Gu X et al (2018) Elevated TREM2 gene dosage reprograms microglia responsivity and ameliorates pathological phenotypes in Alzheimer's disease models. Neuron 97:1032–1048.e5. https://doi.org/10.1016/j.neuron.2018.02.002

Li Z, Zhong L, Gu L et al (2016) Association study of TREM2 polymorphism rs75932628 with leucoaraiosis or Parkinson's disease in the Han Chinese population. BMJ Open 6:e009499. https://doi.org/10.1136/bmjopen-2015-009499

Liao B, Zhao W, Beers DR et al (2012) Transformation from a neuroprotective to a neurotoxic microglial phenotype in a mouse model of ALS. Exp Neurol 237:147–152. https://doi.org/10.1016/j.expneurol.2012.06.011

Lill CM, Rengmark A, Pihlstrøm L et al (2015) The role of TREM2 R47H as a risk factor for Alzheimer's disease, frontotemporal lobar degeneration, amyotrophic lateral sclerosis, and Parkinson's disease. Alzheimers Dement 11:1407–1416. https://doi.org/10.1016/j.jalz.2014.12.009

Liu G, Liu Y, Jiang Q et al (2016) Convergent genetic and expression datasets highlight TREM2 in Parkinson's disease susceptibility. Mol Neurobiol 53:4931–4938. https://doi.org/10.1007/s12035-015-9416-7

Lue LF, Schmitz CT, Serrano G et al (2015) TREM2 protein expression changes correlate with Alzheimer's disease neurodegenerative pathologies in post-mortem temporal cortices. Brain Pathol 25:469–480. https://doi.org/10.1111/bpa.12190

Ma L, Allen M, Sakae N et al (2016) Expression and processing analyses of wild type and p.R47H TREM2 variant in Alzheimer's disease brains. Mol Neurodegener 11:1–9. https://doi.org/10.1186/s13024-016-0137-9

Marek K, Chowdhury S, Siderowf A et al (2018) The Parkinson's progression markers initiative (PPMI) – establishing a PD biomarker cohort. Ann Clin Transl Neurol 5:1460–1477. https://doi.org/10.1002/acn3.644

Matarin M, Salih DA, Yasvoina M et al (2015) A genome-wide gene-expression analysis and database in transgenic mice during development of amyloid or tau pathology. Cell Rep 10:633–644. https://doi.org/10.1016/j.celrep.2014.12.041

Mathys H, Davila-Velderrain J, Peng Z et al (2019) Single-cell transcriptomic analysis of Alzheimer's disease. Nature 570:332–337. https://doi.org/10.1038/s41586-019-1195-2

Meilandt WJ, Ngu H, Gogineni A et al (2020) TREM2 deletion reduces late-stage amyloid plaque accumulation, elevates the Aβ42:Aβ40 ratio, and exacerbates axonal dystrophy and dendritic spine loss in the PS2ApP Alzheimer's mouse model. J Neurosci 40:1956–1974. https://doi.org/10.1523/JNEUROSCI.1871-19.2019

Mengel D, Thelen M, Balzer-Geldsetzer M et al (2016) TREM2 rare variant p.R47H is not associated with Parkinson's disease. Park Relat Disord 23:109–111. https://doi.org/10.1016/j.parkreldis.2015.11.026

Miron VE, Boyd A, Zhao JW et al (2013) M2 microglia and macrophages drive oligodendrocyte differentiation during CNS remyelination. Nat Neurosci 16:1211–1218. https://doi.org/10.1038/nn.3469

Montalbetti L, Ratti MT, Greco B et al (2005) Neuropsychological tests and functional nuclear neuroimaging provide evidence of subclinical impairment in Nasu-Hakola disease heterozygotes. Funct Neurol 20:71–75

Muslimović D, Post B, Speelman JD, Schmand B (2005) Cognitive profile of patients with newly diagnosed Parkinson disease. Neurology 65:1239–1245. https://doi.org/10.1212/01.wnl.0000180516.69442.95

N'Diaye EN, Branda CS, Branda SS et al (2009) TREM-2 (triggering receptor expressed on myeloid cells 2) is a phagocytic receptor for bacteria. J Cell Biol 184:215–223. https://doi.org/10.1083/jcb.200808080

Namba Y, Tomonaga M, Kawasaki H et al (1991) Apolipoprotein E immunoreactivity in cerebral amyloid deposits and neurofibrillary tangles in Alzheimer's disease and kuru plaque amyloid in Creutzfeldt-Jakob disease. Brain Res 541:163–166. https://doi.org/10.1016/0006-8993(91)91092-F

Neumann H, Takahashi K (2007) Essential role of the microglial triggering receptor expressed on myeloid cells-2 (TREM2) for central nervous tissue immune homeostasis. J Neuroimmunol 184:92–99. https://doi.org/10.1016/j.jneuroim.2006.11.032

Ortega-Cubero S, Lorenzo-Betancor O, Lorenzo E et al (2015) TREM2 R47H variant and risk of essential tremor: a cross-sectional international multicenter study. Park Relat Disord 21:306–309. https://doi.org/10.1016/j.parkreldis.2014.12.010

Paloneva J, Manninen T, Christman G et al (2002) Mutations in two genes encoding different subunits of a receptor signaling complex result in an identical disease phenotype. Am J Hum Genet 71:656–662. https://doi.org/10.1086/342259

Paloneva J, Mandelin J, Kiialainen A et al (2003) DAP12/TREM2 deficiency results in impaired osteoclast differentiation and osteoporotic features. J Exp Med 198:669–675. https://doi.org/10.1084/jem.20030027

Pekkarinen P, Hovatta I, Hakola P et al (1998) Assignment of the locus for PLO-SL, a frontal-lobe dementia with bone cysts, to 19q13. Am J Hum Genet 62:362–372. https://doi.org/10.1086/301722

Peng Q, Malhotra S, Torchia JA et al (2010) TREM2- and DAP12-dependent activation of PI3K requires DAP10 and is inhibited by SHIP1. Sci Signal 3:ra38. https://doi.org/10.1126/scisignal.2000500

Perez SE, Nadeem M, He B et al (2017) Neocortical and hippocampal TREM2 protein levels during the progression of Alzheimer's disease. Neurobiol Aging 54:133–143. https://doi.org/10.1016/j.neurobiolaging.2017.02.012

Pesaran T, Karam R, Huether R et al (2016) Beyond DNA: an integrated and functional approach for classifying germline variants in breast cancer genes. Int J Breast Cancer 2016:2469523. https://doi.org/10.1155/2016/2469523

Piccio L, Buonsanti C, Mariani M et al (2007) Blockade of TREM-2 exacerbates experimental autoimmune encephalomyelitis. Eur J Immunol 37:1290–1301. https://doi.org/10.1002/eji.200636837

Piccio L, Deming Y, Del-Águila JL et al (2016) Cerebrospinal fluid soluble TREM2 is higher in Alzheimer disease and associated with mutation status. Acta Neuropathol 131:925–933. https://doi.org/10.1007/s00401-016-1533-5

Poliani PL, Wang Y, Fontana E et al (2015) TREM2 sustains microglial expansion during aging and response to demyelination. J Clin Invest 125:2161–2170. https://doi.org/10.1172/JCI77983

Pottier C, Wallon D, Rousseau S et al (2013) TREM2 R47H variant as a risk factor for early-onset alzheimer's disease. J Alzheimers Dis 35:45–49. https://doi.org/10.3233/JAD-122311

Raha AA, Henderson JW, Stott SRW et al (2016) Neuroprotective effect of TREM-2 in aging and Alzheimer's disease model. J Alzheimers Dis 55:199–217. https://doi.org/10.3233/JAD-160663

Ransohoff RM (2016) How neuroinflammation contributes to neurodegeneration. Science (80-) 353:777–783. https://doi.org/10.1126/science.aag2590

Rayaprolu S, Mullen B, Baker M et al (2013) TREM2 in neurodegeneration: evidence for association of the p.R47H variant with frontotemporal dementia and Parkinson's disease. Mol Neurodegener 8:19. https://doi.org/10.1186/1750-1326-8-19

Redaelli V, Salsano E, Colleoni L et al (2018) Frontotemporal dementia and chorea associated with a compound heterozygous TREM2 mutation. J Alzheimers Dis 63:195–201. https://doi.org/10.3233/JAD-180018

Reitz C, Mayeux R, Alzheimer's Disease Genetics Consortium (2013) TREM2 and neurodegenerative disease. N Engl J Med 369:1564–1565. https://doi.org/10.1056/NEJMc1306509

Robinson JL, Lee EB, Xie SX et al (2018) Neurodegenerative disease concomitant proteinopathies are prevalent, age-related and APOE4-associated. Brain 141:2181–2193. https://doi.org/10.1093/brain/awy146

Satoh J, Tabunoki H, Ishida T et al (2011) Immunohistochemical characterization of microglia in Nasu-Hakola disease brains. Neuropathology 31:363–375. https://doi.org/10.1111/j.1440-1789.2010.01174.x

Satoh JI, Kino Y, Yanaizu M, Saito Y (2018) Alzheimer's disease pathology in Nasu-Hakola disease brains. Intractable Rare Dis Res 7:32–36. https://doi.org/10.5582/irdr.2017.01088

Savage JC, Jay T, Goduni E et al (2015) Nuclear receptors license phagocytosis by trem2+ myeloid cells in mouse models of Alzheimer's disease. J Neurosci 35:6532–6543. https://doi.org/10.1523/JNEUROSCI.4586-14.2015

Schlepckow K, Kleinberger G, Fukumori A et al (2017) An Alzheimer-associated TREM2 variant occurs at the ADAM cleavage site and affects shedding and phagocytic function. EMBO Mol Med 9:1356–1365. https://doi.org/10.15252/emmm.201707672

Schlepckow K, Monroe KM, Kleinberger G et al (2020) Enhancing protective microglial activities with a dual function TREM 2 antibody to the stalk region. EMBO Mol Med 12:e11227. https://doi.org/10.15252/emmm.201911227

Schmid CD, Sautkulis LN, Danielson PE et al (2002) Heterogeneous expression of the triggering receptor expressed on myeloid cells-2 on adult murine microglia. J Neurochem 83:1309–1320. https://doi.org/10.1046/j.1471-4159.2002.01243.x

Sessa G, Podini P, Mariani M et al (2004) Distribution and signaling of TREM2/DAP12, the receptor system mutated in human polycystic lipomembraneous osteodysplasia with sclerosing leukoencephalopathy dementia. Eur J Neurosci 20:2617–2628. https://doi.org/10.1111/j.1460-9568.2004.03729.x

Shi Y, Holtzman DM (2018) Interplay between innate immunity and Alzheimer disease: APOE and TREM2 in the spotlight. Nat Rev Immunol 18:759–772. https://doi.org/10.1038/s41577-018-0051-1

Slattery CF, Beck JA, Harper L et al (2014) R47H TREM2 variant increases risk of typical early-onset Alzheimer's disease but not of prion or frontotemporal dementia. Alzheimers Dement 10:602–608.e4. https://doi.org/10.1016/j.jalz.2014.05.1751

Solje E, Hartikainen P, Valori M et al (2014) The C9ORF72 expansion does not affect the phenotype in Nasu-Hakola disease with the DAP12 mutation. Neurobiol Aging 35:1780.e13–1780.e17. https://doi.org/10.1016/j.neurobiolaging.2014.01.149

Song W, Hooli B, Mullin K et al (2017) Alzheimer's disease-associated TREM2 variants exhibit either decreased or increased ligand-dependent activation. Alzheimers Dement 13:381–387. https://doi.org/10.1016/j.jalz.2016.07.004

Starita LM, Ahituv N, Dunham MJ et al (2017) Variant interpretation: functional assays to the rescue. Am J Hum Genet 101:315–325. https://doi.org/10.1016/j.ajhg.2017.07.014

Strittmatter WJ, Weisgraber KH, Huang DY et al (1993) Binding of human apolipoprotein E to synthetic amyloid β peptide: isoform-specific effects and implications for late-onset Alzheimer disease. Proc Natl Acad Sci U S A 90:8098–8102. https://doi.org/10.1073/pnas.90.17.8098

Suárez-Calvet M, Caballero MÁA, Kleinberger G et al (2016a) Early changes in CSF sTREM2 in dominantly inherited Alzheimer's disease occur after amyloid deposition and neuronal injury. Sci Transl Med 8:369ra178. https://doi.org/10.1126/scitranslmed.aag1767

Suárez-Calvet M, Kleinberger G, Araque Caballero MÁ et al (2016b) sTREM 2 cerebrospinal fluid levels are a potential biomarker for microglia activity in early-stage Alzheimer's disease and associate with neuronal injury markers. EMBO Mol Med 8:466–476. https://doi.org/10.15252/emmm.201506123

Takahashi K, Rochford CDP, Neumann H (2005) Clearance of apoptotic neurons without inflammation by microglial triggering receptor expressed on myeloid cells-2. J Exp Med 201:647–657. https://doi.org/10.1084/jem.20041611

Takahashi K, Prinz M, Stagi M et al (2007) TREM2-transduced myeloid precursors mediate nervous tissue debris clearance and facilitate recovery in an animal model of multiple sclerosis. PLoS Med 4:675–689. https://doi.org/10.1371/journal.pmed.0040124

Tan T, Song Z, Yuan L et al (2016) Genetic analysis of TREM2 variants in Chinese Han patients with sporadic Parkinson's disease. Neurosci Lett 612:189–192. https://doi.org/10.1016/j.neulet.2015.12.029

Thelen M, Razquin C, Hernández I et al (2014) Investigation of the role of rare TREM2 variants in frontotemporal dementia subtypes. Neurobiol Aging 35:2657.e13–2657.e19. https://doi.org/10.1016/j.neurobiolaging.2014.06.018

Tsuang D, Leverenz JB, Lopez OL et al (2013) APOE ε4 increases risk for dementia in pure synucleinopathies. JAMA Neurol 70:223–228. https://doi.org/10.1001/jamaneurol.2013.600

Turnbull IR, Gilfillan S, Cella M et al (2006) Cutting edge: TREM-2 attenuates macrophage activation. J Immunol 177:3520–3524. https://doi.org/10.4049/jimmunol.177.6.3520

Uddin MS, Al Mamun A, Kabir MT et al (2020a) Neuroprotective role of polyphenols against oxidative stress-mediated neurodegeneration. Eur J Pharmacol 886:173412. https://doi.org/10.1016/j.ejphar.2020.173412

Uddin MS, Kabir MT, Al Mamun A et al (2020b) Exploring potential of alkaloidal phytochemicals targeting neuroinflammatory signaling of Alzheimer's disease. Curr Pharm Des 27(3):357–366. https://doi.org/10.2174/1381612826666200531151004

Uddin MS, Kabir MT, Jakaria M et al (2020c) Exploring the potential of neuroproteomics in Alzheimer's disease. Curr Top Med Chem 20:2263–2278. https://doi.org/10.2174/1568026620666200603112030

Uddin MS, Kabir MT, Al Mamun A et al (2020d) Pharmacological approaches to mitigate neuroinflammation in Alzheimer's disease. Int Immunopharmacol 84:106479. https://doi.org/10.1016/j.intimp.2020.106479

Uddin MS, Kabir MT, Rahman MM et al (2020e) TV 3326 for Alzheimer's dementia: a novel multimodal ChE and MAO inhibitors to mitigate Alzheimer's-like neuropathology. J Pharm Pharmacol 72:1001–1012. https://doi.org/10.1111/jphp.13244

Uddin MS, Kabir MT, Rahman MS et al (2020f) Revisiting the amyloid cascade hypothesis: from anti-Aβ therapeutics to auspicious new ways for Alzheimer's disease. Int J Mol Sci 21:5858. https://doi.org/10.3390/ijms21165858

Uddin MS, Tewari D, Al Mamun A et al (2020g) Circadian and sleep dysfunction in Alzheimer's disease. Ageing Res Rev 60:101046. https://doi.org/10.1016/j.arr.2020.101046

Ulland TK, Colonna M (2018) TREM2 — a key player in microglial biology and Alzheimer disease. Nat Rev Neurol 14:667–675. https://doi.org/10.1038/s41582-018-0072-1

Ulland TK, Song WM, Huang SCC et al (2017) TREM2 maintains microglial metabolic fitness in Alzheimer's disease. Cell 170:649–663.e13. https://doi.org/10.1016/j.cell.2017.07.023

Ulrich JD, Ulland TK, Colonna M, Holtzman DM (2017) Elucidating the role of TREM2 in Alzheimer's disease. Neuron 94:237–248. https://doi.org/10.1016/j.neuron.2017.02.042

Verloes A, Maquet P, Sadzot B et al (1997) Nasu-Hakola syndrome: polycystic lipomembranous osteodysplasia with sclerosing leucoencephalopathy and presenile dementia. J Med Genet 34: 753–757. https://doi.org/10.1136/jmg.34.9.753

Voytyuk I, De Strooper B, Chávez-Gutiérrez L (2018) Modulation of γ- and β-secretases as early prevention against Alzheimer's disease. Biol Psychiatry 83:320–327. https://doi.org/10.1016/j.biopsych.2017.08.001

Walton RL, Soto-Ortolaza AI, Murray ME et al (2016) TREM2 p.R47H substitution is not associated with dementia with Lewy bodies. Neurol Genet 2:e85. https://doi.org/10.1212/NXG.0000000000000085

Wang Y, Cella M, Mallinson K et al (2015) TREM2 lipid sensing sustains the microglial response in an Alzheimer's disease model. Cell 160:1061–1071. https://doi.org/10.1016/j.cell.2015.01.049

Wes PD, Sayed FA, Bard F, Gan L (2016) Targeting microglia for the treatment of Alzheimer's disease. Glia 64:1710–1732. https://doi.org/10.1002/glia.22988

Wolfe CM, Fitz NF, Nam KN et al (2019) The role of APOE and TREM2 in Alzheimer's disease—current understanding and perspectives. Int J Mol Sci 20:81. https://doi.org/10.3390/ijms20010081

Yang J, Fu Z, Zhang X et al (2020) TREM2 ectodomain and its soluble form in Alzheimer's disease. J Neuroinflammation 17:204. https://doi.org/10.1186/s12974-020-01878-2

Yeh FL, Wang Y, Tom I et al (2016) TREM2 binds to apolipoproteins, including APOE and CLU/APOJ, and thereby facilitates uptake of amyloid-beta by microglia. Neuron 91:328–340. https://doi.org/10.1016/j.neuron.2016.06.015

Yeh FL, Hansen DV, Sheng M (2017) TREM2, microglia, and neurodegenerative diseases. Trends Mol Med 23:512–533. https://doi.org/10.1016/j.molmed.2017.03.008

Yuan P, Condello C, Keene CD et al (2016) TREM2 haplodeficiency in mice and humans impairs the microglia barrier function leading to decreased amyloid compaction and severe axonal dystrophy. Neuron 90:724–739. https://doi.org/10.1016/j.neuron.2016.05.003

Zheng H, Liu CC, Atagi Y et al (2016) Opposing roles of the triggering receptor expressed on myeloid cells 2 and triggering receptor expressed on myeloid cells-like transcript 2 in microglia activation. Neurobiol Aging 42:132–141. https://doi.org/10.1016/j.neurobiolaging.2016.03.004

Zheng H, Jia L, Liu CC et al (2017) TREM2 promotes microglial survival by activating wnt/β-catenin pathway. J Neurosci 37:1772–1784. https://doi.org/10.1523/JNEUROSCI.2459-16.2017

Zhong L, Chen XF (2019) The emerging roles and therapeutic potential of soluble TREM2 in Alzheimer's disease. Front Aging Neurosci 11:328. https://doi.org/10.3389/fnagi.2019.00328

Zhong L, Chen XF, Wang T et al (2017) Soluble TREM2 induces inflammatory responses and enhances microglial survival. J Exp Med 214:597–607. https://doi.org/10.1084/jem.20160844

Zhong L, Wang Z, Wang D et al (2018) Amyloid-beta modulates microglial responses by binding to the triggering receptor expressed on myeloid cells 2 (TREM2). Mol Neurodegener 13:15. https://doi.org/10.1186/s13024-018-0247-7

Zhou Y, Song WM, Andhey PS et al (2020) Human and mouse single-nucleus transcriptomics reveal TREM2-dependent and TREM2-independent cellular responses in Alzheimer's disease. Nat Med 26:131–142. https://doi.org/10.1038/s41591-019-0695-9

Molecular Mechanisms Underlying the Role of HSPB8 in Neurodegeneration

8

Rupali Patil, Nitu Wankhede, Aman Upaganlawar, and Suvarna Ingale

Abstract

Heat shock proteins (HSPs) regulate protein quality control and are responsible for protein aggregation and disaggregation. Molecular chaperones are members of the small heat shock protein (sHSP) family that maintains cellular homeostasis during unfavorable conditions. The sHSPs due to their chaperone properties avert protein aggregation. The sHSP dysregulation turns out to be an important pathological factor in numerous conditions including neurodegenerative disorders. Recent studies suggest the broad and diversified role of sHSPs in neuroprotection, but the mechanism of sHSPs with the neurodegeneration-promoting signaling pathway is still not clear. Some harmful events like proteasome inhibition induce the chaperone, sHSP-B8 (HSPB8). Misfolded protein toxicity is associated with motor neuron diseases (MNDs) exhibiting expression of HSPB8. Concerning this, HSPs may be considered as a feasible target for the development of drugs that can reduce protein aggregates associated with pathogenic conditions contributing to the development of neurodegenerative disorders. This chapter explores the role of HSPB8 in the regulation of neurodegenerative disorders.

R. Patil (✉)
Department of Pharmacology, GES's Sir Dr. M. S. Gosavi College of Pharmaceutical Education and Research, Nashik, Maharashtra, India

N. Wankhede
Smt. Kishoritai Bhoyar College of Pharmacy, Kamptee, Nagpur, Maharashtra, India

A. Upaganlawar
SNJB's Shriman Sureshdada Jain College of Pharmacy, Neminagar, Chandwad, Nashik, Maharashtra, India

S. Ingale
Department of Pharmacology, Indira College of Pharmacy, Pune, Maharashtra, India

G. M. Ashraf, M. S. Uddin (eds.), *Current Thoughts on Dementia*,
https://doi.org/10.1007/978-981-16-7606-2_8

Keywords

Amyotrophic lateral sclerosis · HSPB8 · Chaperones · Heat shock protein B8 · Motor neuron diseases · Neurodegeneration · Motoneuron diseases

8.1 Introduction

Heat shock proteins (HSPs) are the family of cellular protein, which protects against the stresses responsible for cell injury. As a defense mechanism, organisms significantly increase the synthesis of HSPs against multiple stressors, thus conserving the function of the cell. Based on molecular masses, different families of human HSPs have been identified, and HSP70 superfamily includes HSPA (HSP70) and HSPH (HSP110); the DNAJ family includes HSP40; the HSPB family includes small heat shock protein (sHSP); HSPC family includes HSP90; and human chaperonin families include HSPD/E and CCT (Kampinga et al. 2009). Molecular chaperones are the members of the sHSP family, which participates in cellular homeostasis and maintains cellular functions under unfavorable conditions. The sHSPs provide chaperone specificity and inhibit protein aggregation by binding to misfolded proteins at the hydrophobic domain (Jakob et al. 1993). The different members of the sHSP family may exist in multimeric complexes attributed to variations in subunit numbers (12 to >48) (Candido 2002; McDonald et al. 2012; van Montfort et al. 2001).

In the sHSP family, more attention has been provided to HSPB8, as it is involved in important physiological and pathological conditions. These are the intrinsically disordered proteins (IDPs), which in the process partly retain their structure and are also characterized by structural flexibility via reversible changes in folding (Kazakov et al. 2009). Though other members of this family exist as hetero-oligomers or homo-oligomer, HSPB8 exists mainly as equilibrium mixtures of monomers and dimers (Vos et al. 2008). Numerous studies have presented the involvement of HSPB8 in cellular protein quality control mechanisms, supported by its mutations resulting in the development of motor neuropathy. Besides, it participates in the process of apoptosis, autophagy, and cell proliferation. Based on protein expression level and cell type, HSPB8 also modulates apoptotic signaling (Gober et al. 2003; Hase et al. 2005; Li et al. 2006; Depre et al. 2006). Thus, many diseases, including ischemia, myopathy, diabetes, cataract, and neurodegenerative disorders, may involve sHSP dysregulation due to its involvement in physiological processes (Bakthisaran et al. 2015; Kampinga and Garrido 2012; Sun and MacRae 2005; Kannan et al. 2012).

Native state by substrate refolding may not be achieved after binding of sHSPs (Friedrich et al. 2004; Haslbeck et al. 2005; Mogk et al. 2003). The complex of sHSP and substrate acts as an intermediate, which is processed by the chaperonin family (HSP90) and HSP70 (Lee and Vierling 2000; Nillegoda et al. 2015; Nillegoda and Bukau 2015). An overabundance of HSPB8 in cellular function shows effects on different pathological states, viz. cancers (Modem et al. 2011; Li et al. 2014; Suzuki et al. 2015; Yamamoto et al. 2016), myocardial ischemia (Danan et al. 2007), autoimmune disease (Roelofs et al. 2006; Peferoen et al. 2015), and neurological diseases (Irobi et al. 2004; Rusmini et al. 2015; Crippa et al. 2013; Crippa et al. 2016b; Yang et al. 2015).

The sHSP also provides neuroprotection mediated via diverse mechanisms. Being a chaperone protein, HSPB8 is also highly expressed in the brain and exhibits protection in the neurophysiological state. Also, the levels of HSP8 are elevated during various neuropathological conditions including Alzheimer's disease (AD), Parkinson's disease (PD), and Huntington's disease (HD) (Vicario et al. 2014). Environmental, metabolic, and pathophysiological stress continuously affects both organisms and cells disturbing the integrity of proteome and cell functioning causing the death of the cell. The protective mechanism of the heat shock response pathway mediated via molecular HSP chaperone families helps to counteract the damaging effect of stress produced due to extrinsic and intrinsic factors (Akerfelt et al. 2010; Gomez-Pastor et al. 2018). Within these HSPs, the important ATP-dependent molecular chaperones with small molecular mass (12–42 kDa) (Haslbeck and Vierling 2015) and alpha-crystalline domain are sHSPs (Franck et al. 2004).

This chapter explains the current state of knowledge about the structure and role of HSPB8 in various neurodegenerative diseases, highlighting its involvement in neuropathological conditions, thus presenting a promising novel target in neurodegenerative disease for probing the underlying processes.

8.2 Protein Quality Control System in Neurodegeneration

Maintaining protein balance is essential for normal cellular viability and functioning. Different stressors such as heat, reactive oxygen species (ROS), heavy metals, and mutation can disturb the conformational flexibility of protein required for proper functioning, and even can cause misfolding of existing protein, which ultimately results in dysfunctioning or protein aggregation. Such outcomes are centered on the pathology of several neurological disorders such as AD, PD, and HD. Thus, for minimizing the production of misfolded proteins, protein quality control (PQC) mechanisms have been evolved that maintain normal proteostasis (Hartl et al. 2011). At the post-translational level, PQC involves ingenious mechanisms, which include molecular chaperones that help in maintaining proper protein conformation and/or prevent misfolding and protein aggregation by the ubiquitin–proteasome system (UPS) and autophagy–lysosome pathway (ALP), which destroys proteins that are damaged, irreversibly misfolded, or are no longer required by the cell (Amm et al. 2014). Molecular chaperones and degradative pathways are an integrated part of the PQC system. However, chaperones such as HSPs are constitutively expressed but are elevated during the action of different cell stressors (Morimoto 2006). The co-chaperones such as BCL2-associated athanogene (BAG) family act as nucleotide exchange factors (NEFs) for the molecular chaperones (Takayama and Reed 2001). Neuronal loss is a characteristic feature for chaperone and co-chaperone mutations in

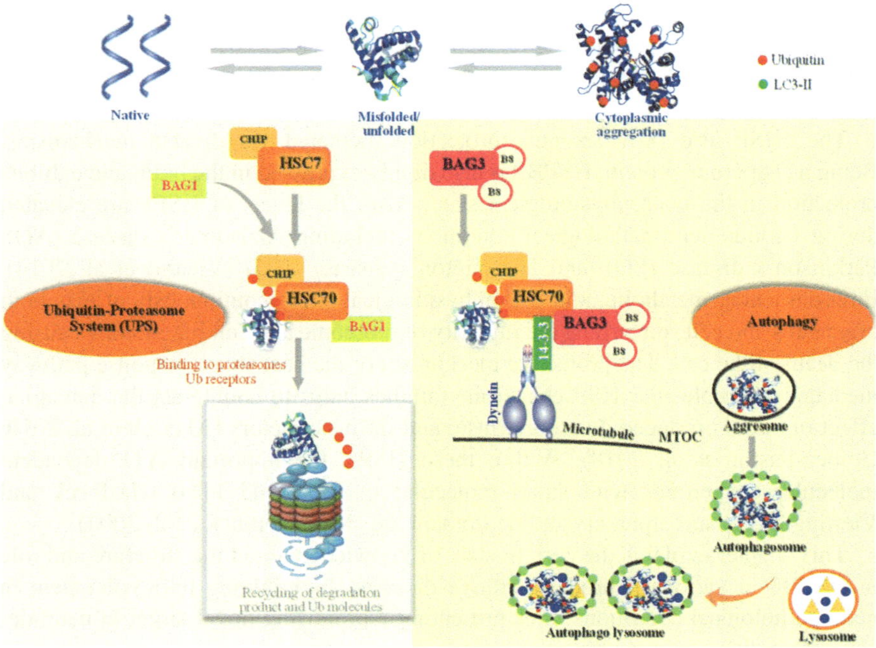

Fig. 8.1 Proteostasis mechanism in neurodegenerative disorders. Molecular chaperones direct misfolded proteins to ubiquitination for degradation. The process is mediated by the HSC70-CHIP complex, which interacts with co-chaperone HSC70 and NEF/BCL2-associated methanogens. BAG1 directs misfolded proteins to polyubiquitination by inhibiting HSP70 and thus promotes degradation via UPS. BAG3 interacts with the protein 14-3-3 and dynein and directs misfolded proteins to autophagosomes for degradation via autophagy (Rusmini et al. 2017). BAG, BCL2-associated athanogene; CHIP, co-chaperone C terminus of Hsc70-interacting protein; HSP, heat shock cognate protein

neurodegenerative diseases (NDs) or other diseases (Smith et al. 2015). This suggests their role in protective mechanism against the degeneration of neurons.

Chaperones control folding of emerging proteins or refolding of existing denatured proteins by directing unfolded, misfolded, or partially folded proteins to degradation via different pathways including UPS, autophagy, and unfolded protein response (UPR). HSPB8 prevents abnormal protein production, which may accumulate in cells escaping degradation during cell death (Cristofani et al. 2017; Minoia et al. 2014; Crippa et al. 2010b) (Fig. 8.1). The endoplasmic reticulum (ER), extensively explored at present, manages the folding and maturation of protein through the UPR signal transduction pathway, which targets the gene specific for ER-associated degradation (ERAD) by translocating the unfolded protein into the cytoplasm to proteasomes, which also participate in PQC (Ron and Walter 2007; Volpi et al. 2017), while specific chaperones and co-chaperones regulate UPS and autophagy pathways (Minoia et al. 2014; Behl 2016; Gamerdinger et al. 2011; Cristofani et al. 2017; Lilienbaum 2013).

UPS has high selectivity and low capacity for misfolded monomeric proteins. The chaperones such as E3-ubiquitin ligase CHIP/STUB1, BAG1, and HSP70 specifically target misfolded proteins to UPS (Fig. 8.1). Autophagy shows low selectivity and high capacity for substrates causing degradation of heteromeric species and damaged organelles (Klionsky et al. 2016). The molecular chaperone actively participates in the autophagy pathway by forming chaperone-assisted selective autophagy complex (CASA), with the target misfolded protein, which is composed of BAG3, CHIP/STUB1, HSP70, and HSPB8. CASA complex activates the receptor, SQSTM1/p62, which binds with LC3 (LC3-II) protein and targets in the direction degradation toward autophagosomes (Klionsky et al. 2016). Several harmful effects in NDs are associated with an imbalance in these two systems of UPS and autophagy (Ciechanover and Kwon 2015; Xilouri and Stefanis 2015; Kakkar et al. 2014; Nikoletopoulou et al. 2015; Senft and Ronai 2015).

The molecular chaperones participate in PQC by activating UPS and thus directing misfolded protein for degradation. The process of degradative pathway involves the interaction of HSC70-CHIP complex along with co-chaperone HSC70, nucleotide exchange factor NEF/BCL2-associated methanogens. BAG1 directs misfolded proteins to polyubiquitination. The interaction of the HSC70-CHIP complex with co-chaperone BAG1 allows misfolded protein degradation via UPS. Alternatively, chaperone-assisted selective autophagy (CASA) involves chaperone HSPB8 and BAG3 complex, which directs the misfolded proteins for degradation via autophagy. HSPB8 helps to recognize misfolded protein and thus acts as a restrictive factor for the formation of a complex. These HSPB8 and BAG3 complexes together interact with HSP70, conjugated with ubiquitin ligase CHIP, whereas BAG3 interacts with the protein 14-3-3 and dynein, which assist delivery of misfolded protein with HSPB8 and BAG3 toward autophagosomes, the microtubule-organizing center (MTOC). The polyubiquitinated protein CHIP in the CASA complex gets recognized by SQSTM1/p62 receptor and inserts the misfolded proteins into autophagosomes (Corti et al. 2020; Rusmini et al. 2017).

Alterations in degradation pathways can result in aggregation of misfolded, which can block PQCS via interfering with autophagy and UPS. The saturation of proteasomes by misfolded protein triggers the expression of HSPB8, which activates the process of autophagy via interaction with HSP70 and BAG3. In the course of failure of dynein-assisted transport and formation of autophagosomes, activation of transcription of BAG1 through unknown factors can stimulate UPS, which is attached to CHIP/HSP70 and leads to misfolded or unfolded proteins to UPS (Fig. 8.2) (Rusmini et al. 2017).

8.3 Distribution of HSPB8

HSPB8 is distributed in various tissues but richly found in the heart, brain, skeletal, and smooth muscle. In human skin, it is present in keratinocytes, which control the growth. The involvement of HSPB8 in cell growth was demonstrated by the study of

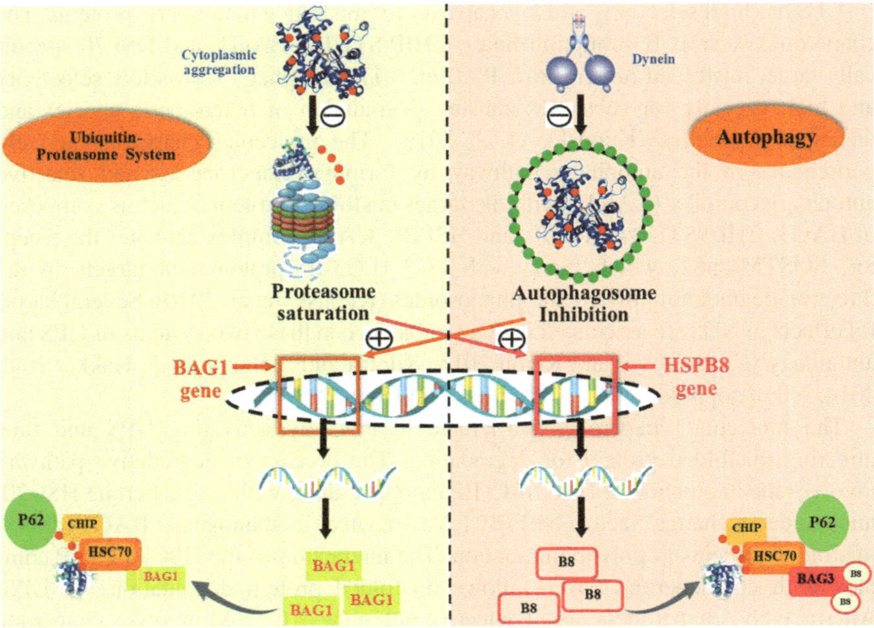

Fig. 8.2 Regulation of protein quality control system. Failure of the degradative system results in the accumulation of misfolded proteins, which blocks UPS and autophagy. As a result, the saturation of proteasomes by misfolded protein triggers the expression of HSPB8, which activates autophagy via HSP70 and BAG3. Concurrently, blockage of autophagosome action activates BAG1, which interacts with CHIP/HSP70 and assists misfolded proteins to UPS (Rusmini et al. 2017). BAG, BCL2-associated athanogene; CHIP, co-chaperone C terminus of Hsc70-interacting protein; HSP, heat shock protein

cultured human keratinocytes in which DNA synthesis and cell proliferation were blocked by inhibition of HSPB8 (Verschuure et al. 2003).

8.3.1 Structure of HSPB8

The sHSPs are made up of two combined sheets of 6–8 β-strands containing conservative α-crystallin domain (ACD) (De Jong et al. 1998). The secondary structure prediction of HSPB8 specifies enrichment of β-strands along with randomly coiled structures (Kim et al. 2006). The unordered structure of HSPB8, evident from the study of far-UV CD, protects against thermal denaturation and proteolysis (Fig. 8.3) (Kazakov et al. 2009).

The approximate molecular mass of HSPB8 is around 22 kDa. It contains a protected amino acid sequence of the α-crystallin domain, which is located on the C-terminal segment (Fig. 8.3). It is also classified as an atypical serine/threonine–protein kinase (Smith et al. 2000). It is an intrinsically disordered protein (IDP) with flexible conformation, which does not have a tertiary structure. It exists in monomer

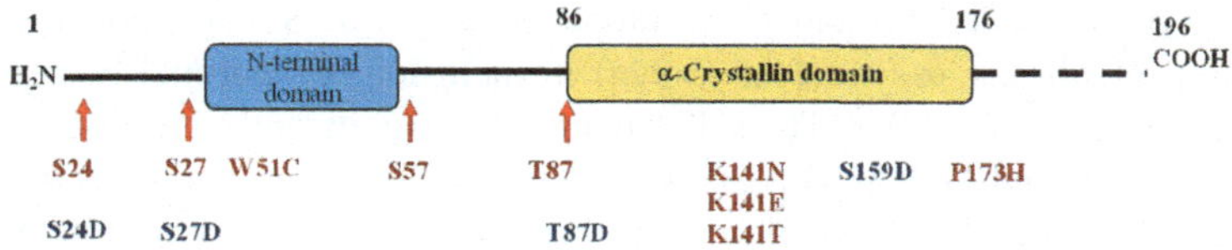

Fig. 8.3 Structure of HSPB8 protein

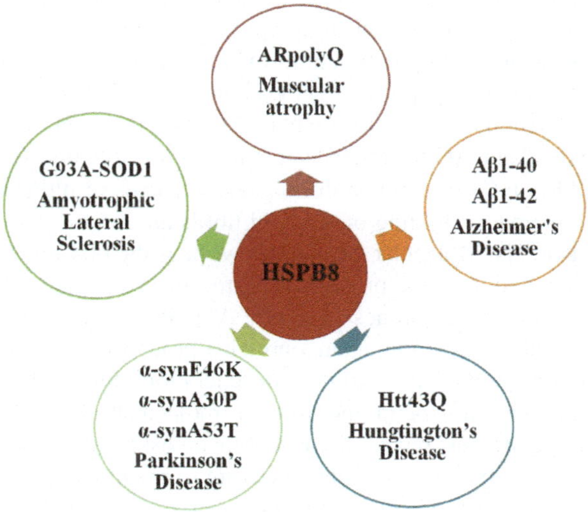

Fig. 8.4 Role of HSPB8 in various neurodegenerative disorders. G93A-SOD1, glycine 93 changed to alanine-superoxide dismutase 1; ARpolyQ, polyglutamine expansion in the androgen receptor; Aβ, amyloid-beta; Htt43Q, huntingtin gene associated with polyglutamine (polyQ) extension of 43 residues; α-syn, α-synuclein variants

form defined by the ultracentrifugation study in solvent glycerol (Chowdary et al. 2004) and differs from others by forming dimers or high-order oligomers. Besides, HSPB8 is enriched in β-strands while lacks β2-strands (Mymrikov et al. 2011).

8.4 sHSP in Neurodegenerative Disorders

Precipitation and aggregation of misfolded proteins are involved in several neurological disorders viz. AD, PD, and amyotrophic lateral sclerosis (ALS) (Fig. 8.4). Also, several studies highlighted the involvement of stress and imbalance in the physiological condition in protein misfolding, which disrupts the proteostasis mechanism. Molecular chaperones such as HSP670, HSP90, and other sHSP families are specialized ATP-dependent chaperone that executes the process of refolding and proteolysis, directing the misfolded protein to UPS and autophagy pathways.

The capability of molecular chaperones to prevent misfolding of protein aggregate formation makes them a novel target in the pathology of many diseases that involves changes in protein conformation (Mogk and Bukau 2017). Mutation and gene alteration in HSPB8 are found to be associated with neurological conditions (Hamouda et al. 2014), whereas enhanced expression of the HSPB8 gene prevents aggregation of HTT43Q in Huntington's disease. Also, HSPB8 facilitates the exclusion of misfolded proteins via autophagy in ALS. Synucleinopathy is the installation of fibrillar α-synuclein (α-syn) in inclusion or neuronal bodies in the processes (Marti et al. 2003). The sHSP expression significantly increases in stress (Bartelt-Kirbach and Golenhofen 2014) and is also found to be co-localized with α-syn in inclusion bodies (Spillantini et al. 1997; Outeiro et al. 2006). Interestingly, HSPB8 removes misfolded proteins that contain elongated polyglutamine chains in other neurodegenerative conditions (Crippa et al. 2010b).

HSPB8 in neurodegenerative diseases may act by preventing the accumulation and aggregation of insoluble proteins. Some of the neurodegenerative conditions associated with protein conformational changes are as follows: AD is characterized by amyloid β-peptides, PD by a mutant α-synuclein forms, ALS by mutant superoxide dismutase 1, HD by mutant huntingtin protein, and muscular dystrophy by an extended CAG tract translated into an polyglutamine (polyQ) tract in the AR protein (ARpolyQ). Protein conformational changes resulting in aggregation or accumulation, and misfolding of amyloid fibrils are responsible for many neurodegenerative disorders. Molecular chaperones work as the first-line defense counter to misfolded, aggregation-prone proteins. The recent investigation suggested the importance of molecular chaperones in ALS, AD, PD, and polyglutamine repeat diseases. It provides protection against proteins prone to aggregation and misfolding and thus acts as potent suppressors of degeneration found in human disease models. Current research has found the role of molecular chaperones in ALS, AD, PD, and polyglutamine repeat diseases (Muchowski and Wacker 2005).

Brain tissues of patients suffering from ailments such as AD, PD, spinocerebellar ataxia type 3 (SCA3), and HD indicate resilient upregulation of HSPB8 in astrocytes along with a minor increase in BAG3. Elevated levels of HSPB8 along with HSPB6 and HSPB1 in multiple sclerosis (MS) are associated with demyelination of white matter (WM) lesion during the active stage of the disease, found entirely in astrocytes but not in oligodendrocytes or microglia. This induction is not detected in the lesions of gray matter (GM) as well (Peferoen et al. 2015). The potential of sHSPs to avert aggregation of α-syn has also been determined by the aggregation process kinetics. The degree of aggregation increases in the presence of gene amplification, macromolecular crowding, and disease-related mutations altering the α-syn aggregation kinetics in cells. It may be associated with the devastation of the protective role due to decreased availability of aggregation contending chaperones (Cox et al. 2016; Rekas et al. 2004). HSPB8 (also known as HSP 22) is associated with the clearance of much-misfolded protein involved in neurodegenerative diseases. Clearance of protein may occur due to the upregulation of autophagy by HSPB8 acting in association with co-chaperone BAG3 (Crippa et al. 2016b). Astrocytes of cerebral areas are the main sites for HSPB8 upregulation in cases of neurodegeneration (Seidel et al. 2012) indicating the importance of astrocytic proteostasis for removal of aggregates in the neuronal microenvironment. HSPB8 restores autophagic flux and removes misfolded aggregates of androgen receptor (AR) poly to promote motor neuron survival of patients suffering from bulbar and spinal muscular atrophy with abnormally long polyQ in mutant AR (Rusmini et al. 2013). Missense mutations in HSPB1 and HSPB8 are mainly involved in the pathogenesis of Charcot-Marie-Tooth (CMT) disease Evgrafov et al. 2004; Irobi et al. 2004; Srivastava et al. 2012).

HSPB8 knockout animals can demonstrate standard locomotor performances. The decrease in HSPB8 aggregates and autophagy were observed in modern knock-in animal models expressing HSPB8 mutant (Bouhy et al. 2018). The cytoprotective role of sHSPs is associated with inhibition of apoptotic machinery by participating in extrinsic and intrinsic apoptotic signaling pathways. HSPB8 also suppresses apoptosis via inhibiting release of cytochrome C from mitochondria (Yang et al. 2015). HSPB8 prevents protein aggregation by getting entombed inside the inclusions of polyglutamine tails with the proteins (Carra et al. 2005). Amyloidosis in the patients of hereditary cerebral hemorrhage revealed the presence of HSPB8 in senile plaques and angiopathy of cerebral amyloid (Carra et al. 2005; Wilhelmus et al. 2006; Wilhelmus et al. 2009).

Carra et al. (2005) studied the involvement of HSPB8 in preventing polyglutamine protein Htt43Q aggregation in the lung fibroblast cell line (CCL39 cells) in Chinese hamster and embryonic kidney 293 cells of humans. Generally, Htt43Q accumulates in perinuclear inclusions consisting of insoluble aggregates of SDS. HSPB8 repressed the gathering of SDS-insoluble Htt43Q. This indicates the role of HSPB8 in sustaining the soluble state of Htt43Q for speedy degradation (Carra et al. 2005; Vos et al. 2010).

8.4.1 Role of HSPB8 in Motor Neuron Diseases

A neurodegenerative disease that affects cortical and/or spinal motor neurons is collectively categorized under motor neuron diseases (MNDs) characterized by progressive muscle weakness and extensor muscle wasting (Irobi et al. 2010). They may be in familial or periodic forms. Pathogenesis of familial MNDs includes altered RNA or protein functions caused by specific gene mutations affecting synthesis or activity of RNA or protein or inducing neurotoxicity and which are specifically involving gain of functions in proteins are ALS and spinal and bulbar muscular atrophy (SBMA). They show unfolding/misfolding due to resistance to folding or conformational instability.

The major proteins affected by missense mutations are HSPB8 and HSPBl that have been reported associated with motor neuropathy (Irobi et al. 2010; Sun et al. 2010). These mutations resulted in alterations mainly at Lysl41 residue in the wild-type HSPB8 protein converting to either Asn (Kl4, NHSPB8) or Glu (K, 41EHSPB8). Numerous studies also show muHSPB8 abnormally interact with HSPB5, HSPBl, and other proteins PASSI, Hic-5 (ARA55), Sam68, BAG3, and TLR4 and act through common signaling pathway for disease progression (Badri et al. 2006; Carra et al. 2009; Fontaine et al. 2006). The sHSPs are implicated in their folding of the protein and additional functions, including protein degradation mediated via the proteasome, RNA processing, redox homeostasis, cell motility, and muscle activity. Thus, mutation in HSPB8 (muHSPB8) may have deleterious effects altering properties and possibly interacting with other proteins. Previous studies also demonstrated that HSPB8 mutations result in protein aggregation along with the reduction in the potential of the mitochondrial membrane in early stage (Irobi et al. 2012).

Impaired cellular functions in misfolded proteins lead to the development of aggregate and cause neurotoxic, which subsequently leads to cell death. PQC system prevents misfolded protein toxicity by reviewing protein folding and clearing damaged substrates. HSPB8 confines to stress granules that get molded after proteotoxic stress and sequester ribonucleoprotein complexes. The intensive accomplishment of the HSPB8-BAG3-HSP70 complex determines the disassembly of stress granules indicating the role of these stress granules in the pathology of ALS. ALS involves the brain motor cortex, brain stem, and anterior horn spinal cord motor neurons (Mateju et al. 2017; Ganassi et al. 2016).

8.4.2 Role of HSPB8 in Amyotrophic Lateral Sclerosis and Muscular Atrophy

Clinically, sporadic ALS (sALS) and familial ALS (fALS) are indistinguishable. The fALS are only 15% of the affected population. It is mainly associated with specific gene mutations involving TAR DNA-binding protein 43 (TDP-43), sequestosome-1 (SQSTM1/p62), superoxide dismutase-1 (SOD-1), fused in sarcoma/translocated in liposarcoma (FUS/TLS), ubiquilin (UBQLN-2), optineurin (OPTN-1), TANK-binding kinase 1 (TBK1), and valosin-containing protein (VCP) (Taylor et al. 2016). These genes play important role in the PQC system and are autophagy-related proteins or mislodge and aggregate applying proteotoxicity (Ju et al. 2009; Taylor et al. 2016; Seguin et al. 2014). The SBMA involves lower motor neurons, neurons of dorsal root ganglia (DRG), distinct androgen target cells in germline tissues, and muscle cells. It varies from ALS by a rate of progression and no involvement of glial or microglia (La Spada et al. 1991; Cortes et al. 2014; Malena et al. 2013; Lieberman et al. 2014; Soraru et al. 2008). The SBMA is associated with the expansion of a CAG repeat in the androgen receptor (AR) gene that leads to elongation of ARpolyQ (La Spada et al. 1991).

ARpolyQ acquires neurotoxic properties by misfolding (Poletti 2004) after binding to testosterone, which acts as its ligand (Katsuno et al. 2002, 2003; Simeoni et al. 2000; Stenoien et al. 1999). Testosterone stimulates conformational changes essential for AR activation possibly damaged by the polyQ. Degradative pathways get altered by the accumulation of misfolded proteins in SBMA, sALS, or fALS. The UPS is possibly flooded by more amount of misfolded/unfolded proteins or inhibited by the poly (Rusmini et al. 2016; Ciechanover and Kwon 2015). Though misfolded protein aggregates could block autophagic flux (Rusmini et al. 2013), the molecular steps that are distorted by the misfolded proteins in these pathways are not tacit. Many chaperones enhance removal of misfolded proteins by aiding proteasomal dilapidation and/or restrictive alterations in autophagic flux (Rusmini et al. 2016; Charmpilas et al. 2017; van Noort et al. 2017).

Chaperone HSPB8 is extensively distributed in many human tissues, at different expression levels. The upregulation of HSPB8 gives protection in the ALS and SBMA (Rusmini et al. 2013; Carra et al. 2013; Crippa et al. 2010b). Mutations in HSPB8 may be responsible for diseases like hereditary motor neuropathy type II (dHMN-II), CMT type 2L, or myopathy, which involve motor neurons and/or muscle cells (Fontaine et al. 2006; Ghaoui et al. 2016). HSPB8 has a vital role in the preservation of motor neuron function and viability, and its mutation impairs the activity of HSPB8 (Kwok et al. 2011). Motor neurons become more susceptible to toxicity induced via misfolded proteins, with age and countenance of HSPB8 in the region of spinal cord that declines with age (Crippa et al. 2010b). HSPB8 mRNA expression is high in the spinal cord sample of ALS patients than in individuals of the same age (Anagnostou et al. 2010). Proteasome impairment is a condition mainly occurring in MNDs and induces HSPB8 expression in cultured motoneurons (Crippa et al. 2010a, b) in the anterior horn spinal cord remaining at end stages of disease in transgenic (Tg) ALS SOD1-G93A mice when compared with wild variety of mice.

Throughout disease development in ALS (Carra et al. 2013) and SBMA (Rusmini et al. 2015), expression of HSPB8 increased drastically in skeletal muscle in mice, contributing to augment the unusual protein removal from muscle to advance cell survival. HSPB8 removes the obstruction of autophagic flux in many NDs. At cellular levels, HSPB8 facilitates misfolded protein autophagic degradation at cellular levels (Rusmini et al. 2013). Recent studies also suggested that enhanced transcription of the C9ORF72 gene results in the expansion of G4C2 hexanucleotide repeats, which form aggregation-prone conformational protein that is difficult to remove via PQC. The molecular chaperones, HSPB8, recognize different peptide repeats (DPRs) generated via transcription alteration and facilitate the degradation of misfolded DPRs responsible for different neurodegenerative diseases (Cristofani et al. 2018). According to increasing genetic and experimental pieces of evidence, translation of ribonucleoprotein complexes and stress granules (SGs) into amyloid-like masses may be responsible for the accumulation of RNA–protein additions in ALS and analogous NDs. Accumulation of misfolded proteins in SGs endorses their transformation into aggregates. The HSPB8-BAG3-HSP70 complex is one of the key factors of granulomatosis (Carra et al. 2017; Ganassi et al. 2016; Mateju et al. 2017).

8.4.3 Role of HSPB8 in Alzheimer's Disease

It is the most common detrimental neurodegenerative condition that leads to dementia and progressive alteration in behavior and learning ability. Pathological factors for disease progression include extracellular protein deposition, and intracellular neurofibrillary tangles (NFTs) result in the formation of senile plaques. With the advancement in research, various pathologies leading to neurodegeneration have been discovered, which majorly includes amyloidal plaque formation and hyperphosphorylation of NFTs (Kumar et al. 2015). Several hypotheses were constructed based on the involvement of causative factors such as the amyloid and tau hypothesis, neurochemical (cholinergic) hypothesis, and inflammation hypothesis (Kurz and Perneczky 2011). In addition to knowing pathology, different studies have also demonstrated the occurrence of α-syn or Lewy related in more than 50% of AD brains, also termed as non-Aβ peptide fragment or non-amyloid-β component of α-syn. Similarly enhanced α-syn levels have been found in cerebrospinal fluid (CSF) of AD patients with cognitive impairment. Further, α-syn also enhances tau hyperphosphorylation. Recent studies suggest that higher α-syn levels are associated with the asymptomatic accumulation of Aβ plaques (Twohig and Nielsen 2019).

One of the distinguishing factors in AD includes senile plaques (SPs) and amyloid angiopathy, which includes deposition protein mainly amyloid-β (Aβ) protein and other proteins such as sHSP and apolipoprotein E, which indirectly interact with them form aggregates. These proteins associated with Aβ result in accumulation and also affect the rate of clearance (Wisniewski and Frangione 1992). The sHSP is involved in the PQC system and thus prevents others from adopting incorrect conformation. Among sHSP, direct interaction between HSP27, actively expressed in astrocytes, and Aβ has been demonstrated (Liang 2000). Furthermore, HSP20 and HSPB2 bind with Aβ and therefore participate in the process of aggregation. HSPB8 has recently gained attention as it contains an α-crystallin domain and it interacts with chaperon HSP27 (Benndorf et al. 2001; Sun et al. 2004). Also, HSPB8 has been demonstrated to prevent protein aggregation during different stress conditions and is expressed in different types of neuronal cells. Furthermore, studies reported that HSPB8 has a higher affinity for Aβ and DAb1–40, and causes reduction in the formation of β-sheet and also inhibits cerebrovascular cytotoxicity mediated by Aβ aggregation (Wilhelmus et al. 2006).

Besides, the HSPB8-BAG3 complex is found to be overexpressed during neurodegenerative conditions and facilitates the clearance of mutated protein prone to aggregation. Postmortem brain studies reported upregulated HSPB8-BAG3 expression in protein conformation disorders such as AD, PDD, and HD. Therefore, the upregulation of HSPB8-BAG3 may contribute to protein homeostasis and in the remodeling of astrocytes during astrogliosis in the above conditions (Seidel et al. 2012). Also, HSP20 and HSP27 prevent Aβ deposition and associated toxicity (Lee et al. 2006). The expression of HSP22 is upregulated with aging and in neurodegenerative conditions like AD, due to deficiency in regulatory mechanism during proteostasis, which can result in misfolded proteins. Numerous studies identified lower expression of HSP22 in excitatory neurons, and also, excitatory glutaminergic neurons are highly susceptible to tau toxicity, thus indicating HSP22 levels are inappropriately being upregulated causing tau activation and making it resistant to proteolytic degradation. In vitro studies have reported that HSP22 significantly reduces tau protein levels, making it a novel target in neurodegenerative conditions (Webster et al. 2020). The miR-425-5p has been recently linked with the pathology of AD and found to be upregulated in AD and also increases tau phosphorylation in HEK293/tau cells. Heat shock protein B8 (HSPB8) has been reported to be targeted by these microRNAs and thus indirectly involved in targeting phosphorylation of tau (Yuan et al. 2020).

8.4.4 Role of HSPB8 in Parkinson's Disease

PD is associated with damage to the dopaminergic network precisely in the substantia nigra. Neuropathologically, PD is characterized by intraneuronal protein aggregates viz. Lewy bodies and Lewy neurites indicating the involvement of alteration in protein handling (Spillantini et al. 1997). The α-syn is an important factor in the pathology of Lewy bodies. Point mutations in α-syn gene is responsible for familial forms of PD, also it can be caused due to an enhanced level of α-syn protein (Olanow and Brundin 2013). Formation of α-syn protein aggregates is a multi-step process that begins with the α-syn misfolding leading to the formation of insoluble oligomers complex and finishes with insoluble fibril formation and aggregates (Ebrahimi-Fakhari et al. 2014). The interaction of HSPB8 with α-syn was found to inhibit the maturation and aggregation of misfolded protein and fibril formation (Bruinsma et al. 2011a).

Based on in vitro studies by Bruinsma et al. (2011a), the most compelling sHSP is HSPB8 in stopping matured fibril development of both mutant and wild-type α-syn (A30P, A53T, and E46K). This study suggests that optimization of the collaboration of α-syn with HSP22 acts as a preparatory point in the expansion of an innovative outcome for involvement in the α-synucleinopathy pathogenesis (Fig. 8.1).

8.4.5 Role of HSPB8 in Huntington's Disease

HD is a common disease inherited by autosomal dominant mutant expansion in the trinucleotide CAG repeats in Htt (huntingtin) gene. HD is associated with polyglutamine (PolyQ). It is characterized by damage in the striatum and cortex neurons resulting in progressive disruption of voluntary motor coordination. The protein that contains a polyglutamine extension of 43 residues (Htt43Q) is unstable. The Htt comprising more than 37 successive glutamines forms insoluble aggregates, a phenotype linked with HD (Ross et al. 2003). In 90% of cells, the development of perinuclear masses takes place as a result of transfection via plasmid encoding an HA-labeled form of Htt43Q and pHDQ43-HA (Wyttenbach et al. 2002). The coexpression of HSPB8 and Htt43Q, intensely reduces aggregation of Htt43Q as >90% of CCL39 cells expressing both Htt43Q and HSPB8 presented with no aggregate in diffuse staining. In the cells wherever inclusion bodies are detected instead of the occurrence of HSPB8, HSPB8 together with the Htt43Q aggregates. HSPB8 actions are similar to that of the chaperones HSP40 and HSP70, which can stop the formation of inclusion of polyglutamine proteins but are often found confined in refractory aggregates (Chai et al. 1999).

8.4.6 Role of HSPB8-BAG3 Induction in Motor Neuron Diseases

HSPB8 is a restrictive element for the autophagic degradation of misfolded proteins. Restoration of autophagy may be achieved by overexpression of HSPB8. The

HSPB8 inducers such as selective estrogen receptor modulators (SERMs) and estrogens (physiological inducers) govern its expression differentially (Piccolella et al. 2017). Doxorubicin and colchicine are powerful HSPB8 inducers. They are autophagy architects of the deduction of insoluble TDP-43 species (Crippa et al. 2016a, b). Trehalose also showed affirmative results in numerous animal models of NDs (Rusmini et al. 2013; Sarkar et al. 2014; He et al. 2016). It also upregulates the expression of BAG3 (Lei et al. 2015). The HSPB2, HSP20 (HSPB6), and HSPB8 are linked to cerebral amyloid angiopathy (CAA) in hereditary cerebral hemorrhage with amyloidosis of the Dutch type (HCHWA-D). Prominently, these sHSPs stimulate interleukin-6 in the cultured neuronal cell, astrocytes, and pericytes, proposing an anti-inflammatory response of sHSPs in HCHWA-D (Wilhelmus et al. 2009).

The majority of AD patients are characterized by HSP20, CAA, HSPB2B3, and HSPB8 that co-localize with CAA and persuade production of intercellular adhesion molecule 1 (ICAM-1), interleukin-8, and monocyte chemoattractant protein by astrocytes in the human brain, strengthening their role in neuroinflammation in AD (Bruinsma et al. 2011b). According to these findings, the exogenous administration of sHSPs shows a defensive role in several diseases having inflammation, protein aggregation, and cell death.

8.5 Recent Development and Future Perspectives

Mechanisms related to the regulation of sHSPs in neurodegeneration by nucleosome remodeling, transcription factor synergy, need to be revealed. Considering the role of HSPB8 in cell physiology, they represent an important target for the treatment of a wide variety of neuronal diseases. The beneficial role of sHSPs in animal models and in clinical trials related to neurodegeneration needs to be explored by interpreting the meticulous regulation and precise targets of these chaperones. Numerous animal models have been developed based on a mutation in the HSPB8 gene in mice to study progressive motor neuropathy via definite neurite degeneration (Bouhy et al. 2018; Ganassi et al. 2016; Irobi et al. 2010).

The protective role of HSPB8 has been explored against TDP43 aggregates in motor neurons, and it also extends survival of hSOD-1G93A mice (Aurelian et al. 2012; Cortese et al. 2018; Rusmini et al. 2017). Trehalose was found to induce HSPB8 expression, thereby reducing ER stress to improve autophagy, delaying disease progression, and prolonging motor neuron survival (Li et al. 2015; Zhang et al. 2014). A potent HSPB8 inducer, colchicine, was found to facilitate autophagy for removal of insoluble TDP-43 in phase II clinical trial in ALS (NCT03693781) (Mandrioli et al. 2019; Rusmini et al. 2017). Recently, surveillance role of HSPB8 in maintaining integrity and dynamism has been explored (Ganassi et al. 2016).

8.6 Conclusion

Though HSPB8 is a type of sHSPs, they differ in many aspects from other sHSPs. The HSPB8 is involved in various neurological disorders including ALS, SBMA, AD, PD, and HD. The chaperone HSPB8 enables the removal of misfolded proteins through autophagy by showing pro-degradative activity and prevents their intracellular accumulation. Activation and recruitment of autophagic machinery in protein folding disorders involve HSPB8 along with co-chaperone BAG3. Astrocytes of cerebral areas undergoing neurodegeneration show upregulation of HSPB8 in the brains of patients with a disease like HD, PD, AD, and SCA3. It also inhibits protein synthesis through the P-eIF2a-stimulating autophagy. The HSPB8 restores autophagic instability and removes misfolded ARpolyQ in spinal and bulbar muscular atrophy to promote motor neuron survival of patients. Induction of HSPB8 in cells affected by MND may be considered as a potential approach to inhibit the onset and progression of the disease.

References

Akerfelt M, Morimoto RI, Sistonen L (2010) Heat shock factors: integrators of cell stress, development and lifespan. Nat Rev Mol Nitul Cell Biol 11(8):545–555

Amm I, Sommer T, Wolf DH (2014) Protein quality control and elimination of protein waste: the role of the ubiquitin-proteasome system. Biochim Biophys Acta, Mol Cell Res 1843(1): 182–196

Anagnostou G, Akbar MT, Paul P, Angelinetta C, Steiner TJ, de Belleroche J (2010) Vesicle-associated membrane protein B (VAPB) is decreased in ALS spinal cord. Neurobiol Aging 31: 969–985

Aurelian L, Laing JM, Lee KS (2012) H11/HSPB8 and its herpes simplex virus type 2 homologue ICP10PK share functions that regulate cell life/death decisions and human disease. Autoim Dis 2012:395329

Badri KR, Modem S, Gerard HC, Khan I, Bagchi M, Hudson AP, Reddy TR (2006) Regulation of Sam68 activity by small heat shock protein 22. J Cell Biochem 99(5):1353–1362

Bakthisaran R, Tangirala R, Rao CM (2015) Small heat shock proteins: role in cellular functions and pathology. Biochim Biophys Acta 1854(4):291–319

Bartelt-Kirbach B, Golenhofen N (2014) Reaction of small heat-shock proteins to different kinds of cellular stress in cultured rat hippocampal neurons. Cell Stress Chaperones 19(1):145–153

Behl C (2016) Breaking BAG: the co-chaperone BAG3 in health and disease. Trends Pharmacol Sci 37:672–688

Benndorf R, Sun X, Gilmont RR, Biederman KJ, Molloy MP, Goodmurphy CW, Cheng H, Andrews PC, Welsh MJ (2001) HSP22, a new member of the small heat shock protein superfamily, interacts with mimic of phosphorylated HSP27 (3DHSP27). J Biol Chem 276(29):26753–26761

Bouhy D, Juneja M, Katona I, Holmgren A, Asselbergh B, De Winter V et al (2018) A knock-in/knock-out mouse model of HSPB8-associated distal hereditary motor neuropathy and myopathy reveals toxic gain-of-function of mutant HSPB8. Acta Neuropathol 135(1):131–148

Bruinsma IB, Bruggink KA, Kinast K, Versleijen AAM, Segers-Nolten IMJ, Subramaniam V, Bea Kuiperij H, Boelens W, de Waal RMW, Verbeek MM (2011a) Inhibition of α-synuclein aggregation by small heat shock proteins. Prot Struct Funct Bioinform 79(10):2956–2967

Bruinsma IB, de Jager M, Carrano A, Versleijen AA, Veerhuis R, Boelens W et al (2011b) Small heat shock proteins induce a cerebral inflammatory reaction. J Neurosci 31(33):11992–12000

Candido EP (2002) The small heat shock proteins of the nematode Caenorhabditis elegans: structure, regulation and biology. Prog Mol Subcell Biol 28:61–78
Carra S et al (2005) HspB8, a small heat shock protein mutated in human neuromuscular disorders, has in vivo chaperone activity in cultured cells. Hum Mol Genet 14(12):1659–1669
Carra S, Brunsting JF, Lambert H, Landry J, Kampinga HH (2009) HSPB8 participates in protein quality control by a non-chaperone-like mechanism that requires eIF2α phosphorylation. J Biol Chem 284(9):5523–5532
Carra S, Rusmini P, Crippa V, Giorgetti E, Boncoraglio A, Cristofani R et al (2013) Different anti-aggregation and pro-degradative functions of the members of the mammalian sHSP family in neurological disorders. Philos Trans Roy Soc Lond Ser B:Biol Sci 368:20110409
Carra S, Alberti S, Arrigo PA, Benesch JL, Benjamin IJ, Boelens W et al (2017) The growing world of small heat shock proteins: from structure to functions. Cell Stress Chaperones 22:601–611
Chai Y, Koppenhafer SL, Bonini NM, Paulson HL (1999) Analysis of the role of heat shock protein (Hsp) molecular chaperones in polyglutamine disease. J Neurosci 19(23):10338–10347
Charmpilas N, Kyriakakis E, Tavernarakis N (2017) Small heat shock proteins in ageing and age-related diseases. Cell Stress Chaperones 22:481–492
Chowdary TK, Raman B, Ramakrishna T, Rao CM (2004) Mammalian Hsp22 is a heat-inducible small heat-shock protein with chaperone-like activity. Biochem J 381(2):379–387
Ciechanover A, Kwon YT (2015) Degradation of misfolded proteins in neurodegenerative diseases: therapeutic targets and strategies. Exp Mol Med 47:e147
Cortes CJ, Ling SC, Guo LT, Hung G, Tsunemi T, Ly L et al (2014) Muscle expression of mutant androgen receptor accounts for systemic and motor neuron disease phenotypes in spinal and bulbar muscular atrophy. Neuron 82:295–307
Cortese A, Laurà M, Casali C, Nishino I, Hayashi YK, Magri S, Taroni F, Stuani C, Saveri P, Moggio M, Ripolone M, Prelle A, Pisciotta C, Sagnelli A, Pichiecchio A, Reilly MM, Buratti E, Pareyson D (2018) Altered TDP-43-dependent splicing in HSPB8-related distal hereditary motor neuropathy and myofibrillar myopathy. Eur J Neurol 25(1):154–163
Corti O, Blomgren K, Poletti A, Beart PM (2020) Autophagy in neurodegeneration: new insights underpinning therapy for neurological diseases. J Neurochem 154(4):354–371
Cox D, Selig E, Griffin MD, Carver JA, Ecroyd H (2016) Small heat-shock proteins prevent alpha-Synuclein aggregation via transient interactions and their efficacy is affected by the rate of aggregation. J Biol Chem 291(43):22618–22629
Crippa V, Carra S, Rusmini P, Sau D, Bolzoni E, Bendotti C et al (2010a) A role of small heat shock protein B8 (HSPB8) in the autophagic removal of misfolded proteins responsible for neurodegenerative diseases. Autophagy 6:958–960
Crippa V, Sau D, Rusmini P, Boncoraglio A, Onesto E, Bolzoni E, Galbiati M, Fontana E, Marino M, Carra S, Bendotti C, de Biasi S, Poletti A (2010b) The small heat shock protein B8 (HSPB8) promotes autophagic removal of misfolded proteins involved in amyotrophic lateral sclerosis (ALS). Hum Mol Genet 19(17):3440–3456
Crippa V et al (2013) Motoneuronal and muscle-selective removal of ALS-related misfolded proteins. Biochem Soc Trans 41(6):1598–1604
Crippa V, Cicardi ME, Ramesh N, Seguin SJ, Ganassi M, Bigi I et al (2016a) The chaperone HSPB8 reduces the accumulation of truncated TDP-43species in cells and protects against TDP-43-mediated toxicity. Hum Mol Genet 25:3908–3924
Crippa V, D'Agostino VG, Cristofani R, Rusmini P, Cicardi ME, Messi E et al (2016b) Transcriptional induction of the heat shock protein B8 mediates the clearance of misfolded proteins responsible for motor neuron diseases. Sci Rep 6:22827
Cristofani R, Crippa V, Rusmini P, Cicardi ME, Meroni M, Licata NV et al (2017) Inhibition of retrograde transport modulates misfolded protein accumulation and clearance in motoneuron diseases. Autophagy 13(8):1280–1303
Cristofani R, Crippa V, Vezzoli G, Rusmini P, Galbiati M, Cicardi ME, Meroni M, Ferrari V, Tedesco B, Piccolella M, Messi E, Carra S, Poletti A (2018) The small heat shock protein B8 (HSPB8) efficiently removes aggregating species of dipeptides produced in C9ORF72-related neurodegenerative diseases. Cell Stress Chaperon 23(1):1–12

Danan IJ, Rashed ER, Depre C (2007) Therapeutic potential of H11 kinase for the ischemic heart. Cardiovasc Drug Rev 25(1):14–29

De Jong WW, Caspers GJ, Leunissen J Am (1998) Genealogy of the α-crystallin—small heat-shock protein superfamily. Int J Biol Macromol 22(3–4):151–162

Depre C et al (2006) H11 kinase prevents myocardial infarction by preemptive preconditioning of the heart. Circ Res 98(2):280–288

Ebrahimi-Fakhari D, Saidi LJ, Wahlster L (2014) Molecular chaperones and protein folding as therapeutic targets in Parkinson's disease and other synucleinopathies. Acta Neuropathol Commun 2(1):1–15

Evgrafov OV, Mersiyanova I, Irobi J, Van Den Bosch L, Dierick I, Leung CL et al (2004) Mutant small heat-shock protein 27 causes axonal Charcot-Marie-Tooth disease and distal hereditary motor neuropathy. Nat Genet 36(6):602–606

Fontaine JM, Sun X, Hoppe AD, Simon S, Vicart P, Welsh MJ et al (2006) Abnormal small heat shock protein interactions involving neuropathy associated HSP22 (HSPB8) mutants. FASEB J 20(12):2168–2170

Franck E, Madsen O, van Rheede T, Ricard G, Huynen MA, de Jong WW (2004) Evolutionary diversity of vertebrate small heat shock proteins. J Mol Evol 59(6):792–805

Friedrich KL, Giese KC, Buan NR, Vierling E (2004) Interactions between small heat shock protein subunits and substrate in small heat shock protein-substrate complexes. J Biol Chem 279(2): 1080–1089

Gamerdinger M, Carra S, Behl C (2011) Emerging roles of molecular chaperones and co-chaperones in selective autophagy: focus on BAG proteins. J Mol Med 89:1175–1182

Ganassi M, Mateju D, Bigi I, Mediani L, Poser I, Lee HO et al (2016) A surveillance function of the HSPB8-BAG3-HSP70 chaperone complex ensures stress granule integrity and dynamism. Mol Cell 63(5):796–810

Ghaoui R, Palmio J, Brewer J, Lek M, Needham M, Evila A et al (2016) Mutations in HSPB8 causing a new phenotype of distal myopathy and motor neuropathy. Neurology 86(4):391–398

Gober MD, Smith CC, Ueda K, Toretsky JA, Aurelian L (2003) Forced expression of the H11 heat shock protein can be regulated by DNA methylation and trigger apoptosis in human cells*. J Biol Chem 278(39):37600–37609. https://doi.org/10.1074/JBC.M303834200

Gomez-Pastor R, Burchfiel ET, Thiele DJ (2018) Regulation of heat shock transcription factors and their roles in physiology and disease. Nat Rev Mol Cell Biol 19(1):4–19

Hamouda MA, Belhacene N, Puissant A, Colosetti P, Robert G, Jacquel A, Mari B, Auberger P, Luciano F (2014) The small heat shock protein B8 (HSPB8) confers resistance to bortezomib by promoting autophagic removal of misfolded proteins in multiple myeloma cells. Oncotarget 5(15):6252–6266

Hartl FU, Bracher A, Hayer-Hartl M (2011) Molecular chaperones in protein folding and proteostasis. Nature 475(7356):324–332

Hase M, Depre C, Vatner SF, Sadoshima J (2005) H11 has dose-dependent and dual hypertrophic and proapoptotic functions in cardiac myocytes. Biochem J 388(2):475–483

Haslbeck M, Vierling E (2015) A first line of stress defense: small heat shock proteins and their function in protein homeostasis. J Mol Biol 427(7):1537–1548

Haslbeck M, Miess A, Stromer T, Walter S, Buchner J (2005) Disassembling protein aggregates in the yeast cytosol. The cooperation of HSP26 with Ssa1 and HSP104. J Biol Chem 280(25): 23861–23868

He Q, Koprich JB, Wang Y, Yu WB, Xiao BG, Brotchie JM et al (2016) Treatment with trehalose prevents behavioral and neurochemical deficits produced in an AAV a-synuclein rat model of Parkinson's disease. Mol Neurobiol 53:2258–2268

Irobi J, Van Impe K, Seeman P, Jordanova A, Dierick I, Verpoorten N et al (2004) Hot-spot residue in small heat-shock protein 22 causes distal motor neuropathy. Nat Genet 36(6):597–601

Irobi J, Almeida-Souza L, Asselbergh B, de Winter V, Goethals S, Dierick I, Krishnan J, Timmermans JP, Robberecht W, de Jonghe P, van den Bosch L, Janssens S, Timmerman V (2010) Mutant HSPB8 causes motor neuron-specific neurite degeneration. Hum Mol Genet 19(16):3254–3265

Irobi J, Holmgren A, De Winter V, Asselbergh B, Gettemans J, Adriaensen D, de Groote CC, Van Coster R, De Jonghe P, Timmerman V (2012) Mutant HSPB8 causes protein aggregates and a reduced mitochondrial membrane potential in dermal fibroblasts from distal hereditary motor neuropathy patients. Neuromuscul Disord 22(8):699–711

Jakob U, Gaestel M, Engel K, Buchner J (1993) Small heat shock proteins are molecular chaperones. J Biol Chem 268(3):1517–1520

Ju JS, Fuentealba RA, Miller SE, Jackson E, Piwnica-Worms D, Baloh RH et al (2009) Valosin-containing protein (VCP) is required for autophagy and is disrupted in VCP disease. J Cell Biol 187:875–888

Kakkar V, Meister-Broekema M, Minoia M, Carra S, Kampinga HH (2014) Barcoding heat shock proteins to human diseases: looking beyond the heat shock response. Dis Model Mech 7:421–434

Kampinga HH, Garrido C (2012) HSPBs: small proteins with big implications in human disease. Int J Biochem Cell Biol 44(10):1706–1710

Kampinga HH, Hageman J, Vos MJ, Kubota H, Tanguay RM, Bruford EA, Cheetham ME, Chen B, Hightower LE (2009) Guidelines for the nomenclature of the human heat shock proteins. Cell Stress Chaperon 14(1):105–111

Kannan R, Sreekumar PG, Hinton DR (2012) Novel roles for alphacrystallins in retinal function and disease. Prog Retin Eye Res 31(6):576–604

Katsuno M, Adachi H, Kume A, Li M, Nakagomi Y, Niwa H et al (2002) Testosterone reduction prevents phenotypic expression in a transgenic mouse model of spinal and bulbar muscular atrophy. Neuron 35:843–854

Katsuno M, Adachi H, Doyu M, Minamiyama M, Sang C, Kobayashi Y et al (2003) Leuprorelin rescues polyglutamine-dependent phenotypes in a transgenic mouse model of spinal and bulbar muscular atrophy. Nat Med 9:768–773

Kazakov AS, Markov DI, Gusev NB, Levitsky DI (2009) Thermally induced structural changes of intrinsically disordered small heat shock protein Hsp22. Biophys Chem 145(2–3):79–85. https://doi.org/10.1016/J.BPC.2009.09.003

Kim MV et al (2006) Structure and properties of K141E mutant of small heat shock protein HSP22 (HspB8, H11) that is expressed in human neuromuscular disorders. Arch Biochem Biophys 454 (1):32–41

Klionsky DJ, Abdelmohsen K, Abe A, Abedin MJ, Abeliovich H, Acevedo Arozena A et al (2016) Guidelines for the use and interpretation of assays for monitoring autophagy (3rd edition). Autophagy 12:1–222

Kumar A, Singh A, Ekavali (2015) A review on Alzheimer's disease pathophysiology and its management: an update. Pharmacol Rep 67(2):195–203

Kurz A, Perneczky R (2011) Novel insights for the treatment of Alzheimer's disease. Prog Neuro-Psychopharmacol Biol Psychiatry 35(2):373–379

Kwok AS, Phadwal K, Turner BJ, Oliver PL, Raw A, Simon AK et al (2011) HSPB8 mutation causing hereditary distal motor neuropathy impairs lysosomal delivery of autophagosomes. J Neurochem 119:1155–1161

La Spada AR, Wilson EM, Lubahn DB, Harding AE, Fischbeck KH (1991) Androgen receptor gene mutations in X-linked spinal and bulbar muscular atrophy. Nature 352:77–79

Lee GJ, Vierling E (2000) A small heat shock protein cooperates with heat shock protein 70 systems to reactivate a heat-denatured protein. Plant Physiol 122(1):189–198

Lee S, Carson K, Rice-Ficht A, Good T (2006) Small heat shock proteins differentially affect Aβ aggregation and toxicity. Biochem Biophys Res Commun 347(2):527–533

Lei Z, Brizzee C, Johnson GV (2015) BAG3 facilitates the clearance of endogenous tau in primary neurons. Neurobiol Aging 36:241–248

Li B et al (2006) Overload of the heat-shock protein H11/HspB8 triggers melanoma cell apoptosis through activation of transforming growth factor-β-activated kinase 1. Oncogene 26(24):3521–3531

Li XS, Qing X, Xiang Yang F, Luo WS (2014) Heat shock protein 22 overexpression is associated with the progression and prognosis in gastric cancer. J Cancer Res Clin Oncol 140(8):1305–1313

Li Y, Guo Y, Wang X, Yu X, Duan W, Hong K, Wang J, Han H, Li C (2015) Trehalose decreases mutant SOD1 expression and alleviates motor deficiency in early but not end-stage amyotrophic lateral sclerosis in a SOD1-G93A mouse model. Neuroscience 298:12–25

Liang JJ-N (2000) Interaction between β-amyloid and lens αB-crystallin. FEBS Lett 484(2):98–101

Lieberman AP, Yu Z, Murray S, Peralta R, Low A, Guo S et al (2014) Peripheral androgen receptor gene suppression rescues disease in mouse models of spinal and bulbar muscular atrophy. Cell Rep 7:774–784

Lilienbaum A (2013) Relationship between the proteasomal system and autophagy. Int J Biochem Mol Biol 4:1–26

Malena A, Pennuto M, Tezze C, Querin G, D'ascenzo C, Silani V et al (2013) Androgen-dependent impairment of myogenesis in spinal and bulbar muscular atrophy. Acta Neuropathol 126:109–121

Mandrioli J, Crippa V, Cereda C, Bonetto V, Zucchi E, Gessani A, Ceroni M, Chio A, D'Amico R, Monsurrò MR, Riva N, Sabatelli M, Silani V, Simone IL, Sorarù G, Provenzani A, D'Agostino VG, Carra S, Poletti A (2019) Proteostasis and ALS: protocol for a phase II, randomised, double-blind, placebo-controlled, multicentre clinical trial for colchicine in ALS (Co-ALS). BMJ Open 9(5):e028486

Marti MJ, Tolosa E, Campdelacreu J (2003) Clinical overview of the synucleinopathies. Move Disord 18(Suppl 6):S21–S27

Mateju D, Franzmann TM, Patel A, Kopach A, Boczek EE, Maharana S et al (2017) An aberrant phase transition of stress granules triggered by misfolded protein and prevented by chaperone function. EMBO J 36(12):1669–1687

McDonald ET, Bortolus M, Koteiche HA, McHaourab HS (2012) Sequence, structure, and dynamic determinants of HSP27 (HSPB1) equilibrium dissociation are encoded by the N-terminal domain. Biochemistry 51(6):1257–1268

Minoia M, Boncoraglio A, Vinet J, Morelli FF, Brunsting JF, Poletti A et al (2014) BAG3 induces the sequestration of proteasomal clients into cytoplasmic puncta: implications for a proteasome-to-autophagy switch. Autophagy 10:1603–1621

Modem S et al (2011) Hsp22 (HspB8/H11) knockdown induces Sam68 expression and stimulates proliferation of glioblastoma cells. J Cell Physiol 226(11):2747–2751

Mogk A, Bukau B (2017) Role of sHSPs in organizing cytosolic protein aggregation and disaggregation. Cell Stress Chaperon 22(4):493–502

Mogk A, Schlieker C, Friedrich KL, Schonfeld HJ, Vierling E, Bukau B (2003) Refolding of substrates bound to small HSPs relies on a disaggregation reaction mediated most efficiently by ClpB/DnaK. J Biol Chem 278(33):31033–31042

Morimoto RI (2006) Stress, aging and neurodegenerative disease. N Engl J Med 355:2254–2255

Muchowski PJ, Wacker JL (2005) Modulation of neurodegeneration by molecular chaperones. Nat Rev Neurosci 6(1):11–22

Mymrikov EV, Seit-Nebi AS, Gusev NB (2011) Large potentials of small heat shock proteins. Physiol Rev 91(4):1123–1159

Nikoletopoulou V, Papandreou ME, Tavernarakis N (2015) Autophagy in the physiology and pathology of the central nervous system. Cell Death Differ 22:398–407

Nillegoda NB, Bukau B (2015) Metazoan HSP70-based protein disaggregases: emergence and mechanisms. Front Mol Biosci 2:57

Nillegoda NB, Kirstein J, Szlachcic A, Berynskyy M, Stank A, Stengel F et al (2015) Crucial HSP70 co-chaperone complex unlocks metazoan protein disaggregation. Nature 524(7564):247–251

Olanow CW, Brundin P (2013) Parkinson's disease and Alpha Synuclein: is Parkinson's disease a prion-like disorder? Mov Disord 28(1):31–40

Outeiro TF, Klucken J, Strathearn KE, Liu F, Nguyen P, Rochet JC et al (2006) Small heat shock proteins protect against alpha-synuclein-induced toxicity and aggregation. Biochem Biophys Res Commun 351(3):631–638

Peferoen LAN et al (2015) Small heat shock proteins are induced during multiple sclerosis lesion development in white but not grey matter. Acta Neuropathol Commun 3(1):87

Piccolella M, Crippa V, Cristofani R, Rusmini P, Galbiati M, Cicardi ME et al (2017) The small heat shock protein B8 (HSPB8) modulates proliferation and migration of breast cancer cells. Oncotarget 8:10400–10415

Poletti A (2004) The polyglutamine tract of androgen receptor: from functions to dysfunctions in motor neurons. Front Neuroendocrinol 25:1–26

Rekas A, Adda CG, Andrew AJ, Barnham KJ, Sunde M, Galatis D et al (2004) Interaction of the molecular chaperone alphaB crystallin with alpha-synuclein: effects on amyloid fibril formation and chaperone activity. J Mol Biol 340(5):1167–1183

Roelofs MF et al (2006) Identification of small heat shock protein B8 (HSP22) as a novel TLR4 ligand and potential involvement in the pathogenesis of rheumatoid arthritis. J Immunol 176 (11):7021–7027

Ron D, Walter P (2007) Signal integration in the endoplasmic reticulum unfolded protein response. Nat Rev Mol Cell Biol 8(7):519–529

Ross CA, Poirier MA, Wanker EE, Amzel M (2003) Polyglutamine fibrillogenesis: the pathway unfolds. Proc Natl Acad Sci 100(1):1–3

Rusmini P, Crippa V, Giorgetti E, Boncoraglio A, Cristofani R, Carra S et al (2013) Clearance of the mutant androgen receptor in motoneuronal models of spinal and bulbar muscular atrophy. Neurobiol Aging 34(11):2585–2603

Rusmini P, Polanco MJ, Cristofani R, Cicardi ME, Meroni M, Galbiati M et al (2015) Aberrant autophagic response in the muscle of a knock-in mouse model of spinal and bulbar muscular atrophy. Sci Rep 5:15174

Rusmini P, Crippa V, Cristofani R, Rinaldi C, Cicardi ME, Galbiati M et al (2016) The role of the protein quality control system in SBMA. J Mol Neurosci 58:348–364

Rusmini P, Cristofani R, Galbiati M, Cicardi ME, Meroni M, Ferrari V, Vezzoli G, Tedesco B, Messi E, Piccolella M, Carra S, Crippa V, Poletti A (2017) The role of the heat shock protein B8 (HSPB8) in motoneuron diseases. Front Mol Neurosci 10:176

Sarkar S, Chigurupati S, Raymick J, Mann D, Bowyer JF, Schmitt T et al (2014) Neuroprotective effect of the chemical chaperone, trehalose in a chronic MPTP-induced Parkinson's disease mouse model. Neurotoxicology 44:250–262

Seguin SJ, Morelli FF, Vinet J, Amore D, De Biasi S, Poletti A et al (2014) Inhibition of autophagy, lysosome and VCP function impairs stress granule assembly. Cell Death Differ 21:1838–1851

Seidel K, Vinet J, den Dunnen WFA, Brunt ER, Meister M, Boncoraglio A et al (2012) The HSPB8-BAG3 chaperone complex is upregulated in astrocytes in the human brain affected by protein aggregation diseases. Neuropathol Appl Neurobiol 38(1):39–53

Senft D, Ronai ZA (2015) UPR, autophagy and mitochondria crosstalk underlies the ER stress response. Trends Biochem Sci 40:141–148

Simeoni S, Mancini MA, Stenoien DL, Marcelli M, Weigel NL, Zanisi M et al (2000) Motoneuronal cell death is not correlated with aggregate formation of androgen receptors containing an elongated polyglutamine tract. Hum Mol Genet 9:133–144

Smith CC, Yu YX, Kulka M, Aurelian L (2000) A novel human gene similar to the protein kinase (PK) coding domain of the large subunit of herpes simplex virus type 2 ribonucleotide reductase (ICP10) codes for a serine-threonine PK and is expressed in melanoma cells. J Biol Chem 275 (33):25690–25699

Smith HL, Li W, Cheetham ME (2015) Molecular chaperones and neuronal proteostasis. Semin Cell Dev Biol 40:142–152

Sorarù G, D'ascenzo C, Polo A, Palmieri A, Baggio L, Vergani L et al (2008) Spinal and bulbar muscular atrophy: skeletal muscle pathology in male patients and heterozygous females. J Neurol Sci 264:100–105

Spillantini MG, Schmidt ML, Lee VM, Trojanowski JQ, Jakes R, Goedert M (1997) Alpha-synuclein in Lewy bodies. Nature 388(6645):839–840

Srivastava AK, Renusch SR, Naiman NE, Gu S, Sneh A, Arnold WD et al (2012) Mutant HSPB1 overexpression in neurons is sufficient to cause age-related motor neuronopathy in mice. Neurobiol Dis 47(2):163–173

Stenoien DL, Cummings CJ, Adams HP, Mancini MG, Patel K, DeMartino G et al (1999) Polyglutamine-expanded androgen receptors form aggregates that sequester heat shock proteins, proteasome components and SRC-1 and are suppressed by the HDJ-2 chaperone. Hum Mol Genet 8:731–741

Sun Y, MacRae TH (2005) The small heat shock proteins and their role in human disease. FEBS J 272(11):2613–2627

Sun X, Fontaine JM, Rest JS, Shelden EA, Welsh MJ, Benndorf R (2004) Interaction of human HSP22 (HSPB8) with other small heat shock proteins. J Biol Chem 279(4):2394–2402

Sun X, Fontaine JM, Hoppe AD, Carra S, Deguzman C, Martin JL, Simon S, Vicart P, Welsh MJ, Landry J, Benndorf R (2010) Abnormal interaction of motor neuropathy-associated mutant HSPB8 (HSP22) forms with the RNA helicase Ddx20 (gemin3). Cell Stress Chaperon 15(5): 567–582

Suzuki M et al (2015) Regulation by heat shock protein 22 (HSPB8) of transforming growth factor-α-induced ovary cancer cell migration. Arch Biochem Biophys 571:40–49

Takayama S, Reed JC (2001) Molecular chaperone targeting and regulation by BAG family proteins. Nat Cell Biol 3:E237–E241

Taylor JP, Brown RH, Cleveland DW (2016) Decoding ALS: from genes to mechanism. Nature 539:197–206

Twohig D, Nielsen HM (2019) α-Synuclein in the pathophysiology of Alzheimer's disease. Mol Neurodegener 14(1):1–19

van Montfort RL, Basha E, Friedrich KL, Slingsby C, Vierling E (2001) Crystal structure and assembly of a eukaryotic small heat shock protein. Nat Struct Biol 8(12):1025–1030

van Noort JM, Bugiani M, Amor S (2017) Heat shock proteins: old and novel roles in neurodegenerative diseases in the central nervous system. CNS Neurol Disord Drug Targets 16:244–256

Verschuure P et al (2003) Expression of small heat shock proteins HspB2, HspB8, Hsp20 and CvHsp in different tissues of the perinatal developing pig. Eur J Cell Biol 82(10):523–530

Vicario M, Skaper D, S., & Negro, A. (2014) The small heat shock protein HSPB8: role in nervous system physiology and pathology. CNS Neurol Disord Drug Targets 13:885–895

Volpi VG, Touvier T, D'antonio M (2017) Endoplasmic reticulum protein quality control failure in myelin disorders. Front Mol Neurosci 9:162

Vos MJ, Hageman J, Carra S, Kampinga HH (2008) Structural and functional diversities between members of the human HSPB, HSPH, HSPA, and DNAJ chaperone families†. Biochemistry 47 (27):7001–7011

Vos MJ et al (2010) HSPB7 is the most potent PolyQ aggregation suppressor within the HSPB family of molecular chaperones. Hum Mol Genet 19(23):4677–4693

Webster JM, Darling AL, Sanders TA, Blazier DM, Vidal-Aguiar Y, Beaulieu-Abdelahad D, Plemmons DG, Hill SE, Uversky VN, Bickford PC, Dickey CA, Blair LJ (2020) HSP22 with an N-terminal domain truncation mediates a reduction in tau protein levels. Int J Mol Sci 21(15): 5442

Wilhelmus MMM, Boelens WC, Otte-Höller I, Kamps B, Kusters B, Maat-Schieman MLC, de Waal RMW, Verbeek MM (2006) Small heat shock protein HSPB8: its distribution in Alzheimer's disease brains and its inhibition of amyloid-β protein aggregation and cerebrovascular amyloid-β toxicity. Acta Neuropathol 111(2):139–149

Wilhelmus MM, Boelens WC, Kox M, Maat-Schieman ML, Veerhuis R, de Waal RM et al (2009) Small heat shock proteins associated with cerebral amyloid angiopathy of hereditary cerebral haemorrhage with amyloidosis (Dutch type) induce interleukin-6 secretion. Neurobiol Aging 30(2):229–240

Wisniewski T, Frangione B (1992) Apolipoprotein E: a pathological chaperone protein in patients with cerebral and systemic amyloid. Neurosci Lett 135(2):235–238

Wyttenbach A et al (2002) Heat shock protein 27 prevents cellular polyglutamine toxicity and suppresses the increase of reactive oxygen species caused by huntingtin. Hum Mol Genet 11 (9):1137–1151

Xilouri M, Stefanis L (2015) Chaperone mediated autophagy to the rescue: a new-fangled target for the treatment of neurodegenerative diseases. Mol Cell Neurosci 66:29–36

Yamamoto N et al (2016) Heat shock protein 22 (HSPB8) limits TGF-β-Stimulated migration of osteoblasts. Mol Cell Endocrinol 436:1–9

Yang B, Zhang H, Mo X, Xiao H, Hu Z (2015) HSPB8 is neuroprotective during oxygen glucose deprivation and reperfusion. Curr Neurovasc Res 12(1):63–72

Yuan J, Wu Y, Li L, Liu C (2020) MicroRNA-425-5p promotes tau phosphorylation and cell apoptosis in Alzheimer's disease by targeting heat shock protein B8. J Neural Transm 127(3): 339–346

Zhang X, Chen S, Song L, Tang Y, Shen Y, Jia L, Le W (2014) MTOR-independent, autophagic enhancer trehalose prolongs motor neuron survival and ameliorates the autophagic flux defect in a mouse model of amyotrophic lateral sclerosis. Autophagy 10(4):588–602

Molecular Insights into the Role of ER Stress in Neurodegenerative Diseases

9

Md. Tanvir Kabir

Abstract

Endoplasmic reticulum (ER) stress can take place due to the disruptions in the structure and functions of ER, along with the changes in calcium homeostasis and buildup of misfolded proteins (MPs). Characteristics of ER response include degradation of MPs, stimulation of ER chaperones, induction of translational attenuation, and alterations in specific proteins. Aggravated or extended ER stress can lead to cell death activation. ER stress is linked with multiple neurodegenerative diseases (NDs), such as Alzheimer's disease, Parkinson's disease, amyotrophic lateral sclerosis, and Huntington's disease. Nonetheless, still the precise roles and causal activities of ER stress in case of the pathogenesis of various disorders are not well known. In this chapter, the probable functions of ER stress in NDs have been summarized. In addition, current understanding of this field has also been highlighted, which might show novel insights regarding disease processes and may help the development of better therapeutic agents for these NDs.

Keywords

Endoplasmic reticulum · ER stress · Misfolded proteins · Alzheimer's disease · Parkinson's disease · Amyotrophic lateral sclerosis · Huntington's disease

M. T. Kabir (✉)
Department of Pharmacy, Brac University, Dhaka, Bangladesh

G. M. Ashraf, M. S. Uddin (eds.), *Current Thoughts on Dementia*,
https://doi.org/10.1007/978-981-16-7606-2_9

9.1 Introduction

Endoplasmic reticulum (ER) is a crucial organelle that plays roles in the production, proper folding, post-translation alteration, and nascent protein delivery to various destinations (Paschen and Frandsen 2001; Breckenridge et al. 2003; Rao et al. 2004). Quality control of proteins via ER ensures proper management of the end product that is essential to maintain normal cellular activities (Paschen and Frandsen 2001; Breckenridge et al. 2003; Rao et al. 2004). Loss of integrity or disruption in the activity of ER can result in ER stress that can take place, for instance, due to the buildup of unfolded proteins and via alterations in calcium homeostasis in ER. Furthermore, ER stress is also responsible for activation of signaling mechanisms including unfolded protein response (UPR) that counters the activities of the main stress containing a genetic or environmental cause (Fig. 9.1) (Paschen and Frandsen 2001; Breckenridge et al. 2003; Rao et al. 2004). UPR has been found to alter the expressions of certain proteins, including those for the ER chaperones, increase degradation of misfolded proteins (MPs), and suppress protein generation to reduce load within the ER (Paschen and Frandsen 2001; Breckenridge et al. 2003; Rao et al. 2004). Nevertheless, if the ER activity is extremely weakened, pathways and genes that cause cell death and/or suppression of survival are also activated (Paschen and Frandsen 2001; Breckenridge et al. 2003; Rao et al. 2004). Cell signaling associated with ER stress-induced apoptosis and cell death is complex and yet to be fully revealed (Breckenridge et al. 2003; Rao et al. 2004). Disrupted activities of the ubiquitin–proteasome system (UPS) are accountable for the degradation of synaptic, ER, and cytosolic proteins, which may eventually play role in ER stress (Imai et al. 2001; Ciechanover and Brundin 2003; Korhonen and Lindholm 2004).

Deposits and accumulation of MPs are the common signs of numerous neurodegenerative diseases (NDs). These MPs influence multiple cell signaling mechanisms, neuronal connectivity, and neuronal death (Bence et al. 2001; Soto 2003). Effects of UPS are reduced in these NDs via protein aggregates or through increased oxidative damage and other toxic products (Bence et al. 2001; Soto 2003). Besides, impairment of the UPS can lead to more buildup of proteins in the cell, which can further result in ER stress and exacerbation of the disorder (Imai et al. 2001; Nishitoh et al. 2002; Ciechanover and Brundin 2003). Environmental toxins, various cell signaling pathways, and reactive oxygen species (ROS) affect mitochondria and result in activation of the caspase family of cysteine proteases leading to cell death (Breckenridge et al. 2003; Degterev et al. 2003; Rao et al. 2004). Indeed, intracellular calcium concentration and calcium secretion from the ER are crucial in interactions between mitochondria and ER and in the regulation of neuronal death (Fig. 9.1) (Verkhratsky 2005).

In this chapter, the studies regarding ER stress and its action in various NDs via studying findings collected regarding molecular and cellular mechanisms associated with ER have been reviewed. Furthermore, how these cellular mechanisms might play a role in the disease process of various NDs has also been discussed.

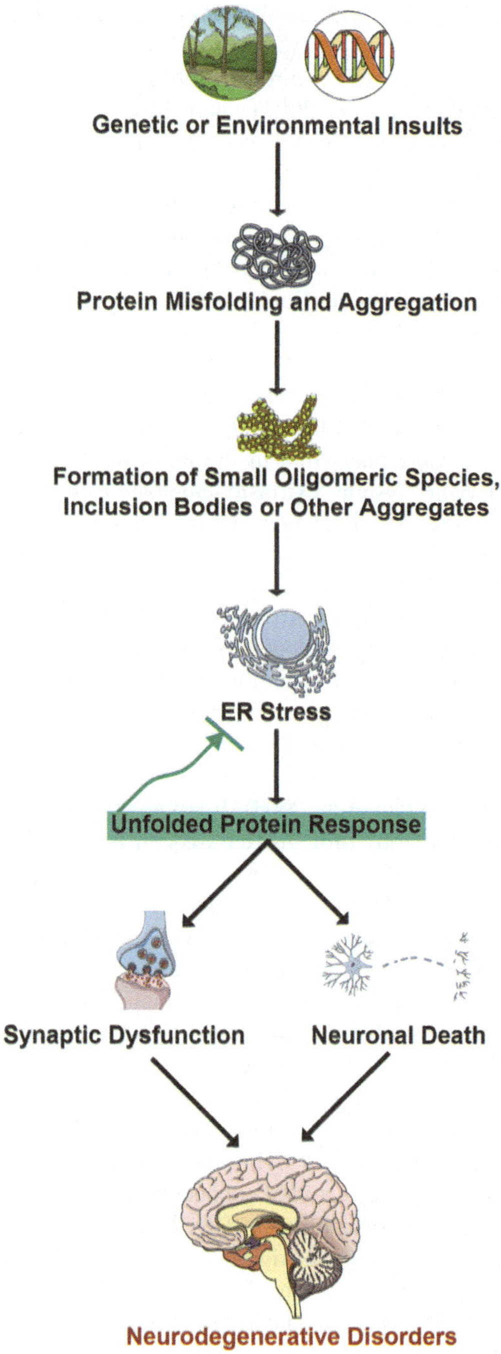

Fig. 9.1 Relationship between ER stress and neurodegenerative disorders. Various genetic or environmental insults might trigger misfolding and aggregation of proteins, which can lead to the generation of various types of aggregates including inclusion bodies, small oligomeric species, or other aggregates, which can further lead to ER stress and eventually unfolded protein response (UPR) activation. UPR is induced to resolve ER stress and endeavors to restore its activities via blocking translation, triggering chaperons, and elevating protein folding in the ER. However, extended ER stress can lead to prolonged UPR activation, which can result in neuronal death, synaptic dysfunction, and eventually neurodegenerative disorders

9.2 ER Stress-Induced Neuronal Cell Death Mechanisms

Like other cells, neuronal death is controlled via two principle pathways (Mattson 2000; Yuan et al. 2003) where the extrinsic pathway takes place due to the action of cell membrane death receptors via caspase-8 activation that causes cleavages of downstream substrates, as well as other caspases (Degterev et al. 2003; Danial and Korsmeyer 2004). Several insults and stressors in the mitochondria can induce the intrinsic pathway. This can induce alterations in membrane permeability transition with the secretion of cytochrome c and other proapoptotic molecules from the inner membrane space (Danial and Korsmeyer 2004). Interestingly, cytochrome c has the ability to bind with apoptotic protease-activating factor 1 and can activate caspase-9 and then caspase-3. Moreover, these 2 pathways can come together on caspase-3 and can induce each other (as demonstrated via truncated B-cell lymphoma 2 (Bcl-2) homology 3 (BH3)-interacting domain death agonist that following caspase-8-cleavage plays role in the mitochondria) (Degterev et al. 2003). In case of the nervous system, extrinsic apoptosis pathway has a less significant contribution as compared to the immune system. Nevertheless, activation of inflammatory responses of death receptors takes place along with increased cleavage of caspase-8 in the brain. Indeed, inflammation is a common feature of multiple NDs and it plays role in the disease process (Wyss-Coray and Mucke 2002). In addition, other than the classical caspase-dependent apoptosis, other kinds of apoptotic processes have been confirmed in multiple cell types (such as neurons) that might or might not include mitochondria (Yuan et al. 2003).

Neurons express various antiapoptotic proteins to counteract both genetic and environmental insult-mediated cell degeneration to fight against cell death (Mattson 2000; Yuan et al. 2003). It is known that multidomain Bcl-2 family involves both antiapoptotic proteins (e.g., B-cell lymphoma extra large (Bcl-xL) and Bcl-2) and proapoptotic proteins (e.g., Bcl-2-interacting killer (Bik), Bak, and bcl-2-like protein 4 (Bax)) (Danial and Korsmeyer 2004). Under normal conditions, Bcl-2 regulates the mitochondrial membrane integrity (Danial and Korsmeyer 2004). It was reported that BH3-only proteins can promote apoptosis via stimulating Bax oligomerization or via suppressing Bcl-2 (Danial and Korsmeyer 2004). Several BH3 proteins including Bik and the cloned Spike are mostly found in ER (Germain et al. 2002; Mund et al. 2003). Furthermore, BH3 protein including Bcl-2-like protein 11 has been found to translocate to the ER membrane and is vital for ER stress-induced cell death mechanisms (Morishima et al. 2004). Nonetheless, more studies are required on the activities of BH3-only proteins in NDs or ER stress-mediated neuronal cell death. There is a vital connection between the mitochondria and ER in the implementation of cell death (Breckenridge et al. 2003; Rao et al. 2004). Bcl-2 family proteins have a significant contribution in interactions between ER and mitochondria, where Bcl-xL and Bcl-2 are linked with ER membrane and mitochondria (Breckenridge et al. 2003; Rao et al. 2004). It has been found that Bcl-2 in association with Bak and Bax, and ER calcium channels including inositol 1,4,5-triphosphate receptor (IP3-R) can control ER calcium concentrations and secretion into the cytosol (Chen et al. 2004).

Indeed, intracellular calcium concentrations are important for numerous neuronal activities such as neuronal death (Verkhratsky 2005). Interestingly, mitochondria's sensitivity to stimulate apoptosis differs from the levels of ROS, calcium, and other metabolites that exist in the cell (Danial and Korsmeyer 2004; Verkhratsky 2005). Except for Bcl-2 proteins, other molecules that play role in the ER–mitochondria interactions have also been detected (Breckenridge et al. 2003; Rao et al. 2004). It has been observed that B-cell receptor-associated protein 31 (Bap31) is crucial for the trafficking of proteins (Breckenridge et al. 2003). Caspase cleavage of BAP31 can result in release of calcium from the ER, which can eventually mediate the cytochrome c release from the mitochondria (Breckenridge et al. 2003). A novel gene, phosphofurin acidic cluster sorting protein 2 (*PACS2*), has been observed to affect several ER activities and cell death by participating with the mitochondria (Simmen et al. 2005). In case of NDs, importance of these proteins in ER stress is still not well known. In addition to this, ER harbors caspase-12 that can be selectively activated/cleaved in the course of ER stress (Nakagawa et al. 2000). Therefore, caspase-12 can cause activation of downstream caspases to induce apoptosis (Rao et al. 2002; Morishima et al. 2002). Caspase-12 involvement with diseases is unclear, since the gene exhibits large deletions in the human genome (Degterev et al. 2003; Obeng and Boise 2005). There is a chance that other caspases take over activities of caspase-12 in humans, as indicated for the human caspase-4 (Hitomi et al. 2004). Nevertheless, still there is an argument regarding the activity of caspase-4 and caspase-12 as initiators of ER stress-mediated apoptosis (Obeng and Boise 2005).

During cell death, caspases can cleave other ER proteins (except Bap31). These involve IP3 receptors, and presenilin-1 (PS1) and presenilin-2 (PS2) that are ER transmembrane proteins, which are crucial in Alzheimer's disease (AD) (Selkoe 2001; Kabir et al. 2020d; Uddin et al. 2020d). Although presenilins (PSs) are antiapoptotic, however, they lose this capacity following caspase cleavage. Several well-known ER proteins and their association with the activity of ER in cell death control are illustrated in Fig. 9.2. Other than these proteins, activation of caspase-12 and ER stress is affected in neurons due to the upstream signals caused via brain-derived neurotrophic factor and due to the presence of hippocalcin (a neuronal calcium-binding protein) (Shimoke et al. 2004; Korhonen et al. 2005). However, more studies are required to discover other proteins and interactions that play role in the ER–mitochondria interaction and regulation of ER-mediated cell death. On the other hand, Bcl-2 along with IP3-R regulates ER calcium stores and release into the cytosol. Increased calcium concentrations sensitize mitochondria to other insults stimulating cell death. Collectively, Bax and the BH3-only proteins can cause suppression of Bcl-2. Bcl-2 can bind with Bap31 that usually leads to antiapoptotic activity in the ER. Several ER chaperones including binding immunoglobulin protein (Bip)/glucose-regulated protein 78 (GRP78) exert protective function and regulate protein folding and UPR components. It is regarded that caspase-12 can directly trigger cell death. It has been observed that calpain (a member of the family of calcium-dependent proteases) can cause caspase-12 activation. Indeed, mutations in PS proteins can play role in AD. Parkin-associated endothelin receptor-like

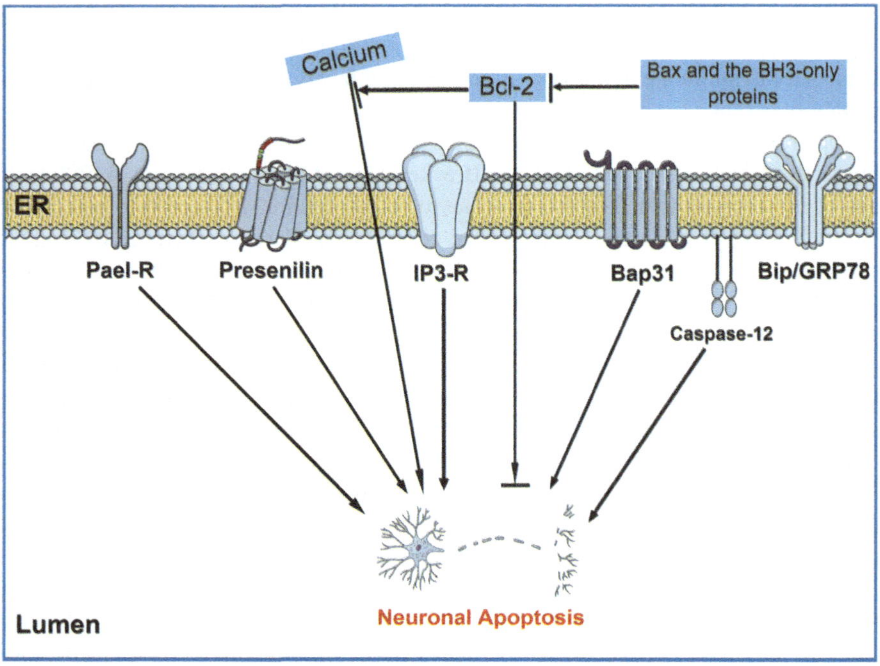

Fig. 9.2 Effects of ER proteins in the regulation of neuronal death (Lindholm et al. 2006). *Bap31*, B-cell receptor-associated protein 31; *Bcl-2*, B-cell lymphoma 2; *BH3*, Bcl-2 homology 3; *Bip*, binding immunoglobulin protein; *ER*, endoplasmic reticulum; *IP3-R*, inositol 1,4,5-triphosphate receptor; *Pael-R*, parkin-associated endothelin receptor-like receptor

receptor (Pael-R) is prone toward unfolding and can induce ER stress. Interestingly, Pael-R acts as a substrate for the parkin (a ubiquitin E3 ligase), which is mutated in case of early-onset PD along with the weakened proteasomal activity. More experiments are required to find out the precise contributions of various ER proteins in regulating cell death in several NDs (Fig. 9.2).

9.3 Role of Endoplasmic Reticulum Stress in Neurodegenerative Diseases

9.3.1 Alzheimer's Disease

Globally, around 24 million people are affected by dementia and most of the cases are AD individuals (Ballard et al. 2011). AD is a big burden for the modern society, and it has been estimated that one new AD case will arise every 33 seconds in the USA, which will eventually result in around one million new AD cases per year by 2050 (Ballard et al. 2011; Cavedo et al. 2014). Characteristics of AD include an irreversible and progressive brain degeneration that further results in a progressive loss of memory and cognitive deficit and even death (Ballard et al. 2011; Kabir et al. 2019b, 2020e). In 1906, Alois Alzheimer characterized AD and observed in the brain that there is a presence of plaques and neurofibrillary tangles (NFTs) (Hippius and Neundörfer 2003; Kabir et al. 2020b; Uddin et al. 2020a). Still, these histopathological hallmarks are used to confirm the diagnosis of the disease in postmortem tissue (Ballard et al. 2011; Scheff et al. 2016). Subsequently, more extensive studies on brains of AD individuals revealed the characteristics of NFTs and senile plaques (also called amyloid plaques) composed of hyperphosphorylated tau (a microtubule assembly protein) and misfolded amyloid-beta (Aβ) protein aggregates, respectively (Brandt et al. 2005; Glenner and Wong 2012). It is known that senile plaques contain a central amyloid core, surrounded by dystrophic neurites and an extreme inflammatory reaction, while NFTs are aberrant filamentous inclusions located in neuronal somata comprised mainly of aberrantly folded tau (Brandt et al. 2005; Oakley et al. 2006; Glenner and Wong 2012). Indeed, the origin of Aβ emerges from the amyloid precursor protein (APP) cleavage, which may go through the non-amyloidogenic pathway (normal processing) or the amyloidogenic pathway (abnormal processing) (Uddin et al. 2020e, f). APP cleavage is induced via signal peptidases during the translocation of APP from the ER to the Golgi apparatus, where it achieves maturation. After the non-amyloidogenic pathway, α-secretase cleaves APP, and this cleavage results in the formation of α-fragments of APPs, which shows no toxicity (Dovey et al. 2001; Luo et al. 2001; Uddin et al. 2020b).

However, in case of amyloidogenic pathway, APP might get alternatively degraded via β-secretase-1 (BACE-1) and γ-secretase, which can further lead to generation of neurotoxic Aβs (Dovey et al. 2001; Luo et al. 2001; Haass 2004; Kabir et al. 2019a). It has been reported that mutations in coding regions of preselinin-1 can mediate the amyloidogenic pathway and are considered as the main causal factor for autosomal-dominant familial AD (FAD) (Berezovska et al. 2005; Uddin et al. 2020c). Disruption in this metabolic cascade can result in the buildup of neurotoxic oligomers that can further form intra- and extra-cellular deposits, which is responsible for the interruption of survival of neurons and synaptic transmission, which can eventually lead to the AD-associated devastating symptoms (Reitz 2012; McGeer and McGeer 2013). Interestingly, rather than the amyloid plaques, small oligomers or intraneuronal Aβ might represent the most related neurotoxic species (Endres and Reinhardt 2013; Kabir et al. 2020c). Histopathological features of AD brains indicate that dysfunction of proteostasis may have a significant contribution to AD pathogenesis (Caughey and Lansbury 2003; Glenner and Wong 2012; Walker et al. 2015; Kabir et al. 2020a). It has been reported that there is an increased level of multiple chaperones including ER stress-related chaperone BiP/GRP78 and heat shock protein-27 in the brains of AD individuals (Hamos et al. 1991). Furthermore, BiP showed an elevated level of expression in nerve cells that did not exhibit deterioration, which indicates that this chaperone might provide neuroprotection even before degeneration in AD. In addition, mutations in presilinin-1 (a gene that is associated with an elevated frequency of FAD) resulted in downregulation of BiP expression (Katayama et al. 1999), which indicates that mutations in PS can alter the ER homeostasis and UPR firing; however, opposite findings revealed that loss of

PS-1 or expression of its variants is not adequate to induce UPR (Sato et al. 2000). In a study, Hoozemans et al. (2005) observed increased BiP level in the hippocampus and temporal cortex regions of AD individuals that was found to be positively linked with elevated proportion of NFTs and Aβ pathology (Hoozemans et al. 2005; Uddin et al. 2019). Immunohistochemistry studies detected that neuronal population were positive for BiP in individuals with AD in comparison with the non-demented elderly people. Furthermore, these researchers also observed that neurons that were positive for BiP did not coexist with NFTs, which indicates that expression of BiP preceded the formation of NFTs (Al Mamun et al. 2020).

Indeed, ER stress was found to be activated during the period of AD; therefore, there is a need to assess the function of other mediators of UPR in AD pathophysiology (Uddin et al. 2020g). Neurons that are located in the hippocampal CA1 area of AD individuals showed an elevated level of phospho-protein kinase RNA-like endoplasmic reticulum kinase (PERK) expression in comparison with the non-demented individuals (controls) (Hoozemans et al. 2009), which was found to be positively linked with Braak stage. These observations are in line with the findings of studies with AD animal models (Page et al. 2006; Abisambra et al. 2013). The observations of elevated level of BiP expression and PERK phosphorylation also suggest that tau can interfere with the ER-associated degradation (ERAD). The eukaryotic initiation factor 2α (eIF2α)/PERK signaling pathway also mediates the UPR that plays a crucial role in AD pathogenesis (Lourenco et al. 2015). Interestingly, its phosphorylation modulated the expression of BACE-1 in AD mouse models (Vassar 2009). It has also been revealed that activating transcription factor 4 (ATF4) regulated the activity of gamma-secretase during amino acid imbalances, which further induces ER stress and mediates amyloidogenic pathway (Mitsuda et al. 2007). Interestingly, Aβ-treated neuronal cultures developed increased level of PERK phosphorylation, which suggests cellular protection, since its silencing elevates the level of neuronal death following treatment with Aβ (Lee et al. 2010a). Thus, activation of the ER stress response takes place in order to maintain neuron homeostasis and to avert AD-related neurodegeneration. In animal models, inositol-requiring enzyme 1 (IRE1) induced c-Jun N-terminal kinase 3 in the Aβ-treated brain, and its deletion was found to restore the translational block stimulated via oligomeric Aβ (Yoon et al. 2012).

X-box-binding protein 1 (XBP1) is also a crucial part of UPR that is also associated with AD pathogenesis. An increased level of XBP1 mRNA splicing was observed in transgenic flies expressing Aβ, which averted toxicity of oligomers (Casas-Tinto et al. 2011). Also, samples obtained from the brains of AD individuals showed increased XBP1 in comparison with age-matched controls, which further suggests the disrupted activity of UPR in AD pathogenesis (Lee et al. 2010b). Splicing of XBP1 mRNA was noticed following Aβ treatment in the temporal cortex region of AD individuals (Katayama et al. 2004) and also in neuronal cell lines (Castillo-Carranza et al. 2012). Genome-wide association studies to detect XBP1 target genes revealed a cluster of AD-related genes (Acosta-Alvear et al. 2007). In the Chinese population, a direct genetic link between an increased rate of AD and polymorphism of XBP1 promoter was revealed, which further indicates that XBP1 might play a role as a UPR etiological element in AD (Liu et al. 2013). Furthermore, the expression of XBP1s downstream target protein disulfide isomerase (PDI) was found to coexist with NFTs in brains of AD individuals (Honjo et al. 2010). In AD mouse models, transporting active form of XBP1 into the hippocampal region induced the restoration of synaptic activity; however, it was also observed that IRE1 removal in the central nervous system (CNS) improved AD pathogenesis (Duran-Aniotz et al. 2017). Interestingly, XBP1s play a similar role in regulating memory and learning-associated mechanisms (Martínez et al. 2016). In case of AD, PERK expression targeting restored synaptic activity via recovering the protein generation of synaptic proteins and ameliorating synaptic transmission and memory capacity (Ma et al. 2013; Duran-Aniotz et al. 2014). In AD models, targeting other eIF2α kinases including general control nonderepressible 2 and protein kinase R affected synaptic activity (Ma et al. 2013; Lourenco et al. 2015). Furthermore, the expression of ATF4 regulated axonal degeneration via a cell nonautonomous process in AD models (Baleriola et al. 2014), which indicates that ATF4 and XBP1s might play opposite roles in AD progression.

In a study, it was suggested that NFTs containing neurons possess elevated phosphorylation of PERK and eIF2a (Culmsee and Landshamer 2006). Moreover, elevated p-eIF2a and p-PERK levels were detected in the hippocampal region of aged P301L mutant tau transgenic mouse models (Ho et al. 2012). In a different study, Abisambra et al. (Van Der Harg et al. 2014) reported that treatment with a specific PERK pathway inhibitor (small molecule) or chemical chaperone tauroursodeoxycholic acid prevented metabolic stress-mediated tau phosphorylation. It was revealed that stimulation of tau phosphorylation can lead to UPR activation (Ho et al. 2012). Indeed, pharmacological stimulation of ER stress might also stimulate tau phosphorylation. Tau hyperphosphorylation and UPR might play roles as self-inducing molecular processes, which can eventually lead to a vicious cycle of neuronal degeneration signaling. In addition to this, unconventional splicing of XBP1 was confirmed in transgenic *Drosophila* tau models (Loewen and Feany 2010). Collectively, growing evidence indicates the presence of multifaceted regulatory molecular networks that induce AD pathogenesis via the ER stress response, even though there is a deficiency in the number of experiments that directly influence the UPR in AD models using mammalian systems (Cornejo and Hetz 2013). Indeed, such techniques will clear its complex relation to confirm marked effect of UPR in the pathogenesis of AD. An outline of the relationship between UPR pharmacological or genetic modification and toxicity/aggregation of classical MPs in different NDs is provided in Table 9.1.

9.3.2 Parkinson's Disease

Characteristics of Parkinson's disease (PD) include severe motor impairments including postural instability, rigidity, slowness of movement, and rest tremors (Jankovic and Aguilar 2008). After AD, PD is considered as the second most frequent age-associated ND that affects around 0.6% of 65- to 69-year-olds and

Table 9.1 A summary of studies linking ER stress with neurodegeneration

Neurodegenerative diseases	Manipulation of UPR	Phenotypic characteristics	Animal model	Reference
Alzheimer's disease	Thapsigargin	Enhanced tau phosphorylation and caspase-3 cleavage	Transgenic (Tg) mouse models	Ho et al. (2012)
	c-Jun N-terminal kinase 3 $(JNK3)^{-/-}$	Reduced amyloid-beta (Aβ), cognitive dysfunction, and neuronal loss	Amyloid precursor protein (APP)/presenilin-1 (PS1) Tg mouse models	Yoon et al. (2012)
	Protein kinase RNA-like endoplasmic reticulum kinase (PERK) central nervous system (CNS) knockout (KO)	Ameliorated memory and learning and long-term potentiation (LTP)		Ma et al. (2013)
	Adeno-associated virus (AAV)-X-box-binding protein 1 (XBP1)	Rescued spine density, memory function, and synaptic plasticity		Cissé et al. (2017)
	Inositol-requiring enzyme 1 (IRE1) CNS KO	Decreased amyloid-beta, ameliorated LTP and learning		Duran-Aniotz et al. (2017)
Parkinson's disease	Salubrinal	Attenuates manifestation of the disease	Overexpression of α-synuclein	Colla et al. (2012a, b)
	AAV-XBP1s	Elevated dopaminergic survival	Neurotoxins	Unterberger et al. (2006)
	Activating transcription factor 6 (ATF6) KO	Elevated neurodegeneration		Egawa et al. (2011), Hashida et al. (2012)
	C/EBP homologous protein (CHOP) KO	Neuroprotection		Silva et al. (2005)
	AAV-binding immunoglobulin protein (BiP)	Dopaminergic survival, reduced aggregation of α-synuclein		Gorbatyuk et al. (2012)
	AAV-XBP1s	Neuroprotection, decreased striatal denervation		Valdés et al. (2014)
Huntington's disease	ATF4 KO	No activity on aggregation	Mutant *Htt* Tg mice	Vidal et al. (2012)
	XBP1s CNS KO			

		Neuroprotection, decreased levels of huntingtin (Htt), ameliorated motor performance		Vidal et al. (2012)
	AAV XBP1s	Reduced aggregation of Htt		Zuleta et al. (2012)
Amyotrophic lateral sclerosis	PERK $^{+/-}$	Enhanced superoxide dismutase 1 (SOD1) aggregation; disease exacerbation	SOD1 overexpressing mouse models	Wang et al. (2011)
	Salubrinal	Enhanced life span		Saxena et al. (2009)
	ATF4 KO	Protection against disease progression; partial embryonic lethality		Matus et al. (2013)
	XBP1 CNS KO	Extended life span; decreased SOD1		Hetz et al. (2009)
	Blockage of apoptosis signal-regulating kinase 1 (ASK-1) binding to SOD1	Death of motor neurons		Nishitoh et al. (2008)
	AAV-SIL-1	Extended survival; delayed muscle denervation		Filézac De L'Etang et al. (2015)
	Guanabenz	Induced disease progression		Vieira et al. (2015)
	Guanabenz	Enhanced survival; delayed disease onset; enhanced motor performance attenuated loss of motor neuron loss; and less accumulation of aggregates		Jiang et al. (2014), Wang et al. (2014)
Prion-related disorders	Salubrinal	Exacerbation of disease	Scrapie prion infection	Moreno et al. (2012)
	XBP1s CNS KO	No effect on disease progression or prion replication		Hetz et al. (2008)

(continued)

Table 9.1 (continued)

Neurodegenerative diseases	Manipulation of UPR	Phenotypic characteristics	Animal model	Reference
	Caspase-12 KO	No activity on disease progression and prion replication		Steele et al. (2007)
	PERK inhibitor	Delays disease progression and decreases neurodegeneration		Moreno et al. (2013)
	AAV-GADD34	Global neuroprotection		Moreno et al. (2012)
	Integrated stress response inhibitor (ISRIB)	Neuroprotection		Halliday et al. (2015)

2.6% of 85- to 89-year-olds (de Lau and Breteler 2006). Symptoms of PD include extensive dopaminergic neuronal loss in the pars compacta portion of the substantia nigra, which eventually leads to a significant reduction in the dopamine level in the brain. Like AD, PD pathogenesis includes the occurrence of neurofibrillary MP deposits called Lewy bodies that are made of MP α-synuclein, which coexists with ubiquitin (Varma and Sen 2015). Although most of the PD cases are sporadic, mutations in multiple genes including parkin, leucine-rich repeat kinase 2 (LRRK2), SNCA (codifies for α-synuclein), PD protein screening system in yeast provide direct evidence of the presence of ER stress in PD. It has been confirmed that α-synuclein specifically targets and suppresses vesicular trafficking from ER to Golgi apparatus via interacting with RAB1 (Cooper et al. 2006). RAB1 overexpression was found to protect from α-synuclein-mediated toxic effects in a fly model and was also found to attenuate the motor decline in a rat model (Coune et al. 2011). Studies with human dopaminergic neurons produced from induced pluripotent stem cells from PD individuals containing α-synuclein mutations also confirmed the significance of ER stress in case of PD (Chung et al. 2013). Furthermore, it has also been demonstrated that α-synuclein disturbs the COPII ER–Golgi traffic, which is a pathway important for activating transcription factor 6 (ATF6) activation and signaling (Credle et al. 2015). In a study, Credle et al. (2015) revealed that α-synuclein suppressed ATF6 processing both indirectly and directly, which further reduced ERAD function.

LRRK2 is the most commonly associated gene that is mutated in case of PD. Indeed, ER stress and LRRK2 pathogenesis are closely associated, as it partly localizes in the ER in dopaminergic neurons of PD individuals (Vitte et al. 2010). It was demonstrated via using *Caenorhabditis elegans* that LRRK2 protected the dopaminergic neurons from 6-hydroxydopamine (6-OHDA) treatment or expression of human α-synuclein, which is commonly linked with elevated BiP expression (Yuan et al. 2011). In addition, LRRK2 homolog-deficient *C. elegans* are more vulnerable toward ER stress (Sämann et al. 2009). Interestingly, E3 ubiquitin ligase Parkin/PARK2 is another familial PD-linked important gene, which is linked with the ERAD and UPS (Mercado et al. 2013). ATF4 transcriptionally controls Parkin, which suggests that it plays a role as an ER stress-inducible protein that induces cytoprotective processes (Bouman et al. 2011) and its subcellular delivery is changed following ER stress (Ledesma et al. 2002). It has been reported that parkin-associated endothelin receptor-like receptor (Pael-R) (a target for proteasome-mediated Parkin-dependent degradation) induced UPR stimulation, which eventually led to neuronal death (a condition that was deteriorated by dysfunctional ER chaperone) (Kitao et al. 2007). In PD models, genetic modification of UPR mediators regulated dopaminergic neuronal death. Following 6-OHDA treatment, genetic ablation of XBP1 during brain development elevated neuronal resistance, along with the increase in various UPR effectors in the substantia nigra region, although there was no activation of proapoptotic UPR markers including CHOP (Valdés et al. 2014). The existence of mild ER stress might be associated with a protective program in hormesis (Matus et al. 2012; Hetz and Mollereau 2014). Furthermore, tunicamycin treatment (non-lethal doses) protected neurodegeneration in mice and fly models of PD (Fouillet et al. 2012). In adult animals, gene therapy to transport XBP1 into the substantia nigra region provided protection against neurodegeneration mediated via PD-mediating neurotoxins (Sado et al. 2009; Valdés et al. 2014). Moreover, this result was also observed in a model where XBP1s were transferred to neuronal stem cells, which were then delivered to animals receiving rotenone treatment (Cui et al. 2012). Following 1-methyl-4-phenyl-1,2,3,6-tetrahydropyridine (MPTP) treatment, animals lacking ATF6 showed exacerbated dopaminergic neuron loss elevated ubiquitin-positive inclusions (Egawa et al. 2011; Hashida et al. 2012). In rats, BiP overexpression utilizing a gene therapy approach provided neuroprotection to dopaminergic neurons following human α-synuclein overexpression (Gorbatyuk et al. 2012). Genetic ablation of proapoptotic growth arrest- and DNA damage-inducible gene 153 (GADD153)/C/EBP homologous protein (CHOP) increased the survivals of dopaminergic neurons following 6-OHDA treatment but not following treatment with MPTP, which indicates that the role of UPR in different PD models might be greatly affected via the selected toxin for dopaminergic degeneration (Silva et al. 2005). Collectively, ER stress contributes significantly to the regulation of survival and physiological functions of dopaminergic neurons (Mercado et al. 2016), therefore suggesting an important target for disease interference (Table 9.1). Nonetheless, interpretations should be carried out carefully because different PD models might cause recruitment of different UPR components, and final outcome of influencing the UPR might result in conflicting findings in a situation-dependent manner.

9.3.3 Amyotrophic Lateral Sclerosis

ALS is a fatal ND, and the characteristics of this disease include chronic degeneration of lower and upper motor neurons from various regions including the spinal cord, cortex, and brainstem, which can further lead to muscle atrophy and paralysis (Maharjan and Saxena 2016). Around 90% of ALS cases are sporadic. Its etiology is not well defined; however, inherited genetic defects in different genes were found to be linked with familial ALS cases (Boylan 2015). Whereas 1–2% of all ALS cases and about 20% of familial ALS cases take place due to superoxide dismutase-1 (*SOD1*) mutations (Dion et al. 2009), around 140 different ALS-associated *SOD1* mutations have been detected (Sreedharan and Brown 2013). Furthermore, various mutated genes have been linked with familial ALS including C9ORF72, ubiquilin-2, fused in sarcoma, and TAR DNA-binding protein 43 (TDP-43). All of these mutated genes are responsible for altering proteostasis and mRNA metabolism (Sreedharan and Brown 2013; Maharjan and Saxena 2016). Indeed, an association of ER stress with ALS is well known and has been confirmed by numerous researchers through using various disease models. It has also been reported that ER stress is one of the initial molecular defects that underlay differential neuronal vulnerability observed in various NDs (Rozas et al. 2017). Mutant SOD1 mouse models and ALS individuals showed changes in ER morphology as individuals showed a detachment of ribosomes in degenerating anterior horn cells, irregular rER cisternae distension, and rough ER fragmentation (Lautenschlaeger et al. 2012; Deitch et al. 2014). An increased level of ER stress was detected in postmortem tissue through the occurrence of various UPR markers including PDI, calnexin, and BiP in the co-existence of mutant SOD1 (Wate et al. 2005; Kikuchi et al. 2006; Atkin et al. 2006). In sALS cases, an increased level of ATF4 and XBP1 expressions was identified in human postmortem spinal cord tissue (Hetz 2012). It was revealed via using three different ALS mouse models that ER stress precisely takes place in vulnerable motor neurons prior to any denervation is seen; subsequently, selective axonal degeneration was also observed (Saxena et al. 2009). Furthermore, proteomic studies of the spinal cord from mutant SOD1 mouse models showed increased levels of ERp57 and PDI (Atkin et al. 2006). PDI also colocalized with SOD1 and TDP-43 in swollen neurites and neuronal cytoplasmic inclusion of ALS patients (Honjo et al. 2011) and single nucleotide polymorphisms in intronic areas of the protein disulfide isomerase A1 (PDIA1) gene were found to play a role as ALS risk factors (Kwok et al. 2013). Moreover, mutations in both ERp57 and PDIA1 are considered risk factors that can lead to ALS development (Gonzalez-Perez et al. 2015).

Animal studies suggested that mutations in PDI can induce motor defects linked with impaired connectivity of motor neurons (Woehlbier et al. 2016). Calreticulin level was decreased in ALS mouse models, which further induced early-stage muscle denervation and muscle weakness (Bernard-Marissal et al. 2015). Furthermore, BiP mutant mice developed the spontaneous motor disease during the aging process, along with aggregation of wild-type endogenous SOD1 and selective motor neuron degeneration (Wang et al. 2010). SIL1 (a BiP cofactor) expression was found to be associated with differential motor neurons vulnerability. Gene therapy for SIL1 delivery to the nervous system provided significant protection from experimental ALS (Filézac De L'Etang et al. 2015). Therefore, alterations in the capacity of ER folding might be linked with the etiology of ALS. Along with the elevated level of PDI and BiP expression (Tobisawa et al. 2003), the expression of mutant SOD1 induces progressive PERK signaling (Atkin et al. 2006; Nagata et al. 2007; Saxena et al. 2009). The existence of proapoptotic UPR downstream mediator GADD153/CHOP has been established in spinal cords of both ALS transgenic mouse models and sporadic ALS individuals (Ito et al. 2009). It was revealed by in vivo studies on motor neurons that chronic ER stress is one of the main pathological features of this ALS model (Sun et al. 2015). Gene expression analyses of ALS brain tissue obtained from individuals with C9orf72 mutations showed a significant change in ER stress-associated genes (Prudencio et al. 2015). In cell cultures, the expression of RAN peptides is obtained from C9orf72 mutation-induced aberrant ER stress levels (Zhang et al. 2014). Phosphorylation of eIF2α was increased via overexpression of TDP-43 in flies. Furthermore, its pharmacological suppression was adequate to weaken its toxic effects (Kim et al. 2014) and targeting one copy of PERK induced the disease process (Wang et al. 2011). Therapeutic approaches to target phosphorylation of eIF2α have also been carried out. Salubrinal (a small molecule that suppresses dephosphorylation of eIF2α treatment) (Boyce et al. 2005) delayed the progression of experimental ALS (Saxena et al. 2009). It was suggested that guanabenz (a specific suppressor of the ER stress-inducible eIF2α phosphatase that suppresses a negative feedback loop of eIF2α phosphorylation) treatment (Tsaytler et al. 2011) induced ALS pathogenesis (Vieira et al. 2015), while various studies also reported protection following treatment with guanabenz (Jiang et al. 2014; Wang et al. 2014). In mutant SOD1 mouse models, sephin-1 (a guanabenz derivate that suppresses the negative feedback signaling for phosphorylation of eIF2α) (Crunkhorn 2015) nearly provided full protection (Das et al. 2015).

It was reported that the expression of ATF4 can modulate ALS pathogenesis (Matus et al. 2013). ATF4 gene deletion was adequate to elevate life span in mutant SOD1 mouse models, which further decreased the expression of proapoptotic genes CHOP and Bcl-2-like 11. Unfortunately, deletion of ATF4 elevated SOD1 aggregation both in vitro and in vivo is probably because of the elevated level of oxidative stress, which confirms its significance in the pathophysiology of ALS (Matus et al. 2013). Increased level of CHOP expression was observed both in microglia, oligodendrocytes, astrocytes, and neurons, which suggests that UPR glial modification may also control ALS pathophysiology (Suzuki and Matsuoka 2012). The role of XBP1 to ALS was also studied. Ablation of XBP1 reduced the severity of experimental ALS, therefore increased life span, which was demonstrated in a mutant SOD1 transgenic mouse model containing XBP1 deletion particularly in the nervous system (Hetz et al. 2009). Interestingly, this phenotype was found to be related to increased activation of macroautophagy machinery in motor neurons lacking XBP1. Still, more studies are required to find out whether direct manipulation of IRE1-XBP1s axis might be effective in ablation of ALS pathogenesis (Rozas et al. 2017). In 2004, a novel autosomal-dominant ALS-causative gene was detected that was reported to encode the vesicle-associated membrane protein-associated protein B (VAPB) (Nishimura et al. 2004). Moreover, this protein interacted with ATF6 through its cytosolic domain, and its malfunction might disturb its function (Gkogkas et al. 2008). It has been suggested that such malfunction might play role in the pathological processes of degenerative motor neuron disease. Transgenic mouse models for ALS-associated mutant VAPB showed signs of ER stress (Suzuki et al. 2009). Collectively, in the ALS field, a significant number of advances in experimental approaches with UPR-based therapeutic potential have been achieved that may prove beneficial in the future (Table 9.1).

9.3.4 Huntington's Disease

HD is a progressive, autosomal-dominant ND, and the characteristics of this ND include behavioral difficulties, cognitive decline, incoordination, dystonia, and chorea (Walker 2007). It has been observed that this disease takes place due to the first exon mutation of the huntingtin (*HTT*) gene, which possesses a tract of glutamine residues (polyQ repeats) that may differ in length among patients that can result in a mutated protein *HTT* with an expanded CAG repeat (Jiang et al. 2016). Prolonged polyglutamine (polyQ) extensions might cause a toxic gain of function (Walker 2007), even though there is a controversy regarding the exact processes that result in neuronal cell death induced via these expansions (Matthias et al. 2015). Expansions of polyQ in distinct proteins are directly associated with the pathology of minimum of 9 different NDs including HD, where each ND contains a different subset of susceptible neuronal populations (Orr and Zoghbi 2007). It is assumed that aggregate generation in the nucleus and cytosol and expression of mutant *HTT* generation overwhelm the proteostasis machinery, which further results in alteration of neuronal physiology (Jiang et al. 2016). It was revealed that mutant *HTT* might disturb various cellular mechanisms that are linked with UPS (Bence et al. 2001; Bennett et al. 2007), transport (Gunawardena and Goldstein 2005), and protein turnover (Bennett et al. 2007).

Stimulation of ER stress in cells expressing polyQ peptides is similar to the mutations seen in *HTT* (Urano et al. 2000; Nishitoh et al. 2002; Kouroku et al. 2002). A process that specifically disrupted via polyQ expanded repeats is ERAD, which resulted in ER stress (Nishitoh et al. 2008; Duennwald and Lindquist 2008). Various downstream targets including CHOP, HERP, and BiP were found to be increased in brain samples obtained from HD individuals (Carnemolla et al. 2009). Similar findings were also seen in an animal model of HD (Cho et al. 2009). A group of genes has been identified that provide an important link between HD and UPR and its association with neurodegeneration (Matthias et al. 2015). Interestingly, in the full-length mutant *HTT* transgenic mouse models, ablation of XBP1 expression was adequate to improve motor performance and decrease neuronal loss in the striatum region. Indeed, these protective activities were similar to reduced accumulation of *HTT* facilitated via the stimulation of autophagy, which perhaps suggests the forkhead box O1 upregulation (Vidal et al. 2012).

However, deficiency of ATF4 did not influence the aggregation of mutant *HTT* (Vidal et al. 2012). Since autophagy is indicated as the preferential pathway for degradation of MP aggregates (including those generated as part of mutant *HTT* (Jiang et al. 2016) and is anticipated to fail during the disease period (Martinez-Vicente et al. 2010), thus XBP1 might play a role as a key *HTT* clearance mediator in case of HD pathogenesis. AAV-XBP1 delivery to the striatum area of an adult mouse model overexpressing a mutant *HTT* decreased its accumulation (Zuleta et al. 2012). In general, various methodological techniques have suggested that the changes in the secretory pathway seen in HD might be associated with disturbances in case of ERAD/protein quality control processes, ER calcium homeostasis, vesicular trafficking, endocytosis, lysosomal/autophagy-induced protein degradation, and ER/Golgi trafficking, which can further lead to ER stress as a usual feature (Vidal et al. 2011). Furthermore, cell-based studies using a mutant *HTT* aggregation assay detected IRE1 as an important mediator of its aggregation (Lee et al. 2012). PERK signaling might also affect mutant *HTT* biology. In a study, Kouroku et al. (2007) revealed that phosphorylation of eIF2α is important for the stimulation of autophagy via *HTT* aggregates (Kouroku et al. 2007). Studies associating ER stress and HD are still poor, and more analyses are required to accurately report the significance of UPR in HD pathogenesis (Table 9.1).

9.3.5 Prion-Related Disorders

Prion-related disorders (PrDs) are also well known as transmissible spongiform encephalopathies. PrDs are a group of related NDs, and the characteristics of PrDs include rapid neurological dysfunction such as psychiatric disturbances, ataxia, and dementia. Etiology of PrDs is divided into familial (inherited in an autosomal-dominant manner), sporadic (spontaneous origin), or infectious (derived from the exposure to material contaminated with infectious prions) PrDs. Indeed, human familial PrDs contain several forms of fatal familial insomnia, Gertmann Straussler-Scheinker syndrome, and Creutzfeldt–Jakob disease (CJD) (Prusiner and Scott 1997). The main molecular mechanism in the pathogenesis of PrDs is the normal cellular prion protein (known as PrPC) conversion into the pathological form known as PrPSc (for scrapie-related PrP) (Prusiner 1998). A probable link between PrDs and ER stress was primarily highlighted through the result of various chaperones including Grp58/ERp57, Grp94, and BiP in the cortex of individuals with sporadic and variant CJD (Hetz et al. 2003). A proteomic screening was revealed that ERp57 was detected in the cerebellum of individuals with sporadic CJD (Yoo et al. 2002). Mouse models infected with scrapie prions exhibited an expression profile in the hippocampus, which suggested that most genes influenced via prion infection are associated with ER stress mediators (Brown et al. 2005).

Nonetheless, other CJD-related studies did not find symptoms of ER stress (Unterberger et al. 2006). Mouse models infected with scrapie prions showed ER stress markers during the pre-symptomatic phase of the disease following the accumulation of PrPSc, such as increased disulfide isomerase Grp58/ERp57 expression levels and increased expression of Grp94 and BiP during the symptomatic phase (Hetz et al. 2005). In the same mouse models, neuronal loss was found to be increased only at the late stage of the disease, which was found to be associated with enhanced pro-caspase-12 processing and decreased expression of ERp57 (Hetz et al. 2003), even though knockout for caspase-12 showed no change in the progression of the disease (Steele et al. 2007). In animal and cellular models, it has been confirmed that ERp57/Grp58 exerts a neuroprotective activity in PrDs (Hetz et al. 2005) and ERp57 is associated with the PrP synthesis and folding (Torres et al. 2015). ER stress took place due to the exposure of cells to purified PrPSc derived from the brains of scrapie-infected mouse models. Targeting of BCL-2 (an anti-apoptotic protein) to the ER membrane might reduce PrPSc toxicity, and prion-infected cells were more vulnerable to ER stress-mediated apoptosis (Hetz and Soto 2006; Hetz et al. 2003; Apodaca et al. 2006). Conversion of PrPSc might influence ER calcium homeostasis at the molecular level (Herms et al. 2000; Mukherjee et al. 2010) and can induce ER stress (Torres et al. 2010), which can eventually lead to cellular death (Ferreiro et al. 2008).

Conflicting findings were observed with the UPR genetic manipulation. In the brain, targeting XBP1 did not influence the replication of prion and its in vivo pathogenesis (Hetz et al. 2008). In a study, Moreno et al. (2012) revealed that continuous protein translational shutdown because of eIF2α phosphorylation induces neuronal death and synaptic dysfunction in prion-infected animals (Moreno et al. 2012). In these models, delivering eIF2α phosphatase via gene therapy provided neuroprotection, while salubrinal treatment worsened the disease progression (Moreno et al. 2012). In prion-infected mouse models, oral administration of a specific PERK inhibitor attenuated disease progression at both symptomatic and preclinical stages (Moreno et al. 2013). In animals, ISRIB (a small compound) treatment blocked the outcomes of eIF2α phosphorylation, which protected PrD (Halliday et al. 2015). Inhibition of PERK did not reduce the misfolding of PrP, which indicates that its protection was particularly associated with enhancements of synaptic functions (Table 9.1).

9.4 ER Stress and Acute Neurodegeneration

Other than the chronic NDs, ER stress is also involved in various acute brain disorders including ischemia. In mice, focal cerebral ischemia took place due to the activation of eIF2α and PERK as well as because of the BiP/Grp78 detachment (Fig. 9.2) (Hayashi et al. 2005). Furthermore, in mice, global ischemia triggered ER stress and activation of ER transcription factors including ATF4 and CHOP/Gadd153 (Hayashi et al. 2005). Hippocampal neurons obtained from CHOP/Gadd153-deficient mouse models showed more resistance against cell death mediated via hypoxia reoxygenation as compared to controls (Tajiri et al. 2004). After ischemia, less neurons are degenerated in the $CHOP^{-/-}$ mouse models, which indicates the significant role of ER stress in stroke/ischemia. ER stress mediates the activation of caspase-12 in case of brain trauma. In the brain, GRP94 (an important hallmark of UPR response) provided protection from ischemia-mediated cell death (Tamatani et al. 2001; Bando et al. 2003). This suggests a direct association of ER response in brain injury. However, more studies are required to find out whether targeting ER stress-linked proteins may be beneficial in the treatment of acute brain disorders.

9.5 Recent Developments and Future Perspectives

As per the current understanding, it is difficult to predict the role of UPR in protein misfolding disorders (PMDs). Therefore, more systematic studies are required in this regard. Interestingly, UPR might exert conflicting and even opposite actions, depending on the context of disease and the analyzed signaling branch. The discovery of new drugs that directly interfere influence of the UPR signaling along with the probability of brain in vivo manipulation of ER stress-linked genes in various animal models is likely to provide evidence about the role of UPR in the progression of PMDs (Cornejo and Hetz 2013; Maly and Papa 2014). Findings of UPR manipulation via genetic or pharmacological techniques and its outcome in disease progression of various models of PMDs are summarized in Table 9.1. In different disease models, UPR manipulation may lead to conflicting outcomes depending on the disease setting and particular studied signaling module. Furthermore, as UPR has a significant contribution to the physiology of multiple organs including the liver, pancreas, and the immune system; thus, severe side effects are likely to occur due to the prolonged administration of UPR-targeting drugs (Cornejo and Hetz 2013; Dufey et al. 2014).

There is a growing interest regarding the use of gene therapy to target UPR locally in certain regions of the brain (Valenzuela et al. 2016). Furthermore, these tools might be effective in clarifying the fundamental link between the occurrence of PMDs and malfunction of ER proteostasis. Indeed, available information indicates that ER stress might influence protein aggregation and directly induce synaptic dysfunction as confirmed in PrDs and AD. A novel function of XBP1s has been identified in enhancing memory and learning-related processes via brain-derived neurotrophic factor regulation (Martínez et al. 2016). Therefore, gene therapies based on XBP1s might act as a dual therapy to target synaptic function, protein aggregation, and proteostasis alteration. It has been reported that ATF4 and phosphorylation of eIF2α can affect cognitive function at the level of neuronal plasticity and synaptic transmission (Costa-Mattioli et al. 2009; Pasini et al. 2015). The probable role of UPR in glial activation and neuroinflammation is still yet to be estimated. In the future, available data regarding the important relationships between UPR and energy control and global proteostasis control might be explored in the setting of NDs. Former studies on ER stress have mostly been carried out by utilizing cultured non-neuronal cells expressing increased levels of exogenous proteins. In future, studies on various mouse models and neuronal systems along with the studies of human samples will eventually raise our understanding regarding the effect of ER stress in various NDs (Lindholm et al. 2006).

9.6 Conclusion

ER organizes various cellular mechanisms through which proteins are generated, properly folded, modified, and delivered to their final destination. Some of these ER-associated cellular mechanisms may go wrong in the case of NDs, which can result in different extents of ER stress that can eventually lead to neuronal death. It is still unclear to what extent cell degeneration takes place because of ER stress or other mechanisms. Considering forthcoming potentials for the treatment of NDs, it is crucial to find out more about which portion of the ER stress response needs to be targeted and at what specific phase of the disease.

References

Abisambra JF, Jinwal UK, Blair LJ et al (2013) Tau accumulation activates the unfolded protein response by impairing endoplasmic reticulum-associated degradation. J Neurosci 33:9498–9507. https://doi.org/10.1523/JNEUROSCI.5397-12.2013

Acosta-Alvear D, Zhou Y, Blais A et al (2007) XBP1 controls diverse cell type- and condition-specific transcriptional regulatory networks. Mol Cell 27:53–66. https://doi.org/10.1016/j.molcel.2007.06.011

Al Mamun A, Uddin MS, Kabir MT et al (2020) Exploring the promise of targeting ubiquitin-proteasome system to combat Alzheimer's disease. Neurotox Res 38:8–17. https://doi.org/10.1007/s12640-020-00185-1

Apodaca J, Kim I, Rao H (2006) Cellular tolerance of prion protein PrP in yeast involves proteolysis and the unfolded protein response. Biochem Biophys Res Commun 347:319–326. https://doi.org/10.1016/j.bbrc.2006.06.078

Atkin JD, Farg MA, Turner BJ et al (2006) Induction of the unfolded protein response in familial amyotrophic lateral sclerosis and association of protein-disulfide isomerase with superoxide dismutase 1. J Biol Chem 281:30152–30165. https://doi.org/10.1074/jbc.M603393200

Baleriola J, Walker CA, Jean YY et al (2014) Axonally synthesized ATF4 transmits a neurodegenerative signal across brain regions. Cell 158:1159–1172. https://doi.org/10.1016/j.cell.2014.07.001

Ballard C, Gauthier S, Corbett A et al (2011) Alzheimer's disease. Lancet 377:1019–1031

Bando Y, Katayama T, Kasai K et al (2003) GRP94 (94 kDa glucose-regulated protein) suppresses ischemic neuronal cell death against ischemia/reperfusion injury. Eur J Neurosci 18:829–840. https://doi.org/10.1046/j.1460-9568.2003.02818.x

Bence NF, Sampat RM, Kopito RR (2001) Impairment of the ubiquitin-proteasome system by protein aggregation. Science (80-) 292:1552–1555. https://doi.org/10.1126/science.292.5521.1552

Bennett EJ, Shaler TA, Woodman B et al (2007) Global changes to the ubiquitin system in Huntington's disease. Nature 448:704–708. https://doi.org/10.1038/nature06022

Berezovska O, Lleo A, Herl LD et al (2005) Familial Alzheimer's disease presenilin 1 mutations cause alterations in the conformation of presenilin and interactions with amyloid precursor protein. J Neurosci 25:3009–3017. https://doi.org/10.1523/JNEUROSCI.0364-05.2005

Bernard-Marissal N, Sunyach C, Marissal T et al (2015) Calreticulin levels determine onset of early muscle denervation by fast motoneurons of ALS model mice. Neurobiol Dis 73:130–136. https://doi.org/10.1016/j.nbd.2014.09.009

Bouman L, Schlierf A, Lutz AK et al (2011) Parkin is transcriptionally regulated by ATF4: evidence for an interconnection between mitochondrial stress and ER stress. Cell Death Differ 18:769–782. https://doi.org/10.1038/cdd.2010.142

Boyce M, Bryant KF, Jousse C et al (2005) A selective inhibitor of eIF2α dephosphorylation protects cells from ER stress. Science (80-) 307:935–939. https://doi.org/10.1126/science.1101902

Boylan K (2015) Familial ALS. Neurol Clin 33:807–830. https://doi.org/10.1016/j.ncl.2015.07.001

Brandt R, Hundelt M, Shahani N (2005) Tau alteration and neuronal degeneration in tauopathies: mechanisms and models. Biochim Biophys Acta Mol basis Dis 1739:331–354

Breckenridge DG, Germain M, Mathai JP et al (2003) Regulation of apoptosis by endoplasmic reticulum pathways. Oncogene 22:8608–8618. https://doi.org/10.1038/sj.onc.1207108

Brown AR, Rebus S, McKimmie CS et al (2005) Gene expression profiling of the preclinical scrapie-infected hippocampus. Biochem Biophys Res Commun 334:86–95. https://doi.org/10.1016/j.bbrc.2005.06.060

Carnemolla A, Fossale E, Agostoni E et al (2009) Rrs1 is involved in endoplasmic reticulum stress response in Huntington disease. J Biol Chem 284:18167–18173. https://doi.org/10.1074/jbc.M109.018325

Casas-Tinto S, Zhang Y, Sanchez-Garcia J et al (2011) The ER stress factor XBP1s prevents amyloid-β neurotoxicity. Hum Mol Genet 20:2144–2160. https://doi.org/10.1093/hmg/ddr100

Castillo-Carranza DL, Zhang Y, Guerrero-Muñoz MJ et al (2012) Differential activation of the ER stress factor XBP1 by oligomeric assemblies. Neurochem Res 37:1707–1717. https://doi.org/10.1007/s11064-012-0780-7

Caughey B, Lansbury PT (2003) Protofibrils, pores, fibrils, and neurodegeneration: separating the responsible protein aggregates from the innocent bystanders. Annu Rev Neurosci 26:267–298. https://doi.org/10.1146/annurev.neuro.26.010302.081142

Cavedo E, Lista S, Khachaturian Z et al (2014) The road ahead to cure Alzheimer's disease: development of biological markers and neuroimaging methods for prevention trials across all stages and target populations. J Prev Alzheimers Dis 1:181–202. https://doi.org/10.14283/jpad.2014.32

Chen R, Valencia I, Zhong F et al (2004) Bcl-2 functionally interacts with inositol 1,4,5-trisphosphate receptors to regulate calcium release from the ER in response to inositol 1,4,5-trisphosphate. J Cell Biol 166:193–203. https://doi.org/10.1083/jcb.200309146

Cho KJ, Lee BI, Cheon SY et al (2009) Inhibition of apoptosis signal-regulating kinase 1 reduces endoplasmic reticulum stress and nuclear huntingtin fragments in a mouse model of Huntington disease. Neuroscience 163:1128–1134. https://doi.org/10.1016/j.neuroscience.2009.07.048

Chung CY, Khurana V, Auluck PK et al (2013) Identification and rescue of α-synuclein toxicity in Parkinson patient-derived neurons. Science (80-) 342:983–987. https://doi.org/10.1126/science.1245296

Ciechanover A, Brundin P (2003) The ubiquitin proteasome system in neurodegenerative diseases: sometimes the chicken, sometimes the egg. Neuron 40:427–446. https://doi.org/10.1016/S0896-6273(03)00606-8

Cissé M, Duplan E, Lorivel T et al (2017) The transcription factor XBP1s restores hippocampal synaptic plasticity and memory by control of the Kalirin-7 pathway in Alzheimer model. Mol Psychiatry 22:1562–1575. https://doi.org/10.1038/mp.2016.152

Colla E, Coune P, Liu Y et al (2012a) Endoplasmic reticulum stress is important for the manifestations of α-synucleinopathy in vivo. J Neurosci 32:3306–3320. https://doi.org/10.1523/JNEUROSCI.5367-11.2012

Colla E, Jensen PH, Pletnikova O et al (2012b) Accumulation of toxic α-synuclein oligomer within endoplasmic reticulum occurs in α-synucleinopathy in vivo. J Neurosci 32:3301–3305. https://doi.org/10.1523/JNEUROSCI.5368-11.2012

Cooper AA, Gitler AD, Cashikar A et al (2006) α-Synuclein blocks ER-Golgi traffic and Rab1 rescues neuron loss in Parkinson's models. Science (80-) 313:324–328. https://doi.org/10.1126/science.1129462

Cornejo VH, Hetz C (2013) The unfolded protein response in Alzheimer's disease. Semin Immunopathol 35:277–292. https://doi.org/10.1007/s00281-013-0373-9

Costa-Mattioli M, Sossin WS, Klann E, Sonenberg N (2009) Translational control of long-lasting synaptic plasticity and memory. Neuron 61:10–26. https://doi.org/10.1016/j.neuron.2008.10.055

Coune PG, Bensadoun JC, Aebischer P, Schneider BL (2011) Rab 1 a over-expression prevents golgi apparatus fragmentation and partially corrects motor deficits in an alpha-synuclein based rat model of Parkinson's disease. J Parkinsons Dis 1:373–387. https://doi.org/10.3233/JPD-2011-11058

Credle JJ, Forcelli PA, Delannoy M et al (2015) α-Synuclein-mediated inhibition of ATF6 processing into COPII vesicles disrupts UPR signaling in Parkinson's disease. Neurobiol Dis 76:112–125. https://doi.org/10.1016/j.nbd.2015.02.005

Crunkhorn S (2015) Neurodegenerative disease: phosphatase inhibitor prevents protein-misfolding diseases. Nat Rev Drug Discov 14:386. https://doi.org/10.1038/nrd4638

Cui M, Si L, Xu T et al (2012) X-box-binding protein 1-modified neural stem cells for treatment of Parkinson's disease. Neural Regen Res 7:736–740. https://doi.org/10.3969/j.issn.1673-5374.2012.10.003

Culmsee C, Landshamer S (2006) Molecular insights into mechanisms of the cell death program: role in the progression of neurodegenerative disorders. Curr Alzheimer Res 3:269–283. https://doi.org/10.2174/156720506778249461

Danial NN, Korsmeyer SJ (2004) Cell death: critical control points. Cell 116:205–219. https://doi.org/10.1016/S0092-8674(04)00046-7

Das I, Krzyzosiak A, Schneider K et al (2015) Preventing proteostasis diseases by selective inhibition of a phosphatase regulatory subunit. Science (80-) 348:239–242. https://doi.org/10.1126/science.aaa4484

de Lau LM, Breteler MM (2006) Epidemiology of Parkinson's disease. Lancet Neurol 5:525–535. https://doi.org/10.1016/S1474-4422(06)70471-9

Degterev A, Boyce M, Yuan J (2003) A decade of caspases. Oncogene 22:8543–8567. https://doi.org/10.1038/sj.onc.1207107

Deitch JS, Alexander GM, Bensinger A et al (2014) Phenotype of transgenic mice carrying a very low copy number of the mutant human G93A superoxide dismutase-1 gene associated with amyotrophic lateral sclerosis. PLoS One 9:e99879. https://doi.org/10.1371/journal.pone.0099879

Dion PA, Daoud H, Rouleau GA (2009) Genetics of motor neuron disorders: new insights into pathogenic mechanisms. Nat Rev Genet 10:769–782. https://doi.org/10.1038/nrg2680

Dovey HF, John V, Anderson JP et al (2001) Functional gamma-secretase inhibitors reduce beta-amyloid peptide levels in brain. J Neurochem 76:173–181. https://doi.org/10.1046/j.1471-4159.2001.00012.x

Duennwald ML, Lindquist S (2008) Impaired ERAD and ER stress are early and specific events in polyglutamine toxicity. Genes Dev 22:3308–3319. https://doi.org/10.1101/gad.1673408

Dufey E, Sepúlveda D, Rojas-Rivera D, Hetz C (2014) Cellular mechanisms of endoplasmic reticulum stress signaling in health and disease. 1. An overview. Am J Physiol - Cell Physiol 307:C582–C594. https://doi.org/10.1152/ajpcell.00258.2014

Duran-Aniotz C, Martínez G, Hetz C (2014) Memory loss in Alzheimer's disease: are the alterations in the UPR network involved in the cognitive impairment? Front Aging Neurosci 6:8. https://doi.org/10.3389/fnagi.2014.00008

Duran-Aniotz C, Cornejo VH, Espinoza S et al (2017) IRE1 signaling exacerbates Alzheimer's disease pathogenesis. Acta Neuropathol 134:489–506. https://doi.org/10.1007/s00401-017-1694-x

Egawa N, Yamamoto K, Inoue H et al (2011) The endoplasmic reticulum stress sensor, ATF6α, protects against neurotoxin-induced dopaminergic neuronal death. J Biol Chem 286:7947–7957. https://doi.org/10.1074/jbc.M110.156430

Endres K, Reinhardt S (2013) ER-stress in Alzheimer's disease: turning the scale? Am J Neurodegener Dis 2:247–265

Ferreiro E, Oliveira CR, Pereira CMF (2008) The release of calcium from the endoplasmic reticulum induced by amyloid-beta and prion peptides activates the mitochondrial apoptotic pathway. Neurobiol Dis 30:331–342. https://doi.org/10.1016/j.nbd.2008.02.003

Filézac De L'Etang A, Maharjan N, Cordeiro Branã M et al (2015) Marinesco-Sjögren syndrome protein SIL1 regulates motor neuron subtype-selective ER stress in ALS. Nat Neurosci 18:227–238. https://doi.org/10.1038/nn.3903

Fouillet A, Levet C, Virgone A et al (2012) ER stress inhibits neuronal death by promoting autophagy. Autophagy 8:915–926. https://doi.org/10.4161/auto.19716

Germain M, Mathai JP, Shore GC (2002) BH-3-only BIK functions at the endoplasmic reticulum to stimulate cytochrome c release from mitochondria. J Biol Chem 277:18053–18060. https://doi.org/10.1074/jbc.M201235200

Gkogkas C, Middleton S, Kremer AM et al (2008) VAPB interacts with and modulates the activity of ATF6. Hum Mol Genet 17:1517–1526. https://doi.org/10.1093/hmg/ddn040

Glenner GG, Wong CW (2012) Alzheimer's disease: initial report of the purification and characterization of a novel cerebrovascular amyloid protein. 1984. Biochem Biophys Res Commun 425:534–539. https://doi.org/10.1016/j.bbrc.2012.08.020

Gonzalez-Perez P, Woehlbier U, Chian RJ et al (2015) Identification of rare protein disulfide isomerase gene variants in amyotrophic lateral sclerosis patients. Gene 566:158–165. https://doi.org/10.1016/j.gene.2015.04.035

Gorbatyuk MS, Shabashvili A, Chen W et al (2012) Glucose regulated protein 78 diminishes α-synuclein neurotoxicity in a rat model of parkinson disease. Mol Ther 20:1327–1337. https://doi.org/10.1038/mt.2012.28

Gunawardena S, Goldstein LSB (2005) Polyglutamine diseases and transport problems: deadly traffic jams on neuronal highways. Arch Neurol 62:46–51. https://doi.org/10.1001/archneur.62.1.46

Haass C (2004) Take five - BACE and the γ-secretase quartet conduct Alzheimer's amyloid β-peptide generation. EMBO J 23:483–488. https://doi.org/10.1038/sj.emboj.7600061

Halliday M, Radford H, Sekine Y et al (2015) Partial restoration of protein synthesis rates by the small molecule ISRIB prevents neurodegeneration without pancreatic toxicity. Cell Death Dis 6: e1672. https://doi.org/10.1038/cddis.2015.49

Hamos JE, Oblas B, Pulaski-Salo D et al (1991) Expression of heat shock proteins in Alzheimer's disease. Neurology 41:345–350. https://doi.org/10.1212/wnl.41.3.345

Hashida K, Kitao Y, Sudo H et al (2012) ATF6alpha promotes astroglial activation and neuronal survival in a chronic mouse model of Parkinson's disease. PLoS One 7:e47950. https://doi.org/10.1371/journal.pone.0047950

Hayashi T, Saito A, Okuno S et al (2005) Damage to the endoplasmic reticulum and activation of apoptotic machinery by oxidative stress in ischemic neurons. J Cereb Blood Flow Metab 25:41–53. https://doi.org/10.1038/sj.jcbfm.9600005

Herms JW, Korte S, Gall S et al (2000) Altered intracellular calcium homeostasis in cerebellar granule cells of prion protein-deficient mice. J Neurochem 75:1487–1492. https://doi.org/10.1046/j.1471-4159.2000.0751487.x

Hetz C (2012) The unfolded protein response: controlling cell fate decisions under ER stress and beyond. Nat Rev Mol Cell Biol 13:89–102. https://doi.org/10.1038/nrm3270

Hetz C, Mollereau B (2014) Disturbance of endoplasmic reticulum proteostasis in neurodegenerative diseases. Nat Rev Neurosci 15:233–249. https://doi.org/10.1038/nrn3689

Hetz CA, Soto C (2006) Stressing out the ER: a role of the unfolded protein response in prion-related disorders. Curr Mol Med 6(1):37–43

Hetz C, Russelakis-Carneiro M, Maundrell K et al (2003) Caspase-12 and endoplasmic reticulum stress mediate neurotoxicity of pathological prion protein. EMBO J 22:5435–5445. https://doi.org/10.1093/emboj/cdg537

Hetz C, Russelakis-Carneiro M, Wälchli S et al (2005) The disulfide isomerase Grp58 is a protective factor against prion neurotoxicity. J Neurosci 25:2793–2802. https://doi.org/10.1523/JNEUROSCI.4090-04.2005

Hetz C, Lee AH, Gonzalez-Romero D et al (2008) Unfolded protein response transcription factor XBP-1 does not influence prion replication or pathogenesis. Proc Natl Acad Sci U S A 105:757–762. https://doi.org/10.1073/pnas.0711094105

Hetz C, Thielen P, Matus S et al (2009) XBP-1 deficiency in the nervous system protects against amyotrophic lateral sclerosis by increasing autophagy. Genes Dev 23:2294–2306. https://doi.org/10.1101/gad.1830709

Hippius H, Neundörfer G (2003) The discovery of Alzheimer's disease. Dialogues Clin Neurosci 5: 101–108

Hitomi J, Katayama T, Eguchi Y et al (2004) Involvement of caspase-4 in endoplasmic reticulum stress-induced apoptosis and Aβ-induced cell death. J Cell Biol 165:347–356. https://doi.org/10.1083/jcb.200310015

Ho YS, Yang X, Lau JCF et al (2012) Endoplasmic reticulum stress induces tau pathology and forms a vicious cycle: implication in Alzheimer's disease pathogenesis. J Alzheimers Dis 28: 839–854. https://doi.org/10.3233/JAD-2011-111037

Honjo Y, Ito H, Horibe T et al (2010) Protein disulfide isomerase-immunopositive inclusions in patients with Alzheimer disease. Brain Res 1349:90–96. https://doi.org/10.1016/j.brainres.2010.06.016

Honjo Y, Kaneko S, Ito H et al (2011) Protein disulfide isomerase-immunopositive inclusions in patients with amyotrophic lateral sclerosis. Amyotroph Lateral Scler 12:444–450. https://doi.org/10.3109/17482968.2011.594055

Hoozemans JJM, Veerhuis R, Van Haastert ES et al (2005) The unfolded protein response is activated in Alzheimer's disease. Acta Neuropathol 110:165–172. https://doi.org/10.1007/s00401-005-1038-0

Hoozemans JJM, Van Haastert ES, Nijholt DAT et al (2009) The unfolded protein response is activated in pretangle neurons in Alzheimer's disease hippocampus. Am J Pathol 174:1241–1251. https://doi.org/10.2353/ajpath.2009.080814

Imai Y, Soda M, Inoue H et al (2001) An unfolded putative transmembrane polypeptide, which can lead to endoplasmic reticulum stress, is a substrate of Parkin. Cell 105:891–902. https://doi.org/10.1016/S0092-8674(01)00407-X

Ito Y, Yamada M, Tanaka H et al (2009) Involvement of CHOP, an ER-stress apoptotic mediator, in both human sporadic ALS and ALS model mice. Neurobiol Dis 36:470–476. https://doi.org/10.1016/j.nbd.2009.08.013

Jankovic J, Aguilar LG (2008) Current approaches to the treatment of Parkinson's disease. Neuropsychiatr Dis Treat 4:743–757. https://doi.org/10.2147/ndt.s2006

Jiang HQ, Ren M, Jiang HZ et al (2014) Guanabenz delays the onset of disease symptoms, extends lifespan, improves motor performance and attenuates motor neuron loss in the SOD1 G93A mouse model of amyotrophic lateral sclerosis. Neuroscience 277:132–138. https://doi.org/10.1016/j.neuroscience.2014.03.047

Jiang Y, Chadwick SR, Lajoie P (2016) Endoplasmic reticulum stress: the cause and solution to Huntington's disease? Brain Res 1648:650–657. https://doi.org/10.1016/j.brainres.2016.03.034

Kabir MT, Sufian MA, Uddin MS et al (2019a) NMDA receptor antagonists: repositioning of memantine as a multitargeting agent for Alzheimer's therapy. Curr Pharm Des 25:3506–3518. https://doi.org/10.2174/1381612825666191011102444

Kabir MT, Uddin MS, Begum MM et al (2019b) Cholinesterase inhibitors for Alzheimer's disease: multitargeting strategy based on anti-Alzheimer's drugs repositioning. Curr Pharm Des 25: 3519–3535. https://doi.org/10.2174/1381612825666191008103141

Kabir MT, Uddin MS, Abdeen A et al (2020a) Evidence linking protein misfolding to quality control in progressive neurodegenerative diseases. Curr Top Med Chem 20:2025–2043. https://doi.org/10.2174/1568026620666200618114924

Kabir MT, Uddin MS, Al Mamun A et al (2020b) Combination drug therapy for the management of Alzheimer's disease. Int J Mol Sci 21:3272. https://doi.org/10.3390/ijms21093272

Kabir MT, Uddin MS, Mathew B et al (2020c) Emerging promise of immunotherapy for Alzheimer's disease: a new hope for the development of Alzheimer's vaccine. Curr Top Med Chem 20:1214–1234. https://doi.org/10.2174/1568026620666200422105156

Kabir MT, Uddin MS, Setu JR et al (2020d) Exploring the role of PSEN mutations in the pathogenesis of Alzheimer's disease. Neurotox Res 38(4):833–849

Kabir MT, Uddin MS, Zaman S et al (2020e) Molecular mechanisms of metal toxicity in the pathogenesis of Alzheimer's disease. Mol Neurobiol 58(1):1–20

Katayama T, Imaizumi K, Sato N et al (1999) Presenilin-1 mutations downregulate the signalling pathway of the unfolded-protein response. Nat Cell Biol 1:479–485. https://doi.org/10.1038/70265

Katayama T, Imaizumi K, Manabe T et al (2004) Induction of neuronal death by ER stress in Alzheimer's disease. J Chem Neuroanat 28:67–78. https://doi.org/10.1016/j.jchemneu.2003.12.004

Kikuchi H, Almer G, Yamashita S et al (2006) Spinal cord endoplasmic reticulum stress associated with a microsomal accumulation of mutant superoxide dismutase-1 in an ALS model. Proc Natl Acad Sci U S A 103:6025–6030. https://doi.org/10.1073/pnas.0509227103

Kim HJ, Raphael AR, Ladow ES et al (2014) Therapeutic modulation of eIF2α phosphorylation rescues TDP-43 toxicity in amyotrophic lateral sclerosis disease models. Nat Genet 46:152–160. https://doi.org/10.1038/ng.2853

Kitao Y, Imai Y, Ozawa K et al (2007) Pael receptor induces death of dopaminergic neurons in the substantia nigra via endoplasmic reticulum stress and dopamine toxicity, which is enhanced under condition of parkin inactivation. Hum Mol Genet 16:50–60. https://doi.org/10.1093/hmg/ddl439

Korhonen L, Lindholm D (2004) The ubiquitin proteasome system in synaptic and axonal degeneration: a new twist to an old cycle. J Cell Biol 165:27–30. https://doi.org/10.1083/jcb.200311091

Korhonen L, Hansson I, Kukkonen JP et al (2005) Hippocalcin protects against caspase-12-induced and age-dependent neuronal degeneration. Mol Cell Neurosci 28:85–95. https://doi.org/10.1016/j.mcn.2004.08.015

Kouroku Y, Fujita E, Jimbo A et al (2002) Polyglutamine aggregates stimulate ER stress signals and caspase-12 activation. Hum Mol Genet 11:1505–1515. https://doi.org/10.1093/hmg/11.13.1505

Kouroku Y, Fujita E, Tanida I et al (2007) ER stress (PERK/eIF2α phosphorylation) mediates the polyglutamine-induced LC3 conversion, an essential step for autophagy formation. Cell Death Differ 14:230–239. https://doi.org/10.1038/sj.cdd.4401984

Kwok CT, Morris AG, Frampton J et al (2013) Association studies indicate that protein disulfide isomerase is a risk factor in amyotrophic lateral sclerosis. Free Radic Biol Med 58:81–86. https://doi.org/10.1016/j.freeradbiomed.2013.01.001

Lautenschlaeger J, Prell T, Grosskreutz J (2012) Endoplasmic reticulum stress and the ER mitochondrial calcium cycle in amyotrophic lateral sclerosis. Amyotroph Lateral Scler 13:166–177. https://doi.org/10.3109/17482968.2011.641569

Ledesma MD, Galvan C, Hellias B et al (2002) Astrocytic but not neuronal increased expression and redistribution of parkin during unfolded protein stress. J Neurochem 83:1431–1440. https://doi.org/10.1046/j.1471-4159.2002.01253.x

Lee DY, Lee KS, Lee HJ et al (2010a) Activation of PERK signaling attenuates Aβ-mediated ER stress. PLoS One 5:e10489. https://doi.org/10.1371/journal.pone.0010489

Lee JH, Won SM, Suh J et al (2010b) Induction of the unfolded protein response and cell death pathway in Alzheimer's disease, but not in aged Tg2576 mice. Exp Mol Med 42:386–394. https://doi.org/10.3858/emm.2010.42.5.040

Lee H, Noh JY, Oh Y et al (2012) IRE1 plays an essential role in ER stress-mediated aggregation of mutant huntingtin via the inhibition of autophagy flux. Hum Mol Genet 21:101–114. https://doi.org/10.1093/hmg/ddr445

Lindholm D, Wootz H, Korhonen L (2006) ER stress and neurodegenerative diseases. Cell Death Differ 13:385–392. https://doi.org/10.1038/sj.cdd.4401778

Liu L, Martin R, Chan C (2013) Palmitate-activated astrocytes via serine palmitoyltransferase increase BACE1 in primary neurons by sphingomyelinases. Neurobiol Aging 34:540–550. https://doi.org/10.1016/j.neurobiolaging.2012.05.017

Loewen CA, Feany MB (2010) The unfolded protein response protects from Tau neurotoxicity in vivo. PLoS One 5:e13084. https://doi.org/10.1371/journal.pone.0013084

Lourenco MV, Ferreira ST, De Felice FG (2015) Neuronal stress signaling and eIF2α phosphorylation as molecular links between Alzheimer's disease and diabetes. Prog Neurobiol 129:37–57. https://doi.org/10.1016/j.pneurobio.2015.03.003

Luo Y, Bolon B, Kahn S et al (2001) Mice deficient in BACE1, the Alzheimer's β-secretase, have normal phenotype and abolished β-amyloid generation. Nat Neurosci 4:231–232. https://doi.org/10.1038/85059

Ma T, Trinh MA, Wexler AJ et al (2013) Suppression of eIF2α kinases alleviates Alzheimer's disease-related plasticity and memory deficits. Nat Neurosci 16:1299–1305. https://doi.org/10.1038/nn.3486

Maharjan N, Saxena S (2016) ER strikes again: proteostasis dysfunction in ALS. EMBO J 35:798–800. https://doi.org/10.15252/embj.201694117

Maly DJ, Papa FR (2014) Druggable sensors of the unfolded protein response. Nat Chem Biol 10:892–901. https://doi.org/10.1038/NCHEMBIO.1664

Martínez G, Vidal RL, Mardones P et al (2016) Regulation of memory formation by the transcription factor XBP1. Cell Rep 14:1382–1394. https://doi.org/10.1016/j.celrep.2016.01.028

Martinez-Vicente M, Talloczy Z, Wong E et al (2010) Cargo recognition failure is responsible for inefficient autophagy in Huntington's disease. Nat Neurosci 13:567–576. https://doi.org/10.1038/nn.2528

Matthias EF, Ravi Kiran Reddy K, Joaquin GL et al (2015) The unfolded protein response and its potential role in Huntington's disease elucidated by a systems biology approach. F1000Research 4:103. https://doi.org/10.12688/f1000research.6358.2

Mattson MP (2000) Apoptosis in neurodegenerative disorders. Nat Rev Mol Cell Biol 1:120–129. https://doi.org/10.1038/35040009

Matus S, Castillo K, Hetz C (2012) Hormesis: protecting neurons against cellular stress in Parkinson disease. Autophagy 8:997–1001. https://doi.org/10.4161/auto.20748

Matus S, Lopez E, Valenzuela V et al (2013) Functional contribution of the transcription factor ATF4 to the pathogenesis of amyotrophic lateral sclerosis. PLoS One 8:e66672. https://doi.org/10.1371/journal.pone.0066672

McGeer PL, McGeer EG (2013) The amyloid cascade-inflammatory hypothesis of Alzheimer disease: implications for therapy. Acta Neuropathol 126:479–497. https://doi.org/10.1007/s00401-013-1177-7

Mercado G, Valdés P, Hetz C (2013) An ERcentric view of Parkinson's disease. Trends Mol Med 19:165–175. https://doi.org/10.1016/j.molmed.2012.12.005

Mercado G, Castillo V, Soto P, Sidhu A (2016) ER stress and Parkinson's disease: pathological inputs that converge into the secretory pathway. Brain Res 1648:626–632. https://doi.org/10.1016/j.brainres.2016.04.042

Mitsuda T, Hayakawa Y, Itoh M et al (2007) ATF4 regulates γ-secretase activity during amino acid imbalance. Biochem Biophys Res Commun 352:722–727. https://doi.org/10.1016/j.bbrc.2006.11.075

Moreno JA, Radford H, Peretti D et al (2012) Sustained translational repression by eIF2α-P mediates prion neurodegeneration. Nature 485:507–511. https://doi.org/10.1038/nature11058

Moreno JA, Halliday M, Molloy C et al (2013) Oral treatment targeting the unfolded protein response prevents neurodegeneration and clinical disease in prion-infected mice. Sci Transl Med 5(206):206ra138. https://doi.org/10.1126/scitranslmed.3006767

Morishima N, Nakanishi K, Takenouchi H et al (2002) An endoplasmic reticulum stress-specific caspase cascade in apoptosis. Cytochrome c-independent activation of caspase-9 by caspase-12. J Biol Chem 277:34287–34294. https://doi.org/10.1074/jbc.M204973200

Morishima N, Nakanishi K, Tsuchiya K et al (2004) Translocation of Bim to the endoplasmic reticulum (ER) mediates ER stress signaling for activation of caspase-12 during ER stress-induced apoptosis. J Biol Chem 279:50375–50381. https://doi.org/10.1074/jbc.M408493200

Mukherjee A, Morales-Scheihing D, Gonzalez-Romero D et al (2010) Calcineurin inhibition at the clinical phase of prion disease reduces neurodegeneration, improves behavioral alterations and increases animal survival. PLoS Pathog 6:e1001138. https://doi.org/10.1371/journal.ppat.1001138

Mund T, Gewies A, Schoenfeld N et al (2003) Spike, a novel BH3-only protein, regulates apoptosis at the endoplasmic reticulum. FASEB J 17:696–698. https://doi.org/10.1096/fj.02-0657fje

Nagata T, Ilieva H, Murakami T et al (2007) Increased ER stress during motor neuron degeneration in a transgenic mouse model of amyotrophic lateral sclerosis. Neurol Res 29:767–771. https://doi.org/10.1179/016164107X229803

Nakagawa T, Zhu H, Morishima N et al (2000) Caspase-12 mediates endoplasmic-reticulum-specific apoptosis and cytotoxicity by amyloid-β. Nature 403:98–103. https://doi.org/10.1038/47513

Nishimura AL, Mitne-Neto M, Silva HCA et al (2004) A mutation in the vesicle-trafficking protein VAPB causes late-onset spinal muscular atrophy and amyotrophic lateral sclerosis. Am J Hum Genet 75:822–831. https://doi.org/10.1086/425287

Nishitoh H, Matsuzawa A, Tobiume K et al (2002) ASK1 is essential for endoplasmic reticulum stress-induced neuronal cell death triggered by expanded polyglutamine repeats. Genes Dev 16:1345–1355. https://doi.org/10.1101/gad.992302

Nishitoh H, Kadowaki H, Nagai A et al (2008) ALS-linked mutant SOD1 induces ER stress- and ASK1-dependent motor neuron death by targeting Derlin-1. Genes Dev 22:1451–1464. https://doi.org/10.1101/gad.1640108

Oakley H, Cole SL, Logan S et al (2006) Intraneuronal β-amyloid aggregates, neurodegeneration, and neuron loss in transgenic mice with five familial Alzheimer's disease mutations: potential factors in amyloid plaque formation. J Neurosci 26:10129–10140. https://doi.org/10.1523/JNEUROSCI.1202-06.2006

Obeng EA, Boise LH (2005) Caspase-12 and caspase-4 are not required for caspase-dependent endoplasmic reticulum stress-induced apoptosis. J Biol Chem 280:29578–29587. https://doi.org/10.1074/jbc.M502685200

Orr HT, Zoghbi HY (2007) Trinucleotide repeat disorders. Annu Rev Neurosci 30:575–621. https://doi.org/10.1146/annurev.neuro.29.051605.113042

Page G, Rioux Bilan A, Ingrand S et al (2006) Activated double-stranded RNA-dependent protein kinase and neuronal death in models of Alzheimer's disease. Neuroscience 139:1343–1354. https://doi.org/10.1016/j.neuroscience.2006.01.047

Paschen W, Frandsen A (2001) Endoplasmic reticulum dysfunction - a common denominator for cell injury in acute and degenerative diseases of the brain? J Neurochem 79:719–725. https://doi.org/10.1046/j.1471-4159.2001.00623.x

Pasini S, Corona C, Liu J et al (2015) Specific downregulation of hippocampal ATF4 reveals a necessary role in synaptic plasticity and memory. Cell Rep 11:183–191. https://doi.org/10.1016/j.celrep.2015.03.025

Prudencio M, Belzil VV, Batra R et al (2015) Distinct brain transcriptome profiles in C9orf72-associated and sporadic ALS. Nat Neurosci 18:1175–1182. https://doi.org/10.1038/nn.4065

Prusiner SB (1998) Prions. Proc Natl Acad Sci U S A 95(23):13363–13383

Prusiner SB, Scott MR (1997) Genetics of prions. Annu Rev Genet 31:139–175. https://doi.org/10.1146/annurev.genet.31.1.139

Rao RV, Castro-Obregon S, Frankowski H et al (2002) Coupling endoplasmic reticulum stress to the cell death program. An Apaf-1-independent intrinsic pathway. J Biol Chem 277:21836–21842. https://doi.org/10.1074/jbc.M202726200

Rao RV, Ellerby HM, Bredesen DE (2004) Coupling endoplasmic reticulum stress to the cell death program. Cell Death Differ 11:372–380. https://doi.org/10.1038/sj.cdd.4401378

Reitz C (2012) Alzheimer's disease and the amyloid cascade hypothesis: a critical review. Int J Alzheimers Dis 2012:369808. https://doi.org/10.1155/2012/369808

Rozas P, Bargsted L, Martínez F et al (2017) The ER proteostasis network in ALS: determining the differential motoneuron vulnerability. Neurosci Lett 636:9–15. https://doi.org/10.1016/j.neulet.2016.04.066

Sado M, Yamasaki Y, Iwanaga T et al (2009) Protective effect against Parkinson's disease-related insults through the activation of XBP1. Brain Res 1257:16–24. https://doi.org/10.1016/j.brainres.2008.11.104

Sämann J, Hegermann J, von Gromoff E et al (2009) Caenorhabditits elegans LRK-1 and PINK-1 act antagonistically in stress response and neurite outgrowth. J Biol Chem 284:16482–16491. https://doi.org/10.1074/jbc.M808255200

Sato N, Urano F, Yoon Leem J et al (2000) Upregulation of BiP and CHOP by the unfolded-protein response independent of presenilin expression. Nat Cell Biol 2:863–870. https://doi.org/10.1038/35046500

Saxena S, Cabuy E, Caroni P (2009) A role for motoneuron subtype-selective ER stress in disease manifestations of FALS mice. Nat Neurosci 12:627–636. https://doi.org/10.1038/nn.2297

Scheff SW, Ansari MA, Mufson EJ (2016) Oxidative stress and hippocampal synaptic protein levels in elderly cognitively intact individuals with Alzheimer's disease pathology. Neurobiol Aging 42:1–12. https://doi.org/10.1016/j.neurobiolaging.2016.02.030

Selkoe DJ (2001) Alzheimer's disease: genes, proteins, and therapy. Physiol Rev 81:741–766. https://doi.org/10.1152/physrev.2001.81.2.741

Shimoke K, Utsumi T, Kishi S et al (2004) Prevention of endoplasmic reticulum stress-induced cell death by brain-derived neurotrophic factor in cultured cerebral cortical neurons. Brain Res 1028: 105–111. https://doi.org/10.1016/j.brainres.2004.09.005

Silva RM, Ries V, Oo TF et al (2005) CHOP/GADD153 is a mediator of apoptotic death in substantia nigra dopamine neurons in an in vivo neurotoxin model of parkinsonism. J Neurochem 95:974–986. https://doi.org/10.1111/j.1471-4159.2005.03428.x

Simmen T, Aslan JE, Blagoveshchenskaya AD et al (2005) PACS-2 controls endoplasmic reticulum–mitochondria communication and Bid-mediated apoptosis. EMBO J 24:717–729. https://doi.org/10.1038/sj.emboj.7600559

Soto C (2003) Unfolding the role of protein misfolding in neurodegenerative diseases. Nat Rev Neurosci 4:49–60. https://doi.org/10.1038/nrn1007

Sreedharan J, Brown RH (2013) Amyotrophic lateral sclerosis: problems and prospects. Ann Neurol 74:309–316. https://doi.org/10.1002/ana.24012

Steele AD, Hetz C, Yi CH et al (2007) Prion pathogenesis is independent of caspase-12. Prion 1: 243–247. https://doi.org/10.4161/pri.1.4.5551

Sun S, Sun Y, Ling SC et al (2015) Translational profiling identifies a cascade of damage initiated in motor neurons and spreading to glia in mutant sod1-mediated ALS. Proc Natl Acad Sci U S A 112:E6993–E7002. https://doi.org/10.1073/pnas.1520639112

Suzuki H, Matsuoka M (2012) TDP-43 toxicity is mediated by the unfolded protein response-unrelated induction of C/EBP homologous protein expression. J Neurosci Res 90:641–647. https://doi.org/10.1002/jnr.22777

Suzuki H, Kanekura K, Levine TP et al (2009) ALS-linked P56S-VAPB, an aggregated loss-of-function mutant of VAPB, predisposes motor neurons to ER stress-related death by inducing aggregation of co-expressed wild-type VAPB. J Neurochem 108:973–985. https://doi.org/10.1111/j.1471-4159.2008.05857.x

Tajiri S, Oyadomari S, Yano S et al (2004) Ischemia-induced neuronal cell death is mediated by the endoplasmic reticulum stress pathway involving CHOP. Cell Death Differ 11:403–415. https://doi.org/10.1038/sj.cdd.4401365

Tamatani M, Matsuyama T, Yamaguchi A et al (2001) ORP150 protects against hypoxia/ischemia-induced neuronal death. Nat Med 7:317–323. https://doi.org/10.1038/85463

Tobisawa S, Hozumi Y, Arawaka S et al (2003) Mutant SOD1 linked to familial amyotrophic lateral sclerosis, but not wild-type SOD1, induces ER stress in COS7 cells and transgenic mice. Biochem Biophys Res Commun 303:496–503. https://doi.org/10.1016/S0006-291X(03)00353-X

Torres M, Castillo K, Armisén R et al (2010) Prion protein misfolding affects calcium homeostasis and sensitizes cells to endoplasmic reticulum stress. PLoS One 5:e15658. https://doi.org/10.1371/journal.pone.0015658

Torres M, Medinas DB, Matamala JM et al (2015) The protein-disulfide isomerase ERp57 regulates the steady-state levels of the prion protein. J Biol Chem 290:23631–23645. https://doi.org/10.1074/jbc.M114.635565

Tsaytler P, Harding HP, Ron D, Bertolotti A (2011) Selective inhibition of a regulatory subunit of protein phosphatase 1 restores proteostasis. Science (80-) 332:91–94. https://doi.org/10.1126/science.1201396

Uddin MS, Kabir MT, Jakaria M et al (2019) Endothelial PPARγ is crucial for averting age-related vascular dysfunction by stalling oxidative stress and ROCK. Neurotox Res 36:583–601. https://doi.org/10.1007/s12640-019-00047-5

Uddin MS, Al Mamun A, Kabir MT et al (2020a) Multi-target drug candidates for multifactorial Alzheimer's disease: AChE and NMDAR as molecular targets. Mol Neurobiol 58(1):281–303. https://doi.org/10.1007/s12035-020-02116-9

Uddin MS, Al Mamun A, Rahman MA et al (2020b) Exploring the promise of flavonoids to combat neuropathic pain: from molecular mechanisms to therapeutic implications. Front Neurosci 14:1–18. https://doi.org/10.3389/fnins.2020.00478

Uddin MS, Kabir MT, Al Mamun A et al (2020c) Pharmacological approaches to mitigate neuroinflammation in Alzheimer's disease. Int Immunopharmacol 84:106479. https://doi.org/10.1016/j.intimp.2020.106479

Uddin MS, Kabir MT, Rahman MM et al (2020d) TV 3326 for Alzheimer's dementia: a novel multimodal ChE and MAO inhibitors to mitigate Alzheimer's-like neuropathology. J Pharm Pharmacol 72:1001–1012. https://doi.org/10.1111/jphp.13244

Uddin MS, Kabir MT, Tewari D et al (2020e) Revisiting the role of brain and peripheral Aβ in the pathogenesis of Alzheimer's disease. J Neurol Sci 416:116974. https://doi.org/10.1016/j.jns.2020.116974

Uddin MS, Rahman MA, Kabir MT et al (2020f) Multifarious roles of mTOR signaling in cognitive aging and cerebrovascular dysfunction of Alzheimer's disease. IUBMB Life 72:1843–1855. https://doi.org/10.1002/iub.2324

Uddin MS, Tewari D, Sharma G et al (2020g) Molecular mechanisms of ER stress and UPR in the pathogenesis of Alzheimer's disease. Mol Neurobiol 57:2902–2919

Unterberger U, Höftberger R, Gelpi E et al (2006) Endoplasmic reticulum stress features are prominent in Alzheimer disease but not in prion diseases in vivo. J Neuropathol Exp Neurol 65:348–357. https://doi.org/10.1097/01.jnen.0000218445.30535.6f

Urano F, Wang XZ, Bertolotti A et al (2000) Coupling of stress in the ER to activation of JNK protein kinases by transmembrane protein kinase IRE1. Science (80-) 287:664–666. https://doi.org/10.1126/science.287.5453.664

Valdés P, Mercado G, Vidal RL et al (2014) Control of dopaminergic neuron survival by the unfolded protein response transcription factor XBP1. Proc Natl Acad Sci U S A 111:6804–6809. https://doi.org/10.1073/pnas.1321845111

Valenzuela V, Martínez G, Duran-Aniotz C, Hetz C (2016) Gene therapy to target ER stress in brain diseases. Brain Res 1648:561–570. https://doi.org/10.1016/j.brainres.2016.04.064

Van Der Harg JM, Nölle A, Zwart R et al (2014) The unfolded protein response mediates reversible tau phosphorylation induced by metabolic stress. Cell Death Dis 5:e1393. https://doi.org/10.1038/cddis.2014.354

Varma D, Sen D (2015) Role of the unfolded protein response in the pathogenesis of Parkinson's disease. Acta Neurobiol Exp (Wars) 75:1–26

Vassar R (2009) O1-03-01: phosphorylation of the translation initiation factor eIF2alpha increases BACE1 levels and promotes amyloidogenesis. Alzheimers Dement 5:P81–P82. https://doi.org/10.1016/j.jalz.2009.05.205

Verkhratsky A (2005) Physiology and pathophysiology of the calcium store in the endoplasmic reticulum of neurons. Physiol Rev 85:201–279. https://doi.org/10.1152/physrev.00004.2004

Vidal R, Caballero B, Couve A, Hetz C (2011) Converging pathways in the occurrence of endoplasmic reticulum (ER) stress in huntingtons disease. Curr Mol Med 11:1–12. https://doi.org/10.2174/156652411794474419

Vidal RL, Figueroa A, Court FA et al (2012) Targeting the UPR transcription factor XBP1 protects against Huntington's disease through the regulation of FoxO1 and autophagy. Hum Mol Genet 21:2245–2262. https://doi.org/10.1093/hmg/dds040

Vieira FG, Ping Q, Moreno AJ et al (2015) Guanabenz treatment accelerates disease in a mutant SOD1 mouse model of ALS. PLoS One 10:e0135570. https://doi.org/10.1371/journal.pone.0135570

Vitte J, Traver S, Maués De Paula A et al (2010) Leucine-rich repeat kinase 2 is associated with the endoplasmic reticulum in dopaminergic neurons and accumulates in the core of lewy bodies in parkinson disease. J Neuropathol Exp Neurol 69:959–972. https://doi.org/10.1097/NEN.0b013e3181efc01c

Walker FO (2007) Huntington's disease. Lancet 369:218–228. https://doi.org/10.1016/S0140-6736(07)60111-1

Walker L, McAleese KE, Thomas AJ et al (2015) Neuropathologically mixed Alzheimer's and Lewy body disease: burden of pathological protein aggregates differs between clinical phenotypes. Acta Neuropathol 129:729–748. https://doi.org/10.1007/s00401-015-1406-3

Wang M, Ye R, Barron E et al (2010) Essential role of the unfolded protein response regulator GRP78/BiP in protection from neuronal apoptosis. Cell Death Differ 17:488–498. https://doi.org/10.1038/cdd.2009.144

Wang L, Popko B, Roos RP (2011) The unfolded protein response in familial amyotrophic lateral sclerosis. Hum Mol Genet 20:1008–1015. https://doi.org/10.1093/hmg/ddq546

Wang L, Popko B, Tixier E, Roos RP (2014) Guanabenz, which enhances the unfolded protein response, ameliorates mutant SOD1-induced amyotrophic lateral sclerosis. Neurobiol Dis 71: 317–324. https://doi.org/10.1016/j.nbd.2014.08.010

Wate R, Ito H, Zhang JH et al (2005) Expression of an endoplasmic reticulum-resident chaperone, glucose-regulated stress protein 78, in the spinal cord of a mouse model of amyotrophic lateral sclerosis. Acta Neuropathol 110:557–562. https://doi.org/10.1007/s00401-005-1080-y

Woehlbier U, Colombo A, Saaranen MJ et al (2016) ALS -linked protein disulfide isomerase variants cause motor dysfunction. EMBO J 35:845–865. https://doi.org/10.15252/embj.201592224

Wyss-Coray T, Mucke L (2002) Inflammation in neurodegenerative disease - a double-edged sword. Neuron 35:419–432. https://doi.org/10.1016/S0896-6273(02)00794-8

Yoo BC, Krapfenbauer K, Cairns N et al (2002) Overexpressed protein disulfide isomerase in brains of patients with sporadic Creutzfeldt-Jakob disease. Neurosci Lett 334:196–200. https://doi.org/10.1016/S0304-3940(02)01071-6

Yoon SO, Park DJ, Ryu JC et al (2012) JNK3 perpetuates metabolic stress induced by Aβ peptides. Neuron 75:824–837. https://doi.org/10.1016/j.neuron.2012.06.024

Yuan J, Lipinski M, Degterev A (2003) Diversity in the mechanisms of neuronal cell death. Neuron 40:401–413. https://doi.org/10.1016/S0896-6273(03)00601-9

Yuan Y, Cao P, Smith MA et al (2011) Dysregulated LRRK2 signaling in response to endoplasmic reticulum stress leads to dopaminergic neuron degeneration in C. elegans. PLoS One 6:e22354. https://doi.org/10.1371/journal.pone.0022354

Zhang YJ, Jansen-West K, Xu YF et al (2014) Aggregation-prone c9FTD/ALS poly (GA) RAN-translated proteins cause neurotoxicity by inducing ER stress. Acta Neuropathol 128:505–524. https://doi.org/10.1007/s00401-014-1336-5

Zuleta A, Vidal RL, Armentano D et al (2012) AAV-mediated delivery of the transcription factor XBP1s into the striatum reduces mutant Huntingtin aggregation in a mouse model of Huntington's disease. Biochem Biophys Res Commun 420:558–563. https://doi.org/10.1016/j.bbrc.2012.03.033

Role of Hypertension and Hyperlipidemia in the Pathogenesis of Dementia

10

Meenakshi Kaira, Vaibhav Walia, and Munish Garg

Abstract

Dementia is not a single ailment, but a term that describes symptoms of impairment in memory, communication, and thinking, which occurs mainly in the elderly person. About 50 million people have dementia worldwide, and about ten million people are reported per day. The intensity of the symptoms varies from patient to patient. Hypertension and hyperlipidemia have been found to increase the risk of the emergence of dementia directly or indirectly. Further, the treatment with the antihypertensive and antihyperlipidemic drugs produces beneficial effects in the treatment of dementia. In the present chapter, the authors described the role of hypertension and hyperlipidemia in the pathogenesis of dementia and described the impact of antihypertensive and antihyperlipidemic drugs for the treatment of dementia.

Keywords

Alzheimer's disease · Dementia · Hyperlipidemia · Hypertension · Lewy body dementia · Lipid · Cholesterol

M. Kaira · M. Garg (✉)
Department of Pharmaceutical Sciences, Maharshi Dayanand University, Rohtak, India

V. Walia
Faculty of Pharmacy, SGT University, Gurugram, India

G. M. Ashraf, M. S. Uddin (eds.), *Current Thoughts on Dementia*,
https://doi.org/10.1007/978-981-16-7606-2_10

10.1 Introduction

Dementia is known to affect 36 million people across the world, 2040 might be about 81.1 million, and approximately 115 million will be afflicted by the year 2050 (Alzheimer's Association 2012; Ferri et al. 2005; Wortmann 2012). Dementia is also referred as neurocognitive disorders often resulted in progressive cognitive impairment (American Psychiatric Association 2013; Chertkow et al. 2013; Sheehan 2012). The patient suffering from dementia experiences impairment related to memory, communication, recognition, and performing tasks. It has been reported that the damages to the cortex and its region may lead to dementia and cognitive impairment (Hildreth and Church 2015). It has also been reported that the individuals suffering from mild cognitive impairment (MCI) cannot be considered under demented (Hildreth and Church 2015). However, the afflicted individuals may face problems related to memory, judgment, and thinking more during aging as compared to normal aging persons (Alzheimer Society of Canada 2016). MCI does not interfere with the independence and daily routine of the individual (Alzheimer Society of Canada 2016) but predisposes the individual toward dementia (Bruscoli and Lovestone 2004; Ganguli et al. 2011; Matthews et al. 2008), but most of them retrieve their normal state or remain unchanged cognitively (Ganguli et al. 2011; Larrieu et al. 2002; Mitchell and Shiri-Feshki 2009).

Alzheimer's disease (AD) is responsible for 60–80% of cases of dementia with manifestations including accelerated cognitive and memory impairment due to the damage of cholinergic neurons and impaired cholinergic signaling (Reitz 2015). AD affects approx. 25 million worldwide, and this value might be quadruple by the year 2050 (Brookmeyer et al. 1998; Ferri et al. 2005; von Strauss et al. 1999). Vascular dementia (VaD) is also known as multi-infarct dementia and is the second most prevalent dementia found in individuals (Alzheimer Society of Canada 2016). Stroke mainly accounts for this form of dementia. VaD accounts for 15–20% of cases of dementia, characterized by the presence of clinical stroke and resulted in cognitive deficits in the patients (Gorelick et al. 2011; Román et al. 1993; Ruitenberg et al. 2001; Thal et al. 2012). Various risk factors responsible for the VaD include hypertension, insulin resistance, and hyperlipidemia (Cohen and Tong 2010; Faraci 2011; Sahathevan et al. 2012; Yates et al. 2012). It was observed that 20–30% of patients are demented after 3 months of stroke and the chances are two- to fivefold of getting dementia in the stroke patients (Desmond et al. 2000; Kokmen et al. 1996; Lindén et al. 2004; Pohjasvaara et al. 1997; Prencipe et al. 1997; Tatemichi et al. 1992, 1994).

However, 50–60% of afflicted patients develop AD later (Kokmen et al. 1996; Korf et al. 2004). Diabetes and hypertension increase the risk of development of VaD, whereas obesity predisposes the patients to AD (Hayden et al. 2006). Lewy body dementia (LBD) accounts for 5–15% cases of dementia (Alzheimer Society of Canada 2016) and is characterized by fluctuating cognition, attention and alertness problems, visual hallucinations, spontaneous parkinsonism, mental status fluctuations, disorganized speech, movement difficulties, sleep disorders, etc. (Alzheimer Society of Canada 2016; Zupancic et al. 2011). About 80% of cases of

LBD patients develop parkinsonism (Muangpaisan 2007). This chapter describes the role of hypertension and hyperlipidemia in the pathogenesis of dementia and potential therapeutic strategy.

10.2 Hypertension in the Emergence of Dementia

The persistent rise in blood pressure beyond the normal limit is known as hypertension (HTN), and individuals with an age greater than 60 years are more prone to develop HTN. HTN increases the risk of dementia and its subtypes in afflicted individuals (Kivipelto et al. 2002; Shah et al. 2012; Tzourio et al. 1999; Whitmer et al. 2005). Blood pressure (BP) is often affected by age, suggesting the possible role of aging also (Korf et al. 2004; Swan et al. 1998). HTN is known to be an emerging cause of dementia in individuals as shown in Fig. 10.1. A study conducted on the Japanese population suggested that the BP > 160/95 mmHg is responsible for the emergence of VaD and AD in elderly patients (Launer et al. 2000). Rise in BP and reduced compliance to BP-lowering drug often led to lower mini-mental state examination (MMSE) scores (Vinyoles et al. 2008). Increased systolic BP (SBP) often worsens memory, reduced MMSE scores, processing speed, and verbal fluency functions (Haan et al. 1999; Singh-Manoux and Marmot 2005). The individuals with the SBP greater than 180 mmHg had a high risk of development of AD and dementia (Kilander et al. 1998). Another study also suggested the same fact that the

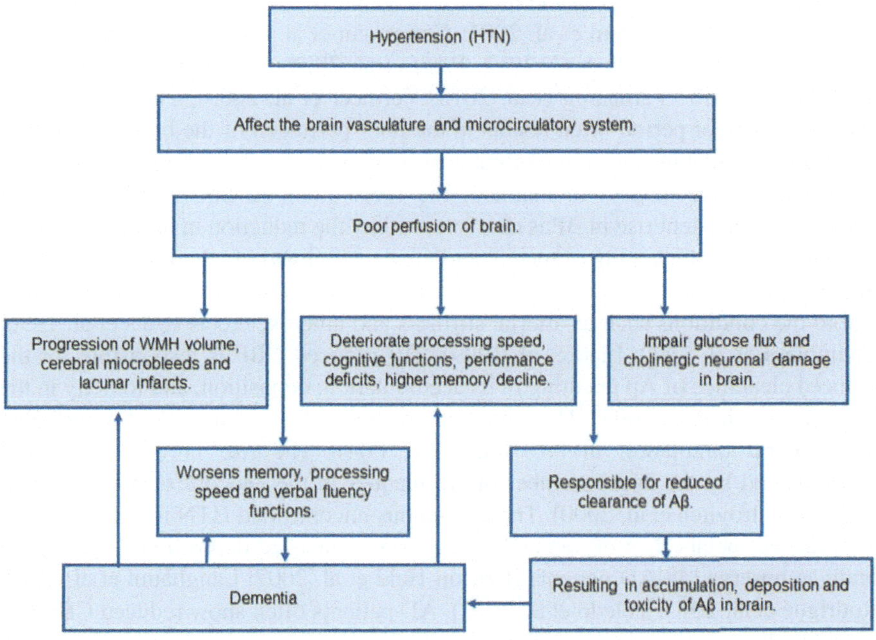

Fig. 10.1 Role of hypertension in the emergence of dementia

SBP > 180 mmHg increases the risk of AD 1.5 times and other related dementia 1.6 times, respectively, and an increase of 10 mmHg in SBP significantly increased the cognitive deficit (Launer et al. 2015). A study performed on 40–60 years individuals suggested that the increase in both SBP and diastolic BP (DBP) resulted in cognitive deficits (Shang et al. 2016). It is suggested that the high SBP resulted in the smaller regional and total brain volumes, whereas the DBP did not found to affect the total brain volume (Glodzik et al. 2012; Harris et al. 2008; Jennings et al. 2012; Leritz et al. 2011; Nagai et al. 2009). But, DBP is often considered as the strong predictor of cognitive deterioration (Cacciatrore et al. 1997) and the increased DBP is responsible for the larger cognitive deficits in later life (Kilander et al. 1998; Tsivgoulis et al. 2009). Besides the HTN, some of the cross-sectional studies also suggested that low BP also results in a higher prevalence of dementia (including AD) (Kennelly et al. 2009a, b). But some studies suggested no link between HTN and cognitive impairment in individuals of age more than 65 years (Scherr et al. 1991). One of the studies also suggested that the BP of about 140–159 mmHg is responsible for better performance, whereas the BP more than 180 mmHg deteriorates performance and memory in the individuals (Andre-Petersson et al. 2001).

Despite this evidence, in general, the untreated HTN often resulted in silent cerebral white matter lesions (WMLs) and deteriorated processing speed, verbal fluency, cognitive functions, performance deficits, higher memory decline, etc., in the afflicted individuals (Elias et al. 2003; Gottesman et al. 2014; Sierra et al. 2004). HTN led to high-grade WML, white matter hyperintensity (WMH) progression and volume, cerebral microbleeds, lacunar infarcts, cerebral ischemia, and ultimately the cognitive impairment responsible for the emergence of VaD (Bezerra et al. 2012; de Leeuw et al. 2002; Dufouil et al. 2001; Gottesman et al. 2010; Harrell et al. 1991; Junqué et al. 1990; Poels et al. 2012; Prins et al. 2004; Steingart et al. 1987; van Sloten et al. 2015; Verhaaren et al. 2013; Vermeer et al. 2003). The uncontrolled HTN for a longer period often results in the poor perfusion of the brain and affects the brain vasculature and microcirculatory system and led to the activation of a compensatory mechanism to maintain the cerebral blood flow (CBF) (Iadecola 2004). The persistent rise in BP is characterized by the reduction in luminal diameter of microvessels, thickening of media, and the development of atheromatous plaques in the larger cerebral arteries. The poor CBF and cerebral hypoperfusion are mainly due to the conditions such as arterial stiffness and atherosclerosis (Bots et al. 1996; Chambless et al. 2002; Liao et al. 1999). The reduced CBF is responsible for the reduced clearance of Aβ resulting in its accumulation, deposition, and toxicity in the brain (Zlokovic et al. 2005). The increased levels and deposition of Aβ further affect the cerebral circulation in AD (Iadecola 2004). The rise in SBP is further accompanied by the larger number of Aβ plaques in the neocortex and the hippocampus (Petrovitch et al. 2000). The continuous uncontrolled HTN is thus related to the reduced metabolism of cerebral glucose and increased deposition of Aβ in the brain as observed in AD patients (Beason-Held et al. 2007; Langbaum et al. 2012; Rodrigue et al. 2013; Toledo et al. 2012). AD patients often show reduced CBF and vascular changes in various brain regions (Farkas and Luiten 2001; Nagahama et al. 2003; Warkentin et al. 2004).

Uncontrolled HTN affects cerebral perfusion. Endothelial nitric oxide synthase (eNOS) is responsible for the maintenance of cerebral perfusion. When cerebral perfusion goes down to a critical point, eNOS upregulates the nitric oxide (NO) to maintain it (de la Torre 1999, 2000a, b, 2002). Chronic HTN causes vascular injury by modulating endothelium and basal NO level (Cooke and Dzau 1997) responsible for the functional vascular changes (McCarron et al. 2006). NO shields the endothelial cell function (Maxwell 2002). HTN-mediated endothelial dysfunction is often linked with the reduced CBF, impair glucose flux in the brain, neuronal cell death and cholinergic neuronal damage, and neurons expressing NOS as observed in the AD patients (Chen et al. 1999; Fernandez-Vizarra et al. 2004; Girouard and Iadecola 2006; Hamel 2004; Selley 2003; Rodrigo et al. 2004a, b; Tong and Hamel 1999). Thus, the poor cerebral perfusion and the reduced CBF are often linked with the abnormal cerebral glucose homeostasis that is considered as an important hallmark of AD.

10.3 Hyperlipidemia in the Emergence of Dementia

Cholesterol is the major lipids of the body and is widely present in the brain. In the brain, cholesterol mainly regulates neuronal function, plasticity, and synaptic remodeling (Pfrieger 2003). Cholesterol is synthesized inside the brain, and the rate-limiting step in the synthesis is the formation of mevalonate from 3-hydroxy-3-methylglutryl-CoA reductase (HMG-CoAR) (Arnaud and Mach 2005). ABCA1 is the ATP-regulated membrane protein responsible for the transportation of cholesterol in neurons (Jurevics and Morell 1995). APOE mRNA is expressed abundantly in the CNS and acts as low-density lipoprotein (LDL) receptor-related proteins (LRP), binds to the cholesterol, transports it through cell surface lipoprotein receptors, and is finally responsible for the internalization of cholesterol in the form of APOE–cholesterol–phospholipid complex. The complex then forms endosome and fuses with lysosomes and ultimately breakdown to release free cholesterol. Free cholesterol produces negative feedback to the enzyme β-hydroxy β-methylglutaryl-CoA (HMG-CoA) and inhibits the endogenous synthesis of cholesterol.

Brain cholesterol is transported in the form of 2,4-hydroxylcholesterol into the systemic circulation through APOE-mediated transport process (Bogdanovic et al. 2001), and thus, the dysfunctioning of APOE increases the brain cholesterol load and subsequent damage (Heverin et al. 2004; Papassotiropoulos et al. 2002). ApoE also mediates the risk of development of AD (Viticchi et al. 2014). Ma et al. (2010) establish the link between lower ApoA-1 level and dementia, and further, it was suggested that the higher ApoB/ApoA-1 ratio is responsible for the cognitive decline in normal individuals (Song et al. 2012). Also, a diet rich in fat and cholesterol promotes blood–brain barrier (BBB) dysfunction resulting in cognitive decline (Hsu and Kanoski 2014). Further, the BBB dysfunction is promoted by aging and inflammatory cytokines, which affect the BBB permeability (Dias et al. 2015). Thus, both aging-mediated neuronal inflammation and high cholesterol level are

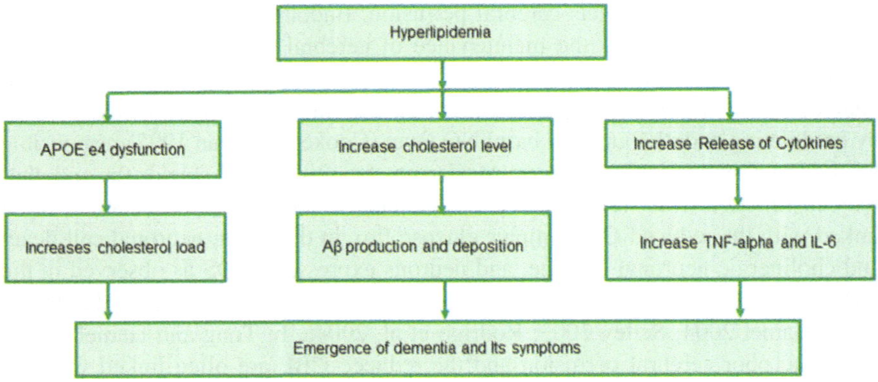

Fig. 10.2 Role of hyperlipidemia in the emergence of dementia

responsible for the BBB dysfunctioning and cognitive impairment. The role of hyperlipidemia and the increased cholesterol level in the emergence of dementia are displayed in Fig. 10.2.

Raised levels of cholesterol thus predispose the individuals to dementia (Solomon et al. 2009). Further, the hypertriglyceridemia, high LDL, and low high-density lipoprotein (HDL) levels increase the emergence of MCI and dementia (Carlsson 2010; Panza et al. 2011). Increased triglyceride level has been observed in almost all cases of dementia (Raffaitin et al. 2009), suggesting triglyceridemia is the major cause of VaD (Panza et al. 2012). Further, the raised levels of triglyceride increase the deposition of Aβ (Burgess et al. 2008). The increased level of LDL cholesterol is responsible for the threefold increase in stroke-associated dementia rather than AD-related dementia (Moroney et al. 1999). Besides this is the small-dense LDL particles that are related to the development of VaD and atherogenic dyslipidemia (Watanabe et al. 2004). A study conducted on 192 adults with AD suggested that LDL level affects the functional performance in subjects (de Oliveira et al. 2017) and implicated in the atherogenic conditions that lead to VaD (Shoji et al. 2009). LDL also promotes the secretions and release of neuroinflammatory cytokines (Dias et al. 2015). Besides this, one study suggested that higher cholesterol concentration lowers the risk of VaD (Mielke et al. 2005). The study conducted on 721 demented patients suggested that high cholesterol increases 20–40% of chances of dementia (Whitmer et al. 2005). The study conducted on 444 men of age above 70 years suggested that the cholesterol level above $\geq$6.5 mmol/L increases the risk of development of AD and is considered as a predictor of later also (Kivipelto et al. 2001; Notkola et al. 1998). Also, midlife obesity raised cholesterol and BP doubled the risk of development of AD and related dementia (Kivipelto et al. 2005). However, the opposite fact has been reported by the study conducted on 392 participants suggesting that the raised cholesterol levels reduce the chances of dementia and suggesting the protective action of cholesterol (Mielke et al. 2005). Another study suggested that the raised levels of total cholesterol decrease the development of AD (Reitz et al. 2004). HDL cholesterol is important for the elimination of cholesterol that inhibits

Table 10.1 Genes that are involved in the regulation of cholesterol and Aβ synthesis, and clearance (Reitz 2012)

Gene	Functions
ABCA1	• Regulate the efflux of cholesterol and levels of ApoE
ABCA2	• Lipid transport and myelination
ABCG1	• Cholesterol efflux
APOA1	• Transport the cholesterol back into the liver
	• Inhibits the aggregation and toxicity of Aβ
APOE (ε2/3/4)	• Transport of cholesterol transfer from astroglia to neurons
CD36	• Microglial binding with Aβ and evoke an immune response
CETP	• Transfer cholesteryl esters from HDL particles to triglyceride-rich lipoproteins
CLU	• Forms insoluble Aβ forms responsible for toxicity
CYP46A1	• Convert cholesterol to 24-hydroxycholesterol;
	• Responsible for the elimination of cholesterol from the brain
DHCR24	• Inhibit Aβ synthetic enzymes;
	• Inhibit the caspase 3 and protect against the Aβ-induced apoptosis
HMGCR	• Cholesterol synthesis
LDLR	• Major APOE receptor and clears the Aβ
LRP1	• Receptor for ApoE and mediates the clearance of Aβ
SOAT1	• Favors the formation of cholesterol esters
SORL1	• Binds to the ApoE lipoproteins and results in their intracellular trafficking
SREBF1	• Regulates lipid homeostasis and cholesterol synthesis pathway

polymerization and aggregation of Aβ, and maintains cognitive abilities, verbal ability, and perceptual speed (Jimenez-Conde et al. 2010; Reynolds et al. 2010). HDL reverses the deleterious effect of oxidized LDL (OxLDL) particles, normalizes the endothelial functions, attenuates the effects of inflammatory cytokines, and protects against cognitive impairment (Appleton et al. 2017; He et al. 2016).

Generally, it is considered as the high serum cholesterol level is an indicator and risk factor for AD (Evans et al. 2000; Jarvik et al. 1995; Notkola et al. 1998; Solomon et al. 2009). Higher cholesterol levels are responsible for the increased production of Aβ further responsible for the learning and memory deficit in experimental animals (Gamba et al. 2015). Various genes and their products have been known to influence the lipid levels, transport, and the formation of Aβ (Table 10.1). A cholesterol-rich environment is also responsible for the increased production and deposition of Aβ in the brain by favoring the amyloidosis (He et al. 2016; Refolo et al. 2000, 2001; Wahrle et al. 2002). Exogenous cholesterol is known to increase the β-cleavage of APP resulting in the increased formation of Aβ40 and Aβ42 (Frears et al. 1999). When the concentration of cholesterol increases, then the internalization and degradation of Aβ decrease resulting in its deposition, intracellular accumulation, and aggregation (Yip et al. 2001). Aβ (5 kDa) in the presence of cholesterol acts as a precursor/seed for aggregation resulting in the formation of fibrillar Aβ peptides (Mizuno et al. 1999). Aβ42 causes the hyperphosphorylation of tau protein (Zou et al. 2003), and the later is promoted by the presence of higher cholesterol levels (McLaurin et al. 2003).

10.4 Relationship Between Aging, Hypertension, Hyperlipidemia, and Dementia

Aging predisposes the individuals to dementia and its types (Corrada et al. 2008), and the risk of dementia increases with the increase in age of individuals, and further this, risk doubles after every 5 years (Corrada et al. 2008). Aging affects the mitochondrial metabolism and process of glycolysis (Brewer 2010) and contributes to the development of insulin resistance (Ford et al. 2002; Kobayashi et al. 2007). Insulin activates the enzyme Akt, which further results in the inactivation of enzyme GSK-3β responsible for the impaired glucose metabolism and antiapoptotic mechanisms (Fang et al. 2000; van der Heide et al. 2006). Further, aging is characterized by reduced insulin signaling (Ford et al. 2002; Kobayashi et al. 2007), reduction in glucose uptake, and insulin resistance (Abdul-Ghani and DeFronzo 2008; Barazzoni et al. 2000; Cortopassi et al. 1992; Kim et al. 2008; Petersen et al. 2003). Also, reduced brain insulin signaling has been observed in the aged mice (Muller et al. 2012). Insulin resistance is characterized by the reduced binding of insulin to insulin receptors (Fulop et al. 2003) resulted in the impaired glucose metabolism, emergence of hypertension and hyperlipidemia (Cornier et al. 2008), increased aggregation of Aβ, and formation of neurofibrillary tangles (Craft et al. 1999; Park et al. 2000; van der Heide et al. 2006) responsible for the development of AD and related dementia (Rivera et al. 2005; Steen et al. 2005). Insulin resistance is thus associated with cognitive decline in dementia (Craft et al. 1999; Vingtdeux et al. 2011). Further, the aging-mediated senescence of adipose tissue is implicated in the emergence of insulin resistance (Minamino et al. 2009) characterized by the increased fat mass and reduced muscle mass (Benbassat et al. 1997; Corpas et al. 1993; Rudman et al. 1990).

Aging interferes with the secretion and functioning of adiponectin (Gulcelik et al. 2013). Adiponectin directly regulates insulin sensitivity and glucose metabolism via activation of AMPK and GLUT4 translocation (Vu et al. 2013). Adiponectin acts as an insulin sensitizer (Cnop et al. 2003; Lim et al. 2008; Pajvani and Scherer 2003), promotes insulin sensitivity (Kadowaki et al. 2006), suppresses liver gluconeogenesis (Yamauchi et al. 2002), and reduces insulin resistance (Stefan et al. 2002; Yamauchi et al. 2001). Adiponectin knockout mice show impaired insulin secretion (Kubota et al. 2002; Okamoto et al. 2008) and insulin resistance (Yamaguchi et al. 2007). Adiponectin promotes NO production and thus maintains the normal functioning of endothelium (Zhu et al. 2008a, b). Further, adiponectin knockout mice are characterized by reduced levels of the NO in endothelium responsible for endothelial dysfunction (Ouedraogo et al. 2007). Adiponectin also regulates microvascular network flow and function (Goldstein et al. 2009; Takaoka et al. 2009) and is responsible for arterial vasodilation (Goldstein and Scalia 2004).

10.5 Drugs Used in the Treatment of Dementia

Cognitive processes are negatively regulated by the renin–angiotensin system (RAS) (Gard 2002; Sharma 1998), and the activation of RAS in the brain promotes neurogenic hypertension (Phillips and de Oliveira 2008). Angiotensin II (AT-II) is the ultimate component of the RAS, and AT-II affects learning and memory (Georgiev and Yonkov 1985). Further, the continuous stimulation of the brain by AT-II causes memory impairment and cognitive decline (Inaba et al. 2009). Also, the increased expression of angiotensin-converting enzyme (ACE) is responsible for cognitive impairment (Markus et al. 1995). Thus, the blockade of the RAS and its signaling component exert beneficial effects in cognitive decline. Further, the inhibition of AT1 receptors and ACE produces improvement in cognition (Fig. 10.3) (Bonini et al. 2006; Costall et al. 1989; de Souza et al. 2004; Kułakowska et al. 1996). Stimulation of RAS is also responsible for the aggregation and deposition of Aβ in the brain of AD patients (Hu et al. 2001), and the treatment with ACE inhibitors reduces the susceptibility of AD and inhibits the aggregation of Aβ aggregation, fibril formation, and Aβ-induced neuronal cell death (Hu et al. 2001). Further, the inhibition of RAS and its downstream signaling components prevents the BBB disruption and microcirculation dysfunction (Ito et al. 2011). Also, the antihypertensives such as RAS inhibitors that can cross the BBB counteract the cognitive decline (Gard 2004). The combined therapy of long-acting ACE inhibitor (i.e., perindopril 4 mg daily) and diuretic (indapamide 2.5 mg daily) reduces the BP by 9/4 mmHg and incidence of dementia by one-third in the stroke-afflicted patients (Tzourio et al. 2003). In another study conducted on 2418 demented patients, 1180 received placebo and 1238 received nitrendipine plus enalapril, and the trial was stopped prematurely. However, the study reported the significant benefit in stroke and treatment reduces the incidence of dementia by 50% (Forette et al. 1998; Staessen et al. 1997).

The lipid-lowering therapy results in the reduction in the dementia onset (Dufouil et al. 2005a, b; Shepardson et al. 2011). The high levels of cholesterol result in the production of oxidative products that results in neuronal inflammation and promotes

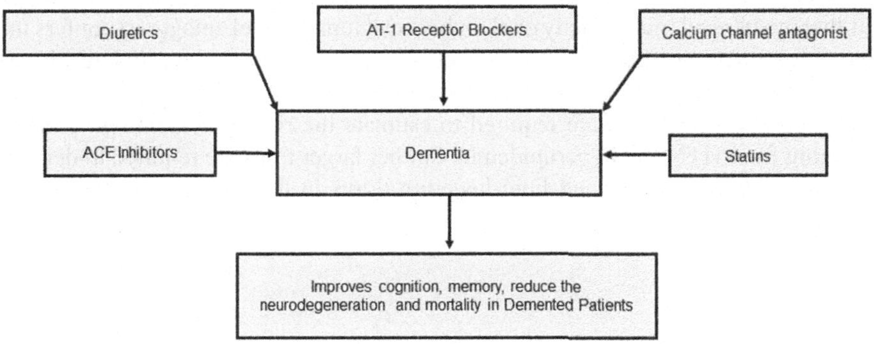

Fig. 10.3 Drugs used in the treatment of dementia

neuronal cell death (Ma et al. 2010; Trousson et al. 2009). Also, the structural changes in the HMG-CoA gene have been reported to be implicated in AD (Leduc et al. 2015). Statins inhibit the enzyme HMG-CoA reductase, reduce the synthesis of cholesterol, improve endothelial function, inhibit thrombus formation, and reduce the risk of AD emergence and dementia by two- to threefold (Beydoun et al. 2011; Furberg 1999; Green et al. 2006; Horsdal et al. 2009; Li et al. 2007, 2010; Ostrowski et al. 2007). Statins improve the CBF, decrease oxidation of LDL, stabilize atherosclerotic plaque, reduce the load of Aβ, and protect the neurons against neuroinflammation and apoptosis (Liao and Laufs 2005; Refolo et al. 2001; Zhang et al. 2013).

10.6 Recent Developments and Future Perspectives

The major limitation is the conflicting results from various studies some, suggesting the importance of SBP or some suggesting the importance of DBP in the prediction of cognitive impairment (Reitz and Luchsinger 2007). Another main limitation is how aging is responsible for the emergence of dementia and its relationship with BP and lipid levels (Langa et al. 2018; Trevisan et al. 2019). Further, no sufficient data are available to estimate the exact threshold of BP and lipid levels increase the risk of dementia. High BP is often responsible for the WML, but data regarding the exact prevalence of WMH in the normal population are still lacking (Abell et al. 2018; Kennelly et al. 2009a, b; Power et al. 2011).

Various studies have reported that the treatment with BP-lowering and lipid-lowering drugs confers beneficial effects in dementia and reduces the emergence of dementia in clinical settings. However, still, the efficacy of these drugs cannot be considered significant in clinical settings. Most of the studies did not use cognition as a primary endpoint, and follow-up time is not so long enough to determine the alterations in cognitive function. Despite the various advancements in medical science, curative and preventive therapy for dementia is currently not available. The type of antihypertensive or the basis for the choice of BP-lowering drug for dementia patients is still not known exactly (Nguyen et al. 2010; Peters et al. 2020). Further, no clinical trials exactly explain how calcium channel antagonist confers the protective effect in dementia (Angeli et al. 2004). The controversial results of antihypertensive therapy in dementia patients (Gorelick et al. 2011). Thus, the larger trials and long-term trials are required to estimate the risk of dementia in patients suffering from HTN and hyperlipidemia. Further larger trials are required to demonstrate the efficacy of BP- and lipid-lowering drugs in the reduction in the risk of development of dementia.

10.7 Conclusion

In conclusion, HTN and hyperlipidemia increase the risk of the development of dementia. The high BP and cholesterol levels result in oxidative damage, neuronal inflammation, and neuronal cell death. Further, the treatment with the antihypertensive and antihyperlipidemic drugs confers the benefits and reduces the dementia risk in afflicted patients. However, no adequate justification has been given in the literature regarding the choice of drugs for the treatment of dementia in afflicted patients. Thus, the efficacy is often doubted due to the lack of adequate and long-term clinical trials. Also, future research should be based upon the determination of the threshold levels of BP and cholesterol beyond which they are responsible for the emergence of dementia. Further clinical trials should be conducted to determine the long-term efficacy and the safety of antihypertensive and antihyperlipidemic drugs in the management of dementia.

References

Abdul-Ghani MA, DeFronzo RA (2008) Mitochondrial dysfunction, insulin resistance, and type 2 diabetes mellitus. Curr Diab Rep 8:173

Abell JG, Kivimäki M, Dugravot A et al (2018) Association between systolic blood pressure and dementia in the Whitehall II cohort study: role of age, duration, and threshold used to define hypertension. Eur Heart J 39(33):3119–3125

Alzheimer Society of Canada (2016) Lewy body dementia. www.alzheimer.ca/en/About-dementia/Dementias/Lewy-Body-Dementia. Accessed 19 Sep 2016

Alzheimer's Association (2012) Alzheimer's disease facts and figures. Alzheimers Dement 2012(8):131–168

American Psychiatric Association (2013) Diagnostic and statistical manual of mental disorders (DSM-5®). American Psychiatric Accusation, Washington, DC

Andre-Petersson L, Hagberg B, Janzon L, Steen G (2001) A comparison of cognitive ability in normotensive and hypertensive 68-year-old men: results from population study "Men born in 1914," in Malmoe, Sweden. Exp Aging Res 27:319–340

Angeli F, Verdecchia P, Reboldi GP, Gattobigio R, Bentivoglio M, Staessen JA, Porcellati C (2004) Calcium channel blockade to prevent stroke in hypertension: a meta-analysis of 13 studies with 103,793 subjects. Am J Hypertens 17:817–822

Appleton JP, Scutt P, Sprigg N, Bath PM (2017) Hypercholesterolaemia and vascular dementia. Clin Sci 131:1561–1578

Arnaud C, Mach F (2005) Pleiotropic effects of statins in atherosclerosis: role on endothelial function, inflammation and immunomodulation. Arch Mal Coeur Vaiss 98:661

Barazzoni R, Short KR, Nair KS (2000) Effects of aging on mitochondrial DNA copy number and cytochromec oxidase gene expression in rat skeletal muscle, liver, and heart. J Biol Chem 275: 3343–3347

Beason-Held LL, Moghekar A, Zonderman AB, Kraut MA, Resnick SM (2007) Longitudinal changes in cerebral blood flow in the older hypertensive brain. Stroke 38:1766–1773

Benbassat CA, Maki KC, Unterman TG (1997) Circulating levels of insulin-like growth factor (IGF) binding protein-1 and-3 in aging men: relationships to insulin, glucose, IGF, and dehydroepiandrosterone sulfate levels and anthropometric measures. J Clin Endocrinol Metab 82:1484–1491

Beydoun MA, Beason-Held LL, Kitner-Triolo MH, Beydoun HA, Ferrucci L, Resnick SM, Zonderman AB (2011) Statins and serum cholesterol's associations with incident dementia and mild cognitive impairment. J Epidemiol Community Health 65:949–957

Bezerra DC, Sharrett AR, Matsushita K, Gottesman RF, Shibata D, Mosley TH, Coresh J, Szklo M, Carvalho MS, Selvin E (2012) Risk factors for lacune subtypes in the Atherosclerosis Risk in Communities (ARIC) Study. Neurology 78:102–108

Bogdanovic N, Bretillon L, Lund EG, Diczfalusy U, Lannfelt L, Winblad B, Russell DW, Björkhem I (2001) On the turnover of brain cholesterol in patients with Alzheimer's disease. Abnormal induction of the cholesterol-catabolic enzyme CYP46 in glial cells. Neurosci Lett 314:45–48

Bonini JS, Bevilaqua LR, Zinn CG, Kerr DS, Medina JH, Izquierdo I, Cammarota M (2006) Angiotensin II disrupts inhibitory avoidance memory retrieval. Horm Behav 50:308–313

Bots ML, Witteman JC, Hofman A, de Jong PT, Grobbee DE (1996) Low diastolic blood pressure and atherosclerosis in elderly subjects: the Rotterdam Study. Arch Intern Med 156:843–848

Brewer GJ (2010) Epigenetic oxidative redox shift (EORS) theory of aging unifies the free radical and insulin signaling theories. Exp Gerontol 45:173–179

Brookmeyer R, Gray S, Kawas C (1998) Projections of Alzheimer's disease in the United States and the public health impact of delaying disease onset. Am J Public Health 88:1337–1342

Bruscoli M, Lovestone S (2004) Is MCI really just early dementia? A systematic review of conversion studies. Int Psychogeriatr 16:129–140

Burgess BL, Parkinson PF, Racke MM, Hirsch-Reinshagen V, Fan J, Wong C, Stukas S, Theroux L, Chan JY, Donkin J, Wilkinson A (2008) ABCG1 influences the brain cholesterol biosynthetic pathway but does not affect amyloid precursor protein or apolipoprotein E metabolism in vivo. J Lipid Res 49:1254–1267

Cacciatrore F, Abet P, Ferrara N, Paolisso G, Amato L, Canonico S, Maggi S, Varricchio M, Rengo F (1997) For the Osservatorio geriatrico campano group: the role of blood pressure in cognitive impairment in an elderly population. J Hypertens 15:135–142

Carlsson CM (2010) Type 2 diabetes mellitus, dyslipidemia, and Alzheimer's disease. J Alzheimers Dis 20:711–722

Chambless LE, Folsom AR, Davis V, Sharrett R, Heiss G, Sorlie P, Szklo M, Howard G, Evans GW (2002) Risk factors for progression of common carotid atherosclerosis: the Atherosclerosis Risk in Communities Study, 1987–1998. Am J Epidemiol 155:38–47

Chen ZY, Su YL, Lau CW, Law WI, Huang Y (1999) Endothelium-dependent contraction and direct relaxation induced by baicalein in rat mesenteric artery. Eur J Pharmacol 374:41–47

Chertkow H, Feldman HH, Jacova C, Massoud F (2013) Definitions of dementia and predementia states in Alzheimer's disease and vascular cognitive impairment: consensus from the Canadian conference on diagnosis of dementia. Alzheimers Res Ther 5:1–8

Cnop M, Havel PJ, Utzschneider KM, Carr DB, Sinha MK, Boyko EJ, Retzlaff BM, Knopp RH, Brunzell JD, Kahn SE (2003) Relationship of adiponectin to body fat distribution, insulin sensitivity and plasma lipoproteins: evidence for independent roles of age and sex. Diabetologia 46:459–469

Cohen RA, Tong X (2010) Vascular oxidative stress: the common link in hypertensive and diabetic vascular disease. J Cardiovasc Pharmacol 55:308

Cooke JP, Dzau VJ (1997) Nitric oxide synthase: role in the genesis of vascular disease. Annu Rev Med 48:489–509

Cornier MA, Dabelea D, Hernandez TL, Lindstrom RC, Steig AJ, Stob NR, Van Pelt RE, Wang H, Eckel RH (2008) The metabolic syndrome. Endocr Rev 29:777–822

Corpas E, Harman SM, Blackman MR (1993) Human growth hormone and human aging. Endocr Rev 14:20–39

Corrada MM, Brookmeyer R, Berlau D, Paganini-Hill A, Kawas CH (2008) Prevalence of dementia after age 90: results from the 90+ study. Neurology 71:337–343

Cortopassi GA, Shibata D, Soong NW, Arnheim N (1992) A pattern of accumulation of a somatic deletion of mitochondrial DNA in aging human tissues. Proc Natl Acad Sci 89:7370–7374

Costall B, Coughlan J, Horovitz ZP, Kelly ME, Naylor RJ, Tomkins DM (1989) The effects of ACE inhibitors captopril and SQ29, 852 in rodent tests of cognition. Pharmacol Biochem Behav 33: 573–579

Craft S, Asthana S, Newcomer JW, Wilkinson CW, Matos IT, Baker LD, Cherrier M, Lofgreen C, Latendresse S, Petrova A, Plymate S (1999) Enhancement of memory in Alzheimer disease with insulin and somatostatin, but not glucose. Arch Gen Psychiatry 56:1135–1140

De la Torre JC (1999) Critical threshold cerebral hypoperfusion causes Alzheimer's disease? Acta Neuropathol 98:1–8

De la Torre JC (2000a) Critically attained threshold of cerebral hypoperfusion: the CATCH hypothesis of Alzheimer's pathogenesis. Neurobiol Aging 21:331–342

De la Torre JC (2000b) Impaired cerebromicrovascular perfusion: summary of evidence in support of its causality in Alzheimer's disease. Ann N Y Acad Sci 924:136–152

De la Torre JC (2002) Alzheimer disease as a vascular disorder: nosological evidence. Stroke 33: 1152–1162

De Leeuw FE, de Groot JC, Oudkerk M, Witteman JC, Hofman A, Van Gijn J, Breteler MM (2002) Hypertension and cerebral white matter lesions in a prospective cohort study. Brain 125:765–772

De Souza FA, Sanchis-Segura C, Fukada SY, De Bortoli VC, Zangrossi H Jr, De Oliveira AM (2004) Intracerebroventricular effects of angiotensin II on a step-through passive avoidance task in rats. Neurobiol Learn Mem 81:100–103

Desmond DW, Moroney JT, Paik MC, Sano M, Mohr JP, Aboumatar S, Tseng CL, Chan S, Williams JB, Remien RH, Hauser WA (2000) Frequency and clinical determinants of dementia after ischemic stroke. Neurology 54:1124–1131

Dias HI, Brown CL, Polidori MC, Lip GY, Griffiths HR (2015) LDL-lipids from patients with hypercholesterolaemia and Alzheimer's disease are inflammatory to microvascular endothelial cells: mitigation by statin intervention. Clin Sci 129:1195–1206

Dufouil C, de Kersaint-Gilly A, Besancon V, Levy C, Auffray E, Brunnereau L, Alperovitch A, Tzourio C (2001) Longitudinal study of blood pressure and white matter hyperintensities: the EVA MRI Cohort. Neurology 56:921–926

Dufouil C, Chalmers J, Coskun O, Besancon V, Bousser MG, Guillon P, Macmahon S, Mazoyer B, Neal B, Woodward M, Tzourio-Mazoyer N (2005a) Effects of blood pressure lowering on cerebral white matter hyperintensities in patients with stroke: the PROGRESS (Perindopril Protection Against Recurrent Stroke Study) Magnetic Resonance Imaging Substudy. Circulation 112:1644–1650

Dufouil C, Richard F, Fievet N, Dartigues JF, Ritchie K, Tzourio C, Amouyel P, Alpérovitch A (2005b) APOE genotype, cholesterol level, lipid-lowering treatment, and dementia: the three-city study. Neurology 64:1531–1538

Elias MF, Elias PK, Sullivan LM, Wolf PA, D'agostino RB (2003) Lower cognitive function in the presence of obesity and hypertension: the Framingham heart study. Int J Obes 27:260–268

Evans RM, Emsley CL, Gao S, Sahota A, Hall KS, Farlow MR, Hendrie H (2000) Serum cholesterol, APOE genotype, and the risk of Alzheimer's disease: a population-based study of African Americans. Neurology 54:240–242

Fang X, Yu SX, Lu Y, Bast RC, Woodgett JR, Mills GB (2000) Phosphorylation and inactivation of glycogen synthase kinase 3 by protein kinase A. Proc Natl Acad Sci 97:11960–11965

Faraci FM (2011) Protecting against vascular disease in brain. Am J Phys Heart Circ Phys 300: H1566–H1582

Farkas E, Luiten PG (2001) Cerebral microvascular pathology in aging and Alzheimer's disease. Prog Neurobiol 64:575–611

Fernandez-Vizarra P, Fernandez AP, Castro-Blanco S, Encinas JM, Serrano J, Bentura ML, Munoz P, Martınez-Murillo R, Rodrigo J (2004) Expression of nitric oxide system in clinically evaluated cases of Alzheimer's disease. Neurobiol Dis 15:287–305

Ferri CP, Prince M, Brayne C, Brodaty H, Fratiglioni L, Ganguli M, Hall K, Hasegawa K, Hendrie H, Huang Y, Jorm A (2005) Global prevalence of dementia: a Delphi consensus study. Lancet 366:2112–2117

Ford ES, Giles WH, Dietz WH (2002) Prevalence of the metabolic syndrome among US adults: findings from the third National Health and Nutrition Examination Survey. JAMA 287:356–359

Forette F, Seux ML, Staessen JA, Thijs L, Birkenhäger WH, Babarskiene MR, Babeanu S, Bossini A, Gil-Extremera B, Girerd X, Laks T (1998) Prevention of dementia in randomised double-blind placebo-controlled Systolic Hypertension in Europe (Syst-Eur) trial. Lancet 352: 1347–1351

Frears ER, Stephens DJ, Walters CE, Davies H, Austen BM (1999) The role of cholesterol in the biosynthesis of β-amyloid. NeuroReport 10:1699–1705

Fulop T, Larbi A, Douziech N (2003) Insulin receptor and ageing. Pathol Biol 51:574–580

Furberg CD (1999) Natural statins and stroke risk. Circulation 99:185–188

Gamba P, Testa G, Gargiulo S, Staurenghi E, Poli G, Leonarduzzi G et al (2015) Oxidized cholesterol as the driving force behind the development of Alzheimer's disease. Front Aging Neurosci 7:119

Ganguli M, Snitz BE, Saxton JA, Chang CC, Lee CW, Vander Bilt J, Hughes TF, Loewenstein DA, Unverzagt FW, Petersen RC (2011) Outcomes of mild cognitive impairment by definition: a population study. Arch Neurol 68:761–767

Gard PR (2002) The role of angiotensin II in cognition and behaviour. Eur J Pharmacol 438:1–4

Gard PR (2004) Angiotensin as a target for the treatment of Alzheimer's disease, anxiety and depression. Expert Opin Ther Targets 8:7–14

Georgiev V, Yonkov D (1985) Participation of angiotensin II in learning and memory. I. Interaction of angiotensin II with saralasin. Methods Find Exp Clin Pharmacol 7:415–418

Girouard H, Iadecola C (2006) Neurovascular coupling in the normal brain and in hypertension, stroke, and Alzheimer disease. J Appl Physiol 100:328–335

Glodzik L, Mosconi L, Tsui W, de Santi S, Zinkowski R, Pirraglia E, Rich KE, McHugh P, Li Y, Williams S, Ali F (2012) Alzheimer's disease markers, hypertension, and gray matter damage in normal elderly. Neurobiol Aging 33:1215–1227

Goldstein BJ, Scalia R (2004) Adiponectin: a novel adipokine linking adipocytes and vascular function. J Clin Endocrinol Metab 89:2563–2568

Goldstein BJ, Scalia RG, Ma XL (2009) Protective vascular and myocardial effects of adiponectin. Nat Clin Pract Cardiovasc Med 6:27–35

Gorelick PB, Scuteri A, Black SE, DeCarli C, Greenberg SM, Iadecola C, Launer LJ, Laurent S, Lopez OL, Nyenhuis D, Petersen RC (2011) Vascular contributions to cognitive impairment and dementia: a statement for healthcare professionals from the American Heart Association/American Stroke Association. Stroke 42:2672–2713

Gottesman RF, Coresh J, Catellier DJ, Sharrett AR, Rose KM, Coker LH, Shibata DK, Knopman DS, Jack CR, Mosley TH Jr (2010) Blood pressure and white-matter disease progression in a biethnic cohort: Atherosclerosis Risk in Communities (ARIC) study. Stroke 41:3–8

Gottesman RF, Schneider AL, Albert M, Alonso A, Bandeen-Roche K, Coker L, Coresh J, Knopman D, Power MC, Rawlings A, Sharrett AR (2014) Midlife hypertension and 20-year cognitive change: the atherosclerosis risk in communities neurocognitive study. JAMA Neurol 71:1218–1227

Green RC, McNagny SE, Jayakumar P, Cupples LA, Benke K, Farrer LA (2006) MIRAGE Study Group. Statin use and the risk of Alzheimer's disease: the MIRAGE study. Alzheimers Dement 2:96–103

Gulcelik NE, Halil M, Ariogul S, Usman A (2013) Adipocytokines and aging: adiponectin and leptin. Minerva Endocrinol 38:203

Haan M, Shemanski L, Jagust WJ, Kuller L, Manolio T, Bryan N (1999) Predictors of cognitive change in the Cardiovascular Health Study: the role of cardiovascular and genetic risk factors. JAMA 282:40–46

Hamel E (2004) Cholinergic modulation of the cortical microvascular bed. Prog Brain Res 145: 171–178

Harrell LE, Duvall E, Folks DG, Duke L, Bartolucci A, Conboy T, Callaway R, Kerns D (1991) The relationship of high-intensity signals on magnetic resonance images to cognitive and psychiatric state in Alzheimer's disease. Arch Neurol 48:1136–1140

Harris P, Alcantara DA, Amenta N, Lopez OL, Eiríksdóttir G, Sigurðsson S, Gudnason V, Madsen S, Thompson PM, Launer LJ, Carmichael OT (2008) Localized measures of callosal atrophy are associated with late-life hypertension: AGES–Reykjavik Study. NeuroImage 43: 489–496

Hayden KM, Zandi PP, Lyketsos CG, Khachaturian AS, Bastian LA, Charoonruk G, Tschanz JT, Norton MC, Pieper CF, Munger RG, Breitner JC (2006) Vascular risk factors for incident Alzheimer disease and vascular dementia: the Cache County study. Alzheimer Dis Assoc Disord 20:93–100

He Q, Li Q, Zhao J, Wu T, Ji L, Huang G, Ma F (2016) Relationship between plasma lipids and mild cognitive impairment in the elderly Chinese: a case-control study. Lipids Health Dis 15:146

van der Heide LP, Ramakers GM, Smidt MP (2006) Insulin signaling in the central nervous system: learning to survive. Prog Neurobiol 79:205–221

Heverin M, Bogdanovic N, Lütjohann D, Bayer T, Pikuleva I, Bretillon L, Diczfalusy U, Winblad B, Björkhem I (2004) Changes in the levels of cerebral and extracerebral sterols in the brain of patients with Alzheimer's disease. J Lipid Res 45:186–193

Hildreth KL, Church S (2015) Evaluation and management of the elderly patient presenting with cognitive complaints. Med Clin 99:311–335

Horsdal HT, Olesen AV, Gasse C, Sørensen HT, Green RC, Johnsen SP (2009) Use of statins and risk of hospitalization with dementia: a Danish population-based case-control study. Alzheimer Dis Assoc Disord 23:18

Hsu TM, Kanoski SE (2014) Blood-brain barrier disruption: mechanistic links between Western diet consumption and dementia. Front Aging Neurosci 6:88

Hu J, Igarashi A, Kamata M, Nakagawa H (2001) Angiotensin-converting enzyme degrades Alzheimer amyloid β-peptide (Aβ); retards Aβ aggregation, deposition, fibril formation; and inhibits cytotoxicity. J Biol Chem 276:47863–47868

Iadecola C (2004) Neurovascular regulation in the normal brain and in Alzheimer's disease. Nat Rev Neurosci 5:347–360

Inaba S, Iwai M, Furuno M, Tomono Y, Kanno H, Senba I, Okayama H, Mogi M, Higaki J, Horiuchi M (2009) Continuous activation of renin-angiotensin system impairs cognitive function in renin/angiotensinogen transgenic mice. Hypertension 53:356–362

Ito U, Hakamata Y, Kawakami E, Oyanagi K (2011) Temporary focal cerebral ischemia results in swollen astrocytic end-feet that compress microvessels and lead to focal cortical infarction. J Cereb Blood Flow Metab 31:328–338

Jarvik GP, Wijsman EM, Kukull WA, Schellenberg GD, Yu C, Larson EB (1995) Interactions of apolipoprotein E genotype, total cholesterol level, age, and sex in prediction of Alzheimer's disease: a case-control study. Neurology 45:1092–1096

Jennings JR, Mendelson DN, Muldoon MF, Ryan CM, Gianaros PJ, Raz N, Aizenstein H (2012) Regional grey matter shrinks in hypertensive individuals despite successful lowering of blood pressure. J Hum Hypertens 26:295–305

Jimenez-Conde J, Biffi A, Rahman R, Kanakis A, Butler C, Sonni S, Massasa E, Cloonan L, Gilson A, Capozzo K, Cortellini L (2010) Hyperlipidemia and reduced white matter hyperintensity volume in patients with ischemic stroke. Stroke 41:437–442

Junqué C, Pujol J, Vendrell P, Bruna O, Jódar M, Ribas JC, Viñas J, Capdevila A, Marti-Vilalta JL (1990) Leuko-araiosis on magnetic resonance imaging and speed of mental processing. Arch Neurol 47:151–156

Jurevics H, Morell P (1995) Cholesterol for synthesis of myelin is made locally, not imported into brain. J Neurochem 64:895–901

Kadowaki T, Yamauchi T, Kubota N, Hara K, Ueki K, Tobe K (2006) Adiponectin and adiponectin receptors in insulin resistance, diabetes, and the metabolic syndrome. J Clin Invest 116:1784–1792

Kennelly SP, Lawlor BA, Kenny RA (2009a) Blood pressure and dementia - a comprehensive review. Ther Adv Neurol Disord 2(4):241–260

Kennelly SP, Lawlor BA, Kenny RA (2009b) Blood pressure and the risk for dementia—a double edged sword. Ageing Res Rev 8:61–70

Kilander L, Nyman H, Boberg M, Hansson L, Lithell H (1998) Hypertension is related to cognitive impairment: a 20-year follow-up of 999 men. Hypertension 31:780–786

Kim JA, Wei Y, Sowers JR (2008) Role of mitochondrial dysfunction in insulin resistance. Circ Res 102:401–414

Kivipelto M, Helkala EL, Laakso MP, Hänninen T, Hallikainen M, Alhainen K, Soininen H, Tuomilehto J, Nissinen A (2001) Midlife vascular risk factors and Alzheimer's disease in later life: longitudinal, population based study. BMJ 322:1447–1451

Kivipelto M, Helkala EL, Laakso MP, Hänninen T, Hallikainen M, Alhainen K, Iivonen S, Mannermaa A, Tuomilehto J, Nissinen A, Soininen H (2002) Apolipoprotein E ε4 allele, elevated midlife total cholesterol level, and high midlife systolic blood pressure are independent risk factors for late-life Alzheimer disease. Ann Intern Med 137:149–155

Kivipelto M, Ngandu T, Fratiglioni L, Viitanen M, Kåreholt I, Winblad B, Helkala EL, Tuomilehto J, Soininen H, Nissinen A (2005) Obesity and vascular risk factors at midlife and the risk of dementia and Alzheimer disease. Arch Neurol 62:1556–1560

Kobayashi J, Nishimura K, Matoba M, Maekawa N, Mabuchi H (2007) Generation and gender differences in the components contributing to the diagnosis of the metabolic syndrome according to the Japanese criteria. Circ J 71:1734–1737

Kokmen E, Whisnant JP, O'fallon WM, Chu CP, Beard CM (1996) Dementia after ischemic stroke: a population-based study in Rochester, Minnesota (1960-1984). Neurology 46:154–159

Korf ES, White LR, Scheltens P, Launer LJ (2004) Midlife blood pressure and the risk of hippocampal atrophy: the Honolulu Asia Aging Study. Hypertension 44:29–34

Kubota N, Terauchi Y, Yamauchi T, Kubota T, Moroi M, Matsui J, Eto K, Yamashita T, Kamon J, Satoh H, Yano W (2002) Disruption of adiponectin causes insulin resistance and neointimal formation. J Biol Chem 277:25863–25866

Kułakowska A, Karwowska W, Wiśniewski K, Braszko JJ (1996) Losartan influences behavioural effects of angiotensin II in rats. Pharmacol Res 34:109–115

Langa KM (2018) Cognitive aging, dementia, and the future of an aging population. In: National Academies of Sciences, Engineering, and Medicine; Division of Behavioral and Social Sciences and Education; Committee on Population, Majmundar MK, Hayward MD (eds) Future Directions for the Demography of Aging: Proceedings of a Workshop. National Academies Press, Washington, DC, p 9

Langbaum JB, Chen K, Launer LJ, Fleisher AS, Lee W, Liu X, Protas HD, Reeder SA, Bandy D, Yu M, Caselli RJ (2012) Blood pressure is associated with higher brain amyloid burden and lower glucose metabolism in healthy late middle-age persons. Neurobiol Aging 33:827–e11

Larrieu S, Letenneur L, Orgogozo JM, Fabrigoule C, Amieva H, Le Carret N, Barberger-Gateau P, Dartigues JF (2002) Incidence and outcome of mild cognitive impairment in a population-based prospective cohort. Neurology 59:1594–1599

Launer LJ, Ross GW, Petrovitch H, Masaki K, Foley D, White LR, Havlik RJ (2000) Midlife blood pressure and dementia: the Honolulu–Asia aging study*. Neurobiol Aging 21:49–55

Launer LJ, Lewis CE, Schreiner PJ, Sidney S, Battapady H, Jacobs DR, Lim KO, D'Esposito M, Zhang Q, Reis J, Davatzikos C (2015) Vascular factors and multiple measures of early brain health: CARDIA brain MRI study. PLoS One 10:e0122138

Leduc V, De Beaumont L, Théroux L, Dea D, Aisen P, Petersen RC, Dufour R, Poirier J (2015) HMGCR is a genetic modifier for risk, age of onset and MCI conversion to Alzheimer's disease in a three cohorts study. Mol Psychiatry 20:867–873

Leritz EC, Salat DH, Williams VJ, Schnyer DM, Rudolph JL, Lipsitz L, Fischl B, McGlinchey RE, Milberg WP (2011) Thickness of the human cerebral cortex is associated with metrics of cerebrovascular health in a normative sample of community dwelling older adults. NeuroImage 54:2659–2671

Li G, Larson EB, Sonnen JA, Shofer JB, Petrie EC, Schantz A, Peskind ER, Raskind MA, Breitner JC, Montine TJ (2007) Statin therapy is associated with reduced neuropathologic changes of Alzheimer disease. Neurology 69:878–885

Li G, Shofer JB, Rhew IC, Kukull WA, Peskind ER, McCormick W, Bowen JD, Schellenberg GD, Crane PK, Breitner JC, Larson EB (2010) Age-varying association between statin use and incident Alzheimer's disease: [See editorial comments by Dr. Mary Hann]. J Am Geriatr Soc 58: 1311–1317

Liao JK, Laufs U (2005) Pleiotropic effects of statins. Annu Rev Pharmacol Toxicol 45:89–118

Liao D, Arnett DK, Tyroler HA, Riley WA, Chambless LE, Szklo M, Heiss G (1999) Arterial stiffness and the development of hypertension: the ARIC study. Hypertension 34:201–206

Lim S, Choi SH, Jeong IK, Kim JH, Moon MK, Park KS, Lee HK, Kim YB, Jang HC (2008) Insulin-sensitizing effects of exercise on adiponectin and retinol-binding protein-4 concentrations in young and middle-aged women. J Clin Endocrinol Metab 93:2263–2268

Lindén T, Skoog I, Fagerberg B, Steen B, Blomstrand C (2004) Cognitive impairment and dementia 20 months after stroke. Neuroepidemiology 23:45–52

Ma MT, Zhang J, Farooqui AA, Chen P, Ong WY (2010) Effects of cholesterol oxidation products on exocytosis. Neurosci Lett 476:36–41

Markus HS, Barley J, Lunt R, Bland JM, Jeffery S, Carter ND, Brown MM (1995) Angiotensin-converting enzyme gene deletion polymorphism: a new risk factor for lacunar stroke but not carotid atheroma. Stroke 26:1329–1333

Matthews FE, Stephan BC, McKeith IG, Bond J, Brayne C (2008) Medical Research Council Cognitive Function and Ageing Study. Two-year progression from mild cognitive impairment to dementia: to what extent do different definitions agree? J Am Geriatr Soc 56:1424–1433

Maxwell AJ (2002) Mechanisms of dysfunction of the nitric oxide pathway in vascular diseases. Nitric Oxide 6:101–124

McCarron RM, Chen Y, Tomori T, Strasser A, Mechoulam R, Shohami E, Spatz M (2006) Endothelial-mediated regulation of cerebral. J Physiol Pharmacol 57:133–144

McLaurin J, Darabie AA, Morrison MR (2003) Cholesterol, a modulator of membrane-associated aβ-fibrillogenesis. Pharmacopsychiatry 36:130–135

Mielke MM, Zandi PP, Sjögren M, Gustafson D, Östling S, Steen B, Skoog I (2005) High total cholesterol levels in late life associated with a reduced risk of dementia. Neurology 64:1689–1695

Minamino T, Orimo M, Shimizu I, Kunieda T, Yokoyama M, Ito T, Nojima A, Nabetani A, Oike Y, Matsubara H, Ishikawa F (2009) A crucial role for adipose tissue p53 in the regulation of insulin resistance. Nat Med 15:1082–1087

Mitchell AJ, Shiri-Feshki M (2009) Rate of progression of mild cognitive impairment to dementia–meta-analysis of 41 robust inception cohort studies. Acta Psychiatr Scand 119:252–265

Mizuno T, Nakata M, Naiki H, Michikawa M, Wang R, Haass C, Yanagisawa K (1999) Cholesterol-dependent generation of a seeding amyloid β-protein in cell culture. J Biol Chem 274:15110–15114

Moroney JT, Tang MX, Berglund L, Small S, Merchant C, Bell K, Stern Y, Mayeux R (1999) Low-density lipoprotein cholesterol and the risk of dementia with stroke. JAMA 282:254–260

Muangpaisan W (2007) Clinical differences among four common dementia syndromes. Geriatr Aging 10:9

Muller AP, Fernandez AM, Haas C, Zimmer E, Portela LV, Torres-Aleman I (2012) Reduced brain insulin-like growth factor I function during aging. Mol Cell Neurosci 49:9–12

Nagahama Y, Nabatame H, Okina T, Yamauchi H, Narita M, Fujimoto N, Murakami M, Fukuyama H, Matsuda M (2003) Cerebral correlates of the progression rate of the cognitive decline in probable Alzheimer's disease. Eur Neurol 50:1–9

Nagai M, Hoshide S, Ishikawa J, Shimada K, Kario K (2009) Insular cortex atrophy as an independent determinant of disrupted diurnal rhythm of ambulatory blood pressure in elderly hypertension. Am J Hypertens 22:723–729

Nguyen Q, Dominguez J, Nguyen L, Gullapalli N (2010) Hypertension management: an update. Am Health Drug Benef 3(1):47–56

Notkola IL, Sulkava R, Pekkanen J, Erkinjuntti T, Ehnholm C, Kivinen P, Tuomilehto J, Nissinen A (1998) Serum total cholesterol, apolipoprotein E {FC12} e4 allele, and Alzheimer's disease. Neuroepidemiology 17:14–20

Okamoto M, Ohara-Imaizumi M, Kubota N, Hashimoto S, Eto K, Kanno T, Kubota T, Wakui M, Nagai R, Noda M, Nagamatsu S (2008) Adiponectin induces insulin secretion in vitro and in vivo at a low glucose concentration. Diabetologia 51:827–835

de Oliveira FF, Chen ES, Smith MC, Bertolucci PH (2017) Longitudinal lipid profile variations and clinical change in Alzheimer's disease dementia. Neurosci Lett 646:36–42

Ostrowski SM, Wilkinson BL, Golde TE, Landreth G (2007) Statins reduce amyloid-β production through inhibition of protein isoprenylation. J Biol Chem 282:26832–26844

Ouedraogo R, Gong Y, Berzins B, Wu X, Mahadev K, Hough K, Chan L, Goldstein BJ, Scalia R (2007) Adiponectin deficiency increases leukocyte-endothelium interactions via upregulation of endothelial cell adhesion molecules in vivo. J Clin Invest 117:1718–1726

Pajvani UB, Scherer PE (2003) Adiponectin: systemic contributor to insulin sensitivity. Curr Diab Rep 3:207–213

Panza F, Frisardi V, Imbimbo PB, Seripa D, Paris F, Santamato A, D'Onofrio G, Logroscino G, Pilotto A, Solfrizzi V (2011) Anti-β-amyloid immunotherapy for Alzheimer's disease: focus on bapineuzumab. Curr Alzheimer Res 8:808–817

Panza F, Solfrizzi V, Logroscino G, Maggi S, Santamato A, Seripa D, Pilotto A (2012) Current epidemiological approaches to the metabolic-cognitive syndrome. J Alzheimers Dis 30:S31–S75

Papassotiropoulos A, Lutjohann D, Bagli M, Locatelli S, Jessen F, Buschfort R, Ptok U, Björkhem I, Von Bergmann K, Heun R (2002) 24S-hydroxycholesterol in cerebrospinal fluid is elevated in early stages of dementia. J Psychiatr Res 36:27–32

Park CR, Seeley RJ, Craft S, Woods SC (2000) Intracerebroventricular insulin enhances memory in a passive-avoidance task. Physiol Behav 68:509–514

Peters R, Yasar S, Anderson CS et al (2020) Investigation of antihypertensive class, dementia, and cognitive decline: a meta-analysis. Neurology 94(3):e267–e281

Petersen KF, Befroy D, Dufour S, Dziura J, Ariyan C, Rothman DL, DiPietro L, Cline GW, Shulman GI (2003) Mitochondrial dysfunction in the elderly: possible role in insulin resistance. Science 300:1140–1142

Petrovitch H, White LR, Izmirilian G, Ross GW, Havlik RJ, Markesbery W, Nelson J, Davis DG, Hardman J, Foley DJ, Launer LJ (2000) Midlife blood pressure and neuritic plaques, neurofibrillary tangles, and brain weight at death: the HAAS*. Neurobiol Aging 21:57–62

Pfrieger FW (2003) Cholesterol homeostasis and function in neurons of the central nervous system. Cell Mol Life Sci 60:1158–1171

Phillips MI, De Oliveira EM (2008) Brain renin angiotensin in disease. J Mol Med 86:715–722

Poels MM, Zaccai K, Verwoert GC, Vernooij MW, Hofman A, van der Lugt A, Witteman JC, Breteler MM, Mattace-Raso FU, Ikram MA (2012) Arterial stiffness and cerebral small vessel disease: the Rotterdam Scan Study. Stroke 43:2637–2642

Pohjasvaara T, Erkinjuntti T, Vataja R, Kaste M (1997) Dementia three months after stroke: baseline frequency and effect of different definitions of dementia in the Helsinki Stroke Aging Memory Study (SAM) cohort. Stroke 28:785–792

Power MC, Weuve J, Gagne JJ, McQueen MB, Viswanathan A, Blacker D (2011) The association between blood pressure and incident Alzheimer disease: a systematic review and meta-analysis. Epidemiology 22(5):646–659

Prencipe M, Ferretti C, Casini AR, Santini M, Giubilei F, Culasso F (1997) Stroke, disability, and dementia: results of a population survey. Stroke 28:531–536

Prins ND, van Dijk EJ, den Heijer T, Vermeer SE, Koudstaal PJ, Oudkerk M, Hofman A, Breteler MM (2004) Cerebral white matter lesions and the risk of dementia. Arch Neurol 61:1531–1534
Raffaitin C, Gin H, Empana JP et al (2009) Metabolic syndrome and risk for incident Alzheimer's disease or vascular dementia: the three-city study. Diabetes Care 32:169–174
Refolo LM, Pappolla MA, Malester B, LaFrancois J, Bryant-Thomas T, Wang R, Tint GS, Sambamurti K, Duff K (2000) Hypercholesterolemia accelerates the Alzheimer's amyloid pathology in a transgenic mouse model. Neurobiol Dis 7:321–331
Refolo LM, Pappolla MA, LaFrancois J, Malester B, Schmidt SD, Thomas-Bryant T, Tint GS, Wang R, Mercken M, Petanceska SS, Duff KE (2001) A cholesterol-lowering drug reduces β-amyloid pathology in a transgenic mouse model of Alzheimer's disease. Neurobiol Dis 8: 890–899
Reitz C (2012) Dyslipidemia and dementia: current epidemiology, genetic evidence, and mechanisms behind the associations. J Alzheimers Dis 30:S127–S145
Reitz C (2015) Genetic diagnosis and prognosis of Alzheimer's disease: challenges and opportunities. Expert Rev Mol Diagn 15:339–348
Reitz C, Luchsinger JA (2007) Relation of blood pressure to cognitive impairment and dementia. Curr Hypertens Rev 3(3):166–176
Reitz C, Tang MX, Luchsinger J, Mayeux R (2004) Relation of plasma lipids to Alzheimer disease and vascular dementia. Arch Neurol 61:705–714
Reynolds CA, Gatz M, Prince JA, Berg S, Pedersen NL (2010) Serum lipid levels and cognitive change in late life. J Am Geriatr Soc 58:501–509
Rivera EJ, Goldin A, Fulmer N, Tavares R, Wands JR, de la Monte SM (2005) Insulin and insulin-like growth factor expression and function deteriorate with progression of Alzheimer's disease: link to brain reductions in acetylcholine. J Alzheimers Dis 8:247–268
Rodrigo J, Fernandez AP, Alonso D et al (2004a) Nitric oxide in the rat cerebellum after hypoxia/ ischemia. Cerebellum 3:194–203
Rodrigo J, Fernandez-Vizarra P, Castro-Blanco S, Bentura ML, Nieto M, Gómez-Isla T, Martínez-Murillo R, MartInez A, Serrano J, Fernandez AP (2004b) Nitric oxide in the cerebral cortex of amyloid-precursor protein (SW) Tg2576 transgenic mice. Neuroscience 128:73–89
Rodrigue KM, Rieck JR, Kennedy KM, Devous MD, Diaz-Arrastia R, Park DC (2013) Risk factors for β-amyloid deposition in healthy aging: vascular and genetic effects. JAMA Neurol 70:600–606
Román GC, Tatemichi TK, Erkinjuntti T, Cummings JL, Masdeu JC, Garcia JH, Amaducci L, Orgogozo JM, Brun A, Hofman A, Moody DM (1993) Vascular dementia: diagnostic criteria for research studies: report of the NINDS-AIREN International Workshop. Neurology 43:250
Rudman D, Feller AG, Nagraj HS, Gergans GA, Lalitha PY, Goldberg AF, Schlenker RA, Cohn L, Rudman IW, Mattson DE (1990) Effects of human growth hormone in men over 60 years old. N Engl J Med 323:1–6
Ruitenberg A, Ott A, van Swieten JC, Hofman A, Breteler MM (2001) Incidence of dementia: does gender make a difference? Neurobiol Aging 22:575–580
Sahathevan R, Brodtmann A, Donnan GA (2012) Dementia, stroke, and vascular risk factors; a review. Int J Stroke 7:61–73
Scherr PA, Hebert LE, Smith LA, Evans DA (1991) Relation of blood pressure to cognitive function in the elderly. Am J Epidemiol 134:1303–1315
Selley ML (2003) Increased concentrations of homocysteine and asymmetric dimethylarginine and decreased concentrations of nitric oxide in the plasma of patients with Alzheimer's disease. Neurobiol Aging 24:903–907
Shah NS, Vidal JS, Masaki K, Petrovitch H, Ross GW, Tilley C, DeMattos RB, Tracy RP, White LR, Launer LJ (2012) Midlife blood pressure, plasma β-amyloid, and the risk for Alzheimer disease: the Honolulu Asia Aging Study. Hypertension 59:780–786
Shang S, Li P, Deng M, Jiang Y, Chen C, Qu Q (2016) The age-dependent relationship between blood pressure and cognitive impairment: a cross-sectional study in a rural area of Xi'an, China. PLoS One 11:e0159485

Sharma P (1998) Meta-analysis of the ACE gene in ischaemic stroke. J Neurol Neurosurg Psychiatry 64:227–230

Sheehan B (2012) Assessment scales in dementia. Ther Adv Neurol Disord 5:349–358

Shepardson NE, Shankar GM, Selkoe DJ (2011) Cholesterol level and statin use in Alzheimer disease: I. Review of epidemiological and preclinical studies. Arch Neurol 68:1239–1244

Shoji T, Hatsuda S, Tsuchikura S, Shinohara K, Kimoto E, Koyama H, Emoto M, Nishizawa Y (2009) Small dense low-density lipoprotein cholesterol concentration and carotid atherosclerosis. Atherosclerosis 202:582–588

Sierra C, De La Sierra A, Salamero M, Sobrino J, Gómez-Angelats E, Coca A (2004) Silent cerebral white matter lesions and cognitive function in middle-aged essential hypertensive patients. Am J Hypertens 17:529–534

Singh-Manoux A, Marmot M (2005) High blood pressure was associated with cognitive function in middle-age in the Whitehall II study. J Clin Epidemiol 58:1308–1315

van Sloten TT, Protogerou AD, Henry RM, Schram MT, Launer LJ, Stehouwer CD (2015) Association between arterial stiffness, cerebral small vessel disease and cognitive impairment: a systematic review and meta-analysis. Neurosci Biobehav Rev 53:121–130

Solomon A, Kåreholt I, Ngandu T, Wolozin B, MacDonald SW, Winblad B, Nissinen A, Tuomilehto J, Soininen H, Kivipelto M (2009) Serum total cholesterol, statins and cognition in non-demented elderly. Neurobiol Aging 30:1006–1009

Song F, Poljak A, Crawford J, Kochan NA, Wen W, Cameron B, Lux O, Brodaty H, Mather K, Smythe GA, Sachdev PS (2012) Plasma apolipoprotein levels are associated with cognitive status and decline in a community cohort of older individuals. PLoS One 7:e34078

Staessen JA, Fagard R, Thijs L, Celis H, Arabidze GG, Birkenhäger WH, Bulpitt CJ, De Leeuw PW, Dollery CT, Fletcher AE, Forette F (1997) Randomised double-blind comparison of placebo and active treatment for older patients with isolated systolic hypertension. Lancet 350:757–764

Steen E, Terry BM, Rivera JE, Cannon JL, Neely TR, Tavares R, Xu XJ, Wands JR, de la Monte SM (2005) Impaired insulin and insulin-like growth factor expression and signaling mechanisms in Alzheimer's disease–is this type 3 diabetes? J Alzheimers Dis 7:63–80

Stefan N, Bunt JC, Salbe AD, Funahashi T, Matsuzawa Y, Tataranni PA (2002) Plasma adiponectin concentrations in children: relationships with obesity and insulinemia. J Clin Endocrinol Metab 87:4652–4656

Steingart A, Hachinski VC, Lau C, Fox AJ, Diaz F, Cape R, Lee D, Inzitari D, Merskey H (1987) Cognitive and neurologic findings in subjects with diffuse white matter lucencies on computed tomographic scan (leuko-araiosis). Arch Neurol 44:32–35

von Strauss E, Viitanen M, De Ronchi D, Winblad B, Fratiglioni L (1999) Aging and the occurrence of dementia: findings from a population-based cohort with a large sample of nonagenarians. Arch Neurol 56:587–592

Swan GE, DeCarli C, Miller BL, Reed T, Wolf PA, Jack LM, Carmelli D (1998) Association of midlife blood pressure to late-life cognitive decline and brain morphology. Neurology 51:986–993

Takaoka M, Nagata D, Kihara S, Shimomura I, Kimura Y, Tabata Y, Saito Y, Nagai R, Sata M (2009) Periadventitial adipose tissue plays a critical role in vascular remodeling. Circ Res 105:906–911

Tatemichi TK, Desmond DW, Mayeux R, Paik M, Stern Y, Sano M, Remien RH, Williams JB, Mohr JP, Hauser WA, Figueroa M (1992) Dementia after stroke: baseline frequency, risks, and clinical features in a hospitalized cohort. Neurology 42:1185–1193

Tatemichi TK, Paik M, Bagiella E, Desmond DW, Stern Y, Sano M, Hauser WA, Mayeux R (1994) Risk of dementia after stroke in a hospitalized cohort: results of a longitudinal study. Neurology 44:1885–1891

Thal DR, Grinberg LT, Attems J (2012) Vascular dementia: different forms of vessel disorders contribute to the development of dementia in the elderly brain. Exp Gerontol 47:816–824

Toledo JB, Toledo E, Weiner MW, Jack CR Jr, Jagust W, Lee VM, Shaw LM, Trojanowski JQ (2012) Alzheimer's Disease Neuroimaging Initiative. Cardiovascular risk factors, cortisol, and amyloid-β deposition in Alzheimer's disease neuroimaging initiative. Alzheimers Dement 8: 483–489

Tong XK, Hamel E (1999) Regional cholinergic denervation of cortical microvessels and nitric oxide synthase-containing neurons in Alzheimer's disease. Neuroscience 92:163–175

Trevisan K, Cristina-Pereira R, Silva-Amaral D, Aversi-Ferreira TA (2019) Theories of aging and the prevalence of Alzheimer's disease. Biomed Res Int 2019:9171424

Trousson A, Bernard S, Petit PX, Liere P, Pianos A, El Hadri K, Lobaccaro JM, Said Ghandour M, Raymondjean M, Schumacher M, Massaad C (2009) 25-hydroxycholesterol provokes oligodendrocyte cell line apoptosis and stimulates the secreted phospholipase A2 type IIA via LXR beta and PXR. J Neurochem 109:945–958

Tsivgoulis G, Alexandrov AV, Wadley VG, Unverzagt FW, Go RC, Moy CS, Kissela B, Howard G (2009) Association of higher diastolic blood pressure levels with cognitive impairment. Neurology 73:589–595

Tzourio C, Dufouil C, Ducimetière P, Alpérovitch A (1999) Cognitive decline in individuals with high blood pressure: a longitudinal study in the elderly. Neurology 53:1948–1952

Tzourio C, Anderson C, Chapman N, Woodward M, Neal B, MacMahon S, Chalmers J (2003) PROGRESS Collaborative Group. Effects of blood pressure lowering with perindopril and indapamide therapy on dementia and cognitive decline in patients with cerebrovascular disease. Arch Intern Med 163:1069–1075

Verhaaren BF, Vernooij MW, de Boer R, Hofman A, Niessen WJ, van der Lugt A, Ikram MA (2013) High blood pressure and cerebral white matter lesion progression in the general population. Hypertension 61:1354–1359

Vermeer SE, Prins ND, den Heijer T, Hofman A, Koudstaal PJ, Breteler MM (2003) Silent brain infarcts and the risk of dementia and cognitive decline. N Engl J Med 348:1215–1222

Vingtdeux V, Davies P, Dickson DW, Marambaud P (2011) AMPK is abnormally activated in tangle-and pre-tangle-bearing neurons in Alzheimer's disease and other tauopathies. Acta Neuropathol 121:337–349

Vinyoles E, De la Figuera M, Gonzalez-Segura D (2008) Cognitive function and blood pressure control in hypertensive patients over 60 years of age: COGNIPRES study. Curr Med Res Opin 24:3331–3340

Viticchi G, Falsetti L, Vernieri F, Altamura C, Altavilla R, Luzzi S, Bartolini M, Provinciali L, Silvestrini M (2014) Apolipoprotein E genotype and cerebrovascular alterations can influence conversion to dementia in patients with mild cognitive impairment. J Alzheimers Dis 41:401–410

Vu V, Bui P, Eguchi M, Xu A, Sweeney G (2013) Globular adiponectin induces LKB1/AMPK-dependent glucose uptake via actin cytoskeleton remodeling. J Mol Endocrinol 51:155–165

Wahrle S, Das P, Nyborg AC, McLendon C, Shoji M, Kawarabayashi T, Younkin LH, Younkin SG, Golde TE (2002) Cholesterol-dependent γ-secretase activity in buoyant cholesterol-rich membrane microdomains. Neurobiol Dis 9:11–23

Warkentin S, Ohlsson M, Wollmer P, Edenbrandt L, Minthon L (2004) Regional cerebral blood flow in Alzheimer's disease: classification and analysis of heterogeneity. Dement Geriatr Cogn Disord 17:207–214

Watanabe T, Koba S, Kawamura M, Itokawa M, Idei T, Nakagawa Y, Iguchi T, Katagiri T (2004) Small dense low-density lipoprotein and carotid atherosclerosis in relation to vascular dementia. Metabolism 53:476–482

Whitmer RA, Sidney S, Selby J, Johnston SC, Yaffe K (2005) Midlife cardiovascular risk factors and risk of dementia in late life. Neurology 64:277–281

Wortmann M (2012) Dementia: a global health priority-highlights from an ADI and World Health Organization report. Alzheimers Res Ther 4:40

Yamaguchi N, Kukita T, Li YJ, Martinez Argueta JG, Saito T, Hanazawa S, Yamashita Y (2007) Adiponectin inhibits osteoclast formation stimulated by lipopolysaccharide from Actinobacillus actinomycetemcomitans. FEMS Immunol Med Microbiol 49:28–34

Yamauchi T, Kamon J, Waki H, Terauchi Y, Kubota N, Hara K, Mori Y, Ide T, Murakami K, Tsuboyama-Kasaoka N, Ezaki O (2001) The fat-derived hormone adiponectin reverses insulin resistance associated with both lipoatrophy and obesity. Nat Med 7:941–946

Yamauchi T, Kamon J, Minokoshi YA, Ito Y, Waki H, Uchida S, Yamashita S, Noda M, Kita S, Ueki K, Eto K (2002) Adiponectin stimulates glucose utilization and fatty-acid oxidation by activating AMP-activated protein kinase. Nat Med 8:1288–1295

Yates KF, Sweat V, Yau PL, Turchiano MM, Convit A (2012) Impact of metabolic syndrome on cognition and brain: a selected review of the literature. Arterioscler Thromb Vasc Biol 32:2060–2067

Yip CM, Elton EA, Darabie AA, Morrison MR, McLaurin J (2001) Cholesterol, a modulator of membrane-associated Aβ-fibrillogenesis and neurotoxicity. J Mol Biol 311:723–734

Zhang YY, Fan YC, Wang M, Wang D, Li XH (2013) Atorvastatin attenuates the production of IL-1β, IL-6, and TNF-α in the hippocampus of an amyloid β1-42-induced rat model of Alzheimer's disease. Clin Interv Aging 8:103

Zhu H, Taylor JW, Bennett DA, Younkin SG, Estus S (2008a) Lack of association of hepatic lipase polymorphisms with late onset Alzheimer's disease. Neurobiol Aging 29:793

Zhu W, Cheng KK, Vanhoutte PM, Lam KS, Xu A (2008b) Vascular effects of adiponectin: molecular mechanisms and potential therapeutic intervention. Clin Sci 114:361–374

Zlokovic BV, Deane R, Sallstrom J, Chow N, Miano JM (2005) Neurovascular Pathways and Alzheimer Amyloid β-peptide. Brain Pathol 15:78–83

Zou K, Kim D, Kakio A, Byun K, Gong JS, Kim J, Kim M, Sawamura N, Nishimoto SI, Matsuzaki K, Lee B (2003) Amyloid β-protein (Aβ) 1–40 protects neurons from damage induced by Aβ1–42 in culture and in rat brain. J Neurochem 87:609–619

Zupancic M, Mahajan A, Handa K (2011) Dementia with Lewy bodies: diagnosis and management for primary care providers. Prim Care Comp CNS Disord 13:PCC.11r01190

Behavioral Abnormalities of Gut Microbiota and Progression of Dementia

11

Mohammad Zubair, Farha Fatima, and Fohad Mabood Husain

Abstract

Several bacteria are found in the human gut, which has a significant impact on the physiological functions of humans and also facilitates communication between brain and gut. These important gut-dwelling bacteria form the intestinal flora. Gut microbiota also establishes the structural and functional relationship with gut mucosa. A number of behavioral and functional aspects of the brain are affected by the activities of gut microbiota. Several diseases such as dementia (Alzheimer's disease), autism spectrum disorder (ASD), and bowel syndrome mostly result from the ineffective functioning of gut microbiota. Dementia describes a group of symptoms associated with cognitive impairment in human beings. Certain conditions such as stress, diabetes, smoking, obesity, and hearing impairment are the risk factors for dementia. Ineffective management of these problems may result in intensive cases of dementia. Certain precautionary measures such as maintenance of physical fitness, management of hearing impairment, and awareness in individuals may help in controlling the increasing prevalence of dementia. The relationship between dementia and gut microbiota is of greater significance since the improper functioning of gut microbiota may lead to impairment in the quality of life. In this chapter, the authors represent the role of gut bacteria and their linkage with Alzheimer's disease.

M. Zubair (✉)
Department of Medical Microbiology, Faculty of Medicine, University of Tabuk, Tabuk, Kingdom of Saudi Arabia

F. Fatima
Department of Zoology, Faculty of Life Science, Aligarh Muslim University, Aligarh, India

F. M. Husain
Department of Food Science and Nutrition, King Saud University, Riyadh, Kingdom of Saudi Arabia

G. M. Ashraf, M. S. Uddin (eds.), *Current Thoughts on Dementia*,
https://doi.org/10.1007/978-981-16-7606-2_11

Keywords

Dementia · Alzheimer's disease · Autism spectrum disorder · Bowel syndrome · Depression · Gut microbiota · Microbiomes

11.1 Introduction

There has been tremendous growth in the fields of microbiology, neuroscience, and gastroenterology in recent years, and these developments have verified the sayings of Hippocrates that all diseases begin through the gut. Scientists have discovered that gut-dwelling bacteria have a greater influence on human physiology such as communication between brain and gut, individual behavior, and brain function (Ignatova 2019). The gut microbiota includes different types of microorganisms such as protozoans, fungi, viruses, and bacteria, which are found in huge number that even exceeds the total cell count of the human body. These microorganisms along with their genomes and substances they produce constitute the gut microbiomes (Dinan and Cryan 2012). These microorganisms play a distinguished role within the human body and form the intestinal flora that is favorable for the host. Probiotic bacteria dominate the intestine of healthy individuals. Matricon et al. (2012) have provided important information regarding probiotics; according to which, the word probiotic is derived from a Greek word, which means "for life." The term, probiotics, further refers to the living microbes that manage the intestinal balance of microbes in the human body. These microbes are capable enough to manage the internal body functions successfully. Scientific researchers on rodents raised in completely hygienic environments suggest that activities of gut microbiota affect the development of emotional behavior, neurotransmitter system of the human brain, and pain and stress modulation systems. However, studies conducted on animals suggest that perturbations of microbiota triggered by probiotics and antibiotics exert a significant impact on adult animals (Bercik et al. 2012).

The findings and theories of the Ancient Greek philosophers identified the association between brain and the gut, which led to the proposition that they are functionally connected (Welcome 2018a, b). Several diseases of the brain including epilepsy are linked with gut dysfunctions. Consequently, these diseases were treated through deprivation of food ingestions (Welcome 2018a). The researches carried out in the late nineteenth century discovered the fundamental relationship between the brain and gut functioning (Thursby and Juge 2017). Consequently, significant researches have been conducted throughout the twentieth century and at the beginning of the twenty-first century that has provided detailed information and knowledge about the functional association between gut microbiota and the brain. Under normal circumstances, gut microbiota is responsible for establishing a functional and structural relationship with gut mucosa and hemostasis of the body. Recent studies have suggested the role of gut microbiota in the prevalence of several diseases within the human body, for example, diseases of the central nervous system such as anxiety, depression, autism spectrum diseases, multiple sclerosis, and attention hypersensitivity disorders (Welcome et al. 2018; Tremlett et al. 2017). Similar researches have also suggested that dysfunctions of the gut microbiota lead to the prevalence of Alzheimer's disease (AD) and also accelerate the speedy progression of diseases (Jiang et al. 2017). Other common diseases associated with the dysfunction of gut microbiota include osteoporosis, cardiovascular disorders, and sarcopenia of muscles.

Bacteria form a greater part of the human microbiota. Recent studies have suggested that the total genome of gut microbiota is 150 times greater than the total human genome (Parashar and Udayabanu 2017; Jin et al. 2017). The recent studies conducted on the prevailing issues suggested that the nature of the relationship between microbiota, gut, and brain is bidirectional. This type of relationship is termed the microbiota–gut–brain axis (Rieder et al. 2017). Other important disorders, such as Parkinson's disease and amyotrophic lateral sclerosis (ALS), are also related to the gut microbiota. The gut microbiota is assumed to be strong enough to induce emotional imbalance in humans by changing the course of human metabolism (Ahmed et al. 2016). In patients suffering from Parkinson's disease, pro-inflammatory gut microbiota dysbiosis is potentially strong enough to promote the abnormal folding of the synuclein that is inflammatory in nature. The occurrence of similar conditions has been observed in patients suffering from Alzheimer's disease (AD), which shares important characteristics with other neurodegenerative disorders including the growth of misfolded proteins, neuroinflammation, and pathological spread through several misfolded proteins (Zhuang et al. 2018).

Since the human body is highly sensitive at the time of birth, the occurrence of bacteria within the human body is quite easy and may affect human growth through the maturity of bacteria during adolescence and childhood. As the periods are found to be critical, the prevalence of such diseases is found with greater probabilities and underlying reasons within the human body (Faust et al. 2012). Therefore, a little change or disruption in the host–microbiota mutualistic association has the potential to alter the signals of the gut–brain axis that results in a negative impact on human health and brain development. Adolescents, who are affected by the gut bacteria during the early phases, develop several neurodevelopment disorders in the later stages of their life. Recent studies have suggested that environmental factors, including the use of antibiotics, infections, stress, and unhealthy diet during the periods of immaturity and instability in microbiota, can significantly impact human growth (Borre et al. 2014). These factors are significant and may lead to the microbial imbalance that causes severe brain disorders.

Adults with matured gut microbiota are likely to have important brain functions such as the synaptic pruning and myelination (Sowell et al. 2003). However, changes that occur in the gut microbiota during adulthood lead to abnormal brain functioning and human behavior. Researchers stress the need for the maintenance of gut microbiota within healthy individuals to refrain from disorders commonly reported in aged people. Studies conducted on the association between aging and brain disorders indicated that aging is a process that is not critically associated with the process of neurodevelopment; however, the body displays the pro-inflammatory progression responses during the process of aging that result in the imbalance of gut microbiota (Franceschi 2007; Claesson et al. 2012).

The abovementioned factors are crucial as they decrease the stability of microbiota within the human body. Other important factors such as an unhealthy diet and low health status, along dietary habits, are highly associated with the increasing occurrence of aging diseases. Additionally, deterioration of gastrointestinal and digestive motility, usage of drugs, and impaired immunity along the malabsorption of nutrients significantly impact the composition of gut microbiota. All factors indicate the decrease in cognitive function and also the reduction in brain weight with the reduced diversity of gut microbiota in older populations. Based on the above facts, this chapter mainly deals with the impact of gut bacteria and the relationship between bacteria and AD.

11.2 Formation of the Normal Gut Microbiota

Previous studies have revealed that the gut microbiota comprises approximately 500–1000 species of microbes. However, recent studies suggested that the overall gut microflora is composed of approximately more than 35,000 microbial species (Ramakrishna 2007; Frank et al. 2007). The perspective proposed by the human microbiome project along with the metagenome of human intestinal tract (MetaHIT) suggests that the prevalence of the gut microbiota includes the human microbiome, which is composed of approximately ten million non-superfluous genes. Among obese and non-obese patients, gut microbiota results in the prevalence of high and low gene count. These factors have significant implications regarding individual health and the occurrence of diseases (Le Chatelier et al. 2013).

Microbiomes with high gene count include specific ratios of *Butyrivibrio crossotus*, *Anaerotruncus colihominis* with a high proportion of *Akkermansia verrucomicrobia*, and *Ruminococcus*. Le Chatelier et al. (2013) revealed that the high gene count microbiome increases the production of organisms that produce butyrate, which helps in increasing the production of methanogenic ecosystems while decreasing the formation of hydrogen sulfide. A significant role is exerted by gut microbiota in the pathogenesis of the metabolic syndrome. Individuals with high gene count usually have a functionally vigorous gut microbiome, which leads to the minimum prevalence of metabolic disorders such as obesity and other health-related issues (Festi et al. 2014).

In contrast to this, individuals with low gene count experience an increasing proportion of pro-inflammatory bacteria including the *Ruminococcus* and *Bacteroides*, which are associated with provocative bowel diseases (Swidsinski et al. 2005). Other participants of low gene count bacteria experience the growth of *Campylobacter*, *Porphyromonas*, *Parabacteroides*, and *Staphylococcus*. Similarly, few key bacterial metabolites in individuals with low gene count include metabolites produced by the degradation of the aromatic amino acids, and β-glucuronide and metabolites produced by the reduction in the dissimilatory nitrite, and all of them cause an adverse impact on human health (Swidsinski et al. 2005).

Healthy gut microbiota among individuals is mostly composed of Bacteroidetes and phyla Firmicutes along with *Verrucomicrobia* and phyla Actinobacteria. Gut microbiota are significant in exhibiting different distributions of the given items at genus and exceeding levels, though the occurrence of the given constituents remains constant (Rinninella et al. 2019). Different organs of the gut such as the esophagus and rectum contain a distant ratio of bacteria, where approximate contents up to 10^{11} per gram are found in stomach and esophagus, while 10^{12} per grams microbial content is located in distal gut and colon (O'Hara and Shanahan 2006). Studies identified *Streptococcus* as the most common genus found in esophagus, duodenum, and jejunum.

Another important genre is *Helicobacter*, which is one of the dominant genera found in the stomach and significantly contributes to molding the overall microbial foundation of the gastric flora. In conditions when *Helicobacter pylori* commensally inhabit the stomach, other associated genres are also found including the *Prevotella*, *Streptococcus*, *Rothia*, and *Veillonella* (Andersson et al. 2008). The diversity of genus disappears when a pathogenic phenotype is obtained by *Helicobacter pylori*. In normal human beings, large intestine hosts approximately 70% of the overall microbiota of the human body.

11.2.1 Gut–Brain Axis

Gut–brain axis acts as the important mode of communication between intestinal microbiota and the central nervous system through various sources including the immune stimulation, neural pathways, hormonal response of gut, bacterial metabolism, and neural pathways (Borre et al. 2014). The gut microbiota is capable enough to influence the synthetic ability of the central nervous system by the involvement of several neuroactive molecules including the catecholamines, histamine, gamma-aminobutyric, acetylcholine, and serotonin. In addition, short-chain fatty acids that are produced during the gastrointestinal process are important in creating an impact over the functions of the central nervous system and cause a significant negative impact on gut–brain axis. Other factors such as intestinal microbiota, complexities with food antigens, and leaky gut syndrome are important in creating a negative influence over gut–brain axis (Gawlik-Kotelnicka and Polguj 2018).

11.2.2 Impact of Gut Microbiota on Behavioral and Functional Aspects of Brain

Colonization of postnatal microbes creates a long-lasting impact on the physiology of host's brain. Several developments have been provided regarding the prevalence, development, and impact of diseases on brain functions in a scientific study (Sudo et al). Sudo et al. (2004) in their study revealed that significant stress response was observed in samples that were kept in a germ-free environment in comparison with the samples that were free from specific pathogens. The following changes were observed in samples with normal gut flora. The behavioral phenotype is, however, due to the low-level expressions of hippocampal and cortical brain-derived neurotrophic factor (BDNF), which serves as the neurotrophin that helps in the growth and neuronal survival (Park and Poo 2013). Samples with single bacterium strains of *Bifidobacterium infantis* were highly responsive in saving the stress responses.

In conditions when the fecal transplant of the microbiome was transferred to samples kept under germ-free conditions during the early age, positive results were observed in the form of restored stress responses to base levels; however, this impact was not significant during the later age (Floch 2010). Other behaviors such as the reduced anxiety among the germ-free samples can only be controlled through the integration of specific pathogen-free microbiota among young samples, since the results are not visible enough in the aged samples. The information suggests that microbiota activities can only be triggered during the early age to reflect greater improvements in the form of smooth neuronal circuits and plasticity during development and the discovery of the involved molecular mechanisms (Tognini 2017).

Another important development includes the genetic expression of the germ-free samples in comparison with the genetic expression of specific pathogen-free samples in frontal cortex, strontium, and hippocampus (Kwon and Chapman 2011). Another important protein, such as the synaptophysin, which is valuable for endocytosis and synaptogenesis, along with *PSD95*, is involved in the plasticity and maturation of synaptic provided a decreasing level of striatum among germ-free samples. This indicates that the changes in synaptic plasticity are highly associated with the deficiency or simply the unavailability of intestinal commensals (Fig. 11.1).

The gut microbiome is further related to the formation of the hippocampal serotonergic system (Clarke et al. 2013). Samples kept in the germ-free environment reflected an increased level of serotonin that is associated with the reduced level of anxiety, although no direct association was found between the behavioral phenotype and changes in neurotransmitters (Clarke et al. 2013). Similarly, a colonization that occurred after weaning indicated a significant failure in restoring the hippocampal and neurochemical dissimilarities in the conventionalized samples. Besides, concerning the germ-free sample, within the amygdala, plasticity-related genes failed to indicate modifications. Furthermore, NMDA receptors and their subunit, i.e., NR2B, indicated downregulations that were parallel to decreased anxiety behaviors (Neufeld et al. 2011). The overall absence of microbes in germ-free samples indicated a visible increase in the volume of the hippocampal and amygdala. Notably, volume of the whole brain remains constant. Besides, alterations within certain areas, such as the morphological aspects of dendritic spines, went through a significant increase in length. The structural differences may be considered in producing the behavioral impact on germ-free samples (Luczynski et al. 2016).

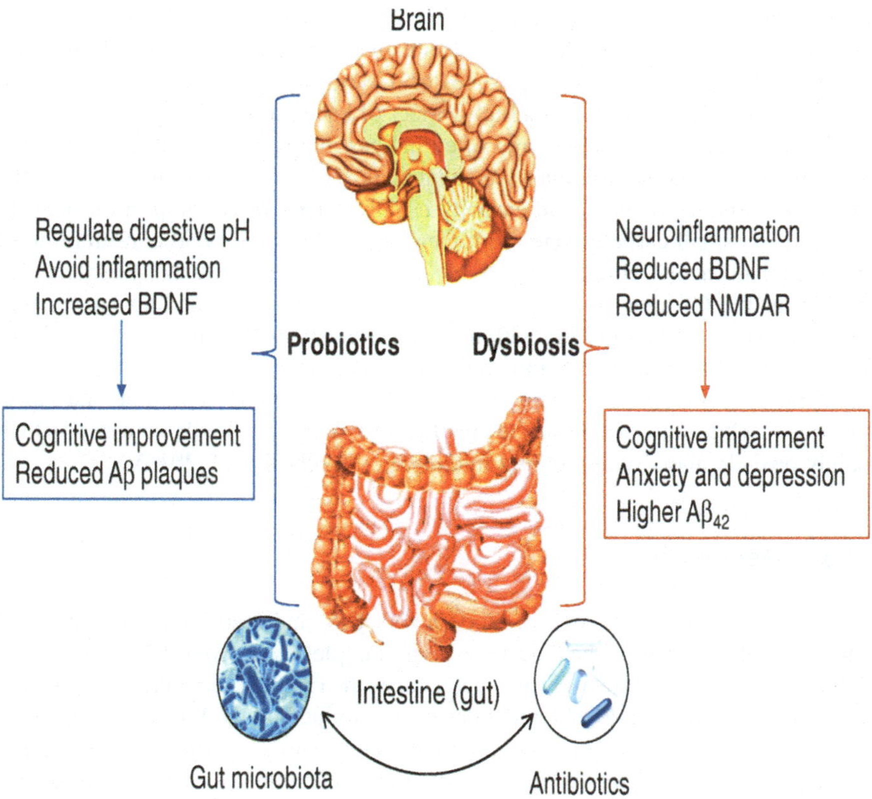

Fig. 11.1 Schematic representation of the role of microbiota–gut–brain axis in Alzheimer's disease. Good bacteria probiotics are capable to stabilize digestive pH, reduce inflammation, and increase neuroprotective molecules, such as brain-derived neurotrophic factor (BDNF). These effects lead to improved cognition and reduced Aβ plaque formation in AD animal models. In contrast, impaired microbiota dysbiosis can induce neuroinflammation and reduce the expression of BDNF and NMDA receptors, leading to cognitive impairment, mood disorders, and higher levels of $A\beta_{42}$. Antibiotics, by affecting gut microbiota composition, interact with this circuit, and produce different effects, depending on their microbiome target. (Adapted from Ref. Angelucci et al. 2019)

11.2.3 Present Understanding Related to Gut Microbiota

The development and understanding related to microbiota have been investigated through several perspectives, including the nasal and vaginal cavities, skin, and other oral cavities (Segata et al. 2012). Abundant information is available regarding the given factors; however, only limited information is available regarding microbiota and its association with the gastrointestinal (GI) tract (Hollister et al. 2014). Presently, gut microbiota is considered as the prime location for bacterial colonies. Most of the information regarding gut microbiota is derived through fecal samples, while only limited studies used collected information through mucosal biopsies. Since the collection of fecal samples is easy, only a limited amount of knowledge is obtained through the given source (Tang et al. 2020). Researchers suggested that a sufficient amount of this knowledge can only be obtained through the small intestine, on account of the presence of huge colonies of bacteria within the small intestine that serves as a significant source of a high variety of bacterial strains.

The degradation of complex and often indigestible carbohydrates promotes the efficiency of colonic microbiota (Zoetendal et al. 2012). The small intestine, on the other hand, due to its shape and capacity usually carries out an import mechanism that converts small carbohydrates and, therefore, reflects the rapid consumption of the overall nutrients. However, faces do not ideally represent the prevalence of bacteria in the GI tract, and they are only responsible for providing information regarding the changes occurring within the large intestine (Nigam et al. 2019). Apart from this, maximum data regarding the problems associated with the gut microbiota are mainly provided by the studies conducted in the regions of Europe and North American, while only a few are available from Asia and South America, which mainly provide biased views on gut microbiota (Marchesi et al. 2016).

11.3 Dementia

Dementia is a major challenge of modern societies, and therefore, it has attracted significant attention from healthcare experts. People aged above 65 are the main victims of the disease; however, this fact reduces the ratio of premature deaths (Lund et al. 1987). Significant development has been provided by the Lancet Commission on Dementia Prevention while outlining the major factors significantly involved in controlling the given disease. Furthermore, a valuable emphasis is provided regarding the prevention and control of dementia by Prince et al. (2016).

In 2015, estimated 47 million cases of dementia were reported worldwide. This ratio is expected to increase thrice in the upcoming years till 2050 (Baumgart et al. 2015). The ratio is expected to increase thrice in the upcoming years, i.e., till 2050. Dementia is usually observed in individuals who are isolated and have lost their functional abilities and depend on other individuals including friends and family members. Individuals going through the following conditions are the major target of the disease, as they require maximum care in terms of health and other social activities. Globally, the overall cost of treatment of dementia is estimated to be up to US$ 18 billion, which is expected to increase in the long run. However, most of the cost is associated with social and family care (Jones et al. 2016). New studies suggested that new methods of medical care are strong enough to reduce the overall costs of care provided to patients with dementia; this problem is inevitable in aged patients who are close to their retirement or in the ninth decade of their life.

Several factors are responsible for the occurrence of dementia. Individual lifestyle is one of the major causes of the occurrence of the problem and, therefore, significantly contributes to the reduction in the problem within the aged population. Dementia is often a delayed observation in certain population, while in others the ratio of the affected people with dementia is increasing day by day. Loss of hearing and social isolation are some consequences of dementia in the affected patients

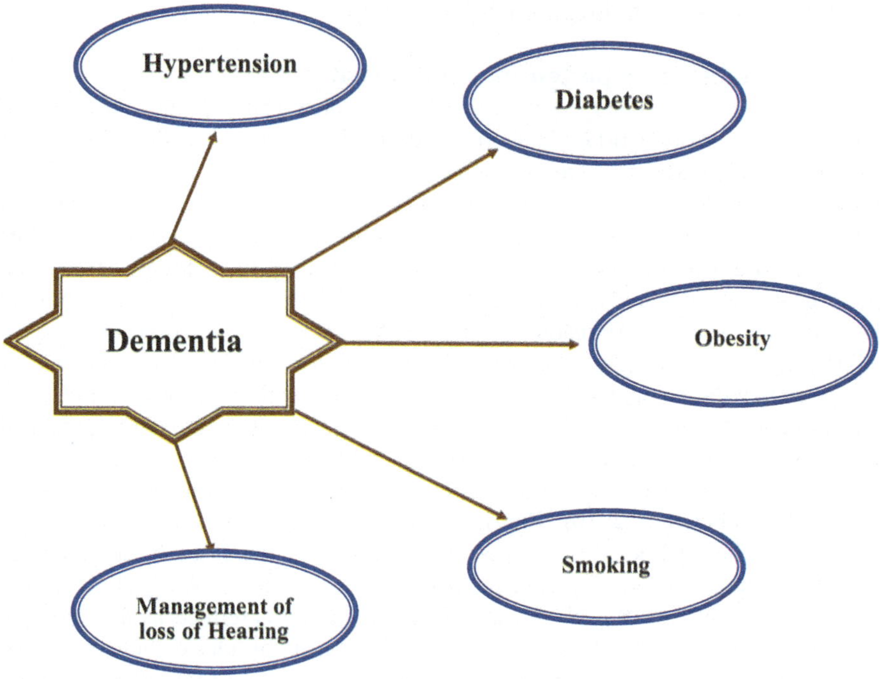

Fig. 11.2 Factors contributing to dementia management

(Matthews et al. 2016). After incorporating the risk factors that are often reversible under different phases of life other than aging, researchers have been able to propose a model that is used to identify several risk factors that one encounters in the individual life. These factors supported the formation of population attributable fractions (PAFs) that are being developed to provide the future consequences of the problem. It further helps in the successful reduction in most of the potential risk factors (Yusuf et al. 2020).

Studies suggested that more than one-third of the cases of dementia can easily be catered through the theoretical knowledge of the given problem. Management of hearing loss, regular exercise and the acquisition of related knowledge, stress management, hypertension, diabetes, obesity, and smoking help in the reduction in the prevalence of the disease (World Health Organization 2019). Thus, individual control regarding the given factors may lead to an escape or prevention of the given factors (Fig. 11.2).

11.3.1 Factors Associated with Dementia

11.3.1.1 Diabetes, Hypertension, and Obesity

Major concerns have emerged regarding high-fat diets and sedentary lifestyles of individuals across the world. Metabolic syndrome is determined as the collection of the morbid conditions that are associated with insulin resistance, obesity, glucose intolerance, dyslipidemia, and hypertension (Halpern et al. 2010). Obesity causes a great economic burden, and several morbidities are associated with it that includes diabetes, cardiovascular disease (hypertension), and fatty liver disease. There is a mechanistic relationship between metabolism of cholesterol taking place within the brain and formation of Aβ plaques leading to AD (Martins et al. 2006).

The risk of developing AD and blood restriction increases on account of increased blood sugar and body fat. Moreover, the gut microbiota, fatty liver disease, and susceptibility to insulin resistance play a significant role in the development of complex metabolic abnormalities (Dumas et al. 2006). It has been shown that there is a significant influence of gut microbiota on cholesterol metabolism and insulin sensitivity (Rabot et al. 2010). The risk of developing cardiovascular disease independent of the traditional cardiovascular risk factors increases because of the increase in the circulating levels of gut microbiota metabolites (Wang et al. 2011).

The important part played by cardiovascular disease is termed as the etiologic hallmark of AD and also highlights the significance of vascular pathology in AD. Considering a variety of processes, gut microbiota has a significant influence on the important physiological processes leading to the development of cardiovascular diseases such as hypertension (Hazen and Smith 2012). It is known that diabetes is of two types, diabetes mellitus type I and type II. Similarly, a decade ago it was proposed that AD might be diabetes mellitus type III (Martins et al. 2006). The development of resistance by the brain toward insulin might be responsible for the progression of AD, which even prevents the appropriate uptake of lipid. This increases stress and inflammation leading to dementia because of increased building up of lipids in the brain (Rani et al. 2016).

The animal models are likely to present a significant and strong association of the development between AD and type 2 diabetes mellitus (Chornenkyy et al. 2019). It has been noted that deterioration of neurons blocks the path of insulin to brain, which results in the development of physical confusion and majority of the signs of AD. Diabetes has a positive impact on AD and vice versa; besides, various disorders are underlined as potential mechanisms for both disorders. Those potential mechanisms include aging, inflammation, Aβ clearance by insulin degradation enzyme (IDE), O-GlcNAcylation, oxidative stress, cerebral vascular insufficiency, glucose metabolism, Aβ aggregation by advanced glycation end products (AGEs), and circulating cortisol (Sandhir and Gupta 2015).

The risk of developing AD increases by twofold in diabetic patients. However, this does not mean that all the individuals suffering from AD are diabetic, and similarly, every diabetic individual does not eventually develop AD. Diabetes mellitus type 2 is considered the major cause of obesity, which is controlled by gut microbiota. It has been observed in a recent study that metabolism of tryptophan, production of microbial metabolites, bacterial cell wall sugars, microbial neurotransmitters, and bile acids produced by gut microbiota control the brain functions (Tremaroli and Bäckhed 2012).

The activity of lipoprotein lipase (LPL) is significantly affected by the gut microbiota. This enzyme is responsible for the release of fatty acids from triglyceride-rich lipoproteins found in muscle, heart, and fat by influencing the expression of fasting-induced adipocyte factor protein (FIAF) (Eckel 1989). FIAF plays a significant role in the prevention of diabetes as it is a key inhibitor of LPL activity. The cellular uptake of fatty acids increases leading to the accumulation of adipocyte triglyceride as a result of upregulation of adipocyte LPL activity. Adiposity is promoted through the suppression of intestinal FIAF by microbes, whereas the accumulation of calories harvested from the diet into fat and their storage in the liver enhances as a result of an increase in hepatic lipogenesis. The innate immune response is activated through the involvement of specific bacterial taxa of the gut microbiota in the energy homeostasis and nutrient uptake leading to low-grade inflammation by lipopolysaccharides (LPSs). There is an association between the presence of low-grade inflammation and constant levels of LPS in the circulation and increased levels of adiposity and insulin resistance. As compared to individuals, who are obesity-resistant, the presence of ileal inflammation, increase in innate immune response, and decrease in LPS activity were observed in the rats, who are susceptible to weight gain (de La Serre et al. 2010).

11.3.1.2 Smoking

The peripheral and central nervous system is modulated as a result of alterations in the gut microbiota, which results in an alteration in the brain functioning. Changes in brain function and subjective reports of mood occur on account of alteration made in microbiota with beneficial bacteria or probiotics. Cigarette smoke contains a complex admixture of 5000 different combustion products, which comprises different cytotoxic and carcinogenic compounds (Durazzo et al. 2014). It contains high concentrations of short- and long-lived ROS, RNS, and other oxidizing agents.

Increasing inheritance and aging of ε4 allele of the *APOE* gene are known as the consistent and strong risk factors leading to the development of AD; however, no proper mechanism has been established, which is responsible for the inception and progression of late onset of AD. A major focus has been placed on the identification of modifiable risk factors for AD, which could be altered for effective reduction in their prevalence during the asymptomatic preclinical stage. This is likely to be responsible for the promotion of a decrease in the prevalence of AD. However, the strength of association between AD and potentially modifiable risk factors is still a major concern (Xu et al. 2015).

Smoking is one of the modifiable risk factors for AD along with hypercholesterolemia, mood disorders, type 2 diabetes mellitus, nutritional supplement intake, hypertension, level of alcohol consumption, level of physical activity, and cognitive engagement. It has been narrated that the risk of AD increases in smokers, who are not the carriers of *APOE ε4*. Moreover, formerly, lifetime smoking was known as a significant risk factor for AD. The increased risk of developing AD was linked with peak years, duration, and measure of cigarette smoking dose. The major source of attrition is the elders, who fail to cope up with premature conditions because they are suffering from smoking-related diseases. This might reduce the number of smokers, who are likely to develop AD in their later life (Durazzo et al. 2014).

There is a significant association between smoking and the development of adverse effects on brain neurobiology in people who have no history of biomedical or psychiatric conditions. However, these individuals might have a history of mild traumatic brain injury (TBI) or a neuropsychiatric disorder such as alcohol abuse or schizophrenia (Durazzo et al. 2014). It is also apparent that there is congruency between neuropathologic and neurocognitive abnormalities characterized as the preclinical stages of AD and pattern of neurobiological and neurocognitive abnormalities observed in the smokers.

Smoking might even relate to multiple neurocognitive and neurobiological abnormalities. The transition from mild cognitive impairment to AD is likely to be observed through progressive regional atrophy and neurocognitive decline in smokers with the increase in age. Smoking even exacerbates neurobiological and neurocognitive sequelae associated with mild TBI and alcohol consumption. There is a direct association of increased free radical concentrations, causing oxidative damage to membrane proteins, DNA, lipids, carbohydrates, and RNA of neuronal, glial, and vascular brain tissue, with smoking (Durazzo et al. 2014). The immune and inflammatory responses in the central and peripheral nervous system are controlled by cytokines including the interferons, chemokines, growth factors, interleukins, and tissue necrosis factor. Cytokine-mediated immune response might trigger the inflammation as a result of oxidative damage imposed by the free radicals of other oxidants. Smoking is also significantly associated with the increased pro-inflammatory cytokine levels through the generation of ROS and other inflammatory mediators in the brain with the help of immune cells, astrocytes, and microglia (Durazzo et al. 2014).

11.3.1.3 Depression

A wide range of neuropsychiatric symptoms and others focusing on specific symptoms (e.g., aggression and agitation) are likely to be assessed in the patients suffering from dementia. The most prevalent behavioral and psychological symptoms of dementia include depression, agitation, apathy, irritability, and anxiety (Mukherjee et al. 2017). However, depression, anxiety, and apathy are known as the most clinically significant symptoms. It has been found that gastrointestinal disease, a high-fat diet, and use of antibiotic are associated with gut microbiota abnormalities. This condition might induce depression-like behavior. Abnormal gut microbiota is considered as the main reason for developing acquired and inborn depression. On the contrary, there is a difference in the microbiota of the individuals suffering from depression (Fan et al. 2018).

The depression-like symptoms are likely to be aggravated by chronic restraint stress along with the involvement of the behavioral, biochemical, and microbiota aberrations as in case of the acquired depression model. The patients suffering from AD possess abnormal gut microbiota, and their hypofunction in the brain and gut–brain can be improved by specific probiotic intervention (Hu et al. 2016). The animal models of behavioral disorders such as anxiety, depression, and cognitive dysfunction are linked with the impact of intestinal microbiota composition on brain function.

The composition of the gut microbiota plays significant role in majority of the metabolic conditions that involve the central nervous system. For instance, the brain might alter the composition of the gut microbiota through the hypothalamic–pituitary–adrenal axis, when it is confronted with stress or depression (Wang and Wang 2016). This might regulate the secretion of cortisol and affect the activity of immune cell in the gut and overall system. The levels of circulating cytokines can be altered through gut microbiota and probiotic agents that have significant impact on brain functions. There is a strong implication of vagus nerve and modulation of systemic tryptophan in relaying signals from the gut to the brain.

The current understanding related to dementia suggests that dementia is neither completely preventable nor it is controllable. Lancet Commission on Dementia Prevention, Intervention and Care is now working for a greater cause, which is how individuals can manage an escape from dementia or may work to control the disease (Orgeta et al. 2019). Most of the manifestations regarding the given problems are now controllable; however, no such measures have been developed until now that may help in the complete prevention and cure of this problem (Orgeta et al. 2019). Available knowledge suggests that care and strong interventions may reduce the speedy impact of dementia on human brain, and the contribution of the patient's family is very vital in this regard. For some people, dementia is similar to mild cognitive impairment. There is a slight difference between dementia and cognitive impairment, as it only affects the daily activities and the social functioning of the affected person (Meyer et al. 2002). As in mild cognitive impairment, patients are capable of performing complex tasks, such as the payment of bills or medications. The difference between dementia and cognitive impairment is often considered as a thin line since for many people dementia is largely associated with cognitive impairment.

Several types of dementia have been recognized, and the most common is in the form of Alzheimer's disease (Román 2002). Next is vascular dementia, which is another commonly recognized form of dementia. Other forms include Lewy bodies and mixed dementia. Similarly, dementias related to brain injury, alcohol abuse, and frontotemporal are least recognized. Therefore, the word dementia itself refers to all sorts of cognitive impairments that are related to brain.

Several beliefs are also associated with this disease as some people associate this disease with a certain form of curse or punishment. These beliefs lead to the avoidance of the necessary treatment (Liew et al. 2015). Consequently, the problem has now been identified as major neurocognitive disorders by the Diagnostic and Statistical Manual of Mental. This represents illnesses that are highly associated with the cognitive abnormalities that are visible during the cognitive brain development of individuals.

11.4 Behavioral Abnormalities and Progression of Dementia

Practitioners have observed that the occurrence of dementia and frontotemporal degeneration that is linked with brain injury, alcohol abuse, and infections are low (Livingston et al. 2017). The word "dementia" here particularly focuses on the various types of dementia. Generally, the word dementia is derived from two words, i.e., Latin words "de" which means "out of" and "mens" which refers to "mind" (Livingston et al. 2017). Both collectively are used to denote a demeaning connotation. It is referred as the declining cognitive ability of the individuals and which poses a detrimental impact on their neurological development. Previous records have highlighted that culturally dementia was stigmatized and was considered to be a curse or a punishment for the individuals suffering from it (Livingston et al. 2017), which further complicates the disease and increases the intensity of the adverse effects. However, now it is considered as a neurological disorder, and its medical treatment is encouraged and promoted.

The diagnostic and statistical manual of mental disorders (DSM) replaced the word dementia with a phrase as "major neurocognitive disorders" to dispel the stigma associated with the term (Livingston et al. 2017). It further defines it as the illness, which causes neural disability and substrate abnormalities leading to a decline in the cognitive ability caused by a disturbance in the normal development of the human brain (Surr et al. 2017). Furthermore, its medical recognition has helped patients of dementia to seek early treatment to overcome the adverse outcomes. Accordingly, the mild neurological disease is also categorized under the heading of dementia, which is equivalent to the World Health Organization (WHO), which had ICD-10 (International Classification of Diseases) as the classifier of the mild cognitive disorders. In other words, dementia is recognized as an irreversible geriatric disorder, which accounts for major morbidity and mortality cause among the aged population.

Commonly recognized symptoms of dementia include the deterioration of self-care, memory, and decline in quality of life. The impact of this disease can be recognized by its high cost of medical treatment, which in the USA alone is approximately $287,000: 57% higher than the costs of other diseases. The recent report issued by WHO has revealed that approximately 47.5 million are suffering from dementia and 90–98% of whom are aged 65 years or above. For developing countries like China, the dementia population is 8.4 million, which costs an expenditure of about US$17.5 billion (Wu et al. 2013). Thereby, understanding its impact as a healthcare expert and practitioner is substantially important, which not only aids in its effective treatment but also helps to improve the techniques, reducing the cost burden and improving the overall quality of life of an individual who is suffering from dementia.

11.4.1 Specific Risk Factors and Mechanisms

Dementia treatment requires the medical team to have a comprehensive evaluation of the risk factors and realize the impact these factors may have collective and individual impacts (Ganguli 2015). Dementia research has identified various factors that contribute to the development of dementia. While some of these risk factors are modifiable, some are not. Developing an adequate understanding of the impact can help devise effective care strategy and contribute to the reduction in the overall percentage of people suffering from dementia. Some of these factors are defined in Fig. 11.3.

11.4.2 Education or Qualification of Individuals

Education is identified as one of the factors that can lead to the development of dementia. Initially, the researcher was debatable against its recognition, though now its prevalence as dementia-producing factors is increasing (Ganguli 2015). The increased rate of education substantially contributes to brain awareness, which eventually reduces the prospects of developing dementia as one becomes older. Surprisingly, education does not impact the pathological development (brain development or changes in the brain itself) of the brain, which occurs over time and is related to dementia (Xu et al. 2016; Gibson et al. 2016). Education reduces the impact caused by these pathologies on the individual thought process, memory, as well as his cognitive ability. In simple words, even though the changes in the brain of

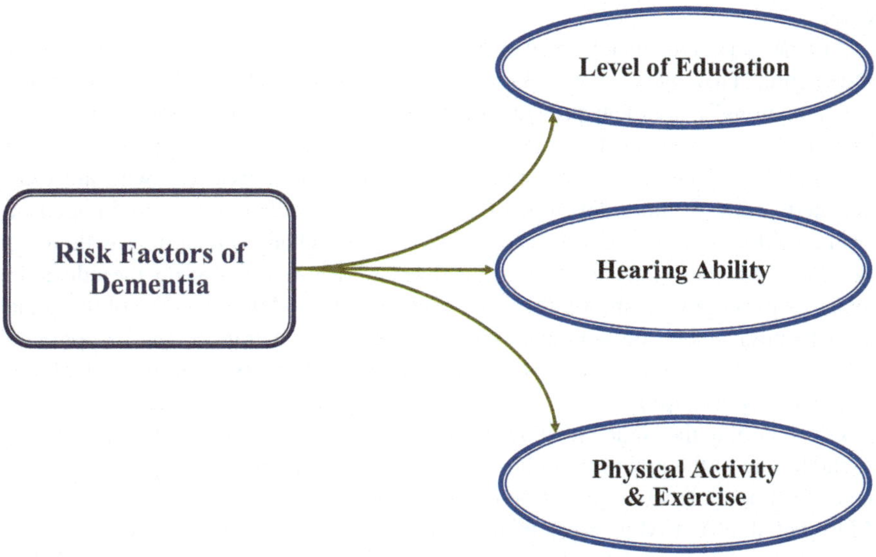

Fig. 11.3 Risks factors associated with dementia

a dementia patient are similar to that in average individual, the changes accompanied with education had a less detrimental impact on the cognitive ability of an individual as compared to others (Skoog et al. 2017).

The recognition of education as a risk factor for dementia is supported by American Journal of Epidemiology, which has also endorsed that higher education level is linked to improved performance of an individual in terms of his cognitive abilities and better scores in the cognitive tests (Weuve et al. 2018). Intriguingly, University of Michigan documented the decline in dementia from the years 2000 to 2012, along with a decreased rate of prevalence of dementia among individuals. The findings of the university research are significant as they point out the correlation that exists between education and dementia; i.e., the advancement in the education level is strongly correlated with the declining rates of dementia among individuals. Likewise, the researcher and medical practitioners have directed their attention to the cause of Alzheimer's disease, a frequently reported cause of dementia that showed that increase in education level could reduce its development prospects (Kenigsberg et al. 2016).

Consequently, it has been observed that prospects of Alzheimer's disease become high when the education level of an individual is low. A comprehensive review of 247 studies has shown that a literacy rate lower than that of 9th grade in an individual leads to the increased risk of dementia. The prime question that arises here is that to what extent the education makes a difference. Research of Annals of Epidemiology has highlighted that the additional year of education acts as an instrumental way to overcome or reduce the prevalence of dementia by 2.1% (Booker et al. 2016). Likewise, another research showed that the prevention against dementia should be initiated at an early stage of an individuals' life, which may continue to the age of 15 accompanied by the education that an individual is required to possess at this level.

Another aspect that needs to be nurtured is how the amount of education is related to dementia. This query answered by the researches showed that higher educational or occupational attainment prevents declining cognitive ability (Stern 2012). The dilemma that how education can affect cognitive ability is based on cognitive reserve. The cognitive reserve points to the idea that individuals who are more educated and have more developed brains can increase their ability to compensate for the decline that emerges as a result of structural changes occurred as their age increases. Another study evaluated the time as to how quickly do the education enhances the cognitive ability; the study findings revealed that the time of the actual impact could not be determined; however, its effective impact on the cognitive ability was observed as a slower decline in the mental ability of an individual over time (Gow et al. 2008).

Even though the understanding of dementia and its types is also undergoing research, one of the previous studies showed that educational and support programs have a positive impact on reduction in the dementia risk through effective education (Marim et al. 2013). More work in academics and accomplishment of a higher level of education is required for developing the required mental and cognitive ability, which helps to overcome the factors that promote the development of dementia. The low educational level makes individuals increasingly vulnerable to cognitive decline as it results in decline or reduced cognitive reserve, which is essential for an individual to sustain or maintain a function regardless of the changes in the brain pathology.

11.4.3 Hearing

An individual's ability to hear is also linked with dementia. Researches have recognized that hearing loss is considered as a major factor contributing to the risk of dementia, which is a relatively novel finding (Stahl 2017). Many questions and queries arise which question the reasons. Generally, the association between the two can be categorized into various aspects. One of the most obvious reasoning is a physiological pathway, which is linked to both hearing loss and a decline in the cognitive ability of an individual (Hardy and Selkoe 2002).

Another possible factor is the cognitive load that the individual has to bear, which causes great strain on the brain (Golub et al. 2017). Let us consider an example when an individual puts in more effort to understand what someone is saying, he directs all his resource toward it which might have been used for other purposes such as encoding what is being said and referring back to the memory. It is observed that treatment of dementia for their hearing loss can aid in optimizing their communication, which positively impacts their everyday life. For example, if a person can hear things easily, he is likely to spend his time in nativities or tasks, which are cognitively more demanding (Golub et al. 2017). This eventually declines the risk of dementia among patients.

Hearing loss can also be grouped based on age as evidence suggests that its detrimental effect on neurological development continues to increase at a later stage of life, particularly in the older population (McCabe et al. 2016). The common effect of dementia and hearing loss is observed in older age people given the increase in their microvascular pathology. For instance, the hearing loss leads to the cognitive load, which makes the brain vulnerable to the changes that make an individual suffer from disengagement, social isolation, and decline in his cognitive abilities. The individual is also required to bear the suffering of increased atrophy, which leads to the decline in cognitive ability, increasing the prospects of dementia development (Thomson et al. 2017). Even though the hearing loss among the individuals accounts for low cognitive ability, so researchers at times have been able to find that individuals with hearing impairment also possess normal baseline cognition.

11.4.4 Exercise and Physical Activity

Dementia is strongly correlated with the physical activity of the individual. Low physical activity increases the prospects of developing dementia (Lamb et al. 2018). Evidence from the literature highlights that indulgence in physical activity leads to the decline of cognitive abilities, particularly in the aged population (Brasure et al.

2018). A growing body of longitudinal study has shown that low indulgence in exercise leads to the development of dementia. It is, therefore, claimed that the increase in the exercise stimulates the brain ability to sustain the old connections while at the same time ascertain the new ones.

Similarly, prospective researches have shown the healthy effect produced by exercising that causes a positive impact on an individual's ability to preserve memory. With the combined results from various researches, it has been found that exercise accounts for 30% decrease in the risk of dementia, where for Alzheimer's disease, this percentage increases up to 45% (Forbes et al. 2015). Results of one study that examined the behavioral pattern of approximately 2000 patients have highlighted that among the five behaviors, not smoking, regular exercises, moderate consumption of alcohol, healthy body weight, and healthy diet and participation in healthy regular exercise, proved to be extremely effective for reducing the dementia prospects (Lamb et al. 2018). In short, medical professionals need to realize that abnormalities of dementia can be overcome by integrating aerobic exercise and some form of physical activity to enhance their performance in terms of their neurological thinking and testing. The outcomes of the improved physical activity can help practitioners to improve their patient's memory, concentration, and processing of the information (Atherton et al. 2016).

Accordingly, it has been observed that not all forms of physical activity prove to be effective. For instance, non-aerobic exercises including stretching and toning, though classified as exercises, are unable to contribute to the declining rate of dementia. The time duration is another aspect to be considered when determining the impact of exercises on dementia patients. The researcher has observed that the aged population must exercise three times a week to reduce the risk of dementia (Brown et al. 2015). Similarly, for the middle-aged population, regular exercises have been found to be fruitful for the patients as it was able to reduce the dementia risk to 52% in these individuals.

The various mechanisms provide an explanation of the impact of physical inactivity on the rate of dementia. For instance, the lack of indulgence in the physical exercise can increase the individual's vascular disease and vascular risk factors. The increase in vascular disease increases the level of inflammatory markers in the blood (such as of C-reactive blood) (Atherton et al. 2016). Since inflammatory markers are associated with the cognitive abilities and their increase poses adverse effect, therefore, this increase leads to the decline in cognitive thinking, escalating the prospects of developing dementia. Exercise is also related and recognized as a neurogenesis stimulator and increases cell proliferation, which can help in protection against the decline in the physical activities, eventually decreasing the dementia prospects (Briones 2006). Thereby, engaging in the regular exercise provides a wide range of health benefits to the human body and its brain, which serve as encouraging factors for regular exercising at all ages.

11.5 Behavioral and Psychological Symptoms of Dementia

It has been assumed that behavioral or psychological symptoms of dementia are prevalent between 50% and 90% of individuals affected with dementia during their lifetime. In addition, these individuals are further observed with long-term residential care facilities and higher prevalence in hospital care facilities as compared to community-dwelling settings (Cerejeira et al. 2012). These symptoms can be led to substantial deterioration of the quality of life as they are a major source of increasing financial burden and caregiver stress. Aggression, calling out continuously, sleep disturbance, lack of motivation and interest, and agitation are included in behavioral or psychological symptoms of dementia (Kales et al. 2015). It has been further reported that behavioral or psychological symptoms of dementia can be a major source of impairment for caregivers and family as well as considered as an imperative determinant of burden, care home admission, and depression of family caregiver.

The presentation and extent of behavioral or psychological symptoms of dementia differ based on brain-damaged state, etiology of the dementia syndrome, and dementia severity. It has been reviewed that the psychological foundation for the behavioral or psychological symptoms of dementia is associated with a disparity of neurotransmitters, which include dopamine, noradrenaline, serotonin, gamma-aminobutyric acid, and acetylcholine. The different neurophysiological modifications related to each subtype explain the variations in the presentation of behavioral or psychological symptoms of dementia (Matsumoto et al. 2007). No modern treatments are available for a number of dementia syndromes; therefore, there is a lack of evidence regarding the treatments that are helpful for managing symptoms. Antipsychotic medications and psychosocial therapies are usually involved in managing symptoms of dementia, and 40% of patients with dementia receive antipsychotic medications in institutional care.

In addition, increased risk of stroke and arrhythmia, and accelerated cognitive decline are considered as serious adverse effects. Increased mortality risk of up to 3.8% is related to the use of antipsychotics in patients with dementia. There is an increased consideration in the role and effectiveness of dietary modifications in offering advantages with a lack of effective and safe treatments for the neuropsychiatric symptoms and in the absence of available cures for advanced dementia syndrome (Ueda et al. 2013). On the other hand, there is a lack of evidence regarding the role of diet and dietary supplements in managing clinical symptoms of dementia. It has been emphasized that the gut–brain axis plays a significant role in behavioral or psychological symptoms of dementia, and emerging evidence recommends an association between neurodegenerative and cognitive aging diseases and the gut microbiota.

A number of contributory factors have been recognized, which can be attributed as constructs associated with the individual with dementia such as environmental factors, and caregiver factors as cognitive reduction alone cannot demonstrate these symptoms and indicators. The association of dementia can be possibly linked with symptoms directly by means of disruption in brain circuitry entailed in emotion and behavior. Progressions in neuroscience have indicated that reciprocal and extensive associations between brain centers exist that regulate cognition and emotion (Hersch and Falzgraf 2007). Three or more frontal–subcortical circuits have been theorized in the circuit model, which have basal ganglia, thalamic, and frontal aspects, that influence human behavior. Synaptic or circuit disconnections in these networks result in the behavioral or psychological symptoms of dementia. Individuals with dementia might be disproportionately affected by undiagnosed illnesses and pain regardless of cognitive impairment.

In the community population of older adults with dementia, 36% had an unidentified illness that was related to behavioral or psychological symptoms of dementia. These symptoms include crying out, hallucinations, repeated questioning, and agitation. In addition, aggressive behavior is related to pain in patients with dementia, and managing pain can dispel such behaviors. These symptoms might emerge as the side effects of drugs or drug–drug interactions (McKeith and Cummings 2005). It has been indicated from clinical experience that long-standing attributes and patterns might influence the progression of behavioral or psychological symptoms of dementia where premorbid personality characteristics are accentuated by the loss of inhibitory control.

The development of these symptoms might be affected by lifelong psychiatric disorders such as schizophrenia, anxiety, bipolar disorder, and depression as well as their management such as mood stabilizers, antipsychotics, antidepressants, and anxiolytics. The special role of family caregivers is associated with one of the difficulties of dementia. Stages of psychological distress, as well as stress, are higher, while subjective well-being, physical health, and self-efficacy are substantially low in the caregivers of dementia patients (Miyamoto et al. 2010). These variations are even greater when compared with non-caregivers. The extent of depression varies from 23% to 85% in individuals who care for patients with dementia and from 16% to 45% in those who care for patients with anxiety in comparison with depression in non-caregivers.

The stress threshold of the individual with dementia becomes reduced and the possibility for higher levels of frustration escalates with reduced ability to respond to stimuli. However, severe agitation and serious anxiety can be developed if it remains persistent. Changes in routine, physical and social environmental changes, lack of stimuli, and too many competing or misleading stimuli can cause stress. A vast range of behavioral, environmental, and caregiver supportive interventions are involved in non-pharmacologic treatments. Medical organizations, expert groups, and numerous guidelines suggest non-pharmacologic strategies as the recommended first-line treatment approach (Fauth and Gibbons 2014). On the other hand, these strategies have not been largely interpreted and implemented in practical clinical management and standard care. There is a lack of consensus regarding categorizing non-pharmacologic interventions; anyhow these are grouped into three categories: the environment, the caregiver, and the individual with dementia.

11.6 Relationship Between Dementia and Gut Microbiota

It has been hypothesized that the gut microbiome is considered as a novel risk factor for dementia considering its negative composition. Similarly, a lower prevalence of *Bacteroides* and a higher prevalence of other bacteria are autonomously and imperatively related to dementia as indicated by multivariable analyses. In addition, these associations are stronger in comparison with those of conventional dementia biomarkers. In previous studies, similar relationships have been discussed regarding patients with carotid stenosis and coronary artery disease (Alkasir et al. 2017). Endothelial cell function is regulated by *Bacteroides* and eradicates inflammation, which is constant with the inverse association between the presence of dementia and population of this genus. On the other hand, it has been emphasized that a smaller population of *Bifidobacterium* and a larger population of *Bacteroidetes* are suggestive of a counter-regulatory influence of *Bacteroidetes* or a repressive influence of *Bifidobacterium* in the gut of patients with Alzheimer's disease (Fig. 11.4).

Due to different variations, this discrepancy might be discussed in the researches, which includes the criteria used for diagnosing dementia, ethnicity, and dietary composition. Therefore, it might be way too early for concluding how to diagnose dementia in the presence of a gut microbiome. However, the associations between several types of gut microbiome and systemic arteriosclerotic diseases indicate common underlying mechanisms involved in the influences of gut microbial composition on multi-organ arteriosclerosis (Vogt et al. 2017). The diversity of the gut microbiome is an imperative potential determinant, but it cannot assess its

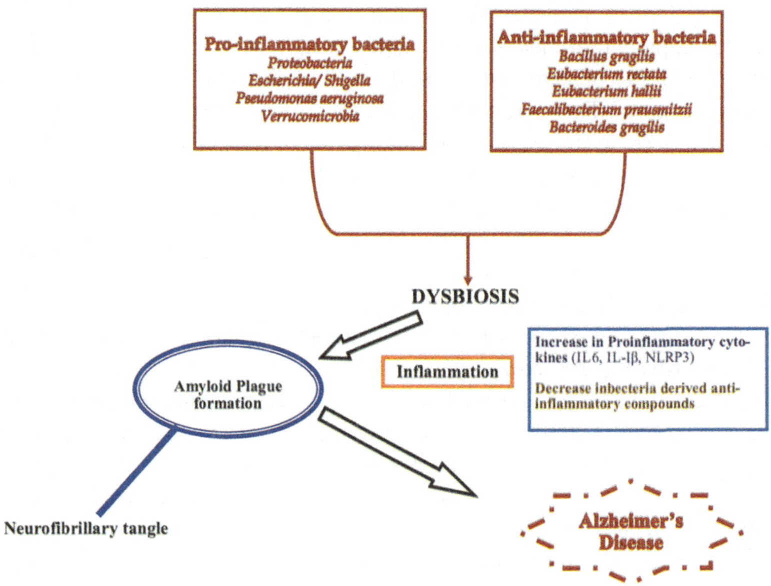

Fig. 11.4 Links between gut microbiota and dementia

significance comprehensively to identify the other bacteria due to lack of effectiveness of the methods. It is speculated that there is a usual microvascular arteriosclerotic and the inflammatory mechanism for cardiovascular and cerebrovascular diseases. For dementia, this mechanism can be accelerated by making several changes in the gut microbiome.

It has been identified that inflammation and microvascular arteriosclerosis are risk factors for such diseases. The cells of the immune system orchestrate the inflammatory response, while infections in the body are responded by inflammation. These cells are orchestrated by the innate branch and the adaptive branch. In the 1990s, the development and implication of inflammation were first observed in Alzheimer's disease pathology with the neuropathological finding of inflammatory proteins, the neurofibrillary tangles, surrounding the amyloid plaques, and activated inflammatory cells (Hu et al. 2016). On the basis of epidemiological findings, it was found that patients suffering from arthritis also had a lower extent of patients affected with Alzheimer's disease (AD). A similar case has been reported for other patients with a high intake of nonsteroidal inflammatory drugs.

The perception of the brain has inaccurately provided the earliest results of an immune-privileged organ, which is not an elicited inflammation for responding to damage or antigens. On the other hand, the presence of activated microglial cells and acute-phase proteins in Aβ plaques stain for components of the complement system in brain tissue of AD patients and inflammatory cytokines. The etiology of AD can be identified through inflammation-related risk factors, which can lead to new strategies to tackle the disease. Studies have diversified to explore for cognitive function at an extent of inflammation-associated risk factors for cognitive decline, dementia, progression in dementia, and AD, since the initial discovery of a potential inflammatory ingredient to the AD combination (Shoemark and Allen 2015).

The development of the gut-associated lymphoid system is affected by the microbiota. Seventy percent of the circulating lymphocytes in the body are located within the intestines. There are several lines of immune cells, which are essential for the host response to microbiota, including dendritic cells, myofibroblasts, and macrophages in the lamina propria. A significant increase in the bacterial addition in the gut is shown by surface and circulating immunoglobulin and gut lymphoid tissue (Moos et al. 2016). Pioneering species in the gut link through surface cell receptors of the immune cells of the gut in the early stages of the human life cycle, which include Toll-like receptors and caspase recruitment domain protein for promoting the expression of host genes that create a mucosal and intraluminal environment for favoring their colonization.

NOD-like receptors are another family of membrane-bound receptors that detect proteins. NOD-like receptors are situated in the cytoplasm and are engaged in the monitoring of bacterial pathogen-related molecular patterns. These receptors are specifically significant in tissues as they are expressed at high levels. The epithelium contains specialized cells, which secrete the protective mucus layer and Paneth cells, and restrict the interaction between bacteria and epithelial cells with respect to the intestinal epithelial cells. When exposed to heat-treated bacteria, these mucosal immune responses are lowered, which suggests that such mechanisms entail the metabolic products of bacterial activity and bacterial cell receptor-mediated sensing (Dinan and Cryan 2012). It has been notified that inflammation cannot be induced by gram-negatives and LPS, while other metabolites and cell components can be entailed such as opportunistic pathogenic bacteria and gram-positive pathogenic bacteria for inducing inflammation.

Probiotics are useful due to their pleiotropic signaling abilities and interfering with the extent of cellular pathways adopted to maintain homeostasis. For example, the serum level of tryptophan is improved by oral probiotic therapy, which is a precursor of serotonin. In addition, due to the important effect of probiotics, there is an increase in the number of beneficial intestinal microbiota. Overall, improved endocrine, immune, neural signaling, and carbohydrate, protein, and lipid metabolism are culminated by the effects of these beneficial microbes, which have a conclusive influence in the neuroinflammation attenuation, thus reducing neurodegeneration.

Prebiotics are ingested by the host, which are nondigestible food components, but they confer an extent of health advantages to the host under the fermentative influence of the intestinal microbiota. Prebiotics specifically oligosaccharides are carbohydrate substances. Resistant starch, insulin, lactulose, xylooligosaccharides, galactooligosaccharides, polydextrose, banana, psyllium, wheat dextrin, whole-grain wheat, and whole-grain corn are included as substances with prebiotic effects (Pistollato et al. 2016). Gut health, memory, anxiety, depression, stress, and cognition are improved by prebiotics similar to probiotics. Improvement in brain mitochondrial function, hippocampal plasticity, and a decrease in microglial activation are related to prebiotics. The result is a reduction in neurodegeneration and attenuation of neuroinflammation. The combination of prebiotics and probiotics is synbiotics, which are beneficial for influencing the central nervous system and gut health.

11.7 Factors Impacting the Normal Gut Microbiota

Several factors are responsible for shaping the healthy gut microbiota; the process, however, is significant in all the stages of the life of an individual (Fig. 11.5).

11.7.1 Child's Age

Though it is clear that the formation of microbes begins right after human birth, evidence suggests that the colonization of the gut begins in the uterus (Rooks and Garrett 2016). Studies related to the sequencing of 16S rRNA indicated that first meconium is found in abundance in the common genera of bacteria such as *Enterococcus*, *Lactococcus*, *Escherichia Shigella*, and *Leuconostoc*. Besides, the mode of delivery is further crucial in shaping the first microbiota in individuals. The maternal vagina also contributes to the development of infants' intestine leading to the colonization of various organisms (Dominguez-Bello et al. 2010). In other cases, colonization takes place through maternal skin flora, characterized by the dominance

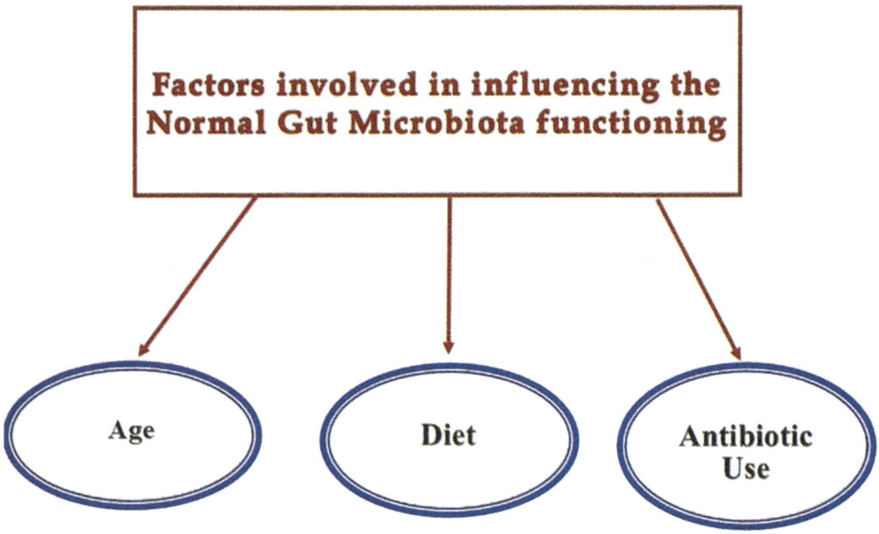

Fig. 11.5 Factors important in influencing the normal gut microbiota

of genera like *Corynebacterium*, *Propionibacterium*, and *Streptococcus* (Biagi et al. 2010; Gosalbes et al. 2013). The initial phase of the development of infant's gut microbiota seems unstable and, therefore, lacks diversity; however, several changes occur as the infant grows and eventually it becomes stable and diverse and indicates 40–60% resemblance with the adult microbiota.

Similarly, the proportions of *Bacteroides* and *Bifidobacterium* are different in adolescents and children. Microbiota is found to be in a stable state between the third and the seventh decade of human life. A significant decrease in the proportion of *Firmicutes* and *Bifidobacteria* is observed with an increase in the populations of *Proteobacterium*, *Staphylococcus*, and *Escherichia coli* (Fig. 11.6).

The following circumstances provide the functional effects in the normal gut flora. These impacts are in the form of the reduced ability of vitamin B12 production, minimum activities in relation to the microbial reductases, immune dysfunctions, and the increased propensity of DNA alterations (Lan et al. 2013).

In the initial stage, gut microbiota is highly affected by the type of feed, i.e., formula or breastfeed; similarly, the temporal allocation is highly affected by the individual lifestyle, dietary patterns, environmental factors, use of antibiotics, and life events. During the preterm phase of the infants, *Lactobacillus* and *Bifidobacterium* are responsible for colonizing the guts of infants. These bacteria are dominant; during breastfeeding, the human milk provides oligosaccharides (HMO) that are easily used by these bacteria. In contrast to this, infants that are fed through the formula feed are likely to have genera like *Enterobacteria*, *Clostridia*, and *Enterococcus*, which dominate the gut microbiota (Groer et al. 2014). Microbiota of the preterm stages is capable of maintaining the gut-associated lymphoid tissue (GALT) and is further involved in the formation of

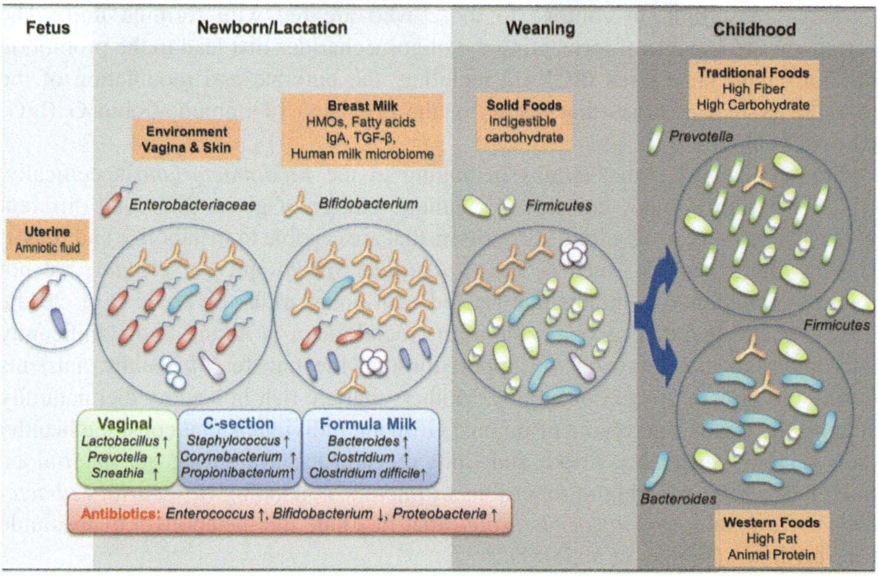

Fig. 11.6 Fetal-to-childhood gut microbiota colonization and important factors affecting this process. The establishment of the gut microbiota may begin in utero and be affected by dynamic shifts in early life. Diversity of the gut microbiota increases with age until it becomes a stable adult microbiota. This process of establishing the gut microbiota is affected by various factors such as delivery mode, methods of milk feeding, the introduction of solid foods, and foods consumed daily in childhood. (Adapted from Ref. Tanaka and Nakayama 2017)

inborn immunity during the development phase. Information suggests that ineffective colonization in the gut microbiota results in the development of several pediatric diseases on account of their poor immune system (Sherman et al. 2015).

11.7.2 Diet

An infant's diet via supplementation with prebiotics or probiotics serves as the primary force of providing a valuable impact over the functioning and colonization of the gut microbiota. This serves as the second crucial factor after delivery (Martin et al. 2010). A significant difference has been suggested by researchers in the formation of the gut microbial when exposed to either breastfeed or formula feeds. Understanding regarding the mode of infant's feed is of substantial value since mothers of the present day are promoting the trend of formula feed. Other than meeting the physiological and nutritional demands of infants, breast milk is highly rich in bioactive compounds that protect against infection (Oddy 2002).

These compounds are highly important in protecting the immune system. Oligosaccharides found in human milk are highly responsive in providing nutrients to infant's colonic bacteria and, therefore, provide an advantage to *Bifidobacterium* (Albenberg and Wu 2014). This phenomenon is very common in infants that are fed through breast milk in contrast to those who are fed with formula milk. The organisms are capable of fermenting the oligosaccharides that lead to the promotion of short-chain fatty acids (SCFAs) including the butyrate and modulation of the immune system of hosts that results in the provision of immunoglobin G (IgG) (Brown et al. 2013).

Studies suggested that strains belonging to the *Bifidobacterium*, specifically, Bifidobacterium *longus*, consists of a unique cluster of genes that code different glycosidases and carbohydrate transporters that are capable to metabolize and import HMOs. Other bacteria, such as the *Clostridium specie* and *Bacteroides*, are not found in abundance in infants that are fed with breast milk (Zivkovic et al. 2011). This indicates that diet serves as the most crucial factor in infants and significantly contributes to the provision of support through the transfer of healthy nutrients during the growth process of infants. Mothers that are rich in healthy diet naturally support the strong microbiota of infants. Growing individuals that consume healthy diets such as vegetables, fibers, and fruits are said to have strong gut microbiota. Such individuals have high proportions of microbes including *Roseburia*, *Eubacterium rectale*, and *Ruminococcus bromii* that facilitate the metabolism of insoluble carbohydrates (Walker et al. 2011).

11.7.3 Antibiotics

Previously, knowledge related to the antibiotics was only limited to their bacteriostatic and bactericidal effects against pathogenic microbes. However, in the modern era, knowledge about the impact of antibiotics on the ecology of the gut microbiota has also become a focus of interest of the researchers. Individuals who prefer regular intake of antibiotics usually experience short- and long-term impacts on the functional behavior of normal gut microbiota. Further, the intestinal flora is capable of affecting the activity of the brain and causes its dysfunctions (Angelucci et al. 2019).

It is evident that human body contains several drug-resistant bacterial genes that lived in the gut way before the discovery of antibiotics. This fact suggests that it is not only antibiotics that negatively affect the growth-related properties, but also other smaller molecules that are highly effective in reducing the growth metabolism in infants (Jakobsson et al. 2010). The factors may be associated with the imbalance within the commensal microbiota that may further impact the growth and formation of the resistance genes (Ng et al. 2013).

Propositions regarding the adverse effect of antibiotics have been discussed several years ago. One of the major effects is associated with the development of the *Salmonella* infection. One of the major impacts of the given mechanism is in the form of the loss of interaction within the species of microbiota, which is responsible for the increase in the formation of the sialic acid. This acid is important in promoting the growth of pathogens including *Clostridium difficile* and *Salmonella typhimurium*. Major changes associated with the gut microbiota are in the form of colonization by foreign microbes (Vogt and Finlay 2017) (Table 11.1).

Table 11.1 The gut bacteria and their metabolites on the nervous system (adapted from Ref. Alkasir et al. 2017)

Gut microbiota	Metabolites product	Effects on the nervous system function
Lactobacillus, *Bifidobacterium*	Gamma-aminobutyric acid (GABA)	Inhibitory neurotransmitter, metabolic disorders can lead to anxiety and depression
Streptococcus, *Escherichia*, *Enterococci*, *Enterococcus*, *Lactococcus*, *Lactobacillus*	Serotonin	Neurotransmitters, regulate emotions
Bacillus	Norepinephrine	Neurotransmitters involved in motor, cognitive, memory, emotion, and other central nervous and endocrine control
Lactobacillus, *Bacillus*	Acetylcholine	Acting on neurotransmitters in the central and peripheral nervous systems, and cognitive function, particularly closely related to learning and memory
Lactobacillus, *Lactococcus*, *Streptococcus*, *Enterococcus*	Histamine	Regulating neurotransmitter; sleep and cognitive function related
Clostridium, *C. sporogenes*	Indole-3-propionic acid (IPA)	Antioxidants, protect neurons
Bacteroides, *Bifidobacterium*, *Propionibacterium*, *Eubacterium*, *Lactobacillus*, *Clostridium*, *Roseburia*, *Prevotella*	Short-chain fatty acids (SCFA)	Carbohydrates (starch, cellulose, etc.), the main products of fermentation, to provide energy for the host, regulate endothelial cell function, promote the synthesis and secretion of neurotransmitters and hormones, reduce inflammation
Blue-green algae (*Cyanobacteria*)	Beta-*N*-methylamino-L-alanine (BMAA)	Neurotoxicity, neuronal damage, and misfolded proteins related
Gram-negative bacteria	Endotoxin	Induced inflammation, release large amounts of inflammatory cytokines (TNF-α, IL-6, and IL-8), obesity, insulin resistance, and diabetes and is closely related to the occurrence of AD
Escherichia, *Bacillus*, *Lactococcus*, *Lactobacillus*, *Streptococcus*	Dopamine	System activity, Parkinson's disease, AD, and depression-related
Spore-forming microbes, *Candida*, *Streptococcus*, *Enterococcus* spp.	Promote 5-HT biosynthesis	Increase the motility of the gut

Table 11.2 Effects of antibiotics in AD rodent models and humans (adapted from Ref. Angelucci et al. 2019)

Antibiotic	Species	Target	Effects
Streptozotocin	• Mice,	• Gram-positive bacteria,	• Memory deficits
	• Rats	• Pancreatic islet cells	
Ampicillin	• Rats	• Gram-positive bacteria,	• Increased serum corticosterone
		• Gram-negative bacteria	• Increased anxiety
			• Memory deficits
Cefepime	• Humans	• Gram-positive bacteria,	• Reduced consciousness, myoclonus, confusion
		• Gram-negative bacteria	
Amoxicillin	• Humans	• Gram-positive bacteria	• Improved cognition
Rifampicin	• Humans,	• Bacterial DNA-dependent, RNA synthesis	• Anti-cholinesterase
	• Rats,		• Antioxidative
	• Mice		• Anti-inflammatory
			• Reduced Aβ
Minocycline	• Mice,	• Gram-positive bacteria,	• Reduced inflammation and microglial activation
	• Rats	• Gram-negative bacteria	• Improved cognition
			• Reduced Aβ
Rapamycin	• Mice,	• Antifungal,	• Improved cognition
	• Rats	• Immunosuppressant,	• Reduced tau
		• mTOR inhibitors	• Reduced Aβ
			• Reduced microglial activation
D-Cycloserine	• Humans,	• Gram-positive bacteria,	• Improved cognition
	• Rats	• Gram-negative bacteria,	
		• NMDA receptor partial agonist	
Doxycycline	• Humans,	• Gram-positive bacteria,	• Improved cognition
	• Mice	• Gram-negative bacteria	• Reduced inflammation

Studies related to the adverse effect of antibiotics suggested that the effects of the short-term (intake of antibiotics for 7 days) antibiotic treatment are powerful enough to last for up to 2 years. Illustrative table summarizes the role of the effects of antibiotics in AD rodent models and humans (Table 11.2). This impedes the recovery of *Bacteroides* (Panda et al. 2014). Likewise, a short-term intake of antibiotics in combination with the clarithromycin induced with triple therapy for the removal of *H. pylori* results in the decrease in changes associated with the *Actino* bacteria along with the increase in folds in the Erm B resistance. The conditions remain similar among several patients whose ages were up to 4 years, while other patients experienced a quick recovery.

Other common antibiotics including ciprofloxacin and beta-lactams provided significant signs of reduction in the microbial diversity (up to 25%). Besides, a significant increase has been detected in the population of *Bacteroidetes*. The results were identified among individuals that have undergone the short-term course of the given drugs. Horizontal gene transfer, in the human intestine, is the most widely known phenomenon associated with the use of long-term antibiotics (Huddleston 2014). The phenomenon is highly useful and is found to be competent with the microbial diversity of the normal gut. Besides, conjugation, natural transformation, and phage transduction are some common mechanisms that are widely used in the process of transferring genetic information into different species. The process of gene transfer may also take place through integrin and transposons. Researchers suggested that within several environmental settings, human gut microbiota is more similar to the horizontal gene transfer (Smillie et al. 2011). This leads to the formation of the reservoir conditions of the resistance genes and, therefore, directs us to maintain the maximum care when using antibiotics for longer durations.

11.8 Recent Developments and Future Perspectives

Since the last three decades, remarkable progress is made by the researchers to understand the disease by recruiting some of the best scientists worldwide. It is possible to delay the onset of AD symptoms; however, there is lack of researches related to this topic because of inadequate funding of research, high cost of clinical studies, better models, antiquated administrative structure of discovery programs, lack of suitable infrastructure, and arcane decision-making systems for selecting and funding innovative ideas. At present, examination of autopsy of their brain tissue after death is the only way to diagnose AD. This highlights the need of understanding the process of neurodegenerative disorders throughout the individual's life. Molecular ecology and metagenomics have played an important role in increasing knowledge of the genetic diversity encoded by microbial life.

Recently, the major focus is on the impact of gut microbiota on brain health by focusing on multiple interrelated systems and organs that involve neural signaling of microbial products and transmitter molecules of the intestinal residents and neurohumoral cells. The transmitter molecules are responsible for maintaining the integrity of the blood–brain barrier. This is the reason gut microbiota-based therapy is considered as a promising potential approach for future therapies of brain diseases. Moreover, there is an increasing focus of scientists toward the application of the microbiota in the diagnosis, treatment, and prognosis of different brain diseases. Therefore, further investigations are required for the manipulation of the gut environment as a preventive and therapeutic tool for providing a good safety profile.

11.9 Conclusion

Dementia affects the daily functioning of an individual, and AD is the most common type of dementia making most of the cases. Recently, it is indicated that gastrointestinal tract microbiota is directly linked to dementia pathogenesis as it triggers metabolic diseases and low-grade inflammation progress. A novel strategy is proposed through the modulation of the microbiota (with the use of probiotics) to manage these disorders and as an adjuvant for psychiatric treatment of dementia and other related diseases. An important role is played by gut microbiota in the postnatal development of endocrine, immune, and neural systems that underpin the signaling of the central nervous system. It is important to understand gut microbiota related to inflammation and metabolic diseases as it is directed associated with AD pathogenesis. Further, the novel vision into the complex biology of AD is enabled through the comparative analysis of gut microbiota, which plays an important role in taking preventive measures such as early diagnosis, identifying new therapeutic targets, and developing novel drugs. Therefore, direct targeting of gut microbiota enzymes through pharmacological inhibitors or activators, or modulation of gut microbiota through probiotics or other dietary intervention holds promise for AD and associated diseases.

References

Ahmed RM, Irish M, Piguet O, Halliday GM, Ittner LM, Farooqi S, Kiernan MC (2016) Amyotrophic lateral sclerosis and frontotemporal dementia: distinct and overlapping changes in eating behaviour and metabolism. Lancet Neurol 15(3):332–342. https://doi.org/10.1016/s1474-4422(15)00380-4

Albenberg LG, Wu GD (2014) Diet and the intestinal microbiome: associations, functions, and implications for health and disease. Gastroenterology 146(6):1564–1572. https://doi.org/10.1053/j.gastro.2014.01.058

Alkasir R, Li J, Li X et al (2017) Human gut microbiota: the links with dementia development. Prot Cell 8:90–102. https://doi.org/10.1007/s13238-016-0338-6

Andersson AF, Lindberg M, Jakobsson H, Bäckhed F, Nyrén P, Engstrand L (2008) Comparative analysis of human gut microbiota by barcoded pyrosequencing. PLoS One 3(7):e2836. https://doi.org/10.1371/journal.pone.0002836

Angelucci F, Cechova K, Amlerova J et al (2019) Antibiotics, gut microbiota, and Alzheimer's disease. J Neuroinflammation 16(1):108. https://doi.org/10.1186/s12974-019-1494-4

Atherton N, Bridle C, Brown D, Collins H, Dosanjh S, Griffiths F, McShane R (2016) Dementia and Physical Activity (DAPA)-an exercise intervention to improve cognition in people with mild to moderate dementia: study protocol for a randomized controlled trial. Trials 17(1):1–12. https://doi.org/10.1186/s13063-016-1288-2

Baumgart M, Snyder HM, Carrillo MC, Fazio S, Kim H, Johns H (2015) Summary of the evidence on modifiable risk factors for cognitive decline and dementia: a population-based perspective. Alzheimers Dement 11(6):718–726. https://doi.org/10.1016/j.jalz.2015.05.016

Bercik P, Collins SM, Verdu EF (2012) Microbes and the gut-brain axis. Neurogastroenterol Motil 24(5):405–413. https://doi.org/10.1111/j.1365-2982.2012.01906.x

Biagi E, Nylund L, Candela M, Ostan R, Bucci L, Pini E, Nikkïla J, Monti D, Satokari R, Franceschi C, Brigidi P (2010) Through ageing, and beyond: gut microbiota and inflammatory

status in seniors and centenarians. PLoS One 5(5):e10667. https://doi.org/10.1371/journal.pone.0010667
Booker A, Jacob LE, Rapp M, Bohlken J, Kostev K (2016) Risk factors for dementia diagnosis in German primary care practices. Int Psychogeriatr 28(7):1059–1065. https://doi.org/10.1017/s1041610215002082
Borre YE, O'Keeffe GW, Clarke G, Stanton C, Dinan TG, Cryan JF (2014) Microbiota and neurodevelopmental windows: implications for brain disorders. Trends Mol Med 20(9): 509–518. https://doi.org/10.1016/j.molmed.2014.05.002
Brasure M, Desai P, Davila H, Nelson VA, Calvert C, Jutkowitz E, McCarten JR (2018) Physical activity interventions in preventing cognitive decline and Alzheimer-type dementia: a systematic review. Ann Intern Med 168(1):30–38. https://doi.org/10.7326/m17-1528
Briones TL (2006) Environment, physical activity, and neurogenesis: implications for prevention and treatment of Alzhemier's disease. Curr Alzheimer Res 3(1):49–54. https://doi.org/10.2174/156720506775697197
Brown EM, Sadarangani M, Finlay BB (2013) The role of the immune system in governing host-microbe interactions in the intestine. Nat Immunol 14(7):660. https://doi.org/10.1038/ni.2611
Brown D, Spanjers K, Atherton N, Lowe J, Stonehewer L, Bridle C, Lamb SE (2015) Development of an exercise intervention to improve cognition in people with mild to moderate dementia: Dementia and Physical Activity (DAPA) Trial, registration ISRCTN32612072. Physiotherapy 101(2):126–134. https://doi.org/10.1016/j.physio.2015.01.002
Cerejeira J, Lagarto L, Mukaetova-Ladinska E (2012) Behavioral and psychological symptoms of dementia. Front Neurol 3:73. https://doi.org/10.3389/fneur.2012.00073
Chornenkyy Y, Wang WX, Wei A, Nelson PT (2019) Alzheimer's disease and type 2 diabetes mellitus are distinct diseases with potential overlapping metabolic dysfunction upstream of observed cognitive decline. Brain Pathol 29(1):3–17. https://doi.org/10.1111/bpa.12655
Claesson MJ, Jeffery IB, Conde S, Power SE, O'connor EM, Cusack S, Harris HM, Coakley M, Lakshminarayanan B, O'Sullivan O, Fitzgerald GF (2012) Gut microbiota composition correlates with diet and health in the elderly. Nature 488(7410):178. https://doi.org/10.1038/nature11319
Clarke G, Grenham S, Scully P, Fitzgerald P, Moloney RD, Shanahan F, Cryan JF (2013) The microbiome-gut-brain axis during early life regulates the hippocampal serotonergic system in a sex-dependent manner. Mol Psychiatry 18(6):666–673. https://doi.org/10.1038/mp.2012.77
Dinan TG, Cryan JF (2012) Regulation of the stress response by the gut microbiota: implications for psychoneuroendocrinology. Psychoneuroendocrinology 37(9):1369–1378. https://doi.org/10.1016/j.psyneuen.2012.03.007
Dominguez-Bello MG, Costello EK, Contreras M, Magris M, Hidalgo G, Fierer N, Knight R (2010) Delivery mode shapes the acquisition and structure of the initial microbiota across multiple body habitats in newborns. Proc Natl Acad Sci 107(26):11971–11975. https://doi.org/10.1073/pnas.1002601107
Dumas ME, Barton RH, Toye A, Cloarec O, Blancher C, Rothwell A, Fearnside J, Tatoud R, Blanc V, Lindon JC, Mitchell SC (2006) Metabolic profiling reveals a contribution of gut microbiota to fatty liver phenotype in insulin-resistant mice. Proc Natl Acad Sci 103(33): 12511–12516. https://doi.org/10.1073/pnas.0601056103
Durazzo TC, Mattsson N, Weiner MW, Alzheimer's Disease Neuroimaging Initiative (2014) Smoking and increased Alzheimer's disease risk: a review of potential mechanisms. Alzheimers Dement 10:S122–S145
Eckel RH (1989) Lipoprotein lipase. N Engl J Med 320(16):1060–1068
Fan C, Song Q, Wang P, Li Y, Yang M, Yu SY (2018) Neuroprotective effects of ginsenoside-Rg1 against depression-like behaviors via suppressing glial activation, synaptic deficits, and neuronal apoptosis in rats. Front Immunol 9:2889. https://doi.org/10.3389/fimmu.2018.02889
Faust K, Sathirapongsasuti JF, Izard J, Segata N, Gevers D, Raes J, Huttenhower C (2012) Microbial co-occurrence relationships in the human microbiome. PLoS Comput Biol 8(7): e1002606. https://doi.org/10.1371/journal.pcbi.1002606

Fauth EB, Gibbons A (2014) Which behavioral and psychological symptoms of dementia are the most problematic? Variability by prevalence, intensity, distress ratings, and associations with caregiver depressive symptoms. Int J Geriatr Psychiatry 29(3):263–271. https://doi.org/10.1002/gps.4002

Festi D, Schiumerini R, Eusebi LH, Marasco G, Taddia M, Colecchia A (2014) Gut microbiota and metabolic syndrome. World J Gastroenterol 20(43):16079. https://doi.org/10.3748/wjg.v20.i43.16079

Floch MH (2010) Fecal bacteriotherapy, fecal transplant, and the microbiome. J Clin Gastroenterol 44(8):529–530. https://doi.org/10.1097/mcg.0b013e3181e1d6e2

Forbes D, Forbes SC, Blake CM, Thiessen EJ, Forbes S (2015) Exercise programs for people with dementia. Cochrane Database Syst Rev (4). https://doi.org/10.1002/14651858.cd006489.pub4

Franceschi C (2007) Inflammaging as a major characteristic of old people: can it be prevented or cured? Nutr Rev 65(Suppl 3):S173–S176. https://doi.org/10.1111/j.1753-4887.2007.tb00358.x

Frank DN, Amand ALS, Feldman RA, Boedeker EC, Harpaz N, Pace NR (2007) Molecular-phylogenetic characterization of microbial community imbalances in human inflammatory bowel diseases. Proc Natl Acad Sci 104(34):13780–13785. https://doi.org/10.1073/pnas.0706625104

Ganguli M (2015) Cancer and dementia: it's complicated. Alzheimer Dis Assoc Disord 29(2):177. https://doi.org/10.1097/wad.0000000000000086

Gawlik-Kotelnicka O, Polguj M (2018) Can microbiology affect psychiatry? A link between gut microbiota and psychiatric disorders. Psychiatr Pol 52(6):1023–1039

Gibson G, Newton L, Pritchard G, Finch T, Brittain K, Robinson L (2016) The provision of assistive technology products and services for people with dementia in the United Kingdom. Dementia 15(4):681–701. https://doi.org/10.1177/1471301214532643

Golub JS, Luchsinger JA, Manly JJ, Stern Y, Mayeux R, Schupf N (2017) Observed hearing loss and incident dementia in a multiethnic cohort. J Am Geriatr Soc 65(8):1691–1697. https://doi.org/10.1111/jgs.14848

Gosalbes MJ, Llop S, Valles Y, Moya A, Ballester F, Francino MP (2013) Meconium microbiota types dominated by lactic acid or enteric bacteria are differentially associated with maternal eczema and respiratory problems in infants. Clin Exp Allergy 43(2):198–211. https://doi.org/10.1111/cea.12063

Gow AJ, Johnson W, Pattie A, Whiteman MC, Starr J, Deary IJ (2008) Mental ability in childhood and cognitive aging. Gerontology 54(3):177–186. https://doi.org/10.1159/000118098

Groer MW, Luciano AA, Dishaw LJ, Ashmeade TL, Miller E, Gilbert JA (2014) Development of the preterm infant gut microbiome: a research priority. Microbiome 2(1):38. https://doi.org/10.1186/2049-2618-2-38

Halpern A, Mancini MC, Magalhães MEC, Fisberg M, Radominski R, Bertolami MC, Nery M (2010) Metabolic syndrome, dyslipidemia, hypertension and type 2 diabetes in youth: from diagnosis to treatment. Diabetol Metab Syndr 2(1):55. https://doi.org/10.1186/1758-5996-2-55

Hardy J, Selkoe DJ (2002) The amyloid hypothesis of Alzheimer's disease: progress and problems on the road to therapeutics. Science 297(5580):353–356

Hazen SL, Smith JD (2012) An antiatherosclerotic signaling cascade involving intestinal microbiota, microRNA-10b, and ABCA1/ABCG1-mediated reverse cholesterol transport. Circ Res 111:948. https://doi.org/10.1161/circresaha.112.277277

Hersch EC, Falzgraf S (2007) Management of the behavioral and psychological symptoms of dementia. Clin Interv Aging 2(4):611

Hollister EB, Gao C, Versalovic J (2014) Compositional and functional features of the gastrointestinal microbiome and their effects on human health. Gastroenterology 146(6):1449–1458. https://doi.org/10.1053/j.gastro.2014.01.052

Hu X, Wang T, Jin F (2016) Alzheimer's disease and gut microbiota. Sci China Life Sci 59(10):1006–1023. https://doi.org/10.1007/s11427-016-5083-9

Huddleston JR (2014) Horizontal gene transfer in the human gastrointestinal tract: potential spread of antibiotic resistance genes. Infect Drug Resist 7:167. https://doi.org/10.2147/idr.s48820

Ignatova V (2019) Influence of gut microbiota on behavior and its disturbances. In: Behavioral neuroscience. IntechOpen, London. https://doi.org/10.5772/intechopen.85317
Jakobsson HE, Jernberg C, Andersson AF, Sjölund-Karlsson M, Jansson JK, Engstrand L (2010) Short-term antibiotic treatment has differing long-term impacts on the human throat and gut microbiome. PLoS One 5(3):e9836. https://doi.org/10.1371/journal.pone.0009836
Jiang C, Li G, Huang P, Liu Z, Zhao B (2017) The gut microbiota and Alzheimer's disease. J Alzheimers Dis 58(1):1–15. https://doi.org/10.3233/jad-161141
Jin Y, Wu S, Zeng Z, Fu Z (2017) Effects of environmental pollutants on gut microbiota. Environ Pollut 222:1–9. https://doi.org/10.1016/j.envpol.2016.11.045
Jones L, Candy B, Davis S, Elliott M, Gola A, Harrington J, Kupeli N, Lord K, Moore K, Scott S, Vickerstaff V (2016) Development of a model for integrated care at the end of life in advanced dementia: a whole systems UK-wide approach. Palliat Med 30(3):279–295. https://doi.org/10.1177/0269216315605447
Kales HC, Gitlin LN, Lyketsos CG (2015) Assessment and management of behavioral and psychological symptoms of dementia. BMJ 350:h369. https://doi.org/10.1136/bmj.h369
Kenigsberg PA, Aquino JP, Berard A, Gzil F, Andrieu S, Banerjee S, Platel H (2016) Dementia beyond 2025: knowledge and uncertainties. Dementia 15(1):6–21. https://doi.org/10.1177/1471301215574785
Kwon SE, Chapman ER (2011) Synaptophysin regulates the kinetics of synaptic vesicle endocytosis in central neurons. Neuron 70(5):847–854. https://doi.org/10.1016/j.neuron.2011.04.001
de La Serre CB, Ellis CL, Lee J, Hartman AL, Rutledge JC, Raybould HE (2010) Propensity to high-fat diet-induced obesity in rats is associated with changes in the gut microbiota and gut inflammation. Am J Physiol Gastrointest Liv Physiol 299(2):G440–G448. https://doi.org/10.1152/ajpgi.00098.2010
Lamb SE, Sheehan B, Atherton N, Nichols V, Collins H, Mistry D, Lall R (2018) Dementia and Physical Activity (DAPA) trial of moderate to high intensity exercise training for people with dementia: randomised controlled trial. BMJ 361:k1675. https://doi.org/10.1136/bmj.k1675
Lan Y, Kriete A, Rosen GL (2013) Selecting age-related functional characteristics in the human gut microbiome. Microbiome 1(1):2. https://doi.org/10.1186/2049-2618-1-2
Le Chatelier E, Nielsen T, Qin J, Prifti E, Hildebrand F, Falony G, Leonard P (2013) Richness of human gut microbiome correlates with metabolic markers. Nature 500(7464):541–546. https://doi.org/10.1038/nature12506
Liew TM, Feng L, Gao Q, Ng TP, Yap P (2015) Diagnostic utility of Montreal cognitive assessment in the fifth edition of diagnostic and statistical manual of mental disorders: major and mild neurocognitive disorders. J Am Med Dir Assoc 16(2):144–148. https://doi.org/10.1016/j.jamda.2014.07.021
Livingston G, Sommerlad A, Orgeta V, Costafreda SG, Huntley J, Ames D, Cooper C (2017) Dementia prevention, intervention, and care. Lancet 390(10113):2673–2734. https://doi.org/10.1016/s0140-6736(17)31363-6
Luczynski P, Whelan SO, O'Sullivan C, Clarke G, Shanahan F, Dinan TG, Cryan JF (2016) Adult microbiota-deficient mice have distinct dendritic morphological changes: differential effects in the amygdala and hippocampus. Eur J Neurosci 44(9):2654–2666. https://doi.org/10.1111/ejn.13291
Lund DA, Pett MA, Caserta MS (1987) Institutionalizing dementia victims: some caregiver considerations. J Gerontol Soc Work 11(1–2):119–135. https://doi.org/10.1300/j083v11n01_10
Marchesi JR, Adams DH, Fava F, Hermes GD, Hirschfield GM, Hold G, Quraishi MN, Kinross J, Smidt H, Tuohy KM, Thomas LV (2016) The gut microbiota and host health: a new clinical frontier. Gut 65(2):330–339. https://doi.org/10.1136/gutjnl-2015-309990
Marim CM, Silva V, Taminato M, Barbosa DA (2013) Effectiveness of educational programs on reducing the burden of caregivers of elderly individuals with dementia: a systematic review. Rev Lat Am Enfermagem 21(SPE):267–275. https://doi.org/10.1590/s0104-11692013000700033

Martin R, Nauta A, Ben Amor K, Knippels L, Knol J, Garssen J (2010) Early life: gut microbiota and immune development in infancy. Benefic Microbes 1(4):367–382. https://doi.org/10.3920/bm2010.0027

Martins IJ, Hone E, Foster JK, Sünram-Lea SI, Gnjec A, Fuller SJ, Nolan D, Gandy SE, Martins RN (2006) Apolipoprotein E, cholesterol metabolism, diabetes, and the convergence of risk factors for Alzheimer's disease and cardiovascular disease. Mol Psychiatry 11(8):721. https://doi.org/10.1038/sj.mp.4001854

Matricon J, Meleine M, Gelot A, Piche T, Dapoigny M, Muller E, Ardid D (2012) Review article: associations between immune activation, intestinal permeability and the irritable bowel syndrome. Aliment Pharmacol Ther 36(11–12):1009–1031. https://doi.org/10.1111/apt.12080

Matsumoto N, Ikeda M, Fukuhara R, Shinagawa S, Ishikawa T, Mori T, Toyota Y, Matsumoto T, Adachi H, Hirono N, Tanabe H (2007) Caregiver burden associated with behavioral and psychological symptoms of dementia in elderly people in the local community. Dement Geriatr Cogn Disord 23(4):219–224. https://doi.org/10.1159/000099472

Matthews FE, Stephan BC, Robinson L, Jagger C, Barnes LE, Arthur A, Brayne C (2016) Cognitive Function and Ageing Studies (CFAS) Collaboration. A two-decade dementia incidence comparison from the Cognitive Function and Ageing Studies I and II. Nat Commun 7:11398. https://doi.org/10.1038/ncomms11398

McCabe M, You E, Tatangelo G (2016) Hearing their voice: a systematic review of dementia family caregivers' needs. The Gerontologist 56(5):e70–e88. https://doi.org/10.1093/geront/gnw078

McKeith I, Cummings J (2005) Behavioural changes and psychological symptoms in dementia disorders. Lancet Neurol 4(11):735–742. https://doi.org/10.1016/s1474-4422(05)70219-2

Meyer JS, Xu G, Thornby J, Chowdhury MH, Quach M (2002) Is mild cognitive impairment prodromal for vascular dementia like Alzheimer's disease? Stroke 33(8):1981–1985. https://doi.org/10.1161/01.str.0000024432.34557.10

Miyamoto Y, Tachimori H, Ito H (2010) Formal caregiver burden in dementia: impact of behavioral and psychological symptoms of dementia and activities of daily living. Geriatr Nurs 31(4):246–253. https://doi.org/10.1016/j.gerinurse.2010.01.002

Moos WH, Faller DV, Harpp DN, Kanara I, Pernokas J, Powers WR, Steliou K (2016) Microbiota and neurological disorders: a gut feeling. BioRes Open Access 5(1):137–145. https://doi.org/10.1089/biores.2016.0010

Mukherjee A, Biswas A, Roy A, Biswas S, Gangopadhyay G, Das SK (2017) Behavioural and psychological symptoms of dementia: correlates and impact on caregiver distress. Dement Geriatr Cogn Disord Extra 7(3):354–365. https://doi.org/10.1159/000481568

Neufeld KM, Kang N, Bienenstock J, Foster JA (2011) Reduced anxiety-like behavior and central neurochemical change in germ-free mice. Neurogastroenterol Motil 23(3):e255–e119. https://doi.org/10.1111/j.1365-2982.2010.01620.x

Ng KM, Ferreyra JA, Higginbottom SK, Lynch JB, Kashyap PC, Gopinath S, Naidu N, Choudhury B, Weimer BC, Monack DM, Sonnenburg JL (2013) Microbiota-liberated host sugars facilitate post-antibiotic expansion of enteric pathogens. Nature 502(7469):96. https://doi.org/10.1038/nature12503

Nigam Y, Knight J, Williams N (2019) Gastrointestinal tract 5: the anatomy and functions of the large intestine. Nurs Times 115(10):50–53

O'Hara AM, Shanahan F (2006) The gut flora as a forgotten organ. EMBO Rep 7(7):688–693. https://doi.org/10.1038/sj.embor.7400731

Oddy WH (2002) The impact of breastmilk on infant and child health. Breastfeed Rev 10(3):5. https://doi.org/10.1016/s0378-3782(07)70044-0

Orgeta V, Mukadam N, Sommerlad A, Livingston G (2019) The lancet commission on dementia prevention, intervention, and care: a call for action. Ir J Psychol Med 36(2):85–88. https://doi.org/10.1017/ipm.2018.4

Panda S, Casellas F, Vivancos JL, Cors MG, Santiago A, Cuenca S, Guarner F, Manichanh C (2014) Short-term effect of antibiotics on human gut microbiota. PLoS One 9(4):e95476. https://doi.org/10.1371/journal.pone.0095476

Parashar A, Udayabanu M (2017) Gut microbiota: implications in Parkinson's disease. Parkinsonism Relat Disord 38:1–7. https://doi.org/10.1016/j.parkreldis.2017.02.002
Park H, Poo MM (2013) Neurotrophin regulation of neural circuit development and function. Nat Rev Neurosci 14(1):7–23. https://doi.org/10.1038/nrn3379
Pistollato F, Sumalla Cano S, Elio I, Masias Vergara M, Giampieri F, Battino M (2016) Role of gut microbiota and nutrients in amyloid formation and pathogenesis of Alzheimer disease. Nutr Rev 74(10):624–634. https://doi.org/10.1093/nutrit/nuw023
Prince M, Ali GC, Guerchet M, Prina AM, Albanese E, Wu YT (2016) Recent global trends in the prevalence and incidence of dementia, and survival with dementia. Alzheimers Res Ther 8(1): 23. https://doi.org/10.1186/s13195-016-0188-8
Rabot S, Membrez M, Bruneau A, Gérard P, Harach T, Moser M, Raymond F, Mansourian R, Chou CJ (2010) Germ-free C57BL/6J mice are resistant to high-fat-diet-induced insulin resistance and have altered cholesterol metabolism. FASEB J 24(12):4948–4959. https://doi.org/10.1096/fj.10-164921
Ramakrishna BS (2007) The normal bacterial flora of the human intestine and its regulation. J Clin Gastroenterol 41:S2–S6. https://doi.org/10.1097/mcg.0b013e31802fba68
Rani V, Deshmukh R, Jaswal P, Kumar P, Bariwal J (2016) Alzheimer's disease: is this a brain specific diabetic condition? Physiol Behav 164:259–267. https://doi.org/10.1016/j.physbeh.2016.05.041
Rieder R, Wisniewski PJ, Alderman BL, Campbell SC (2017) Microbes and mental health: a review. Brain Behav Immun 66:9–17. https://doi.org/10.1016/j.bbi.2017.01.016
Rinninella E, Raoul P, Cintoni M, Franceschi F, Miggiano GAD, Gasbarrini A, Mele MC (2019) What is the healthy gut microbiota composition? A changing ecosystem across age, environment, diet, and diseases. Microorganisms 7(1):14. https://doi.org/10.3390/microorganisms7010014
Román GC (2002) Vascular dementia may be the most common form of dementia in the elderly. J Neurol Sci 203:7–10. https://doi.org/10.1016/s0022-510x(02)00252-6
Rooks MG, Garrett WS (2016) Gut microbiota, metabolites and host immunity. Nat Rev Immunol 16(6):341. https://doi.org/10.1038/nri.2016.42
Sandhir R, Gupta S (2015) Molecular and biochemical trajectories from diabetes to Alzheimer's disease: a critical appraisal. World J Diabetes 6(12):1223. https://doi.org/10.4239/wjd.v6.i12.1223
Segata N, Haake SK, Mannon P, Lemon KP, Waldron L, Gevers D, Huttenhower C, Izard J (2012) Composition of the adult digestive tract bacterial microbiome based on seven mouth surfaces, tonsils, throat and stool samples. Genome Biol 13(6):R42. https://doi.org/10.1186/gb-2012-13-6-r42
Sherman MP, Zaghouani H, Niklas V (2015) Gut microbiota, the immune system, and diet influence the neonatal gut–brain axis. Pediatr Res 77(1–2):127. https://doi.org/10.1038/pr.2014.161
Shoemark DK, Allen SJ (2015) The microbiome and disease: reviewing the links between the oral microbiome, aging, and Alzheimer's disease. J Alzheimers Dis 43(3):725–738. https://doi.org/10.3233/jad-141170
Skoog I, Börjesson-Hanson A, Kern S, Johansson L, Falk H, Sigström R, Östling S (2017) Decreasing prevalence of dementia in 85-year olds examined 22 years apart: the influence of education and stroke. Sci Rep 7(1):1–8. https://doi.org/10.1038/s41598-017-05022-8
Smillie CS, Smith MB, Friedman J, Cordero OX, David LA, Alm EJ (2011) Ecology drives a global network of gene exchange connecting the human microbiome. Nature 480(7376):241. https://doi.org/10.1038/nature10571
Sowell ER, Peterson BS, Thompson PM, Welcome SE, Henkenius AL, Toga AW (2003) Mapping cortical change across the human life span. Nat Neurosci 6(3):309. https://doi.org/10.1038/nn1008

Stahl SM (2017) Does treating hearing loss prevent or slow the progress of dementia? Hearing is not all in the ears, but who's listening? CNS Spectr 22(3):47–250. https://doi.org/10.1017/s1092852917000268

Stern Y (2012) Cognitive reserve in ageing and Alzheimer's disease. Lancet Neurol 11(11): 1006–1012. https://doi.org/10.1016/s1474-4422(12)70191-6

Sudo N, Chida Y, Aiba Y, Sonoda J, Oyama N, Yu XN, Kuboo C, Koga Y (2004) Postnatal microbial colonization programs the hypothalamic-pituitary-adrenal system for stress response in mice. J Physiol 558(1):847–854. https://doi.org/10.1113/jphysiol.2004.063388

Surr CA, Gates C, Irving D, Oyebode J, Smith SJ, Parveen S, Dennison A (2017) Effective dementia education and training for the health and social care workforce: a systematic review of the literature. Rev Educ Res 87(5):966–1002. https://doi.org/10.3102/0034654317723305

Swidsinski A, Weber J, Loening-Baucke V, Hale LP, Lochs H (2005) Spatial organization and composition of the mucosal flora in patients with inflammatory bowel disease. J Clin Microbiol 43(7):3380–3389. https://doi.org/10.1128/jcm.43.7.3380-3389.2005

Tanaka M, Nakayama J (2017) Development of the gut microbiota in infancy and its impact on health in later life. Allergol Int 66(4):515–522

Tang Q, Jin G, Wang G, Liu T, Liu X, Wang B, Cao H (2020) Current sampling methods for gut microbiota: a call for more precise devices. Front Cell Infect Microbiol 10:151. https://doi.org/10.3389/fcimb.2020.00151

Thomson RS, Auduong P, Miller AT, Gurgel RK (2017) Hearing loss as a risk factor for dementia: a systematic review. Laryngosc Investig Otolaryngol 2(2):9–79. https://doi.org/10.1002/lio2.65

Thursby E, Juge N (2017) Introduction to the human gut microbiota. Biochem J 474(11): 1823–1836. https://doi.org/10.1042/bcj20160510

Tognini P (2017) Gut microbiota: a potential regulator of neurodevelopment. Front Cell Neurosci 11:25. https://doi.org/10.3389/fncel.2017.00025

Tremaroli V, Bäckhed F (2012) Functional interactions between the gut microbiota and host metabolism. Nature 489(7415):242. https://doi.org/10.1038/nature11552

Tremlett H, Bauer KC, Appel-Cresswell S, Finlay BB, Waubant E (2017) The gut microbiome in human neurological disease: a review. Ann Neurol 81(3):369–382. https://doi.org/10.1002/ana.24901

Ueda T, Suzukamo Y, Sato M, Izumi SI (2013) Effects of music therapy on behavioral and psychological symptoms of dementia: a systematic review and meta-analysis. Ageing Res Rev 12(2):28–641. https://doi.org/10.1016/j.arr.2013.02.003

Vogt SL, Finlay BB (2017) Gut microbiota-mediated protection against diarrheal infections. J Trav Med 24(Suppl 1):S39–S43. https://doi.org/10.1093/jtm/taw086

Vogt NM, Kerby RL, Dill-McFarland KA, Harding SJ, Merluzzi AP, Johnson SC, Carlsson CM, Asthana S, Zetterberg H, Blennow K, Bendlin BB (2017) Gut microbiome alterations in Alzheimer's disease. Sci Rep 7(1):13537. https://doi.org/10.1038/s41598-017-13601-y

Walker AW, Ince J, Duncan SH, Webster LM, Holtrop G, Ze X, Brown D, Stares MD, Scott P, Bergerat A, Louis P (2011) Dominant and diet-responsive groups of bacteria within the human colonic microbiota. ISME J 5(2):220. https://doi.org/10.1038/ismej.2010.118

Wang HX, Wang YP (2016) Gut microbiota-brain axis. Chin Med J 129(19):2373. https://doi.org/10.4103/0366-6999.190667

Wang LW, Tancredi DJ, Thomas DW (2011) The prevalence of gastrointestinal problems in children across the United States with autism spectrum disorders from families with multiple affected members. J Dev Behav Pediatr 32(5):351–360. https://doi.org/10.1097/dbp.0b013e31821bd06a

Welcome MO (2018a) History of development of gastrointestinal physiology: from antiquity to modern period and the birth of modern digestive physiology. In: Gastrointestinal physiology. Springer, Cham, pp 1–51. https://doi.org/10.1007/978-3-319-91056-7_1

Welcome MO (2018b) The gut microbiota-brain signaling: behavioral abnormalities of the gut microbiota underlie Alzheimer's disease development and progression. Dictatorship or bidirectional relationship? J Res Med Dent Sci 6(5):246–263

Welcome MO, Mastorakis NE, Pereverzev VA (2018) Neuropsychological functions of the gut microbiota. In: Chiappelli F (ed) Advances in psychobiology. Nova Science Publishers, New York, NY, pp 267–279

Weuve J, Barnes LL, de Leon CFM, Rajan KB, Beck T, Aggarwal NT, Evans DA (2018) Cognitive aging in black and white Americans: cognition, cognitive decline, and incidence of Alzheimer disease dementia. Epidemiology 29(1):151. https://doi.org/10.1097/ede.0000000000000747

World Health Organization (2019) Risk reduction of cognitive decline and dementia. WHO guidelines. WHO, Geneva

Wu YT, Lee H, Norton S, Chen C, Chen H, He C, Brayne C (2013) Prevalence studies of dementia in mainland China, Hong Kong and Taiwan: a systematic review and meta-analysis. Alzheimers Dement 9:687–P688. https://doi.org/10.1016/j.jalz.2013.04.353

Xu W, Tan L, Wang HF, Jiang T, Tan MS, Tan L, Yu JT (2015) Meta-analysis of modifiable risk factors for Alzheimer's disease. J Neurol Neurosurg Psychiatry 86(12):1299–1306. https://doi.org/10.1136/jnnp-2015-310548

Xu W, Tan L, Wang HF, Tan MS, Tan L, Li JQ, Yu JT (2016) Education and risk of dementia: dose-response meta-analysis of prospective cohort studies. Mol Neurobiol 53(5):3113–3123. https://doi.org/10.1007/s12035-015-9211-5

Yusuf S, Joseph P, Rangarajan S, Islam S, Mente A, Hystad P, AlHabib KF (2020) Modifiable risk factors, cardiovascular disease, and mortality in 155 722 individuals from 21 high-income, middle-income, and low-income countries (PURE): a prospective cohort study. Lancet 395(10226):795–808. https://doi.org/10.1016/s0140-6736(19)32008-2

Zhuang ZQ, Shen LL, Li WW, Fu X, Zeng F, Gui L, Lü Y, Cai M, Zhu C, Tan YL, Zheng P (2018) Gut microbiota is altered in patients with Alzheimer's disease. J Alzheimers Dis 63(4): 1337–1346. https://doi.org/10.3233/jad-180176

Zivkovic AM, German JB, Lebrilla CB, Mills DA (2011) Human milk glycobiome and its impact on the infant gastrointestinal microbiota. Proc Natl Acad Sci 108(Suppl 1):4653–4658. https://doi.org/10.1073/pnas.1000083107

Zoetendal EG, Raes J, Van Den Bogert B, Arumugam M, Booijink CC, Troost FJ, Bork P, Wels M, De Vos WM, Kleerebezem M (2012) The human small intestinal microbiota is driven by rapid uptake and conversion of simple carbohydrates. ISME J 6(7):1415. https://doi.org/10.1038/ismej.2011.212

Gut–Brain Axis in Alzheimer's Disease: Interplay Between Cholecystokinin, Dysbiosis, and Brain-Derived Neurotrophic Factor

12

Yassamine Ouerdane, Zeinab S. El-Nahas, Fella Ouerdane, Khaled M. Hamam, and Mahmoud Ahmed Ebada

Abstract

The relationship between metabolic profile and the structural integrity of the brain in Alzheimer's disease (AD) patients is of particular interest. Less is known about the impact of dietary changes and lifestyle on the progression of AD. Alteration in eating behavior is one of the observed symptoms in AD with both increased and decreased food intake. Cholecystokinin (CCK) is a satiety hormone and neurotransmitter highly expressed in brain regions like the hippocampus, which is responsible for memory and learning. CCK also regulates the expression of the brain-derived neurotrophic factor (BDNF), which induces the growth and survival of neurons. CCK levels and its receptor expression are both altered in AD patients. Several factors are presumed to affect the CCK release, such as weight gain, weight loss, dysbiosis of gut microbiota, and poor food habits. Thus, CCK is an essential pillar to understand the changes in eating behavior and memory deterioration in AD, and it is crucial to study the interaction induced by these factors in the occurrence and the progression of AD. This chapter aims to analyze the links between CCK, gut microbiota, BDNF, and dietary habits in the pathophysiological mechanism of AD.

Y. Ouerdane · F. Ouerdane
Faculty of Medicine, Saad Dahlab University, Blida, Algeria

Z. S. El-Nahas (✉)
Faculty of Medicine, Menoufia University, Menoufia, Egypt
e-mail: zainab.samy369@med.menofia.edu.eg

K. M. Hamam
Faculty of medicine and surgery, October 6 University, Giza, Egypt

M. A. Ebada
Faculty of Medicine, Zagazig University, Zagazig, Egypt

Resident Physician, Ministry of Health and Population, Cairo, Egypt

G. M. Ashraf, M. S. Uddin (eds.), *Current Thoughts on Dementia*,
https://doi.org/10.1007/978-981-16-7606-2_12

Keywords

Cholecystokinin · Brain-derived neurotrophic factor · Gut microbiota · Food habits · Dysbiosis · Ketone bodies · Alzheimer's disease

12.1 Introduction

Dementia is a clinical syndrome characterized by deterioration in thinking, memory, and other cognitive functions that interfere with the individual's ability to perform everyday activities (Robinson and Tang 2015). The most prevalent subtype of dementia is Alzheimer's disease (AD), corresponding to about 60–70% of all dementia cases (Brookmeyer et al. 2007; Hebert et al. 2003). Other major forms include vascular dementia, dementia with Lewy bodies, mixed dementia, Parkinson's disease dementia, Creutzfeldt–Jakob disease, and normal pressure hydrocephalus (Gaugler et al. 2016). AD is a brain disorder characterized by deterioration of intellectual and behavioral functions causing difficulties in remembering, recalling recent events, language, concentration, and learning new information. The characteristic features of AD are the appearance of extracellular amyloid-beta (Aβ) plaques and neurofibrillary tangles (NFTs) in the intracellular environment and loss of neuronal connection resulting in their destruction and unleashing a series of damaging events to the brain leading to neuroinflammation, mitochondrial dysfunction, and neuronal loss. The exact etiology remains unclear, but it may be a result of the overlap of both genetic and environmental factors. There is no pharmacological treatment for AD; however, available medications help only relieve the symptoms. This disease is considered a burden not only for the affected individuals but also for caregivers dealing with the patient's symptoms, which further results in their stress.

Moreover, the lack of awareness of dementia and stigmatization could delay clinical diagnosis. The current research tries to illustrate a putative pathological model for AD progression. It segregates risk factors from everyday habits and lifestyles to prevent the occurrence of the disease. Interestingly, some factors are believed to have a tremendous impact on the pathophysiological mechanism of AD, including obesity (Pedditzi and Ruth Peters 2016), gut microbial composition (Kowalski and Mulak 2019), changes in brain-derived neurotrophic factor (BDNF) (Song et al. 2015), and cholecystokinin (CCK) levels (Plagman et al. 2019).

Gut microbiota is a complex and dynamic community of microorganisms that resides in the alimentary system. Dysbiosis of gut microbiota can modulate the neurochemical and neuro-metabolic signaling pathways to the brain, resulting in neurodegeneration. It is linked with the genesis of depression, anxiety, multiple sclerosis (Berer et al. 2011; Cryan and Dinan 2012), and AD (Jiang et al. 2017). Besides, it plays a significant role in energy regulation and is considered a causative factor for obesity (Harakeh et al. 2016; Ley et al. 2006; Shabana et al. 2018). The latter is associated with cognitive decline, decreased blood–brain barrier (BBB) integrity, reduced white matter, brain atrophy, and increased risk for late-onset AD

(Droogsma et al. 2015; Elias et al. 2012), pointing out the incumbent role of gut microbiota in energy balance to prevent the deleterious consequences of this imbalance on the brain. Microbial diversity could also provide a glimpse into the potential role gut microbiota plays in modulating metabolic profile since a high microbial diversity was associated with low weight gain and improved metabolic parameters (Menni et al. 2017). Preclinical studies unearth the role of a high-fat diet (HFD) in changing microbial consortia and increasing dementia risk (Nam et al. 2017; Sah et al. 2017; Sanguinetti et al. 2018; Studzinski et al. 2009).

Moreover, these tiny organisms influence several peptides related to cognition and neural plasticity, such as BDNF (Duca et al. 2012) and CCK (Maqsood and Stone 2016). CCK is a neuropeptide and digestive hormone integral for satiety induction and memory function and seems to be altered in the cerebrospinal spinal (CSF) of AD patients (Plagman et al. 2019; Rehfeld 2004). Interestingly, the same peptide enhances the secretion of BDNF and further increases neuronal connections and plasticity (Hwang et al. 2013). Since patients with AD have shown changes in food preferences and food intake, it was necessary to study the role CCK plays along in AD (Morris et al. 1989). This chapter aims to analyze the links between gut microbiota, obesity, CCK, and BDNF in the pathophysiological mechanism of AD to delineate the intertwined relationship of these aspects, which could open avenues toward understanding the mechanisms of the disease and how to alleviate its progression.

12.2 Cholecystokinin Physiology

CCK is a peptide hormone responsible for regulating the digestive process. CCK is synthesized and released by enteroendocrine cells (EECs) located in the proximal small intestine (Larsson and Rehfeld 1978; Rehfeld et al. 1990). Also, it is synthesized in thyroid C cells (Rehfeld et al. 1990), in adrenal medullary cells (Larsson and Rehfeld 1979; Rehfeld 1987), and in the pituitary cells with small amounts (Rehfeld et al. 1987). In the central nervous system (CNS), CCK is present with high concentrations in the amygdala, cortex, hippocampus, and septum (Beinfeld et al. 1981; Crawley and Corwin 1994). Its receptors are expressed mainly in the hippocampus, cerebral cortex, and striatum (Rehfeld et al. 1990).

Concerning the molecular pattern, CCK belongs to a peptide family constituted of a variety of amino acids with various structural forms derived from post-translational modification of a single 115-residue prohormone precursor, pre-pro-cholecystokinin, with a well-preserved C-terminal sequence in which its biological activity resides.

Regarding CCK receptors, there are two identified types of G-protein-coupled CCK receptors, CCKA or (CCK-1) and CCKB or (CCK-2). CCK-1 receptors have a high affinity to sulfated CCK peptides with a negligible affinity to non-sulfated ones. They mediate gall bladder contraction, pancreatic enzyme secretion, and inhibit gastric acid secretion (Chen et al. 2004). CCK-1 receptors are predominantly present in the peripheral alimentary system, anterior pituitary, and areas of the midbrain

(You et al. 1996), whereas CCK-2 receptors are predominantly present in the brain with equal affinities to both sulfated and non-sulfated CCK, alluding to the vital role CCK plays in the brain and the gastrointestinal tract (GIT). There are various forms of CCK. Plasma forms are CCK-58, CCK-33, CCK-22, and CCK-8, while CCK-8 and CCK-5 are mainly released from neurons (Rehfeld 1978; Rehfeld and Hansen 1986).

Ivy and Oldberg first discovered CCK in 1928 in intestinal extracts indicating its role in eliciting gallbladder contraction in cats, dogs, and pigs (Ivy and Oldberg 1928). CCK peptides are released in response to the entry of food into the duodenum, increasing CCK blood levels (Liddle et al. 1985). Its secretion is regulated by nutrients, including ingested fats, proteins, and amino acids, which induce CCK secretion. In contrast, carbohydrates cause only a temporary and transient increase to CCK, deducing that protein- and fat-rich food is considered the most critical stimulus (Liddle et al. 1985; Rehfeld 1998). On the other hand, pancreatic proteases and bile acids (BAs) reduce CCK secretion (Green and Lyman 1972; Ohta et al. 1990; Owyang et al. 1986). The relationship between BAs and CCK is of great interest as they reciprocally affect each other. The demonstrable effect of BAs on CCK secretion is indicated in rats as the administration of taurocholate; its sodium salt is the chief ingredient of BAs, which inhibited the secretion of pancreatic enzymes and CCK (Tomita et al. 1994). In humans, the addition of a mixture of BAs to a test meal in the presence of loxiglumide, a CCK-1 receptor antagonist, prevented the increase in CCK release, so it is suggested that BAs could downregulate CCK secretion (Koop et al. 1996). Also, CCK facilitates bile release into the duodenum as it stimulates gall bladder contraction and relaxes the sphincter of Oddi. Adding to that, infusion of CCK-8 in humans causes a decrease in gall bladder volume by 80% and an increase in bilirubin output by eight- to tenfold, and these effects are mediated by CCK-1 receptors that are located on the smooth muscle layer of the gallbladder and cholinergic nerve terminals (Schjoldager 1994; Schmidt et al. 1991).

Additionally, other molecules affect CCK secretion, such as lipopolysaccharide (LPS) (Bogunovic et al. 2007; Palazzo et al. 2007) in the outer membrane of Gram-negative bacteria and luminal CCK-releasing factor (LCRF) secreted into the intestine (Spannagel et al. 1996). However, the mechanism whereby LCRF affects CCK secretion is not understood. It is also unknown whether LCRF could directly affect CCK cells or stimulate its release in humans. Moreover, diazepam-binding inhibitor (DBI), secreted from the proximal intestine, increases plasma CCK levels, and it is indicated that antiserum to DBI abolishes peptone-induced elevations in plasma CCK (Herzig et al. 1998; Li et al. 2000).

In terms of the physiological functions, it has been found that CCK is not only a gallbladder contraction factor, but also stimulates pancreatic secretion, induces satiety, delays gastric emptying (Rehfeld 2017), triggers insulin and glucagon release (Hermansen 1984; Jensen et al. 1981; Otsuki et al. 1979), and acts as a neurotransmitter in the brain and peripheral neurons (Rehfeld et al. 2007; Rehfeld 2004, 2017). Its receptors mediate CCK signals and contribute to the regulation of memory, anxiety, cognition, analgesia, satiety, and dopamine-mediated behavior

(Bush et al. 1999; Moran et al. 1993). CCK has shown to play a role in delaying gastric emptying in fish, rodents, dogs, and humans (Debas et al. 1975; Olsson et al. 1999), and this is attributable to its effect in the relaxation of the proximal stomach and pyloric contraction (Yamagishi and Debas 1978). Again, loxiglumide has groped its way to determine that the antagonizing effect of the CCK-1 receptor dictates no changes in gastric emptying when physiological levels of CCK-8 are administrated (Konturek et al. 1990; Liddle et al. 1985). Besides, CCK causes a centrally mediated feeling of satiety. It induces satiety through CCK-1 receptors located on vagal afferent nerves, which terminate on dorsal hindbrain neurons of the nucleus of the solitary tract (NST), and thus restricting food amounts (Kerstens et al. 1985; Pupovac and Anderson 2002).

Interestingly, the neuronal expression of CCK outlines its pivotal role. It plays a notable role in memory. CCK sustains memory retention and prevents the degeneration of cholinergic neurons against basal forebrain lesions (Sugaya et al. 1992). In rats, acute intraperitoneal injection of CCK-8S during the Morris water maze task has an evident improvement in spatial memory (Voits et al. 2001). Adding to that, stress-induced CCK fluctuations within the hypothalamus and hippocampus dissect its role in the regulatory process of hypothalamo–pituitary–adrenocortical activity, glucocorticoid concentrations, learning, and remembering (Greisen et al. 2005; Kim et al. 2003).

12.3 Mechanism of Cholecystokinin Signaling from the Gut to the Brain

The regulation of feeding behavior is complex. It comprises hormonal, neural, and nutrient (glucose and protein) signals triggered by food ingestion and absorption to inform the CNS, particularly the hypothalamus, about stored energy levels (Thorens 2008). These signals are transmitted to the lateral hypothalamus via peripheral pathways that implicate vagal afferent fibers, which are the primary neuronal connection relaying visceral sensory information from the periphery (Raybould 2007). Of the hormones controlling feeding behavior, CCK plays a central role (Gibbs et al. 1973; Maclagan 1937). Vagal afferents express CCK1Rs. The peripheral terminals of these neurons lie in both mucosal and muscle layers of the GIT. These terminals are in close apposition to the membrane of EEC in the lamina propria of the mucosal layers (Harrington et al. 2018). Such evidence forms the anatomical basis for recognizing the paracrine mode of the action of CCK, which results in subsequent activation of vagal afferents (Berthoud et al. 2004). Additionally, in CCK1R null mice, the activation of the vagal afferent pathway in the NST was markedly reduced (Whited et al. 2006). This finding suggests that the effect of CCK on the vagus nerve is through the activation of CCK 1 receptors.

Interestingly, CCK interacts with other hormones to induce the sensation of satiety. Evidence points out that CCK interacts with leptin and thus enhances the excitatory responses of vagal afferents to exogenous CCK (Peters et al. 2006). In low plasma CCK levels (e.g., in the fasting state), vagal afferent neurons stimulate the expression of two hormonal receptors responsible for food intake such as cannabinoid CB1 and melanin concentration hormone (MCH)-1 receptors. In contrast, increased CCK concentrations diminish the expression of these receptors. It is essential to point out that in fasting, there is a decrease in the expression of other satiety peptides, including cocaine- and amphetamine-regulated transcript (CART) (Dockray 2009). Administered CCK in fasting rats changes the expression of the genes encoding CART, MCH, and CB1, resulting in an increase in CART and a decrease in MCH and CB1 expression (Burdyga 2004; de Lartigue et al. 2007). The action of CCK in stimulating CART depends on the activation of protein kinase C, phosphorylation of CREB, and activation of MAP kinase. However, the mechanism by which CCK downregulates the expression of MCH and CB1 is not fully understood (Burdyga et al. 2006).

Although this peptide exerts its effects mainly through the activation of peripheral vagal nerves, questions about the direct action of intravenous (IV) CCK administration on the CNS raise much enthusiasm. In a study conducted by Hommer et al., they attempted to examine whether the excitatory effect of CCK on dopamine (DA) cells after its IV administration is peripheral or central (Hommer et al. 1985) since both DA and CCK coexist in many neurons in the substantia nigra (SN) and the ventral tegmental area (Hökfelt et al. 1980). Hommer et al. (1985) performed a series of lesions on the vagal afferents and efferents to the NST. First, the results of acute and chronic supradiaphragmatic lesions of the vagus nerve indicate a remarkable reduction in CCK-induced satiety; however, neither of these lesions was effective in attenuating CCK-induced excitation in the SN. Moreover, high cervical spinal cord transection was similarly without effect.

In contrast, lesions of NST afferents and efferents significantly reduced, but not abolished, CCK-induced DA excitations. Taken together these findings, Hommer et al. (1985) have indicated that peripheral CCK could affect the SN by an activity mediated through CCK receptors in the NST, and a direct effect on DA neurons, suggesting that IV administered CCK may cross the BBB and act directly on nigral DA cells (Hommer et al. 1985). Besides, IV injection of CCK-4 induces panic-like attacks in humans (Bradwejn 1993), which might be in accordance with a transfer in the opposite direction, from the peripheral blood to the CNS. Lundberg et al. (2010) have suggested that CSF levels of immunoreactive CCK-LI concentration can be measured indirectly in the plasma of females. Previous studies reported diurnal and seasonal variations of CCK in CSF and plasma in both genders (Geracioti et al. 1993; Lundberg et al. 2007). The evidence recognizes that CCK peptides are transferable between the CNS and the peripheral blood (Fig. 12.1).

12.4 The Link Between Cholecystokinin and Alzheimer's Disease

Despite relevant preexisting research on AD pathogenesis, the exact mechanism that leads to the occurrence of this disease is not completely clear. Whopping data have agreed that AD pathogenesis is heterogeneous and complex caused by aging and

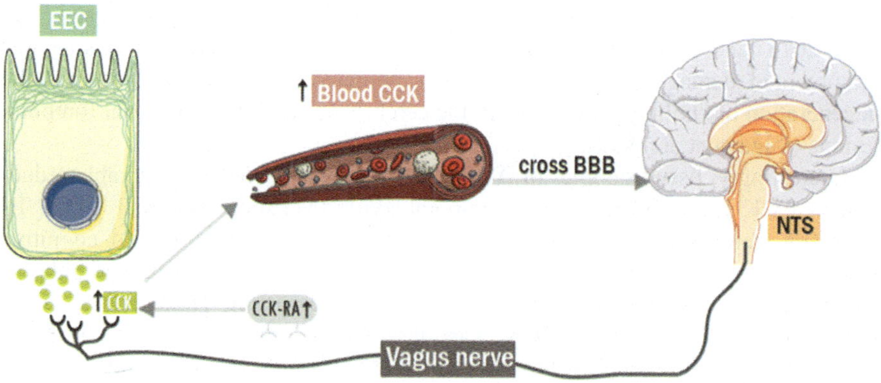

Fig. 12.1 Mechanism of cholecystokinin signaling from the gut to the brain. *CCK* cholecystokinin, *EEC* enteroendocrine cells, *CCK-RA* CCK receptor A, *NTS* nucleus tractus solitarius, *BBB* blood–brain barrier

dual interaction between genetic and environmental risk factors (Lin et al. 2014). Several features characterized the brain of AD patients, including senile plaques and NFTs, thoughtfully described as the cardinal lesions, accompanied by deficits in axonal transport and neuronal loss. The appearance of these lesions involves two peptides, Aβ and a microtubule-associated protein known as tau. The hyperphosphorylation of tau forms the NFTs. Aβ, which is a 40–42 amino acids of the amyloid protein precursor (APP), is the principal constituent of senile plaques. Aβ aggregation could result in oxidative stress, neuroinflammation, and neurotoxicity, all of which can initiate the pathogenic cascade, ultimately leading to apoptosis and deterioration of the neurotransmission systems (Glenner and Wong 1984; Tiwari et al. 2019; Yankner 1996). These hallmarks are observed in specific, vulnerable brain areas, and the hippocampus is one of the earliest areas to be affected (Demars et al. 2010).

In addition to the deleterious effects on cognitive functions, alteration in eating behavior is one of the observed symptoms in AD patients, including both increased and decreased food intakes. Patients with AD have shown changes in food variety preferences (Morris et al. 1989), suggesting instability in weight regulation and hyperphagia, which prevails in a third of all AD patients (Morris et al. 1989). The mechanisms underlying hyperphagia in AD have not been entirely elucidated, but may be due to decreased satiety hormones or reduced responsiveness to these hormones (Adebakin et al. 2012). Several studies underline the implication of CCK in this disease since the presence of CCK in the cerebral cortex is 10 times greater than in the duodenum (Sanders et al. 1982). CCK appeared to have ameliorative influences on hippocampal synaptic plasticity (Wen et al. 2014). Its receptors might play a role in the performance of visual recognition tasks (Sebret et al. 1999) and spatial memory since some studies have displayed decrements in spatial memory performance in the rats lacking CCKA receptors (Matsushita et al. 2003). It has been reported that repeated treatment with CCK-8S has beneficial effects on promoting neuronal network interconnections such as hippocampal excitatory synapses (Wei et al. 2013) and dendritic growth, resulting in protecting the hippocampus against atrophy and damage (Zhang et al. 2013). Moreover, this peptide prevents the degeneration of cholinergic neurons in the cerebral cortex following basal forebrain lesions (Sugaya et al. 1992).

Interestingly, tissue samples of AD brain autopsies showed a remarkable reduction in CCK levels in the middle temporal gyrus (Mazurek and Beal 1991). Its receptors were established to be altered in patients with AD or mild cognitive impairment (MCI) as CCKA, and CCKB receptors have shown to be downregulated, which highlights the role of CCK receptors in AD development (Hokama et al. 2014; Lin et al. 2014). However, CCK was observed in specific sites in the brain surrounding neuritic plaques (Struble et al. 1987). Moreover, the expression level of nutrition-related genes was altered in AD and MCI patients, and it was significantly associated with downregulation in CCKAR expression, suggesting that nutrient supply might be related to AD (Lin et al. 2014). These findings underline the critical recognition of the role this peptide plays in AD pathogenesis.

In a cohort study from Plagman et al. (2019), CSF levels of CCK were associated with better outcomes, which may reflect compensatory protection as AD pathology progresses. Higher CCK levels appeared to be related to better cognitive outcomes since a per ng/mL increase in CCK; there was a roughly 65% decreased likelihood of having MCI or AD compared to cognitively normal (CN) individuals, and a 62% reduced likelihood of MCI progression to AD. Besides, patients with AD had lower levels of CCK than MCI or CN subjects. Higher CSF CCK levels were associated with more regional gray matter (GM) volume in areas such as the parahippocampal gyrus, hippocampus, posterior cingulate cortex, and superior and medial prefrontal gyri; but, they were not related to increased risk for AD diagnosis. However, Plagman et al. (2019) have indicated that higher CCK levels are correlated with higher tau levels, and they attributed this association to a compensatory mechanism CCK plays to enhance memory in AD patients since higher tau levels act as a partial mediator reducing the influence of CCK by nearly 50%. Thus, the enhancing effect of CCK on memory might be reduced as well (Plagman et al. 2019). In addition to the effect of tau on CCK effect (Plagman et al. 2019), other mediators reduce CCK release and thus worsening memory functions. We will focus on them later on in the following sections.

In the pathological cascade of AD, the overlap of various neurotransmitters cannot be neglected since the dysfunction of the dopaminergic system could lead to a decline of cognition in AD animal models, and the opioid system was reported to be related to hyperphosphorylated tau and Aβ production (Anthony et al. 2010; Benamar et al. 2007; Cai and Ratka 2012; Guzman-Ramos et al. 2012; Yakovleva et al. 2007). Several signaling molecules are impaired in AD, including acetylcholine, DA, and opioid system (Cai and Ratka 2012; Pan et al. 2019). CCK family plays a vital role in neuronal functions by interacting with these neurotransmitters (Okubo and Harada 2001). CCK receptors are suggested to have a regulatory role in both the endogenous opioid system and DA release in the brain (Pommier et al. 2002; Wank 1995). Therefore, it is well established to conclude that CCK might be

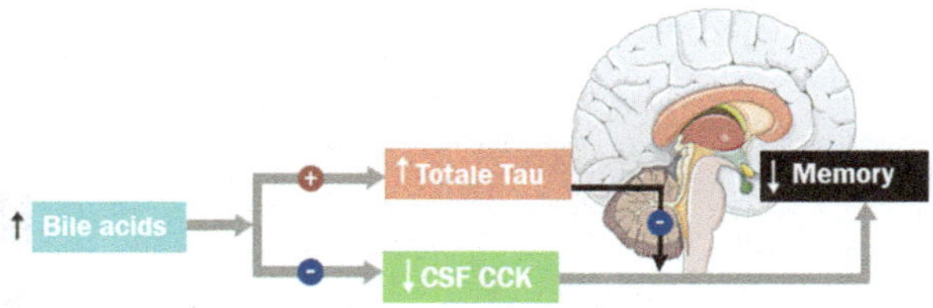

Fig. 12.2 The role of cholecystokinin in Alzheimer's disease. *CCK* cholecystokinin, *CSF* cerebrospinal fluid

involved in the pathogenesis of AD, and it is crucial to determine to what extent this peptide is incriminated.

Neuronal CCK is not alone in shedding light on the pathological cascade of AD. Satiety messages start from the gut and are transmitted to the brain (Hommer et al. 1985). Therefore, a putative role of peripheral CCK might be incriminated in AD trajectory as well. Several other components related to digestion are effective regulators for CCK release in the gut and are closely associated with AD pathogenesis such as BAs, which are known as the potent inhibitor to CCK release. BAs exert inhibitory effects on CCK, independent of the chemical nature of the stimulus such as triglycerides (Gomez et al. 1986), amino acids (Gomez et al. 1988), or mixed liquid test meal containing glucose, long-chain fatty acids, and protein (Gomez et al. 1988; Koop et al. 1989). Both primary and secondary BAs are present in the brains of mice and possibly humans, which indicates that they cross the BBB (MahmoudianDehkordi et al. 2019). In addition, BAs seem to be affected in AD patients (Greenberg et al. 2009; Mapstone et al. 2014; Marksteiner et al. 2018; Olazarán et al. 2015; Pan et al. 2017). Previous studies suggest that BAs are altered in AD and MCI (Kubo et al. 2002). Altered BA profiles are significantly associated with structural and functional changes in the brain, reduced glucose metabolism, and with three CSF biomarkers, including Aβ1–42, t-tau, and p-tau. The altered BA profile was associated with higher tau levels (Nho et al. 2019). Therefore, evidence showing that BAs contribute to AD pathogenesis is overwhelming (Nho et al. 2019).

The authors hypothesized that the decreased levels of CCK in the CSF of AD patients might be related to a potent inhibitory effect exerted by BAs. This inhibitory effect might be the result of gut dysbiosis and poor food habits since both alter BA profiles (Kakimoto et al. 2017; Wang et al. 2019; Yoshitsugu et al. 2019) and consequently lead to deterioration in memory and neuronal plasticity (Fig. 12.2).

12.5 The Link Between Cholecystokinin and Alzheimer's Disease Risk Factors

Several risk factors are implicated in the cascade of AD pathogenesis; they include obesity, a decrease in the BDNF, and dysbiosis in gut microbiota. All of these factors contribute to AD development in later life (Pedditzi and Ruth Peters 2016; Kowalski and Mulak 2019; Song et al. 2015). In Fig. 12.3, we discuss the contribution of these factors in AD and the interplay between CCK and these factors.

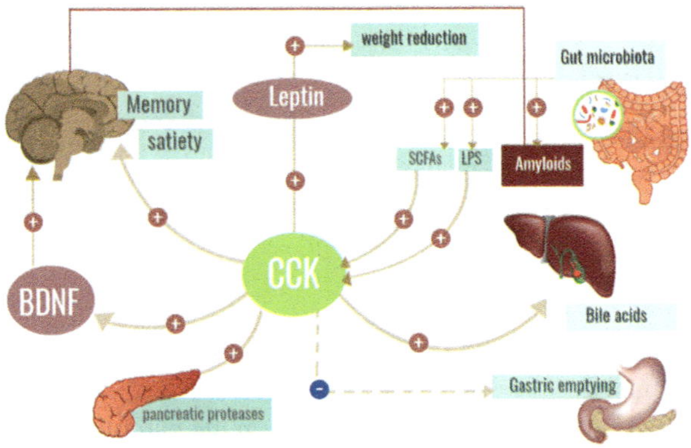

Fig. 12.3 Possible interactions between cholecystokinin, gut microbiota, brain-derived neurotrophic factor, and weight changes in Alzheimer's disease. This figure shows bacterial products in the gut, such as LPS and SCFAs, could trigger CCK release from EEC in the intestinal lumen. CCK secretion delays gastric emptying and induces the sensation of satiety. Furthermore, it enhances the release of bile acids and gallbladder contraction. Also, CCK stimulates pancreatic enzyme secretion. In brain regions, CCK triggers the expression and the release of BDNF, which helps neuronal growth and maturation and improves memory function. Synergic interaction with leptin induces weight reduction. However, in a dysbiosis state, these organisms synthesize an array of proteins referred to as amyloids. These proteins pass through the gastrointestinal barrier and the BBB, and then accumulate in the brain, which in turn unleashes a cascade of damaging events into the brain. *LPS* lipopolysaccharide, *SCFAs* short-chain fatty acids, *CCK* cholecystokinin, *BDNF* brain-derived neurotrophic factor

12.5.1 A Promising Connection Between Brain-Derived Neurotrophic Factor and Alzheimer's Disease

BDNF is an essential protein encoded by the BDNF gene and belongs to neurotrophin family members of growth factors. Its initial synthesis as a precursor protein containing a single peptide takes place in the endoplasmic reticulum. Following endoproteolytic cleavage of the single peptide, pro-BDNF undergoes package and sorting and could further be converted to mature BDNF. The processing of pro-BDNF conversion to BDNF may occur extracellularly or intracellularly (Nagappan et al. 2009; Pang et al. 2004) since pro- and mature forms of BDNF are both colocalized in secretory granules in presynaptic axon terminals (Dieni et al. 2012). This suggests a discrepancy in the exact location where the last cleavage stage occurs. The mature state of BDNF is transported in secretory granules to postsynaptic dendrites or in BDNF-containing vesicles to presynaptic terminals (Leal et al. 2015). Besides, BDNF signaling is elicited through two membrane-bound receptors. BDNF produces its biological response through binding to its low-affinity P75 neurotrophin receptor (P75 NTR) and a tyrosine kinase receptor, referred to as tropomyosin-related kinase B (TrkB) (Chao 2003).

The versatility of BDNF lies in its abilities in promoting neuronal differentiation (maturation), growth, and survival. In addition, it regulates synaptic plasticity and modulates long-term potentiation in the hippocampus sustaining learning and memory formation (Benarroch 2015). Moreover, BDNF has demonstrated an enhanced synaptic transmission in hippocampal cultures (Lemann et al. 1994; Leßmann and Heumann 1998; Levine et al. 1995, 1996, 1998). Also, BDNF regulates the balance between synaptic and extrasynaptic *N*-methyl-D aspartate (NMDA) receptors, which are one of the fundamental neurotransmitter receptors in the brain. Besides, BDNF is conducive to achieve neuroprotection and protect neurons against mitochondrial dysfunction, which is considered the major cause of excitotoxicity, through diminishing extrasynaptic NMDA receptor death signaling (Zito and Scheuss 2009). Consequently, dysregulation of BDNF signaling and excitotoxicity, leading to damaged and killed neurons, is involved in pathological processes, such as brain ischemia (Ferrer et al. 2001; Tejeda et al. 2016; Vidaurre et al. 2012), traumatic brain injury (Rostami et al. 2014; Schober et al. 2012), and neurodegenerative diseases (Nguyen et al. 2016; Plotkin et al. 2014; Porritt et al. 2005), including AD (Jerónimo-Santos et al. 2015). Therefore, AD and other neurodegenerative disorders such as Huntington's disease or Parkinson's disease might be associated with reduced BDNF levels and, consequently, enhanced extrasynaptic NMDA receptor signaling (Lau et al. 2015).

There is an association between excitotoxicity in AD and dysregulation of BDNF, albeit the mechanism is not fully understood. Brain tissue samples of AD patients have shown a decline in the levels of BDNF and its high-affinity receptor TrkB, compared to age-matched controls (Ginsberg et al. 2006; Hock et al. 2000). Weinstein et al. (2014) followed up cognitively healthy older individuals for up to 10 years and reported that higher levels of BDNF were associated with a lower risk of future occurrence of dementia and AD. Additionally, serum BDNF levels reflect behavioral alterations and subtle cognitive changes by the preclinical stage of AD trajectory, which indicates the role of BDNF in shaping the onset of AD neurodegeneration. Several recent studies suggested that plasma BDNF concentration may serve as a potential biomarker for the validation of AD diagnosis (Laske et al. 2006; Leyhe et al. 2008). Alteration in BDNF expression can cause modifications to the serotonergic system, which attributes to degenerative disorders, depression, or mood disorders observed in aged individuals and AD patients (Altar 1999). Moreover, it was demonstrated that BDNF could affect the subcellular trafficking of APP, so interfering BDNF signaling increases Aβ generation (Matrone et al. 2008; Rohe et al. 2009). On the other hand, multiple studies have reported an increase in BDNF concentrations and associated receptor TrkB in the hippocampus and parietal cortex of AD patients and alluded that to a possible association with compensatory mechanisms contributable to the repair by Aβ degradation (Durany et al. 2000; O'Bryant et al. 2009).

Several factors, such as physical activity and diet restriction, are affecting the BDNF expression and release (Lee et al. 2000). Physical exercise induces BDNF expression and its release in the brain and in the peripheral tissues, resulting in promoting body functions (Sleiman et al. 2016). In terms of diet regulatory effect, it has been shown that maintaining a dietary restriction regimen in adult rats results in an increase in the number of newly generated neural cells in the dentate gyrus of the hippocampus and triggers hippocampal BDNF expression (Gomez-Pinilla 2008; Lee et al. 2000, 2002). Thus, research on the regulatory mechanism of BDNF expression by diet and physical exercises could have innovative and interesting therapeutic consequences. However, there is also a close relationship between physiological factors that alter BDNF expression and those affecting neurogenesis in adult rats, which points out a possibility to prevent further neural damage by preventing rapid brain aging, which is related to stress factors (Tapia-Arancibia et al. 2004). Stress and glucocorticoids are responsible for promoting neuronal death, negatively affect cognitive function in experimental systems, and could lead to neuropsychiatric disorders (Landfield et al. 2007; Lupien and McEwen 1997; Porter and Landfield 1998; Sapolsky 2000). According to stimulus intensity, stress can profoundly accelerate the brain aging process (De Kloet et al. 1999; Pardon 2007). It has been reported that chronic stress reduces hippocampal BDNF expression (Smith et al. 1995; Ueyama et al. 1997).

Moreover, it has been shown that glucocorticoids and stress could activate APP degradation, so there is a direct correlation between stress and aggravation of Aβ (Green et al. 2006). This field is still in its infancy. However, if successful, it will contribute to closing the gap between environmental modification factors and neurodegenerative diseases, and could lead to new therapies to cure AD symptoms.

12.5.2 CCK Mediates Brain-Derived Neurotrophic Factor Release

Several studies revealed that CCK-8S dissects a role in stimulating nerve growth factor (NGF) and BDNF synthesis in brain regions, including the cortex, septum, and hippocampus, resulting in neuroprotection (Tirassa et al. 1999; Tirassa and Costa 2007). Endogenous BDNF protein content in the dorsal vagal complex (DVC) increased after peripheral leptin or CCK treatment and decreased after fasting (Bariohay et al. 2005). In another experimental study, the administration of CCK increased the expression of NGF, BDNF, and its receptor TrKB as well as downregulation of TrKA and P75BTR receptors in the hippocampus. A likely suggested hypothesis by the authors is that the stimulatory effect of CCK-8S on BDNF and NGF could result from an indirect action via modulating the cholinergic system and GABAergic afferents (Tirassa and Costa 2007). This is consistent with the finding that CCK-8, parallel to neurotrophin induction, enhances the synthesis of acetylcholine and GABA in the septum and the hippocampus of both healthy and injured brains, and both neurotransmitters play a role in neurotrophin synthesis and release (Tirassa et al. 1999, 2005). This effect of CCK-8S in mediating neurotrophin release was abolished entirely after the administration of the anticholinergic drug (atropine), which raises a particular emphasis for further trials to demonstrate the exact mechanism by which CCK stimulates the secretion of BDNF in the hippocampus (Tirassa et al. 1998).

The effects of CCK-8S on locus coeruleus (LC)-noradrenergic (NA) neuronal cell activity have been studied, and it was revealed that CCK-8S could modulate BDNF expression in LC-NA neurons and induce a significant increase in BDNF mRNA and BDNF protein concomitantly (Hwang et al. 2013). Studies using H_2O_2 as an oxidative stressor showed a decrement in BDNF protein expression (Choi et al. 2010; Son et al. 1999). The administration of CCK-8S induces a delay and prevention of BDNF loss in H_2O_2-treated LC neurons (Hwang et al. 2013). This action is attributed to CCK inhibiting effect on caspase-3 expression, which is caused by the H_2O_2 toxic effect. In the same neurons, CCK-8S activates ERK1/2 and AKT, which are known to control cell proliferation and neural survival positively. These two molecules are upstream regulators of BDNF expression. Thus, it is trustworthy to conclude that CCK-8S stimulates BDNF expression via ERK1/2 and AKT pathways (Hwang et al. 2013).

12.5.3 Obesity as a Risk Factor for Alzheimer's Disease

The last few decades have witnessed a rising prevalence of obesity worldwide (Chooi et al. 2019). Obesity is a chronic progressive disorder characterized by body fat accumulation that promotes as a result of a cluster of associated risk factors leading to positive energy balance and weight gain. The matching of energy consumption and energy expenditure is a cardinal feature to achieve energy balance and body weight homeostasis and consistency, so any bias in this balanced equation could result in obesity. However, the factors contributing to the disequilibrium are not fully understood (Sanmiguel et al. 2015). In clinical practice, obesity is typically defined as a body mass index (BMI) of 30 or higher, which is estimated by the weight in kilogram divided by the square of the height in meters (Khaodhiar et al. 1999). It is a medical issue associated with health problems, including insulin resistance and type 2 diabetes mellitus, hypertension, cardiovascular disease, and gallbladder disease, making it the second leading cause of preventable death (Eckel 1997; Khaodhiar et al. 1999). Obesity is also associated with cognitive decline (Jeong et al. 2005; Whitmer et al. 2005). It is related to a higher risk of AD and vascular dementia in later life (Xu et al. 2011). There is a compelling relationship between higher adiposity in middle age and later impairment in cognitive function (Beydoun et al. 2008; Whitmer et al. 2005, 2008). Obese individuals between 30 and 39 years have a 3.5 increased relative risk ratio for the incidence of AD and later vascular dementia. This relative risk ratio gradually reduced up to the age of 70 years in obese people. However, it is still associated with a high risk of later dementia compared to non-obese controls. Obese individuals who are over 80 years of age showed a reduced risk of subsequent vascular dementia (Wotton and Goldacre 2014).

The total number of obese people with dementia is suggested to be 9–19% higher than the projected number based on aging alone (Loef and Walach 2013; Nepal et al. 2014). Hence, according to these data, midlife obesity is a risk factor for incident dementia in later life, while obesity in older age may reduce this risk. Moreover, recent research has emphasized that rapid weight change, a 10% higher increase or decrease in BMI over 2 years, was associated with a higher risk of dementia compared with a person with stable BMI (Park et al. 2019).

A systematic review and meta-analysis suggested that being overweight/obese <65 years is positively correlated with subsequent dementia; the opposite was seen in those aged ≥65 years (Pedditzi and Ruth Peters 2016). The reason behind this association may be attributable to the causality relationship between obesity and other comorbidities, especially impaired glucose tolerance, subsequent type 2 diabetes mellitus, and low-grade inflammation leading up to metabolic syndrome, all of which contribute to cognitive decline in the short and long term (Gunstad et al. 2010; Lamport et al. 2013, 2014; Yaffe et al. 2004). Also, cardiovascular risk factors, hypertension, and unbalanced insulin levels are associated with a high risk of dementia (Biessels et al. 2006; Pedditzi and Ruth Peters 2016). Additionally, obesity has an appealing correlation with unfavorable changes in brain volume and structure. There is evidence linking midlife obesity with changes in brain structures such as temporal atrophy and gray and white matter volume/integrity changes (Pannacciulli et al. 2006; Taki et al. 2008). Obesity is also linked to reduced brain volume independent of age and morbidity as it has been reported that obesity is related to reduced white matter integrity in many tracts, including the corpus callosum, cerebellar peduncles, cingulum, and corona radiata (Gunstad et al. 2008; Karlsson et al. 2013; Kullmann et al. 2016; Mueller et al. 2011; Papageorgiou et al. 2017; Verstynen et al. 2013; Ward et al. 2005). According to a population-based study, women with temporal lobe atrophy had higher BMIs indicating an association between obesity and brain pathology. Besides, in a longitudinal analysis, BMI is interlinked with temporal lobe atrophy and with gray matter atrophy in the temporal, frontal, and occipital cortices, thalamus, midbrain, and hippocampus (Gustafson et al. 2004; Shefer et al. 2013). Hippocampus has become a warranting interesting area of research. It is implicated in regulating food intake (Davidson et al. 2007), and consequently, any damage in this area can affect food-seeking behavior and body weight regulation as well (Davidson and Jarrard 1993; Higgs et al. 2008; Rozin et al. 1998). Western diets, which are high saturated fat foods, are associated with a small hippocampal volume (Jacka et al. 2015). Obese individuals also displayed a decreased activation of the right parietal cortex during a working memory task (Gonzales et al. 2010).

Additionally, three prospective trials investigating the effects of bariatric surgery on memory function have shown to result in improved memory among bariatric surgical patients in 2 years postoperatively compared to obese controls (Alosco et al. 2014; Gunstad et al. 2011; Miller et al. 2013). Smith et al. (2010) demonstrated that the DASH diet, which is a healthy diet aiming to decrease blood pressure, combined with a weight management program, exhibited potential benefits, including weight reduction and progress in memory function after 4 months. In preclinical trials, surgical lipectomy in mice was associated with normalization of hippocampus-dependent memory and reinstatement of long-term potentiation and dendritic spine density (Erion et al. 2014). Therefore, obesity implication in memory function might be one of the leading causes of memory impairment in AD (McRae-McKee et al. 2019).

From a genetic perspective, a correlation between an increase in BMI in childhood and midlife age and the fat mass and obesity-associated (FTO) gene was reported (Frayling et al. 2007). This gene is located on chromosome 16, and it is ubiquitously expressed in the human body, especially in neurons (Fredriksson et al. 2008; Han et al. 2010). It is copiously expressed in the brains, especially in the regions important for body weight regulation and feeding behavior (Dina et al. 2007). Keller et al. (2011) have suggested that the *FTO AA* genotype potentially increased the risk of dementia, particularly AD. A recent study suggests that the *FTO* gene is involved in the signaling pathway of leptin and its downstream effectors (Tung et al. 2010). Also, leptin might be associated with AD as its concentrations are elevated in the CSF of AD patients compared to controls (Bonda et al. 2014). Elevated *FTO* gene expression resulted in increased mRNA expression of stat3, a transcription factor essential for leptin receptor signaling (Tung et al. 2010). Collectively, these data confirm the implication of obesity as a risk factor in AD. Further work is needed to unravel the specific molecular and cellular mechanism by which obesity-related factors contribute to the pathological cascade, ultimately leading to AD progression.

12.5.4 Cholecystokinin Correlations with Hyperphagia and Obesity

Several lines of evidence suggest the incrimination of CCK patterns in developing obesity in animal and human studies. Abnormalities in CCK could result in obesity, which is consistent with the CCK role in regulating appetite and energy intake since deficiency of CCKA or receptor blockade in rodents could result in hyperphagia and obesity (Moran and Bi 2006). The intraperitoneal administration of CCK dose-dependently has decreased both solid and liquid food intakes in rats and significantly reduced body weight (Gibbs et al. 1973). Nevertheless, after discontinuation in CCK infusion, all meal patterns returned rapidly to normal, and bodyweight immediately recovered (West et al. 1984). Moreover, rats genetically lacking functional CCK1 receptors rapidly develop obesity. OLETF rats, which lack CCK1 receptors as a result of a mutation, become not only hyperphagic and overweight early in life, but also become resistant to the inhibitory effects of exogenous CCK on energy intake (Funakoshi et al. 1995; Moran et al. 1998). However, rather than the other way around, Kopin et al. observed that the administration of CCK in CCK1 receptor knockout mice fails to inhibit food intake, but maintains normal body weight in adult life (Kopin et al. 1999). These findings shed light on the CCK role in long-term maintenance of body weight. CCK has also shown to potentiate appetite and weight reduction by leptin, and the combination would be useful (Matson and Ritter 1999).

Several studies have tended to confirm the presence of altered plasma CCK concentrations after rapid weight loss in obese humans as it was observed that postprandial plasma CCK concentrations are diminished after a mean reduction of 14% in body weight in obese people, which was the result of a low energy diet-induced weight loss. Reductions in CCK levels persisted during the first year weight maintenance period (Sumithran et al. 2011).

Similarly, weight loss of approximately 15% of body weight in obese men ensues with a marked decrease in postprandial CCK concentrations compared with baseline

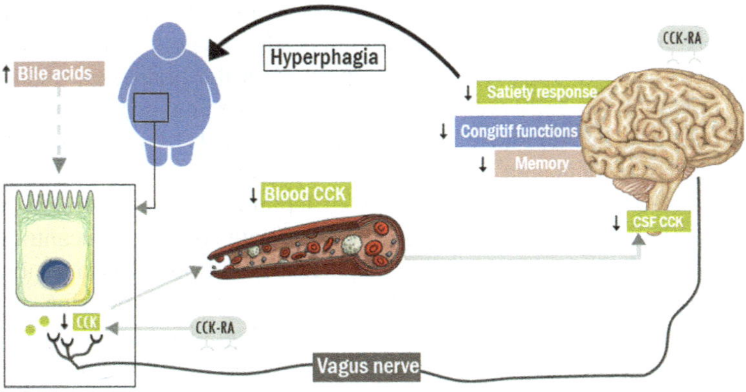

Fig. 12.4 The interplay between obesity and cholecystokinin, and their impact on memory. *CCK* cholecystokinin, *CCKR-A* cholecystokinin receptor A, *CSF* cerebrospinal fluid

(Chearskul et al. 2008). Moreover, in diet-induced obesity, a significant reduction in CCK levels was observed after 10 weeks of diet supplementation (Morris et al. 2008). In addition, basal and postprandial serum CCK levels were significantly reduced in obese women with metabolic syndrome features compared to matched controls (Zwirska-Korczala et al. 2007). These studies suggest that CCK is reduced following weight changes, which goes in line with the observed decline in cognition.

Obese patients display higher BA levels, which are negatively associated with the cognitive restraint of eating (Prinz et al. 2015). As we have discussed before, higher levels of CCK predicted better cognition and more gray matter in specific brain regions. AD patients have reduced CCK levels. Also, weight changes are accompanied by plasma CCK concentrations, which might be reflected in the CSF levels reported in AD patients. This might be illustrated by the inhibitory effect of BAs on CCK release (Fig. 12.4).

12.5.5 Crosstalk of Gut Microbiota and Alzheimer's Disease and Cholecystokinin

Human beings are teeming with microorganisms in and on their bodies, having over 150 times more microbial genes than mammalian genes, which highlight the role these tiny organisms play in our lives. The gut ecosystem harbors a complex and dynamic population of microbiota that is separated from the external environment by the gastric barrier. In terms of taxonomy, the microbiota is classified as phyla, and the prevalent and dominant gut microbial phyla are Firmicutes and Bacteroidetes. Human beings and their gut microbial communities establish a close association in which they coexist in a unique mutualistic scenario (Candela et al. 2012). This unique symbiotic interaction assists in maintaining nutrient metabolism balance and protects against pathogens. Thus, microbiota perturbation has serious consequences on digestive and metabolic disorders. Dysregulation of gut microbiota, caused by stress and some other factors, leads to increasing the permeability of the intestinal mucosa resulting in leaky gut syndrome (La Rosa et al. 2018). This leads to increased immune response and chronic neuroinflammation (Haroon et al. 2012; Mörkl et al. 2018), resulting in cytokine production, which is implicated in the pathophysiology of neurodegenerative disorders, including AD (Grochowska et al. 2019). Thus, it has become crucial to unravel the exact molecular and cellular mechanism by which gut microbiota contributes to the development, course, and prognosis of AD.

Studies on rodents exposed to microbial pathogens during developmental periods have shown anxiety-like behavior and impaired cognitive function (Bilbo et al. 2005; Goehler et al. 2008; Sullivan et al. 2006). The studies on mice grown in germ-free (GF) environment, vulnerable to pathogenic bacterial infections, antibiotics, probiotic agents, or fecal microbiota transplantation emphasized the role gut microbiota plays in host behavioral functions, cognition, and AD pathogenesis (Jiang et al. 2017). However, the extreme complexity and several confounding factors complicate the understanding of the molecular mechanisms by which intestinal microbiota can influence gut–brain communication.

Patients with AD have shown a decrease in the diversity of microbial composition in gut and blood compared with normal controls (Park et al. 2017; Vogt et al. 2017; Xin et al. 2018a, b). Gut bacteria can interact with the host environment and produce amyloids, LPSs, and other immunogenic compounds (Bhattacharjee and Lukiw 2013; Zhao et al. 2015; Hufnagel et al. 2013; Oli et al. 2012; Schwartz and Boles 2013; Syed and Boles 2014). For example, *E. coli* endotoxin contributes to the generation of Aβ fibrils (Asti and Gioglio 2014), which are the regulator for the signaling pathways implicated in neuroinflammation (Friedland 2015). Amyloid generation and dissemination result from the invasion of the intestinal–blood barrier and the BBB when they both become more permeable to small molecules during aging (Hill and Lukiw 2015; Marques et al. 2013; Shoemark and Allen 2015; Tran and Greenwood-Van Meerveld 2013). Amyloid influx across the BBB is also mediated by receptors for advanced glycosylation products (Deane et al. 2003). This amyloid influx depends on amyloid chaperones and apolipoproteins E and J (Zlokovic 1996), while amyloid clearance is governed by low-density lipoprotein receptor-related protein 1 (LRP1) (Deane et al. 2004). These transportation mechanisms rigorously maintain the physiological functions in the brain, and they are known to be altered in AD patients (Weiss et al. 2009). Bacterial-derived amyloids could leak from the GIT to accumulate in the brain (Zhao et al. 2015), causing an increase in reactive oxygen species (ROS) and activation of nuclear factor-jB (NF-jB) signaling, which upregulates the pro-inflammatory microRNA-34a (miRNA-34a). This miRNA-34a, in turn, downregulates the expression of the triggering receptors expressed in microglial/myeloid cells-2 (TREM2), resulting in an impairment of phagocytosis, and consequently, an accumulation of Ab42 peptide (Zhao et al. 2015; Zhao and Lukiw 2013) leading to neuroinflammation, which plays a significant role in the development of AD (Fig. 12.5).

Additionally, recent research has emphasized the role of gut microbiota in modulating the factors implicated in memory function and neural plasticity such as

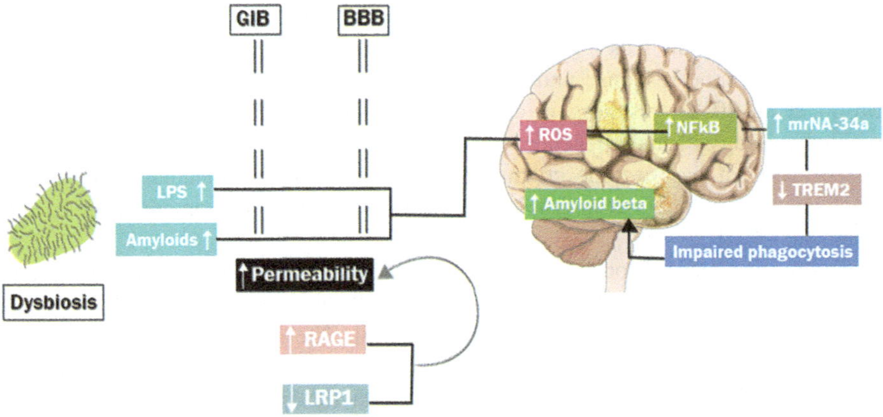

Fig. 12.5 Possible effects of gut microbiota in the brain of Alzheimer's disease patients. *CCK* cholecystokinin, *LPS* lipopolysaccharide, *RAGE* receptor for advanced glycation end products, *ROS* reactive oxygen species, *NFkB* nuclear factor kappa B, *miRNA* microRNA-34a, *TREM2* triggering receptor expressed on myeloid cells 2, *GIB* gastrointestinal barrier, *BBB* blood–brain barrier

CCK and BDNF (Bercik et al. 2011; Duca et al. 2012). Gut microbiota synthesizes an array of active metabolites by the degradation of indigestible carbohydrates such as short-chain fatty acids (SCFAs), acetate, propionate, butyrate (Topping and Clifton 2001), and BAs (Jones et al. 2008). Luminal and circulating SCFAs are the ligands of the receptors located on EEC, and they trigger the release of various metabolically active gut hormones. SCFAs, propionate, and butyrate, and acetate increase the secretion of CCK (Sileikiene et al. 2008). Besides, dissociated cells from the proximal small intestine of GF mice have low CCK expression levels (Duca et al. 2012). Thus, altered (dysbiosis) or unhealthy status of the gut microbiota is usually responsible for the decrease in CCK release, as suggested by previous research (Zhang et al. 2019). Therefore, the reduction in CCK secretion caused by gut microbial dysregulation might lead to a decrease in cognitive functions in the brain. These findings underline the critical role of the gut microbiota as a potent modulator for CCK release. Collectively, these findings raise some questions that have tended to remain in the background such as the factors dysregulating gut microbiota, and the underlying mechanisms by which dysbiosis affects memory, cognitive functions, and CCK hormonal release.

Several lines of evidence suggest that diet could give a state-of-art description of the composition of gut microbiota (Flint et al. 2012; Ley et al. 2006) since changes in the diet appear to affect the composition of gut microbiota within 24 h of the diet change (Ley et al. 2006), and also lead to a reduction in BDNF (Gyorkos et al. 2019). Therefore, it is incumbent to study the interaction of dysbiosis and diet modification on AD pathogenesis.

Diets have revealed their efficacy in enhancing cognitive functions in neurological diseases and irritable bowel syndrome (Lichtwark et al. 2014; Reddel et al.

2019). The disturbance of the gut–brain axis affects the brain functions such as the changes observed in the hippocampus and the frontal lobe in obese human models (Castanon et al. 2014). Diets rich in fat and sugar appeared to have deleterious effects on cognition throughout childhood and during adulthood (Khan et al. 2014). Furthermore, prolonged consumption of an HFD increases Firmicutes/Bacteroidetes ratio (Org et al. 2015; Shi et al. 2014; Ussar et al. 2015) observed in the population of obesity (Ley et al. 2006), and this imbalance results in gut dysbiosis (Fava et al. 2013; Kim et al. 2012; Ussar et al. 2015). This ratio seems to be changeable in people with unrestricted Western diet, a diet with excessive saturated fat and refined sugar (HFS), at different life stages. The ratio increases in the adults, but decreases in the elderly, suggesting a correlation between the decreased ratio and cognitive impairment in the elderly (Mariat et al. 2009).

Further, Western diets increase intestinal permeability and induce BBB damage, which increases the susceptibility to toxin influx from the circulation to the brain (Noble et al. 2017). Strikingly, the implication of gut dysbiosis in cognitive decline in the elderly might be elucidated by its effect on modulating BDNF functions in the CNS via changes in the neurotransmitter functions or by changes in SCFAs availability and actions in the brain (Maqsood and Stone 2016). The hippocampus of the rat-fed HFS diet has appeared to have reduced levels of BDNF, resulting in decreased neuronal plasticity. Therefore, reduced levels of BDNF as a result of the HFS diet influence neuronal and behavioral functions (Molteni et al. 2002).

Moreover, animals maintained on the HSF diet showed decreased levels of synapsin I, cyclic AMP-response element-binding protein (CREB), and growth-associated protein 43 (GAP-43), all of which contribute to BDNF release (Molteni et al. 2002). SCFAs exert a widespread influence on neurodegenerative disorders (Silva et al. 2020). They interfere with protein–protein interactions between Aβs and distribute their assembly into neurotoxic oligomers (Ho et al. 2018). Besides, lower levels of SCFAs could have negative effects on glucose status in the brain, immune responses, energy homeostasis, and epithelial growth factor, affecting the central and peripheral nervous system function (Bienenstock et al. 2015; Canfora et al. 2015). Therefore, lower SCFA levels do not only decrease CCK concentration, and thus worsening memory function in the brain, but also are suggested to directly impact the pathological cascade of AD (Silva et al. 2020). However, the underpinnings of SCFA effect on this disease remain to be determined (Fig. 12.6).

Besides the contributable role of diet to the alterations of microbial composition, it influences the production or aggregation of amyloid proteins (Bieschke et al. 2010; Friedland 2015; Soto 2012). Thus, depicting the underlying mechanisms of the interplay of diet, microbial composition, CCK release, and amyloid production could bring us a significant step forward, elucidating AD pathogenesis and help us to ameliorate its progress by a high-quality diet. However, it is still partially shrouded in mystery and needs further research.

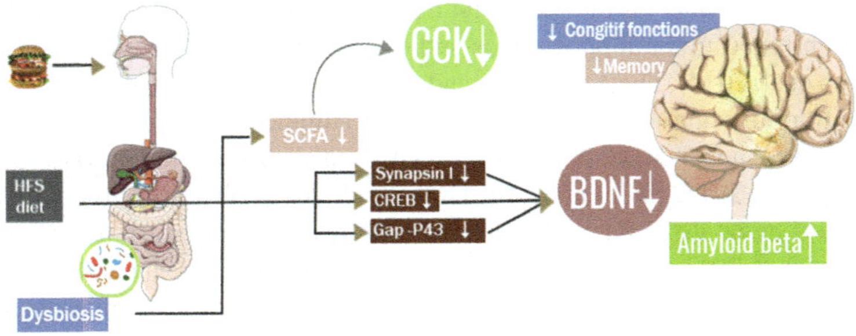

Fig. 12.6 The effect of high-fat sugar diet diet-induced dysbiosis on the brain. *HFS* high-fat sugar diet, *SCFA* short-chain fatty acid, *CCK* cholecystokinin, *CREB* cAMP response element-binding protein, *Gap-P43* growth-associated protein 43, *BDNF* brain-derived neurotrophic factor

12.6 Linkage of Cholecystokinin and Ketone Bodies in Alzheimer's Disease

The term ketone bodies (KBs) encompass three molecules, acetoacetate (AcAc), 3-β-hydroxybutyrate (3HB), and acetone (Laffel 1999). Ketosis refers to a state of elevations of these molecules in the blood or the urine while being used as a fuel in case of reduced glucose uptake or metabolism (Sokoloff 1973). Interestingly, research has found consistent diminutions in regional glucose use in demented patients compared to elderly normal persons (Cunnane et al. 2011). This change in glucose metabolism represents an excellent correlation with clinical disabilities in dementia (Blass 2002). At a molecular level, ROS overproduction and the changes in cellular calcium regulation are the results of brain hypometabolism. The production of ROS and calcium homeostasis are directly altered in the brains of AD patients, which are generally under severe oxidative stress (Lambert et al. 1998). This is a critical mechanism for synapse disruption in AD (Keller et al. 2002). Oxidative stress is associated with plaques and tangles and can plausibly be connected to the clinical aspects of the disease, mainly via disruption of the synaptic activity (Keller et al. 1997, 2002).

In contrast, the metabolism of KB is unaltered, at least, in the early stages of the disease (Castellano et al. 2014). It was postulated that KB ameliorates and improves cognitive functions (Mattson et al. 2018; Xin et al. 2018a, b). Diet supplementation with KB has consistently been found to cause a modest improvement of mental functions in AD patients. Ketone diet, as an initial treatment at preclinical stages, was more pronounced and efficient than after the appearance of AD first symptoms (Hertz et al. 2015). Several animal studies have shown that KB could effectively ameliorate AD symptoms through multiple mechanisms. KB reduces the production of advanced glycation end products (Srikanth et al. 2011; Yao et al. 2011). The accumulation of these products accelerates the progression of AD, as previously described. In clinical research, KB can reduce oxidative stress and inflammation and delay the progression of AD, which is later manifested by improved cognitive function (Henderson et al. 2009).

Additionally, the effect of KB on CCK release was well delineated in previous research. Paoli et al. (2015) stated that ketosis guarantees sustained CCK levels. Moreover, another research is referred to the role of KB in enhancing CCK release (Chearskul et al. 2008). It was well postulated in previous research that KB exerts anorexigenic effects and has an ameliorative influence on memory in the brain (Krikorian et al. 2012; Paoli et al. 2015). Of note, the effect of KB is blocked by the transection of the common hepatic branch of the vagus nerve (Langhans et al. 1985), and the same nerve is associated with memory performance (Sun et al. 2017). Thus, it is suggested that the ketone diet induces satiety and influences memory function through potentiating the CCK release and the vagal signaling pathways. In addition, fenofibrate treatment acts on peroxisome proliferator-activated receptor α (PPAR-α), controls both lipid/glucose metabolism, and reduces body weight gain, adiposity, and food intake (Larsen et al. 2003; Park et al. 2012). Several studies found that fenofibrate treatment is associated with increased CCK production and KB (Grabacka et al. 2016; Park et al. 2012; Reddy and Hashimoto 2001; Srivastava et al. 2006). Decreased food intake after treatment with fenofibrate was correlated with the increase in CCK binding to CCKA receptors in the small intestine. However, in CCK receptor-deficient OLETF rats, this effect was not observed (Park et al. 2012).

Surprisingly, the memory-enhancing effect was observed as well after treatment with fenofibrate (Ancelin et al. 2012; Ouk et al. 2014), which points out a possible role KB plays in enhancing memory and mediating the anorexigenic effect via the same hormonal mechanism of CCK.

12.7 Changes in Cholecystokinin System in Other Dementia Subtypes

Parkinson's disease (PD) is a chronic progressive age-related neurodegenerative disease characterized by so-called movement disorders of bradykinesia, resting tremors, and rigidity. The recognized underlying pathophysiological mechanism of the disease is the loss of DA neurotransmitters. It is associated with loss of dopaminergic neurons in the SN, the presence of Lewy bodies, and non-motor symptoms, which may accompany PD from early stages onward (Hanagasi et al. 2017; Kalia and Lang 2015). Non-motor symptoms can be troublesome and debilitating than motor symptoms, and dementia is regarded as one of which. Its incidence in PD has been increasing (Hanagasi et al. 2017) and is estimated by 24–31% (Aarsland et al. 2005). The cognitive profile of patients with Parkinson's disease dementia describes a constellation of poor planning, impaired episodic memory, abstract reasoning, and mental flexibility, and apathy (Petrova et al. 2012). It is difficult to make an adequate diagnosis at the earliest stages of PD and consequently struggling to manage the symptoms at its later stages (Kalia and Lang 2015). Interestingly, the ketogenic diet has shown benefits for PD patients since it has revealed a significant improvement in non-motor symptoms and the unified Parkinson's disease rating scale; however, further research is needed (Walczyk and Wick 2017).

CCK coexists with DA in dopaminergic neurons and various brain nuclei, so the locus of CCK in the brain is suggested to act as a regulator of the dopaminergic system, modulate its release, and potentiate the DA-mediated behaviors (Crawley et al. 1985). CCKA and CCKB receptors are present in the brain, and both are associated with DA release. CCKA receptor enhances the DA release from the posterior nucleus and mediates the behavioral actions of this neurotransmitter, whereas the CCKB receptor inhibits the DA release from the anterior nucleus accumbens (Crawley 1991). PD results from multifaceted interactions of genetic and environmental factors altering numerous fundamental cellular processes (Fujii et al. 1999). In a case–control study, PD patients showed a significant difference in the gene distribution coding for CCK (Fujii et al. 1999). It is noteworthy to mention that CCK *CT/TT* genotypes correlate with hallucinations, which are a prominent feature for dementia in PD (Wang et al. 2003). However, there are contradictory results as no significant associations were found between PD and any of the genotypes of CCK, CCKAR, and CCKBR. This contradiction might be elucidated by the racial and ethnic differences among species in the studies (Goldman 2011).

Huntington's disease (HD) is an inherited, disabling, progressive neurodegenerative disorder of the CNS characterized clinically by a relentlessly progressive movement disorder and a range of neuropsychiatric problems, including cognitive impairment (Folstein 1989). Clinical diagnosis of HD depends on family history, the development of an extrapyramidal movement disorder, and the presence of a positive genetic test (Paulsen 2011). Despite the progress in understanding the disease pathogenesis, the neuroprotective strategies remain limited. In recent years, searching for biomarkers of disease progression has become increasingly important (Ha and Fung 2012).

In a study conducted by Hays et al. (1981), they found that CCK receptors were significantly decreased in the basal ganglia and the cerebral cortex of postmortem brains of HD patients compared with matched controls suggesting a possible selective loss of CCK receptor-containing neurons in the cerebral cortex of HD patients. However, in another study, it found an increase in immunoreactive CCK-LI concentrations in the prefrontal, premotor, temporal, and occipital cortex of HD postmortem brains. This contradiction might also be elucidated by the racial and ethnic differences among species as well (Goldman 2011), so further research is needed to resolve the contradictory results.

Frontotemporal dementia (FTD) is a heterogeneous group of non-Alzheimer's dementias and is collectively characterized by progressive atrophy that primarily affects frontal or temporal lobes, or both (Warren et al. 2013). As described in previous research, FTD patients frequently described alterations in eating behavior (Ahmed et al. 2014; Ikeda 2002; Rascovsky et al. 2011). A recent study measuring the levels of peptides involved in the regulation of eating behavior shows no differences in CCK concentrations compared with matched controls (Ahmed et al. 2015). To our knowledge, this is the only existing study that compares CCK levels in FTD patients. Further research is warranted to confirm the previous results.

12.8 Recent Developments and Future Perspectives

During the last decades, a frameshift has occurred in the field of AD research and brought a burst of knowledge about genetics, pharmacological, clinical, behavioral, and social research that redefines the disease for better understanding to curtail its progress. Until now, there is no cure for AD and unmet requirements for effective treatments. However, an in-depth look into the significant underlying symptoms, including behavioral functions, mood disturbance, metabolic dysregulation, and inflammation, could lead to novel therapeutics that alleviate the disease progress and guide researchers toward more understanding and searching for effective treatments. The distinctive components present in the brains of AD patients are Aβ and NFTs. Gut microbiota and BDNF are associated with the biological underpinnings of AD pathology, including amyloid generation, inflammatory response, and metabolic disturbances.

Tackling the association between food habits, microbial composition, and AD risk is of considerable interest. It is necessary to characterize the contribution of gut microbiota and dietary patterns on tauopathy and neurodegeneration, so this issue needs much more preclinical studies. Microbial composition and BDNF are both associated with CCK release, which draws a triad association between CCK, microbial composition, and BDNF, since AD risk is highly related to gender (Viña and Lloret 2010) and CSF CCK has shown variations in both genders so that CSF CCK might be a future useful biomarker for AD (Plagman et al. 2019). Therefore, including male and female rats in preclinical studies should be taken into consideration. Tracking the changes in CCK release induced by diet, BA release, and microbial diversity is imperative. Moreover, it is essential to understand the possibility for CCK to cross BBB since contradictions were found (Hommer et al. 1985); thus, more work is needed to conclude in this area.

12.9 Conclusion

The potential implication of CCK in memory, neuronal plasticity, and BDNF release raises much concern. Besides, it has been hypothesized that the trigger for CCK release is through the implication of gut microbial products, and microbial dysbiosis may result in its reduction, indicating a link between CCK release and gut microbiota. Also, diet-induced CCK release affects the microbial composition and BA release. These factors orchestrate a dynamic and complex interplay that might incriminate CCK in the pathophysiology/development of AD. Therefore, preclinical studies are warranted to draw a conclusive mechanism linking dietary patterns, microbial composition, CCK, BAs, and BDNF release all together to achieve a better understanding of the pathology of AD and consequently preventing its progress.

References

Aarsland D, Zaccai J, Brayne C (2005) A systematic review of prevalence studies of dementia in Parkinson's disease. Mov Disord 20:1255. https://doi.org/10.1002/mds.20527

Adebakin A, Bradley J, Gümüsgöz S, Waters EJ, Lawrence CB (2012) Impaired satiation and increased feeding behaviour in the triple-transgenic Alzheimer's disease mouse model. PLoS One 7:e45179. https://doi.org/10.1371/journal.pone.0045179

Ahmed RM, Irish M, Kam J, van Keizerswaard J, Bartley L, Samaras K, Hodges JR, Piguet O (2014) Quantifying the eating abnormalities in frontotemporal dementia. JAMA Neurol 71: 1540. https://doi.org/10.1001/jamaneurol.2014.1931

Ahmed RM, Latheef S, Bartley L, Irish M, Halliday GM, Kiernan MC, Hodges JR, Piguet O (2015) Eating behavior in frontotemporal dementia. Neurology 85:1310–1317. https://doi.org/10.1212/WNL.0000000000002018

Alosco ML, Spitznagel MB, Strain G, Devlin M, Cohen R, Paul R, Crosby RD, Mitchell JE, Gunstad J (2014) Improved memory function two years after bariatric surgery. Obesity 22:32–38. https://doi.org/10.1002/oby.20494

Altar CA (1999) Neurotrophins and depression. Trends Pharmacol Sci 20:59. https://doi.org/10.1016/S0165-6147(99)01309-7

Ancelin M-L, Carrière I, Barberger-Gateau P, Auriacombe S, Rouaud O, Fourlanos S, Berr C, Dupuy A-M, Ritchie K (2012) Lipid lowering agents, cognitive decline, and dementia: the three-city study. J Alzheimers Dis 30:629–637. https://doi.org/10.3233/JAD-2012-120064

Anthony IC, Norrby KE, Dingwall T, Carnie FW, Millar T, Arango JC, Robertson R, Bell JE (2010) Predisposition to accelerated Alzheimer-related changes in the brains of human immunodeficiency virus negative opiate abusers. Brain 133:3685–3698. https://doi.org/10.1093/brain/awq263

Asti A, Gioglio L (2014) Can a bacterial endotoxin be a key factor in the kinetics of amyloid fibril formation? J Alzheimers Dis 39:169–179. https://doi.org/10.3233/JAD-131394

Bariohay B, Lebrun B, Moyse E, Jean A (2005) Brain-derived neurotrophic factor plays a role as an anorexigenic factor in the dorsal vagal complex. Endocrinology 146:5612–5620. https://doi.org/10.1210/en.2005-0419

Beinfeld MC, Meyer DK, Eskay RL, Jensen RT, Brownstein MJ (1981) The distribution of cholecystokinin immunoreactivity in the central nervous system of the rat as determined by radioimmunoassay. Brain Res 212:51–57. https://doi.org/10.1016/0006-8993(81)90031-7

Benamar K, Yondorf M, Barreto VT, Geller EB, Adler MW (2007) Deletion of μ-opioid receptor in mice alters the development of acute neuroinflammation. J Pharmacol Exp Ther 323:990–994. https://doi.org/10.1124/jpet.107.129973

Benarroch EE (2015) Brain-derived neurotrophic factor. Neurology 84:1693–1704. https://doi.org/10.1212/WNL.0000000000001507

Bercik P, Denou E, Collins J, Jackson W, Lu J, Jury J, Deng Y, Blennerhassett P, MacRi J, McCoy KD, Verdu EF, Collins SM (2011) The intestinal microbiota affect central levels of brain-derived neurotropic factor and behavior in mice. Gastroenterology 141:599–609.e3. https://doi.org/10.1053/j.gastro.2011.04.052

Berer K, Mues M, Koutrolos M, AlRasbi Z, Boziki M, Johner C, Wekerle H, Krishnamoorthy G (2011) Commensal microbiota and myelin autoantigen cooperate to trigger autoimmune demyelination. Nature 479:538–541. https://doi.org/10.1038/nature10554

Berthoud HR, Blackshaw LA, Brookes SJH, Grundy D (2004) Neuroanatomy of extrinsic afferents supplying the gastrointestinal tract. Neurogastroenterol Motil 16:28–33. https://doi.org/10.1111/j.1743-3150.2004.00471.x

Beydoun MA, Beydoun HA, Wang Y (2008) Obesity and central obesity as risk factors for incident dementia and its subtypes: a systematic review and meta-analysis. Obes Rev 9:204. https://doi.org/10.1111/j.1467-789X.2008.00473.x

Bhattacharjee S, Lukiw WJ (2013) Alzheimer's disease and the microbiome. Front Cell Neurosci 7: 153. https://doi.org/10.3389/fncel.2013.00153

Bienenstock J et al (2015) Microbiota and the gut-brain axis. Nutr Rev 73:28
Bieschke J, Russ J, Friedrich RP, Ehrnhoefer DE, Wobst H, Neugebauer K, Wanker EE (2010) EGCG remodels mature α-synuclein and amyloid-β fibrils and reduces cellular toxicity. Proc Natl Acad Sci U S A 107:7710–7715. https://doi.org/10.1073/pnas.0910723107
Biessels GJ, Staekenborg S, Brunner E, Brayne C, Scheltens P (2006) Risk of dementia in diabetes mellitus: a systematic review. Lancet Neurol 5:64. https://doi.org/10.1016/S1474-4422(05)70284-2
Bilbo SD, Levkoff LH, Mahoney JH, Watkins LR, Rudy JW, Maier SF (2005) Neonatal infection induces memory impairments following an immune challenge in adulthood. Behav Neurosci 119:293–301. https://doi.org/10.1037/0735-7044.119.1.293
Blass JP (2002) Alzheimer's disease and Alzheimer's dementia: distinct but overlapping entities. Neurobiol Aging 23:1077–1084. https://doi.org/10.1016/S0197-4580(02)00036-2
Bogunovic M, Davé SH, Tilstra JS, Chang DTW, Harpaz N, Xiong H, Mayer LF, Plevy SE (2007) Enteroendocrine cells express functional toll-like receptors. Am J Physiol Liv Physiol 292: G1770–G1783. https://doi.org/10.1152/ajpgi.00249.2006
Bonda DJ, Stone JG, Torres SL, Siedlak SL, Perry G, Kryscio R, Jicha G, Casadesus G, Smith MA, Zhu X, Lee HG (2014) Dysregulation of leptin signaling in Alzheimer disease: evidence for neuronal leptin resistance. J Neurochem 128:162–172. https://doi.org/10.1111/jnc.12380
Bradwejn J (1993) Neurobiological investigations into the role of cholecystokinin in panic disorder. J Psychiatry Neurosci 18:178–188
Brookmeyer R, Johnson E, Ziegler-graham K, Arrighi HM (2007) Forecasting the global burden of Alzheimer's disease. Alzheimers Dement 3:186–191. https://doi.org/10.1016/j.jalz.2007.04.381
Burdyga G (2004) Expression of cannabinoid CB1 receptors by vagal afferent neurons is inhibited by cholecystokinin. J Neurosci 24:2708–2715. https://doi.org/10.1523/JNEUROSCI.5404-03.2004
Burdyga G, Varro A, Dimaline R, Thompson DG, Dockray GJ (2006) Feeding-dependent depression of melanin-concentrating hormone and melanin-concentrating hormone receptor-1 expression in vagal afferent neurones. Neuroscience 137:1405–1415. https://doi.org/10.1016/j.neuroscience.2005.10.057
Bush DEA, DeSousa NJ, Vaccarino FJ (1999) Self-administration of intravenous amphetamine: effect of nucleus accumbens CCK B receptor activation on fixed-ratio responding. Psychopharmacology 147:331–334. https://doi.org/10.1007/s002130051176
Cai Z, Ratka A (2012) Opioid system and Alzheimer's disease. NeuroMolecular Med 14:91–111. https://doi.org/10.1007/s12017-012-8180-3
Candela M, Biagi E, Maccaferri S, Turroni S, Brigidi P (2012) Intestinal microbiota is a plastic factor responding to environmental changes. Trends Microbiol 20:385. https://doi.org/10.1016/j.tim.2012.05.003
Canfora EE, Jocken JW, Blaak EE (2015) Short-chain fatty acids in control of body weight and insulin sensitivity. Nat Rev Endocrinol 11:577. https://doi.org/10.1038/nrendo.2015.128
Castanon N, Lasselin J, Capuron L (2014) Neuropsychiatric comorbidity in obesity: role of inflammatory processes. Front Endocrinol 5:74. https://doi.org/10.3389/fendo.2014.00074
Castellano C-A, Nugent S, Paquet N, Tremblay S, Bocti C, Lacombe G, Imbeault H, Turcotte É, Fulop T, Cunnane SC (2014) Lower brain 18F-fluorodeoxyglucose uptake but normal 11C-acetoacetate metabolism in mild Alzheimer's disease dementia. J Alzheimers Dis 43: 1343–1353. https://doi.org/10.3233/JAD-141074
Chao MV (2003) Neurotrophins and their receptors: a convergence point for many signalling pathways. Nat Rev Neurosci 4:299–309. https://doi.org/10.1038/nrn1078
Chearskul S, Delbridge E, Shulkes A, Proietto J, Kriketos A (2008) Effect of weight loss and ketosis on postprandial cholecystokinin and free fatty acid concentrations. Am J Clin Nutr 87:1238–1246. https://doi.org/10.1093/ajcn/87.5.1238

Chen D, Zhao C-M, Håkanson R, Samuelson LC, Rehfeld JF, Friis-Hansen L (2004) Altered control of gastric acid secretion in gastrin-cholecystokinin double mutant mice. Gastroenterology 126:476–487. https://doi.org/10.1053/j.gastro.2003.11.012

Choi B-S, Sapkota K, Kim S, Lee HJ, Choi H-S, Kim S-J (2010) Antioxidant activity and protective effects of Tripterygium regelii extract on hydrogen peroxide-induced injury in human dopaminergic cells, SH-SY5Y. Neurochem Res 35:1269–1280. https://doi.org/10.1007/s11064-010-0185-4

Chooi YC, Ding C, Magkos F (2019) The epidemiology of obesity. Metabolism 92:6–10. https://doi.org/10.1016/j.metabol.2018.09.005

Crawley JN (1991) Cholecystokinin-dopamine interactions. Trends Pharmacol Sci 12:232. https://doi.org/10.1016/0165-6147(91)90558-A

Crawley JN, Corwin RL (1994) Biological actions of cholecystokinin. Peptides 15:731–755. https://doi.org/10.1016/0196-9781(94)90104-X

Crawley JN, Stivers JA, Blumstein LK, Paul SM (1985) Cholecystokinin potentiates dopamine-mediated behaviors: evidence for modulation specific to a site of coexistence. J Neurosci 5: 1972–1983. https://doi.org/10.1523/jneurosci.05-08-01972.1985

Cryan JF, Dinan TG (2012) Mind-altering microorganisms: the impact of the gut microbiota on brain and behaviour. Nat Rev Neurosci 13:701–712. https://doi.org/10.1038/nrn3346

Cunnane S, Nugent S, Roy M, Courchesne-Loyer A, Croteau E, Tremblay S, Castellano A, Pifferi F, Bocti C, Paquet N, Begdouri H, Bentourkia M, Turcotte E, Allard M, Barberger-Gateau P, Fulop T, Rapoport SI (2011) Brain fuel metabolism, aging, and Alzheimer's disease. Nutrition 27:3–20. https://doi.org/10.1016/j.nut.2010.07.021

Davidson TL, Jarrard LE (1993) A role for hippocampus in the utilization of hunger signals. Behav Neural Biol 59:167–171. https://doi.org/10.1016/0163-1047(93)90925-8

Davidson TL, Kanoski SE, Schier LA, Clegg DJ, Benoit SC (2007) A potential role for the hippocampus in energy intake and body weight regulation. Curr Opin Pharmacol 7:613. https://doi.org/10.1016/j.coph.2007.10.008

De Kloet ER, Oitzl MS, Joëls M (1999) Stress and cognition: are corticosteroids good or bad guys? Trends Neurosci 22:422–426. https://doi.org/10.1016/S0166-2236(99)01438-1

Deane R, Du Yan S, Submamaryan RK, LaRue B, Jovanovic S, Hogg E, Welch D, Manness L, Lin C, Yu J, Zhu H, Ghiso J, Frangione B, Stern A, Schmidt AM, Armstrong DL, Arnold B, Liliensiek B, Nawroth P, Hofman F, Kindy M, Stern D, Zlokovic B (2003) RAGE mediates amyloid-β peptide transport across the blood-brain barrier and accumulation in brain. Nat Med 9:907–913. https://doi.org/10.1038/nm890

Deane R, Wu Z, Sagare A, Davis J, Du Yan S, Hamm K, Xu F, Parisi M, LaRue B, Hu HW, Spijkers P, Guo H, Song X, Lenting PJ, Van Nostrand WE, Zlokovic BV (2004) LRP/amyloid β-peptide interaction mediates differential brain efflux of Aβ isoforms. Neuron 43:333–344. https://doi.org/10.1016/j.neuron.2004.07.017

Debas HT, Farooq O, Grossman MI (1975) Inhibition of gastric emptying is a physiological action of cholecystokinin. Gastroenterology 68:1211–1217. PMID 1126597

Demars M, Hu Y-S, Gadadhar A, Lazarov O (2010) Impaired neurogenesis is an early event in the etiology of familial Alzheimer's disease in transgenic mice. J Neurosci Res 88:2103–2117. https://doi.org/10.1002/jnr.22387

Dieni S, Matsumoto T, Dekkers M, Rauskolb S, Ionescu MS, Deogracias R, Gundelfinger ED, Kojima M, Nestel S, Frotscher M, Barde YA (2012) BDNF and its pro-peptide are stored in presynaptic dense core vesicles in brain neurons. J Cell Biol 196:775–788. https://doi.org/10.1083/jcb.201201038

Dina C, Meyre D, Gallina S, Durand E, Körner A, Jacobson P, Carlsson LMS, Kiess W, Vatin V, Lecoeur C, Delplanque J, Vaillant E, Pattou F, Ruiz J, Weill J, Levy-Marchal C, Horber F, Potoczna N, Hercberg S, Le Stunff C, Bougnères P, Kovacs P, Marre M, Balkau B, Cauchi S, Chèvre JC, Froguel P (2007) Variation in FTO contributes to childhood obesity and severe adult obesity. Nat Genet 39:724–726. https://doi.org/10.1038/ng2048

Dockray GJ (2009) Cholecystokinin and gut–brain signalling. Regul Pept 155:6–10. https://doi.org/10.1016/j.regpep.2009.03.015

Droogsma E, Van Asselt D, Bieze H, Veeger N, De Deyn PP (2015) The relationship of weight change trajectory with medial temporal lobe atrophy in patients with mild Alzheimer's disease: results from a cohort study. Alzheimers Res Ther 7:18. https://doi.org/10.1186/s13195-015-0098-1

Duca FA, Swartz TD, Sakar Y, Covasa M (2012) Increased oral detection, but decreased intestinal signaling for fats in mice lacking gut microbiota. PLoS One 7:e39748. https://doi.org/10.1371/journal.pone.0039748

Durany N, Michel T, Kurt J, Cruz-Sánchez FF, Cervós-Navarro J, Riederer P (2000) Brain-derived neurotrophic factor and neurotrophin-3 levels in Alzheimer's disease brains. Int J Dev Neurosci 18:807–813. https://doi.org/10.1016/S0736-5748(00)00046-0

Eckel RH (1997) Obesity and heart disease: a statement for healthcare professionals from the nutrition committee. Circulation 96:3248. https://doi.org/10.1161/01.CIR.96.9.3248

Elias MF, Goodell AL, Waldstein SR (2012) Obesity, cognitive functioning and dementia: back to the future. J Alzheimers Dis 2:S113. https://doi.org/10.3233/JAD-2011-111175

Erion JR, Wosiski-Kuhn M, Dey A, Hao S, Davis CL, Pollock NK, Stranahan AM (2014) Obesity elicits interleukin 1-mediated deficits in hippocampal synaptic plasticity. J Neurosci 34:2618–2631. https://doi.org/10.1523/JNEUROSCI.4200-13.2014

Fava F, Gitau R, Griffin BA, Gibson GR, Tuohy KM, Lovegrove JA (2013) The type and quantity of dietary fat and carbohydrate alter faecal microbiome and short-chain fatty acid excretion in a metabolic syndrome 'at-risk' population. Int J Obes 37:216–223. https://doi.org/10.1038/ijo.2012.33

Ferrer I, Krupinski J, Goutan E, Martí E, Ambrosio S, Arenas E (2001) Brain-derived neurotrophic factor reduces cortical cell death by ischemia after middle cerebral artery occlusion in the rat. Acta Neuropathol 101:229–238. https://doi.org/10.1007/s004010000268

Flint HJ, Scott KP, Duncan SH, Louis P, Forano E (2012) Microbial degradation of complex carbohydrates in the gut. Gut Microbes 3:289. https://doi.org/10.4161/gmic.19897

Folstein SE (1989) Huntington's disease: a disorder of families. Am J Hum Genet 48:171

Frayling TM, Timpson NJ, Weedon MN, Zeggini E, Freathy RM, Lindgren CM, Perry JRB, Elliott KS, Lango H, Rayner NW, Shields B, Harries LW, Barrett JC, Ellard S, Groves CJ, Knight B, Patch AM, Ness AR, Ebrahim S, Lawlor DA, Ring SM, Ben-Shlomo Y, Jarvelin MR, Sovio U, Bennett AJ, Melzer D, Ferrucci L, Loos RJF, Barroso I, Wareham NJ, Karpe F, Owen KR, Cardon LR, Walker M, Hitman GA, Palmer CNA, Doney ASF, Morris AD, Smith GD, Hattersley AT, McCarthy MI (2007) A common variant in the FTO gene is associated with body mass index and predisposes to childhood and adult obesity. Science 316:889–894. https://doi.org/10.1126/science.1141634

Fredriksson R, Hägglund M, Olszewski PK, Stephansson O, Jacobsson JA, Olszewska AM, Levine AS, Lindblom J, Schiöth HB (2008) The obesity gene, FTO, is of ancient origin, up-regulated during food deprivation and expressed in neurons of feeding-related nuclei of the brain. Endocrinology 149:2062–2071. https://doi.org/10.1210/en.2007-1457

Friedland RP (2015) Mechanisms of molecular mimicry involving the microbiota in neurodegeneration. J Alzheimers Dis 45:349. https://doi.org/10.3233/JAD-142841

Fujii C, Harada S, Ohkoshi N, Hayashi A, Yoshizawa K, Ishizuka C, Nakamura T (1999) Association between polymorphism of the cholecystokinin gene and idiopathic Parkinson's disease. Clin Genet 56:395–400. https://doi.org/10.1034/j.1399-0004.1999.560508.x

Funakoshi A, Miyasaka K, Shinozaki H, Masuda M, Kawanami T, Takata Y, Kono A (1995) An animal model of congenital defect of gene expression of cholecystokinin (CCK)-a receptor. Biochem Biophys Res Commun 210:787–796. https://doi.org/10.1006/bbrc.1995.1728

Gaugler J, James B, Johnson T, Scholz K, Weuve J (2016) 2016 Alzheimer's disease facts and figures. Alzheimers Dement 12:459–509. https://doi.org/10.1016/j.jalz.2016.03.001

Geracioti TD, Nicholson WE, Orth DN, Ekhator NN, Loosen PT (1993) Cholecystokinin in human cerebrospinal fluid: concentrations, dynamics, molecular forms and relationship to fasting and feeding in health, depression and alcoholism. Brain Res 629:260–268. https://doi.org/10.1016/0006-8993(93)91329-Q

Gibbs J, Young RC, Smith GP (1973) Cholecystokinin decreases food intake in rats. J Comp Physiol Psychol 84:488–495. https://doi.org/10.1037/h0034870

Ginsberg SD, Che S, Wuu J, Counts SE, Mufson EJ (2006) Down regulation of trk but not p75NTR gene expression in single cholinergic basal forebrain neurons mark the progression of Alzheimer's disease. J Neurochem 97:475–487. https://doi.org/10.1111/j.1471-4159.2006.03764.x

Glenner GG, Wong CW (1984) Alzheimer's disease: initial report of the purification and characterization of a novel cerebrovascular amyloid protein. Biochem Biophys Res Commun 120: 885–890. https://doi.org/10.1016/S0006-291X(84)80190-4

Goehler LE, Park SM, Opitz N, Lyte M, Gaykema RPA (2008) Campylobacter jejuni infection increases anxiety-like behavior in the holeboard: possible anatomical substrates for viscerosensory modulation of exploratory behavior. Brain Behav Immun 22:354–366. https://doi.org/10.1016/j.bbi.2007.08.009

Goldman JG (2011) New thoughts on thought disorders in Parkinson's disease: review of current research strategies and challenges. Parkinsons Dis 2011:675630. https://doi.org/10.4061/2011/675630

Gomez G, Lluis F, Guo YS, Greeley GH, Townsend CM, Thompson JC (1986) Bile inhibits release of cholecystokinin and neurotensin. Surgery 100:363–368

Gomez G, Upp JR, Lluis F, Alexander RW, Poston GJ, Greeley GH, Thompson JC (1988) Regulation of the release of cholecystokinin by bile salts in dogs and humans. Gastroenterology 94:1036–1046. https://doi.org/10.1016/0016-5085(88)90564-1

Gomez-Pinilla F (2008) The influences of diet and exercise on mental health through hormesis. Ageing Res Rev 7:49. https://doi.org/10.1016/j.arr.2007.04.003

Gonzales MM, Tarumi T, Miles SC, Tanaka H, Shah F, Haley AP (2010) Insulin sensitivity as a mediator of the relationship between BMI and working memory-related brain activation. Obesity 18:2131–2137. https://doi.org/10.1038/oby.2010.183

Grabacka MM, Wilk A, Antonczyk A, Banks P, Walczyk-Tytko E, Dean M, Pierzchalska M, Reiss K (2016) Fenofibrate induces ketone body production in melanoma and glioblastoma cells. Front Endocrinol 7:5. https://doi.org/10.3389/fendo.2016.00005

Green GM, Lyman RL (1972) Feedback regulation of pancreatic enzyme secretion as a mechanism for trypsin inhibitor-induced hypersecretion in rats. Exp Biol Med 140:6–12. https://doi.org/10.3181/00379727-140-36384

Green KN, Billings LM, Roozendaal B, McGaugh JL, LaFerla FM (2006) Glucocorticoids increase amyloid-β and tau pathology in a mouse model of Alzheimer's disease. J Neurosci 26:9047–9056. https://doi.org/10.1523/JNEUROSCI.2797-06.2006

Greenberg N, Grassano A, Thambisetty M, Lovestone S, Legido-Quigley C (2009) A proposed metabolic strategy for monitoring disease progression in Alzheimer's disease. Electrophoresis 30:1235–1239. https://doi.org/10.1002/elps.200800589

Greisen MH, Bolwig TG, Wörtwein G (2005) Cholecystokinin tetrapeptide effects on HPA axis function and elevated plus maze behaviour in maternally separated and handled rats. Behav Brain Res 161:204–212. https://doi.org/10.1016/j.bbr.2005.02.006

Grochowska M, Laskus T, Radkowski M (2019) Gut microbiota in neurological disorders. Arch Immunol Ther Exp 67:375–383. https://doi.org/10.1007/s00005-019-00561-6

Gunstad J, Paul RH, Cohen RA, Tate DF, Spitznagel MB, Grieve S, Gordon E (2008) Relationship between body mass index and brain volume in healthy adults. Int J Neurosci 118:1582–1593. https://doi.org/10.1080/00207450701392282

Gunstad J, Lhotsky A, Wendell CR, Ferrucci L, Zonderman AB (2010) Longitudinal examination of obesity and cognitive function: results from the Baltimore longitudinal study of aging. Neuroepidemiology 34:222–229. https://doi.org/10.1159/000297742

Gunstad J, Strain G, Devlin MJ, Wing R, Cohen RA, Paul RH, Crosby RD, Mitchell JE (2011) Improved memory function 12 weeks after bariatric surgery. Surg Obes Relat Dis 7:465–472. https://doi.org/10.1016/j.soard.2010.09.015

Gustafson D, Lissner L, Bengtsson C, Björkelund C, Skoog I (2004) A 24-year follow-up of body mass index and cerebral atrophy. Neurology 63:1876. https://doi.org/10.1212/01.WNL.0000141850.47773.5F

Guzman-Ramos K, Moreno-Castilla P, Castro-Cruz M, McGaugh JL, Martinez-Coria H, LaFerla FM, Bermudez-Rattoni F (2012) Restoration of dopamine release deficits during object recognition memory acquisition attenuates cognitive impairment in a triple transgenic mice model of Alzheimer's disease. Learn Mem 19:453–460. https://doi.org/10.1101/lm.026070.112

Gyorkos A, Baker MH, Miutz LN, Lown DA, Jones MA, Houghton-Rahrig LD (2019) Carbohydrate-restricted diet and exercise increase brain-derived neurotrophic factor and cognitive function: a randomized crossover trial. Cureus 11:e5604. https://doi.org/10.7759/cureus.5604

Ha AD, Fung VSC (2012) Huntington's disease. Curr Opin Neurol 25:491. https://doi.org/10.1097/WCO.0b013e3283550c97

Han Z, Niu T, Chang J, Lei X, Zhao M, Wang Q, Cheng W, Wang J, Feng Y, Chai J (2010) Crystal structure of the FTO protein reveals basis for its substrate specificity. Nature 464:1205–1209. https://doi.org/10.1038/nature08921

Hanagasi HA, Tufekcioglu Z, Emre M (2017) Dementia in Parkinson's disease. J Neurol Sci 374: 26. https://doi.org/10.1016/j.jns.2017.01.012

Harakeh SM, Khan I, Kumosani T, Barbour E, Almasaudi SB, Bahijri SM, Alfadul SM, Ajabnoor GMA, Azhar EI (2016) Gut microbiota: a contributing factor to obesity. Front Cell Infect Microbiol 6:95. https://doi.org/10.3389/fcimb.2016.00095

Haroon E, Raison CL, Miller AH (2012) Psychoneuroimmunology meets neuropsychopharmacology: translational implications of the impact of inflammation on behavior. Neuropsychopharmacology 37:137. https://doi.org/10.1038/npp.2011.205

Harrington AM, Castro J, Erickson A, Grundy L, Brierley SM (2018) Extrinsic sensory afferent nerves innervating the gastrointestinal tract in health and disease. In: Physiology of the gastrointestinal tract. Elsevier, Amsterdam, pp 387–418. https://doi.org/10.1016/B978-0-12-809954-4.00017-7

Hays SE, Goodwin FK, Paul SM (1981) Cholecystokinin receptors are decreased in basal ganglia and cerebral cortex of Huntington's disease. Brain Res 225:452–456. https://doi.org/10.1016/0006-8993(81)90853-2

Hebert LE, Scherr PA, Bienias JL, Bennett DA, Evans DA (2003) Alzheimer disease in the US population: prevalence estimates using the 2000 census. Arch Neurol 60:1119–1122. https://doi.org/10.1001/archneur.60.8.1119

Henderson ST, Vogel JL, Barr LJ, Garvin F, Jones JJ, Costantini LC (2009) Study of the ketogenic agent AC-1202 in mild to moderate Alzheimer's disease: a randomized, double-blind, placebo-controlled, multicenter trial. Nutr Metab 6:31. https://doi.org/10.1186/1743-7075-6-31

Hermansen K (1984) Effects of Cholecystokinin (CCK)-4, Nonsulfated CCK-8, and sulfated CCK-8 on pancreatic somatostatin, insulin, and glucagon secretion in the dog: studies in vitro*. Endocrinology 114:1770–1775. https://doi.org/10.1210/endo-114-5-1770

Hertz L, Chen Y, Waagepetersen HS (2015) Effects of ketone bodies in Alzheimer's disease in relation to neural hypometabolism, β-amyloid toxicity, and astrocyte function. J Neurochem 134:7–20. https://doi.org/10.1111/jnc.13107

Herzig K-H, Wilgus C, Schön I, Tatemoto K, Fölsch UR (1998) Regulation of the action of the novel cholecystokinin-releasing peptide diazepam binding inhibitor by inhibitory hormones and taurocholate. Regul Pept 74:193–198. https://doi.org/10.1016/S0167-0115(98)00021-4

Higgs S, Williamson AC, Rotshtein P, Humphreys GW (2008) Sensory-specific satiety is intact in amnesics who eat multiple meals: research report. Psychol Sci 19:623–628. https://doi.org/10.1111/j.1467-9280.2008.02132.x

Hill JM, Lukiw W (2015) Microbial-generated amyloids and Alzheimer's disease (AD). Front Aging Neurosci 7:9. https://doi.org/10.3389/fnagi.2015.00009

Ho L, Ono K, Tsuji M, Mazzola P, Singh R, Pasinetti GM (2018) Protective roles of intestinal microbiota derived short chain fatty acids in Alzheimer's disease-type beta-amyloid neuropathological mechanisms. Expert Rev Neurother 18:83–90. https://doi.org/10.1080/14737175.2018.1400909

Hock C, Heese K, Hulette C, Rosenberg C, Otten U (2000) Region-specific neurotrophin imbalances in Alzheimer disease: decreased levels of brain-derived neurotrophic factor and increased levels of nerve growth factor in hippocampus and cortical areas. Arch Neurol 57:846–851. https://doi.org/10.1001/archneur.57.6.846

Hokama M, Oka S, Leon J, Ninomiya T, Honda H, Sasaki K, Iwaki T, Ohara T, Sasaki T, LaFerla FM, Kiyohara Y, Nakabeppu Y (2014) Altered expression of diabetes-related genes in Alzheimer's disease brains: the Hisayama study. Cereb Cortex 24:2476–2488. https://doi.org/10.1093/cercor/bht101

Hökfelt T, Rehfeld JF, Skirboll L, Ivemark B, Goldstein M, Markey K (1980) Evidence for coexistence of dopamine and CCK in meso-limbic neurones. Nature 285:476–478. https://doi.org/10.1038/285476a0

Hommer D, Palkovits M, Crawley J, Paul S, Skirboll L (1985) Cholecystokinin-induced excitation in the substantia nigra: evidence for peripheral and central components. J Neurosci 5:1387–1392. https://doi.org/10.1523/JNEUROSCI.05-06-01387.1985

Hufnagel DA, Tükel Ç, Chapman MR (2013) Disease to dirt: the biology of microbial amyloids. PLoS Pathog 9:1–5. https://doi.org/10.1371/journal.ppat.1003740

Hwang CK, Kim DK, Chun HS (2013) Cholecystokinin-8 induces brain-derived neurotrophic factor expression in noradrenergic neuronal cells. Neuropeptides 47:245–250. https://doi.org/10.1016/j.npep.2013.04.002

Ikeda M (2002) Changes in appetite, food preference, and eating habits in frontotemporal dementia and Alzheimer's disease. J Neurol Neurosurg Psychiatry 73:371–376. https://doi.org/10.1136/jnnp.73.4.371

Ivy AC, Oldberg E (1928) A hormone mechanism for gall-bladder contraction and evacuation. Am J Physiol Content 86:599–613. https://doi.org/10.1152/ajplegacy.1928.86.3.599

Jacka FN, Cherbuin N, Anstey KJ, Sachdev P, Butterworth P (2015) Western diet is associated with a smaller hippocampus: a longitudinal investigation. BMC Med 13:215. https://doi.org/10.1186/s12916-015-0461-x

Jensen SL, Rehfeld JF, Holst JJ, Nielsen OV, Fahrenkrug J, Muckadell OBS (1981) Secretory effects of cholecystokinins on the isolated perfused porcine pancreas. Acta Physiol Scand 111:225–231. https://doi.org/10.1111/j.1748-1716.1981.tb06730.x

Jeong SK, Nam HS, Son MH, Son EJ, Cho KH (2005) Interactive effect of obesity indexes on cognition. Dement Geriatr Cogn Disord 19:91–96. https://doi.org/10.1159/000082659

Jerónimo-Santos A, Vaz SH, Parreira S, Rapaz-Lérias S, Caetano AP, Buée-Scherrer V, Castrén E, Valente CA, Blum D, Ana Maria Sebastião MJD (2015) Dysregulation of TrkB receptors and BDNF function by amyloid-β peptide is mediated by calpain. Cereb Cortex 25:3107

Jiang C, Li G, Huang P, Liu Z, Zhao B (2017) The gut microbiota and Alzheimer's disease. J Alzheimers Dis 58:1–15. https://doi.org/10.3233/JAD-161141

Jones BV, Begley M, Hill C, Gahan CGM, Marchesi JR (2008) Functional and comparative metagenomic analysis of bile salt hydrolase activity in the human gut microbiome. Proc Natl Acad Sci U S A 105:13580–13585. https://doi.org/10.1073/pnas.0804437105

Kakimoto T, Kanemoto H, Fukushima K, Ohno K, Tsujimoto H (2017) Effect of a high-fat–high-cholesterol diet on gallbladder bile acid composition and gallbladder motility in dogs. Am J Vet Res 78:1406–1413. https://doi.org/10.2460/ajvr.78.12.1406

Kalia LV, Lang AE (2015) Parkinson's disease. Lancet 386:986. https://doi.org/10.1016/S0140-6736(14)61393-3

Karlsson HK, Tuulari JJ, Hirvonen J, Lepomäki V, Parkkola R, Hiltunen J, Hannukainen JC, Soinio M, Pham T, Salminen P, Nuutila P, Nummenmaa L (2013) Obesity is associated with white matter atrophy: a combined diffusion tensor imaging and voxel-based morphometric study. Obesity 21:2530–2537. https://doi.org/10.1002/oby.20386

Keller J, Mark R, Bruce A, Blanc E, Rothstein J, Uchida K, Waeg G, Mattson M (1997) 4-Hydroxynonenal, an aldehydic product of membrane lipid peroxidation, impairs glutamate transport and mitochondrial function in synaptosomes. Neuroscience 80:685–696. https://doi.org/10.1016/S0306-4522(97)00065-1

Keller JN, Pang Z, Geddes JW, Begley JG, Germeyer A, Waeg G, Mattson MP (2002) Impairment of glucose and glutamate transport and induction of mitochondrial oxidative stress and dysfunction in synaptosomes by amyloid β-peptide: role of the lipid peroxidation product 4-hydroxynonenal. J Neurochem 69:273–284. https://doi.org/10.1046/j.1471-4159.1997.69010273.x

Keller L, Xu W, Wang HX, Winblad B, Fratiglioni L, Graff C (2011) The obesity related gene, FTO, interacts with APOE, and is associated with Alzheimer's disease risk: a prospective cohort study. J Alzheimers Dis 23:461–469. https://doi.org/10.3233/JAD-2010-101068

Kerstens PJSM, Lamers CBHW, Jansen JBMJ, de Jong AJL, Hessels M, Hafkenscheid JCM (1985) Physiological plasma concentrations of cholecystokinin stimulate pancreatic enzyme secretion and gallbladder contraction in man. Life Sci 36:565–569. https://doi.org/10.1016/0024-3205(85)90638-1

Khan NA, Raine LB, Donovan SM, Hillman CH (2014) IV. The cognitive implications of obesity and nutrition in childhood. Monogr Soc Res Child Dev 79:51–71. https://doi.org/10.1111/mono.12130

Khaodhiar L, McCowen KC, Blackburn GL (1999) Obesity and its comorbid conditions. Clin Cornerstone 2:17–31. https://doi.org/10.1016/S1098-3597(99)90002-9

Kim H, Whang W-W, Kim H-T, Pyun K-H, Cho S-Y, Hahm D-H, Lee H-J, Shim I (2003) Expression of neuropeptide Y and cholecystokinin in the rat brain by chronic mild stress. Brain Res 983:201–208. https://doi.org/10.1016/S0006-8993(03)03087-7

Kim K-A, Gu W, Lee I-A, Joh E-H, Kim D-H (2012) High fat diet-induced gut microbiota exacerbates inflammation and obesity in mice via the TLR4 signaling pathway. PLoS One 7: e47713. https://doi.org/10.1371/journal.pone.0047713

Konturek SJ, Kwiecien N, Obtulowicz W, Kopp B, Oleksy J, Rovati L (1990) Cholecystokinin in the inhibition of gastric secretion and gastric emptying in humans. Digestion 45:1–8. https://doi.org/10.1159/000200218

Koop I, Koop H, Gerhardt C, Schafmayer A, Arnold R (1989) Do bile acids exert a negative feedback control of cholecystokinin release? Scand J Gastroenterol 24:315–320. https://doi.org/10.3109/00365528909093053

Koop I, Schindler M, Bosshammer A, Scheibner J, Stange E, Koop H (1996) Physiological control of cholecystokinin release and pancreatic enzyme secretion by intraduodenal bile acids. Gut 39: 661–667. https://doi.org/10.1136/gut.39.5.661

Kopin AS, Mathes WF, McBride EW, Nguyen M, Al-Haider W, Schmitz F, Bonner-Weir S, Kanarek R, Beinborn M (1999) The cholecystokinin-a receptor mediates inhibition of food intake yet is not essential for the maintenance of body weight. J Clin Invest 103:383–391. https://doi.org/10.1172/JCI4901

Kowalski K, Mulak A (2019) Brain-gut-microbiota axis in Alzheimer's disease. J Neurogastroenterol Motil 25:48–60. https://doi.org/10.5056/jnm18087

Krikorian R, Shidler MD, Dangelo K, Couch SC, Benoit SC, Clegg DJ (2012) Dietary ketosis enhances memory in mild cognitive impairment. Neurobiol Aging 33:425.e19–425.e27. https://doi.org/10.1016/j.neurobiolaging.2010.10.006

Kubo T, Nishimura S, Kumagae Y, Kaneko I (2002) In vivo conversion of racemized?-amyloid ([D-Ser26]A?1-40) to truncated and toxic fragments ([D-Ser26]A?25-35/40) and fragment presence in the brains of Alzheimer's patients. J Neurosci Res 70:474–483. https://doi.org/10.1002/jnr.10391

Kullmann S, Callaghan MF, Heni M, Weiskopf N, Scheffler K, Häring HU, Fritsche A, Veit R, Preissl H (2016) Specific white matter tissue microstructure changes associated with obesity. NeuroImage 125:36–44. https://doi.org/10.1016/j.neuroimage.2015.10.006

La Rosa F, Clerici M, Ratto D, Occhinegro A, Licito A, Romeo M, Carmine Di Lorio PR (2018) The gut-brain axis in Alzheimer's disease and omega-3. A critical overview of clinical trials. Nutrients 10:1267

Laffel L (1999) Ketone bodies: a review of physiology, pathophysiology and application of monitoring to diabetes. Diabetes Metab Res Rev 15:412–426. https://doi.org/10.1002/(SICI)1520-7560(199911/12)15:6<412::AID-DMRR72>3.0.CO;2-8

Lambert MP, Barlow AK, Chromy BA, Edwards C, Freed R, Liosatos M, Morgan TE, Rozovsky I, Trommer B, Viola KL, Wals P, Zhang C, Finch CE, Krafft GA, Klein WL (1998) Diffusible, nonfibrillar ligands derived from A 1-42 are potent central nervous system neurotoxins. Proc Natl Acad Sci 95:6448–6453. https://doi.org/10.1073/pnas.95.11.6448

Lamport DJ, Dye L, Mansfield MW, Lawton CL (2013) Acute glycaemic load breakfast manipulations do not attenuate cognitive impairments in adults with type 2 diabetes. Clin Nutr 32:265–272. https://doi.org/10.1016/j.clnu.2012.07.008

Lamport DJ, Lawton CL, Mansfield MW, Moulin CAJ, Dye L (2014) Type 2 diabetes and impaired glucose tolerance are associated with word memory source monitoring recollection deficits but not simple recognition familiarity deficits following water, low glycaemic load, and high glycaemic load breakfasts. Physiol Behav 124:54–60. https://doi.org/10.1016/j.physbeh.2013.10.033

Landfield PW, Blalock EM, Chen K-C, Porter NM (2007) A new glucocorticoid hypothesis of brain aging: implications for Alzheimer's disease. Curr Alzheimer Res 4:205

Langhans W, Egli G, Scharrer E (1985) Selective hepatic vagotomy eliminates the hypophagic effect of different metabolites. J Auton Nerv Syst 13:255–262. https://doi.org/10.1016/0165-1838(85)90014-1

Larsen PJ, Jensen PB, Sorensen RV, Larsen LK, Vrang N, Wulff EM, Wassermann K (2003) Differential influences of peroxisome proliferator-activated receptors and - on food intake and energy homeostasis. Diabetes 52:2249–2259. https://doi.org/10.2337/diabetes.52.9.2249

Larsson LI, Rehfeld JF (1978) Distribution of gastrin and CCK cells in the rat gastrointestinal tract. Histochemistry 58:23–31. https://doi.org/10.1007/BF00489946

Larsson L-I, Rehfeld JF (1979) Localization and molecular heterogeneity of cholecystokinin in the central and peripheral nervous system. Brain Res 165:201–218. https://doi.org/10.1016/0006-8993(79)90554-7

de Lartigue G, Dimaline R, Varro A, Dockray GJ (2007) Cocaine- and amphetamine-regulated transcript: stimulation of expression in rat vagal afferent neurons by cholecystokinin and suppression by ghrelin. J Neurosci 27:2876–2882. https://doi.org/10.1523/JNEUROSCI.5508-06.2007

Laske C, Stransky E, Leyhe T, Eschweiler GW, Schott K, Langer H, Gawaz M (2006) Decreased brain-derived neurotrophic factor (BDNF)- and β-thromboglobulin (β-TG)-blood levels in Alzheimer's disease [4]. Thromb Haemost 96:102. https://doi.org/10.1160/TH06-03-0173

Lau D, Bengtson CP, Buchthal B, Bading H (2015) BDNF Reduces toxic extrasynaptic NMDA receptor signaling via synaptic NMDA receptors and nuclear-calcium-induced transcription of inhba/activin A. Cell Rep 12:1353–1366. https://doi.org/10.1016/j.celrep.2015.07.038

Leal G, Afonso PM, Salazar IL, Duarte CB (2015) Regulation of hippocampal synaptic plasticity by BDNF. Brain Res 1621:82. https://doi.org/10.1016/j.brainres.2014.10.019

Lee J, Duan W, Long JM, Ingram DK, Mattson MP (2000) Dietary restriction increases the number of newly generated neural cells, and induces BDNF expression, in the dentate gyrus of rats. J Mol Neurosci 15:99–108. https://doi.org/10.1385/JMN:15:2:99

Lee J, Seroogy KB, Mattson MP (2002) Dietary restriction enhances neurotrophin expression and neurogenesis in the hippocampus of adult mice. J Neurochem 80:539–547. https://doi.org/10.1046/j.0022-3042.2001.00747.x

Lemann V, Gottmann K, Heumann R (1994) BDNF, and NT-4/5 enhance glutamatergic synaptic transmission. NeuroReport 6:21

Leßmann V, Heumann R (1998) Modulation of unitary glutamatergic synapses by neurotrophin-4/5 or brain-derived neurotrophic factor in hippocampal microcultures: presynaptic enhancement depends on pre-established paired-pulse facilitation. Neuroscience 86:399–413. https://doi.org/10.1016/S0306-4522(98)00035-9

Levine ES, Dreyfus CF, Black IB, Plummer MR (1995) Brain-derived neurotrophic factor rapidly enhances synaptic transmission in hippocampal neurons via postsynaptic tyrosine kinase receptors. Proc Natl Acad Sci U S A 92:8074–8077. https://doi.org/10.1073/pnas.92.17.8074

Levine ES, Dreyfus CF, Black IB, Plummer MR (1996) Selective role for trkB neurotrophin receptors in rapid modulation of hippocampal synaptic transmission. Mol Brain Res 38:300–303. https://doi.org/10.1016/0169-328X(96)00025-3

Levine ES, Crozier RA, Black IB, Plummer MR (1998) Brain-derived neurotrophic factor modulates hippocampal synaptic transmission by increasing N-methyl-D-aspartic acid receptor activity. Proc Natl Acad Sci U S A 95:10235–10239. https://doi.org/10.1073/pnas.95.17.10235

Ley RE, Turnbaugh PJ, Klein S, Gordon JI (2006) Microbial ecology: human gut microbes associated with obesity. Nature 444:1022–1023. https://doi.org/10.1038/4441022a

Leyhe T, Stransky E, Eschweiler GW, Buchkremer G, Laske C (2008) Increase of BDNF serum concentration during donepezil treatment of patients with early Alzheimer's disease. Eur Arch Psychiatry Clin Neurosci 258:124–128. https://doi.org/10.1007/s00406-007-0764-9

Li Y, Hao Y, Owyang C (2000) Diazepam-binding inhibitor mediates feedback regulation of pancreatic secretion and postprandial release of cholecystokinin. J Clin Invest 105:351–359. https://doi.org/10.1172/JCI7204

Lichtwark IT, Newnham ED, Robinson SR, Shepherd SJ, Hosking P, Gibson PR, Yelland GW (2014) Cognitive impairment in coeliac disease improves on a gluten-free diet and correlates with histological and serological indices of disease severity. Aliment Pharmacol Ther 40:160–170. https://doi.org/10.1111/apt.12809

Liddle RA, Goldfine ID, Rosen MS, Taplitz RA, Williams JA (1985) Cholecystokinin bioactivity in human plasma. Molecular forms, responses to feeding, and relationship to gallbladder contraction. J Clin Invest 75:1144–1152. https://doi.org/10.1172/JCI111809

Lin H, Zhang T, Wu Y, Wang Y, Wang W, Wang Q (2014) Related Genes and Potential Biomarkers for Early Diagnosis of Alzheimer's Disease. Am J Alzheimers Dis Other Dement 29:90–95. https://doi.org/10.1177/1533317513506774

Loef M, Walach H (2013) Midlife obesity and dementia: meta-analysis and adjusted forecast of dementia prevalence in the United States and China. Obesity 21:E51–E55. https://doi.org/10.1002/oby.20037

Lundberg K, Josefsson A, Nordin C (2007) Diurnal and seasonal variation of cholecystokinin peptides in humans. Neuropeptides 41:59–63. https://doi.org/10.1016/j.npep.2006.09.049

Lundberg K, Hilke S, Nordin C, Theodorsson E, Josefsson A (2010) Cholecystokinin in plasma and cerebrospinal fluid—a study in healthy young women. Peptides 31:1625–1628. https://doi.org/10.1016/j.peptides.2010.04.028

Lupien SJ, McEwen BS (1997) The acute effects of corticosteroids on cognition: integration of animal and human model studies. Brain Res Rev 24:1. https://doi.org/10.1016/S0165-0173(97)00004-0

Maclagan NF (1937) The role of appetite in the control of body weight. J Physiol 90:385–394. https://doi.org/10.1113/jphysiol.1937.sp003524

MahmoudianDehkordi S, Arnold M, Nho K, Ahmad S, Jia W, Xie G, Louie G, Kueider-Paisley A, Moseley MA, Thompson JW, St John Williams L, Tenenbaum JD, Blach C, Baillie R, Han X, Bhattacharyya S, Toledo JB, Schafferer S, Klein S, Koal T, Risacher SL, Allan Kling M, Motsinger-Reif A, Rotroff DM, Jack J, Hankemeier T, Bennett DA, De Jager PL, Trojanowski JQ, Shaw LM, Weiner MW, Doraiswamy PM, van Duijn CM, Saykin AJ, Kastenmüller G, Kaddurah-Daouk R (2019) Altered bile acid profile associates with cognitive impairment in

Alzheimer's disease-an emerging role for gut microbiome. Alzheimers Dement 15:76–92. https://doi.org/10.1016/j.jalz.2018.07.217

Mapstone M, Cheema AK, Fiandaca MS, Zhong X, Mhyre TR, MacArthur LH, Hall WJ, Fisher SG, Peterson DR, Haley JM, Nazar MD, Rich SA, Berlau DJ, Peltz CB, Tan MT, Kawas CH, Federoff HJ (2014) Plasma phospholipids identify antecedent memory impairment in older adults. Nat Med 20:415–418. https://doi.org/10.1038/nm.3466

Maqsood R, Stone TW (2016) The gut-brain axis, BDNF, NMDA and CNS disorders. Neurochem Res 41:2819. https://doi.org/10.1007/s11064-016-2039-1

Mariat D, Firmesse O, Levenez F, Guimarães VD, Sokol H, Doré J, Corthier G, Furet JP (2009) The firmicutes/bacteroidetes ratio of the human microbiota changes with age. BMC Microbiol 9:126. https://doi.org/10.1186/1471-2180-9-123

Marksteiner J, Blasko I, Kemmler G, Koal T, Humpel C (2018) Bile acid quantification of 20 plasma metabolites identifies lithocholic acid as a putative biomarker in Alzheimer's disease. Metabolomics 14:1. https://doi.org/10.1007/s11306-017-1297-5

Marques F, Sousa JC, Sousa N, Palha JA (2013) Blood-brain-barriers in aging and in Alzheimer's disease. Mol Neurodegener 8:38. https://doi.org/10.1186/1750-1326-8-38

Matrone C, Ciotti MT, Mercanti D, Marolda R, Calissano P (2008) NGF and BDNF signaling control amyloidogenic route and Aβ production in hippocampal neurons. Proc Natl Acad Sci U S A 105:13139–13144. https://doi.org/10.1073/pnas.0806133105

Matson CA, Ritter RC (1999) Long-term CCK-leptin synergy suggests a role for CCK in the regulation of body weight. Am J Physiol Integr Comp Physiol 276:R1038–R1045. https://doi.org/10.1152/ajpregu.1999.276.4.R1038

Matsushita H, Akiyoshi J, Kai K, Ishii N, Kodama K, Tsutsumi T, Isogawa K, Nagayama H (2003) Spatial memory impairment in OLETF rats without cholecystokinin – a receptor. Neuropeptides 37:271–276. https://doi.org/10.1016/S0143-4179(03)00083-0

Mattson MP, Moehl K, Ghena N, Schmaedick M, Cheng A (2018) Intermittent metabolic switching, neuroplasticity and brain health. Nat Rev Neurosci 19:81–94. https://doi.org/10.1038/nrn.2017.156

Mazurek MF, Beal MF (1991) Cholecystokinin and somatostatin in Alzheimer's disease postmortem cerebral cortex. Neurology 41:716–719. https://doi.org/10.1212/wnl.41.5.716

McRae-McKee K, Evans S, Hadjichrysanthou C, Wong MM, de Wolf F, Anderson RM (2019) Combining hippocampal volume metrics to better understand Alzheimer's disease progression in at-risk individuals. Sci Rep 9:1–9. https://doi.org/10.1038/s41598-019-42632-w

Menni C, Jackson MA, Pallister T, Steves CJ, Spector TD, Valdes AM (2017) Gut microbiome diversity and high-fibre intake are related to lower long-term weight gain. Int J Obes 41:1099–1105. https://doi.org/10.1038/ijo.2017.66

Miller LA, Crosby RD, Galioto R, Strain G, Devlin MJ, Wing R, Cohen RA, Paul RH, Mitchell JE, Gunstad J (2013) Bariatric surgery patients exhibit improved memory function 12 months postoperatively. Obes Surg 23:1527–1535. https://doi.org/10.1007/s11695-013-0970-7

Molteni R, Barnard RJ, Ying Z, Roberts CK, Gómez-Pinilla F (2002) A high-fat, refined sugar diet reduces hippocampal brain-derived neurotrophic factor, neuronal plasticity, and learning. Neuroscience 112:803–814. https://doi.org/10.1016/S0306-4522(02)00123-9

Moran TH, Bi S (2006) Hyperphagia and obesity in OLETF rats lacking CCK-1 receptors. Philos Trans R Soc B Biol Sci 361:1211–1218. https://doi.org/10.1098/rstb.2006.1857

Moran TH, Ameglio PJ, Peyton HJ, Schwartz GJ, McHugh PR (1993) Blockade of type A, but not type B, CCK receptors postpones satiety in rhesus monkeys. Am J Physiol Integr Comp Physiol 265:R620–R624. https://doi.org/10.1152/ajpregu.1993.265.3.R620

Moran TH, Katz LF, Plata-Salaman CR, Schwartz GJ (1998) Disordered food intake and obesity in rats lacking cholecystokinin A receptors. Am J Physiol Integr Comp Physiol 274:R618–R625. https://doi.org/10.1152/ajpregu.1998.274.3.R618

Mörkl S, Lackner S, Meinitzer A, Mangge H, Lehofer M, Halwachs B, Gorkiewicz G, Kashofer K, Painold A, Holl AK, Bengesser SA, Müller W, Holzer P, Holasek SJ (2018) Gut microbiota, dietary intakes and intestinal permeability reflected by serum zonulin in women. Eur J Nutr 57: 2985–2997. https://doi.org/10.1007/s00394-018-1784-0

Morris CH, Hope RA, Fairburn CG (1989) Eating habits in dementia. Br J Psychiatry 154:801–806. https://doi.org/10.1192/bjp.154.6.801

Morris MJ, Chen H, Watts R, Shulkes A, Cameron-Smith D (2008) Brain neuropeptide Y and CCK and peripheral adipokine receptors: temporal response in obesity induced by palatable diet. Int J Obes 32:249–258. https://doi.org/10.1038/sj.ijo.0803716

Mueller K, Anwander A, Möller HE, Horstmann A, Lepsien J, Busse F, Mohammadi S, Schroeter ML, Stumvoll M, Villringer A, Pleger B (2011) Sex-dependent influences of obesity on cerebral white matter investigated by diffusion-tensor imaging. PLoS One 6:e18544. https://doi.org/10.1371/journal.pone.0018544

Nagappan G, Zaitsev E, Senatorov VV, Yang J, Hempstead BL, Lu B (2009) Control of extracellular cleavage of ProBDNF by high frequency neuronal activity. Proc Natl Acad Sci U S A 106: 1267–1272. https://doi.org/10.1073/pnas.0807322106

Nam KN, Mounier A, Wolfe CM, Fitz NF, Carter AY, Castranio EL, Kamboh HI, Reeves VL, Wang J, Han X, Schug J, Lefterov I, Koldamova R (2017) Effect of high fat diet on phenotype, brain transcriptome and lipidome in Alzheimer's model mice. Sci Rep 7:4307–4307. https://doi.org/10.1038/s41598-017-04412-2

Nepal B, Brown LJ, Anstey KJ (2014) Rising midlife obesity will worsen future prevalence of dementia. PLoS One 9:e99305. https://doi.org/10.1371/journal.pone.0099305

Nguyen KQ, Rymar VV, Sadikot AF (2016) Impaired TrkB signaling underlies reduced BDNF-mediated trophic support of striatal neurons in the R6/2 mouse model of Huntington's disease. Front Cell Neurosci 10:37. https://doi.org/10.3389/fncel.2016.00037

Nho K, Kueider-Paisley A, MahmoudianDehkordi S, Arnold M, Risacher SL, Louie G, Blach C, Baillie R, Han X, Kastenmüller G, Jia W, Xie G, Ahmad S, Hankemeier T, van Duijn CM, Trojanowski JQ, Shaw LM, Weiner MW, Doraiswamy PM, Saykin AJ, Kaddurah-Daouk R (2019) Altered bile acid profile in mild cognitive impairment and Alzheimer's disease: relationship to neuroimaging and CSF biomarkers. Alzheimers Dement 15:232–244. https://doi.org/10.1016/j.jalz.2018.08.012

Noble EE, Hsu TM, Kanoski SE (2017) Gut to brain dysbiosis: mechanisms linking western diet consumption, the microbiome, and cognitive impairment. Front Behav Neurosci 11:9. https://doi.org/10.3389/fnbeh.2017.00009

O'Bryant SE, Hobson V, Hall JR, Waring SC, Chan W, Massman P, Lacritz L, Cullum CM, Diaz-Arrastia R (2009) Brain-derived neurotrophic factor levels in Alzheimer's disease. J Alzheimers Dis 17:337–341. https://doi.org/10.3233/JAD-2009-1051

Ohta H, Guan D, Tawil T, Liddle RA, Green GM (1990) Regulation of plasma cholecystokinin levels by bile and bile acids in the rat. Gastroenterology 99:819–825. https://doi.org/10.1016/0016-5085(90)90974-6

Okubo T, Harada S (2001) Polymorphisms of the CCK, CCKAR and CCKBR genes: an association with alcoholism study. J Stud Alcohol 62:413–421. https://doi.org/10.15288/jsa.2001.62.413

Olazarán J, Gil-de-Gómez L, Rodríguez-Martín A, Valentí-Soler M, Frades-Payo B, Marín-Muñoz J, Antúnez C, Frank-García A, Acedo-Jiménez C, Morlán-Gracia L, Petidier-Torregrossa R, Guisasola MC, Bermejo-Pareja F, Sánchez-Ferro Á, Pérez-Martínez DA, Manzano-Palomo S, Farquhar R, Rábano A, Calero M (2015) A blood-based, 7-metabolite signature for the early diagnosis of Alzheimer's disease. J Alzheimers Dis 45:1157–1173. https://doi.org/10.3233/JAD-142925

Oli MW, Otoo HN, Crowley PJ, Heim KP, Nascimento MM, Ramsook CB, Lipke PN, Brady LJ (2012) Functional amyloid formation by Streptococcus mutans. Microbiology 158:2903–2916. https://doi.org/10.1099/mic.0.060855-0

Olsson C, Aldman G, Larsson A, Holmgren S (1999) Cholecystokinin affects gastric emptying and stomach motility in the rainbow trout Oncorhynchus mykiss. J Exp Biol 202:161–170

Org E, Parks BW, Joo JWJ, Emert B, Schwartzman W, Kang EY, Mehrabian M, Pan C, Knight R, Gunsalus R, Drake TA, Eskin E, Lusis AJ (2015) Genetic and environmental control of host-gut microbiota interactions. Genome Res 25:1558–1569. https://doi.org/10.1101/gr.194118.115

Otsuki M, Sakamoto C, Yuu H, Maeda M, Morita S, Ohki A, Kobayashi N, Terashi K, Okano K, Baba S (1979) Discrepancies between the doses of cholecystokinin or caerulein-stimulating exocrine and endocrine responses in perfused isolated rat pancreas. J Clin Invest 63:478–484. https://doi.org/10.1172/JCI109325

Ouk T, Gautier S, Pétrault M, Montaigne D, Maréchal X, Masse I, Devedjian J-C, Deplanque D, Bastide M, Nevière R, Duriez P, Staels B, Pasquier F, Leys D, Bordet R (2014) Effects of the PPAR-α agonist fenofibrate on acute and short-term consequences of brain ischemia. J Cereb Blood Flow Metab 34:542–551. https://doi.org/10.1038/jcbfm.2013.233

Owyang C, May D, Louie DS (1986) Trypsin suppression of pancreatic enzyme secretion. Gastroenterology 91:637–643. https://doi.org/10.1016/0016-5085(86)90633-5

Palazzo M, Balsari A, Rossini A, Selleri S, Calcaterra C, Gariboldi S, Zanobbio L, Arnaboldi F, Shirai YF, Serrao G, Rumio C (2007) Activation of enteroendocrine cells via TLRs induces hormone, chemokine, and defensin secretion. J Immunol 178:4296–4303. https://doi.org/10.4049/jimmunol.178.7.4296

Pan X, Elliott CT, McGuinness B, Passmore P, Kehoe PG, Hölscher C, McClean PL, Graham SF, Green BD (2017) Metabolomic profiling of bile acids in clinical and experimental samples of Alzheimer's disease. Metabolites 7:28. https://doi.org/10.3390/metabo7020028

Pan X, Kaminga AC, Wen SW, Wu X, Acheampong K, Liu A (2019) Dopamine and dopamine receptors in Alzheimer's disease: a systematic review and network meta-analysis. Front Aging Neurosci 11:175. https://doi.org/10.3389/fnagi.2019.00175

Pang PT, Teng HK, Zaitsev E, Woo NT, Sakata K, Zhen S, Teng KK, Yung WH, Hempstead BL, Lu B (2004) Cleavage of proBDNF by tPA/plasmin is essential for long-term hippocampal plasticity. Science 306:487–491. https://doi.org/10.1126/science.1100135

Pannacciulli N, Del Parigi A, Chen K, Le DSNT, Reiman EM, Tataranni PA (2006) Brain abnormalities in human obesity: a voxel-based morphometric study. NeuroImage 31:1419–1425. https://doi.org/10.1016/j.neuroimage.2006.01.047

Paoli A, Bosco G, Camporesi EM, Mangar D (2015) Ketosis, ketogenic diet and food intake control: a complex relationship. Front Psychol 6:27. https://doi.org/10.3389/fpsyg.2015.00027

Papageorgiou I, Astrakas LG, Xydis V, Alexiou GA, Bargiotas P, Tzarouchi L, Zikou AK, Kiortsis DN, Argyropoulou MI (2017) Abnormalities of brain neural circuits related to obesity: a diffusion tensor imaging study. Magn Reson Imaging 37:116–121. https://doi.org/10.1016/j.mri.2016.11.018

Pardon MC (2007) Stress and ageing interactions: a paradox in the context of shared etiological and physiopathological processes. Brain Res Rev 54:251. https://doi.org/10.1016/j.brainresrev.2007.02.007

Park M-K, Han Y, Kim MS, Seo E, Kang S, Park S-Y, Koh H, Kim DK, Lee H-J (2012) Reduction of food intake by fenofibrate is associated with cholecystokinin release in Long-Evans tokushima rats. Kr J Physiol Pharmacol 16:181. https://doi.org/10.4196/kjpp.2012.16.3.181

Park JY, Choi J, Lee Y, Lee JE, Lee EH, Kwon HJ, Yang J, Jeong BR, Kim YK, Han PL (2017) Metagenome analysis of bodily microbiota in a mouse model of Alzheimer disease using bacteria-derived membrane vesicles in blood. Exp Neurobiol 26:369–379. https://doi.org/10.5607/en.2017.26.6.369

Park S, Jeon S-M, Jung S-Y, Hwang J, Kwon J-W (2019) Effect of late-life weight change on dementia incidence: a 10-year cohort study using claim data in Korea. BMJ Open 9:e021739. https://doi.org/10.1136/bmjopen-2018-021739

Paulsen JS (2011) Cognitive impairment in Huntington disease: diagnosis and treatment. Curr Neurol Neurosci Rep 11:474–483. https://doi.org/10.1007/s11910-011-0215-x

Pedditzi E, Ruth Peters NB (2016) The risk of overweight/obesity in mid-life and late life for the development of dementia: a systematic review and meta-analysis of longitudinal studies. Age Ageing 45:14

Peters JH, Simasko SM, Ritter RC (2006) Modulation of vagal afferent excitation and reduction of food intake by leptin and cholecystokinin. Physiol Behav 89:477–485. https://doi.org/10.1016/j.physbeh.2006.06.017

Petrova M, Raycheva M, Traykov L (2012) Cognitive profile of the earliest stage of dementia in Parkinson's disease. Am J Alzheimers Dis Other Dement 27:614–619. https://doi.org/10.1177/1533317512460562

Plagman A, Hoscheidt S, McLimans KE, Klinedinst B, Pappas C, Anantharam V, Kanthasamy A, Willette AA (2019) Cholecystokinin and Alzheimer's disease: a biomarker of metabolic function, neural integrity, and cognitive performance. Neurobiol Aging 76:201–207. https://doi.org/10.1016/j.neurobiolaging.2019.01.002

Plotkin JL, Day M, Peterson JD, Xie Z, Kress GJ, Rafalovich I, Kondapalli J, Gertler TS, Flajolet M, Greengard P, Stavarache M, Kaplitt MG, Rosinski J, Chan CS, Surmeier DJ (2014) Impaired TrkB receptor signaling underlies corticostriatal dysfunction in Huntington's disease. Neuron 83:178–188. https://doi.org/10.1016/j.neuron.2014.05.032

Pommier B, Beslot F, Simon A, Pophillat M, Matsui T, Dauge V, Roques BP, Noble F (2002) Deletion of CCK 2 receptor in mice results in an upregulation of the endogenous opioid system. J Neurosci 22:2005–2011. https://doi.org/10.1523/JNEUROSCI.22-05-02005.2002

Porritt MJ, Batchelor PE, Howells DW (2005) Inhibiting BDNF expression by antisense oligonucleotide infusion causes loss of nigral dopaminergic neurons. Exp Neurol 192:226–234. https://doi.org/10.1016/j.expneurol.2004.11.030

Porter NM, Landfield PW (1998) Stress hormones and brain aging: adding injury to insult? Nat Neurosci 1:3–4. https://doi.org/10.1038/196

Prinz P, Hofmann T, Ahnis A, Elbelt U, Goebel-Stengel M, Klapp BF, Rose M, Stengel A (2015) Plasma bile acids show a positive correlation with body mass index and are negatively associated with cognitive restraint of eating in obese patients. Front Neurosci 9:199. https://doi.org/10.3389/fnins.2015.00199

Pupovac J, Anderson GH (2002) Dietary peptides induce satiety via cholecystokinin-a and peripheral opioid receptors in rats. J Nutr 132:2775–2780. https://doi.org/10.1093/jn/132.9.2775

Rascovsky K, Hodges JR, Knopman D, Mendez MF, Kramer JH, Neuhaus J, van Swieten JC, Seelaar H, Dopper EGP, Onyike CU, Hillis AE, Josephs KA, Boeve BF, Kertesz A, Seeley WW, Rankin KP, Johnson JK, Gorno-Tempini M-L, Rosen H, Prioleau-Latham CE, Lee A, Kipps CM, Lillo P, Piguet O, Rohrer JD, Rossor MN, Warren JD, Fox NC, Galasko D, Salmon DP, Black SE, Mesulam M, Weintraub S, Dickerson BC, Diehl-Schmid J, Pasquier F, Deramecourt V, Lebert F, Pijnenburg Y, Chow TW, Manes F, Grafman J, Cappa SF, Freedman M, Grossman M, Miller BL (2011) Sensitivity of revised diagnostic criteria for the behavioural variant of frontotemporal dementia. Brain 134:2456–2477. https://doi.org/10.1093/brain/awr179

Raybould HE (2007) Mechanisms of CCK signaling from gut to brain. Curr Opin Pharmacol 7: 570–574. https://doi.org/10.1016/j.coph.2007.09.006

Reddel S, Putignani L, Del Chierico F (2019) The impact of low-FODMAPs, Gluten-free, and ketogenic diets on gut microbiota modulation in pathological conditions. Nutrients 11:373. https://doi.org/10.3390/nu11020373

Reddy JK, Hashimoto T (2001) An adaptive metabolic system. Annu Rev Nutr 21:193–230. https://doi.org/10.1146/annurev.nutr.21.1.193

Rehfeld JF (1978) Immunochemical studies on cholecystokinin. II. Distribution and molecular heterogeneity in the central nervous system and small intestine of man and hog. J Biol Chem 253:4022–4030

Rehfeld JF (1987) Preprocholecystokinin processing in the normal human anterior pituitary. Proc Natl Acad Sci 84:3019–3023. https://doi.org/10.1073/pnas.84.9.3019

Rehfeld JF (1998) Accurate measurement of cholecystokinin in plasma. Clin Chem 44:991–1001

Rehfeld JF (2004) Cholecystokinin. Best Pract Res Clin Endocrinol Metab 18:569–586. https://doi.org/10.1016/j.beem.2004.07.002

Rehfeld JF (2017) Cholecystokinin—from local gut hormone to ubiquitous messenger. Front Endocrinol 8:47. https://doi.org/10.3389/fendo.2017.00047

Rehfeld JF, Hansen HF (1986) Characterization of preprocholecystokinin products in the porcine cerebral cortex. Evidence of different processing pathways. J Biol Chem 261:5832–5840

Rehfeld JF, Lindholm J, Andersen BN, Bardram L, Cantor P, Fenger M, Lüdecke DK (1987) Pituitary tumors containing cholecystokinin. N Engl J Med 316:1244–1247. https://doi.org/10.1056/NEJM198705143162004

Rehfeld JF, Johnsen AH, Ødum L, Bardram L, Schifter S, Scopsi L (1990) Non-sulphated cholecystokinin in human medullary thyroid carcinomas. J Endocrinol 124:501–506. https://doi.org/10.1677/joe.0.1240501

Rehfeld JF, Friis-Hansen L, Goetze JP, Hansen TVO (2007) The biology of cholecystokinin and gastrin peptides. Curr Top Med Chem 7:1154–1165. https://doi.org/10.2174/156802607780960483

Robinson L, Tang E (2015) Dementia 3029:1–6. https://doi.org/10.1136/bmj.h3029

Rohe M, Synowitz M, Glass R, Paul SM, Nykjaer A, Willnow TE (2009) Brain-derived neurotrophic factor reduces amyloidogenic processing through control of SORLA gene expression. J Neurosci 29:15472–15478. https://doi.org/10.1523/JNEUROSCI.3960-09.2009

Rostami E, Krueger F, Plantman S, Davidsson J, Agoston D, Grafman J, Risling M (2014) Alteration in BDNF and its receptors, full-length and truncated TrkB and p75NTR following penetrating traumatic brain injury. Brain Res 1542:195–205. https://doi.org/10.1016/j.brainres.2013.10.047

Rozin P, Dow S, Moscovitch M, Rajaram S (1998) What causes humans to begin and end a meal? A role for memory for what has been eaten, as evidenced by a study of multiple meal eating in amnesic patients. Psychol Sci 9:392–396. https://doi.org/10.1111/1467-9280.00073

Sah SK, Lee C, Jang JH, Park GH (2017) Effect of high-fat diet on cognitive impairment in triple-transgenic mice model of Alzheimer's disease. Biochem Biophys Res Commun 493:731–736. https://doi.org/10.1016/j.bbrc.2017.08.122

Sanders DJ, Zahedi-Asl S, Marr AP (1982) Glucagon and CCK in human brain: controls and patients with senile dementia of Alzheimer type. Prog Brain Res 55:465–471. https://doi.org/10.1016/S0079-6123(08)64215-6

Sanguinetti E, Collado MC, Marrachelli VG, Monleon D, Selma-Royo M, Pardo-Tendero MM, Burchielli S, Iozzo P (2018) Microbiome-metabolome signatures in mice genetically prone to develop dementia, fed a normal or fatty diet. Sci Rep 8:4907. https://doi.org/10.1038/s41598-018-23261-1

Sanmiguel C, Gupta A, Mayer EA (2015) Gut microbiome and obesity: a plausible explanation for obesity. Curr Obes Rep 4:250. https://doi.org/10.1007/s13679-015-0152-0

Sapolsky RM (2000) Glucocorticoids and hippocampal atrophy in neuropsychiatric disorders. Arch Gen Psychiatry 57:925. https://doi.org/10.1001/archpsyc.57.10.925

Schjoldager BT-BG (1994) Role of CCK in gallbladder function. Ann N Y Acad Sci 713:207–218. https://doi.org/10.1111/j.1749-6632.1994.tb44067.x

Schmidt WE, Creutzfeldt W, Schleser A, Choudhury AR, Nustede R, Hocker M, Nitsche R, Sostmann H, Rovati LC, Folsch UR (1991) Role of CCK in regulation of pancreaticobiliary functions and GI motility in humans: effects of loxiglumide. Am J Physiol Liv Physiol 260: G197–G206. https://doi.org/10.1152/ajpgi.1991.260.2.G197

Schober ME, Block B, Requena DF, Hale MA, Lane RH (2012) Developmental traumatic brain injury decreased Brain Derived Neurotrophic Factor expression late after injury. Metab Brain Dis 27:167–173. https://doi.org/10.1007/s11011-012-9309-7

Schwartz K, Boles BR (2013) Microbial amyloids - functions and interactions within the host. Curr Opin Microbiol 16:93. https://doi.org/10.1016/j.mib.2012.12.001

Sebret A, Léna I, Crété D, Matsui T, Roques BP, Daugé V (1999) Rat hippocampal neurons are critically involved in physiological improvement of memory processes induced by cholecystokinin-b receptor stimulation. J Neurosci 19:7230–7237. https://doi.org/10.1523/JNEUROSCI.19-16-07230.1999

Shabana, Shahid SU, Irfan U (2018) The gut microbiota and its potential role in obesity. Future Microbiol 13:589. https://doi.org/10.2217/fmb-2017-0179

Shefer G, Marcus Y, Stern N (2013) Is obesity a brain disease? Neurosci Biobehav Rev 37:2489. https://doi.org/10.1016/j.neubiorev.2013.07.015

Shi CZ, Chen HQ, Liang Y, Xia Y, Yang YZ, Yang J, Zhang JD, Wang SH, Liu J, Qin HL (2014) Combined probiotic bacteria promotes intestinal epithelial barrier function in interleukin-10-gene-deficient mice. World J Gastroenterol 20:4636–4647. https://doi.org/10.3748/wjg.v20.i16.4636

Shoemark DK, Allen SJ (2015) The microbiome and disease: reviewing the links between the oral microbiome, aging, and Alzheimer's disease. J Alzheimers Dis 43:725. https://doi.org/10.3233/JAD-141170

Sileikiene V, Mosenthin R, Bauer E, Piepho HP, Tafaj M, Kruszewska D, Weström B, Erlanson-Albertsson C, Pierzynowski SG (2008) Effect of ileal infusion of short-chain fatty acids on pancreatic prandial secretion and gastrointestinal hormones in pigs. Pancreas 37:196–202. https://doi.org/10.1097/MPA.0b013e31816386f4

Silva YP, Bernardi A, Frozza RL (2020) The role of short-chain fatty acids from gut microbiota in gut-brain communication. Front Endocrinol 11:25. https://doi.org/10.3389/fendo.2020.00025

Sleiman SF, Henry J, Al-Haddad R, El Hayek L, Haidar EA, Stringer T, Ulja D, Karuppagounder SS, Holson EB, Ratan RR, Ninan I, Chao MV (2016) Exercise promotes the expression of brain derived neurotrophic factor (BDNF) through the action of the ketone body β-hydroxybutyrate. elife 5:e15092. https://doi.org/10.7554/eLife.15092

Smith MA, Makino S, Kim SY, Kvetnansky R (1995) Stress increases brain-derived neurotropic factor messenger ribonucleic acid in the hypothalamus and pituitary. Endocrinology 136:3743–3750. https://doi.org/10.1210/endo.136.9.7649080

Smith PJ, Blumenthal JA, Babyak MA, Craighead L, Welsh-Bohmer KA, Browndyke JN, Strauman TA, Sherwood A (2010) Effects of the dietary approaches to stop hypertension diet, exercise, and caloric restriction on neurocognition in overweight adults with high blood pressure. Hypertension 55:1331–1338. https://doi.org/10.1161/HYPERTENSIONAHA.109.146795

Sokoloff L (1973) Metabolism of ketone bodies by the brain. Annu Rev Med 24:271–280. https://doi.org/10.1146/annurev.me.24.020173.001415

Son JH, Chun HS, Joh TH, Cho S, Conti B, Lee JW (1999) Neuroprotection and neuronal differentiation studies using substantia nigra dopaminergic cells derived from transgenic mouse embryos. J Neurosci 19:10–20. https://doi.org/10.1523/JNEUROSCI.19-01-00010.1999

Song J-H, Yu J-T, Tan L (2015) Brain-derived neurotrophic factor in Alzheimer's disease: risk, mechanisms, and therapy. Mol Neurobiol 52:1477–1493. https://doi.org/10.1007/s12035-014-8958-4

Soto C (2012) Transmissible proteins: expanding the prion heresy. Cell 149:968. https://doi.org/10.1016/j.cell.2012.05.007

Spannagel AW, Green GM, Guan D, Liddle RA, Faull K, Reeve JR (1996) Purification and characterization of a luminal cholecystokinin-releasing factor from rat intestinal secretion. Proc Natl Acad Sci 93:4415–4420. https://doi.org/10.1073/pnas.93.9.4415

Srikanth V, Maczurek A, Phan T, Steele M, Westcott B, Juskiw D, Münch G (2011) Advanced glycation endproducts and their receptor RAGE in Alzheimer's disease. Neurobiol Aging 32:763–777. https://doi.org/10.1016/j.neurobiolaging.2009.04.016

Srivastava RAK, Jahagirdar R, Azhar S, Sharma S, Bisgaier CL (2006) Peroxisome proliferator-activated receptor-α selective ligand reduces adiposity, improves insulin sensitivity and inhibits atherosclerosis in LDL receptor-deficient mice. Mol Cell Biochem 285:35–50. https://doi.org/10.1007/s11010-005-9053-y

Struble RG, Powers RE, Casanova MF, Kitt CA, Brown EC, Price DL (1987) Neuropeptidergic systems in plaques of Alzheimer's disease. J Neuropathol Exp Neurol 46:567–584. https://doi.org/10.1097/00005072-198709000-00006

Studzinski CM, Li F, Bruce-Keller AJ, Fernandez-Kim SO, Zhang L, Weidner AM, Markesbery WR, Murphy MP, Keller JN (2009) Effects of short-term Western diet on cerebral oxidative stress and diabetes related factors in APP x PS1 knock-in mice. J Neurochem 108:860–866. https://doi.org/10.1111/j.1471-4159.2008.05798.x

Sugaya K, Takahashi M, Kubota K (1992) Cholecystokinin protects cholinergic neurons against basal forebrain lesion. Jpn J Pharmacol 59:125–128. https://doi.org/10.1254/jjp.59.125

Sullivan R, Wilson DA, Feldon J, Yee BK, Meyer U, Richter-Levin G, Avi T, Michael T, Gruss M, Bock J, Helmeke C, Braun K (2006) The International Society For Developmental Psychobiology Annual Meeting Symposium: impact of early life experiences on brain and behavioral development. Dev Psychobiol 48:583–602. https://doi.org/10.1002/dev.20170

Sumithran P, Prendergast LA, Delbridge E, Purcell K, Shulkes A, Kriketos A, Proietto J (2011) Long-term persistence of hormonal adaptations to weight loss. N Engl J Med 365:1597–1604. https://doi.org/10.1056/NEJMoa1105816

Sun L, Peräkylä J, Holm K, Haapasalo J, Lehtimäki K, Ogawa KH, Peltola J, Hartikainen KM (2017) Vagus nerve stimulation improves working memory performance. J Clin Exp Neuropsychol 39:954–964. https://doi.org/10.1080/13803395.2017.1285869

Syed AK, Boles BR (2014) Fold modulating function: bacterial toxins to functional amyloids. Front Microbiol 5:401. https://doi.org/10.3389/fmicb.2014.00401

Taki Y, Kinomura S, Sato K, Inoue K, Goto R, Okada K, Uchida S, Kawashima R, Fukuda H (2008) Relationship between body mass index and Gray Matter Volume in 1,428 healthy individuals. Obesity 16:119–124. https://doi.org/10.1038/oby.2007.4

Tapia-Arancibia L, Rage F, Givalois L, Arancibia S (2004) Physiology of BDNF: focus on hypothalamic function. Front Neuroendocrinol 25:77–107. https://doi.org/10.1016/j.yfrne.2004.04.001

Tejeda GS, Ayuso-Dolado S, Arbeteta R, Esteban-Ortega GM, Vidaurre OG, Díaz-Guerra M (2016) Brain ischaemia induces shedding of a BDNF-scavenger ectodomain from TrkB receptors by excitotoxicity activation of metalloproteinases and γ-secretases. J Pathol 238: 627–640. https://doi.org/10.1002/path.4684

Thorens B (2008) Glucose sensing and the pathogenesis of obesity and type 2 diabetes. Int J Obes 32:S62–S71. https://doi.org/10.1038/ijo.2008.208

Tirassa P, Costa N (2007) CCK-8 induces NGF and BDNF synthesis and modulates TrkA and TrkB expression in the rat hippocampus and septum: effects on kindling development. Neurochem Int 50:130–138. https://doi.org/10.1016/j.neuint.2006.07.008

Tirassa P, Stenfors C, Lundeberg T, Aloe L (1998) Cholecystokinin-8 regulation of NGF concentrations in adult mouse brain through a mechanism involving CCK A and CCK B receptors. Br J Pharmacol 123:1230–1236. https://doi.org/10.1038/sj.bjp.0701718

Tirassa P, Aloe L, Stenfors C, Turrini P, Lundeberg T (1999) Cholecystokinin-8 protects central cholinergic neurons against fimbria-fornix lesion through the up-regulation of nerve growth factor synthesis. Proc Natl Acad Sci 96:6473–6477. https://doi.org/10.1073/pnas.96.11.6473

Tirassa P, Costa N, Aloe L (2005) CCK-8 prevents the development of kindling and regulates the GABA and NPY expression in the hippocampus of pentylenetetrazole (PTZ)-treated adult rats. Neuropharmacology 48:732–742. https://doi.org/10.1016/j.neuropharm.2004.12.001

Tiwari S, Atluri V, Kaushik A, Yndart A, Nair M (2019) Alzheimer's disease: pathogenesis, diagnostics, and therapeutics. Int J Nanomedicine 14:5541–5554. https://doi.org/10.2147/IJN.S200490

Tomita H, Miyasaka K, Matsumoto M, Funakoshi A (1994) Direct, concentration-dependent inhibition by taurocholate of pancreatic exocrine secretion and CCK release in conscious rats. Dig Dis Sci 39:1544–1549. https://doi.org/10.1007/BF02088062

Topping DL, Clifton PM (2001) Short-chain fatty acids and human colonic function: roles of resistant starch and nonstarch polysaccharides. Physiol Rev 81:1031. https://doi.org/10.1152/physrev.2001.81.3.1031

Tran L, Greenwood-Van Meerveld B (2013) Age-associated remodeling of the intestinal epithelial barrier. J Gerontol A Biol Sci Med Sci 68:1045

Tung YCL, Ayuso E, Shan X, Bosch F, O'Rahilly S, Coll AP, Yeo GSH (2010) Hypothalamic-specific manipulation of Fto, the ortholog of the human obesity gene FTO, affects food intake in rats. PLoS One 5:e8771. https://doi.org/10.1371/journal.pone.0008771

Ueyama T, Kawai Y, Nemoto K, Sekimoto M, Toné S, Senba E (1997) Immobilization stress reduced the expression of neurotrophins and their receptors in the rat brain. Neurosci Res 28: 103–110. https://doi.org/10.1016/S0168-0102(97)00030-8

Ussar S, Griffin NW, Bezy O, Fujisaka S, Vienberg S, Softic S, Deng L, Bry L, Gordon JI, Kahn CR (2015) Interactions between gut microbiota, host genetics and diet modulate the predisposition to obesity and metabolic syndrome. Cell Metab 22:516–530. https://doi.org/10.1016/j.cmet.2015.07.007

Verstynen TD, Weinstein A, Erickson KI, Sheu LK, Marsland AL, Gianaros PJ (2013) Competing physiological pathways link individual differences in weight and abdominal adiposity to white matter microstructure. NeuroImage 79:129–137. https://doi.org/10.1016/j.neuroimage.2013.04.075

Vidaurre OG, Gascón S, Deogracias R, Sobrado M, Cuadrado E, Montaner J, Rodríguez-Peña A, Díaz-Guerra M (2012) Imbalance of neurotrophin receptor isoforms TrkB-FL/TrkB-T1 induces neuronal death in excitotoxicity. Cell Death Dis 3:e256. https://doi.org/10.1038/cddis.2011.143

Viña J, Lloret A (2010) Why women have more Alzheimer's disease than men: gender and mitochondrial toxicity of amyloid-β peptide. J Alzheimers Dis 20:S527. https://doi.org/10.3233/JAD-2010-100501

Vogt NM, Kerby RL, Dill-McFarland KA, Harding SJ, Merluzzi AP, Johnson SC, Carlsson CM, Asthana S, Zetterberg H, Blennow K, Bendlin BB, Rey FE (2017) Gut microbiome alterations in Alzheimer's disease. Sci Rep 7:13537. https://doi.org/10.1038/s41598-017-13601-y

Voits M, Hasenöhrl RU, Huston JP, Fink H (2001) Repeated treatment with cholecystokinin octapeptide improves maze performance in aged Fischer 344 rats. Peptides 22:1325–1330. https://doi.org/10.1016/S0196-9781(01)00459-4

Walczyk T, Wick JY (2017) The ketogenic diet: making a comeback. Consult Pharm 32:388. https://doi.org/10.4140/TCP.n.2017.388

Wang J, Si Y-M, Liu Z-L, Yu L (2003) Cholecystokinin, cholecystokinin-A receptor and cholecystokinin-B receptor gene polymorphisms in Parkinson's disease. Pharmacogenetics 13:365–369. https://doi.org/10.1097/00008571-200306000-00008

Wang Y, Gao X, Zhang X, Xiao Y, Huang J, Yu D, Li X, Hu H, Ge T, Li D, Zhang T (2019) Gut microbiota dysbiosis is associated with altered bile acid metabolism in infantile cholestasis. mSystems 4:e00463. https://doi.org/10.1128/mSystems.00463-19

Wank SA (1995) Cholecystokinin receptors. Am J Physiol Liv Physiol 269:G628–G646. https://doi.org/10.1152/ajpgi.1995.269.5.G628

Ward MA, Carlsson CM, Trivedi MA, Sager MA, Johnson SC (2005) The effect of body mass index on global brain volume in middle-aged adults: a cross sectional study. BMC Neurol 5:23. https://doi.org/10.1186/1471-2377-5-23

Warren JD, Rohrer JD, Rossor MN (2013) Frontotemporal dementia. BMJ 347:f4827–f4827. https://doi.org/10.1136/bmj.f4827

Wei X, Zhang Y, Zhang L, Xu S, Chen X, Wang C, Si J, Ma K, Wang Q (2013) CCK-8S increases the firing frequency of CCK-positive neurons and facilitates excitatory synaptic transmission in primary rat hippocampal neurons. Neurosci Lett 549:34–39. https://doi.org/10.1016/j.neulet.2013.06.043

Weinstein G, Beiser AS, Choi SH, Preis SR, Chen TC, Vorgas D, Au R, Pikula A, Wolf PA, DeStefano AL, Vasan RS, Seshadri S (2014) Serum brain-derived neurotrophic factor and the risk for dementia: the Framingham heart study. JAMA Neurol 71:55–61. https://doi.org/10.1001/jamaneurol.2013.4781

Weiss N, Miller F, Cazaubon S, Couraud PO (2009) The blood-brain barrier in brain homeostasis and neurological diseases. Biochim Biophys Acta Biomembr 1788:842. https://doi.org/10.1016/j.bbamem.2008.10.022

Wen D, Sun D, Zang G, Hao L, Liu X, Yu F, Ma C, Cong B (2014) Cholecystokinin octapeptide induces endogenous opioid-dependent anxiolytic effects in morphine-withdrawal rats. Neuroscience 277:14–25. https://doi.org/10.1016/j.neuroscience.2014.06.048
West DB, Fey D, Woods SC (1984) Cholecystokinin persistently suppresses meal size but not food intake in free-feeding rats. Am J Physiol Integr Comp Physiol 246:R776–R787. https://doi.org/10.1152/ajpregu.1984.246.5.R776
Whited KL, Thao D, Lloyd KCK, Kopin AS, Raybould HE (2006) Targeted disruption of the murine CCK 1 receptor gene reduces intestinal lipid-induced feedback inhibition of gastric function. Am J Physiol Liv Physiol 291:G156–G162. https://doi.org/10.1152/ajpgi.00569.2005
Whitmer RA, Gunderson EP, Barrett-Connor E, Quesenberry CP, Yaffe K (2005) Obesity in middle age and future risk of dementia: a 27 year longitudinal population based study. Br Med J 330: 1360–1362. https://doi.org/10.1136/bmj.38446.466238.E0
Whitmer RA, Gustafson DR, Barrett-Connor E, Haan MN, Gunderson EP, Yaffe K (2008) Central obesity and increased risk of dementia more than three decades later. Neurology 71:1057–1064. https://doi.org/10.1212/01.wnl.0000306313.89165.ef
Wotton CJ, Goldacre MJ (2014) Age at obesity and association with subsequent dementia: record linkage study. Postgrad Med J 90:547–551. https://doi.org/10.1136/postgradmedj-2014-132571
Xin L, Ipek Ö, Beaumont M, Shevlyakova M, Christinat N, Masoodi M, Greenberg N, Gruetter R, Cuenoud B (2018a) Nutritional ketosis increases NAD+/NADH ratio in healthy human brain: an in vivo study by 31P-MRS. Front Nutr 5:62. https://doi.org/10.3389/fnut.2018.00062
Xin Y, Diling C, Jian Y, Ting L, Guoyan H, Hualun L, Xiaocui T, Guoxiao L, Ou S, Chaoqun Z, Jun Z, Yizhen X (2018b) Effects of oligosaccharides from morinda officinalis on gut microbiota and metabolome of APP/PS1 transgenic mice. Front Neurol 9:412. https://doi.org/10.3389/fneur.2018.00412
Xu WL, Atti AR, Gatz M, Pedersen NL, Johansson B, Fratiglioni L (2011) Midlife overweight and obesity increase late-life dementia risk: a population-based twin study. Neurology 76:1568–1574. https://doi.org/10.1212/WNL.0b013e3182190d09
Yaffe K, Kanaya A, Lindquist K, Simonsick EM, Harris T, Shorr RI, Tylavsky FA, Newman AB (2004) The metabolic syndrome, inflammation, and risk of cognitive decline. J Am Med Assoc 292:2237–2242. https://doi.org/10.1001/jama.292.18.2237
Yakovleva T, Marinova Z, Kuzmin A, Seidah NG, Haroutunian V, Terenius L, Bakalkin G (2007) Dysregulation of dynorphins in Alzheimer disease. Neurobiol Aging 28:1700–1708. https://doi.org/10.1016/j.neurobiolaging.2006.07.002
Yamagishi T, Debas HT (1978) Cholecystokinin inhibits gastric emptying by acting on both proximal stomach and pylorus. Am J Physiol Metab 234:E375. https://doi.org/10.1152/ajpendo.1978.234.4.E375
Yankner BA (1996) Mechanisms of neuronal degeneration in Alzheimer's disease. Neuron 16:921–932. https://doi.org/10.1016/S0896-6273(00)80115-4
Yao J, Chen S, Mao Z, Cadenas E, Brinton RD (2011) 2-Deoxy-D-glucose treatment induces ketogenesis, sustains mitochondrial function, and reduces pathology in female mouse model of Alzheimer's disease. PLoS One 6:e21788. https://doi.org/10.1371/journal.pone.0021788
Yoshitsugu R, Kikuchi K, Iwaya H, Fujii N, Hori S, Lee DG, Ishizuka S (2019) Alteration of bile acid metabolism by a high-fat diet is associated with plasma transaminase activities and glucose intolerance in rats. J Nutr Sci Vitaminol (Tokyo) 65:45–51. https://doi.org/10.3177/jnsv.65.45
You Z-B, Herrera-Marschitz M, Pettersson E, Nylander I, Goiny M, Shou H-Z, Kehr J, Godukhin O, Hökfelt T, Terenius L, Ungerstedt U (1996) Modulation of neurotransmitter release by cholecystokinin in the neostriatum and substantia nigra of the rat: regional and receptor specificity. Neuroscience 74:793–804. https://doi.org/10.1016/0306-4522(96)00149-2
Zhang L, Wei X, Zhang Y, Xu S, Chen X, Wang C, Wang Q (2013) CCK-8S increased the filopodia and spines density in cultured hippocampal neurons of APP/PS1 and wild-type mice. Neurosci Lett 542:47–52. https://doi.org/10.1016/j.neulet.2013.03.023
Zhang X, Grosfeld A, Williams E, Vasiliauskas D, Barretto S, Smith L, Mariadassou M, Philippe C, Devime F, Melchior C, Gourcerol G, Dourmap N, Lapaque N, Larraufie P, Blottière HM,

Herberden C, Gerard P, Rehfeld JF, Ferraris RP, Fritton JC, Ellero-Simatos S, Douard V (2019) Fructose malabsorption induces cholecystokinin expression in the ileum and cecum by changing microbiota composition and metabolism. FASEB J 33:7126–7142. https://doi.org/10.1096/fj.201801526RR

Zhao Y, Lukiw WJ (2013) TREM2 signaling, miRNA-34a and the extinction of phagocytosis. Front Cell Neurosci 7:131. https://doi.org/10.3389/fncel.2013.00131

Zhao Y, Dua P, Lukiw WJ (2015) Microbial sources of amyloid and relevance to amyloidogenesis and Alzheimer's disease (AD). J Alzheimers Dis Parkinson 5:177. https://doi.org/10.4172/2161-0460.1000177

Zito K, Scheuss V (2009) NMDA receptor function and physiological modulation. In: Encyclopedia of neuroscience. Elsevier Ltd, Amsterdam, pp 1157–1164. https://doi.org/10.1016/B978-008045046-9.01225-0

Zlokovic BV (1996) Cerebrovascular transport of Alzheimer's amyloid β and apolipoproteins J and E: possible anti-amyloidogenic role of the blood-brain barrier. Life Sci 59:1483. https://doi.org/10.1016/0024-3205(96)00310-4

Zwirska-Korczala K, Konturek SJ, Sodowski M, Wylezol M, Kuka D, Sowa P, Adamczyk-Sowa M, Kukla M, Berdowska A, Rehfeld JF, Bielanski W, Brzozowski T (2007) Basal and postprandial plasma levels of PYY, ghrelin, cholecystokinin, gastrin and insulin in women with moderate and morbid obesity and metabolic syndrome. J Physiol Pharmacol 58(Suppl 1):13–35

Part II

Therapeutic Interventions for Dementia

Cholesterol and Dementia: A Possible Therapeutic Approach

13

Jessica Sarahi Alavez-Rubio and Teresa Juárez-Cedillo

Abstract

Dementia is a syndrome characterized by deterioration in memory, behavior, and the ability to perform everyday activities. Currently available drugs have no impact on progression and do not offer a cure, so identifying and studying new therapeutic approaches is important. There is evidence that indicates disturbances in cholesterol homeostasis can be related to dementia. The present chapter aims to address the relationship that exists between the metabolism of cholesterol and dementia, especially in Alzheimer's disease and vascular dementia, and how this relationship has been used for therapeutic purposes, from the use of statins to the development of new drugs, such as ACAT inhibitors, highlighting the importance of cholesterol in the proper functioning of the brain and how this can represent a useful path for designing new therapeutic approaches.

Keywords

ACAT1 inhibitors · Alzheimer's disease · Cholesterol metabolism · Statins · Vascular dementia · Amyloid beta · Tau pathology · Dementia

J. S. Alavez-Rubio
Universidad Nacional Autónoma de México, Mexico City, Mexico

T. Juárez-Cedillo (✉)
Unidad de Investigación Epidemiológica y en Servicios de Salud, Área Envejecimiento, Centro Médico Nacional Siglo XXI, Instituto Mexicano del Seguro Social (Actualmente comisionada en la Unidad de Investigación en Epidemiológica Clínica, Hospital General Regional No. 1 Dr. Carlos Mac Gregor Sánchez Navarro, Instituto Mexicano del Seguro Social), Mexico City, Mexico
e-mail: terezillo@exalumno.unam.mx; teresa.juarez@imss.gob.mx

G. M. Ashraf, M. S. Uddin (eds.), *Current Thoughts on Dementia*,
https://doi.org/10.1007/978-981-16-7606-2_13

13.1 Introduction

Dementia is the main cause of disability and dependency among older adults worldwide, has implications in cognitive abilities, behavior, and memory, interfering with daily life activities. The impact of dementia is not only significant economically, but also represents important human costs to countries, societies, families, and individuals (WHO, ADI 2012). Alzheimer's disease (AD) contributes in 60–70% of cases, being the main cause of dementia (WHO, ADI 2012); the second most common type is vascular dementia (VaD), with 17–25% of cases (Appleton et al. 2017). AD is a multifactorial disease with modifiable lifestyle-related risk factors and non-modifiable genetic risk factors. One of these factors is altered cholesterol homeostasis, which has been related to brain disorders and significant neurodegenerative diseases (Arenas et al. 2017). Any change in lipid metabolism causes an altered lipid composition of intracellular membrane compartments, which is a common biomarker in many neuronal disorders (Hussain et al. 2019).

Cholesterol plays a fundamental role in physiology and brain function. Studies in vitro and in vivo have shown that high cholesterol levels favor the production of Aβ (Refolo et al. 2000; Fassbender et al. 2001) and it has also been related to tau pathology (Boimel et al. 2009; Glöckner et al. 2011; Glöckner and Ohm 2014; Gratuze et al. 2016; van der Kant et al. 2019, 2020), the two main pathological characteristics of AD. In addition, in epidemiological studies, high plasma cholesterol levels have been associated with an increased risk of developing AD (Kivipelto et al. 2001; Pappolla et al. 2003; Whitmer et al. 2005) and VaD (Solomon et al. 2009b) at older ages; however, the results are still controversial (Mielke et al. 2005).

The proposed hypotheses try to explain the origin and progress, focused on understanding its pathogenesis and finding new therapeutic strategies. But despite the relationship between lipid metabolism and the pathogenesis of dementia, there are few therapeutic approaches focused on this connection. Statins and acyl-CoA: cholesterol acyltransferase 1 (ACAT1) inhibitors are within this type of treatment. Statins are currently the first-line pharmacological therapy for the treatment of hyperlipidemia in the primary prevention of coronary heart disease (Schultz et al. 2018). Additionally, statins have been shown to reduce the risk of AD and to improve cognitive impairment in some cases. However, studies have obtained conflicting results (Jick et al. 2000; Wolozin 2000; Rockwood et al. 2002; Arvanitakis et al. 2008; Cramer et al. 2008; Sparks et al. 2008; Haag et al. 2009; Swiger et al. 2013; Mcguinness et al. 2016).

On the other hand, ACAT inhibitors seem to be a promising option; however, there is still a long way to go. The enzyme ACAT1 has the main role in cholesterol homeostasis, preventing over-accumulation of free cholesterol in the cells, forming cholesteryl esters, and it has been noted that the balance between the free cholesterol and esters is an important point in the control of amyloidogenesis by blocking ACAT1 activity, where an important reduction of cholesteryl esters, amyloid level, and brain amyloid plaques has been reported (Puglielli et al. 2001; Hutter-Paier et al. 2004; Huttunen et al. 2007, 2009, 2010; Bryleva et al. 2010; Murphy et al. 2013; Shibuya et al. 2014, 2015b). Although these results have not been conclusive,

ACAT1 has been identified as a potential therapeutic target (Wollmer et al. 2003; Bertram et al. 2005; Zhao et al. 2005; Lämsä et al. 2007; Chen et al. 2018).

The present chapter aims to address the relationship that exists between the metabolism of cholesterol and dementia, especially in AD and VaD, and how this relationship has been used for therapeutic purposes, from the use of statins, which has been widely explored, with results that are not sufficiently defined, to the development of new drugs, such as ACAT inhibitors; however, these are still in preclinical stages, and there are still many unknowns to be resolved. Undoubtedly, studies in in vitro and animal models have shown promising results. So, given the importance of cholesterol in the proper functioning of the brain, it is believed that it can be a strategy to which attention should be paid and that it may be useful in designing therapeutic approaches for dementia.

13.2 Dementia and Its Relationship with Cholesterol

AD and VaD are the main forms of dementia. Patients with AD present loss of synapses and neurons, as well as senile extracellular plaques and neurofibrillary tangles (NFTs). Senile plaques are an aggregate of Aβ, whereas tau protein is hyperphosphorylated under pathological conditions, which causes its separation from microtubules and promotes the production of insoluble aggregates, which leads to the appearance of helical filaments and the presence of NFTs in the brain (Di Paolo and Kim 2012). In the hypothesis of the amyloid cascade, the pathogenesis of AD, the altered tau protein, and the formation of NFTs are considered consequences after Aβ toxicity (Hardy and Higgins 1992). Although alterations in tau are not considered as the first event in the AD pathogenesis, have a key role in this process (Dickson et al. 2013). On the other hand, the VaD is the result of general and local effects of vascular disease, which causes lesions due to vascular accident and other changes in tissue perfusion (Kalaria 2016). The etiology of VaD includes damage to the large and small vessels, with brain disease of small vessels being the most common cause (Appleton et al. 2017); clinical symptoms vary widely depending on the causes and location of the damage (Panza et al. 2006).

VaD and AD share pathology evident in brain imaging and post-mortem brain. Over half of individuals with AD have vascular pathologies, the mixed etiology becoming more common as age increases, and this combination has a synergistic effect and leads to increased likelihood of dementia (Attems and Jellinger 2014; Kapasi and Schneider 2016; Vijayan and Reddy 2016).

Finally, although dementia can be caused in multiple ways, growing evidence suggests one interrelationship between dementia and lifestyle-related risk factors. These risk factors include a sedentary lifestyle, obesity, unbalanced diets, tobacco use, harmful use of alcohol, diabetes mellitus, midlife hypertension, and hypercholesterolemia, in addition to other risk factors such as cognitive inactivity, social isolation, low educational level, and depression (de la Torre 2002; WHO 2017).

Thus, to understand its pathogenesis and find new therapeutic approaches, several hypotheses to explain the origin and progress of dementia have been proposed. One of them is related to cholesterol metabolism, which plays an important role in physiology and brain function (Arenas et al. 2017). Cholesterol homeostasis is regulated by a dynamic balance of absorption, de novo synthesis, esterification, catabolism, and release (Canevari and Clark 2007); thus, when a process fails, it may result in homeostatic disruption, and this has been associated with neurodegenerative diseases (Holtzman et al. 2011; Lütjohann et al. 2012).

13.3 Cholesterol Metabolism and Homeostasis in the Brain

The brain is highly enriched in lipid content; lipids contribute 50–60% of its dry weight. Those found in the brain are grouped as sphingolipids, glycerophospholipids, and cholesterol (Hussain et al. 2019). The latter is the most important component of the mammalian cell membranes (Dietschy 2009). The brain is the organ with the highest cholesterol content; it has a concentration ten times higher compared to other tissues, and contains 23% of total cholesterol, although it is only 2.1% of body weight (Dietschy and Turley 2001). In the central nervous system (CNS), cholesterol has several essential functions, as it is required for cellular processes such as glial cell proliferation, neurite outgrowth, microtubules stability, and it is abundantly present in the synaptic membranes to aid in nerve signal transmission and increasing nerve conduction velocity (Petrov et al. 2016; Hussain et al. 2019).

Its homeostasis is highly regulated by various control mechanisms which influence the biosynthesis, transport, storage, and elimination of cholesterol in brain (Sun et al. 2015b). Because the blood-brain barrier (BBB) prevents the uptake of low-density lipoprotein-cholesterol (LDL-c) from circulation, cholesterol present in the brain is generated almost entirely by de novo synthesis (Canevari and Clark 2007). In the CNS, more than 99.5% of total cholesterol is in its non-esterified form (Björkhem et al. 2004), and it is found in two main stores: about 70% is in myelin, and the rest is made up of the plasma membranes of glial cells and neurons (Björkhem et al. 2004; Petrov et al. 2016).

Cellular cholesterol is synthesized mainly in the endoplasmic reticulum (ER) (Canevari and Clark 2007; Martin et al. 2014). Through membrane-bound transcription factors known as sterol regulatory element-binding proteins (SREBPs) cells can modulate their cholesterol level; when levels are low, SREBPs promote the generation of transcription factors that bind to the sterol regulatory element (SRE-1), which controls the transcription of 3-hydroxy-3-methyl-glutaryl-CoA reductase (HMGR) and other genes related to the metabolism and transport of cholesterol and other lipids (Zhang and Liu 2015). High cholesterol levels exert control by feedback inhibition and by stimulating HMGR ubiquitination and degradation by the proteasome (Canevari and Clark 2007; Zhang and Liu 2015).

Cholesterol synthesis in the developing brain is relatively high, while mature neurons decrease their endogenous cholesterol synthesis, so that the supply of cholesterol in the adult brain relies mainly on the cholesterol production of the glial cells, especially the astrocytes and oligodendrocytes (Sun et al. 2015b). In the

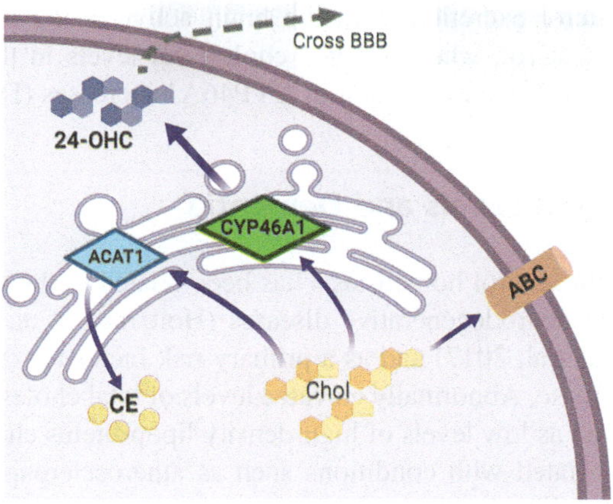

Fig. 13.1 The neurons have different pathways to regulate excess cholesterol, such as esterification by ACAT1 and its storage in lipid droplets within the cell, direct excretion through ABC transporters or conversion to 24-OHC through CYP46A1. *24-OHC* 24-hydroxycholesterol, *ABC* ATP-binding cassette, *ACAT1* acyl-CoA:cholesterol acyltransferase 1, *BBB* blood-brain barrier, *CE* cholesterol esters, *Chol* cholesterol

brain, cholesterol is transported by apolipoproteins, mainly apolipoprotein E (APOE); astrocytes are the major source of APOE, which forms lipoprotein complexes (Petrov et al. 2016). These complexes will be taken up by LDL, VLDL, APOE, receptors expressed on the neuron (Canevari and Clark 2007), and it will be these lipoprotein complexes that, when internalized in the cells, will hydrolyze with the lysosomes to release cholesterol (Sun et al. 2015b).

Neurons are more capable of responding to an excess of cholesterol than astrocytes. They can handle excess cholesterol in different ways (Fig. 13.1), such as esterification and subsequent intracellular storage in lipid droplets (LDs), direct excretion through ATP-binding cassette (ABC) transporters (Bogdanovic et al. 2001; Moutinho et al. 2016) or conversion to 24(S)-hydroxycholesterol (24-OHC) through CYP46A1 (Lund et al. 2003).

Cholesterol esters represent about 1% of the total cholesterol present in the adult brain (Chan et al. 2012). The esterification of excess free cholesterol takes place primarily in the endoplasmic reticulum, it is catalyzed by the enzyme acyl-Coenzyme A:cholesterol acyltransferase 1 (ACAT1), also named sterol *O*-acyltransferase 1 (SOAT1) (Chang et al. 1995).

CYP46A1 is responsible for the hydroxylation of cholesterol to 24-OHC, representing the major elimination pathway (40–50% of brain cholesterol) (Lund et al. 2003; Xie et al. 2003). The 24-OHC is transported to the BBB by APOE, where it fluxes out of the brain via diffusion or by the organic anion transporter across the barrier (Lütjohann et al. 2012). When 24-OHC is in the general circulation, it binds to LDL and is excreted in bile salts. The expression of the proteins needed for cholesterol biosynthesis and transport is stimulated by the nuclear liver X receptor, by the action of 24-OHC and other oxysterols derived in the brain. Therefore, an increase in cholesterol excretion from the brain activates the de novo synthesis mechanism of cholesterol, whereas when cholesterol levels in the ER membrane pass the threshold level, the expression of CYP46A1 increases (Dietschy 2009).

13.4 Cholesterol Levels and Dementia

Dysregulation of cholesterol homeostasis has been related to the presence of major brain disorders and neurodegenerative diseases (Holtzman et al. 2011; Lütjohann et al. 2012; Arenas et al. 2017) and as a primary risk factor for cardiovascular and cerebrovascular disease. Abnormally elevated levels of total cholesterol, LDL-c, and triglycerides, as well as low levels of high-density lipoproteins cholesterol (HDL-c) have all been associated with conditions such as atherosclerosis, coronary artery disease, and hyperlipidemia; they can cause vascular-related brain changes, from small-vessel ischemic infarcts to serious events such as stroke, and have also been related with the development of pathological disorders such as AD and VaD (Solomon et al. 2009a, b; Leritz et al. 2016). Nevertheless, the association between hyperlipidemia and dementia remains controversial because studies examining the association between cholesterol and dementia or cognitive functions have reported conflicting results (West et al. 2008; Leritz et al. 2016; Wendell et al. 2016; Zhao et al. 2019), as summarized in Table 13.1.

In 1906, Alois Alzheimer described the first case of AD-related dementia (Alzheimer 1906), from that moment, the idea of an altered lipid metabolism arose because the findings included not only the presence of Aβ-plaques and neurofibrillary tangles but also the presence of "lipoidic granules" or "adipose inclusions" (Lütjohann et al. 2012). Nevertheless, the relationship between AD and lipid metabolism was established when the ε4 allele of *APOE* (*APOE**ε4) was pointed out as the main genetic risk factor for sporadic AD (Corder et al. 1993; Saunders et al. 1993; Strittmatter et al. 1993; St Clair et al. 1995), although this relationship has not been established for VaD (Yin et al. 2012; Rohn 2014; Sun et al. 2015a; Skrobot et al. 2016; Ikram et al. 2017). The importance of *APOE**ε4 in AD is based on evidence that it modulates the aggregation and elimination of Aβ by binding to it (Shibuya et al. 2015a).

One of the main factors associated with dementia is age, with which alterations in cholesterol levels have also been reported, as Eckert et al. (2001) showed that brains of adult mice exhibited high cholesterol concentrations, as well as an increase in the flow of cholesterol through the CNS in aging and in early onset AD has been reported. In humans, high plasma cholesterol levels at midlife have been linked to an increased risk of developing AD in old age (Kivipelto et al. 2001; Pappolla et al. 2003; Whitmer et al. 2005). Specifically, high cholesterol levels have been shown to promote Aβ formation in in vitro and animal models (Refolo et al. 2000; Fassbender et al. 2001) and to decrease the secretion of soluble amyloid (Lütjohann and von Bergmann 2003; Posse 2012; Maulik et al. 2013; Wood et al. 2014). The Aβ cascade hypothesis (Hardy and Higgins 1992) is the most widely accepted hypothesis for explaining AD development, which proposes that alteration in Aβ precedes and

Table 13.1 Relationship of cholesterol levels with dementia

Population	Results	References
1449 subjects aged 65–79 years with follow-up of 21 years	Midlife elevated serum TC level (≥6.5 mmol/L) was a risk factor for MCI (OR = 1.9, 95% CI = 1.2–3.0)	Kivipelto et al. (2001)
185 non-demented individuals, ≥85 years	High TC and high LDL-c were associated with better memory scores in non-carriers of the *APOE*ε4* allele	West et al. (2008)
1382 subjects who were not demented 21 years after the baseline examination	In participants without dementia, high midlife TC may be a determinant of poorer cognitive performance later in life	Solomon et al. (2009a)
9844 subjects, 40–45 years	Midlife serum TC (≥240 mg/dL) was associated with an elevated risk of AD (HR = 1.57, 95% CI = 1.23–2.01) and VaD (HR = 1.26, 95% CI = 0.82–1.96)	Solomon et al. (2009b)
220 participants with mild to moderate AD, 75.31 years (SD = 8.4)	High TC levels are associated with lower MMSE and more neuropsychiatric symptoms	Hall et al. (2014)
120 individuals, 43–85 years	The memory, executive function, and language were negatively associated with age and TG levels, while LDL-c levels presented a positive relationship	Leritz et al. (2016)
190 participants without neurologic and psychiatric disease, 54–83 years	There is nonlinear age-modified associations between TC and LDL-c levels and performance on measures of memory and speeded executive functioning	Wendell et al. (2016)
3565 participants, 50–89 years	Patients with low (120 mg/dL) (HR = 1.29, 95% CI = 1.04–1.61) and high (210 mg/dL) (HR = 1.16, 95% CI = 1.01–1.33) levels of non-HDL-c during their 60s and 70s had higher risk for AD compared to patients with intermediate levels. Showing a potential U-Shaped relationship	Marcum et al. (2018)
Patients with AD 117 and controls = 117, ≥55 years	Higher serum TC and LDL-c levels and lower HDL-c levels were associated with the risk of AD in a cross-sectional study	Chen et al. (2019)
52 individuals, 75–93 years	Serum HDL-c may be a biomarker of both memory function and cortical structure. High serum level of HDL-c was associated with preserved memory function	Kinno et al. (2019)
1762 subjects, 40–85 years	High serum TC may be a risk factor of cognitive impairment in the elderly male, high serum LDL-c for female subjects and high serum TG may be a protector of cognitive impairment in the middle-aged male participants	Zhao et al. (2019)

HDL-c high-density lipoprotein cholesterol, *HR* hazard ratio, *LDL-c* low-density lipoprotein cholesterol, *MCI* mild cognitive impairment, *OR* odds ratio, *TC* total cholesterol, *TG* triglyceride, *VaD* vascular dementia triggers tau pathology, along with other neurodegenerative and compensatory sequelae (Walker et al. 2018).

The two-hit vascular hypothesis, another hypothesis for the AD development, states that cerebrovascular damage (hit 1) is a first insult that is self-sufficient to star neuronal injury and neurodegeneration, but can also stimulate Aβ accumulation in the brain (hit 2) (Nelson et al. 2016).

In VaD there are also a wide variety of pathological mechanisms related to its origin, where lipids are vitally important, such as lipid oxidation, to which the brain is particularly susceptible, due to the high content of polyunsaturated fatty acids (Appleton et al. 2017). Both high levels of LDL-c and low levels of HDL-c have known risk factors for carotid atherosclerosis and coronary artery disease, which can result in cognitive impairment secondary to cerebral hypoperfusion or embolism (Appleton et al. 2017). But the relationship with VaD is still unclear, on one other hand it has been reported that no association was found between levels of LDL-c and image markers obtained by nuclear magnetic resonance (Gouw et al. 2008; van Dijk et al. 2008), and, on the other hand; unexpectedly, an association of increased HDL-c and decreased LDL-c levels with an elevated risk of worsening of white matter grade on serial magnetic resonance imaging scan has also been reported (Longstreth et al. 2005). White matter findings are common in the elderly and can be related to cognitive impairments, increased risk of future stroke, and development of AD (Provenzano et al. 2013).

Meanwhile, the relationship between serum total cholesterol (TC), cognition, and VaD is currently a matter of debate, as in persons without dementia, high levels of TC during midlife may be a determinant of worse cognitive functioning later in life, but declining levels of TC after midlife may lead to poorer cognitive status (Solomon et al. 2009a), and increasing TC levels are associated with an increased risk of developing VaD (Solomon et al. 2009b); moreover, other studies have reported results with an inverse association (Mielke et al. 2005).

13.4.1 Role of Cholesterol in Amyloid Plaques

The accumulations of Aβ are one of the hallmarks of AD. Soluble Aβ is a normal component of plasma and cerebrospinal fluid. It is a product of amyloid precursor protein (APP) processing (Suh 2002) and then is catabolized by Aβ-degrading peptidases or by other ways (Araki and Tamaoka 2015), when there is deregulation in these processes, a pathological accumulation of Aβ occurs and toxic oligomers are formed (Lütjohann et al. 2012). β- and γ-secretases participate in the formation of the Aβ peptide, which is generated from APP. β-secretase (also named BACE-1) cleaves at the N-terminal end of the Aβ region, and then γ-secretase cleavage of the membrane-bound C-terminal fragment to release Aβ (Cordy et al. 2006; Saxena 2009). In this way two main variants of Aβ are generated, one with 42 and the other with 40 amino-acid peptides ($A\beta_{42}$ and $A\beta_{40}$), being $A\beta_{40}$ the main in cerebrospinal fluid (Suh 2002), and $A\beta_{42}$ the major component of amyloid plaques. $A\beta_{42}$ is more hydrophobic and less soluble, so it tends to aggregate and deposit, it is the cause of neurotoxicity and synaptic loss (Saxena 2009). The appearance of these amyloid plaques have been related to the triggering of other pathological processes present in AD, and it has been suggested that this may be due directly to the binding of Aβ to cell membranes through the binding of Aβ to receptors present in microglial and neuronal cells (Saxena 2009; Khan et al. 2017).

Cholesterol has been proposed as a factor capable of regulating APP cleavage and Aβ production (Posse 2012). Lipid rafts are involved in this relationship, there is evidence that suggests that lipid rafts participate in the production, metabolism, aggregation, and neurotoxicity of Aβ (Cordy et al. 2006; Cossec et al. 2010; Araki and Tamaoka 2015). The lipid rafts are distinct membrane domains with high concentrations of cholesterol and glycosphingolipids; they are involved in different cellular processes (Araki and Tamaoka 2015). β- and γ-secretases, the enzymes responsible for the Aβ production, are mainly located in lipid rafts (Araki and Tamaoka 2015; Kim et al. 2016), it is thought that the increase of cholesterol in these domains can favor the recruitment of APP and these necessary enzymes and thereby increase the Aβ production (Vetrivel and Thinakaran 2006).

13.4.2 Role of Cholesterol in Tau Protein

Neurofibrillary tangles are also a hallmark in Alzheimer's disease. In general, it has been considered that Aβ accumulation is the first pathological feature to appear and that this leads to the accumulation of tau; however, there is also evidence that the tau pathology can progress independently of the accumulation of Aβ, which has raised the interest in identifying the Aβ-independent factors that regulate tau pathology (van der Kant et al. 2020), and one of the factors that have highlighted is the cholesterol metabolism. The relationship between altered cholesterol metabolism with Aβ has been explored more widely but between tau pathology and cholesterol is still unclear. In the search for evidence of this relationship between cholesterol and pathological tau, studies in animal models have also been carried out, testing the effect of a high-cholesterol diet and observing the effect on tau levels, but contradictory results have been obtained (Glöckner et al. 2011; Gratuze et al. 2016).

However, there is also evidence that important molecules in cholesterol metabolism are related to tau levels, such as CYP46A1 (Burlot et al. 2015), APOE, 24-OHC (Leoni et al. 2010), and cholesteryl esters (van der Kant et al. 2019). This idea also is supported by the results of a study where tau and Aβ positron emission tomography (PET) were used (Sepulcre et al. 2018), in that study, different pathways for the accumulation of tau and Aβ were identified; however, the propagation patterns of both are associated with a common genetic profile connected to lipid metabolism, in which APOE has a fundamental role. Besides, it has also been shown that excess of cholesteryl esters increase the accumulation of phosphorylated tau (p-tau) due to the inhibition of its proteasomal degradation and decreased levels of cholesteryl esters also cause decreased levels of p-tau in human isogenic induced pluripotent stem cell (hiPSC) lines with mutations in the cholesterol-binding domain of APP or APP null alleles (Shibuya et al. 2015b; van der Kant et al. 2019).

And when using drugs that inhibit cholesterol synthesis, such as statins, decreased levels of tau have been reported, in a tau transgenic mice model (Boimel et al. 2009), where also this effect is observed despite the absence of Aβ pathology, which might suggest that the path by which tau pathology reduce occurs is independent of Aβ. In clinical trials, this effect on tau levels has also been observed after the administration of statins (Riekse et al. 2006; Li et al. 2017). The evidence suggests that cholesterol modulates both tau and amyloid pathology through related but independent mechanisms (van der Kant et al. 2019, 2020). And a bidirectional mode of action has been suggested (Glöckner and Ohm 2014): altered cholesterol metabolism may promote tau pathology, but tau pathology may also modify cholesterol homeostasis, causing a kind of vicious circle in the accumulation of pathological tau.

13.5 Cholesterol Metabolism as a Therapeutic Strategy for Dementia

Even though cholesterol itself is only one piece of the larger puzzle of factors related to dementia, it has been proposed that strategies that interfere with or modify its metabolism could be interesting and useful approaches to treating dementia. Although there is a large amount of evidence in favor of the connection between lipid metabolism and VaD and AD pathogenesis, few therapeutic strategies have been developed related to it. In this vein, enzymes, receptors, and other components involved in cholesterol metabolism have been pointed out as possible therapeutic targets. Such is the case of cholesterol 24-hydroxylase encoded by the *CYP46A1* gene, APOE, ABCA1, Liver X Receptor (LXR), LRP1, ACAT1, and others. Also, using cholesterol-lowering drugs, such as statins, have been observed to show beneficial effects (Bhattacharyya and Kovacs 2010; Koldamova et al. 2010; Zlokovic et al. 2010; Di Paolo and Kim 2012; Sodhi and Singh 2013; Burlot et al. 2015; Mast et al. 2017; Suidan and Ramaswamy 2019).

Here we focus on statins and ACAT inhibitors (Table 13.2). Statins are drugs that help in the prevention of coronary artery disease and stroke; statins mainly lower serum cholesterol levels; however, they have other effects (Daneschvar et al. 2015). On the other hand; it has been proposed that the balance between cholesterol esters and free cholesterol has a relevant role in the control of amyloidogenesis (Bhattacharyya and Kovacs 2010). Blocking of enzyme activity of ACAT1 has revealed favorable effects such as reduction of cholesteryl esters, amyloid level, and brain amyloid plaques; for these reasons, they have been pointed out as therapeutic targets (Puglielli et al. 2001; Hutter-Paier et al. 2004; Huttunen et al. 2007, 2009, 2010; Bryleva et al. 2010; Murphy et al. 2013; Shibuya et al. 2014, 2015b).

Table 13.2 Statins and ACAT1 inhibitors possibly applied in the treatment of dementia

Status	Statins	ACAT inhibitors
Clinical study	Statins are a class of prescription drugs used to reduce blood levels of LDL-c.	ACAT1 inhibitors were tested in phase III studies for atherosclerosis (Nicholls et al. 2006; Meuwese et al. 2009)
Pharmacological effects	Their primary mechanism of action is the lowering of serum cholesterol but they have other pharmacological effects, such as their anti-inflammatory, antioxidant, inhibitor of platelet aggregation, and neuroprotector activity (Daneschvar et al. 2015).	ACAT1 inhibitors decrease Aβ production by reducing APP, and increase autophagy, which increases Aβ and tau clearance.
Prospects for use in dementia	Due to conflicting results, an agreement has not been reached and it is still a matter of debate. A study with simvastatin is currently ongoing, which is planned to be completed in 2021 (US National Library of Medicine 2009).	The results in preclinical studies are promising but the design of new, more specific and smaller molecules is necessary.

13.5.1 ACAT1/SOAT1 as a Therapeutic Target

The enzyme ACAT1/SOAT1, encoded by *ACAT1* gene, esterifies excess free cholesterol. ACAT1 had been extensively investigated as a possible therapeutic target for atherosclerosis, recently beneficial effects have been noted in animal models of AD and also cancer (Antalis et al. 2010; Bemlih et al. 2010; Lee et al. 2015; Li et al. 2016). Cholesteryl esters form small cytoplasmic lipids droplets in the aqueous medium, which can be stored inside or go through the membrane and get out of the cell (Lämsä et al. 2007; Shibuya et al. 2015a). Excess free cholesterol can be cytotoxic, the reason why a high enzymatic activity of ACAT1 can protect the cell (Rogers et al. 2015).

ACAT is a part of the membrane-bound *O*-acyltransferase (MBOAT) enzyme family, and it is present in two isoforms, ACAT1 and ACAT2. Both enzymes use long-chain fatty acyl-coenzyme A to transform free cholesterol (their main substrate) to cholesteryl esters (Rogers et al. 2015). Each isoform has its intracellular location and its metabolic function. ACAT1 is found in the endoplasmic reticulum and is widely expressed, found in almost all tissues, such as the kidney, adrenal glands, macrophages, and the brain. ACAT2 is also expressed in various tissues, but at a lower level than ACAT1, and is mainly present in the liver and the intestine, in the latter it regulates the absorption of cholesterol in the cells (Rogers et al. 2015).

In animal models and brains of AD patients, in the regions with high amyloid load, a higher level of cholesterol esters is observed. Aβ production has been proposed to be regulated by compartmentation of intracellular cholesterol and cytoplasmic cholesterol esters formed by ACAT1 (Chan et al. 2012; Tajima et al. 2013). A study conducted on a Chinese hamster ovary (CHO) cell model and in a neuronal cell model (Puglielli et al. 2001) showed that cholesteryl esters levels were correlated with production of Aβ, these results were part of the first experimental evidence linking ACAT activity with AD. In cultured cells, high levels of cholesterol esters favor the release of Aβ, whereas pharmacological inhibition of ACAT1 reduces the level of Aβ cholesteryl esters (Lämsä et al. 2007; Shibuya et al. 2015a). In addition, it has been reported that ACAT1 also uses long-chain unsaturated fatty acid and 24-OHC as substrates to produce 24-OHC ester. The accumulation of this favors the formation of atypical lipid droplet-like structures together with an enlarged membrane that is similar to a swollen ER structure. 24-OHC esterification mediated by ACAT1 and subsequent LDs storage play a key role in 24S-OHC apoptosis and necrosis (Yamanaka et al. 2014; Urano et al. 2019). Just as atypical lipid droplet-like structures have a function in the death of cells, the LDs also have important implications.

As mentioned, one of the ways in which excess cholesterol is metabolized is its esterification by ACAT1, and its subsequent intracellular storage in LDs (Chang et al. 1995; Bogdanovic et al. 2001; Moutinho et al. 2016). LDs are not simply passive lipid storage vessels; they participate in many complex cellular functions (Cingolani and Czaja 2017). LDs are very small subcellular organelles found in various cells, and they are the main storage site of neutral lipids, such as triglycerides and cholesteryl esters, which form the lipid center that is surrounded by a phospholipid monolayer (Walther and Farese 2012). LDs arise from the ER (Geng and Guo 2017) and their number and size can change rapidly and vary significantly in each cell type, their diameter is between 1 and 10 μm, although in white adipocytes it can be over 50 μm (Geng and Guo 2017). LDs play fundamental roles in lipid metabolism and energy homeostasis; the breakdown of stored lipids from LDs into free fatty acids (FFAs) supplies energy to the cell, and regulates other cellular processes as FFAs can, for example, activate cell signaling pathways or trigger cellular injury (Cingolani and Czaja 2017); in addition, they contribute to protein storage, folding and degradation (Pennetta and Welte 2019).

In many cancerous tissues and different metabolic diseases, as fatty liver, obesity, and atherosclerosis, dysregulation of LDs metabolism has been observed (Geng and Guo 2017). ACAT1 is highly expressed in glioblastoma (GBM) tumor tissues, and its level of expression has correlated with the LDs presence. Data has shown that genetic or pharmacological inhibition of ACAT1 significantly decreases CE synthesis and LDs formation in GBM cells, and suppressed GBM growth both in vitro and in mouse models (Geng et al. 2017). Once turned incorporated into LDs, they are stable, relatively inert, and harmless. This protective function is probably the reason for the abundant accumulation of LDs in many diseases, states characterized by altered lipid supply and metabolism (Welte 2015). In the nervous system, it is now established that both neurons and glia can accumulate LDs under certain disease conditions. But it is not yet known what their function is under these conditions or if they are present under normal conditions in the nervous system, although their altered functioning is neurons and glia is related to neurodegeneration (Liu et al. 2015; Welte 2015).

13.5.1.1 ACAT1 Inhibitors

Atherosclerosis is a chronic syndrome in which plaque builds up inside the arteries. Plaque is made up of cholesterol, fatty deposits, and other blood-derived substances (Nelson et al. 2016). It is the leading cause of strokes, heart attacks, and peripheral vascular disease because atherosclerosis alters the functioning of the blood vessels and therefore it could impair neurovascular coupling and cerebral blood flow (Shabir et al. 2018).

Initially, the ACAT inhibitor drugs were designed for this pathology. But these exhibited poor absorption, they acted locally by blocking cholesterol absorption in the intestine (Hainer et al. 1994; Roth 1998). After failed results of the ACAT inhibitors (Hainer et al. 1994; Roth 1998), they were ruled out as a viable option to reduce plasma cholesterol levels in humans (Llaverias and Alegret 2004). However the search for other molecules continued and the CI-976 compound was synthesized (Bocan et al. 1991), which seemed to exert its ACAT inhibition activity directly in arterial wall (Llaverias and Alegret 2004), but as a side effect, it had adrenal toxicity. So the development of more water-soluble inhibitors was sought, thus synthesizing avasimibe (CI-1011) (Lee et al. 1996), which was evaluated in a phase III clinical study with 509 patients, unfortunately there were no favorable results in coronary atherosclerosis measured by intravascular ultrasound (Tardif et al. 2004).

Subsequently, a potent ACAT1 and ACAT2 inhibitor, pactimibe, was developed, and after showing positive results in the preclinical phase, two studies, ACTIVATE (Nicholls et al. 2006) and CAPTIVATE (Meuwese et al. 2009), were carried out to evaluate its efficacy and safety; unfortunately, no changes were observed in the progression of the atheroma (Nissen et al. 2006). CAPTIVATE was suspended prematurely but in the preliminary report, no significant differences were reported in the results measured by ultrasound carotid intima-media-thickness (CIMT). Even in the pactimibe group a 7.3% increase in LDL-c levels was observed compared with 1.4% in the placebo group ($p = 0.001$) (Meuwese et al. 2009).

Atherosclerosis has been related to AD and dementia pathogenesis, and they share various risk factors, including hypercholesterolemia, diabetes, and aging (Nelson et al. 2016). ACAT1 inhibitors have been tested on AD models and have shown favorable results. For example, the administration of CP-113,818 for 60 days in a transgenic mouse model of AD decreases significantly the number of Aβ plaques by 88% and the cognitive deficit (Hutter-Paier et al. 2004). Avasimibe was also evaluated in hAPP transgenic mice overexpressing human APP_{751} and a decrease in amyloid plaque, limitation in generation, and an increase in Aβ clearance were reported (Huttunen et al. 2010). Table 13.3 summarizes the studies conducted with pharmacology or genetic inhibition and their most important contribution as evidence of their positive effects or their possible mechanism. ACAT inhibitors could be part of another class of drugs that indirectly regulate the generation of Aβ (Hutter-Paier et al. 2004).

The mechanism that has been proposed (Fig. 13.2a) to explain the reduction of Aβ pathology implies that ACAT1 inhibition causes an excess of free cholesterol, leading to a negative APP regulation. The cholesterol increases, so it is converted by

Table 13.3 Principal contributions on ACAT1 inhibitors and its possible mechanism

Model and therapy	Main findings	Reference
In vitro model with CP-113,818 and Dup128	Both reduced $A\beta_{42}$ and $A\beta_{total}$ secretion in a concentration-dependent manner. CP-113,818 at 10 μM generates maximum increment in FC (~42%) and maximum reduction in CEs (~45%).	Puglielli et al. (2001)
	CEs levels are correlated with Aβ levels and ACAT activity directly regulates Aβ production through the regulation of CEs production.	
CP-113,818 by systemic administration (60 days) via implantable slow-release biopolymer pellets. In vitro and hAPP transgenic mice	The pellets reduced the aggregation of plaques of Aβ (88–99%) and membrane/insoluble Aβ levels (83–96%), also reducing brain CEs (86%). Spatial learning was enhanced and correlated with reduced Aβ levels.	Hutter-Paier et al. (2004)
	It is suggested that the administration of ACAT1 inhibitors by slow-release biopolymer could be considered as a possible preventive and therapeutic strategy.	
Knockdown of ACAT1 expression by RNAi	siRNA (3 μg), ACAT1 protein levels were reduced by 54.4 ± 11.0% and also CEs levels 21.6 ± 4.4%. Reductions were also observed in the levels of APP-C99 in 48.4 ± 5.2%, APP-C83 in 27.4 ± 4.1%, Aβ in 39.2%, and $A\beta_{42}$ in 27.8%.	Huttunen et al. (2007)
	It is proposed that ACAT1 RNAi follows the same or similar pathways as the pharmacological inhibition of ACAT1, altering the levels of secretase (α-, β-, and γ-), Aβ, APP (-C99 and -C83).	
CI-1011 and CP-113-818 in hAPP transgenic mice	The traffic of APP depends on sterols, it is suggested that ACAT1 inhibitors modulate the traffic of APP in the early secretory pathway, limiting the availability of mature APP for β- and γ-secretases, enzymes needed to generate Aβ.	Huttunen et al. (2009)
ACAT1 gene ablation in triple transgenic (3XTg-AD) mice	A model is proposed to explain what occurs with the ACAT1 inhibition, increases cholesterol, which is a substrate for CYP46A1 and causes an increase in 24-OHC biosynthesis in neurons, which causes a down-regulation of hAPP protein level, probably due to an increase in its rate of degradation at the ER. There is also a	Bryleva et al. (2010)

(continued)

Table 13.3 (continued)

Model and therapy	Main findings	Reference
	decrease in HMGR and cholesterol biosynthesis due to the increase in 24-OHC. As a consequence, the total cholesterol level decreases in the 3XTg-AD mouse brains.	
Young (6.5 months old) and aged (16 months old) hAPP transgenic mice treated with CI-1011 via implantable slow-release biopolymer pellets (2 months)	CI-1011, in young animals, reduced levels of $A\beta_{40}$ in 14%, $A\beta_{42}$ in 26%, and C-terminal fragments of APP. In old mice, it seemed to reduce the diffuse amyloid load, but there was no effect on plasma Aβ levels.	Huttunen et al. (2010)
	It is proposed that by decreasing the generation and increasing the Aβ clearance, the inhibition of ACAT1 can reduce Aβ level.	
Adeno-associated virus (AAV)-mediated *ACAT1* gene knockdown in AD mice	The release of AVV in the correct area of the brain affects ACAT activity and the AD-like phenotype, lowering Aβ and hAPP levels in the brain to similar levels compared to complete genetic inhibition of *ACAT1*. Thus, it is proposed that *ACAT1* knockdown gene therapy may be a suitable approach for the treatment of AD.	Murphy et al. (2013)
K604 in vitro and mice model with *ACAT1* gene knock-out	Like pharmacological inhibition, genetic inhibition of *ACAT1* promotes autophagy of $A\beta_{42}$ and its lysosomal proteolysis, independently of mTOR to activate lysosomal expression and modulate network, therefore it is a strategy to benefit various neurodegenerative disorders.	Shibuya et al. (2014)
K604 in in vitro model and triple transgenic mice (3XTg-AD)	In addition to the increase in autophagy, a reduction in P301L-tau level was also shown in a model of neuronal cells and also in the brains of young mice, but not in those of old mice.	Shibuya et al. (2015b)
Genetic inhibition in APP/PSN double-transgenic AD mice and human neuroblastoma SH-SY5Y	Reduction of ACAT1 attenuated the cytotoxicity induced by Aβ, suggesting that ACAT1 may prove to be a meaningful target for the prevention and treatment of AD.	Chen et al. (2018)
	Using genome-wide association, it was established that *PTGS2* and *ACAT1* showed a very strong correlation in the gene regulatory network established in this study. *ACAT1* knockdown in SHSY5Y cells reduced the expression of *PTGS2* significantly and ACAT1	

(continued)

Table 13.3 (continued)

Model and therapy	Main findings	Reference
	over-expression upregulated the expression of *PTGS2*.	
	PTGS2 is a fundamental neuroinflammatory factor and is extensively related to neuronal damage, inflammatory response, and neurodegenerative disorders.	

Aβ amyloid beta, *ACAT1* acyl-Coenzyme A:cholesterol acyltransferase, *APP* amyloid protein precursor, *CEs* cholesteryl esters, *FC* free cholesterol, *PTGS2* prostaglandin-endoperoxide synthase 2

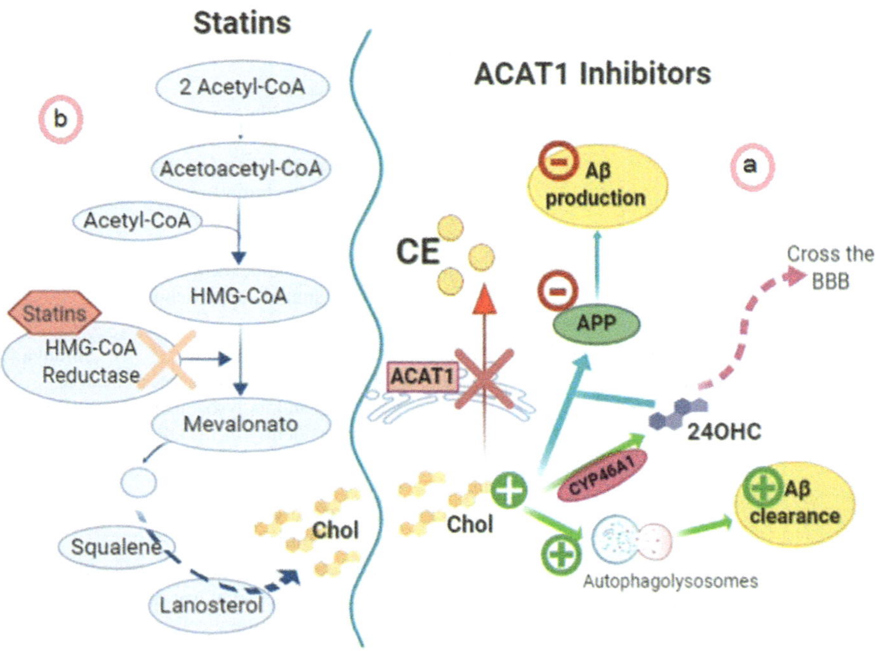

Fig. 13.2 (**a**) The inhibition of ACAT1 increases the level of cholesterol and the 24-OHC, as a consequence, decreases the level of APP, decreasing the Aβ production. The increase in cholesterol level may promote the fusion of membranes to form autophagolysosomes and improve clearance of Aβ. (**b**) Statins inhibit the enzyme HMG-CoA reductase, blocking the conversion of HMG-CoA to mevalonate, thereby interfering in subsequent steps for de novo cholesterol synthesis. *24-OHC* 24-hydroxycholesterol, *Aβ* amyloid beta, *ACAT1* acyl-CoA: cholesterol acyltransferase 1 *CE* cholesterol esters, *APP* amyloid precursor protein, *BBB* blood-brain barrier, *Chol* cholesterol, *HMG-CoA* 3-hydroxy-3-methyl-glutaryl-CoA

CYP46A1 to 24-OHC; subsequently, 24-OHC crosses the BBB to the periphery, so this lowers the level of cholesterol in the brain (Bhattacharyya and Kovacs 2010). With the blocking of ACAT1 the Aβ production is decreased and it is also likely that its clearance to be stimulated (Shibuya et al. 2014).

Microglia, which is a macrophage tissue present in the CNS, not only has a special role in the Aβ clearance, but also in the interaction, internalization, and regulation of pathological tau (Hopp et al. 2018; Perea et al. 2018). However, tau clearance by microglia has received less attention, but microglial phagocytosis is very important in the clearance of tau, and therefore in its spread and in the progression of AD (Luo et al. 2015; Bolós et al. 2016). There is also evidence that impaired microglial clearance of Aβ contributes to AD pathogenesis (Hickman et al. 2008; Griciuc et al. 2013). In mice models, Shibuya et al. (2014) reported an increase in $A\beta_{42}$ phagocytosis and its lysosomal degradation after ACAT1 inhibition.

Another mechanism involved in the decrease of Aβ levels is macroautophagy, also called autophagy, it is a cell degradation process where an autophagosome, double-membrane structure, traps cytosolic components, including peptides or denatured and/or aggregation-prone proteins, such as Aβ. Later it forms autophagolysosomes, uniting with lysosomes to degrade its content (Mizushima 2007). This process of autophagy can be initiated by inhibition of the mammalian target of rapamycin (mTOR), a serine-threonine kinase that modulates cell growth, proliferation, and cellular homeostasis (Caccamo et al. 2010). However, an increase was also observed in lysosome biogenesis and autophagy with genetic inhibition and with inhibitor K-604, independently of mTOR (Shibuya et al. 2014), by means of the transcription factor EB (TFEB), which is related to the transcription of genes involved to lysosomal expression and regulation, modulating lysosome synthesis and autophagy (Settembre et al. 2011). Compared to mitochondria and ER, ACAT1 is mostly localized at the mitochondria-associated ER membrane (MAM) (Area-Gomez et al. 2012). It is thought that MAM participates in the formation of autophagolysosomes and that it is possible that by blocking ACAT1, cholesterol levels in MAM are altered, which could lead to an increase in the formation of autophagosomes and lysosomal volume. Furthermore, increased cholesterol levels in autophagosomes could favor the union of membranes between lysosomes and autophagosomes.

There are still parts to be elucidated, and it is possible that there are other pathways involved; for example, a genetic genomics approach has revealed that *ACAT1* and prostaglandin-endoperoxide synthase 2 (*PTGS2*) have a very high correlation in the gene regulatory network (Chen et al. 2018). PTGS2 is a significantly neuroinflammatory factor and it is widely related to neuronal damage, inflammatory response, and neurodegenerative disorders. Definitely, PTGS2 inhibitors by themselves cannot prevent the occurrence of AD, but their participation in AD should not be ignored. This type of research reminds us that dementia is a complex disease and therefore requires a comprehensive approach.

13.5.2 Statin Therapy as a Possible Therapeutic Strategy

Statins are undoubtedly useful in preventing coronary heart disease events. They are a group of drugs that inhibit HMGR, which plays a relevant role in the production of cholesterol. Statins block the conversion of HMRG to mevalonate (Fig. 13.2b), thereby reducing de novo synthesis of cholesterol. They decrease the formation and release of LCL-c into the circulation and increase the activity of the LDL receptor, which lowers the levels of triglycerides and LDL-c and augments HDL-c levels (Mcguinness et al. 2016). In addition to the aforementioned effects, statins also exert independent effects of lowering cholesterol levels, these are called pleiotropic effects. These include decreased endothelial dysfunction, increased expression of endothelial nitric oxide synthase, improved nitric oxide bioavailability, and strong anti-inflammatory and antioxidant properties. Their pleiotropic properties are involved in various organs and systems, so they are beneficial for a wide variety of diseases (Rohilla et al. 2011; Oesterle et al. 2017).

There are reports that the diagnosis of AD and VaD is less frequent in people who consume statins (Jick et al. 2000; Wolozin 2000; Rockwood et al. 2002). In the case of AD, their use decreases the APP generation, the Aβ neuronal secretion (Lütjohann and von Bergmann 2003), NFT burden (Boimel et al. 2009), and p-tau (Riekse et al. 2006; Li et al. 2017). And for VaD, considering stroke as its major risk factor, reducing levels of cholesterol using statins decreases the risk of stroke in high-risk populations and people with non-cardioembolic stroke or transient ischemic attack. Hence, by decreasing the risk of stroke, statins may also act to decrease the incidence of post-stroke dementia (Mcguinness et al. 2016). Several studies have been conducted to observe or test the use of statins in the prevention or treatment of dementia, obtaining contradictory results (Jick et al. 2000; Wolozin 2000; Rockwood et al. 2002; Arvanitakis et al. 2008; Cramer et al. 2008; Sparks et al. 2008; Haag et al. 2009; Swiger et al. 2013; Mcguinness et al. 2016). Thus, to date, there is not enough evidence that statins are beneficial in the prevention or treatment of dementia (Mcguinness et al. 2014, 2016; Schultz et al. 2018); however, this hypothesis continues to be tested in current clinical studies (US National Library of Medicine 2009).

Statins have also, paradoxically, been related to a reversible cognitive impairing effect in some patients. Even the Food and Drug Administration (FDA), in 2012 posted a safety statement announcing important safety label changes to statins, in which it included information about the potential for generally non-serious and reversible cognitive side effects, such as confusion, memory loss, among others (FDA 2012). Thus, the question has emerged as to how statins could exert both effects, prevention of dementia, and reversible cognitive impairment. Possible explanations have emerged, include the decrease of synthesis of coenzyme Q10 with increasing oxidative stress and reduction of cerebral energy production, and depletion of central nervous system myelin due to cholesterol reduction (Schultz et al. 2018; Tan et al. 2019).

Nevertheless, it has also been reported that evidence is not sufficient to affirm that statins cause cognitive impairment to a significant degree and that reported cases seem to be rare (Bitzur 2016). More recently, it has been recognized that there is growing evidence for reversible cognitive impairment for a small percentage of the population of statin users, and it has been suggested that this fact needs to be recognized to be approached in a better way (Schultz et al. 2018; Tan et al. 2019).

13.6 Recent Developments and Future Perspectives

Despite the great impact of dementia, there are currently only two classes of drugs available to treat it such as cholinesterase inhibitors (ChEIs) and *N*-methyl-D-aspartate (NMDA) receptor antagonists (Szeto and Lewis 2016; Kim and Factora 2018). Nevertheless, they do not provide a cure, as they are only focused on improving cognitive and behavioral symptoms. The prescribed pharmacotherapy should be periodically evaluated and discontinued if it is considered that dementia has advanced to the point that the patient is totally dependent for all activities of daily living and the preservation of cognitive and functional status is no longer possible, or when the adverse effects outweigh the beneficial effects (Kim and Factora 2018). Although there is a lot of evidence linking lipid metabolism with AD and VaD, it is still pending to clarify some aspects. In the case of AD, studies suggest that cholesterol influences the regulation of Aβ production and metabolism, but the relationship with plasma cholesterol has not yet been fully clarified, given the doubt concerning how plasma cholesterol could influence Aβ production and accumulation in the brain if cholesterol cannot freely cross of the BBB and the cholesterol in brain is practically produced by de novo synthesis. A possible explanation is that this effect is mediated by oxysterols, especially 27-OHC, and it has been proposed that cholesterol-enriched diets induce augmented Aβ production in plasma and brain, and that this effect may also be mediated by oxysterols, considering that they can pass through the BBB into the CNS and exert their cytotoxic and pro-apoptotic impact on neurons (Liu et al. 2018).

With VaD, even the genetic question is less clear, but what is without discussion is the undoubted role that cholesterol plays in the main risk factors for developing VaD. Regarding the dilemma of total cholesterol levels, evidence suggests that more important than the total cholesterol level would be the ratio of free cholesterol to cholesterol esters, equilibrium where the enzyme ACAT1 has a fundamental role. In this sense, some treatments for dementia involved with cholesterol metabolism have been proposed as possible therapeutic alternatives in the treatment of dementia. In the case of statins, it is a drug already approved and marketed, as an additional indication, and in the case of ACAT1 inhibitors, it has been proposed to develop drugs. Since statins are a type of drug commercialized some time ago, we have more information about their safety and adverse effects. And biologically, it seems possible that statins could prevent dementia due to their participation in lowering cholesterol levels. However, possible biases in studies to date could interfere with obtaining strong results, so currently they are not indicated to improve cognitive status.

With ACAT1 inhibitors, in which several lines of evidence from cell models and animal studies have shown that ACAT1 inhibition is an effective way of decreasing cerebral Aβ, one consideration is whether the inhibitors will be able to cross the BBB, and reach significant levels in the brain to have an impact on ACAT1 inhibition and regulate AD pathology (Murphy et al. 2013). There is also concern related to the specificity of the enzyme blockage, and if ACAT1 inhibitors could be trapped within the lipid bilayer, causing a high concentration in this area, causing some unwanted effect (Shibuya et al. 2015a). Considering the above, it is necessary to design new, more specific, and smaller molecules, which are capable of crossing the BBB and avoiding unwanted side effects (Shibuya et al. 2015b).

Although the results of preclinical studies are favorable for inhibitors of ACAT1, it is required to go from studies in animal models to clinical studies. In some ways, previous studies with failed results in atherosclerosis provide us information about the safety of its use in humans, but it would be needed to prove the effectiveness and the safety of the new molecules. In general, because cholesterol is a fundamental compound related to the modulates of membrane structure and functioning, it is not surprising that disturbances in cholesterol homeostasis are associated with pathogenesis of dementia; keeping its homeostasis is vital to maintaining its proper functioning. Moreover, in the brains of patients with AD, abnormally high levels of cholesterol esters have been reported, and some pharmacological and genetic modifications have presented multiple benefits. Nevertheless, there are still gaps to be resolved, with evidence suggesting that it may be a therapeutic target.

13.7 Conclusion

The information contained in this chapter allows us to understand that dementia is a chronic disease with a great economic impact and society. And what despite efforts to find an adequate treatment, to date there is none, so the search for new strategies continues to be a priority issue and a challenge to humanity. Given that the altered cholesterol metabolism has been associated with the development of dementia, it has been noted as a possible therapeutic strategy for this disease. Being the use of statins and ACAT1 inhibitors the first approach both in different stages of development. Although there are still doubts and aspects to be resolved, they stand out as a possible alternative. The advance in the understanding pathological mechanisms of dementia allows directing new research toward the search for new therapeutic strategies and alerts us to the importance of a comprehensive treatment for this complex pathology.

References

Alzheimer A (1906) Über einen eigenartigen schweren Erkrankungsprozeß der Hirnrinde. Neurol Zentralblatt 23:1129–1136

Antalis CJ, Arnold T, Rasool T et al (2010) High ACAT1 expression in estrogen receptor negative basal-like breast cancer cells is associated with LDL-induced proliferation. Breast Cancer Res Treat 122:661–670. https://doi.org/10.1007/s10549-009-0594-8

Appleton JP, Scutt P, Sprigg N, Bath PM (2017) Hypercholesterolaemia and vascular dementia. Clin Sci 131:1561–1578. https://doi.org/10.1042/CS20160382

Araki W, Tamaoka A (2015) Amyloid beta-protein and lipid rafts: focused on biogenesis and catabolism. Front Biosci 20:314–324. https://doi.org/10.2741/4311

Area-Gomez E, Lara M, Tambini MD et al (2012) Upregulated function of mitochondria-associated ER membranes in Alzheimer disease. EMBO J 31:4106–4123. https://doi.org/10.1038/emboj.2012.202

Arenas F, Garcia-Ruiz C, Fernandez-Checa JC (2017) Intracellular cholesterol trafficking and impact in neurodegeneration. Front Mol Neurosci 10:382. https://doi.org/10.3389/fnmol.2017.00382

Arvanitakis Z, Schneider J, Wilson R et al (2008) Statins, incident Alzheimer disease, change in cognitive function, and neuropathology. Neurology 70:1795–1802

Attems J, Jellinger KA (2014) The overlap between vascular disease and Alzheimer's disease – lessons from pathology. BMC Med 12:1–12

Bemlih S, Poirier MD, El Andaloussi A (2010) Acyl-coenzyme A: Cholesterol acyltransferase inhibitor avasimibe affect survival and proliferation of glioma tumor cell lines. Cancer Biol Ther 9:1025–1032. https://doi.org/10.4161/cbt.9.12.11875

Bertram L, Hsiao M, Mullin K et al (2005) ACAT1 is not associated with Alzheimer's disease in two independent family-based samples. Mol Psychiatry 10:522–524. https://doi.org/10.1038/sj.mp.4001646

Bhattacharyya R, Kovacs DM (2010) ACAT inhibition and amyloid beta reduction. Biochim Biophys Acta 1801:960–965. https://doi.org/10.1016/j.bbalip.2010.04.003

Bitzur R (2016) Remembering statins: do statins have adverse cognitive effects? Diabetes Care 39: 253–259. https://doi.org/10.2337/dcS15-3022

Björkhem I, Meaney S, Fogelman AM (2004) Brain cholesterol: long secret life behind a barrier. Arterioscler Thromb Vasc Biol 24:806–815. https://doi.org/10.1161/01.ATV.0000120374.59826.1b

Bocan TM, Mueller SB, Uhlendorf PD et al (1991) Comparison of CI-976, an ACAT inhibitor, and selected lipid-lowering agents for antiatherosclerotic activity in iliac-femoral and thoracic aortic lesions. A biochemical, morphological, and morphometric evaluation. Arterioscler Thromb 11: 1830–1843

Bogdanovic N, Bretillon L, Lund EG et al (2001) On the turnover of brain cholesterol in patients with Alzheimer's disease. Abnormal induction of the cholesterol-catabolic enzyme CYP46 in glial cells. Neurosci Lett 314:45–48. https://doi.org/10.1016/S0304-3940(01)02277-7

Boimel M, Grigoriadis N, Lourbopoulos A et al (2009) Statins reduce the neurofibrillary tangle burden in a mouse model of tauopathy. J Neuropathol Exp Neurol 68:314–325. https://doi.org/10.1097/NEN.0b013e31819ac3cb

Bolós M, Llorens-Martín M, Jurado-Arjona J et al (2016) Direct evidence of internalization of tau by microglia in vitro and in vivo. J Alzheimers Dis 50:77–87. https://doi.org/10.3233/JAD-150704

Bryleva EY, Rogers MA, Chang CCY et al (2010) ACAT1 gene ablation increases 24(S)-hydroxycholesterol content in the brain and ameliorates amyloid pathology in mice with AD. Proc Natl Acad Sci 107:3081–3086. https://doi.org/10.1073/pnas.0913828107

Burlot MA, Braudeau J, Michaelsen-Preusse K et al (2015) Cholesterol 24-hydroxylase defect is implicated in memory impairments associated with Alzheimer-like Tau pathology. Hum Mol Genet 24:5965–5976. https://doi.org/10.1093/hmg/ddv268

Caccamo A, Majumder S, Richardson A et al (2010) Molecular interplay between mammalian target of rapamycin (mTOR), amyloid-β, and Tau: effects on cognitive impairments. J Biol Chem 285:13107–13120. https://doi.org/10.1074/jbc.M110.100420

Canevari L, Clark JB (2007) Alzheimer's disease and cholesterol: the fat connection. Neurochem Res 32:739–750. https://doi.org/10.1007/s11064-006-9200-1

Chan RB, Oliveira TG, Cortes EP et al (2012) Comparative lipidomic analysis of mouse and human brain with Alzheimer disease. J Biol Chem 287:2678–2688. https://doi.org/10.1074/jbc.M111.274142

Chang CCY, Chen J, Thomas M et al (1995) Regulation and immunolocalization of acyl-coenzyme A:cholesterol acyltransferase in mammalian cells as studied with specific antibodies. J Biol Chem 270:29532–29540. https://doi.org/10.1074/jbc.270.49.29532

Chen Y, Zhu L, Ji L et al (2018) Silencing the ACAT1 gene in human SH-SY5Y neuroblastoma cells inhibits the expression of cyclo-oxygenase 2 (COX2) and reduces β-amyloid-induced toxicity due to activation of protein kinase C (PKC) and ERK. Med Sci Monit 24:9007–9018. https://doi.org/10.12659/msm.912862

Chen H, Du Y, Liu S et al (2019) Association between serum cholesterol levels and Alzheimer's disease in China: a case-control study. Int J Food Sci Nutr 70:405–411. https://doi.org/10.1080/09637486.2018.1508426

Cingolani F, Czaja MJ (2017) Regulation and functions of autophagic lipolysis. Trends Endocrinol Metab 27:696–705. https://doi.org/10.1016/j.tem.2016.06.003.Regulation

Corder EH, Saunders AM, Strittmatter WJ et al (1993) Gene dose of Apolipoprotein E type 4 allele and risk of Alzheimer's disease in late onset families. Science 261:921–923

Cordy JM, Hooper NM, Turner AJ (2006) The involvement of lipid rafts in Alzheimer's disease. Mol Membr Biol 23:111–122. https://doi.org/10.1080/09687860500496417

Cossec JC, Simon A, Marquer C et al (2010) Clathrin-dependent APP endocytosis and Aβ secretion are highly sensitive to the level of plasma membrane cholesterol. Biochim Biophys Acta Mol Cell Biol Lipids 1801:846–852. https://doi.org/10.1016/j.bbalip.2010.05.010

Cramer C, Haan MN, Galea S et al (2008) Use of statins and incidence of dementia and cognitive impairment without dementia in a cohort study. Neurology 71:344–350. https://doi.org/10.1212/01.wnl.0000319647.15752.7b

Daneschvar HL, Aronson MD, Smetana GW (2015) Do statins prevent Alzheimer's disease? A narrative review. Eur J Int Med 26:666–669. https://doi.org/10.1016/j.ejim.2015.08.012

Di Paolo G, Kim T (2012) Linking Lipids to Alzheimer's disease: cholesterol and beyond. Nat Rev Neurosci 12:284–296. https://doi.org/10.1038/nrn3012.Linking

Dickson J, Kruse C, Montagna DR et al (2013) Alternative polyadenylation and miR-34 family members regulate tau expression. J Neurochem 127:1–7. https://doi.org/10.1038/jid.2014.371

Dietschy JM (2009) Central nervous system: cholesterol turnover, brain development and neurodegeneration. Biol Chem 390:287–293. https://doi.org/10.1515/BC.2009.035

Dietschy JM, Turley SD (2001) Cholesterol metabolism in the brain. Curr Opin Lipidol 12:105–112. https://doi.org/10.1194/jlr.r400004-jlr200

van Dijk EJ, Prins ND, Vrooman HA et al (2008) Progression of cerebral small vessel disease in relation to risk factors and cognitive consequences: Rotterdam scan study. Stroke 39:2712–2719. https://doi.org/10.1161/STROKEAHA.107.513176

Eckert GP, Wood WG, Müller WE (2001) Effects of aging and beta-amyloid on the properties of brain synaptic and mitochondrial membranes. J Neural Transm 108:1051–1064. https://doi.org/10.1007/s007020170024

Fassbender K, Simons M, Bergmann C et al (2001) Simvastatin strongly reduces levels of Alzheimer's disease β-amyloid peptides Aβ42 and Aβ40 in vitro and in vivo. PNAS 98:5856–5861

FDA (2012) Drug safety communication: important safety label changes to cholesterol-lowering statin drugs. In: FDA Consum. Heal. Inf., Expand. Advice statin risks. FDA, Rome. www.fda.gov/consumer

Geng F, Guo D (2017) Lipid droplets, potential biomarker and metabolic target in glioblastoma. Int Med Rev 3:1–9

Geng F, Cheng X, Wu X et al (2017) Inhibition of SOAT1 suppresses glioblastoma growth via blocking SREBP-1-mediated lipogenesis. Clin Cancer Res 22:5337–5348. https://doi.org/10.1158/1078-0432.CCR-15-2973.Inhibition

Glöckner F, Ohm TG (2014) Tau pathology induces intraneuronal cholesterol accumulation. J Neuropathol Exp Neurol 73:846–854. https://doi.org/10.1097/NEN.0000000000000103

Glöckner F, Meske V, Lütjohann D, Ohm TG (2011) Dietary cholesterol and its effect on tau protein: a study in apolipoprotein e-deficient and P301L human tau mice. J Neuropathol Exp Neurol 70:292–301. https://doi.org/10.1097/NEN.0b013e318212f185

Gouw AA, van der Flier WM, Fazekas F et al (2008) Progression of white matter hyperintensities and incidence of new lacunes over a 3-year period: the leukoaraiosis and disability study. Stroke 39:1414–1420. https://doi.org/10.1161/STROKEAHA.107.498535

Gratuze M, Julien J, Morin F et al (2016) High-fat, high-sugar, and high-cholesterol consumption does not impact tau pathogenesis in a mouse model of Alzheimer's disease-like tau pathology. Neurobiol Aging 47:71–73. https://doi.org/10.1016/j.neurobiolaging.2016.07.016

Griciuc A, Serrano A, Parrado AR et al (2013) Alzheimer's disease risk gene CD33 inhibits microglial uptake of amyloid beta. Neuron 78:631–643. https://doi.org/10.1016/j.neuron.2013.04.014.Alzheimer

Haag MDM, Hofman A, Koudstaal PJ et al (2009) Statins are associated with a reduced risk of Alzheimer disease regardless of lipophilicity. The Rotterdam study. J Neurol Neurosurg Psychiatry 80:13–17. https://doi.org/10.1136/jnnp.2008.150433

Hainer JW, Terry JG, Connell JM et al (1994) Effect of the acyl-CoA: cholesterol acyltransferase inhibitor DuP 128 on cholesterol absorption and serum cholesterol in humans. Clin Pharmacol Ther 56:65–74. https://doi.org/10.1038/clpt.1994.102

Hall JR, Wiechmann AR, Johnson LA et al (2014) Total cholesterol and neuropsychiatric symptoms in Alzheimer's disease: the impact of total cholesterol level and gender. Dement Geriatr Cogn Disord 38:300–309. https://doi.org/10.1159/000361043

Hardy J, Higgins G (1992) Alzheimer's disease: the amyloid cascade hypothesis. Science 256:184–185. https://doi.org/10.1177/875512250001600415

Hickman SE, Allison EK, Khoury J (2008) Microglial dysfunction and defective β-amyloid clearance pathways in aging Alzheimer's disease mice. J Neurosci 28:8354–8360. https://doi.org/10.1523/JNEUROSCI.0616-08.2008.Microglial

Holtzman DM, Morris JC, Goate A (2011) Alzheimer's disease: the challenge of the second century. Sci Transl Med 3:1–35. https://doi.org/10.1126/scitranslmed.3002369.Alzheimer

Hopp SC, Lin Y, Oakley D et al (2018) The role of microglia in processing and spreading of bioactive tau seeds in Alzheimer's disease. J Neuroinflammation 15:1–15. https://doi.org/10.1186/s12974-018-1309-z

Hussain G, Wang J, Rasul A et al (2019) Role of cholesterol and sphingolipids in brain development and neurological diseases. Lipids Health Dis 18:1–12

Hutter-Paier B, Huttunen HJ, Puglielli L et al (2004) The ACAT inhibitor CP-113,818 markedly reduces amyloid pathology in a mouse model of Alzheimer's disease. Neuron 44:227–238. https://doi.org/10.1016/j.neuron.2004.08.043

Huttunen HJ, Greco C, Kovacs DM (2007) Knockdown of ACAT-1 reduces amyloidogenic processing of APP. FEBS Lett 581:1688–1692. https://doi.org/10.1016/j.febslet.2007.03.056

Huttunen HJ, Peach C, Bhattacharyya R et al (2009) Inhibition of acyl-coenzyme A: cholesterol acyl transferase modulates amyloid precursor protein trafficking in the early secretory pathway. FASEB J 23:3819–3828. https://doi.org/10.1096/fj.09-134999

Huttunen HJ, Havas D, Peach C et al (2010) The acyl-coenzyme A:cholesterol acyltransferase inhibitor CI-1011 reverses diffuse brain amyloid pathology in aged amyloid precursor protein transgenic mice. J Neuropathol Exp Neurol 69:777–788. https://doi.org/10.1097/NEN.0b013e3181e77ed9.The

Ikram MA, Bersano A, Manso R et al (2017) Genetics of vascular dementia – review from the ICVD working group. BMC Med 15:1–7. https://doi.org/10.1186/s12916-017-0813-9

Jick H, Zornberg GL, Jick SS et al (2000) Statins and the risk of dementia. Lancet 356:1627–1631

Kalaria RN (2016) Neuropathological diagnosis of vascular cognitive impairment and vascular dementia with implications for Alzheimer's disease. Acta Neuropathol 131:659–685. https://doi.org/10.1007/s00401-016-1571-z

van der Kant R, Langness VF, Herrera CM et al (2019) Cholesterol metabolism is a druggable axis that independently regulates tau and amyloid-β in iPSC-derived Alzheimer's disease neurons. Cell Stem Cell 24:363–375. https://doi.org/10.1016/j.stem.2018.12.013

van der Kant R, Goldstein LSB, Ossenkoppele R (2020) Amyloid-β-independent regulators of tau pathology in Alzheimer disease. Nat Rev Neurosci 21:21–35. https://doi.org/10.1038/s41583-019-0240-3

Kapasi A, Schneider JA (2016) Vascular contributions to cognitive impairment, clinical Alzheimer's disease, and dementia in older persons. Biochim Biophys Acta 1862:878–886. https://doi.org/10.1016/j.bbadis.2015.12.023

Khan A, Corbett A, Ballard C (2017) Emerging amyloid and tau targeting treatments for Alzheimer's disease. Expert Rev Neurother 17:697–711. https://doi.org/10.1080/14737175.2017.1326819

Kim LD, Factora RM (2018) Alzheimer dementia: starting, stopping drug therapy. Cleve Clin J Med 85:209–214. https://doi.org/10.3949/ccjm.85a.16080

Kim Y, Kim C, Jang HY, Mook-Jung I (2016) Inhibition of cholesterol biosynthesis reduces γ-secretase activity and amyloid-β generation. J Alzheimers Dis 51:1057–1068. https://doi.org/10.3233/JAD-150982

Kinno R, Mori Y, Kubota S et al (2019) High serum high-density lipoprotein-cholesterol is associated with memory function and gyrification of insular and frontal opercular cortex in an elderly memory-clinic population. NeuroImage Clin 22:1–9. https://doi.org/10.1016/j.nicl.2019.101746

Kivipelto M, Helkala EL, Hänninen T et al (2001) Midlife vascular risk factors and late-life mild cognitive impairment: a population-based study. Neurology 56:1683–1689. https://doi.org/10.1212/WNL.56.12.1683

Koldamova R, Fitz NF, Lefterov I (2010) The role of ATP-binding cassette transporter A1 in Alzheimer's disease and neurodegeneration. Biochim Biophys Acta 1801:824–830. https://doi.org/10.1016/j.bbalip.2010.02.010.The

Lämsä R, Helisalmi S, Herukka SK et al (2007) Study on the association between SOAT1 polymorphisms, Alzheimer's disease risk and the level of CSF biomarkers. Dement Geriatr Cogn Disord 24:146–150. https://doi.org/10.1159/000105164

Lee HT, Sliskovic DR, Picard JA et al (1996) Inhibitors of Acyl-CoA:cholesterol O-Acyl transferase (ACAT) as hypocholesterolemic agents. CI-1011: an acyl sulfamate with unique cholesterol-lowering activity in animals fed noncholesterol-supplemented diets. J Med Chem 39:5031–5034. https://doi.org/10.1021/jm960674d

Lee SSY, Li J, Tai JN et al (2015) Avasimibe encapsulated in human serum albumin blocks cholesterol esterification for selective cancer treatment. ACS Nano 9:2420–2432. https://doi.org/10.1021/nn504025a

Leoni V, Solomon A, Kivipelto M (2010) Links between ApoE, brain cholesterol metabolism, tau and amyloid β-peptide in patients with cognitive impairment. Biochem Soc Trans 38:1021–1025. https://doi.org/10.1042/BST0381021

Leritz EC, Mcglinchey RE, Salat DH, Milberg WP (2016) Elevated levels of serum cholesterol are associated with better performance on tasks of episodic memory. Metab Brain Dis 31:465–473. https://doi.org/10.1007/s11011-016-9797-y.Elevated

Li J, Gu D, Lee SSY et al (2016) Abrogating cholesterol esterification suppresses growth and metastasis of pancreatic cancer. Oncogene 35:6378–6388. https://doi.org/10.1038/onc.2016.168

Li G, Mayer CL, Morelli D et al (2017) Effect of simvastatin on CSF Alzheimer disease biomarkers in cognitively normal adults. Neurology 89:1251–1255. https://doi.org/10.1212/WNL.0000000000004392

Liu L, Zhang K, Sandoval H et al (2015) Glial lipid drolets and ROS induced by mitochondrial defects promote neurodegeneration. Cell 160:177–190. https://doi.org/10.1016/j.cell.2014.12.019.Glial

Liu Q, An Y, Ma W et al (2018) High-cholesterol diet results in elevated amyloid-β and oxysterols in rats. Mol Med Rep 17:1235–1240. https://doi.org/10.3892/mmr.2017.8003

Llaverias G, Alegret M (2004) Inhibidores de la acil coenzima A:colesterol aciltransferasa (ACAT): mecanismos y perspectivas terapéuticas. Clin Invest Arter 16:250–261. https://doi.org/10.1016/S0214-9168(04)79002-6

Longstreth WT, Arnold AM, Beauchamp NJ et al (2005) Incidence, manifestations, and predictors of worsening white matter on serial cranial magnetic resonance: the cardiovascular health study. Stroke 36:56–61. https://doi.org/10.1161/01.STR.0000149625.99732.69

Lund EG, Xie C, Kotti T et al (2003) Knockout of the cholesterol 24-hydroxylase gene in mice reveals a brain-specific mechanism of cholesterol turnover. J Biol Chem 278:22980–22988. https://doi.org/10.1074/jbc.M303415200

Luo W, Liu W, Hu X et al (2015) Microglial internalization and degradation of pathological tau is enhanced by an anti-tau monoclonal antibody. Sci Rep 5:1–12. https://doi.org/10.1038/srep11161

Lütjohann D, von Bergmann K (2003) 24S-hydroxycholesterol: a marker of brain cholesterol metabolism. Pharmacopsychiatry 36:102–106. https://doi.org/10.1055/s-2003-43053

Lütjohann D, Meichsner S, Pettersson H (2012) Lipids in Alzheimer's disease and their potential for therapy. Clin Lipidol 7:65–78. https://doi.org/10.2217/clp.11.74

Marcum ZA, Walker R, Bobb JF et al (2018) Serum cholesterol and incident Alzheimer's disease: findings from the adult changes in thought study. J Am Geriatr Soc 66:2344–2352. https://doi.org/10.1111/jgs.15581.Serum

Martin MG, Pfrieger F, Dotti CG (2014) Cholesterol in brain disease: sometimes determinant and frequently implicated. EMBO Rep 15:1036–1052. https://doi.org/10.15252/embr.201439225

Mast N, Saadane A, Valencia-olvera A et al (2017) Cholesterol-metabolizing enzyme cytochrome P450 46A1 as a pharmacologic target for Alzheimer's disease. Neuropharmacology 123:465–476. https://doi.org/10.1016/j.neuropharm.2017.06.026.Cholesterol-Metabolizing

Maulik M, Westaway D, Jhamandas JH, Kar S (2013) Role of cholesterol in APP metabolism and its significance in Alzheimer's disease pathogenesis. Mol Neurobiol 47:37–63. https://doi.org/10.1007/s12035-012-8337-y

Mcguinness B, Craig D, Bullock R et al (2014) Statins for the treatment of dementia. Cochrane Database Syst Rev (7):CD007514

Mcguinness B, Craig D, Bullock R, Passmore P (2016) Statins for the prevention of dementia. Cochrane Database Syst Rev (1):CD003160. https://doi.org/10.1002/14651858.CD003160.pub3. www.cochranelibrary.com

Meuwese M, Groot E, Duivenvoorden R et al (2009) ACAT inhibition and progression of carotid atherosclerosis in patients with familial hypercholesterolemia. JAMA 301:1131–1139

Mielke M, Zandi P, Sjögren M et al (2005) High total cholesterol levels in late life associated with a reduced risk of dementia. Neurology 64:1689–1695

Mizushima N (2007) Autophagy: process and function. Genes Dev 21:2861–2873. https://doi.org/10.1101/gad.1599207

Moutinho M, Nunes MJ, Rodrigues E (2016) Cholesterol 24-hydroxylase: brain cholesterol metabolism and beyond. Biochim Biophys Acta Mol Cell Biol Lipids 1861:1911–1920. https://doi.org/10.1016/j.bbalip.2016.09.011

Murphy SR, Chang CCY, Dogbevia G et al (2013) Acat1 knockdown gene therapy decreases amyloid-β in a mouse model of Alzheimer's disease. Mol Ther 21:1497–1506. https://doi.org/10.1038/mt.2013.118

Nelson AR, Sweeney MD, Sagare AP, Zlokovic BV (2016) Neurovascular dysfunction and neurodegeneration in dementia and Alzheimer's disease. Biochim Biophys Acta 1862:887–900. https://doi.org/10.1016/j.bbadis.2015.12.016.Neurovascular

Nicholls SJ, Sipahi I, Schoenhagen P et al (2006) Intravascular ultrasound assessment of novel antiatherosclerotic therapies: rationale and design of the Acyl-CoA:Cholesterol Acyltransferase Intravascular Atherosclerosis Treatment Evaluation (ACTIVATE) Study. Am Heart J 152:67–74. https://doi.org/10.1016/j.ahj.2005.10.025

Nissen SE, Tardif J-C, Nicholls SJ et al (2006) Effect of ACAT inhibition on the progression of coronary atherosclerosis. N Engl J Med 354:1253–1263. https://doi.org/10.1056/NEJMoa070635

Oesterle A, Laufs U, Liao JK (2017) Pleiotropic effects of statins on the cardiovascular system. Circ Res 120:229–243. https://doi.org/10.1161/CIRCRESAHA.116.308537

Panza F, D'Introno A, Colacicco AM et al (2006) Lipid metabolism in cognitive decline and dementia. Brain Res Rev 51:275–292. https://doi.org/10.1016/j.brainresrev.2005.11.007

Pappolla MA, Bryant-Thomas TK, Herbert D et al (2003) Mild hypercholesterolemia is an early risk factor for the development of Alzheimer amyloid pathology. Neurology 61:199–205. https://doi.org/10.1212/01.WNL.0000070182.02537.84

Pennetta G, Welte M (2019) Emerging links between lipid droplets and motor neuron diseases. Dev Cell 45:427–432. https://doi.org/10.1016/j.devcel.2018.05.002.Emerging

Perea JR, Llorens-Martín M, Ávila J, Bolós M (2018) The role of microglia in the spread of Tau: relevance for tauopathies. Front Cell Neurosci 12:1–8. https://doi.org/10.3389/fncel.2018.00172

Petrov AM, Kasimov MR, Zefirov AL (2016) Brain cholesterol metabolism and its defects: linkage to neurodegenerative diseases and synaptic dysfunction. Acta Nat 8:58–73

Posse E (2012) Reciprocal regulation of cholesterol and beta amyloid at the subcellular level in Alzheimer's disease. Can J Physiol Pharmacol 90:753–764. https://doi.org/10.1139/y2012-076

Provenzano FA, Muraskin J, Tosto G et al (2013) White matter hyperintensities and cerebral amyloidosis: necessary and sufficient for clinical expression of Alzheimer disease? JAMA Neurol 70:455–461. https://doi.org/10.1001/jamaneurol.2013.1321

Puglielli L, Konopka G, Pack-Chung E et al (2001) Acyl-coenzyme A: Cholesterol acyltransferase modulates the generation of the amyloid β-peptide. Nat Cell Biol 3:905–912. https://doi.org/10.1038/ncb1001-905

Refolo LM, Pappolla MA, Malester B et al (2000) Hypercholesterolemia accelerates the Alzheimer's amyloid pathology in a transgenic mouse model. Neurobiol Dis 7:321–331. https://doi.org/10.1006/nbdi.2000.0304

Riekse RG, Li G, Petrie EC et al (2006) Effect of statins on Alzheimer's disease biomarkers in cerebrospinal fluid. J Alzheimers Dis 10:399–406. https://doi.org/10.3233/JAD-2006-10408

Rockwood K, Kirkland S, Hogan DB et al (2002) Use of lipid-lowering agents, indication bias, and the risk of dementia in community-dwelling elderly people. Arch Neurol 59:223–227. https://doi.org/10.1001/archneur.59.2.223

Rogers MA, Liu J, Song B et al (2015) Acyl-CoA:cholesterol acyltransferases (ACATs/SOATs): enzymes with multiple sterols as substrates and as activators. J Steroid Biochem Mol Biol 151:102–107. https://doi.org/10.1016/j.jsbmb.2014.09.008.Acyl-CoA

Rohilla A, Rohilla S, Kumar A et al (2011) Pleiotropic effects of statins: a boulevard to cardioprotection. Arab J Chem 9:21–27. https://doi.org/10.1016/j.arabjc.2011.06.025

Rohn TT (2014) Is apolipoprotein E4 an important risk factor for vascular dementia? Int J Clin Exp Pathol 7:3504–3511

Roth BD (1998) ACAT inhibitors: evolution from cholesterol-absorption inhibitors to antiatherosclerotic agents. Drug Discov Today 3:19–25

Saunders AM, Strittmatter WJ, Schmechel D et al (1993) Association of apolipoprotein E allele epsilon 4 with late-onset familial and sporadic Alzheimer's disease. Neurology 43:1467–1472. https://doi.org/10.1212/WNL.43.8.1467

Saxena U (2009) Lipid metabolism and Alzheimer's disease: pathways and possibilities. Expert Opin Ther Targets 13:331–338. https://doi.org/10.1517/14728220902738720

Schultz BG, Patten DK, Berlau DJ (2018) The role of statins in both cognitive impairment and protection against dementia: a tale of two mechanisms. Transl Neurodegener 7:1–11. https://doi.org/10.1186/s40035-018-0110-3

Sepulcre J, Grothe MJ, d'Oleire Uquillas F et al (2018) Neurogenetic contributions to amyloid beta and tau spreading in the human cortex. Nat Med 24:1910–1918. https://doi.org/10.1038/s41591-018-0206-4

Settembre C, Di Malta C, Polito VA et al (2011) TFEB links autophagy to lysosomal biogenesis. Science 332:1429–1433. https://doi.org/10.1126/science.1204592.TFEB

Shabir O, Berwick J, Francis SE (2018) Neurovascular dysfunction in vascular dementia, Alzheimer's and atherosclerosis. BMC Neurosci 19:1–16. https://doi.org/10.1186/s12868-018-0465-5

Shibuya Y, Chang CCY, Huang L-H et al (2014) Inhibiting ACAT1/SOAT1 in microglia stimulates autophagy-mediated lysosomal proteolysis and increases A 1-42 clearance. J Neurosci 34: 14484–14501. https://doi.org/10.1523/JNEUROSCI.2567-14.2014

Shibuya Y, Chang CC, Chang T-Y (2015a) ACAT1/SOAT1 as a therapeutic target for Alzheimer's disease. Future Med Chem 7:2451–2467. https://doi.org/10.4155/fmc.15.161

Shibuya Y, Niu Z, Bryleva EY et al (2015b) Acyl-CoA:cholesterol acyltransferase 1 blockage enhances autophagy in the neurons of triple transgenic Alzheimer's disease mouse and reduces human P301L-tau content at the pre-symptomatic stage. Neurobiol Aging 36:2248–2259. https://doi.org/10.1016/j.neurobiolaging.2015.04.002.Acyl-CoA

Skrobot OA, McKnight AJ, Passmore PA et al (2016) A validation study of vascular cognitive impairment genetics meta-analysis findings in an independent collaborative cohort. J Alzheimers Dis 53:981–989. https://doi.org/10.3233/jad-150862

Sodhi RK, Singh N (2013) Liver X receptors: emerging therapeutic targets for Alzheimer's disease. Pharmacol Res 72:45–51. https://doi.org/10.1016/j.phrs.2013.03.008

Solomon A, Kåreholt I, Ngandu T et al (2009a) Serum total cholesterol, statins and cognition in non-demented elderly. Neurobiol Aging 30:1006–1009. https://doi.org/10.1016/j.neurobiolaging.2007.09.012

Solomon A, Kivipelto M, Wolozin B et al (2009b) Midlife serum cholesterol and increased risk of Alzheimer's and vascular dementia three decades later. Dement Geriatr Cogn Disord 28:75–80. https://doi.org/10.1159/000231980

Sparks L, Kryscio RJ, Sabbagh MN et al (2008) Reduced risk of incident AD with elective statin use in a clinical trial cohort. Curr Alzheimer Res 5:416–421. https://doi.org/10.2174/156720508785132316

St Clair D, Rennie M, Slorach E et al (1995) Apolipoprotein E ε4 allele is a risk factor for familial and sporadic presenile Alzheimer's disease in both homozygote and heterozygote carriers. J Med Genet 32:642–644. https://doi.org/10.1136/jmg.32.8.642

Strittmatter WJ, Saunders AM, Schmechel D et al (1993) Apolipoprotein E: high-avidity binding to beta-amyloid and increased frequency of type 4 allele in late-onset familial Alzheimer disease. Proc Natl Acad Sci 90:1977–1981. https://doi.org/10.1073/pnas.90.5.1977

Suh Y-H (2002) Amyloid precursor protein, presenilins, and alpha-synuclein: molecular pathogenesis and pharmacological applications in Alzheimer's disease. Pharmacol Rev 54:469–525. https://doi.org/10.1124/pr.54.3.469

Suidan GL, Ramaswamy G (2019) Targeting apolipoprotein E for Alzheimer's disease: an industry perspective. Int J Mol Sci 20:1–14

Sun JH, Tan L, Wang HF et al (2015a) Genetics of vascular dementia: systematic review and meta-analysis. J Alzheimers Dis 46:611–629. https://doi.org/10.3233/JAD-143102

Sun JH, Yu JT, Tan L (2015b) The role of cholesterol metabolism in Alzheimer's disease. Mol Neurobiol 51:947–965. https://doi.org/10.1007/s12035-014-8749-y

Swiger KJ, Manalac RJ, Blumenthal RS et al (2013) Statins and cognition: a systematic review and meta-analysis of short- and long-term cognitive effects. Mayo Clin Proc 88:1213–1221. https://doi.org/10.1016/j.mayocp.2013.07.013

Szeto J, Lewis S (2016) Current treatment options for Alzheimer's disease and Parkinson's disease dementia. Curr Neuropharmacol 14:326–338. https://doi.org/10.2174/1570159X14666151208112754
Tajima Y, Ishikawa M, Maekawa K et al (2013) Lipidomic analysis of brain tissues and plasma in a mouse model expressing mutated human amyloid precursor protein/tau for Alzheimer's disease. Lipids Health Dis 12:1–14
Tan B, Rosenfeldt F, Ou R et al (2019) Evidence and mechanisms for statin-induced cognitive decline. Expert Rev Clin Pharmacol 12:397–406. https://doi.org/10.1080/17512433.2019.1606711
Tardif JC, Grégoire J, L'Allier PL et al (2004) Effects of the acyl coenzyme A:cholesterol acyltransferase inhibitor avasimibe on human atherosclerotic lesions. Circulation 110:3372–3377. https://doi.org/10.1161/01.CIR.0000147777.12010.EF
de la Torre J (2002) Alzheimer disease as a vascular disorder nosological. Stroke 33:1152–1162
Urano Y, Vo DH, Hirofumi A, Noguchi N (2019) 24 (S)-hydroxycholesterol induces ER dysfunction-mediated unconventional cell death. Cell Death Dis 5:113. https://doi.org/10.1038/s41420-019-0192-4
US National Library of Medicine (2009) ClinicalTrials.gov. In: Trial Simvastatin Amnestic Mild Cogn. Impair. Patients. https://clinicaltrials.gov/ct2/show/NCT00842920
Vetrivel KS, Thinakaran G (2006) Amyloidogenic processing of β-amyloid precursor protein in intracellular compartments. Neurology 66:S69–S73. https://doi.org/10.1212/01.wnl.0000192107.17175.39
Vijayan M, Reddy PH (2016) Stroke and vascular dementia and Alzheimer's disease-molecular links. J Alzheimers Dis 54:427–443. https://doi.org/10.3233/JAD-160527.Stroke
Walker LC, Lynn DG, Chernoff YO (2018) A standard model of Alzheimer's disease? Prion 12:261–265. https://doi.org/10.1080/19336896.2018.1525256
Walther T, Farese R (2012) Lipid droplets and cellular lipid metabolism. Annu Rev Biochem 81:687–714. https://doi.org/10.1146/annurev-biochem-061009-102430.Lipid
Welte MA (2015) Expanding roles for lipid droplets. Curr Biol 25:470–481. https://doi.org/10.1016/j.cub.2015.04.004.Expanding
Wendell C, Zonderman A, Katzel L et al (2016) Nonlinear associations between plasma cholesterol levels and neuropsychological function. Neuropsychology 30:980–987. https://doi.org/10.1037/neu0000298.Nonlinear
West R, Beeri MS, Schmeidler J et al (2008) Better memory functioning associated with higher total and LDL cholesterol levels in very elderly subjects without the APOE4 allele. Am J Geriatr Psychiatry 16:781–785. https://doi.org/10.1097/JGP.0b013e3181812790.Better
Whitmer RA, Sidney S, Selby J et al (2005) Midlife cardiovascular risk factors and risk of dementia in late life. Neurology 64:277–281. https://doi.org/10.1212/01.WNL.0000149519.47454.F2
WHO (2017) Global action plan on the public health response to dementia 2017 - 2025. WHO, Geneva. https://www.who.int/mental_health/neurology/dementia/action_plan_2017_2025/en/
WHO, ADI (2012) Dementia: a public health priority. https://www.who.int/mental_health/publications/dementia_report_2012/en/
Wollmer MA, Streffer JR, Tsolaki M et al (2003) Genetic association of acyl-coenzyme A: cholesterol acyltransferase with cerebrospinal fluid cholesterol levels, brain amyloid load, and risk for Alzheimer's disease. Mol Psychiatry 8:635–638. https://doi.org/10.1038/sj.mp.4001296
Wolozin B (2000) Decreased prevalence of alzheimer disease associated with 3-hydroxy-3-methyglutaryl coenzyme A reductase inhibitors. Arch Neurol 57:1439. https://doi.org/10.1001/archneur.57.10.1439
Wood GW, Ling L, Muller W, Eckert G (2014) Cholesterol as a causative agent in Alzheimer disease: a debatable hypothesis. J Neurochem 129:559–572. https://doi.org/10.1111/jnc.12637.Cholesterol
Xie C, Lund EG, Turley SD et al (2003) Quantitation of two pathways for cholesterol excretion from the brain in normal mice and mice with neurodegeneration. J Lipid Res 44:1780–1789. https://doi.org/10.1194/jlr.M300164-JLR200

Yamanaka K, Urano Y, Takabe W et al (2014) Induction of apoptosis and necroptosis by 24 (S)-hydroxycholesterol is dependent on activity of acyl-CoA: cholesterol acyltransferase 1. Cell Death Dis 5:990–999. https://doi.org/10.1038/cddis.2013.524

Yin YW, Li JC, Wang JZ et al (2012) Association between apolipoprotein E gene polymorphism and the risk of vascular dementia: a meta-analysis. Neurosci Lett 514:6–11. https://doi.org/10.1016/j.neulet.2012.02.031

Zhang J, Liu Q (2015) Cholesterol metabolism and homeostasis in the brain. Prot Cell 6:254–264. https://doi.org/10.1007/s13238-014-0131-3

Zhao FG, Wang YH, Yang JF et al (2005) Association between acyl-coenzyme A: Cholesterol acyltransferase gene and risk for Alzheimer's disease in Chinese. Neurosci Lett 388:17–20. https://doi.org/10.1016/j.neulet.2005.06.020

Zhao B, Shang S, Li P et al (2019) The gender- and age-dependent relationships between serum lipids and cognitive impairment: a cross-sectional study in a rural area of Xi'an, China. Lipids Health Dis 18:1–11

Zlokovic BV, Deane R, Sagare AP et al (2010) Low-density lipoprotein receptor-related protein-1: a serial clearance homeostatic mechanism controlling Alzheimer's amyloid β-peptide elimination from the brain. J Neurochem 115:1077–1089. https://doi.org/10.1111/j.1471-4159.2010.07002.x.Low-density

Therapeutic Potential of PPARs in Alzheimer's Disease

14

Aman Upaganlawar, Chandrashekhar Upasani, and Stella Amarachi Ihim

Abstract

Alzheimer's disease (AD) is the most leading cause of dementia, characterized by the accumulation of amyloid beta (Aβ) plaques in the brain, leading to neuronal degeneration and loss. Advances in recent years have highlighted the role of new targets explaining the cause of the underlying process which includes tau protein's hyperphosphorylation leading to the development of neurofibrillary tangles (NFTs), affecting synaptic function, inflammatory responses, lipid, and energy metabolism. In addition, neuroinflammation has been found to play a significant role in the pathophysiology and progression of AD. However, the current treatment targets the later aspects of the pathophysiological state and thus has minimal impact on the disease. Peroxisome proliferator-activated receptor (PPAR) is a ligand-activated transcription factor with function, centered on the modulation of glucose utilization, lipid breakdown, and the expression of the gene. Furthermore, numerous in vivo and in vitro studies have shown that PPARγ activation has been found to modulate Aβ homeostasis by regulating Aβ production, downregulate inflammatory gene expression, suppress and functionally inactivate inflammation transcription factors, and improve mitochondria function. Hence, PPARγ and its agonists might exercise their potentials in treating AD by modifying the diverse aspects involved in the pathophysiology of AD. Thus signify a pivotal pharmacological therapeutic mark for AD novel drug discovery.

A. Upaganlawar
SNJB's Shriman Sureshdada Jain College of Pharmacy, Neminagar, Chandwad, Nashik, Maharashtra, India

C. Upasani
Department of Pharmacology, Shriman Sureshdada Jain College of Pharmacy, Chandwad, India

S. A. Ihim (✉)
Department of Pharmacology and Toxicology, University of Nigeria, Nsukka, Enugu State, Nigeria

G. M. Ashraf, M. S. Uddin (eds.), *Current Thoughts on Dementia*,
https://doi.org/10.1007/978-981-16-7606-2_14

Here, we represent the complex role played by this potential target in the treatment of AD, describing preclinical and clinical studies using agonist of this receptor, signaling, and the pathophysiological role played by neuroinflammation in the disease process.

Keywords

Amyloid plaques · Neurofibrillary tangles · Peroxisome proliferator-activated receptor · Alzheimer's disease · PPAR agonists · PPARγ

14.1 Introduction

Alzheimer's disease (AD) is a prevalent dementia form, widely branded by advancement in lapse of memory and diminished cognitive functions. Nearly about 47 million individuals suffer from dementia all over the world and it accounts for roughly 60–80% of the entire dementia cases (Jeffrey et al. 2019; Alzheimer's association 2016). It is often identified in adults not less than 65 years of age and its prevalence upsurges after each decade with women being the most commonly affected. To date, there is scanty knowledge of the molecular mechanisms related to its pathogenesis (Maria et al. 2016). It is thought the AD develops from buildup of Aβ plaques in certain brain areas leading to extensive loss of neurons and neurofibrillary tangles (Maria et al. 2016). Pathologically, it is associated with mitochondrial damage, neuroinflammation, and glutamate excitotoxicity, among others ultimately resulting in synaptic and neuronal death. Despite the huge volume of preclinical and clinical investigations, currently available therapies only provide symptomatic relief to most patients (Khan et al. 2019). Modification of the disease process of AD, therefore, offers a lifeline for novel drug therapeutics in AD. Currently available therapeutic options targeting the cholinergic system, glutamatergic system, and the Aβ plaques mostly address the symptomatic presentations of the disease (Maria et al. 2016) and do not impede the rate of disease progression, hence the worldwide quest to access novel treatments to halt, slow, or avert AD.

The nuclear receptor PPARs is a recently documented goal for the treatment of AD (Manoj et al. 2018) and holds a promising future. PPARs are lipid sensor nuclear receptors that modify inflammatory response and Aβ clearance (Mandrekar-Colucci and Landreth 2011). Investigation in the field of AD continues to shed illumination on the possible new roles for these receptors. Reports show they perform a significant part in the modulation of inflammation and energy stability and lipid metabolism (Heneka et al. 2011; Spiegelman 1998). They seem to play a role in the modulation of glucose, lipid metabolism, and gene expression linked to inflammation (Mandrekar-Colucci and Landreth 2011). PPARs modulate the expression of genes that are necessary for cell differentiation and other metabolic processes (Spiegelman 1998).

Therefore, these receptor agonists epitomize an attractive beneficial target for AD. The PPARγ is the most studied isoform of PPARs and holds promising potential as an anti-AD therapy. In this chapter, the history of PPARs, their molecular structure, tissue distribution, isoforms, functions, molecular signaling pathway, and the therapeutic potentials of the utilization of PPARs in the treatment of AD are discussed. The cellular targets of the PPAR agonists and its potential as anti-AD therapy are explored.

14.2 Historical Outlook of the Peroxisomes

Between 1954 and 1964, peroxisomes were identified and described as small organelle present in the cytoplasm of many cells especially eukaryotic cells, containing enzymes such as catalase and some oxidases and usually involved in the catabolism of increasingly extended fatty acid chains, split fatty acid chains (Roy et al. 2013), D-amino acids, polyamines, reduction of reactive oxygen species (mostly H_2O_2), and the synthesis of plasmogens (phospholipids used in the production of myelin and necessary for the functioning of the mammalian brain and lungs). Shortly after this, the PPARs were revealed (Heneka et al. 2011).

PPARs, which are part of the steroid hormone superfamily, were discovered in 1990. They are ligand-activated transcription factors, which are part of the nuclear hormone receptor super family, regulating the expression of genes (Roy et al. 2013). Other associates of the class of nuclear family receptors range from the retinoid receptors, estrogen receptors, FXR, vitamin D receptors, LXR, thyroid hormone receptors, etc.

Briefly, in the 1960s, a group of researchers in Switzerland noted that when hypolipidemic drugs were administered in rodents, there was resultant hepatomegaly as well as a remarkable rise in the figures and size of peroxisomes in the liver of such animals (Mathivat and Cottet 1953; Cottet et al. 1953; Thorp and Waring 1962). These agents capable of initiating such effects were termed 'microbodies' at that time.

After a decade, other groups of researchers in the same field noted that aside from the activities of microbodies previously described, there were a group of unrelated structural compounds that were able to reproduce the same pharmacological effects as microbodies—hepatomegaly and upsurge in the dimensions and quantity of liver peroxisomes. Hence, they coined the term 'peroxisome proliferators' to describe these structurally different compounds capable of producing such effects (Svoboda et al. 1969; Azarnoff and Svoboda 1969; Jihan and Mostafa 2015). Within a short while, they further probed and discovered that these agents, peroxisome proliferators, upon chronic administration to animals such as the rats and mice, caused an increased incidence of adenoma and carcinoma in the liver of these rodents, in addition to the proliferation of peroxisomes (Jihan and Mostafa 2015). The mechanism behind this was not understood, however it was hypothesized that they are likely to be receptor-mediated (Jihan and Mostafa 2015). Years later, another group of scientists in England discovered the hypothesized receptors using estrogen receptor DNA binding domain as a tool in rat livers treated with the peroxisome proliferators and named the receptors, PPARs (Issemann and Green

1990). Only subsequently were they called later PPARα (Huang et al. 1994; Wang-Soo and Jaetaek 2015). Subsequent to the detection of mouse receptor PPARα, PPARs were discovered in additional animal kinds including rats and humans (Dreyer et al. 1992; Schmidt et al. 1992).

14.3 Molecular Structure of PPARs

PPARs (which are present in three isoforms α, β or δ, and γ) share similar three-dimensional structure with other nuclear receptors members and contain functional domains (Elrod and Sun 2008; Bugge and Mandrup 2010; Usuda and Kanda 2014) which include the N-terminal section (A/B domain), DNA binding domain (DBD, C-domain), a flexible hinge section (D-domain), carboxyl-terminal (C-terminal, E/F region) which contains the LiBD as well as a ligand-dependent activation domain (AF-2) (Fig. 14.1).

14.3.1 N-Terminal (A/B Domain)

This NH_2-terminal region is a domain encompassing the ligand-independent transcription activation function 1 (AF-1) (Jihan and Mostafa 2015). (AF-1) motif is a focal point of kinase phosphorylation. N-terminal is also accountable for the constitutive transcriptional activity of PPAR-responsive gene once a ligand is absent. In addition, it plays a role in subtype target gene expression.

14.3.2 DNA Binding Domain (C-domain, DBD)

This area has a helix-loop assembly which is stabilized by two zinc atoms (zinc finger binding motifs). It is accountable for the coupling of PPAR to peroxisome proliferator response elements (PPRE) in the promoter section of target genes (Elrod and Sun 2008; Usuda and Kanda 2014). It also participates in the dimerization of PPAR with RXRs (retinoid X receptor) as this process is important for the activation

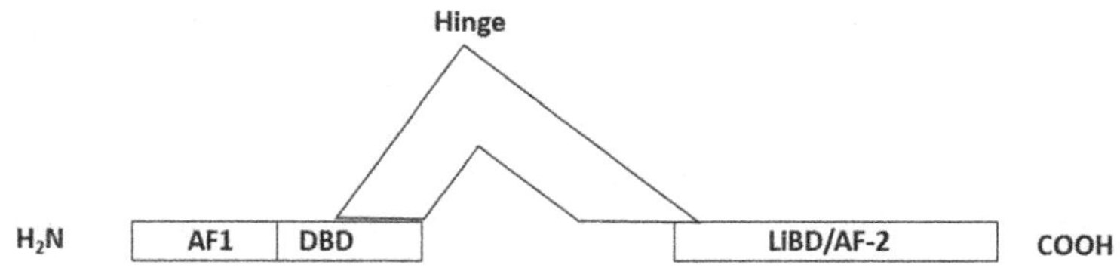

Fig. 14.1 Structural representation of PPARs with domains including the N-terminal section (A/B domain), DNA binding domain (DBD, C-domain), a flexible hinge section (D-domain), carboxyl-terminal (C-terminal, E/F region) which contains the ligand-binding domain (LiBD) as well as a ligand-dependent activation domain (AF-2). (Adapted from Usuda and Kanda 2014)

of transcription (Feige et al. 2005) as well as for the binding of co-activators (Tomaru et al. 2006) to the PPAR-RXR protein complex (Wang-Soo and Jaetaek 2015).

14.3.3 Hinge Region

This region couples the DBD to the LiBD and also represents a docking domain or territory for coactivators and cofactors. Besides, this region can modify the DNA binding capability of PPAR and is also involved in the binding of corepressors (Li et al. 2007; Gray et al. 2006).

14.3.4 Carboxyl Terminal (E/F Domain)

This has a ligand-binding function and is the principal domain in the receptor. This region of PPARs houses the LiBD and another segment, the ligand-dependent activation domain (AF-2), responsible for the engagement of PPAR cofactors (Issemann and Green 1990; Hauser et al. 2000; Zoete et al. 2007).

14.3.5 Ligand Binding Domain

This section is a hydrophobic pocket with Y shape to which ligands connect either to stimulate or inhibit the receptor transactivation, located within the region of the carboxyl terminal (Schmidt et al. 1992). Basically, it facilitates the heterodimerization of PPAR with the 9-cis-retinoic acid receptor (retinoid X receptor or RXR), causing a conformational alteration including nuclear rearrangement. This PPAR-RXR heterodimer complex once formed in its active state fixes to the PPRE in the promoter region of the target gene (Schmidt et al. 1992), associates with and alters coactivators/corepressors dynamics, and subsequently modulates the expression of target genes. Nuclear receptor family members and the different PPAR (α, β or δ and γ) may have some differences only in their LiBD as each can only be activated by diverse agonists that are specific. For instance, the vitamin D receptors are activated by vitamin D and the FXR is activated by bile acids (Usuda and Kanda 2014; Hauser et al. 2000).

14.4 Tissue Distribution and Expression of PPARS and Their Target Genes

There are three known PPAR isotypes such as PPARα, PPARβ/δ, and PPARγ. Tissue distribution, specificity of ligands, and physiological roles differentiate the isoforms. There is only a small overlap in activity as each isoform activates or suppresses different genes.

14.4.1 PPARα

This receptor isoform is highly expressed and found in abundance in most tissues with rapid oxidation of fatty acid and catabolism especially the liver (hepatocytes) and heart (cardiomyocytes) cells (Dong et al. 2015). They are also highly distributed in tissues with high fatty acid storage capacity. In addition, they are found with high expression and distribution in the kidney cortex, white and brown adipose tissues, adipocytes, large intestine, spleen, non-neuronal cells (microglia and astroglia), adrenal gland (Dong et al. 2015), vascular endothelium, and vascular smooth muscles. It is less commonly expressed in immunity-related cells such as monocytes, macrophages, Peyer's patch of the digestive system, skeletal muscles, colon mucosa, cecum, and the placenta (Azhar and Kellwy 2007).

14.4.2 PPARβ/δ

This isotype of PPAR is ubiquitously distributed especially in the liver, intestine, kidney, esophagus, and skeletal muscle (Barish et al. 2006; Kilgore and Billin 2008).

14.4.3 PPARγ

This isoform of PPAR is less distributed within the cardiac cells and skeletal muscles but most largely found in the adipose tissues (the white and brown tissues mainly responsible for the storage of large amounts of fatty acids) (Alexandre and Yves 2016). They are also found in the vascular endothelium, vascular smooth muscles, brain cells, immunity cells/immunologic system (such as macrophages, monocytes, Peyer's patches, bone marrow, and lymphocytes), liver, intestines, kidney, and retina but generally in traces (Miller and Etgen 2003; Wang-Soo and Jaetaek 2015). It should be noted that two splice variants of PPARγ exist, namely PPARγ1 and PPARγ2 (Medina-Gomez et al. 2007). These two differ only by 30 amino acids in the N9 terminal end. While PPARγ1 is extensively found in tissues (skeletal muscle heart, liver) at small levels, both are vastly distributed in adipose tissue (Table 14.1).

14.5 PPAR During Development

During fetal development, PPARβ/δ is expressed, unlike PPARα and PPARγ transcripts which appear late at Day 13.5 of gestation in rats and mice (Markus and Michael 2008). Only a little information is understood regarding the distribution of the PPARs during the development of humans. However, it is known that diverse PPAR isoforms originate from a uniform *PPAR* gene and express tissue-reliant forms during fetal development. Based on target genes and tissue localization, these isoforms exhibit diverse pharmacological, physiological, and biological roles as well as diverse ligand specificities (Keller et al. 2000; Heneka et al. 2011).

Table 14.1 Tissue distributions and primary functions of PPAR subfamily (adapted from Toyohiko 2007)

PPAR subfamily	Distribution in tissues	Primary function
PPARα	High concentrations within the kidney cortex, adipose tissues, adipocytes, intestine, spleen, non-neuronal cells, immune cells, and vessels	Fatty acid oxidation and metabolism of lipids
PPARβ/δ	Highly ubiquitous and present in many tissues, the liver, intestine, kidney, esophagus, and skeletal muscle	Lipid metabolism
PPARγ	Abundant in white and brown adipose tissues	Adipogenesis, lipid metabolism, inflammation, glucose regulation

14.6 Isoforms of PPARs and Their Ligands

PPARs are transcription factors that fit into the superfamily of ligand-activated nuclear receptors. The isoforms of PPARs when activated by distinct and precise ligands heterodimerize with the nuclear retinoic acid receptor (RXR) to modulate gene expression (Markus and Michael 2008). All PPARs are stimulated by natural ligands like fatty acids and lipid metabolites/lipid-derived substrates such as eicosanoids and other numerous xenobiotics. There are numerous natural and artificial agonists/ligands of PPARs useful in the handling of various diseases. These ligands possess different properties, distinct gene expression profiles, specificity for distinct PPAR receptors (Keller et al. 2000), as well as absorption profiles which ultimately results in eliciting diverse clinical response when such receptors are activated. One of the major characteristics of PPAR ligand-binding cavity which differentiates it from other nuclear receptors is the size as it is 3–4 times larger. These agonists on binding to PPAR cause gene transcription involved in glucose and lipid homeostasis.

Isoforms of PPAR are generally concerned with energy homeostasis and lipid metabolism and are encoded by diverse genes. PPARβ/δ is known to regulate glucose utilization, cell inflammation, and differentiation while PPARα has been concerned with the modulation of lipid metabolism and inflammatory processes. PPARγ is associated with glucose metabolism, inflammatory pathways, and adipocyte differentiation. The activity of the receptor is changed posttranslationally by phosphorylation, sumoylation, presence, and the activity of coregulators (Diradourian et al. 2005; Rajnish and Flint 2008).

14.6.1 PPARα

This isoform is encoded by chromosome 22q12-13.1 (Roy et al. 2013). They were the first set of PPAR isoforms to be discovered hence the name. When PPARα is activated by endogenous ligands such as fatty acids and their derivatives, fatty acid catabolism, gluconeogenesis, and ketone acid synthesis are often mediated. However, in rodents, their activation mediates immune modulation, amino acid metabolism, reduction in muscle and liver steatosis, plasma triglycerides including insulin resistance. Several genetic variants and polymorphs of PPARα have been found, differing at their exon or codons. It is postulated that this variant may account for differences in species response to PPARα activators. Polymorphism of PPARα has also been seen among diverse populations (Gervois et al. 1999), e.g. a variant PPARα V227A is a key polymorphism in the Japanese and a link has been noted amid this polymorph, the development of non-alcoholic fatty liver disease and protection against obesity (Naito et al. 2006).

Primarily, PPARα regulates energy homeostasis by enhanced stimulation of the breakdown of fatty acid and cholesterol thereby improving and increasing gluconeogenesis and the reduction of the levels of serum triglycerides. It is a significant controller of lipid metabolism and is often activated in settings associated with energy deprivation. Recent research has shown the role in the treatment of dyslipidemia. They have also portrayed anti-inflammatory and antithrombotic activities in vessels, hence a place in the management of vascular inflammation (Bajaj et al. 2007; Markus and Michael 2008).

14.6.2 PPARγ

In humans, PPARγ is positioned on chromosome 3, band 3P 25, and consists of at least 11 exons that yield 9 transcript variants. It is unique as it comprises three isoforms namely PPARγ1, PPARγ 2, and PPARγ 3 (Tontonoz et al. 1994; Meirhaeghe et al. 2003). PPARγ 2 differs from PPARγ 1 and PPARγ 3 only in the ligand non-dependent region at the N-terminal. Polymorphism has also been associated with PPARγ. Alternate transcription and splicing are accountable for the production of different PPARγ mRNA; PPAR1, 2, 3, and 4, although PPARγ 1, 3, and 4 may be translated into similar proteins. Polymorphism of PPARγ has been linked with a part in the progress of certain diseases (Jihan and Mostafa 2015). For instance, research proposes that PPARγ Pro 12 Ala expression may be involved in the progression of dementia at an early age in life and may strengthen the danger of cognitive dysfunction and dementia when diabetes progresses (Jihan and Mostafa 2015).

PPARγ functions predominantly to stimulate and enhance the regulation of glucose homeostasis as its activation is related to a drop in the levels of serum glucose probably associated with its capacity to control endocrine factors hence the advancement of specific PPARγ agonist for type II diabetes mellitus (Willson et al. 2001). It is also concerned with lipid metabolism, inflammation, insulin sensitivity, and cardiovascular function by regulating the transcription of certain genes concerned with these metabolic processes (Maria et al. 2016; Khan et al. 2019). They also modulate genes concerned with the signaling of insulin and the distribution of proinflammatory cytokines such as tumor necrosis factor (TNFα), hence a key target for thiazolidinediones (TZDs) which are antidiabetic medications that sensitize cells to insulin. Since PPARγ distribution has been noted in monocytes/macrophages, dendritic cells, granulocytes, mast cells, T and B cells they are suggested to perform a crucial part in immune regulatory functions.

14.6.3 PPARβ/δ

The PPARβ/δ has been linked with the existence of polymorphism located on chromosome 6p21.2-21.1 (Roy et al. 2013). The presence of PPARδ + 294T/C polymorphism is suggested to be related to raised levels of low-density lipoproteins (LDL) and apolipoproteins B, low levels of high-density lipoprotein (HDL), and complex likelihood of coronary heart disease. Unlike PPARγ and PPARα, PPARβ/δ is not an easy and frequent target by currently available drugs because they are ubiquitously expressed (Skogsberg et al. 2003; Aberle et al. 2006). Hence its physiological function and role is less studied and understood.

PPARβ/δ seems to be concerned with oxidation of fatty acid and lipid catabolism particularly in the skeletal muscles during binding and also responds to VLDL-derived fatty acids, eicosanoids as well as prostaglandins (Skogsberg et al. 2003). Its stimulation has been demonstrated to encourage proliferation and differentiation of cells, limiting gain in weight, promoting anti-inflammatory effects in vessel walls via deactivation of the expression of vascular cell adhesion molecules (VCAM)1 and monocyte chemoattractant protein (MCP)1 (Barish and Evans 2004). It has also been demonstrated to enhance catabolism of lipid in adipose tissues and the cardiac system hence improving HDL cholesterol levels and insulin resistance (Choi et al. 2012).

14.6.4 Ligand Binding of PPARs

The transcriptional activity of PPARs is mediated by ligand binding. To date, numerous ligands have been recognized that bind to stimulate and modify PPAR activity (Kliewer et al. 2001). Generally, the unique binding property of PPAR ligands is that its binding is more than 3 times larger than other nuclear receptors promoting their potential to link a heterogeneous set of artificial and natural lipophilic molecules like essential fatty acids (EFA) and their metabolites including arachidonic acid. PPARαs are known to bind natural ligands like unsaturated fatty acids, omega-3-fatty acids, prostaglandins (PG), leukotriene B4, 8-hydroxyeicosatetraenoic acid, and artificial ligands such as fibrates like clofibrates, fenofibrate, and bezafibrate (Heneka et al. 2011; Markus and Michael 2008).

PPARβ/δ has been shown to bind such natural ligands as unsaturated fatty acids, carbaprostacyclin l, and VLDL components. A typical synthetic ligand for this receptor is GW501516, GW0742 (Markus and Michael 2008).

The ligands bound by PPARγ include unsaturated fatty acids and derivatives from fatty acid found in diet or intracellular pathways such as cysophosphatidic acid, nitrolinoleic acid, PGD2, and 15d-PGJ2. Others include 15-hydroxyeicosatetranoic acid, 9-hydoxy and 13-hydroxyoctadecadienoic acid. Synthetic ligands of PPARγ include drugs of the TZD family such as rosiglitazone, pioglitazone, troglitazone, and ciglitazone which are potent insulin sensitizers and are useful in the management of diabetes mellitus. Non-TZD synthetic compounds such as L-tyrosine-based GW-7845, diinololymethane analogs, some non-steroidal anti-inflammatory drugs (NSAIDs), the novel synthetic triterpenoid, 2-cyano-3,12-dioxooleana-1,9-dien-28-oic acid (CDDO), and its derivatives also function as PPARγ agonists (Heneka et al. 2011; Markus and Michael 2008). 15d-PGJ2 which is obtained from arachidonic acid by the catalytic actions of cyclooxygenase 2 (Cox 2) and prostaglandin D synthase is believed to be the utmost effective intrinsic ligand for PPARγ as it activates the receptor even at low micromolar concentrations. It is worthy of note that all isotypes of PPAR may be stimulated by polyunsaturated fatty acids with diverse affinities and capabilities (Bernando and Minghetti 2006).

14.6.5 PPAR Signaling

The modulation of the expression of various genes by PPAR occurs through a complex mechanism. Mechanistically, PPARγ and PPARα, PPARβ/δ heterodimerize with a class of nuclear receptor called retinoid-X-receptor, resulting in a complex. The LiBD facilitates this heterodimerization process. In situations where ligands are absent and heterodimers are unstimulated, the heterodimers are associated with certain molecules called corepressors such as N-CoR and SMRT, that overwhelm and block the transcription of genes (Desvergne and Wahli 1999; Fan et al. 2018). The corepressor complex preserves the chromatin and prevents the employment of coactivator complex and transcription activation. The precise DNA sections of target genes that bind with PPAR-RXR dimers are labeled PPAR hormone response elements (PPREs) in the promoter region of target genes. They are originated in the promoter region of PPAR responsive genes like fatty acid-binding protein (aP2). Upon the binding of ligand to the hydrophobic pocket of nuclear receptor—PPAR receptor, a conformational change is induced in the LiBD core subsequently its activation (Mandrekar-Colucci and Landreth 2011). Modification in the conformation causes the release and displacement of corepressors—NCoR/SMRT or Notl. Recruitment of ligand and the displacement of corepressors results in the LiBD stabilization and the binding of cofactors/coactivators such as CBP/P300, P160/SRC-1, vitamin D receptor linked protein (DRIP), or thyroid hormone receptor-associated protein (TRAP) complex, which causes the initiation of the PPAR molecule. Ligands that initiate PPAR-RXR cause an interchange of corepressors for coactivators (Desvergne and Wahli 1999; Fan et al. 2018) (Fig. 14.2).

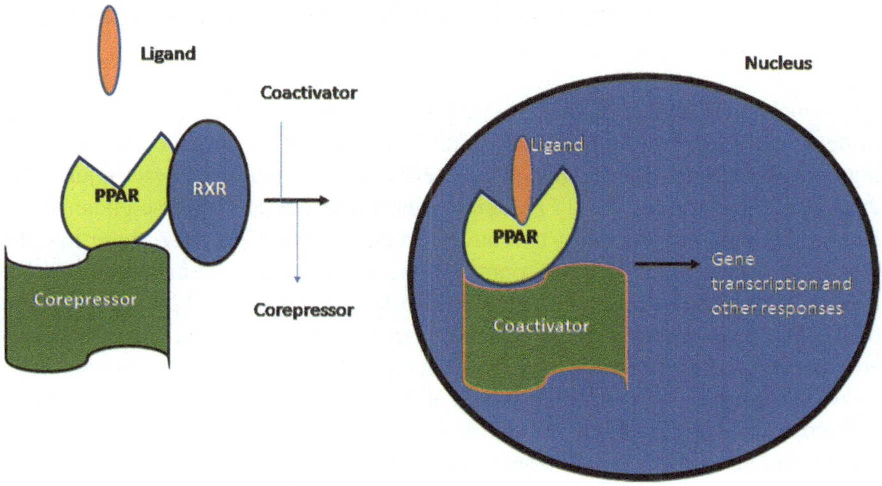

Fig. 14.2 Structural representation of PPARγ activation mechanism. PPARγ ligand binds to the PPAR-RXR heterodimer recruiting corepressors and coactivators eventually leading to gene transcription modulation, insulin secretion, lipid metabolism, and inflation mediating response. (Adapted from Mandrekar-Colucci and Landreth 2011)

Coactivators are potentiators of transcription and core repressors that inhibit gene expression of nuclear receptors are the subdivisions of transcriptional coregulators (Alzheimer's association 2016; Maria et al. 2016; Jeffrey et al. 2019). PPAR transcriptional coregulators play a vital part in disturbing the physiological and pathological role of the receptors and therefore serve as possible therapeutic targets for novel drug delivery. Transcriptional coactivators engaged to the PPAR receptor from coactivator receptor complexes subsequently cause the creation of a greater transcriptional complex (Querfurth and Laferla 2010; Khan et al. 2019) which afterward initiates the transcription of specific target gene concerned with different physiological and pathological progressions. More specifically PPAR-RXR heterodimer stimulates PPRE in the promoter region of target genes. This complex scaffold recruits histone acetyltransferases and RNA polymerase complex which collectively induce chromatin relaxation to authorize gene transcription (Fan et al. 2018). Aside from the above-stated actions of PPARs involving gene transcription, non-genomic actions associated with PPARs in the cytoplasm have been noted. For instance, nongenomic modulation of PPAR is facilitated by interface with cytosolic second messengers including kinases in phosphatases. In general, the activities of PPARs are controlled by posttranslational modulation like phosphorylation and sumoylation.

14.7 Pathophysiology of Neuroinflammation in Alzheimer's Disease

AD, which is a complex progressive neurodegenerative disorder, is a principal root of dementia. Time of life, history of family, apolipoprotein E4 genotype, diabetes, high blood pressure, obesity, hypercholesterolemia, and traumatic brain injury (Lin et al. 2005; Reitz et al. 2011) are among the numerous and diverse endangering factors connected to the onset and advancement of AD. Generally, these may be classified as genetic factors, environmental factors, and impaired metabolic activity (Ferrer 2012). The pathophysiological hallmark of AD is branded by the formation of amyloid plaques, intracellular neurofibrillary tangles resulting in nerve cell death and the advancement of memory loss and a rapid decline in cognitive function.

Aβ protein precursor (AβPP) is a type of integral protein that helps neurons grow and repair itself. Just like other proteins, AβPPs are broken down and recycled after progressive cleaving by α- and β-secretases (Hicks et al. 2013). Amyloid plaques are molded by the accumulation of Aβ, a cleavage result of AβPP when AβPP is gradually cleaved by β-secretase (BACE1) and γ-secretase respectively, resulting in the development of Aβ peptides and a soluble ectodomain sAPPβ (Hicks et al. 2013; Nalivaeva and Turner 2013; Zhang et al. 2012). Aβ monomers aggregate gradually into oligomers, fibrils, and non-soluble amyloid plaques.

Conversely, tangles are formed when tau, which promotes the stabilization of neuronal microtubules, is hyperphosphorylated, therefore accumulating into tangles comprised of paired helical filaments called neurofibrillary tangles (NFTs) (Tanzi and Bertram 2005). It is postulated that deposition of Aβ and its accumulation impairs synaptic and neuronal activities which provide the opportunity for the development of neurofibrillary tangles, subsequently resulting in loss of neurotransmitter function and then neuron loss. The elimination of cholinergic neurons and signaling in the basal forebrain is thought to generate a cholinergic shortage which leads to short-term memory loss in AD (Akiyama et al. 2000). Although there is a suggestion for the pivotal roles played by plaques and tangles in the pathophysiology of AD, the basis of the sporadic class of AD is still unclear. Genetically, the metabolism of Aβ is related to the metabolism of lipid as some allele of ApoE, the lipid carrier protein, is connected with expressively amplified risk for AD (Markus and Michael 2008).

Another pivotal hallmark in AD is the existence of chronic neuroinflammation within the brain without any sign of leucocyte infiltration while amyloid plaques are crowded by abundant microglia and astrocytes (Akiyama et al. 2000; Markus and Michael 2008). The inflammatory process of AD involves the Aβ initiation of glial cells (microglia and astrocytes) which produces proinflammatory substances as a powerful force for neurodegeneration. Glial cells are a highly heterogeneous populace of non-excitable cells of the central nervous system (CNS) involved in numerous brain functions (Verkhratsky et al. 2015). Microglial are the foremost form of brain immune defense and astrocytes are an indispensable neurosupportive cell type. Microglial and astrocytes are specialized glial cells and a chief component of the CNS innate immune system which plays a diverse dominant part in the elimination

(Markus and Michael 2008; Manoj et al. 2018) of trashes and toxins from the cerebrospinal fluid (CSF), release of neuroprotective factors, contributing prominently to synaptogenesis and the modulation of data processing, signal transmission, regulation of neural and synaptic plasticity and providing metabolic and trophic support to neurons (Manoj et al. 2018). Studies reveal that NFTs and plaques cause immunological responses in the brain and exist close to activated glial cells. The initiation of glial cells, astrocytes, and microglial is accompanied by its binding to Aβ oligomers via numerous receptors such as MARCO, scavenger receptor class B member I, CD36, and Toll-like receptors (TLR) (Yu and Ye 2015; Cai et al. 2016). Cyclin-dependent kinase 5 (CDK5) has been implicated in the initiation of microglia and astrocytes.

The result of the binding between Aβ oligomers and the receptor is the induction and release of inflammatory cytokines, chemokines, interleukin, monocyte chemotactic protein I (MCP-1), and TNFα (Westin et al. 2012; Yu et al. 2016). Elevated production of cytokines results in the initiation of nuclear factor-kappa B (NF-kB) pathway and the consequent initiation of mitogen-activated protein kinase (MAPK) pathways which have its proinflammatory gene expression dependent on Aβ. It is hypothesized that these proinflammatory mediators upsurge the activation and products of amyloidogenic pathway especially Aβ. Some studies also demonstrate that NF-kB signaling initiated by TNFα causes increase in Aβ synthesis propelled by β-secretase (BACE-I) transcription (Chen et al. 2012; Liao et al. 2004). Cytokines and MAPK cause the activation of the complement cascade, proinflammatory enzymes such as cyclooxygenase-2 (CoX-2) (Fattahi and Mirshafiey 2014). The enzymes may be related to the excessive secretion of S100B, a neurotrophin expressed by stimulated astrocytes that is capable of inducing NF-kB and encouraging tauopathy. Other proteins from the S100 family: S100A9 and S100A12 fashioned by stimulated microglia and macrophages have also been found to be increased in AD in the brain. S100A9 has also been found to surround Aβ deposits surrounding blood vessels in the brain (Vogl et al. 2012). A robust connection amid late-onset AD (LOAD) and over 20 genomic loci has also been discovered, with the strongest risk factor for LOAD being ApoE. The existence of other *ApoE* alleles is linked with amplified buildup of Aβ (Payton et al. 2016; Gottschalk and Mihovilovic 2016).

It is hypothesized that microglial activation and the simultaneous creation of mediators may hasten the progression of AD and neuronal loss (Gottschalk and Mihovilovic 2016). Such activation is a neuroprotective response primarily aimed at eliminating harmful stimuli, by modifying Aβ-related neurotoxicity, degrading and eliminating Aβ therefore generating a defensive barrier that surrounds plaques. Neuronal cell death and loss are also facilitated by the secretion of reactive oxygen species, nitric oxide (NO) (Vogl et al. 2012), proteolytic enzymes, complement factors, and excitatory amino acids. However unrestrained and protracted activation progresses past physiological control, overriding the beneficial effect.

14.7.1 Basis for the Use of PPAR Agonists in Alzheimer's Disease

Agonist of PPAR signifies a striking therapeutic target for AD. Increasing evidence suggests that anti-inflammatory therapies in AD disease patients may be beneficial for its treatment (Fan et al. 2018). In recent years, a wide build of evidence has established the effectiveness of PPARγ agonists in improving the disease pathology, memory, and learning in diverse modes.

PPARs have been observed to have the capacity to suppress the inflammatory response in peripheral macrophages hence the impression that they may be valuable for CNS disorders (De'la Monte and Wands 2006) having a component of inflammation. Following the use of NSAIDs to delay the onset of neuroinflammatory response, triggered by the deposition of plaques and tau hyperpolarization in AD and reduce the associated risks to develop AD, NSAIDs were observed to directly activate PPARγ. Several other lines of evidence support the hypothesis regarding the role PPARγ and its agonist may play in the treatment of AD (Corbett et al. 2015). PPARα has not been investigated considerably in AD and its mRNA and protein levels have been discovered to be small in AD brain tissues (De'la Monte and Wands 2006) though some recent studies are beginning to highlight the need to study PPARα signaling in AD (Corbett et al. 2015). Therefore, PPARγ was spotlighted and investigated as a striking beneficial target for the AD therapy. An inequality between Aβ manufacture and elimination leads to Aβ buildup, tau hyperpolarization, and neuronal degeneration and apoptosis (Corbett et al. 2015). PPARγ has been found to reduce Aβ levels, highlighting the role played by PPARγ agonists in decreasing Aβ accumulation.

In specialized glial cells, PPARγ is the leading PPAR isoform in microglia while astrocytes have all three PPAR isotypes (Fan et al. 2018). PPARγ expressed on microglia when activated has been demonstrated to diminish neuroinflammation and neurotoxicity in AD animal models. PPARγ agonists have also been useful as remedies for diabetes mellitus which increases the chance of developing AD. Other PPAR receptors and their agonists have not yet received promising results or still under examination for use in the management of AD (Corbett et al. 2015). Multiple evidence abounds that in AD, many physiological functions are altered. Therefore, PPARγ and its agonists might exercise their potentials in treating AD by modifying the diverse aspects involved in the pathophysiology of AD. This includes genetics, insulin sensitivity and signaling, energy metabolism, lipid metabolism, inflammation, and others.

14.8 PPARγ, Genetics, and Dyslipidemia in Alzheimer's Disease

Apolipoprotein E (ApoE) is a lipoprotein ubiquitously distributed in the brain and improves the destruction of Aβ including its engulfment by microglia and astrocytes (Barage and Sonawane 2015). ApoE is also known to play a significant part in the transport of lipid around the body, metabolism and cholesterol homeostasis, axonal growth, and synaptogenesis (Gottschalk and Mihovilovic 2016). In the brain, it is chiefly produced and released with HDL particles by astrocytes, facilitating cholesterol movement and reutilization within the CNS. ApoE E2, E3, and E4 are three naturally existing variants of ApoE identified in humans that modify the risks and age of onset of LOAD. However, the accumulation of ApoE4 is associated with an accumulation of Aβ. ApoE-mRNA levels have also been observed to be increased in AD patients especially LOAD (Payton et al. 2016; Gottschalk and Mihovilovic 2016). Genetically, the chromosome 19, band 13P 32, gene-rich section comprising of TOMM40, ApoE, and APOCI-mRNAs genes which have been implicated in AD exhibits an enriched potential of housing PPARγ binding sites (Subramanian et al. 2017). PPARγ agonists are demonstrated to diminish levels of TOMMO40, ApoE, and APOCI-mRNA through transcriptional regulation especially in ApoE-mRNA which has been linked with an amplified risk for AD.

PPARγ has also been shown to upregulate six LOAD-associated genes (ABCA7, ApoE, CASS4, CELF1, PTK2B, and ZCWPW) and downregulate DSG2 gene in PPARγ knockout experiment (Barrera et al. 2018). This has drawn attention to the possibility of using PPARγ as a drug class for the prevention or delay of the onset of the development of AD.

Rosiglitazone and pioglitazone which are PPARγ agonists have been demonstrated to initiate Aβ deterioration by promoting ApoE concentration within the brain (Fan et al. 2018). ApoE is observed to upsurge the amount of Aβ degrading enzyme neprilysin and insulin in astrocytes and microglia promoting Aβ degradation (Jiang et al. 2008). It is also worthy of note that the lipidation of ApoE is mediated predominantly by ATP bindings cassette A1 (ABCA1) by facilitating the loading of ApoE with phospholipids and cholesterol as ApoE serves as a scaffold for the development of HDL particles. Therefore, the role of ABCA1 is to modulate ApoE function in the CNS (Wahrle et al. 2008). A study demonstrated that the excessive expression of human ABCA1 in mice led to lowered Aβ levels and plaque burden. Of utmost importance is the observation that ABCA1 gene expression is transcriptionally modulated by the nuclear receptors, liver X receptor (LXRs), PPARγ, and RXR. PPARγ agonists, pioglitazone and rosiglitazone, increase ABCA1 and ApoE levels, hence reducing the levels of Aβ by half and this effect is considered to be minor to its induction of LXRα expression (Calkin et al. 2005; Llaverias et al. 2006; Barrera et al. 2018). A clinical study of greater than 500 patients experiencing mild to moderate AD who were managed for 6 months using rosiglitazone showed significant improvement in patients without ApoE4 allele (Risner et al. 2006). However, patients with ApoE4 were unresponsive to the drug and presented no enhancement in a cognitive test. The outcome of this research was consistent with the effect of the influence of *ApoE4* genotype (Risner et al. 2006; Sato et al. 2009) but the foundation of the diverse effects of rosiglitazone in patients with respect to *ApoE* genotype is unsolved. Another study using pioglitazone in first-time type II diabetic AD patients displayed noteworthy enhancement in neuropsychological tests, regional cerebral blood flow, and plasma Aβ levels during pioglitazone treatment (Sato et al. 2009). These studies demonstrate that the amyloid clearance pathway reliant on TZDs is also dependent on the distribution of functional ApoE4.

Aβ is cleared in the brain by the enzymatic action of key enzymes like insulin-degrading enzyme (IDE), neprilysin (NEP), and matrix metalloproteinase (MMP) 9 or non-enzymatic pathways including drainage by perivascular basement membranes (Mandrekar-Colucci and Landreth 2011), engulfment by microglia or astrocytes and elimination facilitated by receptors like LDL receptor-related protein 1 (LRP1) and P-glycoprotein located chiefly on the abluminal side of the cerebral endothelium. Low-density lipoprotein LRP1 is the transmembrane receptor that participates in facilitating the endocytosis of certain proteins including ApoE and Aβ (Mandrekar-Colucci and Landreth 2011). In AD patients, LRP1 levels are decreased symbolizing their part in facilitating Aβ metabolism. Studies show that PPARγ transcriptionally controls LRP1 gene because of the existence of PPRE on the LRP1 promoter section (Gauthier et al. 2003). In vitro studies performed in HepG2 cells reveal that in a concentration-dependent manner, rosiglitazone transcriptionally activated LRP1 gene (Rondón-Ortiz et al. 2017).

Another hypothesis suggests that PPARγ can modulate Aβ homeostasis by regulating Aβ production (Landreth et al. 2008). Studies showed that when neuroblastoma cells (stably transfected with human APP) were stirred with inflammatory cytokines, they stimulated the APP processing enzyme beta secretase (BACE 1) resultant in increased secretion of Aβ. However, PPARγ activation reduced Aβ secretion through its ability to suppress BACE 1 transcription (Jiang et al. 1998; Gauthier et al. 2003; Sastre et al. 2003; Landreth et al. 2008; Rondón-Ortiz et al. 2017).

14.9 PPARγ and Neuroinflammation in Alzheimer's Disease

Numerous clinical AD clinical trial failures have motivated scientists to explore the management of AD in the pre-symptomatic phase, wherein inflammation plays a key part in the development of neurodegeneration (Mandrekar-Colucci and Landreth 2011). Initial studies which explored the activities of PPARγ in AD were founded upon the capability of NSAID to trigger the PPARγ receptor. Several investigations have shown that NSAID use decreases the risk of AD by up to 80% and that these effects may originate from the potential of NSAIDs to activate PPARγ (Mandrekar-Colucci and Landreth 2011), inhibit brain inflammations responses in AD (Kielian and Drew 2003; Landreth and Heneka 2001; Lehmann et al. 1997) as well as reduce the expression of iNOS by initiation of PPARγ in vitro and in vivo (Heneka et al. 1999). In immune cells the ability of PPARγ to downregulate inflammatory gene expression represents a potential beneficial effect of PPARs. Many studies propose that the anti-inflammatory property of TZDs may involve the regulation of microglia and astrocyte inflammation, possibly including CDK5 and downregulation of proinflammatory cytokines (Heneka et al. 1999).

Firstly, PPARγ stimulation has been demonstrated to modify the feedback response of microglia to Aβ deposition thereby increasing Aβ phagocytosis and decreasing the release of cytokines. In vitro, PPARγ ligands were able to inhibit Aβ-mediated activation of microglia and suppress neuronal death (Luna-Medina et al. 2005). In rat in vivo models, co-injected ciglitazone and ibuprofen or pioglitazone or rosiglitazone administered orally suppressed Aβ-induced microglia cytokine generation (Heneka et al. 2003). In A/T mice expressing elevated levels of Aβ and TGF-β1, pioglitazone use decreased astrocyte and microglial initiation in the mouse cortex and hippocampus. These studies and others hypothesize that PPARγ activation may be associated with enhanced uptake and removal of Aβ from the medium and modify the activation of specialized glial cells and inflammatory cytokines and this may be a potential therapeutic pharmacological goal for AD drug development (Calhoun et al. 1999; Camacho et al. 2004).

Moreso, physiological distribution of PPARγ in the brain is fairly less than its distribution in AD patients as measured by mRNA levels, suggesting a pathophysiological involvement by an anti-inflammatory role for PPARγ in AD (Calhoun et al. 1999; Camacho et al. 2004). Following this discovery, more studies were conducted which showed that the PPARγ activation suppressed and functionally inactivated inflammation transcription factors such as nuclear factor-kB (NF-kB), STAT-1 and transcription factor activator protein-1 as well as cyclooxygenase 2 (COX-2), MMP-9, iNOS, proinflammatory cytokines, chemokines and interleukins (IL) (Lenglet et al. 2015). The mechanism by which PPARγ inactivates NF-kB-dependent promoters is not clear but may be associated with the ability of PPARγ to block NF-kB-dependent genes expression through corepressor interference (Pascual et al. 2005). Studies showing the potential of PPARγ natural agonists such as 15d-PGJ2 and synthetic agonists especially TZDs to inhibit IL-6, TNFα, and COX-2 have been demonstrated in monocytic and microglial cell cultures stimulated with Aβ (Combs et al. 2000).

It is also hypothesized that the effect of PPARγ on AD through the inflammatory component may result from the stimulation of PPARγ which causes the polarization of circulating monocytes to macrophages from the M1 microglia phenotype (a classic activated phenotype possessing proinflammatory and neurotoxic activities via the release of inflammatory cytokines) to the activated microglia M2 phenotype (which exhibit anti-inflammatory and neurotrophic effects and degrades toxic aggregates resulting from its anti-inflammatory interleukin production such as IL-1α, 1L-1β, TNF, and NO) (Combs et al. 2000; Calhoun et al. 1999; Camacho et al. 2004; Pascual et al. 2005; Bouhlel et al. 2007; Michelucci et al. 2009; Lenglet et al. 2015; Choi et al. 2017). Schematic representation of the cascade of molecular signaling neuroinflammatory involvement in AD is shown in Fig. 14.3.

14.10 PPARγ and Energy Metabolism in the Mitochondria of Alzheimer's Disease

Cumulative evidence advocates that PPARγ ligands play a pivotal role in mitochondria regulation in adipose tissues (Qiang et al. 2012) and other organs indicative of a likely benefit against mitochondrial dysfunction in AD patients. The mitochondria may be important in cerebral hypometabolism noted in AD patients as it has a serious role in energy metabolism and neuronal apoptosis (Obulesu and

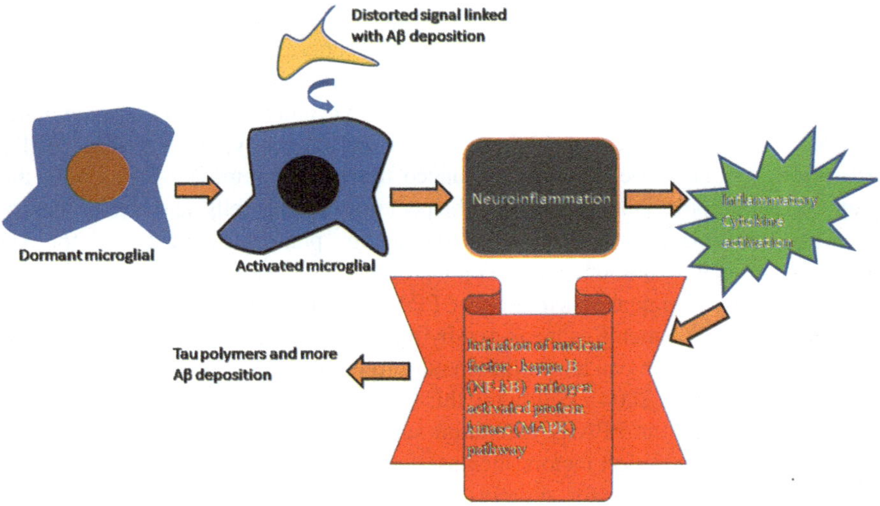

Fig. 14.3 Schematic representation of the cascade of molecular signaling neuroinflammatory involvement in AD. (Adapted from Yu et al. 2016)

Lakshmi 2014). Studies reveal that AD patients show impairment in glucose use and metabolism in areas of the brain controlling memory and cognition as well as an ApoE4-dependent imbalance in glucose use in these brain sections affected by the disease (Bookheimer et al. 2000; Thorp and Waring 1962). In AD patients, the number of neuronal mitochondria is less and the remaining has pronounced morphological changes. It has also been postulated that PPARγ ligands may improve mitochondria function as the basis for their benefits in improving recollection and cognition in AD patients (Roses et al. 2007). Another study also shows that activation of PPARγ by pioglitazone yielded significant elevations in mitochondria DNA and genes involved in mitochondria biogenesis in fat tissues.

In addition, PPAR gamma coactivator 1 alpha (PGC1-α) which is a transcriptional coactivator for mitochondria biogenesis and cellular energy metabolism, predominantly expressed in the brain has been observed to be diminished in postmortem analysis of AD patients (Bogacka et al. 2005). Reports reveal that PPARγ activation by agonists stimulated brain mitochondria biogenesis and or respiration and/or in muscle cells and prevented mitochondria dimensions decrease possibly by enhanced PGC1-α distribution in cultured hippocampal neurons or muscle cells (Hondares et al. 2006; Zolezzi et al. 2013; Puigserver et al. 1998). In vivo mouse model of AD in APP23 shows that PGC1-α gene delivery enhances spatial and recollection memory as well as a decrease in Aβ levels through a reduction in BACE 1 activity (Katsouri et al. 2016). However, it should be noted that excess PGC1-α may have deleterious effects through mitochondria proliferation producing heart, brain, and muscle toxicity leading to impairment of cognitive function (Clark et al. 2012; Ciron et al. 2012).

14.11 PPARγ, Insulin Signaling, and Metabolism in Alzheimer's Disease

Several studies suggest that insulin signaling and metabolism are associated with AD (Razay et al. 2007; Craft et al. 1999) as insulin receptors are predominantly found in the olfactory bulb, hypothalamus, hippocampus, cerebral cortex, striatum, and cerebellum brain areas and affect the process of memory formation (Unger and Betz 1998; Anna et al. 2009).

PPARγ which are useful agents in type 2 diabetic patients are believed to mediate insulin sensitization through a series of complex processes and unclear mechanisms (Anna et al. 2009; Landreth et al. 2008). Insulin binds to its tyrosine kinase receptor, eliciting phosphorylation of insulin receptor substrate protein (IRS) as well as phosphatidylinositol-3-kinase activation leading to numerous biological effects. Other downstream kinases are also activated (Anna et al. 2009). The result is the promotion of glucose uptake, lipid metabolism, gene transcription modulation, and other effects (Sugden and Holness 2004). Insulin is necessary to sustain neuronal homeostasis and survival, encouraging memory and learning especially in the hippocampus (Bloemer et al. 2014) brain area. Post-mortem research in AD patients' brains has also revealed decreased amounts of insulin, insulin-like growth factor (IGF-1 and IGF-2), and insulin receptor substrate 1 (IRS-1). Insulin resistance induced by diet has also been observed to be linked with improved Aβ levels and plaque formation in the brain as well as inflammatory markers. Type II diabetes mellitus patients show that they are more prone to develop AD compared to other individuals, indicating a correlation between insulin resistance, hyperinsulinemia, and memory impairment and the risk for AD (Landreth et al. 2008).

These common pathological similarities between AD and diabetes mellitus lead to growing interest in using PPARγ agonists which are insulin-sensitizing agents such as rosiglitazone and pioglitazone as potential treatment for AD (Nicolakakis et al. 2008). For instance, in high-fat diet rats, rosiglitazone, a PPARγ ligand has been observed to improve neuronal insulin resistance, attenuate brain mitochondria dysfunction and oxidative stress.

14.12 PPARγ and Canonical Wnt/β-Catenin Signaling in Alzheimer's Disease

The Wnt/β-catenin pathway has a significant place in cell fate and embryonic development (Azhar and Kelley 2007; Clevers and Nusse 2012). It is also important in mediating functions in the CNS such as synaptic plasticity and neuroprotection. When Wnt is activated, Wnt ligands link with both Frizzled and low-density lipoprotein receptor-related protein 5/6 (LRP5/60). Disheveled binds to axin, which inhibits the glycogen synthase kinase-3beta (GSK-3β) phosphorylation of β-catenin. Then, β-catenin amasses in the cytosol, translocates to the nucleus and subsequently binds to lymphoid enhancer factor/T-cell co-transcription factors (LEF/TCF) leading to the transcription of Wnt-responsive genes including PPAR beta/delta, Axin-2, cyclin D1, CD44, c-myc, and COX-2 (Angers and Moon 2009; Alexandre and Yves 2016).

In AD, Wnt/β-catenin signaling is downregulated while PPARγ is upregulated. Dysfunctional Wnt signaling has been associated with AD (Inestrosa and Varela-Nallar 2014; Tapia-Rojas and Inestrosa 2018) resulting in Aβ deposition, tau hyperphosphorylation, and cognitive impairment. PPARγ activation in many tissues and pathological states promotes the repression of the Wnt/β-catenin pathway (Liu et al. 2006). Stimulation of Wnt signaling is able to defend against Aβ deposition and upgrade cognitive performance in AD patients as some recent studies show that PPARγ facilitated defense of hippocampal neurons against Aβ induced toxicity directly correlated with β-catenin levels, suppression of GSK 3β activity, and elevated levels of Wnt target genes.

14.13 PPARγ and Neurofibrillary Tangles

Studies recently reveal that PPARγ can affect tau pathology. Tau proteins are microtubule proteins responsible for mediating microtubule stabilization by interacting with tubulin. In AD, tau is hyperphosphorylated resulting in neurotoxicity (Sodhi et al. 2011). Kinases that modulate tau phosphorylation include cyclin-dependent kinases (CDK2 and CDK5), GSK-3β, mitogen-activated protein kinase (MAPK), extracellular signal-regulated protein kinase 1/2 (ERK1/2), c-Jun N-terminal kinase (JNK), Akt, protein kinase a (PKA) (d'abramo et al. 2006), and calcium-calmodulin protein kinase 2 (CaMKII), while PP1, PP2A, PP2B, and PP2C are kinases that promote tau dephosphorylation (d'abramo et al. 2006). Experimental studies performed in vitro in tau transfected cell model involving GSK 3β using PPARγ agonists-troglitazone and pioglitazone resulted in reduced hyperphosphorylation of tau at Ser 202, Ser 396, and Ser 404. The mechanism for this protection against tau hyperphosphorylation is still unclear.

14.14 Recent Developments and Future Perspectives

Although numerous clinical researches have examined the efficacy of PPARγ agonists in AD have shown some benefits, many other trials using pioglitazone or rosiglitazone (Geldmacher et al. 2011) have not shown consistent, statistically significant, and sufficient benefits to support their use in AD patients with or without comorbidities or other peculiarities (Cheng et al. 2015). Concerns on the safety of the utilization of these agents also persist as regards their adverse effects including edema and cardiomegaly (Miller et al. 2011). Therefore, the development of improved and specific selective PPARγ modulators or agonists with improved therapeutic benefit is necessary. This led to the advancement of selective PPARγ modulators. Selective PPARγ modulators which are partial agonists of PPARγ also called SPARMS or SPPARγMs by analogy to selective estrogen receptor modulators (SERMS) continue to attract attention (Yan et al. 2003).

SPPPARγMs selectively link to the LiBD of PPARγ through an activation function 2 motifs (AF2). SPPPARγMs, unlike traditional PPARγ full agonists—TZDs, only form unstable bonding with the Tyr473 site (Tyr473 site is involved in activating transcriptional coactivator binding pocket of the receptor LiBD) in the helix 12 of human PPPARγ receptor LiBD resulting in diminished receptor conformational stability. This results in altered transcriptional activity receptor coactivator interaction and diminished harmful effects on the heart and body as observed with full PPARγ ligands (Cheng et al. 2015). Therefore, SPPARγMs may favor the finding of new pharmacological therapy for AD.

There is an urgent need for the development of newer anti-AD drugs. In preclinical studies, ibuprofen was reported to drastically reduce inflammation and the number of amyloid plaques deposited and this response may be due to the ability of ibuprofen to activate PPARγ (Yan et al. 2003; Miller et al. 2011). When 6 months old Tg 2576 animals with amyloid plaques were treated with pioglitazone and ibuprofen for about 4 months unlike pioglitazone, ibuprofen reduced the plaque deposition as well as the number of activated microglia (Lim et al. 2000; Morihara et al. 2005). In another complementary study, NSAIDS were also found to play a role in improving memory and synaptic plasticity through cyclooxygenase (COX)-related actions. A novel 2-(benzylidene) hexanoic acid-containing molecule was synthesized by Flesch et al. (2015) and was shown to possess PPARγ agonistic action, γ-secretase modulatory effect and can inhibit 5-lipooxygenase, hence a potential therapy for AD (Firuzi et al. 2008). T3D-959, an orally active PPARγ agonist, was recently found to improve cognitive functions in AD experimental models (Tong et al. 2016; De la Monte et al. 2017).

In clinical studies, more than 105 agents have been developed for the treatment of AD. Many of these agents are in different stages. Up to 60% of the drugs in development for AD are disease-modifying therapies (DMT's), 14% are symptomatic cognition improvers while less than 13% address changes in behavior (Watson et al. 2004; Cummings et al. 2014, 2016, 2017, 2018). In clinical studies, although rosiglitazone exhibited anti-AD effect in mild to moderate AD patients, in a clinical trial, results were not as expected in patients carrying *ApoE4* allele but in patients with *ApoE4* allele, there were cognitive improvements (Cummings and Fox 2017). In a smaller study, insulin resistance was enhanced by rosiglitazone but exhibited no improvements in cognitive functions in patients with mild cognitive impairment and insulin resistance (Angelopoulou and Pipperi 2018). Unfortunately, the effects of the currently available therapies for AD, *N*-methyl-D-aspartate receptor antagonist and cholinesterase inhibitors, and several other combination therapies are not able to prevent disease progression or limited due to adverse effects. Research which is now focused on PPARγ agonist in the development of new AD therapies seem to be safer with fewer side effects and significant progress has emerged from its use as an antidiabetic therapy thus far (Kuusisto et al. 1997; Craft et al. 2000; Watson et al. 2004; Cheng et al. 2015; Alexandre and Yves 2016; Angelopoulou and Pipperi 2018; Khan et al. 2019).

14.15 Conclusion

AD represents the most highly neurodegenerative form of dementia. To date, research has been limited in ascertaining the pathological basis for the disease. This has also limited the discovery of potential novel novels to effectively treat AD. Hence it remains needful to search for newer molecules to effectively treat AD. PPARs, especially PPARγ, have been shown from research to be potential candidates as agonists of this receptor modulate gene transcription, induce Aβ clearance, inhibit neuroinflammation. Therefore, this has drawn attention to its potential role in AD. This chapter emphasizes the possible therapeutic role PPAR could play as a therapeutic target in the treatment of AD. Of particular interest is the functional role played by PPARγ in suppressing the inflammatory component of AD, thereby improving learning and memory. However, more molecular mechanistic studies are necessary for the exact action of PPARs in AD.

References

Aberle IH, Beil FU, Seedorf U (2006) Association of the T+294C polymorphism in PPAR delta with low HDL cholesterol and coronary heart disease risk in women. Int J Med Sci 3:108–111

Akiyama H, Barger S, Barnum S et al (2000) Inflammation and Alzheimer's disease. Neurobiol Aging 21(3):383–421

Alexandre V, Yves L (2016) Alzheimer disease: cross talk between the canonical Wnt/Beta-catenin pathway and PPARs alpha and gamma. Front Neurosci 10:459. https://doi.org/10.3389/fnins.2016.00459

Alzheimer's Association (2016) Alzheimer's disease facts and figures. Alzheimers Dement 12:459–509

Angelopoulou E, Pipperi C (2018) DPP-4 inhibitors: a promising therapeutic approach against Alzheimer's disease. Ann Transl Med 6(12):255

Angers S, Moon RT (2009) Proximal events in Wnt signal transduction. Nat Rev Mol Cell Biol 10: 468–477

Anna L, Luigi L, Sebastio P et al (2009) Cross talk between PPARγ and insulin signaling and modulation of insulin sensitivity. PPAR Res 2009:818945

Azarnoff DL, Svoboda DJ (1969) Microbodies in experimentally altered cells. VI. Thyroxine displacement from plasma proteins and clofibrate effect. Arch Int Pharmacodyn Ther 181: 386–393

Azhar S, Kelley G (2007) PPARα: its role in the human metabolic syndrome. Futur Lipidol 2(1): 31–53

Bajaj M, Suraamornkul S, Hardies LJ et al (2007) Effects of peroxisome proliferator-activated receptor (PPAR)-alpha and PPAR-gamma agonists on glucose and lipid metabolism in patients with type 2 diabetes mellitus. Diabetologia 50:1723–1731

Barage SH, Sonawane KD (2015) Amyloid cascade hypothesis: pathogenesis and therapeutic strategies in Alzheimer's disease. Neuropeptides 52:1–18

Barish GD, Evans RM (2004) PPARs and LXRs: atherosclerosis goes nuclear. Trends Endocrinol Metab 15(4):158–165

Barish GD, Narkar VA, Evans RM (2006) PPAR delta: a dagger in the heart of the metabolic syndrome. J Clin Inv 116(3):590–597

Barrera J, Subramanian S, Chiba-Falek O et al (2018) Probing the role of PPARγ in the regulation of late-onset Alzheimer's disease-associated genes. PLoS One 13(5):e0196943

Bernando A, Minghetti L (2006) PPAR-γ agonists as regulators of microglial activation and brain inflammation. Curr Pharm Design 12(1):93–109

Bloemer J, Bhattacharya S, Amin R et al (2014) Impaired insulin signaling and mechanisms of memory loss. Prog Mol Biol Transl Sci 121:413–449

Bogacka I, Xie H, Bray GA et al (2005) Pioglitazone induces mitochondrial biogenesis in human subcutaneous adipose tissue in vivo. Diabetes 54:1392–1399

Bookheimer SY, Strojwas MH, Cohen MS et al (2000) Patterns of brain activation in people at risk for Alzheimer's disease. N Engl J Med 343:450–456

Bouhlel MA, Derudas B, Rigamonti E et al (2007) PPARγ activation primes human monocytes into alternative M2 macrophages with anti-inflammatory properties. Cell Metab 6(2):137–143

Bugge A, Mandrup S (2010) Molecular mechanisms and genome-wide aspects of PPAR subtype specific transactivation. PPAR Res 2010:169506. 1–12

Cai Z, Liu N, Wang C et al (2016) Role of RAGE in Alzheimer's disease. Cell Mol Neurobiol 36(4):483–495

Calhoun ME, Burgermeister P, Phinney AL et al (1999) Neuronal overexpression of mutant amyloid precursor protein results in prominent deposition of cerebrovascular amyloid. Proc Natl Acad Sci USA 96(24):14088–14093

Calkin AC, Forbes JM, Smith CM et al (2005) Rosiglitazone attenuates atherosclerosis in a model of insulin insufficiency independent of its metabolic effects. Arterioscler Thromb Vasc Biol 25: 1903–1909

Camacho IE, Serneels L, Spittaels K et al (2004) Peroxisome-proliferator-activated receptor gamma induces a clearance mechanism for the amyloid-beta peptide. J Neurosci 24:10908–10917

Chen CH, Zhou W, Liu S et al (2012) Increased NF-kappaB signalling up-regulates BACE1 expression and its therapeutic potential in Alzheimer's disease. Int J Neuropsychopharmacol 15(1):77–90

Cheng H, Shang Y, Jiang L (2015) The peroxisome proliferators activated receptor-gamma agonists as therapeutics for the treatment of Alzheimer's disease and mild-to moderate Alzheimer's disease: a meta-analysis. Int J Neurosci 126(4):299–307

Choi YJ, Roberts BK, Wang X et al (2012) Effects of the PPAR-d agonist MBX-8025 on atherogenic dyslipidemia. Atherosclerosis 220:470–476

Choi JY, Kim JY, Kim JY et al (2017) M2 phenotype microglia-derived cytokine stimulates proliferation and neuronal differentiation of endogenous stem cells in ischemic brain. Exp Neurobiol 26(1):33

Ciron C, Lengacher S, Dusonchet J et al (2012) Sustained expression of PGC-1α in the rat nigrostriatal system selectively impairs dopaminergic function. Hum Mol Gene 21(8): 1861–1876

Clark J, Silvaggi JM, Kiselak T et al (2012) Pgc-1α overexpression downregulates Pitx3 and increases susceptibility to MPTP toxicity associated with decreased BDNF. PLoS One 7(11): e48925

Clevers H, Nusse R (2012) Wnt/beta-catenin signaling and disease. Cell 149:1192–1205. https://doi.org/10.1016/j.cell.2012.05.012

Combs CK, Johnson DE, Karlo JC et al (2000) Inflammatory mechanisms in Alzheimer's disease: inhibition of beta-amyloid-stimulated proinflammatory responses and neurotoxicity by PPAR gamma agonists. J Neurosci 20:558–567

Corbett GT, Gonzalez FJ, Pahan K (2015) Activation of peroxisome proliferator-activated receptor alpha stimulates ADAM10-mediated proteolysis of APP. Proc Natl Acad Sci U S A 112:8445–8450. https://doi.org/10.1073/pnas.1504890112

Cottet J, Redel J, Krumm-Heller C et al (1953) Hypocholesterolemic property of sodium phenylethylacetate (22 TH) in the rat. Bull Acad Natl Med 137:441–442

Craft S, Asthana S, Schellenberg G et al (1999) Insulin metabolism in Alzheimer's disease differs according to apolipoprotein E genotype and gender. Neuroendocrinology 70:146–152

Craft S, Asthana S, Schellenberg G et al (2000) Insulin effects on glucose metabolism, memory, and plasma amyloid precursor protein in Alzheimer's disease differ according to apolipoprotein-E genotype. Ann N Y Acad Sci 903:222–228

Cummings JL, Fox N (2017) Defining disease modification for Alzheimer's disease clinical trials. J Prev Alzheimers Dis 4:109–115

Cummings JL, Morstorf T, Zhong K (2014) Alzheimer's disease drug-development pipeline: few candidates, frequent failures. Alzheimers Res Ther 6:37–43

Cummings J, Morstorf T, Lee G (2016) Alzheimer's drug development pipeline: 2016. Alzheimers Dement 2:222–232

Cummings J, Lee G, Mortsdorf T et al (2017) Alzheimer's disease drug development pipeline: 2017. Alzheimers Dement 3:367–384

Cummings J, Ritter A, Zhong K (2018) Clinical trials for disease-modifying therapies in Alzheimer's disease: a primer, lessons learned and a blueprint for the future. J Alzheimers Dis 64(Suppl 1):S3–S22

d'abramo C, Ricciarelli R, Pronzato MA et al (2006) Troglitazone, a peroxisome proliferator-activated receptor gamma agonist, decreases tau phosphorylation in CHOtau4R cells. J Neurochem 98(4):1068–1077

De la Monte SM, Tong M, Schiano I et al (2017) Improved brain insulin/IGF signaling and reduced neuroinflammation with T3D-959 in an experimental model of sporadic Alzheimer's disease. J Alzheimers Dis 55(2):849–864

De'la Monte SM, Wands JR (2006) Molecular indices of oxidative stress and mitochondrial dysfunction occur early and often progress with severity of Alzheimer's disease. J Alzheimers Dis 9:167–181

Desvergne B, Wahli W (1999) Peroxisome proliferator activated receptors: nuclear control of metabolism. Endocr Rev 20(5):649–688

Diradourian C, Girard J, Pégorier JP (2005) Phosphorylation of PPARs: from molecular characterization to physiological relevance. Biochimie 87(1):33–38

Dong C, Zhou H, Shen C et al (2015) Role of peroxisome proliferator-activated receptors gene polymorphisms in type 2 diabetes and metabolic syndrome. World J Diabetes 6(4):654–661

Dreyer C, Krey G, Keller H et al (1992) Control of the peroxisomal beta-oxidation pathway by a novel family of nuclear hormone receptors. Cell 68:879–887

Elrod HA, Sun SY (2008) PPARγ and apoptosis in cancer. PPAR Res 2008:704165

Fan H, Pengfei X, Yonggong Z (2018) The opportunities and challenges of peroxisome proliferator-activated receptors ligands in clinical drug discovery and development. Int J Mol Sci 19:2189. https://doi.org/10.3390/ijms19082189

Fattahi MJ, Mirshafiey A (2014) Positive and negative effects of prostaglandins in Alzheimer's disease. Psychiatry Clin Neurosci 68(1):50–60

Feige JN, Gelman L, Tudor C et al (2005) Fluorescence imaging reveals the nuclear behavior of peroxisome proliferator-activated receptor/retinoid X receptor heterodimers in the absence and presence of ligand. J Biol Chem 280:17880–17890

Ferrer I (2012) Defining Alzheimer as a common age-related neurodegenerative process not inevitably leading to dementia. Prog Neurobiol 97:38–51

Firuzi O, Zhuo J, Chinnici CM et al (2008) 5-Lipoxygenase gene disruption reduces amyloid-β pathology in a mouse model of Alzheimer's disease. FASEB J 22:1169

Flesch D, Ness J, Lamers C et al (2015) SAR-studies of γ-secretase modulators with PPAR-γ-agonistic and 5-lipoxygenase-inhibitory activity for Alzheimer's disease. Bioorg Med Chem Lett 25(4):841–846

Gauthier A, Vassiliou G, Benoist F et al (2003) Adipocyte low density lipoprotein receptor-related protein gene expression and function is regulated by peroxisome proliferator-activated receptor γ. J Biol Chem 278(14):11945–11953

Geldmacher DS, Fritsch T, McClendon MJ et al (2011) A randomized pilot clinical trial of the safety of pioglitazone in treatment of patients with Alzheimer disease. JAMA Neurol 68(1): 45–50

Gervois P, Torra IP, Chinetti G et al (1999) A truncated human peroxisome proliferator-activated receptor alpha splice variant with dominant negative activity. Mol Endocrinol 13:1535–1549

Gottschalk WK, Mihovilovic M (2016) The role of upregulated APOE in Alzheimer's disease etiology. J Alzheimers Dis Parkinson 6(1):209

Gray JP, Davis JW, Gopinathan L et al (2006) The ribosomal protein rpL11 associates with and inhibits the transcriptional activity of peroxisome proliferator-activated receptor-α. Toxicol Sci 89:535–546

Hauser S, Adelmant G, Sarraf G et al (2000) Degradation of the peroxisome proliferator-activated receptor gamma is linked to ligand dependent activation. J Biol Chem 275:18527–18533

Heneka MT, Feinstein DL, Galea E et al (1999) Peroxisome proliferator-activated receptor gamma agonists protect cerebellar granule cells from cytokine induced apoptotic cell death by inhibition of inducible nitric oxide synthase. J Neuroimmunol 100:156–168

Heneka MT, Gavrilyuk V, Landreth GE et al (2003) Noradrenergic depletion increases inflammatory responses in brain: effects on kappaB and HSP70 expression. J Neurochem 85:387–398

Heneka MT, Elisabet R, Michael H et al (2011) Impact and therapeutic potential of PPARs in Alzheimer's disease. Curr Neuropharmacol 9:643–650

Hicks DA, Makova NZ, Gough M et al (2013) The amyloid precursor protein represses expression of acetylcholinesterase in neuronal cell lines. J Biol Chem 288:26039–26051. https://doi.org/10.1074/jbc.M113.461269

Hondares E, Mora O, Yubero P et al (2006) Thiazolidinediones and rexinoids induce peroxisome proliferator-activated receptor-coactivator (PGC)-1alpha gene transcription: an autoregulatory loop controls PGC-1alpha expression in adipocytes via peroxisome proliferator-activated receptor-gamma coactivation. Endocrinology 147:2829–2838

Huang Q, Alvares K, Chu R et al (1994) Association of peroxisome proliferator-activated receptor and Hsp72. J Biol Chem 269:8493–8497

Inestrosa NC, Varela-Nallar L (2014) Wnt signaling in the nervous system and in Alzheimer's disease. J Mol Cell Biol 6(1):64–74

Issemann I, Green S (1990) Activation of a member of the steroid hormone receptor superfamily by peroxisome proliferators. Nature 347:645–650

Jeffrey LC, Gary T, Ballard C (2019) Treatment combinations for Alzheimer's disease: current and future pharmacotherapy options. J Alzheimers Dis 67:779–794

Jiang C, Ting AT, Seed B (1998) PPAR-gamma agonists inhibit production of monocyte inflammatory cytokines. Nature 391:82–86

Jiang Q, Lee CYD, Mandrekar S et al (2008) ApoE promotes the proteolytic degradation of Aβ. Neuron 58(5):681–693

Jihan Y, Mostafa B (2015) Peroxisome proliferator-activated receptors: features, functions, and future. Nuclear Recep Res 2:101188. https://doi.org/10.11131/2015/101188

Katsouri L, Lim YM, Blondrath K et al (2016) PPARγ-coactivator-1α gene transfer reduces neuronal loss and amyloid-β generation by reducing β-secretase in an Alzheimer's disease model. Proc Natl Acad Sci U S A 113(43):12292–12297

Keller JM, Collet P, Bianchi A et al (2000) Implications of peroxisome proliferator-activated receptors (PPARS) in development, cell life status and disease. Int J Dev Biol 44(5):429–442

Khan MA, Alam Q, Haque A (2019) Current progress on peroxisome proliferator-activated receptor gamma agonist as an emerging therapeutic approach for the treatment of Alzheimer's disease: an update. Curr Neuropharmacol 17(3):232–246. https://doi.org/10.2174/1570159x16666180828100002

Kielian T, Drew PD (2003) Effects of peroxisome proliferator activated receptor-gamma agonists on central nervous system inflammation. J Neurosci Res 71:315–325

Kilgore KS, Billin AN (2008) PPARβ/δ ligands as modulators of the inflammatory response. Curr Opin Investig Drugs 9(5):463–469

Kliewer SA, Xu HE, Lambert MH et al (2001) Peroxisome proliferator activated receptors: from genes to physiology. Recent Prog Horm Res 56:239–265

Kuusisto J, Koivisto K, Mykkanen L et al (1997) Association between features of the insulin resistance syndrome and Alzheimer's disease independently of apolipoprotein E4 phenotype: cross sectional population-based study. BMJ 315:1045–1049
Landreth GE, Heneka MT (2001) Anti-inflammatory actions of peroxisome proliferator-activated receptor gamma agonists in Alzheimer's disease. Neurobiol Aging 22:937–944
Landreth G, Qingguang J, Shweta M et al (2008) PPAR agonists as therapeutics for the treatment of Alzheimer's disease. Neurotherapeut J Am Soc Exp Neur 5:481–489
Lehmann JM, Lenhard JM, Oliver BB et al (1997) Peroxisome proliferator-activated receptors alpha and gamma are activated by indomethacin and other non-steroidal anti-inflammatory drugs. J Biol Chem 272:3406–3410
Lenglet S, Montecucco F, Mach F (2015) Role of matrix metalloproteinases in animal models of ischemic stroke. Curr Vasc Pharmacol 13(2):161–166
Li D, Kang Q, Wang DM (2007) Constitutive coactivator of peroxisome proliferator-activated receptor (PPARγ), a novel coactivator of PPARγ that promotes adipogenesis. Mol Endocrinol 21:2320–2333
Liao YF, Wang BJ, Cheng HT et al (2004) Tumor necrosis factor-alpha, interleukin-1beta, and interferon-gamma stimulate gamma-secretase-mediated cleavage of amyloid precursor protein through a JNK-dependent MAPK pathway. J Biol Chem 279(47):49523–44932
Lim GP, Yang F, Chu T et al (2000) Ibuprofen suppresses plaque pathology and inflammation in a mouse model for Alzheimer's disease. J Neurosci 20(15):5709–5714
Lin JD, Handschin C, Spiegelman BM (2005) Metabolic control through the PGC-1 family of transcription coactivators. Cell Metab 1(6):361–370
Liu J, Wang H, Zuo Y et al (2006) Functional interaction between peroxisome proliferator-activated receptor gamma and beta-catenin. Mol Cell Biol 26:5827–5837. https://doi.org/10.1128/MCB.00441-06
Llaverias G, Rebollo A, Pou J et al (2006) Effects of rosiglitazone and atorvastatin on the expression of genes that control cholesterol homeostasis in differentiating monocytes. Biochem Pharmacol 71:605–614
Luna-Medina R, Cortes-Canteli M, Alonso M et al (2005) Regulation of inflammatory response in neural cells in vitro by thiadiazolidinones derivatives through peroxisome proliferator-activated receptor gamma activation. J Biol Chem 280:21453–21462
Mandrekar-Colucci S, Landreth GE (2011) Nuclear receptors as therapeutic targets for Alzheimer's disease. Expert Opin Ther Targets 15(9):1085–1097
Manoj G, Priyanka DP, Jenna B et al (2018) Signaling mechanisms of selective PPARγ modulators in Alzheimer's disease. PPAR Res 2018:2010675
Maria RB, Aniello I, Luca S et al (2016) Targeting neuroinflammation in Alzheimer's disease. J Inflamm Res 9:199–208
Markus PK, Michael H (2008) PPARs in Alzheimer's disease. PPAR Res 2008:403896. https://doi.org/10.1155/2008/403896
Mathivat A, Cottet J (1953) Clinical trials on the hypocholesterolemia-producing effect of 2-phenylbutyric acid. Bull Mem Soc Med Hop Paris 69:1030–1048
Medina-Gomez G, Gray SL, Yetukuri L et al (2007) PPAR gamma prevents lipotoxicity by controlling adipose tissue expandability and peripheral lipid metabolism. PLoS Genet 3(4):e64
Meirhaeghe A, Fajas L, Gouilleux F (2003) A functional polymorphism in a STAT5B site of the human PPAR gamma 3 gene promoter affects height and lipid metabolism in a French population. Arterioscler Thromb Vasc Biol 23:289–294
Michelucci A, Heurtaux T, Grandbarbe L et al (2009) Characterization of the microglial phenotype under specific pro-inflammatory and anti-inflammatory conditions: effects of oligomeric and fibrillar amyloid-β. J Neuroimmunol 210(12):3–12
Miller AR, Etgen GJ (2003) Novel peroxisome proliferator activated receptor ligands for type 2 diabetes and the metabolic syndrome. Expert Opin Investig Drugs 12(9):1489–1500
Miller BW, Willett KC, Desilets AR (2011) Rosiglitazone and pioglitazone for the treatment of Alzheimer's disease. Ann Pharmacother 45(11):1416–1424

Morihara T, Teter B, Yang F et al (2005) Ibuprofen suppresses interleukin-1beta induction of pro-amyloidogenic alpha1-antichymotrpsin to ameliorate beta-amyloid pathology in Alzheimer's models. Neuropsychopharmacology 30(6):1111–1120

Naito H, Yamanoshita O, Kamijima M et al (2006) Association of V227A PPAR alpha polymorphism with altered serum biochemistry and alcohol drinking in Japanese men. Pharmacogenet Genomics 16:569–577

Nalivaeva NN, Turner AJ (2013) The amyloid precursor protein: a biochemical being main brain development, function and disease. FEBS Lett 587:2046–2054. https://doi.org/10.1016/j.febslet.2013.05.010

Nicolakakis N, Aboulkassim T, Ongali B et al (2008) Complete rescue of cerebrovascular function in aged Alzheimer's disease transgenic mice by antioxidants and pioglitazone, a peroxisome proliferator-activated receptor γ agonist. J Neurosci 28(37):9287–9296

Obulesu M, Lakshmi MJ (2014) Apoptosis in Alzheimer's disease: an understanding of the physiology, pathology and therapeutic avenues. Neurochem Res 39(12):2301–2312

Pascual G, Fong AL, Ogawa S et al (2005) A SUMOylation-dependent pathway mediates transrepression of inflammatory response genes by PPAR-gamma. Nature 437:759–763

Payton A, Sindrewicz P, Pessoa V et al (2016) TOMM40 poly-T variant modulates gene expression and is associated with vocabulary ability and decline in nonpathologic aging. Neurobiol Aging 39:217.e1–217.e7

Puigserver P, Wu Z, Park CW et al (1998) A cold-inducible coactivator of nuclear receptors linked to adaptive thermogenesis. Cell 92:829–839

Qiang L, Wang L, Kon N et al (2012) Brown remodeling of white adipose tissue by SirT1-dependent deacetylation of PPARγ. Cell 150(3):620–632

Querfurth HW, LaFerla FM (2010) Alzheimer's disease. N Engl J Med 362(4):329–344

Rajnish KC, Flint BM (2008) PPARs: a therapeutic target in Parkinson's disease. J Neurochem 106:506–518

Razay G, Vreugdenhil A, Wilcock G (2007) The metabolic syndrome and Alzheimer disease. Arch Neurol 64:93–96

Reitz C, Brayne C, Mayeux R (2011) Epidemiology of Alzheimer disease. Nat Rev Neurol 7:137–152

Risner ME, Saunders AM, Altman JF et al (2006) Efficacy of rosiglitazone in a genetically defined population with mild-to-moderate Alzheimer's disease. Pharmacogenom J 6:246–254

Rondón-Ortiz AN, Lino CL, Martínez-Málaga J et al (2017) High concentrations of rosiglitazone reduce mRNA and protein levels of LRP1 in HepG2 cells. Front Pharmacol 8:772

Roses AD, Saunders AM, Huang Y et al (2007) Complex disease associated pharmacogenetics: drug efficacy, drug safety, and confirmation of a pathogenetic hypothesis (Alzheimer's disease). Pharmacogenom J 7:10–28

Roy E, Ralph AD, Muhammad A (2013) In vivo actions of peroxisome proliferator–activated receptors. Diab Ther 36(2):S162–S174

Sastre M, Dewachter I, Landreth GE et al (2003) Nonsteroidal anti-inflammatory drugs and peroxisome proliferator-activated receptor-gamma agonists modulate immunostimulated processing of amyloid precursor protein through regulation of beta-secretase. J Neurosci 23:9796–9804

Sato T, Hanyu H, Hirao K et al (2009) Efficacy of PPAR-gamma agonist pioglitazone in mild Alzheimer disease. Neurobiol Aging 32(9):1626–1633

Schmidt A, Endo N, Rutledge SJ et al (1992) Identification of a new member of the steroid hormone receptor superfamily that is activated by a peroxisome proliferator and fatty acids. Mol Endocrinol 6:1634–1641

Skogsberg J, McMahon AD, Karpe F et al (2003) Peroxisome proliferator activated receptor delta genotype in relation to cardiovascular risk factors and risk of coronary heart disease in hypercholesterolaemic men. J Intern Med 254:597–604

Sodhi RK, Singh N, Jaggi AS (2011) Neuroprotective mechanisms of peroxisome proliferator-activated receptor agonists in Alzheimer's disease. Naunyn Schmiedeberg's Arch Pharmacol 384:115–124

Spiegelman BM (1998) PPAR-gamma: adipogenic regulator and thiazolidinedione receptor. Diabetes 47:507–514

Subramanian S, Gottschalk WK, Kim SY (2017) The effects of PPARγ on the regulation of the TOMM40-APOE-C1 genes cluster. Biochim Biophys Acta (BBA) - Mol Basis Dis 1863(3): 810–816

Sugden M, Holness M (2004) Potential role of peroxisome proliferator-activated receptor-α in the modulation of glucose-stimulated insulin secretion. Diabetes 53(suppl 1):S71–S81

Svoboda D, Azarnoff D, Reddy J (1969) Microbodies in experimentally altered cells. II. The relationship of microbody proliferation to endocrine glands. J Cell Biol 40:734–746

Tanzi RE, Bertram L (2005) Twenty years of the Alzheimer's disease amyloid hypothesis: a genetic perspective. Cell 120(4):545–555

Tapia-Rojas C, Inestrosa N (2018) Loss of canonical Wnt signaling is involved in the pathogenesis of Alzheimer's disease. Neural Regen Res 13(10):1705

Thorp JM, Waring WS (1962) Modification of metabolism and distribution of lipids by ethyl chlorophenoxyisobutyrate. Nature 194:948–949

Tomaru T, Satoh T, Yoshino S et al (2006) Isolation and characterization of a transcriptional cofactor and its novel isoform that bind the deoxyribonucleic acid-binding domain of peroxisome proliferator-activated receptor-gamma. Endocrinology 147:377–388

Tong M, Dominguez C, Didsbury J et al (2016) Targeting Alzheimer's disease neuro-metabolic dysfunction with a small molecule nuclear receptor agonist (T3D-959) reverses disease pathologies. J Alzheimers Dis Parkinson 6(3):238

Tontonoz P, Hu E, Graves RA et al (1994) PPAR gamma 2: tissue-specific regulator of an adipocyte enhancer. Genes Dev 8:1224–1234

Toyohiko A (2007) Current status of carcinogenicity assessment of peroxisome proliferator-activated receptor agonists by the US FDA and a mode-of-action approach to the carcinogenic potential. J Toxicol Pathol 20(4):197–202

Unger JW, Betz M (1998) Insulin receptors and signal transduction proteins in the hypothalamo-hypophyseal system: a review on morphological findings and functional implications. Histol Histopathol J13(4):1215–1224

Usuda D, Kanda T (2014) Peroxisome proliferator-activated receptors for hypertension. World J Cardiol 6(8):744–754

Verkhratsky A, Nedergaard M, Hertz L (2015) Why are astrocytes important? Neurochem Res 40(2):389–401

Vogl T, Gharibyan AL, Morozova-Roche LA (2012) Proinflammatory S100A8 and S100A9 proteins: self-assembly into multifunctional native and amyloid complexes. Int J Mol Sci 13(3):2893–2917

Wahrle SE, Jiang H, Parsadanian M et al (2008) Overexpression of ABCA1 reduces amyloid deposition in the PDAPP mouse model of Alzheimer disease. J Clin Invest 118:671–682

Wang-Soo L, Jaetaek K (2015) Peroxisome proliferator-activated receptors and the heart: lessons from the past and future directions. PPAR Res 2015:271983

Watson JL, Ryan L, Silverberg N et al (2004) Obstacles and opportunities in Alzheimer's clinical trial recruitment. Health Aff 33(4):574–579

Westin K, Buchhave P, Nielsen H et al (2012) CCL2 is associated with a faster rate of cognitive decline during early stages of Alzheimer's disease. PLoS One 7:e30525

Willson TM, Lambert MH, Kliewer SA (2001) Peroxisome proliferator-activated receptor γ and metabolic disease. Annu Rev Biochem 70:341–367

Yan Q, Zhang J, Liu H et al (2003) Anti-inflammatory drug therapy alters beta-amyloid processing and deposition in an animal model of Alzheimer's disease. J Neurosci 23(20):7504–7509

Yu Y, Ye RD (2015) Microglial Aβ receptors in Alzheimer's disease. Cell Mol Neurobiol 35(1): 71–83

Yu S, Liu YP, Liu YH et al (2016) Diagnostic utility of VEGF and soluble CD40L levels in serum of Alzheimer's patients. Yu Clin Chim Acta 453:154–159

Zhang H, Ma Q, Zhang YW et al (2012) Proteolytic processing of Alzheimer's beta-amyloid precursor protein. J Neurochem 120:9–21. https://doi.org/10.1111/j.1471-4159.2011.07519.x

Zoete V, Grosdidier A, Michielin O (2007) Peroxisome proliferator-activated receptor structures: ligand specificity, molecular switch and interactions with regulators. Biochim Biophys Acta 1771:915–925

Zolezzi JM, Silva-Alvarez C, Ordenes D et al (2013) Peroxisome proliferator-activated receptor (PPAR) γ and PPARα agonists modulate mitochondrial fusion-fission dynamics: relevance to reactive oxygen species (ROS)-related neurodegenerative disorders? PLoS One 8(5):e64019

Exploring the Role of Statins in Reversing the Cognitive and Neurovascular Dysfunctions in Dementia

15

Manisha Singh, Vinayak Agarwal, Pranav Pancham, Shriya Agarwal, Siddhi Bhardwaj, and Shalini Mani

Abstract

Statins are an important class of drugs to treat dyslipidemia and have a lot of pleiotropic effects that aid in improving vascular flow, reducing inflammation by decreasing the production of reactive oxygen species (ROS), decreasing the risk of dementia, and improving cognitive functions too. Their extensive utilization on the global scale has made them the most prescribed drugs for preventing coronary heart diseases (CHD). They inhibit the rate-limiting enzyme, β-hydroxy-β-methylglutaryl-CoA reductase (HMGR) of mevalonate pathway that plays an important role in multiple cellular processes by forming various sterol isoprenoids. Further, statins were reported to inhibit the prevalence of Alzheimer's disease (AD) by inhibiting α- and β-secretases. Therefore, this chapter highlights the recent insights about the probable role of statins in reversing or limiting the cognitive decline in dementia and the possible mechanisms related to this.

Keywords

Coronary heart diseases · Alzheimer's disease · β-hydroxy β-methylglutaryl-CoA · HMG-CoA · Cerebral amyloid angiopathy · Neurovascular dysregulation

15.1 Introduction

The occurrence and prevalence of dementia are escalating with around 50 million existing cases and ten million add every year (Villemagne et al. 2013; Shaw et al. 2018). The exponential rise in dementia cases is more reported in developing nations

M. Singh (✉) · V. Agarwal · P. Pancham · S. Agarwal · S. Bhardwaj · S. Mani
Department of Biotechnology, Jaypee Institute of Information Technology, Noida, India

G. M. Ashraf, M. S. Uddin (eds.), *Current Thoughts on Dementia*,
https://doi.org/10.1007/978-981-16-7606-2_15

with 58% of dementia-affected population residing in low- and middle-income countries (Rajkumar et al. 2020). So, understanding the pathological barriers and therapeutic limitations to develop an efficient treatment in this regard, thereby becoming of utmost importance in the current scenario (James et al. 2012). As per the most recently published data by the Alzheimer's Association, Alzheimer's disease (AD) is the most common cause of dementia, contributing 60–70% of dementia cases (Jellinger 2007). There are no efficient preventative as well as treatment options available according to the date and the frequency to acquire dementia doubles every 5 years in people over 60 years and 40% of people over 85 years and older having AD (Albert et al. 2013). It is anticipated that the occurrence of AD will increase three-folds higher by 2050 and it would approximate cost 100 billion dollars to US economy annually to care for AD patients – both direct and indirect cost (Lewczuk et al. 2018; Zalar et al. 2018).

The pathological features that characterize AD are the deposition of 39–42 amino acid peptide called Aβ and the presence of neurofibrillary tangles (NFTs) in cerebral cortex and hippocampus regions of brain (Zalar et al. 2018; Reitz et al. 2013). Also, deposition of beta amyloid (Aβ) in the form of senile plaques (SPs) and intracellular NFTs in soluble deposits of hyperphosphorylated tau protein is another cause (James et al. 2012; Baumgart et al. 2015). Further, the endoproteolytic cleavage of amyloid precursor protein (APP), a transmembrane protein by β- and γ-secretases results in the formation of neurotoxic Aβ peptides and α-secretase, moreover a third enzyme cleaves APP and prevents the formation of Aβ (Vickrey et al. 2006). It is a well-known fact that statins (HMG-CoA [β-hydroxy-β-methylglutaryl-CoA]) have vast multiple effects on balancing many physiological processes in humans like accelerating vascular flow, repressing inflammation and restricting the ROS (Zaleta et al. 2012; Fitzpatrick et al. 2009). So based on these key features, the role of statins in limiting dementia symptoms was further explored. Almost all the cholesterols are synthesized in the brain and it escapes the brain through the blood-brain barrier (BBB) in the form of apolipoprotein (ApoE) or by converting to a more polar compound (24 (S)-hydroxycholesterol), that increases in individuals with AD (Reitz et al. 2013; Choi et al. 2018). Also, the role of cholesterol in modulating the processing of APP to Aβ was explored and it was observed that patients taking cholesterol-lowering statins were at lower risk of acquiring AD, whereas higher cholesterol levels implicated the higher risk of AD (Livingston et al. 2005; Ogino et al. 2019).

The positive impact of statins on reducing the risk of acquiring dementia or reversing the AD-related pathological events is still not very clearly explored and has remained more as a debatable issue. The chapter highlights the synergistic interaction of vascular and neurodegenerative pathologies which is initiated by cognitive decline. Further, the role of statins in reducing vascular dementia and the mechanism for regulating tau and Aβ by statins has also been discussed and the chapter ends with the possible adversities associated with statins and it's future perspectives.

15.2 Synergistic Interaction of Vascular and Neurodegenerative Pathologies

One of the distinctive attributes associated with neurodegenerative disorders (NDDs)-related pathological events is a significant decline in cognitive ability and several correlated competencies in the patients (Zhao et al. 2016). Recent studies have revealed that malfunctioning in neurovascular coupling could also potentially aid in developing the risk of dementia due to impairing cerebral blood flow in response to conventional neuronal functions only (Soni et al. 2019; Griffith et al. 2007). Prevalence and co-existence of cerebral amyloid angiopathy (CAA) have been primarily exhibited in AD-affected subjects and found to additionally degrade the cerebral vascular flow (Zhao et al. 2018; Yang et al. 2020). Thus, their involvement evades development as well as the progression of neurodegenerative pathologies besides eliciting the synergistic effects with over-expression of tumor necrosis factor-alpha (in microglia) and amyloid-β protein (Iadecola and Gottesman 2019). These cascades of synergistic events lead to tremendous neuronal damage in the cortical region, eventually resulting in dementia (Michalicova et al. 2017). Subsequently, many clinical evidence have indicated the role of statin in preventing CVA and dementia both together due to its progressive role in lowering lipid content and regulating the isoprenoid pathway (Park et al. 2020). Setting neurovascular dysregulation as a reference, several studies observed that restricted blood flow causing ischemia is a chief architect in causing further ailments including neurological defects along with several tissue injuries due to hypoxic environment in enclosed brain anatomy (Santos et al. 2017).

Furthermore, the general homeostasis mechanics states that the supply of oxygen is directly proportional to the uniform blood flow across capillaries and any morphological or functional degeneration of capillaries will severely influence the oxygenation process across the organ system (Spires-Jones et al. 2017; Hachinski et al. 2019). Thus, a shunted blood flow in cortical regions results in hypoxic tissue and cognitive damage (see Fig. 15.1). Lately, it has been demonstrated that there is a disruption in the micro-vascular blood flow in AD-affected subjects that straight away corresponds to the degree of accompanying cognitive impairment in the diseased patients (Power et al. 2018). Besides this correlation, elevating neurovascular unit (NVU) dysregulation is observed as a decisive characteristic condition common in both dementia and stroke-affected subjects, also simultaneous detrition of BBB is distinctively marked as an early condition in cognition impairment in AD (see Fig. 15.1). Cerebral small vessel diseases (CSVD), BBB breakdown, and CAA along neural inflammation are some of the characteristics of neurodegenerative pathologies that can be specifically selected as a novel therapeutic target for preventing synergistic expression of vascular and neurodegenerative pathologies (Yan et al. 2020).

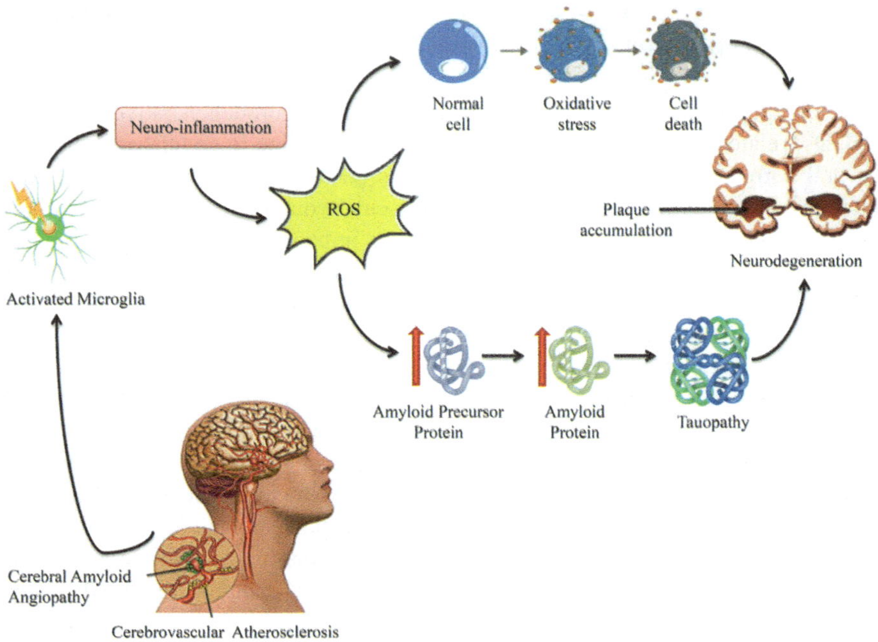

Fig. 15.1 Therapeutic expression of statin in suppressing synergistic effect of vascular and neurodegenerative pathologies in affected subjects

15.3 Neurovascular Dysregulation Initiated Cognitive Impairment

Conditions hypothesizing the relationship of cerebral vasculature in the etiology of neurodegeneration have cemented its place in most of the associated theories of CAA (Liao and Laufs 2005). Moreover, several links between cognition and aberrant changes in cerebral microvasculature that cognates with glucose-oxygen transport are gaining momentum in research findings (Bankstahl et al. 2018). Further, it has been observed that the central nervous system (CNS) lacks energy reserves and is heavily dependent upon a vital continuous supply of oxygen & glucose via cerebral blood flow and thus, NVUs play a pivotal role in facilitating this channel and evading potential cognitive impairments (Yarchoan et al. 2012). Any minutest discrepancy in supply perfusion even of 60–90 seconds could trigger an ischemic shock environment leading to cognitive decline, e.g., malfunctioning or deficiencies in glucose transporter 3 (GLUT3) proteins (Toth et al. 2017).

Also, deprivation in constant oxygen and glucose supply lasting beyond three hours could result in the release of lactic acid in response to shifting of anaerobic metabolism pathways, additionally hindering the ionic strength within the neurons (Levit et al. 2020). These events occur due to the inability to generate ATPs, which initiates the cascade of further failure in downstream energy-dependent processes leading to activation of apoptotic pathways (Tarantini et al. 2017; Costea et al. 2019). NVU comprises vasculature cells (pericytes, endothelium, and smooth muscles cells) that cater to the demand and supply of oxygen & glucose to the brain and are also responsible for molecular exchange between BBB and maintain a clearance of metabolic by-products, trafficking immune cells to support brain cells (Caruso et al. 2019; Yu et al. 2020). Therefore, any slightest anomaly in NVU would compromise the cerebral blood flow rate and eventually adds to the development of dementia and ultimately alters the homeostasis in CNS microenvironment's causing cognition impairment (Kisler et al. 2017; Sweeney et al. 2018).

15.4 Effect of Pharmacodynamic Parameters of Statins in Central Nervous System

Almost all the statins have decent absorbability with attaining the T_{max} in 4–5 hours and the various categories of statins are lovastatin, atorvastatin, fluvastatin, pravastatin, simvastatin, and rosuvastatin (Kokkinos et al. 2013; Geybels et al. 2013; Pocobelli et al. 2008). They are being further categorized as hydrophilic (rosuvastatin and pravastatin) and lipophilic (lovastatin, atorvastatin, and simvastatin) in solubility nature; however, fluvastatin was found to be comparatively lesser lipophilic in nature (McFarland et al. 2014; Shitara and Sugiyama 2006). Owing to their enormous pleiotropic effects, their permeability through BBB to brain is also commendable specifically in case of simvastatin and lovastatin (Alsheikh-Ali et al. 2007). Therefore, their outreach to the neuronal cells (neurons, glial cells) is much higher and so is their neuroprotective effects attained by reducing Aβ deposition (Butterfield et al. 2011; Fassbender et al. 2001; Simons et al. 2002), though the benefits of this property is still a debatable issue.

Among many hypotheses, there was evidence shown by Bettermann et al. (Bettermann et al. 2012) that propensity and probability of lipophilic statins to control or reverse the pathological dimensions of dementia are much appreciable than the hydrophilic ones. On the contrary, some reports had shown higher rates of cognitive dysfunction associated with lipophilic statins as compared to the other classes (Schultz et al. 2018). So, to date the exact mechanics related to statins is still under exploration and would need evaluation at various levels (Rockwood et al. 2002).

15.5 Statin and Its Influence in Modulating Neurodegenerative Pathologies

Being a hydroxymethylglutaryl-CoA reductase inhibitor, its usage in treating CVDs is quite widespread but now it has equally shown a beneficial effect in attenuating the ROS production along with inflammatory responses in APP (Yarchoan et al. 2012; Fracassi et al. 2019). Drugs like simvastatin or fluvastatin have been reported for resolving the neuronal degenerative processes (Schultz et al. 2018). Several research

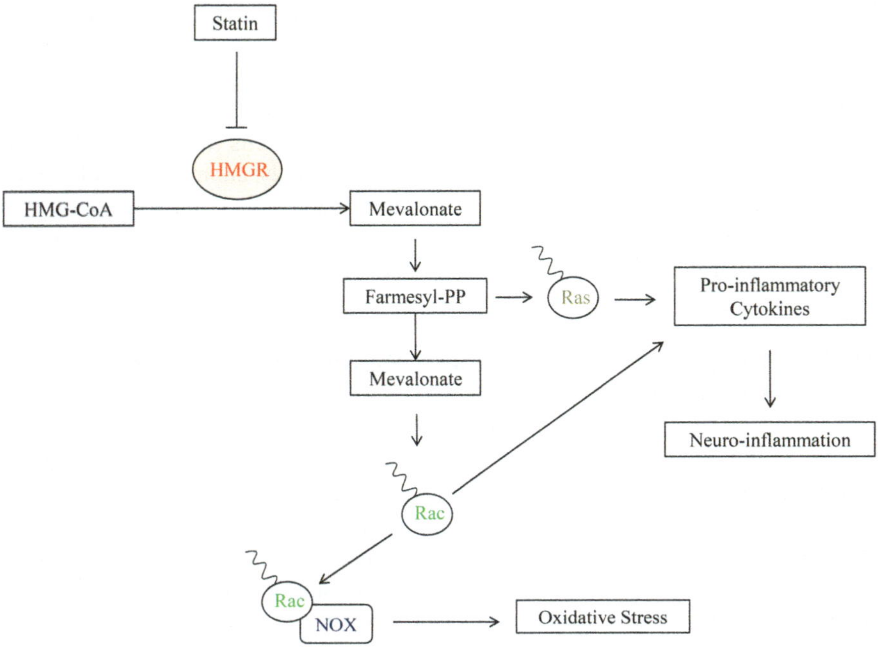

Fig. 15.2 Statin inhibiting HMGR to suppress neuroinflammation and oxidative stress in neurodegenerative pathologies. HMGR, β-hydroxy-β-methylglutaryl-CoA reductase; NOX, NADHP oxidase; HMG-CoA, β-hydroxy-β-methylglutaryl-CoA

studies has shown statins in aiding dyslipidemia as an efficient and potentially prescribed medicine that primarily exhibits inhibitory expression with respect to HMGR, a rate-limiting enzyme of mevalonate (MVL) pathway (Bilgel et al. 2018). Consequently, MVL pathway has observed to mediate various essential biochemical reactions in the biological system including CNS (see Fig. 15.2), apart from being involved in the biogenesis of cholesterol it is also commonly referred to as cholesterol/isoprenoid pathway (Chu et al. 2018; Yang et al. 2020). Statins express an extremely intense inhibition toward HMGR that further hampers the synthesis of cholesterol (see Fig. 15.2) and several other by-products (Smith et al. 2017).

It has been reported to hold attributes that facilitate pleiotropic responses and thus, making it an ideal contender to function as a modulator in diseased CNS environment, especially in neurodegenerative pathologies. Besides this, statins have also been reported to exhibit strong suppression toward anti-inflammation and oxidative stress in the distinct physiopathological environment (Li et al. 2018b; Yang and Williamson 2019). However, after retrospection, it is found that the HMGR inhibition by statin results in decreased prenylation further leading the way toward NOX activity suppression and ultimately reducing pro-inflammatory cytokines production (see Fig. 15.2). Despite statins'influence in modulating neurodegenerative pathologies, it has also been associated with being an effective neurogenesis inducer in adults (Sato and Morishita 2015; Marchetti and Abbracchio

2005; Lukiw 2008). Statin belongs to a class of drugs that tends to prompt differentiation as well as the proliferation of neuronal precursor cells by utilizing Wnt, Akt, and RhoA pathways (Hammad et al. 2019). Therefore, employment of statin as a therapeutic modulator in such NDDs becomes much more imminent (Bilgel et al. 2018; Bitzur 2016; Roy et al. 2017).

15.6 Statins and its Their Related Mechanisms in Neurodegenerative Disorders

Various biochemical mechanisms can be explored based on studying the relationship between statin and its application in reducing dementia. Some recent studies have highlighted the roles of certain key mechanisms associated with statin's expression in lowering cognitive impairments (see Fig. 15.3) (Chuang et al. 2015). Statin primarily acts as a potent anti-inflammatory compound and is actively involved in inducing endothelial nitric oxide synthase enzyme that plays a crucial role in vascular maintenance and functions (Poly et al. 2020). Also, augmentation of endothelial nitric oxide synthase is directly related to cerebral blood flow, cerebral vasomotor reactivity and neurovascular dysfunction escalations (Bagheri et al. 2020; Oesterle et al. 2017).

However, if we observe its detailed background then these attributes are associated with improvement in cognition functioning under an effective treatment regime (Wanleenuwat et al. 2019; Hassanabad and McBride 2019). Secondly, the role of protein ApoE in the brain controls key biochemical roles like lipid transportation as HDL in cholesterol metabolism. Rise in ApoE-ε4 allele levels of ApoE additionally results in the accumulation of SPs in affected subjects (especially AD)

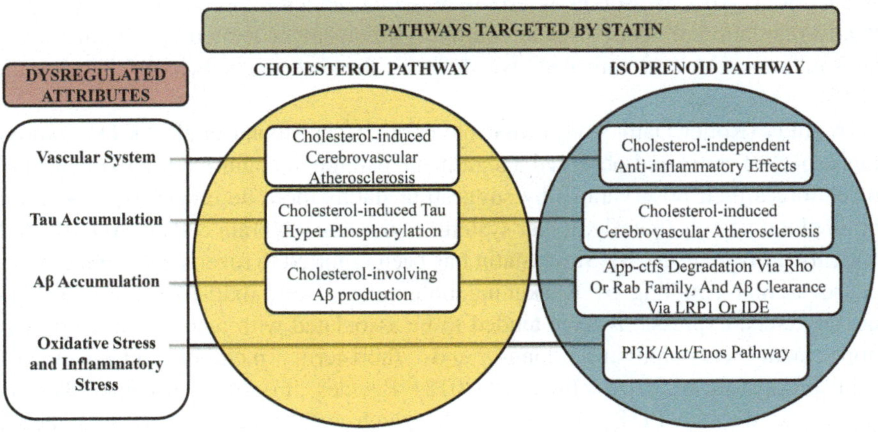

Fig. 15.3 Statin targeted pathways to modulate several pathologies associated with neurodegeneration. Akt, v-akt murine thymoma viral oncogene homolog; eNOS, endothelial nitric oxide synthase; IDE, insulin-degrading enzyme; APP-CTFs, C-terminal amyloid precursor protein; LRP1, low-density lipid-related protein 1; Aβ, amyloid beta

(Iannelli et al. 2018). Therefore, it is imperative to regulate low lipid lipoprotein oxidation by statins, a vital mechanism that also supplements in the prevention and reduction of atherosclerosis. Statin tends to increase the levels of low-density lipid-related protein-1 (LRP1 acts as a receptor of ApoE) that facilitates clearance of Aβ (Ahmadi et al. 2018). Thus, it in turn evades Aβ accumulation and tauopathy (increased levels to tau protein and NFTs) in the affected subjects too.

Besides the biochemical expression of statin in mediating several processes, it is necessary to investigate the chemical profile of the drug and comprehend it holistically. It is primarily an HMG-CoA reductase inhibitor, showcasing pleiotropic effects facilitating ROS suppression, improving vascular flow (see Fig. 15.3), and inhibiting neural inflammation (Schonewille et al. 2016; du Souich et al. 2017). Also, simvastatin has been reported to significantly attenuate the neural inflammation and oxidative stress in dementia-affected mice model, but fluvastatin exhibits somewhat similar yet novel expressions with respect to simvastatin to reduce the oxidative stress along with obliterating cognitive impairment and neural degeneration (Bagheri et al. 2020).

15.7 Role of Statins in Reducing Vascular Dementia

Clinical studies in the recent past have reported several statins-modulated biochemical functions in subjects affected with vascular dementia; however, it is imperative to first explore the chemical profile and mechanism of action for an accurate examination of the drug (Raz et al. 2016). Atorvastatin, simvastatin, and pitavastatin are the most potent and frequently employed statin-based drugs for lipid-lowering purposes; moreover, all of these drugs operate via CYP3A4 pathway for metabolic activities (Kisler et al. 2017; Nelson et al. 2016). Vascular dementia primarily occurs due to the dysregulation of neurovascular coupling, which can be treated by administrating anti-inflammatory and anti-oxidative drugs in the diseased subjects (Zhang et al. 2018).

A study (Rotter Dam Study) (Scheele et al. 2012; Slooter et al. 2001) explored this aspect of statin and observed a significant decline in cognitive impairments and it reinforces their observations by suggesting statins more dominant expression on brain blood vessels (neurovascular system) rather than on brain cells in AD-affected patients (Poly et al. 2020). Simvastatin has been reported to reverse dysregulation in neurovascular coupling by weakening inflammation and oxidative stress. Subsequently, such expressions have tended to be associated with recovery in cognition impairment encompassing long- and short-term memory improvements (Sinyavskaya et al. 2018; Zhu et al. 2018). Besides, this statin also modulate its expression through PI3K/Akt pathway in which an elevated level of endothelial nitric oxide synthase (eNOS) is observed, commonly referred to as isoprenoid effect maintaining the balanced vascular functions (Poly et al. 2020; Fahim et al. 2019). Simvastatin has been observed to raise the levels of basal Akt, phosphorylated Akt, and eNOS which tend to repair cognition impairments in TG2576 mice.

Additionally, drugs such as pitavastatin and atorvastatin have shown protective attributes in response to NVU degeneration and cognition impairments in TG2576 mice as reported by Kurata et al. (2012) (Kurata et al. 2012; Uemura et al. 2020). Statin controls a pivotal role in preventing high plasma cholesterol levels in systemic circulation which could cause cerebral atherosclerosis resulting in CAA and SP formation, a characteristic AD pathology (see Fig. 15.3). Henceforth, statin modulates key functions in the neurovascular system to prevent and reduce the risk of developing AD (Zhang et al. 2018).

15.8 Role of Statins in Reducing Alzheimer's Disease Pathology

As mentioned above, the characteristics of AD are the presence of Aβ peptide and NFTs composed of hyperphosphorylated tau protein. Impairment of memory and cognitive functions are linked with the accumulation of Aβ, increase in oxidative stress and impairment in lipid metabolism altogether (Burns et al. 2006). There is considerable attention given to the cholesterol homeostasis, especially *APOE*, required for the transportation of cholesterol (Fracassi et al. 2019; Maki 2017). Particularly, the *APOE4* allele of the *APOE* gene is linked to the increased cholesterol levels and a higher risk of AD. It was well established that the *APOE* lipoproteins, despite their interaction amidst various cell receptors to deliver lipids, eventually gets associated with Aβ peptide. Thereafter, they either promotes its aggregation or reduces its SPs clearance or both (Liu et al. 2012; Noureddine et al. 2020; Jeske et al. 2020). All these events initiate the progression toward toxic pathways leading to synaptic dysfunction and neurodegeneration. Additionally, various studies indicate that the isoprenoids/protein prenylation and the GTPases have an effect on various aspects of AD, suggesting the role of prenylated proteins on the disease pathogenesis (Fassbender et al. 2001).

It has been observed that the statin-induced isoprenoid depletion causes decreased prenylation of protein and promotes the non-amyloidogenic processing of APP and reduced formation of Aβ. All these evidence suggest that statins act as a neuroprotective agent in the treatment and prevention of AD (Goetzl 2020; Nakagomi et al. 2020). Also, various studies have demonstrated that statin drugs like simvastatin or pravastatin have decreased the level of Aβ in the cerebrospinal fluid (CSF) and brain. Another type, atorvastatin has been shown to reduce the Aβ aggregation by curtailing down the APP content, β-secretase, and oxidative stress in animal models of AD (Pedrini et al. 2005; Kemp et al. 2020).

Furthermore, lovastatin decreases the formation of the components of Aβ plaques and protected the neurons in an Aβ toxicity experimental model by increasing Wnt signaling and decreasing the activity of GSK-3β. In another study, transgenic mice model with APP mutation (APP-Tg), pitavastatin, and atorvastatin have demonstrated to exert the neuroprotective action against the inflammatory cytokines (IL-6, TNF-α, IL-1β), reducing the formation of Aβ, tau phosphorylation and thus, improving the cognitive functions (Ostrowski et al. 2016).

15.9 Role of Statins in Reducing Parkinson's Disease Pathology

Parkinson's disease (PD) is the second most common NDDs after the AD, characterized by the combination of both motor and non-motor symptoms wherein motor symptoms indicated the rigidity of muscle, postural instability, bradykinesia, resting tremor and gait disturbances, and non-motor manifestations constitute sleep disorders, olfactory dysfunction, depression, pain, psychiatric disorders, and cognitive loss (Alecu and Bennett 2019; Tuttle et al. 2016). The degeneration of dopaminergic neurons causes the dysfunction of motor functions in PD patients and at the cellular level, neuronal loss is associated with neurite degeneration and the presence of Lewy body (LBs – cytoplasmic inclusions) that involves the aggregation of α-synuclein (Liu et al. 2012; Guerrero-Ferreira et al. 2018). Different genes are linked to the familiar form of PD, with altered mitochondrial homeostasis, making it one of the pathological causes besides the other concerns like – inflammation, free radicals formation, excitotoxicity, autophagy, and mitochondrial dysfunction leading to cellular dysfunctioning in PD subjects (Jangula and Murphy 2013; Li et al. 2018c). PD-associated mutated genes include α-synuclein, PINK, DJ-1, LRRK2, and parkin which are either directly or indirectly related to mitochondrial dysfunctioning (Hijaz and Volpicelli-Daley 2020).

The role of statins in deducing PD-related dementing process is again backed by its prevalent characteristic of exhibiting neuroprotective actions through various mechanisms including apoptosis regulation, anti-inflammatory effects, and alleviation of oxidative stress (Bartels 2011). Also, statins can present protective effects by modulation of autophagy where the research group showed that the statins reduce down the neurotoxicity induced by rotenone in SH-SY5Y cells by modulating the autophagy signaling (Eriksson et al. 2017). They determined that statins upregulate the autophagic markers like AMP-activated protein kinase (AMPK) and Beclin-1. Also, the expression of mTOR was increased by treatment of cells with rotenone (Bagheri et al. 2020; Braak and Del Tredici 2008). This study explains that statin-induced neuroprotection is mediated by an increase in autophagy. Apart from AMPK and Beclin-1-mediated statin expression, alpha-synuclein has also been reported to be one of the primary factors which lead to the progression of several neurodegenerative diseases including PD. Alpha-synuclein, being a terminal nerve protein, plays a pivotal role in facilitating dopaminergic release and plasticity. However, a significant decline in dopaminergic neurons has been marked as a crucial initial stride toward PD pathogenesis which is due to the intra-cytoplasmic inclusion of Lewy bodies through substantia nigra par compacta (SNpc). Moreover, this inclusion further leads to the synthesis of DA from tyrosine hydroxylase which is followed by series of distinctive misfolding and accumulation of Lewy bodies in the SNpc paramount ultimately leading to the progression of PD (Cheng et al. 2010).

Statins have been observed to attenuate glial activation, inhibit oxidative stress, protect dopaminergic neurons, and suppress aggregation of α synuclein along with the ability to reduce the levels of α synuclein conglomeration in the detergent-insoluble fractions of the transfected cells. Statins can turn down the risk of PD by impeding the degeneration of dopaminergic neurons which may be associated with the ability of statins to directly stimulate SREB (Sterol Regulatory Element Binding Proteins) translocation. SREBP involves transcriptional activity that may restore the expression of presynaptic dopamine markers and come up with neuroprotection of dopaminergic neurons (Bieschke et al. 2006). High cholesterol levels seeding PD has a dual role as a protector against lysosomal membrane permeabilization as well as a stimulator of α synuclein accumulation (Bosco et al. 2006). It has been reported that simvastatin and lovastatin could improve the function of the SNpc by promoting neuronal repair & regeneration, inhibiting oxidative stress, facilitating the immune system, increasing the expression of neurotrophic factors, and downregulating cholesterol. Thereby in the process, it tapper down all the neurodegenerative pathologies and in turn prevent PD progression. However, to conclude, the current research demonstrates that the role of statins in PD is not yet completely conclusive and well-defended one rather, it further needs elaborative experimental data to infer about its probable role in the treatment of PD subjects (Bar-On et al. 2006).

15.10 Role of Statins in Reducing Huntington's Disease Pathology

Huntington's disease (HD) is an autosomal inherited dominant NDD, described by psychiatric and motor disturbances along with dementia. It is caused by the repetition of CAG nucleotides in the gene encoding for huntingtin (HTT) protein and many studies have shown that there is an alteration of cholesterol homeostasis in HD. Despite some research indicating that the cholesterol levels are downregulated in the HD brain (Chen et al. 2016; Shankaran et al. 2017), there are other opinions too, that says there is a gradual reduction of cholesterol synthesis and a constant steady-state of cholesterol levels is maintained for longer time suggesting the compensatory mechanisms (Valenza et al. 2007).

On the contrary, there is also a viewpoint that the cholesterol accumulation in HD brain causes neurotoxicity (Trushina et al. 2006; Del Toro et al. 2010). According to the research in 2016 by Chen et al., the decrease in cholesterol synthesis is caused by altered cholesterol homeostasis and the dysregulation in the clearance of cholesterol which is caused by decreased expression of CYP46A1 (rate-limiting enzyme for the degradation of cholesterol) (Chen et al. 2016). The findings that cholesterol perturbations are involved in the HD pathogenesis directed the hypothesis that the pharmacological interventions on the metabolism of sterols could lead to a reduction in the symptomatology of HD (Saeedi et al. 2020).

Therefore, simvastatin was explored for the same and it has been observed that it improved the synaptic phenotype of R6/2 strain of transgenic mouse model of HD (Chen et al. 2016). Some studies also hypothesized that cholesterol is accumulated in the cell membrane (Trushina et al. 2006) and simvastatin could increase the redistribution, mobilization, and/or efflux of accumulated cholesterol (Burns et al. 2006) additionally showing the increase in BDNF production, membrane fluidity, and exerting anti-inflammatory effects (Smith et al. 2017; Wang et al. 2020).

15.11 Regulation of tau and Aβ by Statins

Statin plays a distinct role in managing AD in diseased subjects by regulating tau and Aβ expression and one of the few primary targets for them in AD pathology is NFTs, a hyper-phosphorylated and filamentous form of tau protein. This state is distinctly termed as tauopathy and is characterized by severe neuronal loss accompanied by dementia due to NFTs over-expression (Blanchard and Tsai 2019). As studies suggest NFTs over-expression is triggered due to the amalgamation of several distinct factors such as high plasma cholesterol levels, inflammatory, and oxidative stress (Shakour et al. 2019; Geifman et al. 2017). Thus, statin drugs such as atorvastatin and simvastatin at a certain optimum dosage, like - 40 mg/day of simvastatin, referred to as optimum dose have shown to effectively repress the tau and NFTs hyperphosphorylation in dyslipidemia helping in achieving improved cognitive state (see Table 15.1) in the affected subjects (Iba et al. 2020). Besides from tau protein, Aβ and its metabolism are strongly associated with AD progression; thus, maintaining a steady state of production and clearance is eminent for proper functioning. And in mitigating this concern, statins have been observed to utilize both isoprenoid-dependent as well as cholesterol pathways to target affected sites for impeding AD progression and development (Shakour et al. 2019).

Furthermore, it has been known that the cerebral vascular system facilitates Aβ clearance with the help of LRP1 protein, therefore, in diseased AD subjects fluvastatin is administered, which further utilizes isoprenoid-dependent pathway for up-regulating LRP1 levels and subsequently achieving normal Aβ clearance. It has been also reported that there is a notable rise in lysosomal degradation of APP-CFTs in the brains of juvenile C57BL/6 mice models and expression was reported to be mediated via an isoprenoid-dependent pathway which is more or less homologs to LRP1 up-regulation. Thereby, evading Aβ accumulation which has been reported the root cause in several AD associated pathologies (Geifman et al. 2017). Table 15.2 provides some of the natural-based statins that can be utilized for the treatment of NDDs. Apart from this, statin has also been observed to regulate Aβ biogenesis by a systematic and periodic discharge of α/β/γ-secretase to facilitate APP-CFTs degradation along with APP-trafficking.

15.12 Neurovascular Unit Dysregulation Initiated Cognitive Impairment

Condition hypothesizing the relationship of cerebral vasculature in the etiology of neurodegeneration has cemented its place in most of the associated theories of CAA. Moreover, several links between cognition and aberrant changes in cerebral microvasculature that cognates with glucose-oxygen transport are gaining momentum in research findings (Levit et al. 2020; Humpel 2008; Chakraborty et al. 2017). Further, it has been observed that central nervous system (CNS) lacks energy reserves and is heavily dependent upon a vital continuous supply of oxygen & glucose via cerebral blood flow, and thus, NVUs play a pivotal role in facilitating this channel and

Table 15.1 Pharmacological aspects of statin-based drugs

Statin-based drugs	Pharmacological attributes					References
	Metabolic channel	Half-life (h)	Solubility	Bioavailability	Absorption	
Simvastatin	CYP3PA4	1.9–3	Lipophilic	< 5%	65–85%	(Kumar et al. 2018)
Atorvastatin	CYP3PA4	11–30	Lipophilic	12%	30%	(Xiang et al. 2020)
Lovastatin	CYP3PA4	2.5–3	Lipophilic	< 5%	31%	(Cicero et al. 2018)
Pravastatin	CYP3PA4	11	Hydrophilic	18%	37%	(Kumar et al. 2019)
Rosuvastatin	CYP2C9	20	Hydrophilic	20%	50%	(Farkouh and Baumgärtel 2019)
Pitavastatin	CYP2C9	11	Lipophilic	> 60%	80%	(Groll et al. 2019)
Fluvastatin	CYP2C9	0.5–2.3	Lipophilic	10–35%	98%	(Choi et al. 2017)

Table 15.2 Natural alternative for statin-based drugs

Biologics	Decline in lipoprotein expression			References
	Low-density lipids (%)	High-density lipids (%)	Triglycerides (%)	
Niacin	5–25	15–35	20–50	(Wełniak et al. 2020)
Fibrates	5–20	15–35	25–50	(Kostis et al. 2019)
Bile acid sequestrants	15–30	3–5	–	(Yu et al. 2020)
Red yeast rice	5–20	3–5	–	(Tarafdar and Pula 2018)
Plant sterol/ stanol	1.2–8	–	1.1–15	(Sweeney et al. 2018)
Policosanol	16.7–20.2	–	–	(Barthold et al. 2020)

evading potential cognitive impairments. Any minutest discrepancy in supply perfusion even of 60–90 seconds could trigger an ischemic shock environment leading to cognitive decline (Cipollini et al. 2019; Levit 2018), for example, malfunctioning or deficiencies in GLUT3 protein. Also, deprivation in constant oxygen and glucose supply lasting beyond 3 h could result in the release of lactic acid in response to shifting of anaerobic metabolism pathways, further hindering the ionic strength within the neurons (Yan et al. 2020). These events occur due to the inability to generate ATP that initiates the cascade of further failure in downstream energy-dependent processes leading to activation of apoptotic pathways. NVU comprising of vasculature cells (pericytes, endothelium, and smooth muscle cells) that cater to the demand and supply of oxygen & glucose to the brain (Stanimirovic and Friedman 2012). NVUs are also responsible for molecular exchange between BBB and maintain clearance of metabolic by-products, trafficking immune cells to support brain cells (Kisler et al. 2017; Cai et al. 2017; McConnell et al. 2019; Minter et al. 2016).

NVU has a crucial role in the regulation of CBF (cerebral blood flow) as well as for the proper functioning of the brain. CBF dysregulation is due to changes in vascular innervation caused by neuronal loss. Various studies have demonstrated the expression of α synuclein which is associated with the increase in permeability of the blood-brain barrier (McNaull et al. 2010; Holmes 2017). NVU disruption diminishes the supply of oxygen and nutrients to the brain inclusively clearance of neurotoxic substances such as β amyloid and α synuclein reduction in parenchyma takes place which leads to an increase in Aβ deposition further reducing the levels of α synuclein within the parenchyma. Increases in age, senile, astrocytes, and microglia produce various cytokines chemokines, and ROS that disrupts the integrity of BBB causing rearrangement of tight junctions leads to an increase in Aβ deposition (Keaney and Campbell 2015; Morrison and Filosa 2019). The reduced expression of P-gp (Pre-glycoprotein) in the midbrain examined in PD is related to dysfunction of

BBB. The major reason for the reduction in levels of P-gp in the midbrain is due to the accumulation of α synuclein and other neurotoxic substances in the brain. Harmful substance accumulation causes brain damage that is because of decreased efflux in membrane transport through P-gp. Therefore, any slightest of anomaly in NVU would compromise on the cerebral blood flow rate and eventually adds to development of dementia and ultimately alters the CNS microenvironment's homeostasis causing cognitive impairment (Sagare et al. 2012; Hachinski et al. 2019; Mauro et al. 2015).

15.13 Adversities with the Use of Statins

Many related studies are supporting the fact that the efficacy of statin administration has surely reduced the mortality rates in cardiovascular disorders but then it does have its related controversies and conflicts for its longer use (Schultz et al. 2018; Li et al. 2018a). It has been reported that its long-standing usage may lead to myalgia, rhabdomyolysis, etc. However, more recently many scientific literature have supported that the statins may exhibit potential neuroprotective properties and can prove to be an efficacious approach in ameliorating the neurovascular and neurodegenerative issues further resolving the issues of vascular and Alzheimer's dementia (Chu et al. 2018). Through its post-administration neurocognitive consequences and related side effects after long use cannot be denied either (Li et al. 2018b).

15.14 Status of Statin-Based Clinical Studies

Much later, after evident consumption of statins to an extent, that it became one of the most prescribed drugs globally, it has been observed that it has the potential to reduce cognitive impairments as well (Ott et al. 2015). However, in clinical trials (Phase III) and then in the re-evaluation of the facts too, it did not report any such observations but in 2012, FDA (Food and Drug Administration) announced the label change for statins claiming that statins may pose some cognitive compromise (Swiger et al. 2013; Richardson et al. 2013). Although the FDA advisory has explained the benefits of statins for treating cardiovascular concerns, however the risk expectancy of 1.33% way to higher (Power et al. 2018). To re-evaluate the same, randomized controlled trials (RCTs) meta-analysis was also done and it was found that there were almost negligible reports about the cognitive decline in the treated subjects with multifaceted dimensional analysis (Strom et al. 2015; Rojas-Fernandez et al. 2014). This was again validated by the American College of Cardiology/ American Heart Association (ACC/AHA) in their guidelines in 2013 that statins didn't exhibit any adverse effects so based on the same, researchers requested FDA to re-check on their label change announcement related with side effects of statins on cognition (Trompet et al. 2010).

15.15 Recent Developments and Future Perspectives

A considerable number of research studies have been carried out on the applications of statin and the associated risk of dementia over the past decade (Barthold et al. 2020). The need for improvement in the common vascular or neurodegenerative pathophysiological mechanisms persists between the use of statins and dementia risk reduction (Petek et al. 2018). Therefore, more translational research and clinical studies need to be carried out to establish a fundamental biological explanation for the observational studies (Janson 2016).

There should be more emphasis on high-quality longitudinal studies with longer follow-up periods, including both the new as well as the chronic statin users, which will help in decreasing the selection biases (Šubic 2019). Some of the potential variables, for example, medication dose, exposure, duration, and the date of initiation, should be given much reputation as this will help to measure the exposure levels in an improved manner. Similarly, epidemiological and clinical research, insights for the informed treatments, patient compliance, and the substitution of more tolerable statins will provide a direction for future diagnosis (Turner 2020).

15.16 Conclusion

The specific role of statins in regulating biological activities in CNS has gained as a novel and potential modulator in the brain processes, specifically in dementia. Currently, there is no successful therapy for dementia and its socioeconomic burden is massively escalating with the aging population; more so, when there is a persistent decrease in stroke and cardiovascular morbidity and mortality rates, providing high-risk population for dementia. The preventative effect of statin on AD includes regulation of inflammatory and oxidative stress, tau, cholesterol metabolism, vascular integrity, and plasticity, which are interdependently/independently involved in the AD pathogenesis. The preclinical/clinical data also exhibit that it alleviates dementia symptomatology. So, it may be possible that statin therapy clears out the Aβ plaques, further decreasing the neuronal loss, hence treating dementia. Despite these promising data, there are many contradictions too, that highlight the controversial role of statin in AD, raising concerns for its pharmacological efficacy. In conclusion, even though efficient neuropharmacological effects are exerted by the administration of statin, more detailed analysis is needed to confirm its therapeutic role in dementia.

References

Ahmadi Y, Ghorbanihaghjo A, Naghi-Zadeh M, Yagin NL (2018) Oxidative stress as a possible mechanism of statin-induced myopathy. Inflammopharmacology 26(3):667–674

Albert MS, DeKosky ST, Dickson D, Dubois B, Feldman HH, Fox NC et al (2013) The diagnosis of mild cognitive impairment due to Alzheimer's disease: recommendations from the National

Institute on Aging-Alzheimer's association workgroups on diagnostic guidelines for Alzheimer's disease. Focus 11(1):96–106

Alecu I, Bennett SA (2019) Dysregulated lipid metabolism and its role in α-Synucleinopathy in Parkinson's disease. Front Neurosci 13:328

Alsheikh-Ali AA, Maddukuri PV, Han H, Karas RH (2007) Effect of the magnitude of lipid lowering on risk of elevated liver enzymes, rhabdomyolysis, and cancer: insights from large randomized statin trials. J Am Coll Cardiol 50(5):409–418

Bagheri H, Ghasemi F, Barreto GE, Sathyapalan T, Jamialahmadi T, Sahebkar A (2020) The effects of statins on microglial cells to protect against neurodegenerative disorders: a mechanistic review. Biofactors 46(3):309–325

Bankstahl M, Breuer H, Leiter I, Märkel M, Bascuñana P, Michalski D et al (2018) Blood–brain barrier leakage during early epileptogenesis is associated with rapid remodeling of the neurovascular unit. ENeuro 5(3). https://doi.org/10.1523/ENEURO.0123-18.2018

Bar-On P, Rockenstein E, Adame A, Ho G, Hashimoto M, Masliah E (2006) Effects of the cholesterol-lowering compound methyl-β-cyclodextrin in models of α-synucleinopathy. J Neurochem 98(4):1032–1045

Bartels A (2011) Blood-brain barrier P-glycoprotein function in neurodegenerative disease. Curr Pharm Des 17(26):2771–2777

Barthold D, Joyce G, Brinton RD, Wharton W, Kehoe PG, Zissimopoulos J (2020) Association of combination statin and antihypertensive therapy with reduced Alzheimer's disease and related dementia risk. PLoS One 15(3):e0229541

Baumgart M, Snyder HM, Carrillo MC, Fazio S, Kim H, Johns H (2015) Summary of the evidence on modifiable risk factors for cognitive decline and dementia: a population-based perspective. Alzheimers Dement 11(6):718–726

Bettermann K, Arnold AM, Williamson J, Rapp S, Sink K, Toole JF et al (2012) Statins, risk of dementia, and cognitive function: secondary analysis of the ginkgo evaluation of memory study. J Stroke Cerebrovasc Dis 21(6):436–444

Bieschke J, Zhang Q, Bosco DA, Lerner RA, Powers ET, Wentworth P et al (2006) Small molecule oxidation products trigger disease-associated protein misfolding. Acc Chem Res 39(9):611–619

Bilgel M, An Y, Helphrey J, Elkins W, Gomez G, Wong DF et al (2018) Effects of amyloid pathology and neurodegeneration on cognitive change in cognitively normal adults. Brain 141(8):2475–2485

Bitzur R (2016) Remembering statins: do statins have adverse cognitive effects? Diabetes Care 39 (Supplement 2):S253–S2S9

Blanchard JW, Tsai LH (2019) Unraveling the paradox of statins with human neurons: new leads in Alzheimer's disease. Cell Stem Cell 24(3):347–349

Bosco DA, Fowler DM, Zhang Q, Nieva J, Powers ET, Wentworth P et al (2006) Elevated levels of oxidized cholesterol metabolites in Lewy body disease brains accelerate α-synuclein fibrilization. Nat Chem Biol 2(5):249–253

Braak H, Del Tredici K (2008) Invited article: nervous system pathology in sporadic Parkinson disease. Neurology 70(20):1916–1925

Burns MP, Igbavboa U, Wang L, Wood WG (2006) Cholesterol distribution, not total levels, correlate with altered amyloid precursor protein processing in statin-treated mice. Neuromolecular Med 8(3):319–328

Butterfield DA, Barone E, Mancuso C (2011) Cholesterol-independent neuroprotective and neurotoxic activities of statins: perspectives for statin use in Alzheimer disease and other age-related neurodegenerative disorders. Pharmacol Res 64(3):180–186

Cai W, Zhang K, Li P, Zhu L, Xu J, Yang B et al (2017) Dysfunction of the neurovascular unit in ischemic stroke and neurodegenerative diseases: an aging effect. Ageing Res Rev 34:77–87

Caruso P, Signori R, Moretti R (2019) Small vessel disease to subcortical dementia: A dynamic model, which interfaces aging, cholinergic dysregulation and the neurovascular unit. Vasc Health Risk Manag 15:259

Chakraborty A, De Wit N, Van Der Flier W, De Vries H (2017) The blood brain barrier in Alzheimer's disease. Vasc Pharmacol 89:12–18

Chen JY, Tran C, Hwang L, Deng G, Jung ME, Faull KF et al (2016) Partial amelioration of peripheral and central symptoms of huntington's disease via modulation of lipid metabolism. J Huntingtons Dis 5(1):65–81

Cheng HC, Ulane CM, Burke RE (2010) Clinical progression in Parkinson disease and the neurobiology of axons. Ann Neurol 67(6):715–725

Choi D, Choi S, Park SM (2018) Effect of smoking cessation on the risk of dementia: a longitudinal study. Ann Clin Trans Neurol 5(10):1192–1199

Choi S, Oh D-S, Jerng UM (2017) A systematic review of the pharmacokinetic and pharmacodynamic interactions of herbal medicine with warfarin. PLoS One 12(8):e0182794

Chu C-S, Tseng P-T, Stubbs B, Chen T-Y, Tang C-H, Li D-J et al (2018) Use of statins and the risk of dementia and mild cognitive impairment: A systematic review and meta-analysis. Sci Rep 8(1):1–12

Chuang CS, Lin CL, Lin MC, Sung FC, Kao CH (2015) Decreased prevalence of dementia associated with statins: a national population-based study. Eur J Neurol 22(6):912–918

Cicero AF, Bove M, Borghi C (2018) Pharmacokinetics, pharmacodynamics and clinical efficacy of non-statin treatments for hypercholesterolemia. Expert Opin Drug Metab Toxicol 14(1):9–15

Cipollini V, Anrather J, Orzi F, Iadecola C (2019) Th17 and cognitive impairment: possible mechanisms of action. Front Neuroanat 13:95

Costea L, Mészáros Á, Bauer H, Bauer H-C, Traweger A, Wilhelm I et al (2019) The blood–brain barrier and its intercellular junctions in age-related brain disorders. Int J Mol Sci 20(21):5472

Del Toro D, Xifró X, Pol A, Humbert S, Saudou F, Canals JM et al (2010) Altered cholesterol homeostasis contributes to enhanced excitotoxicity in Huntington's disease. J Neurochem 115(1):153–167

du Souich P, Roederer G, Dufour R (2017) Myotoxicity of statins: mechanism of action. Pharmacol Ther 175:1–16

Eriksson I, Nath S, Bornefall P, Giraldo AMV, Öllinger K (2017) Impact of high cholesterol in a Parkinson's disease model: prevention of lysosomal leakage versus stimulation of α-synuclein aggregation. Eur J Cell Biol 96(2):99–109

Fahim VF, Wadie W, Shafik AN, Attallah MI (2019) Role of simvastatin and insulin in memory protection in a rat model of diabetes mellitus and dementia. Brain Res Bull 144:21–27

Farkouh A, Baumgärtel C (2019) Mini-review: medication safety of red yeast rice products. Int J Gen Med 12:167

Fassbender K, Simons M, Bergmann C, Stroick M, Lütjohann D, Keller P et al (2001) Simvastatin strongly reduces levels of Alzheimer's disease β-amyloid peptides Aβ42 and Aβ40 in vitro and in vivo. Proc Natl Acad Sci 98(10):5856–5861

Fitzpatrick AL, Kuller LH, Lopez OL, Diehr P, O'Meara ES, Longstreth W et al (2009) Midlife and late-life obesity and the risk of dementia: cardiovascular health study. Arch Neurol 66(3): 336–342

Fracassi A, Marangoni M, Rosso P, Pallottini V, Fioramonti M, Siteni S et al (2019) Statins and the brain: more than lipid lowering agents? Curr Neuropharmacol 17(1):59–83

Geifman N, Brinton RD, Kennedy RE, Schneider LS, Butte AJ (2017) Evidence for benefit of statins to modify cognitive decline and risk in Alzheimer's disease. Alzheimers Res Ther. 9(1): 1–10

Geybels MS, Wright JL, Holt SK, Kolb S, Feng Z, Stanford JL (2013) Statin use in relation to prostate cancer outcomes in a population-based patient cohort study. Prostate 73(11): 1214–1222

Goetzl EJ (2020) Advancing medicine for Alzheimer's disease: A plasma neural exosome platform. FASEB J 34(10):13079–13084

Griffith M, Tajik M, Wing S (2007) Patterns of agricultural pesticide use in relation to socioeconomic characteristics of the population in the rural US South. Int J Health Serv 37(2):259–277

Groll AH, Rijnders BJ, Walsh TJ, Adler-Moore J, Lewis RE, Brüggemann RJ (2019) Clinical pharmacokinetics, pharmacodynamics, safety and efficacy of liposomal amphotericin B. Clin Infect Dis 68(Supplement_4):S260–SS74

Guerrero-Ferreira R, Taylor NM, Mona D, Ringler P, Lauer ME, Riek R et al (2018) Cryo-EM structure of alpha-synuclein fibrils. Elife 7:e36402

Hachinski V, Einhäupl K, Ganten D, Alladi S, Brayne C, Stephan BC et al (2019) Special topic section: linkages among cerebrovascular, cardiovascular, and cognitive disorders: preventing dementia by preventing stroke: the Berlin manifesto. Int J Stroke:1747493019871915

Hammad MA, Sulaiman SAS, Aziz NA, Noor DAM (2019) Evaluation of statins impacts on cognitive function among diabetic patients. Diabetes Metab Syndr Clin Res Rev 13(3): 1797–1803

Hassanabad AF, McBride SA (2019) Statins as potential therapeutics for lung cancer: molecular mechanisms and clinical outcomes. Am J Clin Oncol 42(9):732–736

Hijaz BA, Volpicelli-Daley LA (2020) Initiation and propagation of α-synuclein aggregation in the nervous system. Mol Neurodegener 15(1):1–12

Holmes KR. Progressive slowing of the cerebrovascular response in the temporal and parietal cortex of patients with Alzheimer's disease and mild cognitive impairment 2017

Humpel C (2008) Alzheimer's disease—is it caused by cerebrovascular dysfunction? In: Neurovascular Medicine Pursuing Cellular Longevity for Healthy Aging, p 369

Iadecola C, Gottesman RF (2019) Neurovascular and cognitive dysfunction in hypertension: epidemiology, pathobiology, and treatment. Circ Res 124(7):1025–1044

Iannelli F, Lombardi R, Milone MR, Pucci B, De Rienzo S, Budillon A et al (2018) Targeting mevalonate pathway in cancer treatment: repurposing of statins. Recent Pat Anticancer Drug Discov 13(2):184–200

Iba M, Kim C, Florio J, Mante M, Adame A, Rockenstein E et al (2020) Role of alterations in protein kinase p38γ in the pathogenesis of the synaptic pathology in dementia with Lewy bodies and α-Synuclein transgenic models. Front Neurosci 14:286

James BD, Bennett DA, Boyle PA, Leurgans S, Schneider JA (2012) Dementia from Alzheimer disease and mixed pathologies in the oldest old. JAMA 307(17):1798–1800

Jangula A, Murphy EJ (2013) Lipopolysaccharide-induced blood brain barrier permeability is enhanced by alpha-synuclein expression. Neurosci Lett 551:23–27

Janson CG (2016) Diverse effects of statins in Alzheimer's disease. Sci Transl Med 8(330)):330ec44

Jellinger KA (2007) The enigma of mixed dementia. Alzheimers Dement 3(1):40–53

Jeske R, Albo J, Marzano M, Bejoy J, Li Y (2020) Engineering brain-specific Pericytes from human pluripotent stem cells. Tissue Eng Part B Rev 26(4):367–382

Keaney J, Campbell M (2015) The dynamic blood–brain barrier. FEBS J 282(21):4067–4079

Kemp SS, Aguera KN, Cha B, Davis GE (2020) Defining endothelial cell-derived factors that promote pericyte recruitment and capillary network assembly. Arterioscler Thromb Vasc Biol 120:314948

Kisler K, Nelson AR, Montagne A, Zlokovic BV (2017) Cerebral blood flow regulation and neurovascular dysfunction in Alzheimer disease. Nat Rev Neurosci 18(7):419

Kokkinos PF, Faselis C, Myers J, Panagiotakos D, Doumas M (2013) Interactive effects of fitness and statin treatment on mortality risk in veterans with dyslipidaemia: a cohort study. Lancet 381(9864):394–399

Kostis WJ, Dobrzynski JM, Kostis JB (2019) Advantages and disadvantages in clinical trials. In: Management of hypertension. Springer, Cham, pp 215–238

Kumar P, Mangla B, Singh S (2018) Pitavastatin: A potent drug. Int J Pharma Res Health Sci 6(1): 2070–2074

Kumar AB, Umashankar M, Lakshmi K, Sriram V (2019) Pharmacodynamic and pharmacokinetic overview on statins in the management of hyperlipidemic complications in clinical practice. Drug Invention Today 11(5):1063–1070

Kurata T, Kawai H, Miyazaki K, Kozuki M, Morimoto N, Ohta Y et al (2012) Statins have therapeutic potential for the treatment of Alzheimer's disease, likely via protection of the neurovascular unit in the AD brain. J Neurol Sci 322(1–2):59–63

Levit A. White matter inflammation and executive dysfunction: implications for Alzheimer disease and vascular cognitive impairment. 2018

Levit A, Hachinski V, Whitehead SN (2020) Neurovascular unit dysregulation, white matter disease, and executive dysfunction: the shared triad of vascular cognitive impairment and Alzheimer disease. GeroScience 42:1–21

Lewczuk P, Riederer P, O'Bryant SE, Verbeek MM, Dubois B, Visser PJ et al (2018) Cerebrospinal fluid and blood biomarkers for neurodegenerative dementias: an update of the consensus of the task force on biological markers in psychiatry of the world federation of societies of biological psychiatry. World J Biol Psychiatry 19(4):244–328

Li H-H, Lin C-L, Huang C-N (2018a) Neuroprotective effects of statins against amyloid β-induced neurotoxicity. Neural Regen Res 13(2):198

Li R, Wang T-J, Lyu P-Y, Liu Y, Chen W-H, Fan M-Y et al (2018b) Effects of plasma lipids and statins on cognitive function. Chin Med J 131(4):471

Li Y, Zhao C, Luo F, Liu Z, Gui X, Luo Z et al (2018c) Amyloid fibril structure of α-synuclein determined by cryo-electron microscopy. Cell Res 28(9):897–903

Liao JK, Laufs U (2005) Pleiotropic effects of statins. Annu Rev Pharmacol Toxicol 45:89–118

Liu Y-H, Giunta B, Zhou H-D, Tan J, Wang Y-J (2012) Immunotherapy for Alzheimer disease—the challenge of adverse effects. Nat Rev Neurol 8(8):465–469

Livingston G, Johnston K, Katona C, Paton J, Lyketsos CG, Old Age Task Force of the World Federation of Biological Psychiatry (2005) Systematic review of psychological approaches to the management of neuropsychiatric symptoms of dementia. Am J Psychiatr 162(11): 1996–2021

Lukiw WJ (2008) Emerging amyloid beta (ab) peptide modulators for the treatment of Alzheimer's disease (AD). Expert Opin Emerg Drugs 13(2):255–271

Maki T (2017) Novel roles of oligodendrocyte precursor cells in the developing and damaged brain. Clin Exp Neuroimmunol 8(1):33–42

Marchetti B, Abbracchio MP (2005) To be or not to be (inflamed)–is that the question in anti-inflammatory drug therapy of neurodegenerative disorders? Trends Pharmacol Sci 26(10): 517–525

Mauro C, De Rosa V, Marelli-Berg F, Solito E (2015) Metabolic syndrome and the immunological affair with the blood–brain barrier. Front Immunol 5:677

McConnell HL, Li Z, Woltjer RL, Mishra A (2019) Astrocyte dysfunction and neurovascular impairment in neurological disorders: correlation or causation? Neurochem Int 128:70–84

McFarland AJ, Anoopkumar-Dukie S, Arora DS, Grant GD, McDermott CM, Perkins AV et al (2014) Molecular mechanisms underlying the effects of statins in the central nervous system. Int J Mol Sci 15(11):20607–20637

McNaull BBA, Todd S, McGuinness B, Passmore AP (2010) Inflammation and anti-inflammatory strategies for Alzheimer's disease–a mini-review. Gerontology 56(1):3–14

Michalicova A, Banks A, W, Legath J, Kovac A. (2017) Tauopathies-focus on changes at the neurovascular unit. Curr Alzheimer Res 14(7):790–801

Minter MR, Taylor JM, Crack PJ (2016) The contribution of neuroinflammation to amyloid toxicity in Alzheimer's disease. J Neurochem 136(3):457–474

Morrison HW, Filosa JA (2019) Stroke and the neurovascular unit: glial cells, sex differences, and hypertension. Am J Phys Cell Phys 316(3):C325–CC39

Nakagomi T, Tanaka Y, Nakagomi N, Matsuyama T, Yoshimura S (2020) How long are reperfusion therapies beneficial for patients after stroke onset? Lessons from lethal ischemia following early reperfusion in a mouse model of stroke. Int J Mol Sci 21(17):6360

Nelson AR, Sweeney MD, Sagare AP, Zlokovic BV (2016) Neurovascular dysfunction and neurodegeneration in dementia and Alzheimer's disease. Biochim Biophys Acta-Mol Basis Dis 1862(5):887–900

Noureddine FY, Altara R, Fan F, Yabluchanskiy A, Booz GW, Zouein FA (2020) Impact of the renin–angiotensin system on the endothelium in vascular dementia: unresolved issues and future perspectives. Int J Mol Sci 21(12):4268

Oesterle A, Laufs U, Liao JK (2017) Pleiotropic effects of statins on the cardiovascular system. Circ Res 120(1):229–243

Ogino E, Manly JJ, Schupf N, Mayeux R, Gu Y (2019) Current and past leisure time physical activity in relation to risk of Alzheimer's disease in older adults. Alzheimers Dement 15(12): 1603–1611

Ostrowski SM, Johnson K, Siefert M, Shank S, Sironi L, Wolozin B et al (2016) Simvastatin inhibits protein isoprenylation in the brain. Neuroscience 329:264–274

Ott BR, Daiello LA, Dahabreh IJ, Springate BA, Bixby K, Murali M et al (2015) Do statins impair cognition? A systematic review and meta-analysis of randomized controlled trials. J Gen Intern Med 30(3):348–358

Park L, Hochrainer K, Hattori Y, Ahn SJ, Anfray A, Wang G et al (2020) Tau induces PSD95–neuronal NOS uncoupling and neurovascular dysfunction independent of neurodegeneration. Nat Neurosci 23(9):1079–1089

Pedrini S, Carter TL, Prendergast G, Petanceska S, Ehrlich ME, Gandy S (2005) Modulation of statin-activated shedding of Alzheimer APP ectodomain by ROCK. PLoS Med 2(1):e18

Petek B, Villa-Lopez M, Loera-Valencia R, Gerenu G, Winblad B, Kramberger M et al (2018) Connecting the brain cholesterol and renin–angiotensin systems: potential role of statins and RAS-modifying medications in dementia. J Intern Med 284(6):620–642

Pocobelli G, Newcomb PA, Trentham-Dietz A, Titus-Ernstoff L, Hampton JM, Egan KM (2008) Statin use and risk of breast cancer. Cancer: interdisciplinary. Int J Am Cancer Society 112(1): 27–33

Poly TN, Islam MM, Walther BA, Yang H-C, Wu C-C, Lin M-C et al (2020) Association between use of statin and risk of dementia: a meta-analysis of observational studies. Neuroepidemiology 54(3):214–226

Power MC, Rawlings A, Sharrett AR, Bandeen-Roche K, Coresh J, Ballantyne CM et al (2018) Association of midlife lipids with 20-year cognitive change: a cohort study. Alzheimers Dement 14(2):167–177

Rajkumar AP, Aarsland D, Hu M, Baig F, Margolis RL, Byars JA et al (2020) Delirium, dementia, and other cognitive disorders. In: New Oxford textbook of psychiatry. Oxford University Press, Oxford, p 371

Raz L, Knoefel J, Bhaskar K (2016) The neuropathology and cerebrovascular mechanisms of dementia. J Cereb Blood Flow Metab 36(1):172–186

Reitz C, Jun G, Naj A, Rajbhandary R, Vardarajan BN, Wang L-S et al (2013) Variants in the ATP-binding cassette transporter (ABCA7), apolipoprotein E ϵ4, and the risk of late-onset Alzheimer disease in African Americans. JAMA 309(14):1483–1492

Richardson K, Schoen M, French B, Umscheid CA, Mitchell MD, Arnold SE et al (2013) Statins and cognitive function: a systematic review. Ann Intern Med 159(10):688–697

Rockwood K, Kirkland S, Hogan DB, MacKnight C, Merry H, Verreault R et al (2002) Use of lipid-lowering agents, indication bias, and the risk of dementia in community-dwelling elderly people. Arch Neurol 59(2):223–227

Rojas-Fernandez CH, Goldstein LB, Levey AI, Taylor BA, Bittner V (2014) An assessment by the statin cognitive safety task force: 2014 update. J Clin Lipidol 8(3):S5–S16

Roy S, Weinstock JL, Ishino AS, Benites JF, Pop SR, Perez CD et al (2017) Association of cognitive impairment in patients on 3-hydroxy-3-methyl-glutaryl-CoA reductase inhibitors. J Clin Med Res 9(7):638

Saeedi M, Rastegari A, Hariri R, Mirfazli SS, Mahdavi M, Edraki N et al (2020) Design and synthesis of novel arylisoxazole-chromenone carboxamides: investigation of biological activities associated with Alzheimer's disease. Chem Biodivers 17(5):e1900746

Sagare AP, Bell RD, Zlokovic BV (2012) Neurovascular dysfunction and faulty amyloid β-peptide clearance in Alzheimer disease. Cold Spring Harb Perspect Med 2(10):a011452

Santos CY, Snyder PJ, Wu W-C, Zhang M, Echeverria A, Alber J (2017) Pathophysiologic relationship between Alzheimer's disease, cerebrovascular disease, and cardiovascular risk: a review and synthesis. Alzheimer's Dement Diagn Assess Dis Monit 7:69–87

Sato N, Morishita R (2015) The roles of lipid and glucose metabolism in modulation of β-amyloid, tau, and neurodegeneration in the pathogenesis of Alzheimer disease. Front Aging Neurosci 7: 199

Scheele J, De Schepper E, Van Meurs J, Hofman A, Koes B, Luijsterburg P et al (2012) Association between spinal morning stiffness and lumbar disc degeneration: the Rotterdam study. Osteoarthr Cartil 20(9):982–987

Schonewille M, de Boer JF, Mele L, Wolters H, Bloks VW, Wolters JC et al (2016) Statins increase hepatic cholesterol synthesis and stimulate fecal cholesterol elimination in mice. J Lipid Res 57(8):1455–1464

Schultz BG, Patten DK, Berlau DJ (2018) The role of statins in both cognitive impairment and protection against dementia: a tale of two mechanisms. Trans Neurodegen 7(1):5

Shakour N, Bianconi V, Pirro M, Barreto GE, Hadizadeh F, Sahebkar AJ (2019) In silico evidence of direct interaction between statins and β-amyloid. J Cell Biochem 120(3):4710–4715

Shankaran M, Di Paolo E, Leoni V, Caccia C, Bardile CF, Mohammed H et al (2017) Early and brain region-specific decrease of de novo cholesterol biosynthesis in Huntington's disease: a cross-validation study in Q175 knock-in mice 98:66–76

Shaw LM, Arias J, Blennow K, Galasko D, Molinuevo JL, Salloway S et al (2018) Appropriate use criteria for lumbar puncture and cerebrospinal fluid testing in the diagnosis of Alzheimer's disease. Alzheimers Dement 14(11):1505–1521

Shitara Y, Sugiyama Y (2006) Pharmacokinetic and pharmacodynamic alterations of 3-hydroxy-3-methylglutaryl coenzyme A (HMG-CoA) reductase inhibitors: drug–drug interactions and interindividual differences in transporter and metabolic enzyme functions. Pharmacol Ther 112(1):71–105

Simons M, Schwärzler F, Lütjohann D, Von Bergmann K, Beyreuther K, Dichgans J et al (2002) Treatment with simvastatin in normocholesterolemic patients with Alzheimer's disease: a 26-week randomized, placebo-controlled, double-blind trial. Ann Neurol 52(3):346–350

Sinyavskaya L, Gauthier S, Renoux C, Dell'Aniello S, Suissa S, Brassard P (2018) Comparative effect of statins on the risk of incident Alzheimer disease. Neurology 90(3):e179–ee87

Slooter A, Bots M, Havekes L, del Sol AI, Cruts M, Grobbee D et al (2001) Apolipoprotein E and carotid artery atherosclerosis: the Rotterdam study. Stroke 32(9):1947–1952

Smith KB, Kang P, Sabbagh MN, AsDN I (2017) The effect of statins on rate of cognitive decline in mild cognitive impairment. Alzheimer's Dementia Trans Res Clin Intervent 3(2):149–156

Soni M, Orrell M, Bandelow S, Steptoe A, Rafnsson S, d'Orsi E et al (2019) Physical activity pre-and post-dementia: English longitudinal study of ageing. Aging Ment Health 23(1):15–21

Spires-Jones TL, Attems J, Thal DR (2017) Interactions of pathological proteins in neurodegenerative diseases. Acta Neuropathol 134(2):187–205

Stanimirovic DB, Friedman A (2012) Pathophysiology of the neurovascular unit: disease cause or consequence? J Cereb Blood Flow Metab 32(7):1207–1221

Strom BL, Schinnar R, Karlawish J, Hennessy S, Teal V, Bilker WB (2015) Statin therapy and risk of acute memory impairment. JAMA Intern Med 175(8):1399–1405

Šubic A (2019) Vpliv Demence Na Preventivo in Zdravljenje Bolnikov Z Ishemično Možgansko Kapjo: Univerza v Ljubljani. Medicinska fakulteta

Sweeney MD, Kisler K, Montagne A, Toga AW, Zlokovic BV (2018) The role of brain vasculature in neurodegenerative disorders. Nat Neurosci 21(10):1318–1331

Swiger KJ, Manalac RJ, Blumenthal RS, Blaha MJ, Martin SS (2013) Statins and cognition: a systematic review and meta-analysis of short-and long-term cognitive effects. In: Mayo clinic proceedings. Elsevier, Rochester

Tarafdar A, Pula G (2018) The role of NADPH oxidases and oxidative stress in neurodegenerative disorders. Int J Mol Sci 19(12):3824

Tarantini S, Tran CHT, Gordon GR, Ungvari Z, Csiszar A (2017) Impaired neurovascular coupling in aging and Alzheimer's disease: contribution of astrocyte dysfunction and endothelial impairment to cognitive decline. Exp Gerontol 94:52–58

Toth P, Tarantini S, Csiszar A, Ungvari Z (2017) Functional vascular contributions to cognitive impairment and dementia: mechanisms and consequences of cerebral autoregulatory dysfunction, endothelial impairment, and neurovascular uncoupling in aging. Am J Phys Heart Circ Phys 312(1):H1–H20

Trompet S, van Vliet P, de Craen AJ, Jolles J, Buckley BM, Murphy MB et al (2010) Pravastatin and cognitive function in the elderly. Results of the PROSPER study. J Neurol 257(1):85

Trushina E, Singh RD, Dyer RB, Cao S, Shah VH, Parton RG et al (2006) Mutant huntingtin inhibits clathrin-independent endocytosis and causes accumulation of cholesterol in vitro and in vivo. Hum Mol Genet 15(24):3578–3591

Turner AJ (2020) Lipid rafts and amyloid metabolism: role in pathogenesis of Alzheimer's disease. Российский физиологический журнал им ИМ Сеченова 106(5):539–562

Tuttle MD, Comellas G, Nieuwkoop AJ, Covell DJ, Berthold DA, Kloepper KD et al (2016) Solid-state NMR structure of a pathogenic fibril of full-length human α-synuclein. Nat Struct Mol Biol 23(5):409–415

Uemura MT, Maki T, Ihara M, Lee VM, Trojanowski JQ (2020) Brain microvascular pericytes in vascular cognitive impairment and dementia. Front Aging Neurosci 12:80

Valenza M, Leoni V, Tarditi A, Mariotti C, Björkhem I, Di Donato S et al (2007) Progressive dysfunction of the cholesterol biosynthesis pathway in the R6/2 mouse model of Huntington's disease. Neurobiol Dis 28(1):133–142

Vickrey BG, Mittman BS, Connor KI, Pearson ML, Della Penna RD, Ganiats TG et al (2006) The effect of a disease management intervention on quality and outcomes of dementia care: a randomized, controlled trial. Ann Intern Med 145(10):713–726

Villemagne VL, Burnham S, Bourgeat P, Brown B, Ellis KA, Salvado O et al (2013) Amyloid β deposition, neurodegeneration, and cognitive decline in sporadic Alzheimer's disease: a prospective cohort study. Lancet Neurol 12(4):357–367

Wang P, Velagapudi R, Kong C, Rodriguiz RM, Wetsel WC, Yang T et al (2020) Neurovascular and immune mechanisms that regulate postoperative delirium superimposed on dementia. Alzheimers Dement 16(5):734–749

Wanleenuwat P, Iwanowski P, Kozubski W (2019) Alzheimer's dementia: pathogenesis and impact of cardiovascular risk factors on cognitive decline. Postgrad Med 131(7):415–422

Wełniak A, Koszykowska P, Kędziora-Kornatowska K (2020) Statins and Alzheimer—current stand and further development directions. Med Res J 5(3):191–200

Xiang Q, Wu W, Zhao N, Li C, Xu J, Ma L et al (2020) The influence of genetic polymorphisms in drug metabolism enzymes and transporters on the pharmacokinetics of different fluvastatin formulations. Asian J Pharm Sci 15(2):264–272

Yan C, Zhou Y, Chen Q, Luo Y, Zhang JH, Huang H et al (2020) Dysfunction of the neurovascular unit in diabetes-related neurodegeneration. Biomed Pharmacother 131:110656

Yang M, Williamson J (2019) Blood pressure and statin effects on cognition: a review. Curr Hypertens Rep 21(9):70

Yang Z, Wang H, Edwards D, Ding C, Yan L, Brayne C et al (2020) Association of blood lipids, atherosclerosis and statin use with dementia and cognitive impairment after stroke: a systematic review and meta-analysis. Ageing Res Rev 57:100962

Yarchoan M, Xie SX, Kling MA, Toledo JB, Wolk DA, Lee EB et al (2012) Cerebrovascular atherosclerosis correlates with Alzheimer pathology in neurodegenerative dementias. Brain 135(12):3749–3756

Yu X, Ji C, Shao A (2020) Neurovascular unit dysfunction and neurodegenerative disorders. Front Neurosci 14:334

Zalar B, Maver A, Kovanda A, Peterlin A, Peterlin B (2018) Clinical exome sequencing in dementias: a preliminary study. Psychiatr Danub 30(2):216–219

Zaleta AK, Carpenter BD, Porensky EK, Xiong C, Morris JC (2012) Agreement about diagnosis among patients, companions, and professionals following a dementia evaluation. Alzheimer Dis Assoc Disord 26(3):232

Zhang X, Wen J, Zhang Z (2018) Statins use and risk of dementia: A dose–response meta analysis. Medicine 97(30):e11304

Zhao A, Tohidkia MR, Siegel DL, Coukos G, Omidi Y (2016) Phage antibody display libraries: a powerful antibody discovery platform for immunotherapy. Crit Rev Biotechnol 36(2):276–289

Zhao X, Eyo UB, Murugan M, Wu LJ (2018) Microglial interactions with the neurovascular system in physiology and pathology. Dev Neurobiol 78(6):604–617

Zhu X-C, Dai W-Z, Ma T (2018) Overview the effect of statin therapy on dementia risk, cognitive changes and its pathologic change: a systematic review and meta-analysis. Ann Trans Med 6(22):435

Role of Immunotherapy in Ameliorating Proteopathic Dementia

16

Manisha Singh, Pranav Pancham, Shriya Agarwal, Harleen Kaur, Vinayak Agarwal, Raj Kumar Tiwari, Shalini Mani, and Rachana

Abstract

Dementia is a syndrome that marks a significant cognitive decline with an estimated 50 million populations being affected by this globally and addition of ten million cases every year predicting a compelling threat to society. It commonly involves the neuronal accumulation of proteins leading to protein toxicity, transmission interruptions, cognitive dysfunction, and eventually neuronal death. Currently, novel techniques using different protocols for early theragnostic and prognosis of this age-related disorder have been analyzed to develop efficient and reproducible combinatorial and biologically viable options. Subsequently, immunotherapies have gained much importance among the researchers with promising leads to control and avert the dementing process, while specifically targeting the senile plaques and protein accumulation to control neurotoxicity in neurodegenerative disorders (NDDs). Therefore, in this chapter authors have explored all the reported and possible immunologically applicable options for improving the further cognitive decline in dementia.

Keywords

Neurodegenerative disorders · Alzheimer's disease · Proteopathy · Neuroinflammation · Active immunization · Passive immunization · Synaptic transmission

M. Singh (✉) · P. Pancham · S. Agarwal · V. Agarwal · S. Mani · Rachana
Department of Biotechnology, Jaypee Institute of Information Technology, Noida, India

H. Kaur
Amity Institute of Biotechnology, Amity University, Noida, India

R. K. Tiwari
Department of Pharmaceutical Sciences, School of Health Sciences, University of Petroleum and Energy Studies, Dehradun, Uttarakhand, India

G. M. Ashraf, M. S. Uddin (eds.), *Current Thoughts on Dementia*,
https://doi.org/10.1007/978-981-16-7606-2_16

16.1 Introduction

Dementia is defined as progressive loss of intellectual acuity and enhanced cognitive decline significantly eroding out the quality and functioning of life creating biological stress on individuals and institution of their existence (Valera et al. 2016). It is caused due to the existence of extraneuronal systemic diseases, neuronal toxicity, and cerebral degeneration (Velayudhan et al. 2017). And most of the countries now have a higher percentage of aged population therefore, dementia has become a public health emergency, which has encouraged many countries to develop various national plans and research outcomes targeting this cause. The most prevalent pathophysiological mechanism governing the process of dementia arises due to the abnormal aggregation of functional proteins (Ross and Poirier 2005). And the hydrophobic amino acids buried deep within the folds of proteins cause protein folding/misfolding, prompting the polypeptide chains to unveil their hydrophobic regions on the surface (Oddo et al. 2006). These exposed hydrophobic moieties exhibit a higher propensity to attract other such groups, causing proteins aggregation and they travel from one cell to another, dictating other non-native polypeptide chains to come together to increase aggregates size and causing the normal protein machinery to go erratic in the process (Pooler et al. 2014). Further, these pathological events are the commonest sites in Alzheimer's disease (AD) where aggregation of amyloid beta (Aβ) plaques triggers the cascade of interruptions via synaptic failures, neuronal injuries, and finally resulting in cognitive impairments (Kwon et al. 2020; Wenning and Jellinger 2005). Tau and Aβ plaques further reduce the systemic solubility and result in sequestration and agglomeration of neurofibrillary tangles (NFTs) (Sigurdsson 2008). Thereafter, similar advancements are marked in Parkinson's disease (PD), where the characteristic deposit of Lewy bodies as main constituent and α-synuclein (α-syn) is a major associated factor (Boche et al. 2010).

In addition to these, there are other genetic factors like *PARK-3, −4, −5, −6,* and *−7* known to have a possible involvement in the pathogenesis of PD and elicitation of dementia symptoms. Owing to the major functional contributions of α-syn gene in the survival of dopaminergic neurons, many extensive research studies are aimed at understanding its role in the development of dementia characteristics and further conclusive pathophysiological mechanics still need to be elucidated (Kayed and Jackson 2009). Moreover, other rare incidents where protein misfolding happens due to the interaction between native and infectious proteins involves the PrPc (exogenous prion species) proteins, to undergo misfolding that causes initiation of prion's disease (PrD) (Neff et al. 2008). Hence, these molecular deficits lead to neurodegeneration, and then dementia eventually triggering enormous inflammatory responses. And most of the therapeutic strategies are directed toward clearing out these protein aggregates (tau, Aβ, α-syn) (Wang et al. 2019).

However, most recently immunotherapy-based neuroprotective approaches have attracted the attention of the scientific community in particular due to its specificity toward targeting particular protein types for clearance, neutralization, reduced inflammation apart from controlling its spread and synaptic damage (Takamatsu et al. 2017). Consequently, the passive immunization has exhibited a significant clearing out of tau and Aβ clogs in rodent models, suggesting some early success in translational research in this regard (Hanger et al. 2009). Also, active immunization targeting the tau clearance to enhance cognition and limit cognitive deficits was witnessed in the tau transgenic models wherein this approach proves to be efficient enough in controlling the tauopathies (Tran et al. 2014); (Cunningham and Skelly 2012). Hence, this chapter provides an overview of the problem of dementia, highlighting the pathological mechanics of NDDs, suitability of immunotherapeutic approaches, and the latest updates with its therapeutic efficacy.

16.2 Global Scenario of Dementia and Its Types

According to the Global Burden of Diseases (GBD) data (2016), the cumulative number of dementia patients increased from 20.2 million in 1990 to 43.8 million (Rosset and Roriz-Cruz 2014). Despite many efforts planned in this regard, there is still non-availability of standardized theranostic protocols for dementia, and the only available, currently accepted protocol is based on exponential age-based release in incidence and prevalence of dementia (Chatterjee et al. 2016). However, though it has strong relativity with age progression and is more common in old age this is not the only cause of dementia and it has been observed that population of almost every stratum, age, and gender are affected equally (Heaton et al. 1996). Furthermore, it has been noted that there are almost around 100 types of existing dementia subtypes and highest shareholding is by AD, also known as dementia of Alzheimer's type (DAT) followed by vascular dementia (VaD), Lewy bodies-initiated dementia (DLB), frontotemporal dementia (FTD), dementia of Parkinson's type (DPT), Huntington's type (DHT), and inherited dementias (IDs) (Laine et al. 1998).

As discussed, AD is a well-known commonest cause leading to dementia and resulting due to the exorbitant and aberrant deposition of protein plaques in the neuronal transmission pathway (Davis et al. 1990). These plaques or tangles leads to seizers in specific areas of brain which are responsible for storing memories like the amygdala, hippocampus, cerebellum and prefrontal cortex. These seizures hinder the normal network of neurotransmitters together with causing the trigger in memory loss known as DAT (Hart and Semple 1990). Similarly, in case of DLB/DPT, there is a deposition of spherical proteins known as Lewy bodies blocking the neurotransmitters to source dementia (Biessels et al. 2014). Further, the circulatory compromise in the cerebral vascular system due to stroke, hypertension, infarction, higher cholesterol, and other secondary causes induces interruptions in oxygen supply to the cortical areas of the brain resulting in VaD. However, FTD is recognized as the rarer form of dementia, covering various types of neuronal disorders such as pick's disease and motor neuron diseases (MNDs) that specifically affect the frontal and temporal lobes of the brain hence, is more related to the behavioral symptoms rather than memory (Rubin et al. 1987). All these types of dementia lead to irreversible progressive changes and to date there is no promising therapeutic mediation found therefore, it is all the way more important to understand their involved pathological mechanics and design the relevant medication (Prinz et al. 1982).

16.3 Mechanisms of Protein Toxicity in Neurodegenerative Dementias

Abnormal inter- and extracellular accumulation of misfolded protein aggregates sets the primary platform for NDDs further causing dementia (Chung et al. 2018). Though these protein aggregates are not interrelated they do exhibit many common features like dense fibrillar morphology, mostly ubiquitinated, resistant to proteolytic degradation, and have secondary β sheet structures, all marked as neurotoxic elements (Wong and Krainc 2017). Subsequently, the primary component of AD-linked amyloid plaques, Aβ accumulation instigates the protein toxicity and neurodegeneration by formation of a variety of oligomeric species from monomeric Aβ protein aggregates, the Aβ oligomer (oAβ) then further instigates Aβ protein aggregation to form short protofibrils that are flexible and asymmetrical, the protofibrils eventually elongate into insoluble symmetric fibrillar assemblies consisting of β-strand repeats, perpendicularly oriented to the fiber axis (Núñez et al. 2012). This fibrillar form of Aβ protein is impervious to hydrolytic degradation and is considered neurotoxic (Table 16.1). These Aβ fibrils are also reported to elevate abnormal synaptic neurotransmitter glutamate and excitotoxicity through N-methyl-D aspartic acid receptors (NMDARs) (Sharma et al. 2015).

Additionally, oAβ has been directly linked to neuronal apoptosis due to the disruption of impaired mitochondrial dysfunction, intracellular calcium balance, and promote reactive oxygen species (ROS) and their accumulation-mediated

Table 16.1 Types of protein accumulation-based toxicity causing dementia

S. No.	Proteins involved in NDDs	Intermediate products formed	Imbalance caused	Related signs and symptoms	References
1	Tau	Tau species aggregates	Respiratory, ATP production reduced	Cognitive dysfunction	(Duggal et al. 2019; Collin et al. 2014; Watanabe et al. 2005)
2	Aβ	Aβ aggregates	Insoluble senile plaques	Cognitive dysfunction	(Núñez et al. 2012; Sharma et al. 2015)
3	α-Synuclein	Lewy body and Lewy neurite	Disrupts cellular homeostasis	Motor, cognitive, and psychiatric	(Boland et al. 2018)
4	Huntingtin	CAG repeats	Toxicity	Motor, cognitive, and psychiatric	(Kim et al. 2018; Collin et al. 2014)

synaptic dysfunctions have also been known to interact with several receptors (Warrick et al. 2005). Another receptor ephrin type-B receptor 2 (EPhB2) also plays a critical role in the development and maturation of the CNS and its interaction with oAβ has shown to reduce its levels in hippocampal neurons. The interactions between oAβ and PirB (paired immunoglobulin-like receptor B) would enlist cofilin-signaling components that further instigate depolymerization of actin from microtubules and eventually resulting in cognitive deficits through synaptic dysfunction (Guo et al. 2020). Subsequently, the histological changes in patients suffering from DAT are prevalent in the frontal and temporal lobes of the brain, marked with accumulated senile plaques (SPs) (Yamada et al. 1987; Morris et al. 1996). These SPs are composed of extracellular Aβ protein aggregates with mutations in the Aβ sequence, abnormal neuronal degradation, and accumulation of NFTs in the intracellular region (Lendon et al. 1998).

Besides this, there is another integral protein related to AD, tau, also known as proteins of tangles and is one of the major components of NFTs (Steiner et al. 2018). It is a microtubular paired helical filamentous proteins that work in balancing the genomic and proteomic levels along with fastening up the axonal transport (Duggal et al. 2019; Boutajangout et al. 2010). When they get hyperphosphorylated, they fail to bind with the microtubules and get overexpressed in the path, further triggering cell apoptosis and cognitive dysfunction (Salminen et al. 2011; Wang and Colonna 2019). However, there are many other types of proteins too, that are non-amyloidogenic in nature like synuclein, representing a membrane-bound class of presynaptic protein (Boland et al. 2018). Synuclein binding with the fibrillar inclusion bodies in DLB/DPT and its aggregation in the glial inclusion bodies makes it a toxic protein and is known as synucleinopathies (Jankovic et al. 2018). Correspondingly, in DHT the aggregation of ubiquitinated huntingtin protein fragments in neuronal cytoplasmic or nuclear regions causes intraneuronal/perineuronal lesions eventually leading to cellular apoptosis and cognitive decline (Kim et al. 2018; Collin et al. 2014) (Table 16.1).

16.4 Mechanisms of Protein Toxicity in Dementia of Alzheimer's Type

16.4.1 Amyloid Cascade Hypothesis

Identification of accumulated monomeric Aβ protein aggregates in the form of amyloid plaques in intracellular neurons is considered a key component in the diagnosis of AD-associated amyloid plaques (Watanabe et al. 2005). Aβ peptide is a product formed by cleavage of type 1 transmembrane amyloid precursor protein (APP) by protease, APPs are synthesized from *AAP* encoding gene that resides on chromosome 21 (Lehmensiek et al. 2002). The protease-mediated cleavage of *APP* is carried out by mainly three proteolytic secretase enzymes α-, β-, and γ-secretases wherein α-secretase composed of *ADAM* (a disintegrin and metalloprotease domain) *9,10* and *17*; β-secretase in the brain is called *BACE1* and γ-secretase has four core

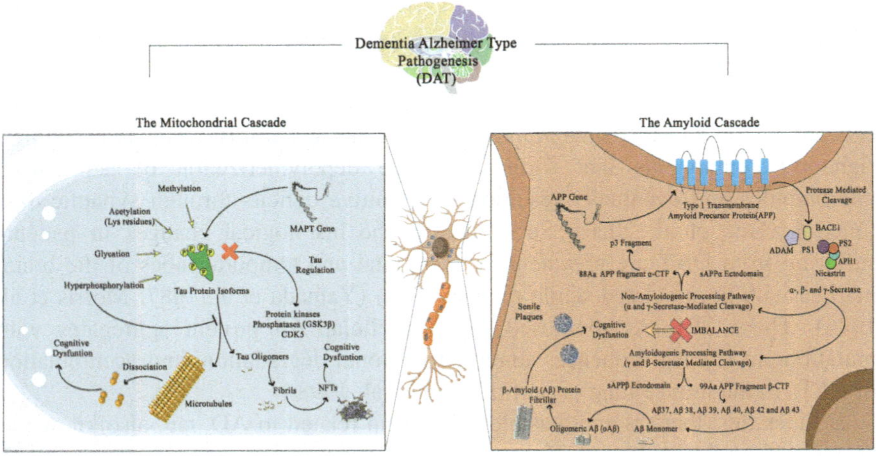

Fig. 16.1 Pathogenesis of protein accumulation and toxicity through mitochondrial and amyloid cascade hypothesis. *NFTs* neurofibrillary tangles, *GSK3β* glycogen synthase kinase, *3β* MAPT microtubule-associated protein tau, *CDK5* cyclin-dependent kinase, *APP* amyloid precursor protein, *sAPP* α-soluble amyloid precursor protein, *ADAM* disintegrin and metalloproteinases, *BACE1* β-site amyloid precursor protein cleaving, *PS1* presenilin 1, *PS2* presenilin 2, *APHI* anterior pharynx defective 1, *αCTF* α C terminal fragment

components nicastrin, *APH1*, PEN2, and presenilin's (PS1 and PS2) (Sorkina et al. 2005; Stephans et al. 2002). These proteolytic secretase enzymes can cleave *APP* to form 8–11 isoforms through the mechanism of alternate transcriptional splicing (Foyer et al. 1997; Sorkina et al. 2005). However, 2 isoforms (*APP751*, *APP770*) are commonly expressed in neurons and glial cells, and 1 isoform (*APP695*) in neurons only (Foyer et al. 1997; Mohamed et al. 2016). The APP cleavage pathway based on the isoforms can be categorized into non-amyloidogenic and amyloidogenic processing pathways and imbalance in the pathway is hypothesized as one of the major causes of DAT (Pimplikar 2009) (Fig. 16.1). Remarkably, both pathways have been shown to compete and can be further studied to derive possible strategies for the reduction of Aβ protein by enhancing both the pathways and reducing the amyloidogenic pathway (Pimplikar 2009; Tiiman et al. 2013).

α- and γ-secretases-mediated cleavage is observed in the non-amyloidogenic pathway involving full cleavage of APP by α-secretase, further release of sAPPα ectodomain outside the cell membrane, and retaining of APP fragment of 83 amino acids with C terminal (α-CTF or C83) inside the plasma membrane that additionally cleaves the γ-secretase releasing small p3 fragment in extracellular space (Kawahara and Kato-Negishi 2011; Stancu et al. 2014). In contrast to the non-amyloidogenic pathway, the amyloidogenic pathway follows subsequent cleavage of APP by γ-secretase and β-secretase complexes (Stancu et al. 2014). As upon cleavage of APP by β-secretase releases sAPPβ ectodomain and APP fragment of 99 amino acids with C-terminal (β-CTF or C99) which is further cleaved by γ-secretase releasing peptides consisting of various chain length such as Aβ37, Aβ38, Aβ39, Aβ40, Aβ42, and Aβ43 (Potter and Wisniewski 2012; Verma et al. 2015). The solubility of Aβ42 is not as abundant as Aβ40 and hence possesses a higher tendency to aggregate due to its hydrophobic nature and is also considered the main component of the SPs and hypnotized the formation and accumulation leading to AD pathogenesis (Casadesus et al. 2004).

16.4.2 Mitochondrial Cascade Hypothesis

Many research studies have attested to the role of Aβ accumulation in the mitochondrial set up of AD patient's brains and its structural alteration leading to mitochondrial dysfunction in systemic AD pathogenesis (Swerdlow et al. 2010; Berg et al. 2010). This alteration reduces the mitochondrial capacity to perform its elementary task during respiration which gets compromised in cerebral regions associated with amyloid plaques (Jellinger 2010). ATP production reduced by the increase in proteins which regulate fission and fusion in mitochondrial, therefore, impairing the mitochondrial dynamics and elevating mitochondrial linked oxidative stress, which has been observed in early manifestation sign in AD, suggesting the potential to drive Aβ-induced AD pathogenesis by inducing ROS (Manke et al. 2013). Amyloid-binding alcohol dehydrogenase, cyclophilin D, cytochrome c, and apoptosis-inducing factors have been also linked to mitochondrial dysfunction, cell apoptosis, and neuronal loss (Cardoso 2011). Another aspect to progressive dementia is the existence of tau protein, a microtubule-binding component, translated from *MAPT* gene on chromosome 17 (Bartley et al. 2012). Alternative splicing results in six different isoforms variants. Isoforms can be distinguished through the composition of 29 N-terminal amino acids insert and 3R and 4R microtubules binding domain repeats, where 4R has 3R ratio are maintained in 1:1 ratio (Fig. 16.1).

However, 4R has a greater affinity for microtubule synthesis. Alteration in *MAPT* pre-mRNA splicing could lead to an imbalance in ratio. However, complete inactivation of tau protein doesn't influence axonal transport, hence, suggesting the association of *MAP1* and *MAP2* to compensate for tau protein-mediated microtubule assembly (Kaplowitz 2002). The physiological functions of tau as a microtubule-binding component to the C-terminals repeats promote polymerization and stability. In neurons, tau isoforms functions are mediated through post-translational modification activity such as phosphorylation and proteolytic cleavage (Sikriwal et al. 2008). Tau phosphorylation modulates microtubule binding and hyperphosphorylation promotes dissociation from microtubules and increased aggregation (Swerdlow et al. 2010). Therefore, more biologically oriented strategies are needed that include gene therapy approaches that regulate protein degradation and clearance by various mechanisms through proteolysis and autophagy routes along with the immunotherapeutic approaches. Both active and passive types of immunization approaches can be employed to modulate inflammatory responses and oligomerization activities (Chagkutip et al. 2003; Hoozemans et al. 2012).

16.4.3 Lipid Peroxidation Malfunctioning

As the brain and blood-brain barrier (BBB) requires a significantly high amount of oxygen and has the highest rate of lipid metabolism that is essential for cell membrane lipid replacement, therefore, the brain's reliance on mitochondria to maintain homeostasis backfires when an accumulation of tau or amyloid protein occurs (Halliwell and Chirico 1993; Kappus and Sies 1981). This results in mitochondrial dysfunction and generation of ROS in brain through leakage of electrons from electron transport chain, cyclooxygenases (COXs), cytochrome p450s (CYPs), and lipoxygenases and reduced antioxidant mediated defense mechanism, thereby attacking surrounding membrane proteins (Horton et al. 1987).

ROS-mediated interaction with cell lipids bound arachidonic acid (AA) leads to lipid peroxidation through a non-specific method by instigating a chain reaction of hydrogen dissociation from lipid molecules to produce the aldehydic product in oxidized lipid membranes (free 4-hydroxy-2-trans-nonenal (HNE)), which is highly lethal and can interact through bond formation with other protein such as glucose transport in hippocampal, *BACE1* expression, thereby increasing A*β* production and accumulation that ultimately causes cell wall instability and cell death (Dix and Aikens 1993). Lipid peroxidation has been known to target redox metals in brain, which leads to iron-mediated necrosis (ferroptosis) and inflammation (Cheeseman 1993; Gutteridge 1995; Chatterjee et al. 2016). Polyunsaturated fatty acids (PUFAs) such as linoleic acid, AA, and docosahexaenoic acid (DHA) causing NDDs and cognitive dysfunction are also considered preferential targets of lipid peroxidation (Halliwell 1978).

16.5 Genetic Mechanisms Involved in Dementia

The most prevalent factor in 60–80% cases of all dementia-related pathogenesis is heritability and genetics related. Recent studies implicate various genetic factors among which apolipoprotein E (*APOE*) gene on chromosome 19, which produces three allelic variants of glycoproteins (APOE2, APOE3, and APOE4), under neuronal dysfunction accumulates in microglia and astrocytes, which further advocates and enhances Aβ seeding and oligomerization and its accumulation in the brain (Canessa et al. 1993; Lesch and Mössner 1998). Some studies suggest the downregulation of Aβ accumulation through *APOE2* and *APOE3*; however, *APOE4* has been associated with Aβ accumulation and tau pathogenesis-mediated cognitive dysfunction such as memory loss or impairment (Chu et al. 2009).

Unbalanced APOE2 levels have been also linked to tauopathy (Kretzschmar 2005). Overexpression of genetic factors such as siglec-3 (CD33), bridging integrator 1 (BIN1), and sortilin-related receptor 1 (SORLA) has been associated with neuronal degeneration and AD pathogenesis leading to cognitive deficits (Zecca et al. 2004). Downregulation of PU.1 a myeloid transcription factor and its expression is fundamental to microglia development, phagocytosis function and is considered a key regulatory mechanism to genes (ABCA7, CD33, TREM2, MS4A4A,

MS4A6A, TYROBP, Aif1, and MYBPC3) associated with AD pathogenesis (Durrenberger et al. 2015; Nakano et al. 2017).

16.6 RNS–ROS Generated Stress in Dementia

Production of ROS along with reactive nitrogen species (RNS) are considered as the most crucial biochemical anomaly that play part in developing neurodegenerative pathologies (Fransen et al. 2012). Their biosynthesis also leads to a significant rise in the oxidative stress level as well as creation of hypoxic environment which further guides the diseased subject toward severe tissue damage and cognitive impairment (Di Meo et al. 2016). First and foremost, RNS production is primarily mediated through nitric oxide (NO) as a major contributor. Despite having a pivotal role in RNS synthesis NO is highly essential for neuronal development, synaptic plasticity, and facilitating neurotransmitter release (Tamir et al. 2002). Predominantly NO synthesis in human system is regulated by three nitric oxide synthase enzymes (NOS), which are endothelial NOS (eNOS), inducible NOS (iNOS), and neuronal NOS (nNOS) (Berg et al. 2011). However, RNS biogenesis also involves combinations of several superoxide ($O2^-$) and nitric oxide radicals (NO^-), dinitrogen trioxide (N_2O_3), nitrogen dioxide, and peroxynitrite anion ($ONOO^-$) (Wang et al. 2006). RNS distress is seen as a characteristic attribute in identifying cognitive defects as it simply affects the intercellular moiety of neuronal cells which leads to associated synaptic malfunction along with microglial activation which subsequently initiates the path for ROS production (Bhat et al. 2015). Various research studies have validated the presence of ROS as a key to pathogenesis of NDDs such as AD (Jomova et al. 2010). Surging ROS and RNS build-up is accompanied by elevated metal ion levels and a compromised anti-oxidant system making the brain extremely vulnerable to protein oxidation and lipid peroxidation (Fenster et al. 2002). Recently concluded studies have displayed ROS targeting protein moieties leading to the formation of Michael adduct between cysteine, histidine, and lysine residues. Protein oxidation generally results in the protein carbonyl formation as end-products which are seen as biomarkers for the presence of oxidative stress along with 42% and 37% increment in hippocampus and inferior parietal lobule respectively in AD pathologies (Agostinho et al. 2010).

16.7 Neuroinflammatory Complex Generated Stress in Dementia

ROS-mediated sequential inflammatory response due to infection, neuronal dysfunction, protein accumulation, or toxicity in neuronal cells such as microglia and astrocytes has been observed to be associated with generation of stress and proinflammatory characteristics (Taylor et al. 2013; Mosley et al. 2006). Prolong activation of such neuronal cells in central nervous system (CNS) is the primary cause of inflammatory response through neuroinflammatory mediator complex, which are mainly found in cerebrospinal fluid (CSF), neurons, blood, brain, and serum in AD patients (Agostinho et al. 2010; García-Bueno et al. 2008). Tumor necrosis factor-α (TNF-α), interleukin-2, -6 and -1β, and cyclooxygenase-2 (COX-2), transforming growth factor (TGF) β1-β2 and caspase-3/7/8-mediated inflammation can cause imbalance in regulatory and homeostasis function which further accelerates neuronal degeneration, Aβ accumulation, and cell apoptosis (Nerurkar et al. 2011). Therefore, systemic inflammation leading to cognitive dysfunction through upregulation of COX-2, caspases activation, and stimulation of astrocytes-mediated interleukin complexes are major causes of APP-derived Aβ aggerates and are considered as a biomarker in AD diagnosis (Liu et al. 2015).

16.8 Therapeutic Approaches for Dementia

16.8.1 Pharmacological Interventions

The treatment approaches can be broadly divided into two categories: non-pharmacological and pharmacological (Schulz et al. 2002). However, the most prominent treatment methods against DAT are pharmacological ones and have been used for targeting a reduction in brain amyloid levels by reducing the amyloid production and removal of Aβ accumulates. Targeting γ-secretase directly has been seen to generate side effects but certain non-steroidal anti-inflammatory drugs like ibuprofen and flurbiprofen have shown to partly inhibit γ-secretase, therefore altering the splicing mechanism of APP (Hosie et al. 2019). However, they have minimal benefits. Similarly, the majority of FDA-approved drugs for the treatment of AD such as donepezil, galantamine, and rivastigmine have been seen to counterbalance the neurotransmitter imbalance by acting as acetylcholinesterase inhibitors (AChEIs) and increase acetylcholine availability at synapses and ultimately delaying cognitive deterioration but these drug intakes possess certain potential side effects (diarrhea, nausea, and vomiting) (Bullock et al. 2019).

Another drug that has shown promising results is memantine, a low-affinity, non-competitive, NMDA receptor binding drug. It can block calcium channels thereby affecting glutamatergic transmission and may improve cognitive and behavioral outcomes (Gitlin et al. 2010). Memantine in combination with AChEIs have shown synergistic effect without adverse side effects (Cooke et al. 2010; Qaseem et al. 2018). Apart from these drugs, there haven't been any new FDA-approved drug since 2003 as we know of Aβ plague that has been observed to be associated with abnormal accumulation of metal ions such as copper, iron, and zinc; in a clinical trial phase II, deferiprone, an iron-chelating agent has been observed to reduce CSF Aβ in early AD (Kleinstäuber et al. 2015).

16.8.2 Non-Pharmacological Intervention

The non-pharmacological approaches against DAT have been an extensive area of research and have elaborate and wide range of approaches and techniques (Bowen et al. 2011; Tripathi and Tiwari 2009). Another approach that is widely accepted and researched is immunotherapeutic approaches that can be broadly categorized into active and passive immunotherapy and have been utilized to enhance the removal of Aβ accumulates in transgenic mouse models and prevention of DAT and PrD (Hoffmann et al. 2010; Neal and Wright 2003).

However, there are other approaches too that have shown efficacy up to a certain limit and generally include exercises, motor, and cognitive rehabilitation techniques (Douglas et al. 2004; Cohen-Mansfield 2004). These approaches require a lot of resource allocation and many professionals such as psychologists, therapists, and caregivers, who offer comprehensive and precise individualized management. Such activities have been shown to enhance cognitive function and reduce the risk of DAT (Gitlin et al. 2007; Buchanan et al. 2007). The advancement of technology has allowed application of other multidimensional strategies such as aromatherapy, art therapy, music therapy, telemedicine, virtual reality, and gaming (Fonseca et al. 2015; Remington et al. 2006).

16.9 Targeting Protein Accumulation by Immunotherapy for Dementia

Immunotherapies including immune checkpoint blockades that increase adaptive immune responses systemically and are efficient enough to combat NDDs specifically AD though current studies fail in explaining any potent effect of such checkpoints in transgenic mouse models (Ishihara et al. 2019). The immunotherapeutic approaches comprise of manipulation in immune responses by various experimentation on active, passive, and cellular immunizations along with their route of administration (Valera et al. 2016). However, findings from the literature of experimental models and human genetic studies propose that the innate immunity plays a major role in treating various NDDs by displaying an effective toxic protein clearance in cortical regions (Troquier et al. 2012; Gao et al. 2019). Further, by blocking the activation of T-cell, cytotoxic T lymphocyte-associated antigen-4 (CTLA-4), programmed cell death-1 (PD-1) lymphocyte activation gene-3 (*LAG-3*) immune checkpoints can be inhibited thus, developing suppression of immune system known as tolerance toward the host cells (Noman et al. 2019; Livingston et al. 2005). Also, LAG-3 is known to be responsible for enabling the spread of α-syn gene in transgenic mouse models of PD and therapeutic involvement of PD-1 specific antibodies, increases the infiltrations of myeloid cells in nervous systems, and decreases the Aβ plaques in cortex and hippocampus (Ishihara et al. 2019; Turner 2005). Though in more advanced researches it was been noted that inhibition of PD-1 alone is not sufficient to influence the entry of myeloid cells in CNS therefore, development in this field can be promoted by establishing inflammasome activation in AD, PD, and MS disease models (Vanneman and Dranoff 2012). Inflammasome complexes that are activated in microglia have shown promising results in innate immunity, secretion, and maturation of pro-inflammatory cytokines like interleukins (IL-β) (Seimetz et al. 2010). This large complex comprises of sensor molecule (*NLRP3*, NLR family pyrin domains-containing protein 3), adaptor protein (ASC), and caspases-1, together form a functional unit upon protein misfolding, damage of tissues, and other infections (Valera and Masliah 2013).

Moreover, the hallmark point in AD is plaque formation in afflicted subjects thus, targeting them through vaccination is required as this approach may prove to be more efficacious than other drug-based approaches (Hong et al. 2020). Active immunization encompasses vaccines involving a particular antigen, i.e., Aβ peptide that focuses on eliciting immune responses against antigen in the host body (Yang et al. 2019; Guetin et al. 2009). Another strategy involves the use of affitopes-short chains of amino acid that mimic segments of native $A\beta_{1-42}$ without identifying the sequence (Yang et al. 2019). AD-01 and AD-02 affitopes target N-terminal of Aβ fragments and modify AD in animal models, whereas passive immunization involves consistent administration of monoclonal or polyclonal antibodies against its antigen (anti-Aβ antibodies) by preventing oligomer formation. Currently, monoclonal antibodies are administered intravenously in AD patients (solanezumab [LY-2062430], GSK-933776, MABT-5102A and bapineuzumab [AAB-001]) (Keu et al. 2017; Zhai et al. 2015). Both these immunizations are responsible for successful inhibition or clearance of toxic Aβ deposits and cognitive decline from transgenic mouse models (prior human trials) (de Haas et al. 2016). Nevertheless, removal of Aβ aggregates can be done by various immune-mediated mechanisms such as extraction of plaques by plasma antibodies from the brain; solubilization binding of antibody and plaque; phagocytosis of aggregates by microglia (Fig. 16.2) (Pittock and Balice-Gordon 2012). Equivalently, it has been also reported that cellular immunotherapies play a vital role in eliciting microglial cell phagocytosis in AD subjects by regulating cellular T-cell, inflammatory, and Aβ immune responses (Lacroix et al. 2018; Lall and Baloh 2017).

The advantage of active immunization is that it ensures constant high concentrations of antibodies thus limiting follow-ups and costs (Iba et al. 2020). With passive immunizations, particular Aβ epitopes can be targeted more effortlessly, and therefore, rapid control of antibodies level is possible (Pahnke et al. 2009). In elderly patients, passive immunotherapy is more promising than active immunotherapy because these patients do not respond to vaccines, anymore (Kortylewski et al. 2009). Nevertheless, administering antibodies is costly as well as time-consuming but the major risk associated with this type of immunization is the development of cerebral amyloid angiopathy and vasogenic edema with microhemorrhages which is observed to be higher in passive immunizations than overactive immunizations. Hence, there is a compelling need of combinatorial and alternative approaches to be considered for determining effectors and antigen targets such as tau rather than antibodies (Panza et al. 2016).

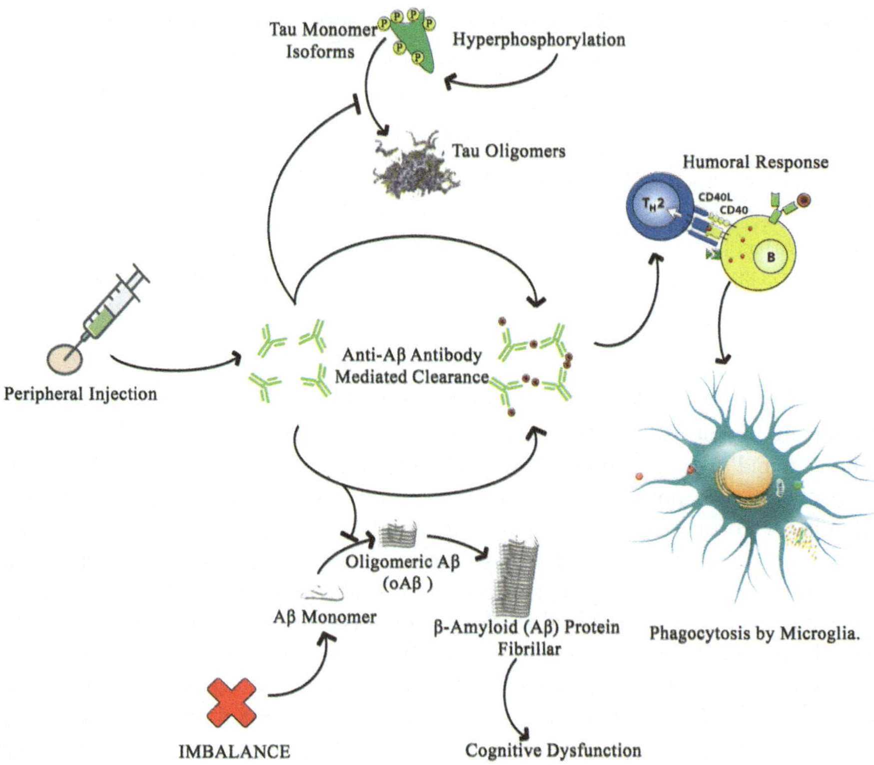

Fig. 16.2 Mechanism for active and passive immunotherapy goals for reducing dementia and its pathological causes

16.10 Immunotherapeutic Approaches for Dementia

Contemporary times have witnessed several clinical trials operating distinctly to develop an efficient and efficacious immunotherapeutic regime to clear protein aggregates (tau and amyloid aggregates) accompanied by other immunotherapies in AD patients eliciting immune system to treat dementia and cognitive impairments. Immunotherapeutic approaches for targeting amyloid have been in trials since 2000, however, for targeting tau aggregates trials have been started for the last 5 years. Thus, for devising a thorough and reproducible therapeutic approach active and passive clinical trials are planned to attain the desired target (Arasi et al. 2018).

As discussed above, active immunization is considered more potent and efficacious than passive form since the former provides less administration of induced autologous antibody and the effect of treatment is not constrained by the production of antidrug antibody. Also, anti-Aβ-antibodies (Abs) have been found to avert Aβ fibrilization, neurotoxicity, and neuronal cell death in in vitro cultures. Such preliminary results further pushed the research's direction toward in vivo models' systems

Table 16.2 Clinical studies aimed at designing Aβ-Abs-based active immunotherapy

Company/sponsor	Therapeutic	Clinical trial stage	References
Axon neuroscience SE	AADvac-1	Phase I	(Ceyzériat et al. 2020)
		Phase II	
Novartis	CAD106	Phase 1	(Arasi et al. 2018)
AC immune SA, Janssen	ACI-35	Phase I	(Pfaar et al. 2018)
		Phase I	
Janssen/Pfizer	ACC-001	Phase II	(Pfaar et al. 2018)
Affinis AG/GSK	AFFITOME	Phase II	(Wagenmann et al. 2019)
AC immune	ACI-24	Phase I/IIa	(Verma et al. 2018)

for a better understanding of Aβ-mediated AD pathology. These trials involve the administration of $A\beta_{1-42}$/Aβ homologous peptide along with specific adjuvants (alum/Freund's) as a form of active immunization. This combinational Ab has been observed to not only ward off any cognitive impairment but also avert the Aβ plaque formation. Furthermore, immuno-histo-chemical assay explored the intricate expression of the Ab in the human system to indicate the prospects of achieving labeled amyloid plaque region in AD subjects (Pfaar et al. 2018). Peripheral injections are the preferred mode for administrating anti-Aβ monoclonal Abs directly into the systemic circulation and their therapeutic expression is facilitated by humoral response (controlled by TH2 cells) instead of cell-mediated response (controlled by TH1 cells) as it leads to cytotoxicity in neuronal cells in the diseased subjects. One of the trials using the active immunotherapy approach is AN1792 that resulted in fatal meningoencephalitis that occurred in 6% of patients due to mediation of T-cell which led to the termination of the trial. Also, inflammation and tau aggregates in the mouse model were observed due to the use of full-length human tau proteins, further ceasing the trial. The first vaccine to be tested for tau aggregation was AADvac1 (Table 16.2), produced by Axon Neuroscience SE and used truncated tau protein of 151–391/4R that was assumed to be pathological fragment for eliciting aggregation and misfolding. Monoclonal antibodies (mAbs) named DC8E8 were also used against this fragment and studies showed that it disrupted tau-tau interactions, responsible for tau aggregation. This epitope was found in MTBR repeat region and vaccine was developed by functionalizing peptide fragment to a carrier protein that elicited B-cell-mediated immunity. Another vaccine, named CAD106 (Table 16.2) developed by Novartis Pharmaceuticals aimed to target B-cell epitopes only by using adjuvant carrier derived from many copies of bacteriophage Qβ (coat protein) and directed amino terminals of Aβ fragment ($A\beta_{1-6}$). In the category of ongoing trials of phase II, it is ACC-001 that uses the same Aβ fragment attached with carrier protein and adjuvant QS-21, a surface-active saponin (Verma et al. 2018).

Consequently, FUNDAMANT (72-week trials) vaccine showed continuous immune response by providing booster doses for 48 and 72 weeks, and its high levels of antibodies were associated with reducing hippocampal atrophy (Vander Zanden and Chi 2020). Also, in a phase II trial for 24 months, ADAMANT was analyzed, and statement released in September 2019 announced no adverse events associated with this immunization (Table 16.2). Apart from active immunization, artificially fabricated Abs were administered in the diseased subject's system as a form of passive immunization (Table 16.3) to elicit specific immune responses (Ceyzériat et al. 2020). This is considered as one of the fastest and the easiest ways to facilitate monoclonal anti-Aβ-Abs expression without hindering the cell-mediated immune response as the Abs are passively transported exogenously (as represented in Fig. 16.2). Several passive clinical trials have recorded a significant decline in Aβ levels and improved cognitive attributes by implementing this methodology (Table 16.3). It is quite fascinating that several research studies have indicated toward Aβ plaque disassembly, activation of microglial that in turn leads to clearance of tau protein due to Abs expression (Plotkin and Cashman 2020). Besides these, Ab functions to block the Aβ toxicity in CNS, also referred to as "Peripheral Sink Effect", triggers the sudden fall in Aβ levels in the neurovascular system, primarily due to Ab binding to soluble Aβ (sAβ) in circulation leading to sAβ being flushed out of CNS.

Despite quality results, passive immunization has been closely associated with several limitations such as crossing BBB, periodic and systematic injection regimes for chronic diseases (such as dementia), antigen (Ag) targeting, risk of hemorrhage, and immunogenic response against Abs itself. Till now, passive immunotherapies trials have not been much reported. Solanezumab and bapineuzumab are phase III trials that did not show any disease-modification and clinical improvement results (Schilling et al. 2018). Also, the new mAbs against the N-terminal region which will enter soon in a clinical trial is LY3303560 but its precise epitope has not been out yet (Loureiro et al. 2020). In publicly available data, passive immunization with LY3303560 has shown to bind with tau proteins in animal models and also pertains acceptable pharmacokinetics (Wagenmann et al. 2019).

16.11 Antibody Engineering for Optimized Immunotherapy against Dementia

Among several therapeutic alternatives antibody-mediated immunotherapy is regarded as the most promising of all to achieve an optimized result against neurodegeneration (Sumner et al. 2018). Designed Abs must be highly sensitive toward a specific Ag to elicit targeted immune response thus it is imperative to identify as well as develop such a mechanism (Chanier and Chames 2019). However, there are several second-generation Abs that are still in developmental stages and one of the key aspects of engineering Ab is its highly specific epitopes which allows eliciting a strong and specific immune response (Nyakatura et al. 2017). There are couple of antibodies such as aducanumab and BAN2401 which account for different APP gene mutations prevalent in AD subjects such as arctic APP mutation protein (E693G) (Table 16.3). Thus, enabling them to be potent enough to prevent Aβ aggregation, neurotoxicity, and in the process prevents cognitive defects and progression of dementia (Zhao et al. 2016).

Table 16.3 Clinical studies aimed at designing Aβ-Abs-based passive immunotherapy

Potential antibodies	Clinical trial stage	Current status	Epitope	Isotype	ARIA-E Safety (%)	References
Solanezumab	III	Failed to meet clinical endpoints	Soluble; aa 16–24	IgG1	0.9	(Schilling et al. 2018)
Bapineuzumab	III		N-term; aa 16–24	IgG1	10	(Schilling et al. 2018)
Crenezumab	III		Conformational; aa 12–24	IgG4	0.3	(Plotkin and Cashman 2020)
Gantenerumab	III		Aa 3–11 & 19–25	IgG1	10	(Vander Zanden and Chi 2020)
Aducanumab	III	Active	Conformational; aa 3–6	IgG1	37–41	(Loureiro et al. 2020)
BAN2401	III		Conformational artic mutation	IgG1	< 1	(Vander Zanden and Chi 2020)

16.12 Recent Developments and Future Perspectives

The major challenge for the delivery of monoclonal antibodies either by active or passive immunization is their transport to the CNS either by crossing the corridors of BBB or through various other targeted routes (Kwon et al. 2020). Many past reports have shown that the treatment of AD through antibody injection that crossed the BBB shows hardly 0.1% of the success rate, where this low rate of transport was either due to their early metabolization in the liver or due to elimination through the kidneys (Medina 2011). Therefore, this accounts for the lower success rate of the anti-Aβ antibody therapies (Liu et al. 2012). Contrary to this, some studies target the BBB receptors "cargo" for the transcytosis of the proteins and help in increasing the penetration of the administered dose through the BBB by 2–3% (Ghaffar and Feinstein 2007). So, this approach can be navigated and studied further for developing an efficient transport medium for immunotherapeutic vaccines (Yu et al. 2019). Also, we need to equally look into the clearance mechanism of these antibodies from the CNS after they bind to the target protein, as they may lead to the release of some inflammatory mediators that may affect the normal functioning of healthy cells, so we need to keep a check on the adversities caused due to this too (Vincent et al. 2011).

16.13 Conclusion

The overview of immunotherapeutic strategies to date has shown promising results, specifically in case of DAT wherein monoclonal antibodies are targeted for Aβ aggregations. However, we still have to go long way and research community has not yet reported the viable options after the inception of the first trial with monoclonal Abs reported in 1977. Also, administering them before or after the onset of symptoms is been extensively explored so that maximum benefits can be derived after the vaccine administration. Moreover, the entire biological activity initiated due to these vaccinations exhibited very effective suppression of the neuroinflammatory responses by controlling the microglial activation and eliciting the immunomodulatory benefits through regulation of the complementary system apart from laying the foundation for the neurogenesis process. Also, the initiation of pre-clinical studies to seek innate immunity stimulation is another approach that can be relied upon. Besides this, the approaches that target antibody peptides need to be focused to control the pathological events in the early onset stage. Equally, it is important to ensure and predict that there is no toxicity or other risks involved while deigning the same as it happened in case of tau protein aggregation.

References

Agostinho P, Cunha A, R, Oliveira C. (2010) Neuroinflammation, oxidative stress and the pathogenesis of Alzheimer's disease. Curr Pharm Des 16(25):2766–2778

Arasi S, Caminiti L, Crisafulli G, Pajno GB (2018) A general strategy for de novo immunotherapy design: the active treatment of food allergy. Expert Rev Clin Immunol 14(8):665–671

Bartley MG, Marquardt K, Kirchhof D, Wilkins HM, Patterson D, Linseman DA (2012) Overexpression of amyloid-β protein precursor induces mitochondrial oxidative stress and activates the intrinsic apoptotic cascade. J Alzheimers Dis 28(4):855–868

Berg EL, Yang J, Melrose J, Nguyen D, Privat S, Rosler E et al (2010) Chemical target and pathway toxicity mechanisms defined in primary human cell systems. J Pharmacol Toxicol Methods 61(1):3–15

Berg RM, Møller K, Bailey DM (2011) Neuro-oxidative-nitrosative stress in sepsis. J Cereb Blood Flow Metab 31(7):1532–1544

Bhat AH, Dar KB, Anees S, Zargar MA, Masood A, Sofi MA et al (2015) Oxidative stress, mitochondrial dysfunction and neurodegenerative diseases; a mechanistic insight. Biomed Pharmacother 74:101–110

Biessels GJ, Strachan MW, Visseren FL, Kappelle LJ, Whitmer RA (2014) Dementia and cognitive decline in type 2 diabetes and prediabetic stages: towards targeted interventions. Lancet Diabet Endocrinol 2(3):246–255

Boche D, Denham N, Holmes C, Nicoll JA (2010) Neuropathology after active Aβ42 immunotherapy: implications for Alzheimer's disease pathogenesis. Acta Neuropathol 120(3):369–384

Boland B, Yu WH, Corti O, Mollereau B, Henriques A, Bezard E et al (2018) Promoting the clearance of neurotoxic proteins in neurodegenerative disorders of ageing. Nat Rev Drug Discov 17(9):660–688

Boutajangout A, Quartermain D, Sigurdsson EM (2010) Immunotherapy targeting pathological tau prevents cognitive decline in a new tangle mouse model. J Neurosci 30(49):16559–16566

Bowen A, Knapp P, Gillespie D, Nicolson DJ, Vail A (2011) Non-pharmacological interventions for perceptual disorders following stroke and other adult-acquired, non-progressive brain injury. Cochrane Database Syst Rev 2011(4):CD007039

Buchanan JA, Christenson AM, Ostrom C, Hofman N (2007) Non-pharmacological interventions for aggression in persons with dementia: a review of the literature. Behav Anal Today 8(4):413

Bullock L, Bedson J, Jordan JL, Bartlam B, Chew-Graham CA, Campbell P (2019) Pain assessment and pain treatment for community-dwelling people with dementia: a systematic review and narrative synthesis. Int J Geriatr Psychiatry 34(6):807–821

Canessa CM, Horisberger J-D, Rossier BC (1993) Epithelial sodium channel related to proteins involved in neurodegeneration. Nature 361(6411):467–470

Casadesus G, Takeda A, Perry G (2004) Challenging the amyloid cascade hypothesis: senile plaques and amyloid-β as protective adaptations to Alzheimer disease. Ann N Y Acad Sci 1019:1–4

Ceyzériat K, Zilli T, Millet P, Frisoni GB, Garibotto V, Tournier BB (2020) Learning from the past: a review of clinical trials targeting amyloid, tau and Neuroinflammation in Alzheimer's disease. Curr Alzheimer Res 17(2):112–125

Chagkutip J, Vaughan RA, Govitrapong P, Ebadi M (2003) 1-Methyl-4-phenylpyridinium-induced down-regulation of dopamine transporter function correlates with a reduction in dopamine transporter cell surface expression. Biochem Biophys Res Commun 311(1):49–54

Chanier T, Chames P (2019) Nanobody engineering: toward next generation immunotherapies and immunoimaging of cancer. Antibodies 8(1):13

Chatterjee S, Peters SA, Woodward M, Arango SM, Batty GD, Beckett N et al (2016) Type 2 diabetes as a risk factor for dementia in women compared with men: a pooled analysis of 2.3 million people comprising more than 100,000 cases of dementia. Diabetes Care 39(2): 300–307

Cheeseman K (1993) Mechanisms and effects of lipid peroxidation. Mol Asp Med 14(3):191–197

Chu J, Hong NA, Masuda CA, Jenkins BV, Nelms KA, Goodnow CC et al (2009) A mouse forward genetics screen identifies LISTERIN as an E3 ubiquitin ligase involved in neurodegeneration. Proc Natl Acad Sci 106(7):2097–2103

Chung CG, Lee H, Lee SB (2018) Mechanisms of protein toxicity in neurodegenerative diseases. Cell Mol Life Sci 75(17):3159–3180

Cohen-Mansfield J (2004) Nonpharmacologic interventions for inappropriate behaviors in dementia: a review, summary, and critique. Focus 9(2):361–308

Collin L, Bohrmann B, Göpfert U, Oroszlan-Szovik K, Ozmen L, Grüninger F (2014) Neuronal uptake of tau/pS422 antibody and reduced progression of tau pathology in a mouse model of Alzheimer's disease. Brain 137(10):2834–2846

Cooke ML, Moyle W, Shum DH, Harrison SD, Murfield JE (2010) A randomized controlled trial exploring the effect of music on agitated behaviours and anxiety in older people with dementia. Aging Ment Health 14(8):905–916

Cunningham C, Skelly DT (2012) Non-steroidal anti-inflammatory drugs and cognitive function: are prostaglandins at the heart of cognitive impairment in dementia and delirium? J Neuroimmune Pharmacol 7(1):60–73

Davis PB, Morris JC, Grant E (1990) Brief screening tests versus clinical staging in senile dementia of the Alzheimer type. J Am Geriatr Soc 38(2):129–135

de Haas N, de Koning C, Spilgies L, de Vries IJM, Hato SV (2016) Improving cancer immunotherapy by targeting the STATe of MDSCs. Onco Targets Ther 5(7):e1196312

Di Meo S, Reed TT, Venditti P, Victor VM (2016) Role of ROS and RNS sources in physiological and pathological conditions. Oxidative Med Cell Longev 2016:1245049

Dix TA, Aikens J (1993) Mechanisms and biological relevance of lipid peroxidation initiation. Chem Res Toxicol 6(1):2–18

Douglas S, James I, Ballard C (2004) Non-pharmacological interventions in dementia. Adv Psychiatr Treat 10(3):171–177

Duggal NA, Niemiro G, Harridge SD, Simpson RJ, Lord JM (2019) Can physical activity ameliorate immunosenescence and thereby reduce age-related multi-morbidity? Nat Rev Immunol 19(9):563–572

Durrenberger PF, Fernando FS, Kashefi SN, Bonnert TP, Seilhean D, Nait-Oumesmar B et al (2015) Common mechanisms in neurodegeneration and neuroinflammation: a BrainNet Europe gene expression microarray study. J Neural Transm 122(7):1055–1068

Fenster CP, Weinsier RL, Darley-Usmar VM, Patel RP (2002) Obesity, aerobic exercise, and vascular disease: the role of oxidant stress. Obes Res 10(9):964–968

Fonseca LM, Navatta ACR, Bottino CM, Miotto EC (2015) Cognitive rehabilitation of dementia in adults with down syndrome: a review of non-pharmacological interventions. Dementia Geriatric Cognitive Disorders Extra 5(3):330–340

Foyer CH, Lopez-Delgado H, Dat JF, Scott IM (1997) Hydrogen peroxide-and glutathione-associated mechanisms of acclimatory stress tolerance and signalling. Physiol Plant 100(2): 241–254

Fransen M, Nordgren M, Wang B, Apanasets O (2012) Role of peroxisomes in ROS/RNS-metabolism: implications for human disease. Biochimica et Biophysica Acta (BBA)-Molecul Basis Dis 1822(9):1363–1373

Gao S, Yang D, Fang Y, Lin X, Jin X, Wang Q et al (2019) Engineering nanoparticles for targeted remodeling of the tumor microenvironment to improve cancer immunotherapy. Theranostics 9(1):126

García-Bueno B, Caso JR, Leza JC (2008) Stress as a neuroinflammatory condition in brain: damaging and protective mechanisms. Neurosci Biobehav Rev 32(6):1136–1151

Ghaffar O, Feinstein A (2007) The neuropsychiatry of multiple sclerosis: a review of recent developments. Curr Opin Psychiatry 20(3):278–285

Gitlin LN, Winter L, Dennis MP, Hauck WW (2007) A non-pharmacological intervention to manage behavioral and psychological symptoms of dementia and reduce caregiver distress: design and methods of project ACT3. Clin Interv Aging 2(4):695

Gitlin LN, Winter L, Dennis MP, Hodgson N, Hauck WW (2010) A biobehavioral home-based intervention and the Well-being of patients with dementia and their caregivers: the COPE randomized trial. JAMA 304(9):983–991

Guetin S, Portet F, Picot M, Pommié C, Messaoudi M, Djabelkir L et al (2009) Effect of music therapy on anxiety and depression in patients with Alzheimer's type dementia: randomised, controlled study. Dement Geriatr Cogn Disord 28(1):36–46

Guo T, Zhang D, Zeng Y, Huang TY, Xu H, Zhao Y (2020) Molecular and cellular mechanisms underlying the pathogenesis of Alzheimer's disease. Mol Neurodegener 15(1):1–37

Gutteridge J (1995) Lipid peroxidation and antioxidants as biomarkers of tissue damage. Clin Chem 41(12):1819–1828

Halliwell B (1978) Biochemical mechanisms accounting for the toxic action of oxygen on living organisms: the key role of superoxide dismutase. Cell Biol Int Rep 2(2):113–128

Halliwell B, Chirico S (1993) Lipid peroxidation: its mechanism, measurement, and significance. Am J Clin Nutr 57(5):715S–725S

Hanger DP, Seereeram A, Noble W (2009) Mediators of tau phosphorylation in the pathogenesis of Alzheimer's disease. Expert Rev Neurother 9(11):1647–1666

Hart S, Semple JM (1990) Neuropsychology and the dementias. Psychology Press, London

Heaton RK, Ryan L, Grant I, Matthews CG (1996) Demographic influences on neuropsychological test performance. Neuropsychol Assess Neuropsychiatr Disord 2:141–163

Hoffmann T, Bennett S, Koh CL, McKenna KT (2010) Occupational therapy for cognitive impairment in stroke patients. Cochrane Database Syst Rev 2010(9):CD006430

Hong WX, Haebe S, Lee AS, Westphalen CB, Norton JA, Jiang W et al (2020) Intratumoral immunotherapy for early-stage solid tumors. Clin Cancer Res 26(13):3091–3099

Hoozemans JJ, Van Haastert ES, Nijholt DA, Rozemuller AJ, Scheper W (2012) Activation of the unfolded protein response is an early event in Alzheimer's and Parkinson's disease. Neurodegener Dis 10(1–4):212–215

Horton A, Fairhurst S, Bus JS (1987) Lipid peroxidation and mechanisms of toxicity. CRC Crit Rev Toxicol 18(1):27–79

Hosie A, Siddiqi N, Featherstone I, Johnson M, Lawlor PG, Bush SH et al (2019) Inclusion, characteristics and outcomes of people requiring palliative care in studies of non-pharmacological interventions for delirium: a systematic review. Palliat Med 33(8): 878–899

Iba M, Kim C, Florio J, Mante M, Adame A, Rockenstein E et al (2020) Role of alterations in protein kinase p38γ in the pathogenesis of the synaptic pathology in dementia with Lewy bodies and α-Synuclein transgenic models. Front Neurosci 14:286

Ishihara J, Ishihara A, Sasaki K, Lee SS-Y, Williford J-M, Yasui M et al (2019) Targeted antibody and cytokine cancer immunotherapies through collagen affinity. Sci Transl Med 11(487): eaau3259

Jankovic J, Goodman I, Safirstein B, Marmon TK, Schenk DB, Koller M et al (2018) Safety and tolerability of multiple ascending doses of PRX002/RG7935, an anti–α-synuclein monoclonal antibody, in patients with Parkinson disease: a randomized clinical trial. JAMA Neurol 75(10): 1206–1214

Jellinger KA (2010) Basic mechanisms of neurodegeneration: a critical update. J Cell Mol Med 14(3):457–487

Jomova K, Vondrakova D, Lawson M, Valko M (2010) Metals, oxidative stress and neurodegenerative disorders. Mol Cell Biochem 345(1–2):91–104

Kaplowitz N (2002) Biochemical and cellular mechanisms of toxic liver injury. Seminars Liver Dis 22(2):137–144. 2002: Copyright© 2002 by Thieme Medical Publishers, Inc., 333 Seventh Avenue, New . . .

Kappus H, Sies H (1981) Toxic drug effects associated with oxygen metabolism: redox cycling and lipid peroxidation. Experientia 37(12):1233–1241

Kawahara M, Kato-Negishi M (2011) Link between aluminum and the pathogenesis of Alzheimer's disease: the integration of the aluminum and amyloid cascade hypotheses. Int J Alzheimers Dis 2011:276393

Kayed R, Jackson GR (2009) Prefilament tau species as potential targets for immunotherapy for Alzheimer disease and related disorders. Curr Opin Immunol 21(3):359–363

Keu KV, Witney TH, Yaghoubi S, Rosenberg J, Kurien A, Magnusson R et al (2017) Reporter gene imaging of targeted T cell immunotherapy in recurrent glioma. Sci Transl Med 9(373):eaag2196

Kim C, Spencer B, Rockenstein E, Yamakado H, Mante M, Adame A et al (2018) Immunotherapy targeting toll-like receptor 2 alleviates neurodegeneration in models of synucleinopathy by modulating α-synuclein transmission and neuroinflammation. Mol Neurodegener 13(1):1–18

Kleinstäuber M, Witthöft M, Steffanowski A, Van Marwijk H, Hiller W, Lambert M (2015) Pharmacological interventions for somatoform disorders in adults, a Cochrane systematic review. J Psychosom Res 6(78):606–607

Kortylewski M, Kujawski M, Herrmann A, Yang C, Wang L, Liu Y et al (2009) Toll-like receptor 9 activation of signal transducer and activator of transcription 3 constrains its agonist-based immunotherapy. Cancer Res 69(6):2497–2505

Kretzschmar D (2005) Neurodegenerative mutants in drosophila: a means to identify genes and mechanisms involved in human diseases? Invertebr Neurosci 5(3–4):97–109

Kwon S, Iba M, Kim C, Masliah E (2020) Immunotherapies for aging-related neurodegenerative diseases—emerging perspectives and new targets. Neurotherapeutics 17(3):935–954

Lacroix R, Rozeman EA, Kreutz M, Renner K, Blank CU (2018) Targeting tumor-associated acidity in cancer immunotherapy. Cancer Immunol Immunother 67(9):1331–1348

Laine M, Laakso M, Vuorinen E, Rinne J (1998) Coherence and informativeness of discourse in two dementia types. J Neurolinguistics 11(1–2):79–87

Lall D, Baloh RH (2017) Microglia and C9orf72 in neuroinflammation and ALS and frontotemporal dementia. J Clin Invest 127(9):3250–3258

Lehmensiek V, Tan E-M, Schwarz J, Storch A (2002) Expression of mutant α-synucleins enhances dopamine transporter-mediated MPP+ toxicity in vitro. Neuroreport 13(10):1279–1283

Lendon CL, Lynch T, Norton J, McKeel D, Busfield F, Craddock N et al (1998) Hereditary dysphasic disinhibition dementia A frontotemporal dementia linked to 17 q21--22. Neurology 50(6):1546–1555

Lesch K-P, Mössner R (1998) Genetically driven variation in serotonin uptake: is there a link to affective spectrum, neurodevelopmental, and neurodegenerative disorders? Biol Psychiatry 44(3):179–192

Liu B, Xu C, Wu X, Liu F, Du Y, Sun J et al (2015) Icariin exerts an antidepressant effect in an unpredictable chronic mild stress model of depression in rats and is associated with the regulation of hippocampal neuroinflammation. Neuroscience 294:193–205

Liu Y-H, Giunta B, Zhou H-D, Tan J, Wang Y-J (2012) Immunotherapy for Alzheimer disease—the challenge of adverse effects. Nat Rev Neurol 8(8):465–469

Livingston G, Johnston K, Katona C, Paton J, Lyketsos CG, Psychiatry OATFotWFoB. (2005) Systematic review of psychological approaches to the management of neuropsychiatric symptoms of dementia. Am J Psychiatr 162(11):1996–2021

Loureiro JC, Pais MV, Stella F, Radanovic M, Teixeira AL, Forlenza OV et al (2020) Passive antiamyloid immunotherapy for Alzheimer's disease. Curr Opin Psychiatry 33(3):284–291

Cardoso SM (2011) The mitochondrial cascade hypothesis for Parkinson's disease. Curr Pharm Des 17(31):3390–3397

Manke A, Wang L, Rojanasakul Y (2013) Mechanisms of nanoparticle-induced oxidative stress and toxicity. Biomed Res Int 2013:942916

Medina M (2011) Recent developments in tau-based therapeutics for neurodegenerative diseases. Recent Pat CNS Drug Discov 6(1):20–30

Mohamed T, Shakeri A, Rao PP (2016) Amyloid cascade in Alzheimer's disease: recent advances in medicinal chemistry. Eur J Med Chem 113:258–272

Morris J, Storandt M, McKeel D, Rubin E, Price J, Grant E et al (1996) Cerebral amyloid deposition and diffuse plaques in "normal" aging: Evidence for presymptomatic and very mild Alzheimer's disease. Neurology 46(3):707–719

Mosley RL, Benner EJ, Kadiu I, Thomas M, Boska MD, Hasan K et al (2006) Neuroinflammation, oxidative stress, and the pathogenesis of Parkinson's disease. Clin Neurosci Res 6(5):261–281

Nakano N, Matsuda S, Ichimura M, Minami A, Ogino M, Murai T et al (2017) PI3K/AKT signaling mediated by G protein-coupled receptors is involved in neurodegenerative Parkinson's disease. Int J Mol Med 39(2):253–260
Neal M, Wright PB (2003) Validation therapy for dementia. Cochrane Database Syst Rev 3: CD001394
Neff F, Wei X, Nölker C, Bacher M, Du Y, Dodel R (2008) Immunotherapy and naturally occurring autoantibodies in neurodegenerative disorders. Autoimmun Rev 7(6):501–507
Nerurkar PV, Johns LM, Buesa LM, Kipyakwai G, Volper E, Sato R et al (2011) Momordica charantia (bitter melon) attenuates high-fat diet-associated oxidative stress and neuroinflammation. J Neuroinflammation 8(1):64
Noman MZ, Hasmim M, Lequeux A, Xiao M, Duhem C, Chouaib S et al (2019) Improving cancer immunotherapy by targeting the hypoxic tumor microenvironment: new opportunities and challenges. Cell 8(9):1083
Núñez MT, Urrutia P, Mena N, Aguirre P, Tapia V, Salazar J (2012) Iron toxicity in neurodegeneration. Biometals 25(4):761–776
Nyakatura EK, Soare AY, Lai JR (2017) Bispecific antibodies for viral immunotherapy. Hum Vaccin Immunother 13(4):836–842
Oddo S, Vasilevko V, Caccamo A, Kitazawa M, Cribbs DH, LaFerla FM (2006) Reduction of soluble Aβ and tau, but not soluble Aβ alone, ameliorates cognitive decline in transgenic mice with plaques and tangles. J Biol Chem 281(51):39413–39423
Pahnke J, Krohn M, Scheffler K (2009) The role of blood-brain barrier in the pathogenesis of Alzheimer dementia--implications for immunological therapies for plaque dissolution. Fortschr Neurol Psychiatr 77:S21–S24
Panza F, Solfrizzi V, Seripa D, Imbimbo BP, Lozupone M, Santamato A et al (2016) Tau-based therapeutics for Alzheimer's disease: active and passive immunotherapy. Immunotherapy 8(9): 1119–1134
Pfaar O, Alvaro M, Cardona V, Hamelmann E, Mösges R, Kleine-Tebbe J (2018) Clinical trials in allergen immunotherapy: current concepts and future needs. Allergy 73(9):1775–1783
Pimplikar SW (2009) Reassessing the amyloid cascade hypothesis of Alzheimer's disease. Int J Biochem Cell Biol 41(6):1261–1268
Pittock SJ, Balice-Gordon R (2012) Emerging synaptic biomarkers in autoimmune dementia: NMDARs targeted by IgA. Neurology 78(22):1710–1711. AAN Enterprises
Plotkin SS, Cashman NR (2020) Passive immunotherapies targeting Aβ and tau in Alzheimer's disease. Neurobiol Dis 144:105010
Pooler AM, Noble W, Hanger DP (2014) A role for tau at the synapse in Alzheimer's disease pathogenesis. Neuropharmacology 76:1–8
Potter H, Wisniewski T (2012) Apolipoprotein E: Essential catalyst of the Alzheimer amyloid cascade. Int J Alzheimers Dis 2012:489428
Prinz PN, Vitaliano PP, Vitiello MV, Bokan J, Raskind M, Peskind E et al (1982) Sleep, EEG and mental function changes in senile dementia of the Alzheimer's type. Neurobiol Aging 3(4): 361–370
Qaseem A, Wilt TJ, Kansagara D, Horwitch C, Barry MJ, Forciea MA (2018) Hemoglobin A1c targets for glycemic control with pharmacologic therapy for nonpregnant adults with type 2 diabetes mellitus: a guidance statement update from the American College of Physicians. Ann Intern Med 168(8):569–576
Remington R, Abdallah L, Melillo KD, Flanagan J (2006) Managing problem behaviors associated with dementia. Rehabil Nurs 31(5):186–192
Rizzi L, Rosset I, Roriz-Cruz M (2014) Global epidemiology of dementia: Alzheimer's and vascular types. BioMed Res Int 2014:908915
Ross CA, Poirier MA (2005) What is the role of protein aggregation in neurodegeneration? Nat Rev Mol Cell Biol 6(11):891–898
Rubin EH, Morris JC, Berg L (1987) The progression of personality changes in senile dementia of the Alzheimer's type. J Am Geriatr Soc 35(8):721–725

Salminen A, Ojala J, Kaarniranta K, Hiltunen M, Soininen H (2011) Hsp90 regulates tau pathology through co-chaperone complexes in Alzheimer's disease. Prog Neurobiol 93(1):99–110

Schilling S, Rahfeld J-U, Lues I, Lemere CA (2018) Passive Aβ immunotherapy: current achievements and future perspectives. Molecules 23(5):1068

Schulz R, O'Brien A, Czaja S, Ory M, Norris R, Martire LM et al (2002) Dementia caregiver intervention research: in search of clinical significance. The Gerontologist 42(5):589–602

Seimetz D, Lindhofer H, Bokemeyer C (2010) Development and approval of the trifunctional antibody catumaxomab (anti-EpCAM× anti-CD3) as a targeted cancer immunotherapy. Cancer Treat Rev 36(6):458–467

Sharma GS, Kumar T, Dar TA, Singh LR (2015) Protein N-homocysteinylation: from cellular toxicity to neurodegeneration. Biochimica et Biophysica Acta (BBA)-General Subjects 1850(11):2239–2245

Sigurdsson EM (2008) Immunotherapy targeting pathological tau protein in Alzheimer's disease and related tauopathies. J Alzheimers Dis 15(2):157–168

Sikriwal D, Ghosh P, Batra JK (2008) Ribosome inactivating protein saporin induces apoptosis through mitochondrial cascade, independent of translation inhibition. Int J Biochem Cell Biol 40(12):2880–2888

Sorkina T, Hoover BR, Zahniser NR, Sorkin A (2005) Constitutive and protein kinase C-induced internalization of the dopamine transporter is mediated by a clathrin-dependent mechanism. Traffic 6(2):157–170

Stancu I-C, Vasconcelos B, Terwel D, Dewachter I (2014) Models of β-amyloid induced tau-pathology: the long and "folded" road to understand the mechanism. Mol Neurodegener 9(1):1–14

Steiner JA, Quansah E, Brundin P (2018) The concept of alpha-synuclein as a prion-like protein: ten years after. Cell Tissue Res 373(1):161–173

Stephans SE, Miller GW, Levey AI, Greenamyre JT (2002) Acute mitochondrial and chronic toxicological effects of 1-methyl-4-phenylpyridinium in human neuroblastoma cells. Neurotoxicology 23(4–5):569–580

Sumner IL, Edwards RA, Asuni AA, Teeling JL (2018) Antibody engineering for optimized immunotherapy in Alzheimer's disease. Front Neurosci 12:254

Swerdlow RH, Burns JM, Khan SM (2010) The Alzheimer's disease mitochondrial cascade hypothesis. J Alzheimers Dis 20(s2):S265–SS79

Takamatsu Y, Ho G, Koike W, Sugama S, Takenouchi T, Waragai M et al (2017) Combined immunotherapy with "anti-insulin resistance" therapy as a novel therapeutic strategy against neurodegenerative diseases. NPJ Parkinson's Dis 3(1):1–10

Tamir S, Izrael S, Vaya J (2002) The effect of oxidative stress on ERα and ERβ expression. J Steroid Biochem Mol Biol 81(4–5):327–332

Taylor JM, Main BS, Crack PJ (2013) Neuroinflammation and oxidative stress: co-conspirators in the pathology of Parkinson's disease. Neurochem Int 62(5):803–819

Tiiman A, Palumaa P, Tougu V (2013) The missing link in the amyloid cascade of Alzheimer's disease–metal ions. Neurochem Int 62(4):367–378

Tran HT, Chung CH-Y, Iba M, Zhang B, Trojanowski JQ, Luk KC et al (2014) α-Synuclein immunotherapy blocks uptake and templated propagation of misfolded α-synuclein and neurodegeneration. Cell Rep 7(6):2054–2065

Tripathi RK, Tiwari SC (2009) Psychotherapeutic approaches in the management of elderlies with dementia an overview. Delhi Psychiatry Jl 12(1):31–41

Troquier L, Caillierez R, Burnouf S, F. J Fernandez-Gomez, Grosjean M-E, Zommer N et al (2012) Targeting phospho-Ser422 by active tau immunotherapy in the THYTau22 mouse model: a suitable therapeutic approach. Curr Alzheimer Res 9(4):397–405

Turner S (2005) Behavioural symptoms of dementia in residential settings: a selective review of non-pharmacological interventions. Aging Ment Health 9(2):93–104

Valera E, Masliah E (2013) Immunotherapy for neurodegenerative diseases: focus on α-synucleinopathies. Pharmacol Ther 138(3):311–322

Valera E, Spencer B, Masliah E (2016) Immunotherapeutic approaches targeting amyloid-β, α-synuclein, and tau for the treatment of neurodegenerative disorders. Neurotherapeutics 13(1):179–189

Vander Zanden CM, Chi EY (2020) Passive immunotherapies targeting amyloid beta and tau oligomers in Alzheimer's disease. J Pharm Sci 109(1):68–73

Vanneman M, Dranoff G (2012) Combining immunotherapy and targeted therapies in cancer treatment. Nat Rev Cancer 12(4):237–251

Velayudhan L, Ffytche D, Ballard C, Aarsland D (2017) New therapeutic strategies for Lewy body dementias. Curr Neurol Neurosci Rep 17(9):68

Verma M, Vats A, Taneja V (2015) Toxic species in amyloid disorders: oligomers or mature fibrils. Ann Indian Acad Neurol 18(2):138

Verma V, Cushman TR, Selek U, Tang C, Welsh JW (2018) Safety of combined immunotherapy and thoracic radiation therapy: analysis of 3 single-institutional phase I/II trials. Int J Radiat Oncol Biol Phys 101(5):1141–1148

Vincent A, Bien CG, Irani SR, Waters P (2011) Autoantibodies associated with diseases of the CNS: new developments and future challenges. Lancet Neurol 10(8):759–772

Wagenmann M, Worm M, Akboga Y, Karjalainen M, Hohlfeld JM (2019) Randomized immunotherapy trial in dual-allergic patients using "active allergen placebo" as control. Allergy 74(8): 1480–1489

Wang J-Y, Wen L-L, Huang Y-N, Chen Y-T, Ku M-C (2006) Dual effects of antioxidants in neurodegeneration: direct neuroprotection against oxidative stress and indirect protection via suppression of gliamediated inflammation. Curr Pharm Des 12(27):3521–3533

Wang S, Colonna M (2019) Microglia in Alzheimer's disease: a target for immunotherapy. J Leukoc Biol 106(1):219–227

Wang Z, Gao G, Duan C, Yang H (2019) Progress of immunotherapy of anti-α-synuclein in Parkinson's disease. Biomed Pharmacother 115:108843

Warrick JM, Morabito LM, Bilen J, Gordesky-Gold B, Faust LZ, Paulson HL et al (2005) Ataxin-3 suppresses polyglutamine neurodegeneration in drosophila by a ubiquitin-associated mechanism. Mol Cell 18(1):37–48

Watanabe Y, Himeda T, Araki T (2005) Mechanisms of MPTP toxicity and their implications for therapy of Parkinson's disease. Med Sci Monit 11(1):RA17–RA23

Wenning GK, Jellinger KA (2005) The role of α-synuclein and tau in neurodegenerative movement disorders. Curr Opin Neurol 18(4):357–362

Wong YC, Krainc D (2017) α-Synuclein toxicity in neurodegeneration: mechanism and therapeutic strategies. Nat Med 23(2):1–13

Yamada M, Tsukagoshi H, Otomo E, Hayakawa M (1987) Cerebral amyloid angiopathy in the aged. J Neurol 234(6):371–376

Yang Q, Peng J, Shi K, Xiao Y, Liu Q, Han R et al (2019) Rationally designed peptide-conjugated gold/platinum nanosystem with active tumor-targeting for enhancing tumor photothermal-immunotherapy. J Control Release 308:29–43

Yu D, Yan H, Zhou J, Yang X, Lu Y, Han Y (2019) A circuit view of deep brain stimulation in Alzheimer's disease and the possible mechanisms. Mol Neurodegener 14(1):1–12

Zecca L, Youdim MB, Riederer P, Connor JR, Crichton RR (2004) Iron, brain ageing and neurodegenerative disorders. Nat Rev Neurosci 5(11):863–873

Zhai L, Spranger S, Binder DC, Gritsina G, Lauing KL, Giles FJ et al (2015) Molecular pathways: targeting IDO1 and other tryptophan dioxygenases for cancer immunotherapy. Clin Cancer Res 21(24):5427–5433

Zhao A, Tohidkia MR, Siegel DL, Coukos G, Omidi Y (2016) Phage antibody display libraries: a powerful antibody discovery platform for immunotherapy. Crit Rev Biotechnol 36(2):276–289

Hypoxic-Hyperoxic Training in Dementia 17

Christos Tsagkaris, Rehab A. Rayan, Eleni Konstantara, Lolita Matiashova, and Valeriia Danilchenko

Abstract

The treatment of dementia spectrum disorders has been long debated. Pharmacological, cognitive, and behavioral interventions have strived to change the landscape of dementia treatment, while a considerable amount of research has unraveled novel aspects of the pathophysiology of dementia. Recent studies have indicated the involvement of hypoxia in the metabolism of amyloid plaques, one of the key pathophysiological mechanisms of dementia. Hypoxic-hyperoxic training (HHT) has been developed under the notion of inducing tolerance against hypoxia. Evidence regarding the efficacy of this intervention derives from preclinical and clinical studies. In both cases the results are promising in terms of efficacy and safety, although there are concerns regarding the design of the existing studies. Moreover, the role of HHT in the context of multimodal treatment of dementia oughts to be precisely described. The purpose of this chapter is to provide an overview of HHT. In this frame, the authors summarize the role of

C. Tsagkaris (✉)
University of Crete, Faculty of Medicine, Heraklion, Greece

R. A. Rayan
Department of epidemiology, High Institute of Public Health, Alexandria University, Alexandria, Egypt

E. Konstantara
Medical University of Sofia, Faculty of Medicine, Sofia, Bulgaria

L. Matiashova
"L.T. Malaya Therapy National Institute of the National Academy of Medical Sciences of Ukraine", Kharkiv, Ukraine

V. Danilchenko
Kharkiv Regional Perinatal Center, Department of Post-Intensive Care, Rehabilitation and Nursing of Premature Newborns, Kharkiv, Ukraine

G. M. Ashraf, M. S. Uddin (eds.), *Current Thoughts on Dementia*,
https://doi.org/10.1007/978-981-16-7606-2_17

hypoxia in the pathophysiology of dementia and the existing preclinical and clinical studies and discuss HHT with regard to therapeutics and public health.

Keywords

Intermittent hypoxia · Hypoxic-hyperoxic training · Cognitive performance · Dementia

17.1 Introduction

In particular, hypoxia seems to regulate the metabolism of the amyloid-beta peptide leading to the formation of amyloid plaques on CNS tissue (Hugo and Ganguli 2014). Amyloid plaques are prominent pathological features of degeneration, while many disorders and comorbidities of elderly individuals are linked to hypoxia (Murphy et al. 2013). Although inhibiting or reversing the deteriorating effect of hypoxia on the CNS is not feasible, hypoxic-hyperoxic training (HHT) projects to make the CNS tissue more tolerant to hypoxia and the subsequent neurodegenerative alterations (Murphy et al. 2013). Variants of HHT training have been developed and tested in the preclinical and clinical context ('2020 Alzheimer's disease facts and figures' 2020). These studies have reported that HHT can be safe and effective, and draw a red line, where HHT might be harmful.

However, the number of studies as well as their inherent characteristics, such as the number of subjects and the follow-up periods have given birth to concerns. A preliminary bibliometric analysis in PubMed suggests that the presence of HHT in dementia research is limited not only in numbers of publications but also in time. With multimodal dementia treatment strategies being on the rise, incorporating HHT to a multitarget approach might upscale its potential. Finally, yet importantly, a meta-analysis could shed light on the qualitative aspects of the existing evidence. The authors review the existing studies and discuss their contribution to the body of evidence as well as the concerns arising from their results and design. The bibliometric footprint of HHT research in the realm of dementia research, the potential position of HHT in multimodal dementia management strategies, and the relevant implications in current therapeutics and health policy are also discussed.

This chapter aims to provide an evidence-based overview of HHT in the treatment of dementia. The authors present the spectrum and pathophysiology of dementia highlighting the involvement of hypoxia.

17.2 Epidemiology, Clinical Spectrum, and Burden of Dementia

Neurocognitive disorders, especially major neurocognitive disorders (dementias), have serious effects on patients, families, the health system, and the economy (Hugo and Ganguli 2014). Alzheimer's disease (AD) is a major risk for mortality (Murphy et al. 2013), admission to hospitals and nursing facilities, and home healthcare in the

United States (US). The expanses of health services and the informal expanses of non-paid caregiving of dementia patients are high and escalating. Caregivers from the family suffer high affective pressure, depression among other health issues ('2020 Alzheimer's disease facts and figures', 2020). Globally, In 2010, almost 35.6 million individuals were assumed to be surviving with dementia, a number anticipated to grow to about 115.4 million individuals by 2050 (Prince et al. 2013).

Dementia prevalence grows significantly with growing age and duplicating every 5 years following age 65 years. In high-income countries, prevalence is 5% to 10% in persons aged 65 years and older, and is often higher in females than males, mostly since females survive longer than males. Higher prevalence has been documented in African American and Latino/Hispanic peoples than in white non-Hispanic ones in the US. Literature worldwide reports that dementia prevalence is lower in sub-Saharan Africa and higher in Latin America than in the remaining regions globally. Now, evaluating mild cognitive impairment (MCI) prevalence is challenging since it relies on the accurate descriptions and subcategories of the investigated MCI (Ward et al. 2012; Uddin et al. 2020). Universally, life expectancy is growing with the aging population growing more quickly in low-income and middle-income countries, where dementia prevalence is estimated to grow (Luo et al. 2012). Recent investigations show that prevalence might be plateauing or declining in high-income countries (Matthews et al. 2013; Rocca et al. 2011).

17.3 Pathophysiology of Dementia with a Focus on Aβ Plaques

Amyloid plaques and neurofibrillary tangles are featured abnormalities, which determine AD. Amyloid plaques comprise mainly a 40–42 amino acid peptide called amyloid-β (Aβ), which is accumulated in fibrils, including a high β-sheet structure. Plaques turn insoluble and sediment in the brain's outside cell spaces. Amyloid plaques are usually linked to distended, dystrophic neurites, astrogliosis, and activated microglia that make a neuritic plaque. However, amyloid plaques and neurofibrillary tangles aggregate inside the cell in neurons. Aβ is naturally created by neurons inside the brain and released in the brain's outside cell spaces where during the pathogenicity of AD it shifts configuration, turns insoluble, and sediment as plaques. Aβ does not have an identified, physiologic role, yet increasing evidence has shown that under specific testing circumstances, Aβ could regulate synaptic transmission. Yet, the function of Aβ in a natural synaptic role or the disease's context is unknown (Cirrito and Holtzman 2008).

Table 17.1 shows hypoxia alters phosphorylating tau and forming Aβ at several levels. Molecular and cellular mechanisms are the foundation of hypoxic-triggered neurotoxicity and cellular death is complicated and many-sided, comprising several interlinked pathways and leading to the formation and release of proteins causing neurons' death as shown in Fig. 17.1 (Bhatia et al. 2017).

Prion pathways seem to be included in almost all neurodegenerative disorders that first appear in the elderly people. A natural cell protein periodically or through a familial transformation misfolds into an unhealthful pattern or prion. Next, the prion

Table 17.1 The pathways of hypoxia toward the initiation and progression of AD (Bhatia et al. 2017)

Increased phosphorylating tau	Increased Aβ formation
Increased cisplatin p35, p25	Increased protein phosphatase A (PPA)
Increased asparaginyl endopeptidase	Increased β/γ-secretase
Increased inhibitor of protein phosphatase 2A (I2 PP2A)	Increased presenilin-1
Decreased protein phosphatase 2A (PP2A)	Decreased neprilysin
	Increased locating γ-secretase from the cellular body to axon

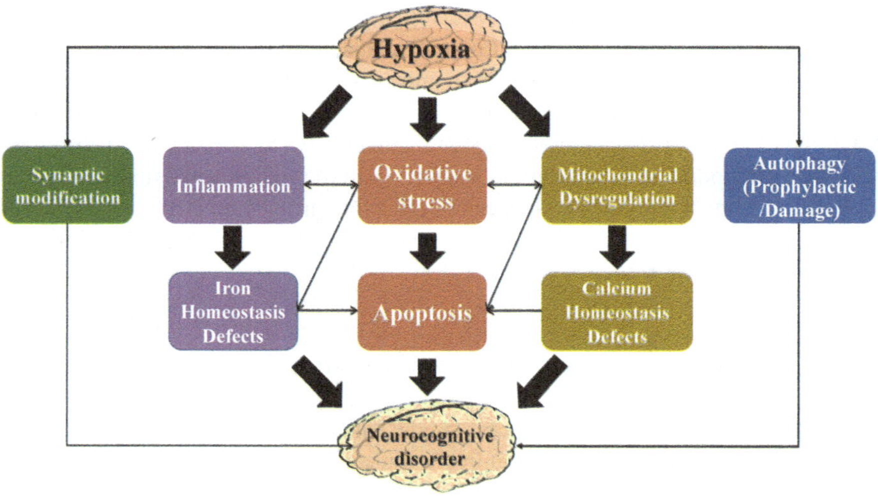

Fig. 17.1 Pathophysiological mechanisms relating hypoxia to neurodegenerative conditions

works as a model, forcing likely other proteins to misfold. This procedure happens through the years and in several regions of the central nervous system (CNS). Several prions turn insoluble, and such as amyloid, cannot be easily removed via the cell. Research infers prion or likewise pathways in Alzheimer's disease (firmly), and in Parkinson's disease, Huntington disease, frontotemporal dementia, and amyotrophic lateral sclerosis. Such prions are not infectious like those in Creutzfeldt-Jakob disease, however they could be heritable (Juebin 2019).

17.4 The Role of Hypoxia in Neurodegeneration and Dementia

Hypoxia regulates the metabolism of amyloid plaque protein (APP), causing a growing production of Aβ through the amyloidogenic mechanism. Time-reliant hypoxic upregulation of APP has been further proven at the mRNA and protein levels, following 10–180 mins of ischemia, which might function as a guarding pathway to raise the levels of neuroprotective soluble APPα (Shi et al. 2000). Yet, most times the growing APP leads to more levels of Aβ, not soluble APPα since hypoxia prefers metabolizing APP through the amyloidogenic mechanism (Lall et al. 2019).

Hypoxia could cause calcium-homeostasis dysregulation via fostering the aggregation of Aβ, which was first displayed in PC12 cells (Green and Peers 2001). Cells were subjected to 5% oxygen over 24 h prior to the test. It was then replicated in primary cultures of central neurons subjected to 2.5% oxygen over 24 h (Webster et al. 2006), where hypoxia raised the levels of calcium inside the cell through upregulating the L-type calcium channels, an outcome, which was replicated on applying externally aggregated Aβ under normoxic status. Adding inhibitors of β- or γ-secretases eliminated such a hypoxic surge of calcium currents, hence showing that such impact of hypoxia was associated with Aβ (Green and Peers 2001; Webster et al. 2006) that could change trafficking of the channel by interacting closely with the α-subunit of the L-type channel (Scragg et al. 2005).

Deregulations of metabolism have been observed in many areas of the patient's brain in AD, resulting in adaptive modifications to sustain metabolic homeostasis and healthy aging. Lower concentrations of the principal glucose transporters in the brain were linked to a higher level of tau phosphorylation and subsequent production of neurofilaments. The disturbed metabolism in hypoxia state could make Aß and produce neurofibrillary tangles. In contrast to conventional techniques that emphasized one target/mechanism, a cluster of targets would be needed to regain the disease-hindered many proteins to delay or manage the pathogenicity of AD. These elements comprise stabilizers of hypoxia-inducible factor (HIF-1α) as seen in Table 17.2. Yet, there are disputes considering the benefits of these elements in AD. However, the function of HIFs in acute hypoxia, hypoxia stimulates beta-secretase enzyme that uses APP to produce Aß peptide – was mainly triggered by HIF-1α (Zhang et al. 2007). A critical decrease of such an enzyme concentration in the cortex and the hippocampus was observed in the HIF-1α conditional knock-out mice. Activating HIFs displayed side effects in autoimmune conditions and carcinogenicity as shown in Table 17.2. Hence, considering the good and bad impacts of HIFs in hypoxia in nerve and non-nerve cells, the therapeutic success against the neurodegenerative conditions might require unique brain states with targeted delivery of elements in the upcoming comprehensive research (Ashok et al. 2017).

Research has reported that chronic hypoxia elevates β-secretase (BACE1) expression, hence inducing the amyloidogenic mechanism (Sun et al. 2006) in both in vitro and animal-stroke models (Peers et al. 2009). Decreased BACE1 expression was observed in mice lacking HIF1-a, a cardinal hypoxia associated transcriptional factor. This finding suggests that HIF1-a serves as a promoter of BACE1 (Sun et al. 2006). Elevated BACE1 levels in combination with other HIF1-triggered genes, such as APH1 – a subunit of the γ-secretase complex, which in turn catalyses the intramembrane Notch proteolysis, contribute to an amyloidogenic cascade (Lall et al. 2019).

Table 17.2 Neuroprophylactic and harmful effects of hypoxia-inducible factor (HIF-1α)

Neuroprophylactic effects	Studies	Harmful effects	Studies
Reduced expression of hypoxia-induced factor 1α (HIF-1α) over a long time, hence, a decreased glycolysis rate was shown in astrocytes for both rats and mice	(Schubert et al. 2009)	In brain stroke and ischemia, HIF-1α overexpression elevated beta-secretase expression leading to the amyloid precursor protein (APP) amyloidogenic rendering	(Zhang et al. 2007)
HIF-I could weaken the Aβ protein-triggered apoptosis of the hippocampal neurons	(Sun et al. 2014)	HIF is the key transcription factor to induce the development of cancer	(Hanahan and Weinberg 2011)
In an AD rat model, an intracerebroventricular injection of a recombinant adeno-associated virus (rAAV) vector expressing the human HIF-1α gene, rAAV-HIF-1α suppressed the hippocampal neuronal apoptosis	(Sun et al. 2010)	Better metabolic power of glucose could enable cancer cell growth	(Brahimi-Horn and Pouysségur 2007)
The neuro-prophylactic impact of a brain permeable iron chelator was induced via elevating the expression of HIF-1α and the HIF-1α -related genes like vascular endothelial growth factor (VEGF), erythropoietin, enolase-l, and p21	(Avramovich-Tirosh et al. 2010)	HIF-1α stimulated VEGFs expression and their receptors, causing neo-angiogenesis	(Jiang et al. 2016)
HIF-1α stimulated expression of aldolase A, enolase-l, and glucose transporter-I concentration in the frontal cortex of transgenic mice	(Mechlovich et al. 2014)	HIF-1α expression knock-down, halted hypoxia effect on the human ß-site APP cleavage enzyme 1 gene transcription	(Guglielmotto et al. 2009; Sun et al. 2006)
Iron chelators could improve the DNA linking of HIF-I to the hypoxic-response elements in cortical cultures and the hippocampal neuronal cell line HI9–7, hence, the neurophylactic effect	(Zaman et al. 1999)	In ischemia/hypoxia, reactive oxygen species stimulate the inflammatory responses in the cell and decrease genes expression for keeping synaptic morphology and performance	(Bazan et al. 2002)
Cobalt chloride stimulated HIF-I in the cortical cultures and inhibited the oxidative stress-triggered neuron death	(Zaman et al. 1999)		
HIF-prolyl hydroxylase domain enzymes (PHDs)	(Siddiq et al. 2005)		

(continued)

Table 17.2 (continued)

Neuroprophylactic effects	Studies	Harmful effects	Studies
inhibitors were prophylactic agents of cortical neurons from oxidative or ischemic insult			
Cobalt chloride or iron chelators stimulated the HIF-I and had prophylactic effects on cells from mitochondrial dysfunction in glial cells	(Yang et al. 2005)		

Hypoxia is evidently linked to AD. Because of the modified protective breakdown of APP, the deposited Aβ is the primary pathogenic inducer of AD. A major pathway where Aβ results in the observed AD pathophysiology is deregulating calcium homeostasis in both neurons and astrocytes, leading to neuron cell death. Aβ could stimulate microglia, causing a pathological neuroinflammatory reaction that adds more to the pathogenesis of AD. Yet, such a reaction in AD brains could be induced separately from Aβ. Hence, neuroinflammation might be another inducer of pathology in the context of AD as shown in Fig. 17.2 (Lall et al. 2019).

17.5 Hypoxic-Hyperoxic Training on Cognitive Performance

Using novel therapies could prevent and control dementia maintaining or even improving cognitive performance. Short repeated and controlled sessions of exposures to mild hypoxia rates, known as intermittent hypoxia-hyperoxia training (IHHT), or intermittent hypoxia-normoxia training (IHT) might enhance cerebral vascular performance and mitochondrial energy formation, like that have been shown in training sports and/or medical rehabilitations (Serebrovska et al. 2019). For example, IHT might benefit elderly health (Dudnik et al. 2018). Previously, an altered intermittent hypoxia training technique, IHHT, was investigated in many clinical trials in many European nations (Dudnik et al. 2018; Glazachev et al. 2017; Susta et al. 2017). The causing IHHT or IHT cellular pathways might involve enhanced mitochondrial performance, regulated glucose transport, stimulated antioxidant enzyme system, and anti-inflammatory responses. Such impacts could prevent and control cardiovascular and cerebral conditions, like AD (Mallet et al. 2018; Manukhina et al. 2016). Adding IHT to exercise improved exercise performance and brain function in the elderly (Bayer et al. 2019). IHT enhanced walking following chronic spinal cord insult (Branitzki-Heinemann et al. 2016). Hypoxia resulted in a critical decrease of spontaneous and phorbol myristate acetate (PMA)-induced NET production in cultured neutrophils (Dougherty et al. 2018).

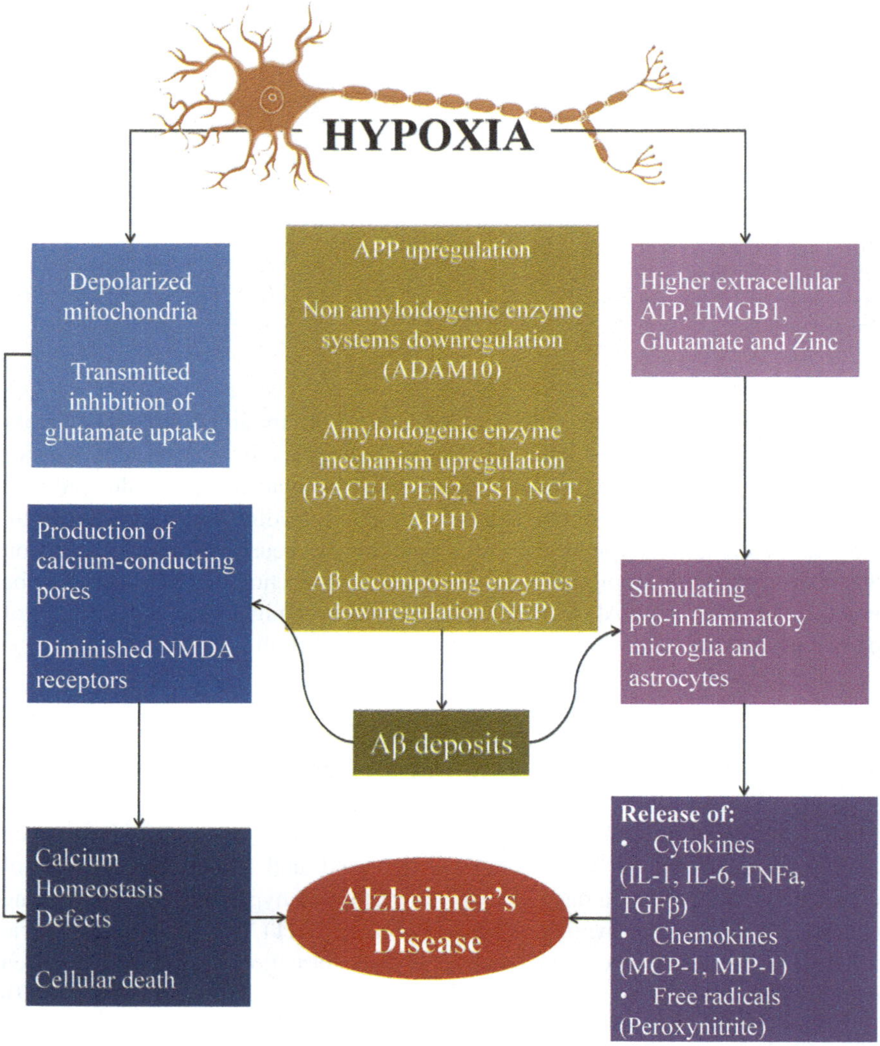

Fig. 17.2 An overview of the different relations between hypoxia and Alzheimer's disease, where hypoxia is a major risk factor for scattered cases of Alzheimer's disease

17.6 Evidence from Preclinical Studies for Hypoxia-Hyperoxia Treatment

Preclinical studies have approached hypoxia-hyperoxia treatment in the context of dementia's pathophysiology. The majority of them have focused on AD investigating the role of induced hypoxia as a countermeasure to AD's pathogenetic mechanisms such as brain hypoperfusion, Aβ deposition, and neuritic plaques formation, and specific metabolites levels in the CNS.

Gao et al. (2011) focused on the pathological evidence concerning the deposition of amyloid plaques on neurons. The researchers took into account clinical data demonstrating the reduction of cerebral perfusion prior to memory and cognitive impairment in patients with AD. It has been established that reduced cerebral perfusion leads to hypoxia in the context of AD (Greene and Lee 2012). Furthermore, studies have shown that the improvement of the oxygen supply in the brain might have a positive impact on AD pathology. Normobaric hyperoxia (NBO) not only provides more oxygen but was also found to be protective in recent studies. Morris water maze tests showed that NBO treatment improved spatial learning and memory problems in AβPP/PS1 transgenic mice (Gao et al. 2011). Immunohistochemical and thioflavin S staining revealed that administering NBO treatment could greatly reduce the deposition of Aβ and the formation of neuritic plaques in the cortex and hippocampus of these mice. Immunoblotting and ELISA assay revealed that NBO treatment reduced Aβ production by inhibiting γ-secretase cleavage of AβPP. From the above, the study suggested the potential therapeutic effect of NBO at the early stages of AD (Gao et al. 2011).

Watanabe et al. (2019) conducted an animal study based on quantitative proton magnetic resonance spectroscopy. This study aimed to evaluate the focal metabolic alterations, following a 24-h hypoxic or hyperoxic exposure with a history of ischemic brain insult, in 60 female Wistar rats which were divided into two groups of equal number according to age, young-3 months old and aged-24 months old. Researchers further divided each group into three respective subgroups including 10 rats each. Nembutal (30 mg·kg-1) was used for the sedation of two of these groups, after overnight fast, and cerebral ischemia was induced to them by ligation of the right common carotid artery. After that, the researchers measured the concentrations of eight metabolites (alanine, choline-containing compounds, total creatine, γ-aminobutyric acid, glutamate, lactate, myo-inositol, and N-acetylaspartate) in extracts of three distinct brain regions (frontoparietal and occipital cortices and the hippocampus) from both hemispheres. On the grounds of these findings, it appears that the regulation of normoxic condition was associated with significant increases in lactate and myoinositol concentrations in the hippocampus of the aged rats. This was observed at a lesser scale in the young ones. In the ischemia-hypoxic condition, the most prevailing alterations of brain metabolites were spotted in the hippocampal regions of both young and aged rats, but the effects were more evident in the aged animals than among the young ones. Ischemia-hyperoxic stress induced less changes in terms of cerebral metabolites, which might indicate a lower level of tissue damage (Watanabe et al. 2019).

17.7 Evidence from Clinical Studies for Hypoxia-Hyperoxia Treatment

Clinical studies assessed the contribution of various hypoxia-hyperoxia treatment modalities in the prevention or treatment of dementia, focusing on AD. A study conducted by Malle et al. (2016) focused on the effects that normobaric hypoxia (NH) exposure has on memory and physiology of the human body. They also studied the physiological and cognitive effects of oxygen breathing before and after the NH exposure. For this study 86 healthy men were randomly allocated in four groups. The groups were: the normoxia-air group ($N = 23$), whose subjects were breathing air, the hypoxia-air group ($N = 22$), whose subjects underwent NH and after that to air-breathing, the normoxia-O_2 group ($N = 21$), where subjects were treated similarly to the normoxia-air group, but with the addition of 100% O_2 breathing periods and the hypoxia-O_2 group ($N = 20$), where subjects were exposed to 100% O_2 prior and after NH exposure. The researchers used the paced auditory serial addition test to assess the memory of the subjects. Moreover, peripheral oxygen saturation (Spo_2), heart rate (HR), and electroencephalogram (EEG) were documented (Malle et al. 2016).

This study suggested that acute NH exposure triggered a physiological-adaptive response decreasing Spo_2 and increasing HR, but not the same as the physiological response to acute hypobaric hypoxia. Impairment in working memory was also caused by acute NH. Oxygen breathing after NH exposure caused a slowing in the electroencephalogram (EEG) which is associated with making working memory ability worse. For this reason, NH is suggested to be followed by air-breathing (Malle et al. 2016).

As it is already demonstrated, chronic hypoxia stimulates angiogenesis in the brain (Luo et al. 2012). According to their findings, therapeutic intermittent hypoxic training (IHT) can improve the vascularization of the cerebral tissue and prevent AD. When in hypoxia, cerebral angiogenesis starts by the transcription factor, HIF-1 when genes with promoter regions containing hypoxic response elements, including the VEGF gene, are activated (Watanabe et al. 2019).

Other clinical studies have reported similar findings. In particular, Bayer et al. (2017) in a clinical study in 2017 studied 34 patients from the Geriatric Day Clinic aged between 64 and 92 years old who participated in a controlled trial. These patients received randomly multimodal training programs (MTP) and IHHT (experimental group-EG) or MTP and placebo-breathing with machine face mask (experimental group-CG) in a double-blind manner. Before and after the 5- to 7-week intervention period (multimodal training intervention, MTI + IHHT vs. MTI + ambient air), their cognitive function was evaluated by the dementia-detection test (DemTect), the Sunderland clock-drawing test (CDT), and functional exercise capacity by the total distance of the 6-minute walk test (6MWT) (Bayer et al. 2017).

Results from other studies manifested that after MTI + IHHT was administered, DemTect showed important improvement (+16.7% vs. -0.39%, $P < 0{,}001$) and similar were the findings of the 6MWT with a more marked increase in EG than CG (+24.1% vs. +10.8%, $P = 0.021$). Furthermore, the CDT showed similar results with DemTect with an increase in EG but decrease in CG (+10.7% vs. −8%, P = 0,031). Also, a relation between the changes of the 6MWT, the DemTect, and the CDT was detected. The studies concluded that IHHT is easy in terms of application, well-tolerated among elderly patients up to 92 years and effective with regard to cognitive function and exercise capacity in geriatric patients after MTI (Cirrito and Holtzman 2008).

Bayer et al. in 2019 updated the previous study by performing some additional tests and including new results. Like before, she recruited 34 patients (from 64 to 92 years old) in a 5–7 weeks long double-blind clinical trial. During this time MTI was performed, where variants such as strength, endurance, balance, reaction, flexibility, coordination, and cognitive function were measured through exercises as well as IHHT in the hypoxic group (HG) which was subjected to 10–14% oxygen breathing for 4–7 min followed by 2–4 min 30–40% oxygen whereas the Normoxic Group (NG) was subjected to placebo treatment with the surrounding air. Before and after all treatments, mobility was evaluated by the Tinetti Mobility Test (TMT), the Timed-Up-and-Go Test (TUG) and Barthel-Index, while overall health status was evaluated by the EQ visual analog scale (EQ VAS) which is a part of the EQ-5D Test. These tests showed that after the MTI plus IHHT or normoxia sessions, results of the TMT, TUG, Barthel Index, and EQ-VAS did not reveal any important differences between HG and NG.

Another study indicated that IHHT added to MTI did not cause any additional improvements in the patient's health and mobility compared to MTI alone (Pichiule and Lamanna 2002). Serebrovska et al. (2019) also conducted a study in 2019 which examined the effects of intermittent hypoxic-hyperoxic training (IHHT) on aged patients who had a precursor of AD known as mild cognitive impairment (MCI). The study used twenty-one participants between 51 and 74 years of age which were divided into three groups: The first one was the Healthy Control (n = 7), the second one the MCI+Sham (n = 6), and the last one the MCI+IHHT (n = 8). IHHT was performed four times of repeated 5-min hypoxia (12% FIO_2) and 3-min hyperoxia sessions (33% FIO_2), five times per week for a total of three weeks. In total, this adds up to a sum of 15 sessions. After that cognitive parameters, Aβ and APP expression, microRNA 29, and long non-coding RNA in isolated platelets, as well as NETs in peripheral blood, were measured (Serebrovska et al. 2019).

The study found an initial decrease in cognitive function indices in both groups 2 (MCI+Sham) and 3 (MCI+IHHT) and important connections between intellectual test scores and the levels of circulating biomarkers of AD. IHHT caused the advancement in intellectual test scores, as well as a significant boost in APP ratio and decline in Aβ expression and NETs formation the next day after the completion of the three-week IHHT, which became more evident 1 month after IHHT. In conclusion, the results from this trial study suggested a possible usage of IHHT as a new therapy to improve cognitive function in patients who have pre-AD symptoms and to even slow down its progress (Serebrovska et al. 2019) (Table 17.3).

Table 17.3 Clinical studies investigating the therapeutic effects of hypoxic-hyperoxic training on dementia

Studies	Number of participants	Intervention	Outcomes
(Malle et al. 2016)	86	Normobaric hypoxia (NH) exposure	Acute NH exposure caused a typical physiological-normal response as well as an impairment in working memory. Oxygen breathing after NH exposure caused a slowing in electroencephalography (EEG) making working memory ability worse
(Bayer et al. 2017)	34	Multimodal training programs (MTP) and intermittent hypoxic-hyperoxic training (IHHT) or MTP and placebo-breathing	IHHT helped in the improvement in cognitive function and exercise capacity in geriatric patients after multimodal training intervention (MTI)
(Bayer et al. 2019)	34	MTP and IHHT or MTP and placebo-breathing + additional tests	After the MTI + IHHT or normoxia sessions, results of the Tinetti mobility test (TMT), Timed-Up-and-Go Test (TUG), Barthel index and the emotional quotient visual analog scale (EQ-VAS) did not reveal any important differences between the hypoxic and the normoxic group
(Serebrovska et al. 2019)	21	IHHT was performed	Potential usage of IHHT as a new therapy to improve cognitive function in patients with pre-AD symptoms and to even slow down its progress

17.8 Hypoxia-Hyperoxia Treatment: Assessment of the Evidence

In this chapter, we elaborated on the mechanisms and the evidence supporting HHT in the treatment of dementia. The authors retrieved information from original studies spanning from preclinical to clinical research. Evidence seems promising regarding HHT. The effects of hypoxia on dementia and the nervous system appear ambiguous and highly dependent on the design of each study. In a previous review of the relevant literature, the authors concluded that hypoxia can prevent and treat AD (Lall et al. 2019). This chapter bends toward the same conclusion with regard to ameliorating the condition of patients with AD, although it is currently impossible to conclude that hypoxia-hyperoxia treatment can prevent AD in healthy individuals (Korczyn 2012). Currently, there is one pilot study indicating the beneficial effects of IHT on individuals suffering from MCI. In this study (Wang et al. 2020), it was reported that IHT sessions enhanced cerebral vasodilation and improved the individuals' performance on mini-mental state examination and digit span scores. However, although the California verbal learning test score tended to improve, trail-making test-B and controlled oral word association test scores remained unchanged (Wang et al. 2020).

Another review highlighted the variability of beneficial effects of intermittent hypoxia training on patients with AD. The variability has been associated with genetic factors (Manukhina et al. 2016), which is consistent with our findings (Salih et al. 2019; Grozeva et al. 2019).

HHT reveals a significant interplay between preclinical and clinical studies, although one would argue that conducting clinical studies, when there is still a lot to be studied at a preclinical level, can be futile (Hsu and Marshall 2016; Insel et al. 2019). Not only preclinical studies pave the way for clinical interventions, but they also report key features supporting the personalization of HHT (Navarrete-Opazo and Mitchell 2014). A study investigating 40 male mice with the administration of 40% oxygen under normal atmospheric pressure was performed at the early stages of AD. The researchers reported significant improvement in AβPP/PS1 transgenic mice after 4–8 weeks of treatment and no effect in wild-type mice. This evidence adds up to the hypothesis of genetic interference that was suspected in clinical studies, urging for the search of genetic biomarkers to explore the eligibility of subjects for HHT (Gao et al. 2011).

The same study reported that NH reduced γ-secretase cleavage of AβPP and Aβ in mice. This finding indicated that HHT can hinder the progression of the disease (Gao et al. 2011). Again this finding brings up the quest for translational research and experimental models mimicking human physiology and pathophysiology. So far the inadequacy between models and humans has backfired on multiple occasions in dementia research (Al Dahhan et al. 2019; Meshalkina et al. 2017; Bolker 2019) As long as large-scale clinical studies verify these outcomes, HHT can be offered to stabilize the disease, while cognitive training can improve the quality of life of the patients (Couch et al. 2020; Whitehouse et al. 1997).

Other studies have identified further biomarkers in patients with AD. Exercise capacity, cognitive performance, and safety in geriatric patients have been correlated with an increase in APP130 and APP110 fractions in platelets, decrease in Aβ expression, and downregulation of lncRNA BACE-AS and NETs formation (Bayer et al. 2017; Serebrovska et al. 2019).

Experience from oncology and rheumatology has shown that genetic biomarkers can define groups of patients with optimal outcomes to specific regimens or treatment modalities (*'P and T : a peer-reviewed Journal for Formulary Management'* 2011, Miteva-Marcheva et al. 2020).

On a different note, the study of Marci et al. (Malle et al. 2016) emphasized safety. Their results suggested that hypoxia can result in cerebral ischemia and subsequent damage of the hippocampus-controlled functions (Macri et al. 2010).

One more study on healthy young men verified that NH and hypobaric hypoxia (HH) can have adverse effects on memory. In both cases, the damage was associated with the physiological reaction toward acute NH and HH. These studies urge for cautiousness and more rigorous research to define the therapeutic range and parameters of HHT drawing a clear line between projected benefit and adverse effects. Furthermore, participants of existing and future clinical studies need to be followed up accordingly (Malle et al. 2016).

17.9 The Bibliometric Footprint of Hypoxic-Hyperoxic Training Research

A preliminary analysis of findings related to HHT and dementia or specific dementia – spectrum conditions such as Alzheimer's indicates that the number of available studies of all genres is small. We can find hardly more than 20 studies investigating therapeutic variants of HHT in dementia out of more than 700 studies focusing on the role of hypoxia in dementia. Even these studies account for about 3% of PubMed indexed dementia studies. In terms of time, dementia studies display an increasing trend since the middle of the twentieth centuries with the first published and indexed studies tracing back to the nineteenth century. Hypoxia-related studies in the context of dementia appeared in the 1950s, although a significant increase in the number of these studies appeared at the beginning of the twenty-first century. Variants of HHT have been scarcely used in the twentieth century. The majority of these studies were published between 2005 and 2020. However, their numbers are not sufficient to describe a clear upward trend (Pubmed–NIH 2020).

17.10 Hypoxia-Hyperoxia Treatment in the Context of Contemporary Healthcare

So far, HHT has been used for the management or treatment of various conditions. HHT represents a training technique in sports aiming to increase endurance and enhance sympathetic stimulation during physical activity (Bonetti et al. 2006). In clinical settings, HHT has been predominantly used in neurology and cardiology. Chronic intermittent hypoxic preconditioning has been applied for the prevention of seizures and subsequent neurodegeneration (Zhen et al. 2014). At the same time, there are reports of the use of HHT in the context of sleep apnea based on its antioxidant potential (Turnbull 2018). The progress and incorporation of HHT in these fields can be a paradigm for HHT in dementia. However, it is important to discuss the clinical implications of HHT in dementia with a focus on the distinct features of dementia spectrum conditions and their therapeutics (Turnbull 2018).

The early diagnosis of dementia spectrum diseases is a challenge for modern biomedicine (Fox et al. 2013). The first challenge lies in ruling out systemic diseases such as hypothyroidism or vitamin B12 deficiency that may lead to a dementia-like syndrome. Once excluded, defining the exact type of dementia can also be complicated. ICD-10 clinical criteria or DSM-5 criteria may be used for the diagnosis of dementia and subtyping. Given that ICD-10 does not include criteria for the presence of Lewy Bodies, many clinicians rely on the DSM-5 criteria or the Consensus criteria (Shaji et al. 2018). Data-driven diagnostic modalities may ease the diagnostic procedure; however, merging conventional diagnostics will novelties can take time (Tolonen et al. 2018).

A detailed assessment usually allows the clinician to determine the cause of the particular dementia syndrome. Clinical recognition of dementia's subtypes is easier in the early stages of the disease. At later stages, even distinguishing AD from vascular dementia can be difficult. Currently, evidence from clinical studies on hypoxia-hyperoxia treatment is mainly related to AD. Future studies will need to assess thoroughly this treatment in the context of other dementing conditions (Mathuranath et al. 2000).

Moreover, researchers and clinicians ought to pay attention to the integration of hypoxia-hyperoxia treatment to existing therapeutic schemes (Verges et al. 2015; Zucchella et al. 2018). To this date, the treatment of dementia can be either pharmacological or non-pharmacological. When it comes to pharmacological treatment, it is notable that the currently available options are rather symptomatic treatments with limited effectiveness and questionable cost-effectiveness features (Massoud 2009; Versijpt 2014). A comprehensive evaluation of the possible subtype, associated behavioral and psychological symptoms, and comorbidities are necessary to determine the appropriate pharmacological treatment. However, the need for disease-modifying agents remains unmet to this date (Shaji et al. 2018).

On the other hand, behavioral and psychosocial interventions can be effective without putting patients with dementia at risk of adverse events, drug interactions, or polypharmacy (Barus et al. 2019). This is the basis of cognitive training and rehabilitation and other multimodal management schemes in dementia (Theodoros and Scordilis Dorothy Martha 2019). Apart from these, depending on the skills and the abilities of the patient, multisensory stimulation, music/dancing therapy, massage and animal-assisted therapy are available. The combination of pharmacological and non-pharmacological modalities is decided on an individual basis taking into account the psychosocial traits of the patient and the needs of their caregivers as well (Table 17.4) (Kales et al. 2015).

When it comes to HHT, rigorous assessment and follow-up of patients or healthy individuals undergoing such interventions will require improved imaging techniques. Watanabe et al. (2019) have suggested the visualization of A2 noradrenergic neurons with MRI based on the detection of noradrenaline groups of cells in the brain by T1-weighted MRI with magnetization transfer.

HHT would face cost and implementation issues. So far, HHT has been praised as a potent research modality with a favorable cost analysis and promising therapeutic features (Mateika et al. 2015). However, HHT sessions still represent a significant financial burden at an individual level, since the majority of people receiving such sessions for other conditions are supported by funding sources such as sports teams (Park et al. 2017). With small studies, absence of long-term results, a potential need for expensive additional genetic testing, and absence of robust pharmacoeconomics

Table 17.4 Various treatment strategies in dementia

Cognitive-emotion-oriented interventions
• Cognitive rehabilitation
• Simulated presence therapy (SPT)
• Reminiscence therapy
• Reality orientation therapy (ROT)
Sensory stimulation-oriented interventions
• Light therapy
• Music therapy • Drama therapy • Aromatherapy
• Transcutaneous electrical nerve stimulation (TENS)
• Snoezelen (multisensory orientation) therapy.
Behavior management oriented therapy
Other psychosocial interventions
• Animal therapy
• Therapeutic exercise/physical activity

studies, HHT has a long way till state or private insurance providers agree to cover its cost (Scharlach et al. 2005). Regulatory and legal parameters are also implicated. For example, the European Medicines Agency (EMA) has special procedures for non-pharmaceutical products, while general regulatory measures such as the general data protection regulation (GDPR) can pose additional obstacles ('Legal framework' n.d.) (EMA 2020). Overall, clinical research is more likely to make HHT accessible to a number of carefully selected patients. In case robust results validate the existing evidence, insurance coverage may be applied and HHT will become accessible to a larger number of individuals. The incorporation of HHT in multimodal experimental schemes can not only increase its accessibility, but it can also generate extra licensing procedure (Tsagkaris et al. 2020; Griffiths et al. 2019).

17.11 Recent Developments and Future Perspectives

The current scarcity of preclinical and clinical research on HHT is a call for action. Our search identified 4 preclinical and 4 clinical studies that were published in indexed journals. The subjects of these studies – either animals for preclinical or humans for clinical – did not exceed $n = 100$ in most cases. As a result of this, systematic approaches on such studies with small populations or short follow-up time would be difficult to lead to credible conclusions. Future studies ought to investigate the effect of hypoxia-hyperoxia treatment in larger population sets. On the other hand, the indication that hereditary traits may be implicated in the efficacy of HHT offers an opportunity of tracking genetic biomarkers and providing HHT in the frame of precision medicine in the future. Finally, yet importantly, provided that HHT is incorporated into multimodal interventions, it will be important to assess its effectiveness in combination and comparison to other pharmaceutical and non-pharmaceutical approaches (Chalfont et al. 2020; Lee et al. 2019).

17.12 Conclusion

HHT appears like a beneficial additional treatment for dementia. It is based on documented aspects of the pathophysiology of the disease. The existing evidence cannot be conclusive but it indicates the potential efficacy of HHT in the prevention of dementia in healthy individuals. Genetic factors can lead to variability in the outcomes of the treatment, while NH and HH need to be handled with cautiousness, due to their potential effect on memory impairment. The presence of HHT variants in bibliography is small but growing. The extent to which this intervention can reach depends not only on inherent factors but also on external factors, including funding and support from stakeholders and integration in AD MTP strategies.

References

'2020 Alzheimer's disease facts and figures' (2020) Alzheimer's and Dementia. https://doi.org/10.1002/alz.12068.

Angelopoulos T, Scordilis Dorothy Martha, T. C. (2019) 'Cognitive Impairment and Rehabilitation in Alzheimer's Disease No Title'.

Ashok BS, Ajith TA, Sivanesan S (2017) Hypoxia-inducible factors as neuroprotective agent in Alzheimer's disease. Clin Exp Pharmacol Physiol 44(3):327–334. https://doi.org/10.1111/1440-1681.12717

Avramovich-Tirosh Y et al (2010) Up-Regulation of Hypoxia-Inducible Factor (HIF)-1α and HIF-Target Genes in Cortical Neurons by the Novel Multifunctional Iron Chelator Anti-Alzheimer Drug, M30. Curr Alzheimer Res. https://doi.org/10.2174/156720510791162403

Barus R et al (2019) Drug interactions with dementia-related pathophysiological pathways worsen or prevent dementia. Br J Pharmacol 176(18):3413–3434. https://doi.org/10.1111/bph.14607

Bayer U et al (2017) Intermittent hypoxic–hyperoxic training on cognitive performance in geriatric patients. Alzheimer's Dement Translat Res Clin Intervent. https://doi.org/10.1016/j.trci.2017.01.002

Bayer U et al (2019) Effects of intermittent hypoxia-hyperoxia on mobility and perceived health in geriatric patients performing a multimodal training intervention: a randomized controlled trial. BMC Geriatr. https://doi.org/10.1186/s12877-019-1184-1

Bazan NG, Palacios-Pelaez R, Lukiw WJ (2002) Hypoxia signaling to genes: significance in Alzheimer's disease. Mol Neurobiol. https://doi.org/10.1385/MN:26:2-3:283

Bhatia D et al (2017) 'Hypoxia and its emerging therapeutics in neurodegenerative, inflammatory and renal diseases. Hypoxia Hum Dis. https://doi.org/10.5772/66089

Bolker JA (2019) Selection of models: evolution and the choice of species for translational research. Brain Behav Evol Brain. https://doi.org/10.1159/000500317

Bonetti DL, Hopkins WG, Kilding AE (2006) High-intensity kayak performance after adaptation to intermittent hypoxia. Int J Sports Physiol Perform 1(3):246–260. https://doi.org/10.1123/ijspp.1.3.246

Brahimi-Horn MC, Pouysségur J (2007) Hypoxia in cancer cell metabolism and pH regulation. Essays Biochem. https://doi.org/10.1042/BSE0430165

Branitzki-Heinemann K et al (2016) Formation of neutrophil extracellular traps under low oxygen level. Front Immunol. https://doi.org/10.3389/fimmu.2016.00518

Chalfont G, Milligan C, Simpson J (2020) A mixed methods systematic review of multimodal non-pharmacological interventions to improve cognition for people with dementia. Dementia 19(4):1086–1130. https://doi.org/10.1177/1471301218795289

Cirrito JR, Holtzman DM (2008) Anti-amyloid-β immunotherapy as a treatment for Alzheimer's disease. In: CNS regeneration. Elsevier, Amsterdam, pp 295–318. https://doi.org/10.1016/B978-012373994-0.50014-2

Couch E et al (2020) Outcomes tested in non-pharmacological interventions in mild cognitive impairment and mild dementia: a scoping review. BMJ Open. https://doi.org/10.1136/bmjopen-2019-035980

Al Dahhan NZ, De Felice FG, Munoz DP (2019) Potentials and pitfalls of cross-translational models of cognitive impairment. Front Behav Neurosci 13. https://doi.org/10.3389/fnbeh.2019.00048

Dougherty BJ et al (2018) Daily acute intermittent hypoxia improves breathing function with acute and chronic spinal injury via distinct mechanisms. Respirat Physiol Neurobiol. https://doi.org/10.1016/j.resp.2017.05.004

Dudnik E et al (2018) Intermittent hypoxia-hyperoxia conditioning improves cardiorespiratory fitness in older comorbid cardiac outpatients without hematological changes: a randomized controlled trial. High Alt Med Biol. https://doi.org/10.1089/ham.2018.0014

European Medicines Agency. Legal framework. Available at: https://www.ema.europa.eu/en/about-us/what-we-do/legal-framework Accesseded 12 Dec 2020

Fox C et al (2013) The pros and cons of early diagnosis in dementia. Br J Gen Pract 63(612):e510–e512. https://doi.org/10.3399/bjgp13X669374

Gao B et al (2011) Effect of normobaric hyperoxia on behavioral deficits and neuropathology in Alzheimer's disease mouse model. J Alzheimers Dis. https://doi.org/10.3233/JAD-2011-110308

Glazachev O et al (2017) Adaptations following an intermittent hypoxia-hyperoxia training in coronary artery disease patients: a controlled study. Clin Cardiol. https://doi.org/10.1002/clc.22670

Green KN, Peers C (2001) Amyloid β peptides mediate hypoxic augmentation of Ca2+ channels. J Neurochem 77(3):953–956. https://doi.org/10.1046/j.1471-4159.2001.00338.x

Greene NH, Lee LA (2012) Modern and evolving understanding of cerebral perfusion and autoregulation. Adv Anesth. https://doi.org/10.1016/j.aan.2012.08.003

Griffiths AW et al (2019) Barriers and facilitators to implementing dementia care mapping in care homes: results from the DCMTM™ EPIC trial process evaluation. BMC Geriatr. https://doi.org/10.1186/s12877-019-1045-y

Grozeva D et al (2019) Benefits and challenges of rare genetic variation in Alzheimer's disease. Curr Genet Med Rep. https://doi.org/10.1007/s40142-019-0161-5

Guglielmotto M et al (2009) The up-regulation of BACE1 mediated by hypoxia and ischemic injury: role of oxidative stress and HIF1α. J Neurochem. https://doi.org/10.1111/j.1471-4159.2008.05858.x

Hanahan D, Weinberg RA (2011) Hallmarks of cancer: the next generation. Cell. https://doi.org/10.1016/j.cell.2011.02.013

Hsu D, Marshall GA (2016) 'Primary and secondary prevention trials in Alzheimer disease: looking back moving forward. Curr Alzheimer Res. https://doi.org/10.2174/1567205013666160930112125

Huang Juebin (2019) 'Dementia - Neurologic Disorders', MSD Manual Professional Edition. Available at: https://www.msdmanuals.com/professional/neurologic-disorders/delirium-and-dementia/dementia.

Hugo J, Ganguli M (2014) Dementia and cognitive impairment. epidemiology, diagnosis, and treatment. Clin Geriatr Med. https://doi.org/10.1016/j.cger.2014.04.001

Insel PS et al (2019) Determining clinically meaningful decline in preclinical Alzheimer disease. Neurology 93(4):e322–e333. https://doi.org/10.1212/WNL.0000000000007831

Jiang M, Qin C, Han M (2016) Primary breast cancer induces pulmonary vascular hyperpermeability and promotes metastasis via the VEGF-PKC pathway. Mol Carcinog. https://doi.org/10.1002/mc.22352

Kales HC, Gitlin LN, Lyketsos CG (2015) Assessment and management of behavioral and psychological symptoms of dementia. BMJ (Online). https://doi.org/10.1136/bmj.h369

Korczyn AD (2012) Why have we failed to cure alzheimer's disease? J Alzheimers Dis. https://doi.org/10.3233/JAD-2011-110359

Lall R, Mohammed R, Ojha U (2019) What are the links between hypoxia and alzheimer's disease? Neuropsychiatr Dis Treat. https://doi.org/10.2147/NDT.S203103

Lee G et al (2019) Predicting Alzheimer's disease progression using multi-modal deep learning approach. Sci Rep. https://doi.org/10.1038/s41598-018-37769-z

'Legal Framework' (n.d.). Available at: https://www.ema.europa.eu/en/about-us/what-we-do/legal-framework

Luo J et al (2012) Hypoxia induces angiogenic factors in brain microvascular endothelial cells. Microvasc Res. https://doi.org/10.1016/j.mvr.2011.11.004

Macri MA et al (2010) Region-specific effects on brain metabolites of hypoxia and hyperoxia overlaid on cerebral ischemia in young and old rats: a quantitative proton magnetic resonance spectroscopy study. J Biomed Sci. https://doi.org/10.1186/1423-0127-17-14

Malle C et al (2016) Physiological and cognitive effects of acute normobaric hypoxia and modulations from oxygen breathing. Aerosp Med Human Perform. https://doi.org/10.3357/AMHP.4335.2016

Mallet RT et al (2018) Cardioprotection by intermittent hypoxia conditioning: evidence, mechanisms, and therapeutic potential. Am J Physiol Heart Circ Physiol. https://doi.org/10.1152/ajpheart.00060.2018

Manukhina EB et al (2016) Intermittent hypoxia training protects cerebrovascular function in Alzheimer's disease. Exp Biol Med. https://doi.org/10.1177/1535370216649060

Massoud F (2009) Cost-effectiveness of pharmacological treatment in vascular dementia. Canad J Neurol Sci 36(6):673–674. https://doi.org/10.1017/S031716710000826X

Mateika JH et al (2015) Intermittent hypoxia: a low-risk research tool with therapeutic value in humans. J Appl Physiol 118(5):520–532. https://doi.org/10.1152/japplphysiol.00564.2014

Mathuranath PS et al (2000) A brief cognitive test battery to differentiate Alzheimer's disease and frontotemporal dementia. Neurology. https://doi.org/10.1212/01.wnl.0000434309.85312.19

Matthews FE et al (2013) A two-decade comparison of prevalence of dementia in individuals aged 65 years and older from three geographical areas of England: results of the cognitive function and ageing study I and II. Lancet 382(9902):1405–1412. https://doi.org/10.1016/S0140-6736(13)61570-6

Mechlovich D et al (2014) The novel multi-target iron chelator, M30 modulates HIF-1α-related glycolytic genes and insulin signaling pathway in the frontal cortex of APP/PS1 Alzheimer's disease mice. Curr Alzheimer Res. https://doi.org/10.2174/1567205010666131212112529

Meshalkina DA, Song C, Kalueff AV (2017) Better lab animal models for translational neuroscience research and CNS drug development. Lab Anim. https://doi.org/10.1038/laban.1236

Miteva-Marcheva NN et al (2020) Application of pharmacogenetics in oncology. Biomarker Res 8(1):32. https://doi.org/10.1186/s40364-020-00213-4

Murphy SL, Xu J, Kochanek KD (2013) 'Deaths: final data for 2010.', National vital statistics reports: from the Centers for Disease Control and Prevention. Natl Vital Statist Syst 61(4):1–117. Available at: https://pubmed.ncbi.nlm.nih.gov/24979972/

Navarrete-Opazo A, Mitchell GS (2014) Therapeutic potential of intermittent hypoxia: a matter of dose. Am J Phys Regul Integr Comp Phys 307(10):R1181–R1197. https://doi.org/10.1152/ajpregu.00208.2014

Park H-Y, Shin C, Lim K (2017) Intermittent hypoxic training for 6 weeks in 3000 m hypobaric hypoxia conditions enhances exercise economy and aerobic exercise performance in moderately trained swimmers. Biol Sport. https://doi.org/10.5114/biolsport.2018.70751

Peers C et al (2009) Hypoxia and neurodegeneration. Ann New York Acad Sci. https://doi.org/10.1111/j.1749-6632.2009.05026.x

P&T (2011) Personalized medicine: identifying the appropriate patient through biomarkers in oncology. P & T 36(7 Suppl 1):3–10. Available at: https://www.ncbi.nlm.nih.gov/pmc/articles/PMC3171807/

Pichiule P, Lamanna JC (2002) Angiopoietin-2 and rat brain capillary remodeling during adaptation and deadaptation to prolonged mild hypoxia. J Appl Physiol. https://doi.org/10.1152/japplphysiol.00318.2002

Prince M et al (2013) The global prevalence of dementia: a systematic review and metaanalysis. Alzheimer's Dementia. https://doi.org/10.1016/j.jalz.2012.11.007

Pubmed - National Institute of Health, Search Ierms [Dementia AND Hypoxia], Available: https://pubmed.ncbi.nlm.nih.gov/?term=dementia+AND+hypoxia&sort=date. Accessed 12 December 2020

Rocca WA et al (2011) Trends in the incidence and prevalence of Alzheimer's disease, dementia, and cognitive impairment in the United States. Alzheimer's Dementia. https://doi.org/10.1016/j.jalz.2010.11.002

Salih DA et al (2019) Genetic variability in response to amyloid beta deposition influences Alzheimer's disease risk. Brain Commun. https://doi.org/10.1093/braincomms/fcz022

Scharlach A, Dal Santo TS, Mills-Dick K (2005) Assuring the quality of long-term care insurance benefits through care management. J Aging Soc Policy 17(3):39–59. https://doi.org/10.1300/J031v17n03_03

Schubert D, Soucek T, Blouw B (2009) The induction of HIF-1 reduces astrocyte activatio... [Eur J Neurosci. 2009] - PubMed result. Europ J Neurosci 7(4):751

Scragg JL et al (2005) Alzheimer's amyloid peptides mediate hypoxic up-regulation of L-type Ca 2+ channels. FASEB. https://doi.org/10.1096/fj.04-2659fje

Serebrovska ZO et al (2019) Intermittent hypoxia-hyperoxia training improves cognitive function and decreases circulating biomarkers of Alzheimer's disease in patients with mild cognitive impairment: a pilot study. Int J Mol Sci. https://doi.org/10.3390/ijms20215405

Shaji KS et al (2018) Clinical practice guidelines for management of dementia. Indian J Psychiatry. https://doi.org/10.4103/0019-5545.224472

Shi J et al (2000) Hypoperfusion induces overexpression of β-amyloid precursor protein mRNA in a focal ischemic rodent model. Brain Res. https://doi.org/10.1016/S0006-8993(99)02113-7

Siddiq A et al (2005) Hypoxia-inducible factor prolyl 4-hydroxylase inhibition. J Biol Chem 280(50):41732–41743. https://doi.org/10.1074/jbc.M504963200

Sun X et al (2006) Hypoxia facilitates Alzheimer's disease pathogenesis by up-regulating BACE1 gene expession. Proc Natl Acad Sci U S A. https://doi.org/10.1073/pnas.0606298103

Sun Y et al (2014) A viral vector expressing hypoxia-inducible factor 1 alpha inhibits hippocampal neuronal apoptosis. Neural Regen Res 9(11):1145. https://doi.org/10.4103/1673-5374.135317

Sun YM et al (2010) Effect of intracerebroventricular injection of rAAV-HIF-1α on hippocampal neuronal apoptosis in a rat model of Alzheimer disease. Nan fang yi ke da xue xue bao J Southern Med Univ 30(12):2711–2714

Susta D, Dudnik E, Glazachev OS (2017) A programme based on repeated hypoxia–hyperoxia exposure and light exercise enhances performance in athletes with overtraining syndrome: a pilot study. Clin Physiol Funct Imaging. https://doi.org/10.1111/cpf.12296

Tolonen A et al (2018) Data-driven differential diagnosis of dementia using multiclass disease state index classifier. Front Aging Neurosci 10. https://doi.org/10.3389/fnagi.2018.00111

Tsagkaris C et al (2020) Mithridatism for dementia? Hypoxic - Hyperoxic training in dementia. Mental Health Global Challeng J 3(1):21–26. https://doi.org/10.32437/mhgcj.v3i1.82

Turnbull CD (2018) Intermittent hypoxia, cardiovascular disease and obstructive sleep apnoea. J Thorac Dis 10(S1):S33–S39. https://doi.org/10.21037/jtd.2017.10.33

Uddin MS et al (2020) Revisiting the role of brain and peripheral Aβ in the pathogenesis of Alzheimer's disease. J Neurol Sci 416:116974. https://doi.org/10.1016/j.jns.2020.116974

Verges S et al (2015) Hypoxic conditioning as a new therapeutic modality. Front Pediatr 3. https://doi.org/10.3389/fped.2015.00058

Versijpt, J. (2014) Effectiveness and cost-effectiveness of the pharmacological treatment of Alzheimer's disease and vascular dementia , J Alzheimer's Dis. Edited by A. D. Korczyn, 42 (s3), pp. S19–S25. doi: https://doi.org/10.3233/JAD-132639.

Wang H et al (2020) Intermittent hypoxia training for treating mild cognitive impairment: a pilot study. Am J Alzheimers Dis Other Dement. https://doi.org/10.1177/1533317519896725

Ward A et al (2012) Mild cognitive impairment: disparity of incidence and prevalence estimates. Alzheimers Dement 8(1):14–21. https://doi.org/10.1016/j.jalz.2011.01.002

Watanabe T et al (2019) Magnetic resonance imaging of noradrenergic neurons. Brain Struct Funct. https://doi.org/10.1007/s00429-019-01858-0

Webster NJ et al (2006) Amyloid peptides mediate hypoxic increase of L-type Ca2+ channels in central neurones. Neurobiol Aging 27(3):439–445. https://doi.org/10.1016/j.neurobiolaging.2005.02.002

Whitehouse PJ et al (1997) Quality-of-life assessment in dementia drug development: position paper from the International Working Group on Harmonization of Dementia Drug Guidelines. Alzheimer Dis Assoc Disord. 11(3):56–60

Yang YT, Ju TC, Yang DI (2005) Induction of hypoxia inducible factor-1 attenuates metabolic insults induced by 3-nitropropionic acid in rat C6 glioma cells. J Neurochem. https://doi.org/10.1111/j.1471-4159.2005.03032.x

Zaman K et al (1999) Protection from oxidative stress-induced apoptosis in cortical neuronal cultures by iron chelators is associated with enhanced DNA binding of hypoxia-inducible factor-1 and ATF-1/CREB and increased expression of glycolytic enzymes, p21(waf1/cip1), and erythropoietin. J Neurosci. https://doi.org/10.1523/jneurosci.19-22-09821.1999

Zhang X et al (2007) Hypoxia-inducible factor 1α (HIF-1α)-mediated hypoxia increases BACE1 expression and β-amyloid generation. J Biol Chem 282(15):10873–10880. https://doi.org/10.1074/jbc.M608856200

Zhen J et al (2014) Chronic intermittent hypoxic preconditioning suppresses pilocarpine-induced seizures and associated hippocampal neurodegeneration. Brain Res 1563:122–130. https://doi.org/10.1016/j.brainres.2014.03.032

Zucchella C et al (2018) The multidisciplinary approach to Alzheimer's disease and Dementia. A narrative review of non-pharmacological treatment. Front Neurol 9. https://doi.org/10.3389/fneur.2018.01058

Music Therapy in Dementia

18

Shreya Bhargava, Isha Jain, Manisha Singh, and Rachana

Abstract

Dementia is not a single condition; it describes a set of symptoms that may include impairment in memory, communication, and thinking. Alzheimer's disease is the most common form of dementia and may contribute to 60 to 80% of cases. Alzheimer's dementia results in permanent impairment of social or occupational functioning which includes difficulty in recognizing close ones or solving critical problems. Out of various researched remedies for treating patients suffering from dementia, music therapy is one such technique that is being practiced across the world. The impact of various types of sounds causes positive changes over neural synapses and plays a significant role in improving the severe symptoms in patients. Music therapy and the personal/private carers and nurses appointed to give such treatments to the patients have been shown to play an important role in curing dementia. The present chapter describes the basics of dementia and the mechanism of action of music therapy. Evidence from various case studies regarding the success of direct and indirect music therapy have been discussed in curing dementia.

Keywords

Dementia · Music therapy · Occupational functioning · Alzheimer's disease · Vascular dementia · Direct music therapy · Indirect music therapy

S. Bhargava · I. Jain · M. Singh · Rachana (✉)
Department of Biotechnology, Jaypee Institute of Information Technology, Noida, India
e-mail: rachana.dr@iitbombay.org

G. M. Ashraf, M. S. Uddin (eds.), *Current Thoughts on Dementia*,
https://doi.org/10.1007/978-981-16-7606-2_18

18.1 Introduction

Dementia is described as a group of syndrome that occurs as a result of various other diseases and disorders of the brain, which is generally progressive and chronic. It mainly comprises impairment of the higher-order cortical functions, which includes comprehension, memory, the process of the discerning, learning process, calculative ability, language recognition, and social judgment (Grande et al. 2019). Usually, dementia can arise in two different age groups, before or after 65 years.

When the onset of dementia occurs before the age of 65 years, it is said to be early-onset dementia whereas, when people develop this disease after the age of 65, it is said to be late-onset dementia (Balin et al. 2018). As studied by several scientists, it is majorly caused by chemical and structural changes in the human brain which leads to loss of neurons and overall shrinking of the volume of the brain (Tan et al. 2017). Several medicines have been used for the treatment of dementia and now music therapy (MT) is also been given to such patients and it has been shown to have a positive impact in improving the conditions (Umbrello et al. 2019).

In the present chapter, the authors have discussed the basics about dementia, mechanism of action of MT, and the evidence of the success of this non-pharmacological treatment by describing a series of researches where listening to music resulted in enhanced memory, decreased feelings of anxiety and agitation, and improved social skills.

18.2 Risk Factors of Dementia

Risk factors are generally defined as visible or invisible causes responsible for either increasing or decreasing the chances of developing that particular condition. In case of dementia age, sex, genetics, and vascular risks are believed to be the major risk factors (Booker et al. 2016).

Age is one of the most consistent and prominent risk factors for dementia. The occurrence rate of this disease doubles after a gap of every 5 years as reported by several researchers. Nevertheless, the age of a person is said to be the biggest risk factor for dementia (Osler et al. 2019). In certain cases, vascular risk factors are also involved in causing dementia which includes blood pressure, diabetes mellitus, stroke, and heart disease (Takeda et al. 2020). These types of conditions usually cause dementia when they occur or affect the human body in different combinations with other causing factors (Kloppenborg et al. 2008; Savva and Stephan et al. 2010; Viswanathan et al. 2009). Vascular dementia is said to be more common in the case of males as compared to females whereas, the cases of Alzheimer's disease (AD) are more common in women (Savva et al. 2009; Sung et al. 2012).

Another risk factor involved in causing dementia is genetics. A minimum of around 20 different genes are observed to be linked with dementia or AD (Lourida et al. 2019). In case of AD, three genes, coding for different proteins (presenilin 1 and 2, amyloid precursor protein) were found to be linked with the early onset of dementia (Hunter and Brayne 2018). The amyloid precursor protein is a protein integrated in the plasma membrane, usually found concentrated specifically in the synapses of neurons and even in other several tissues (Tang 2019). The presenilin class of proteins are special transmembrane proteins responsible for the cleavage of amyloid protein (Galla et al. 2020). Multiple mutations in this protein are said to be associated with occurrence of fronto-temporal dementia. These peculiar combinations of genes which generally cause this condition are rare in nature (Cacabelos 2008; Verghese et al. 2011). Down syndrome is another condition responsible for dementia (Bayen et al. 2018; Lott and Head 2019) which occurs due to the presence of three copies of chromosome number 21 and carries those genes which are majorly associated with amyloid production which in turn might be responsible in some cases of dementia especially in middle age dementia (Coppus et al. 2006; Coskun et al. 2010).

Besides the above-discussed factors, there are many other lifestyle factors responsible for causing dementia. These factors include smoking and drinking alcohol, etc. (Gupta and Warner 2008). Smoking, in general, affects the blood vessels which reach the different parts of the brain and can increase the risk of vascular dementia (which will be covered later in the chapter) (Meng et al. 2020). Also, the prolonged consumption of alcohol can lead to vascular changes in the brain and can lead to a high risk of developing the above-mentioned disease (Hulsegge et al. 2014; Shimada et al. 2018). Other than this, various scientists have observed many other plausible risk factors for the conditions such as consumption of non-steroidal anti-inflammatory agents, depression, hormone replacement therapy, and exposure to toxins such as aluminum (Bakulski et al. 2020; Cantón-Habas et al. 2020). Head trauma is also said to be a plausible risk in the early onset of dementia (Chen et al. 2009; Kristman et al. 2014).

18.3 Stages of Dementia

Dementia is classified into four progressive stages which have been formulated based on mini-mental state examination (MMSE) scores.

18.3.1 Mild Cognitive Impairment

In this stage, the signs and symptoms are not severe and do not affect the daily routine of patients. However, 70% of such patients go-ahead to develop clear symptoms of dementia in the later age range of life (Fymat 2019).

18.3.2 Early Stage Dementia

In this stage symptoms like difficulties with recalling certain moments, personality change, and social withdrawal start to get significantly visible (Fink et al. 2018; Steeman et al. 2006; Martyr and Clare 2018).

18.3.3 Middle Stage Dementia

In this stage, severe signs are: difficulties in problem solving and impairment in societal judgment. The patient might require assistance for personal hygiene and overall care as well (Kerpershoek et al. 2018; Lin et al. 2013).

18.3.4 Late Stage Dementia

Patients in this stage observe drastic changes in their overall personality and recognizing skills and assistance are needed for their personal safety, hygiene, and overall care (van den Dungen et al. 2012; Tekok-Kilic et al. 2007).

18.4 Types of Dementia

18.4.1 Alzheimer's Disease

Alzheimer's disease (AD) is known to be the most prominent type of dementia. It accounts for the major proportion of the cases related to dementia, i.e., up to 70% of the total cases (Klassen and Ahlskog 2011; Sacktor and Robertson 2014). The clinical features for AD include memory loss and word-finding difficulties (Yang et al. 2018; Zucchella et al. 2018). With the advancement of the disease, memory loss and language difficulties become common and severe. This causes difficulty in everyday activities such as shopping, handling money, and navigating routes. Other features include anxiety and lack of motivation and related emotions (Allen et al. 2013; Ewers et al. 2012; Rocca et al. 2011).

18.4.2 Vascular Dementia

Vascular dementia (VD) is the second most prominent type of dementia. This condition arises when the blood supply toward the brain is compromised by arterial diseases (Smith 2017). This results in the reduction of neuronal functions which usually occur due to axonal and cerebral abnormalities and leads to the gradual death of brain cells (Kalaria 2016). The clinical features of VD include a stroke which might usually occur progressively (Anor et al. 2017; Smith 2017). In addition to the features associated with AD, apathy and slowing down of cognitive processes is also common in this type of dementia (Jackson et al. 2013; Iadecola 2013).

18.4.3 Dementia with Lewy Bodies

This form of dementia generally accounts for approximately 10% cases of dementia worldwide (Jellinger and Korczyn 2018). It is said to be a combination of AD and

Parkinson's disease (PD) as it shares various characteristic features in common (McKeith et al. 2017; Mueller et al. 2017). Lewy bodies are said to be the combination of proteins called alpha-synuclein which commonly occurs in few parts of brain, including the cerebral cortex. The clinical features for this type include difficulty in maintaining alertness, difficulty in planning out things, and disorientation of the entire space (Tsunoda et al. 2018). Even hallucinations and recurring falls are said to be the most prominent features of dementia caused by Lewy bodies (Jellinger and Attems 2013; McKeith 2007).

18.4.4 Fronto-Temporal Dementia

Fronto-temporal dementia is one of the rarest forms of dementia and this type of condition mostly affects the front region of the brain which is responsible for emotions, language recognition, and planning out things (Bright et al. 2019; Olney et al. 2017). This condition is characterized by the diminished dynamic fluidity inside the brain and narrowing of the meta-state distance caused by the dynamic state of connectivity (Premi et al. 2019). The clinical features include two broadly classified categories having behavioral and language changes (Convery et al. 2019). These types of features might result in multiple problems associated with the normal functioning of the brain (Bang et al. 2015; Warren et al. 2013).

18.4.5 Huntington's and Parkinson's Diseases

Huntington's and Parkinson's diseases generally cause abnormal movements and difficulty in coordinating with the other parts of the body along with cognitive issues. These cognitive changes occur initially, resulting in becoming one of the common causes in advanced stage dementia (~50%) (Stopa et al. 2018; Zarowitz et al. 2014).

18.4.6 Corticobasal Degeneration

In corticobasal degeneration, there is damage caused and significant shrinkage of the brain due to the abnormal protein's (tau) deposition in the brain (Zhang et al. 2020). Additional symptoms include loss of balance and movement difficulties (Armstrong et al. 2013; Luzzi et al. 2007).

18.4.7 Creutzfeldt-Jacob Disease

Creutzfeldt-Jacob disease is caused due to the presence of infectious protein particles in the brain called prions (Groveman et al. 2019). It is a rare form of the disease which affects one individual in a million and also might take quite a lot of years for an already infected person to develop visible symptoms. This disease usually starts with lethargy, severe changes in overall mood, and delays in recalling different events (Bougard et al. 2016). It may develop various psychiatric forms including dementia as well (Abudy et al. 2014; Riemenschneider et al. Riemenschneider et al. 2003; Tschampa et al. 2001; Zerr and Parchi 2018).

18.4.8 Mixed Dementia

Mixed dementia occurs as a result of more than one type of dementia-related symptoms usually in the late advanced stages of patients suffering from this disease (where age is more than 80 years) (Custodio et al. 2017; Davies et al. 2018). This condition is characterized by the presence of macroscopically visible lesions and increased frequency of a condition known as cerebral amyloid angiopathy (De Reuck et al. 2018). Generally, a mixture of AD and vascular changes are seen in such patients (Abudy et al. 2014; Jellinger and Attems 2007).

18.5 Music Therapy: A Healer in Disguise

Music makes a unique connection among the organisms conveying emotions (Reybrouck and Eerola 2017). It is a language which need not necessarily be expressed only through words, rather can be felt. Like any other form of art, music holds an aesthetic perspective and has tremendous healing powers which have been recognized as well as, documented worldwide. The shift in focus of the music from a simple societal model toward a more specific neuroscientific model has broadened the horizons for research in this field. The therapy involving music and its elements is being studied for its effects on language-related, cognitive, and sensorimotor functions (Matthews 2015; Zarowitz et al. 2014). It has been found that musical memories are usually often longer preserved than non-musical memories inside the brain (Armstrong et al. 2013).

Music is said to comprise of various small elements like singing, playing different types of instruments, moving or dancing to the beats of music, writing songs or sharing memories associated with any type of music which altogether helps meeting their long-lost requirements like a sense of achieving a particular goal, happiness including elements of success or attainment of the meaning of life (Héroux et al. 2020).

The active participation encouraged by music interventions can enhance a person's life through various types of biophysiological responses and also through self-awareness, self-discovery, and increased self-esteem, and complete satisfaction. Several findings of researchers working in this field revealed that feelings of anxiety, shyness, and stubbornness decreased and the positive social behavioral aspects such as eye contact, smiling, and handshaking increased following every successful MT. These findings reflect that this therapy can work wonders in providing a sense of familiarity, comfort like nothing else can which was observed even in the late stages of dementia where the internal stimuli were almost non-existent (Abudy et al. 2014; Low et al. 2013).

18.6 Mechanisms of Action of Music Therapy

18.6.1 Regeneration Mechanism of the Neurons

There exists a prominent study which provided proof that MT causes positive effects on the cranial nerves from fetal to adult human being. Researchers have observed that musical elements cause a strong impact on neuronal responses also, the overall cell count changed (Fang et al. 2017). Besides, this specialized clinical research performed by another scientist depicted that listening to music can promote significant recovery of neurons and cognitive reservation during the early recovery stage of post-stroke trauma (Choi et al. 2010). Several research-related evidence showed that steroids regulated the processes like neurogenesis, cognition, and neuroprotection (Trimble 2007), and a strong connection between music-related activity and steroid hormones was also discovered (Fang et al. 2017; Kuhn et al. 2007). The effect of music on steroid hormones such as cortisol (Fig. 18.1) and estrogen and its production via the auditory pathway particularly through the neural pathway (circuits related to emotions inside the brain) is mediated by the cerebral limbic system. This system comprises the hypothalamic-pituitary-adrenal axis and amygdaloid complex (Fusani and Gahr 2006; Khalfa et al. 2003).

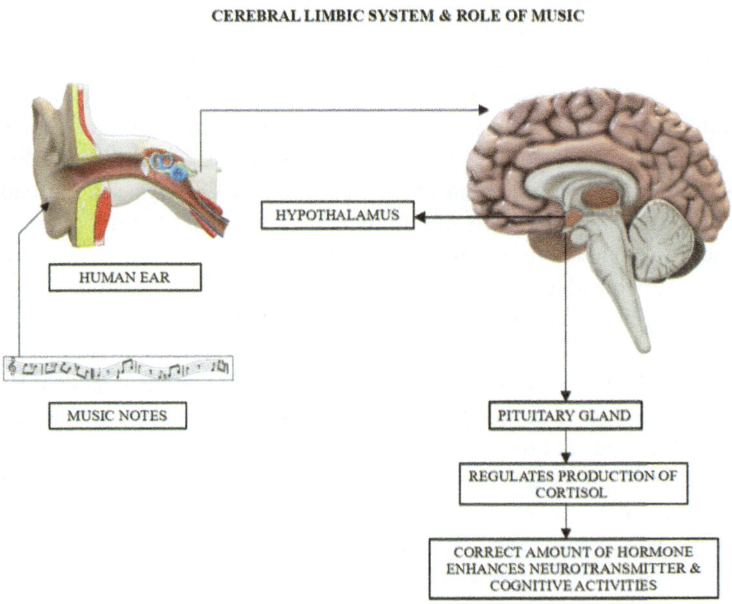

Fig. 18.1 Induction of different hormones during music therapy

In a study, a special therapy related to music was given to juvenile rats for 2 h daily. The controls were also considered and were exposed to the background noise which was considered as the main factor for them. This therapy lasted for 3 weeks with 60 days old rats who were exposed to the training in fear extinction, auditory fear conditioning. Anterior cingulate cortex (ACC) brain-derived neurotrophic factor (BDNF) assays were performed to analyze the impact of MT. BDNS protein helps in the survival of existing neurons, and it also promotes the growth as well as differentiation of newly formed neurons. During fear extinction, rats were showing less behavior of freezing as compared with control rats. Juvenile rats were showing results like a decreasing rate in anxiety behaviors, increases fear extinction, and the rapid increase of BDNF levels in ACC in adult rats (Chen et al. 2019).

18.6.2 Involvement of Neuroendocrine Pathway

MT has been shown to influence the level of hormones including cortisol (Fig. 18.1), testosterone, and estrogen (Trimble 2007). Fukui et al. enlisted patients suffering from AD to listen to his chosen playlist and songs with more frequent verbal contact with the music therapist. The treatment was continued for 1 month with which fugue decreased with a significant secretion of compounds like testosterone and 17-β-estradiol (Abbott 2002; Ménard et al. 2017), which meant that hormones had a preventive impact on patients battling AD through this therapy. This way MT has been found to be better than hormone replacement therapy as it is non-invasive, safe without having long-lasting side effects (Herman et al. 2019; Trimble 2007).

In the same line, to analyze the neurobiological effects of the Chinese traditional five elements music therapy, the experiment was carried with male Wistar rats (Särkämö et al. 2008). They were randomly assigned different experimental groups (powerful, sad, gentle, joyful, and music groups) and control groups. Experimental groups were exposed to mild sound pressure levels (between 50 and 60 db) for 2 hours/day which comprised an overall duration of 28 days. At the end of the session, the concentration levels of Glu (Glutamic acid) and GABA (gamma amino butyric acid) were noted down and matched with the different types of music which were taken into consideration. The levels of amino acids like aspartate, glycine, and glutamine were also measured afterward. As a result of MT, in the case of powerful music the levels of GABA dropped, whereas the levels of glutamic acid increased. In the case of a sad type of music, the levels of glutamic acid dropped and GABA levels increased. In the case of gentle music, the levels of both types of molecules remained constant and remained at the 0 levels. In the case of joyful music, the glutamic acid levels increased and GABA levels dropped significantly. It was concluded that different types of melodies were responsible for producing different effects over amino acids and the related neurotransmitter levels (Hao et al. 2020). Music seems to promote the secretion of several neurotransmitters, neuropeptides, and other biochemicals like endocannabinoids, endorphins, nitric oxide, and dopamine (Särkämö et al. 2008). It was suggested that music took part in the reward, stress and arousal, immunity, and social affiliation-related emotions of human beings (Armstrong et al. 2013; Davidsson et al. 2002).

18.6.3 Neuropsychiatric Mechanism

One different opinion raised by several scientists and researchers is that it is the emotional competencies that significantly influenced the cognitive test scores of dementia patients rather than music therapy (Belchev et al. 2017). Captivatingly, almost all the researches in the years that passed by have suggested that MT had therapeutic effects on neuropsychiatric symptoms along with cognition effectiveness (Fukui and Toyoshima 2008; Nowrangi et al. 2015). Irish et al. (2006) used "spring" created by vivaldi from "the four seasons" as background music for verifying the recollecting power of the autobiographical memory of AD patients in musical conditions, which was found to be improved. As feelings of anxiety decreased gradually with time, it was concluded that anxiety reduction might become one of the promising mechanisms for enhancing autobiographical memory recall with music (Fukui et al. 2012; Vik-Mo 2019). As a development, another scientist found that sad music was the most effective in autobiographical memory. So, it was pointed out that music itself could not conjure memory, instead, the neuropsychiatric symptoms linked with music had a great result on semantic memory (Boso et al. 2006; Woodward 2005).

18.6.4 Neuroplasticity Mechanism

A functional magnetic resonance imaging (fMRI) was used by several scientists to detect the change in the overall functioning of the brain while the patients suffering from dementia were engaged in different types of singing activities generally considering the case of karaoke devices (Jung et al. 2019). A 6 months long music training increased the neural activities in the right portion of the affected brain, which indicated that MT improved the neural efficacy of the AD or dementia-affected patients. As a result, the activation of different regions of the brain during the karaoke task, i.e., left lingual and right angular gyrus, the site of multimodal sensory integration might have been caused by music and reading-related processing in the brain. According to one more study, the participation of the parietal lobe in pitch processing while listening to music and playing instruments is crucial (Satoh et al. 2003; Ashford 2015). This also echoed that music might play a pivotal role in neuroplasticity-related mechanisms in the brain which were depicted by the resonance imaging results of the scientist (Gallego and García 2017; Muresanu 2007).

As discussed above, it is clear that there is a scientific basis for the improvement of neurological conditions during MT. There have been many studies conducted by several scientists throughout the world over this, though the number of studies is not that high and we haven't reached up to the molecular level to derive the clear-cut results from these studies. There are several cases where medicines like antidepressants, tranquilizers, and antipsychotics could not be tolerated (Kessing et al. 2009; Schneider et al. 2006), so in such cases, MT was chosen to treat dementia which consists of destructiveness and anxiety as a whole (O'Connor and Gray 2014; O'Connor 2016).

18.7 Case Studies Associated with Music Therapy and Its Effect on Brain Function and Dementia

18.7.1 Effect of Music Therapy on Brain Function

18.7.1.1 Effect of Music Therapy on Preterm Babies Mental and Overall Development

Preterm birth carries several risk factors for the baby including neurological impairment and gaps in motor function, differences in cognition, and problems related to behavioral development. Here in this study the author specially designed MT for preterm infants which are known as creative music therapy (CMT). The study included the selection of 60 medically stable premature infants less than 32 weeks of gestational age. The CMT sessions were conducted for 20 minutes thrice a week. The music included lullabies and soft rhythms which can calm down the babies and were also adjusted according to the individual requirement of the baby. The non-invasive type magnetic resonance imaging (MRI) was conducted at the corrected age for all the infants and the developmental changes in the brain were observed. The analysis of the result concluded that MT could positively affect the growth of premature infants (Haslbeck et al. 2017).

18.7.1.2 Effect of Music Therapy on Brain Function

Popescu et al. (2004) worked over the magnetoencephalographic signals, while the selected group of people listened to music. The study aimed to correlate various formulated studies, which explained the relation between the time course of a particular region of the brain's activation and the dynamic nature of music elements. It was depicted that the front region of the brain responded mainly toward the slow-paced music and the motor-related regions showed the smoother or transient types of responses which were later mapped with the temporal scales of music. The different types of rhythms and pieces of music were considered in this study. The final results confirmed that the overall activity in the motor regions and related structures especially, the lateral and supplementary motor areas were directly related to the extent of rhythmicity, which was derived from elements of music. This further concluded how music expression affected the overall performance of the human brain (Popescu et al. 2004).

In a study by Large and Snyder (2009) brain process during the musical performances was understood and, it was observed that high-frequency neural activity results from the high-frequency oscillatory activity which further leads to the communication between the neural areas while the rhythm of the music going on (Large and Snyder 2009).

A study conducted by Chen et al. (2008) was based on the listening of the predicted tapping and then again tapping with the musical rhythms. They conducted 2 fMRI studies involving rhythm perception and production. Mid-pre-motor cortex (PMC) and cerebellum were observed during predicted tapping. The second activity was conducted to check whether the previous activity was motor planning or rehearsal and, in this activity, there was no tapping along with the musical rhythm. It was proven that the motor regions were the same which were engaged in both activities. The ventral and dorsal parts of the PMC acted differently as ventral PMC occurred during the action-coupled process, whereas dorsal part was working only on actions, which were based on higher rules of temporal organization. These activities had shown that the action-perception process is responsible for the link between auditory and motor processes during musical rhythm (Chen et al. 2008).

18.7.2 Case Studies Based on the Detailed Response of Dementia Patients

18.7.2.1 Effects of Music on Agitated -Type Behavior

In this study, nurses opted MT for AD patients and it was used twice a week (Baker et al. 2008). It was given to the patients when their caregiver activities were going on to avoid unnecessary disruptive behavior of patients. CNAs (certified nursing assistants) conducted their activities based on certain formulated rules like the requirement of a basic overview of dementia and its relation with the irritable behavior of patients, ability to use I-pods while imparting this type of therapy and the last one included obeying the key elements of a successful music program always. Music therapists created different playlists according to the caregiver activities like for bathing there was a separate type of music, for dancing, there was a separate type of music and for a sing-along, there was a different music type. No headphones were allowed during music therapies. This method was found to improve certain behaviors like agitation, anger issues, etc., in patients suffering from dementia (Ridder et al. 2013).

Another study was conducted as a non-randomized quantitative study under the guidance of several authors in Australia. It was initially aimed to investigate the long-term effects of the grouped MT on the patients suffering from AD. The convenient type of sampling was conducted to select patients exhibiting the moderate symptoms of the above-mentioned disease. These patients were selected from a total of 13 different types of nursing homes. Weekly MT sessions were provided to them to see the significant changes in their anger-related emotions. In total, the agitation levels were completely assessed five times in an entire year. The detailed model-based study showed no significant difference between the two groups which were formed (one reference model and one group with high agitation levels). But the music therapist's observation stated that the patients felt relaxed or less agitated during or immediately after the therapy session. The limitations of the study conducted came out to be the very small sample size (19 participants) and the fact that only five elaborative assessments were conducted in an entire year (Wall and Duffy 2010).

In another study, several criteria were adopted for the patients to be enrolled in MT like NPI-C (neuropsychiatric index) score, MMSE score, and CDI (children's depression inventory) score. Fifty-nine patients were then enrolled who matched certain criteria. There were 3 rounds of MT therapy which were 30 minutes each and the control group did several activities like playing with cards, and newspaper reading, etc., according to the patient preference. The first session was started with two factors which were empathetic behavior and non-empathetic behavior in which they assessed whether the patients made compatible relationships with them or not. The second session was based on the reactions of the patients to the MT therapy where they got to know about the patient's interest by their certain reactions like how they laugh, how they are enjoying music, and what are the body movements involved in that process. It was found that NPI scores were getting improved in some activities or behavior like irritation, anger, and how they behaved at night with other people. Reduction in non-EB (non-empathetic behavior) was also taking place (Gogoularadja and Bakshi 2020).

In another experiment (Prince et al. 2013) about 40 patients suffering from AD of mild to moderate form had undergone MT for 6 weeks and many factors were studied. Patients were selected from 2 geriatric residencies in the region of Murcia. Patients with dementia had shown to improve neuropsychiatric symptoms and cognitive functions to a much lesser extent (Gallego and García 2017). Patients were divided into two groups in which there were less than 12 patients in each group. Headphones were not allowed while listening to the song. Patients attended this therapy for 2 weeks lasting for 45 minutes each. Patients were examined for a series of parameters like neuropsychiatric, cognitive, and functional assessment after 3 weeks (6 sessions) and at the end of the study period (12 sessions). It was observed that depression did not improve after a total of 6 sessions of therapy ($p > 0.05$) and MT did not have any significant effect on BI scores but, it had an optimistic impact on feelings related to anxiety & depression according to the HADS (hospital anxiety and depression scale) scores. In totality, MT helped in stimulating cognitive function, improved mood, and reduced behavior problems which are triggered by stressful conditions (García et al. 2012; Satoh et al. 2015).

MT was getting famous in Japan also for the individual as well as group-oriented therapy (Eguchi 2018). Yukiko was a lady in Japan who was identified with AD-type dementia and after that, her family agreed to participate in home-based MT because Yukiko had a keen interest in music. There were several steps which were adopted to conduct MT for her such as Music therapists usually come and interact with the patient and used to take an interview of the patient and her family, as well. Sessions were being held at Yukiko's home with her Music therapist. After some sessions some positive responses could be seen in her behavior like, she used to recollect her memories for which she was nostalgic, she used to sing her favorite songs as a part of the therapy. This case study concluded that MT can help in social competencies as it helped her in building her social connections, communications skills, etc. This therapy did not improve dementia symptoms much but really helped Yukiko on a personal level (Otera et al. 2020).

18.7.2.2 Indian Classical Music and Dementia

The roots of one of the oldest forms of music reside in India (dates back to 5000–2000 BC). The ancient manuscripts, called the "vedas", are said to be the source of inspiration for classical music. The samaveda, one of the 4 main vedas is said to be the main originating source of this form of music (Hegde et al. 2012). There exist different verses in classical music which are used for chanting. Ragas, meaning the "one which induces an emotion in mind" and taal (a rhythmic structure) form the main body of Indian classical music (Zarowitz et al. 2014). These terms form the basis of different compositions and melodic structure. The ragas are said to evoke a combination of various emotions like sadness, anger, devotion, passion, romance, etc. (Roy et al. 2017). The expression of these structures of Indian classical music is intended to vary during the main performance which is a result of the complex rhythmic cycle that evolved with the due course of time (Hegde 2017).

In a study conducted by several scientists, 20 musically untrained individuals were made to listen to the North Indian classical music (NICM) tunes and it showed high positive results which were comparable to the relaxed states found after a meditation session (Bardekar and Gurjar 2016). Moreover, the systolic and diastolic blood pressure showed a significant decrease, sensations related to stress, anxiety, and depression dropped and thereby increased the feelings related to ultimate satisfaction, hope, optimism, and harmony (Bardekar and Gurjar 2016).

Ragas are considered to possess healing powers and can enhance the overall well-being of an individual (Zarowitz et al. 2014. In the case of a patient named Harbhajan Singh who took the assistance of his daughter, Prabhjot Panwar for overcoming his illness temporarily (Parmar and Puwar 2019). Singh was fond of Hindustani classical ragas and it became the basis of his therapy later on. From the information collected based on his entire treatment during the sessions, it was concluded that he showed visible signs of improvement like moving his head and hands on the specific tunes of ragas, and in the next half of the session he even started recognizing the song and connected it with his memories related to his college days (Parmar and Puwar 2019).

18.7.3 Case Studies Including Carer/Nurse Training

18.7.3.1 Carer Training (Ridder, Denmark)

In this case authors focused mainly on the indirect and direct music therapy practices which are prevalent in various homecare centers in Denmark (Stige 2018). The concept of indirect music therapy included the sharing of knowledge related to music-related healing in patients suffering from dementia and AD. The training happened more in a teaching-specific manner. Direct music therapy included the art of process learning with real-life examples. An experienced music ambassador, Marie Munk Madsen taught the participants the basic model of therapy related to music. The goals of the training which was provided included the ideas for implementing the music activities. The tools required to enforce the same in daily living were the theoretical knowledge related to the same and the courage which is required to use the body dynamically along with the voice to put the best effort in for treating the patients. Altogether, 20 caregivers from 11 different nursing homes or daycare centers were a part of this learning initiative. They weren't expected to have musical skills to participate. The training process required the participants to be active at all times and be able to cope with the pressure related to time and different situations. Adequate time was devoted to interacting with several patients from these 11 nursing homes from where the caregivers were selected. As a result, the carers learned the art of understanding and curing the patients at the same time. This training strategy stated that a music therapist isn't just required to act as a facilitator but also, as a supervisor and knowledge distributor for other carers (Schneider et al. 2006).

18.7.3.2 Training Social Workers and Caregivers in the Family (Wosch, Germany)

The University of Applied Sciences Wurzburg-Schweinfurt, Germany, formulated three main streams of music therapy skill sharing. First is the counseling which was required for the family caregivers of patients with dementia who are staying at home. Second is instructing the therapy techniques to the social workers and the third is forming projects in social work using several techniques in music therapy. After looking into the 3-year long calculated statistics, it was concluded by the authors that most family caregivers do not require special training for learning the art of empathizing and tackling patients with dementia. There were two main areas which emerged as a result of a one-day workshop for family caregivers and social workers. One area was about the importance of sharing the experiences of being a family caregiver and how that closeness affects the behavioral aspects of patients. Second area was to develop an individualistic approach toward music in a more informal way being the family caregivers (Wood and Ansdell 2018). It was also found that elements of group music therapy and therapeutic songwriting when integrated for the social work students increase their self-confidence with a significant rise in social competencies and behavior with social-communicative orientation (Baker and Yeates 2018). A group of carefully selected patients were treated with both direct and indirect therapeutic techniques and it was concluded that both direct and indirect forms of music therapy work as driving forces for teaching the paid/family caregivers the art of handling people with dementia and can together improve the behavioral changes in the patients (Raglio et al. 2008).

18.7.3.3 Polyphonic Partnerships (Stige, Norway)

This approach involves the connection between music and nature and also the surrounding environment while getting musical therapy into implementation (Stige 2018). This indirect type of MT is based on the ecological perspective of music which was initially developed in the presence of biological elements. Ecological perspectives mean that the goals and their practices of therapy focus mainly on fostering the health-promoting relationships between several individuals including both carers and patients (Morell and Shoemark 2018). MT is generally viewed as a group initiative which can impact the larger audience and at a given time and can also run the economic activities related to the same domain (referring to nursing homes). In Norway, there are few guidelines formulated keeping in mind both the ecological and economical perspectives. They suggest the individualistic approach for handling the patients with dementia if needed in severe cases. The main objective for the carers was to improvise/teach other carers the ability to build emotional connections with their family patients or the patients in general. The implementation of the indirect mode of music therapy mainly requires the skill of staying occupied in polyphonic partnerships which permit semi-professional environments and promote organizational change. Both professionalism and collaboration can foster interactive sessions and make them easy between the carers and patients. It was concluded from the study that indirect music therapy requires cooperation from both sides and it mainly requires assistance from the side of professionals working in the same field. It also includes the mutual sharing of knowledge for the larger benefit of carers and patients (McDermott et al. 2018).

18.7.4 Case Studies Based on Song-Writing

18.7.4.1 Therapeutic Songwriting for the Family Caregivers (Baker, Australia)

Family carers play an important role in shaping the behavior of the people around with dementia as they impact them emotionally, physically, and enhance their overall well-being (Stige 2018). Therapeutic songwriting comes with several advantages which include the opportunity to tell the patients their own story, keeping a real-time track of their progress, allowing the people to process/re-process their own emotions and recollect the long-lost memories. The songwriting in this special way also encourages the pairing of the emotional content with the lyrics. The series of songs that were presented in front of the family carers instilled in them a sense of compassion, increased the level of understanding toward the patients, gave them time to introspect their role as a caregiver and reflect on their identity. This particular support program differed from the others in its way of preserving the emotions of both the carers and the people with dementia. This technique of therapeutic songwriting is kept under the category of direct music therapy and it also benefits the patients indirectly as the emotional competencies of carers are increased and this was tried over a small group of patients (small sample size) in a nursing home in Australia which gave positive results and enhanced the emotions competencies of the patients at large (Bunt and Stige 2014).

18.7.4.2 Songwriting by the Patient

In another study, one of the patients involved in this therapy was Margaret who was 94-year-old and her therapy of writing songs lasted for 18 months for the treatment of dementia. This songwriting technique has helped Margaret to indulge herself in expressing her feelings fully. Her rhythmic attention was improving day by day although her cognition power was not so good but, as the days were passing, she was able to compose songs efficiently. Margaret was able to experience her life moments; she was able to recognize her past gradually. MT has developed her creativity too much (Ahessy 2017).

18.8 Recent Developments and Future Perspectives

Future research with MT should aim to indulge in high-quality research with large sample sizes. There should be proper planning of such experiments and there should be consistency in using the set protocols and assessing the parameters to evaluate the same. Studies at the molecular level should also be conducted along with measuring psychological behavior, physiological parameters to make out some conclusion at the molecular level so that we can further move into the deeper planning to treat such patients with this harmless therapy, which seems to be promising with these primitive studies conducted so far.

18.8.1 Implications of Music Therapy for Clinical Practice

Many health professionals and music therapists are taking part in musical intervention for the treatment of several diseases. Several authors have described and discussed various effects of MT on various diseases including dementia. Music therapists need to take any action and provide some basic knowledge and guidance, through education and specific medical personnel training to the medical staff or carers, to improve music-related medicinal practices. Now the therapeutic goals must be combined with the active singing experts to improve and accomplish the long-term setting of the people with dementia. Further, it will be best for the music therapists to train the family members/caregivers to incorporate the singing for daily caregiving with people having dementia (Cho 2018).

18.8.1.1 Musical Intervention in Improving the Quality of Life

The musical intervention helped to the loss of depressive symptoms, higher self-esteem, and a better quality of life (Cheung et al. 2019; Pongan et al. 2017). A case study was conducted where MT was given to about 60 patients of age 8–10 years old, suffering from a pediatric brain tumor (Payrau et al. 2017). Pre- and post-treatment results were observed for these subjects and were compared with the control group. The experimental group has received this therapy 45 minutes weekly for 52 weeks. Placebo intervention was given to the control group and three factors: depressive symptoms, self-esteem, and the quality of life were studied. Results were collected at baseline, 6 months, and 12 months after the intervention began. After the 12 months follow-up, changes like loss of depressive symptoms ($P < 0.001$), higher levels of self-esteem ($P < 0.001$), and improved quality of life ($P < 0.001$) were observed.

18.8.1.2 Musical Intervention in the Depressive State

Depression is one of the very common psychological disorders which leads a person into more anger issues, reduced social gatherings, and more toward isolation (Kok and Reynolds 2017). The musical intervention is useful to cure depressive states but very limited research has been performed in this field (Leubner and Hinterberger 2017). In their study, 28 different studies were analyzed and several factors like the length of the trial, patient's age, active or passive singing, and the type of sessions (individual or group) were considered. The main focus of the researchers was on only one type of music genre (classical, western, instrumental, or vocal). Patients had shown tremendous improvement in cognitive and emotional benefits when they sang or listened to any of these song types. MT also helped in several therapeutic approaches like it had shown positive effects on the patients before heart surgery and gives relaxation during angiography. Musical interventions have shown to help in improving the quality of sleep in old age people and it has also helped in improving memory in children. The immune system has also been reported to be strengthened with MT (Leubner and Hinterberger 2017).

The impact of music on the cognitive functioning of the brain is explained through this example where the authors analyzed that sudden behavioral or cognitive dysfunction leads to nerve damage especially in old age (Wang et al. 2018). They performed an elaborated literature search over platforms like PubMed and EMbase. A total of 34 studies were included out of which most of their quality checks were based on PEDro (physiotherapy evidence database) and CASP scale scores. These studies were divided into several subgroups based on the factors they took into consideration and later on the meta-analysis was performed. As the results of all of these studies being analyzed, it was found that in the majority of cases music therapy was successful in decreasing the behavioral symptoms related to depression, stress and at the same time lowered down the risks attached to cognitive dysfunction. The positive trends in these studies and their impact on the overall working of the brain reassure the researchers in this field to explore this non-pharmacological medium of treatment for reducing several risk factors attached to dementia, especially in elderly patients (Zhang et al. 2017).

18.8.1.3 Impact of Music on Memory

Memory loss which generally refers to an unusual type of forgetfulness is counted as one of the early symptoms of dementia and other neurodegenerative diseases (Gluck et al. 2016). In one study the strategy to treat this condition included the use of various types of MTs for encoding the verbal or written information about the patient (Abrahan et al. 2020). The biggest role in making the strategy successful was played by emotional competencies. In this study, 30 older adults (OA) and 24 young adults (YA) were made to upskill themselves in various sets and subsets of music which was either positively or negatively balanced. Both immediate and delayed memory flashbacks were recorded. The results depicted that the performance of OAs was less efficient than the YAs in the case of immediate recalling of memories which were directly associated with positive tracks of music. Lyrics that were sung were better retained when compared with spoken words in music in the case of OAs. The time duration in recalling memories is independent of the type of music which was played in front of the participants. The analysis of these results shows that music intervention for recollecting the forgotten memories was beneficial for people of all age groups especially the aged adults. The research is yet to unfold the several hypothetical understandings of the mechanisms which are responsible for encoding music and its advantage (Ratovohery et al. 2018).

18.9 Conclusion

MT has been found to improve the severity of dementia and behavior issues of the patients and it can be the best method to cure dementia as it is a kind of home therapy with no harm. This therapy also helped in working as a facilitator and a supervisor. It helped in improving the harmony within patients as well. Songwriting techniques when integrated with some social work behaviors helped in increasing self-confidence with a gradual rise in social competencies. Direct or indirect type of therapy helped in improving the art of handling people with dementia and improved the emotional competencies. So, this therapy aids in improving the cognitive function, mood and also reduces the issues/behaviors which were activated by stressful conditions in the patients suffering from dementia. In most of the cases only behavioral changes have been studied for patients with dementia. In the future, the researchers can even study many other changes like physical, cognitive, and physiological at the molecular level for unfolding the mysteries of the human brain and its relation with music and its components.

References

Abbott A (2002) Music, maestro, please! Nature 416(6876):12–14

Abudy A, Juven-Wetzler A, Zohar J (2014) The different faces of Creutzfeldt-Jacob disease CJD in psychiatry. Gen Hosp Psychiatry 36(3):245–248

Abrahan VD, Shifres F, Justel N (2020 Sep) Impact of music-based intervention on verbal memory: an experimental behavioral study with older adults. Cogn Process 21:1–4

Ahessy B (2017 Sep) Song writing with clients who have dementia: a case study. Arts Psychother 1(55):23–31

Allen GR, Erdmann MV, Randall JE, Ching P, Rauzon MJ, Hayashi LA, Thomas MD, Robertson DR, Taylor L, Coste M (2013) Reef fishes of the East Indies. University of Hawai'i Press, Honolulu

Anor CJ, O'Connor S, Saund A, Tang-Wai DF, Keren R, Tartaglia MC (2017) Neuropsychiatric symptoms in Alzheimer disease, vascular dementia, and mixed dementia. Neurodegener Dis 17 (4–5):127–134

Armstrong MJ, Litvan I, Lang AE, Bak TH, Bhatia KP, Borroni B, Boxer AL, Dickson DW, Grossman M, Hallett M, Josephs KA (2013) Criteria for the diagnosis of corticobasal degeneration. Neurology 80(5):496–503

Ashford JW (2015) Treatment of Alzheimer's disease: the legacy of the cholinergic hypothesis, neuroplasticity, and future directions. J Alzheimers Dis 47(1):149–156

Baker F, Wigram T, Stott D, McFerran K (2008) Therapeutic songwriting in music therapy: part I: who are the therapists, who are the clients, and why is songwriting used? Nord J Music Ther 17(2):105–123

Baker FA, Yeates S (2018) Carers' experiences of group therapeutic songwriting: An interpretive phenomenological analysis. Br J Music Ther 32(1):8–17

Bakulski KM, Seo YA, Hickman RC, Brandt D, Vadari HS, Hu H, KyunPark S (2020) Heavy metals exposure and Alzheimer's disease and related dementias. J Alzheimers Dis 76(4):1–28

Balin BJ, Hammond CJ, Little CS, Hingley ST, Al-Atrache Z, Appelt DM, Whittum-Hudson JA, Hudson AP (2018 Oct) Chlamydia pneumoniae: An etiologic agent for late-onset dementia. Front Aging Neurosci 9(10):302

Bang J, Spina S, Miller BL (2015) Frontotemporal dementia. Lancet 386(10004):1672–1682

Bardekar AA, Gurjar AA (2016) Study of Indian classical ragas structure and its influence on human body for music therapy. In: 2nd International Conference on Applied and Theoretical Computing and Communication Technology (iCATccT) 2016 Jul 21. IEEE, Piscataway, pp 119–123

Bayen E, Possin KL, Chen Y, de Langavant LC, Yaffe K (2018) Prevalence of aging, dementia, and multimorbidity in older adults with down syndrome. JAMA Neurol 75(11):1399–1406

Belchev Z, Levy N, Berman I, Levinzon H, Hoofien D, Gilboa A (2017) Psychological traits predict impaired awareness of deficits independently of neuropsychological factors in chronic traumatic brain injury. Br J Clin Psychol 56(3):213–234

Booker A, Jacob LE, Rapp M, Bohlken J, Kostev K (2016) Risk factors for dementia diagnosis in German primary care practices. Int Psychogeriatr 28(7):1059–1065

Boso M, Politi P, Barale F, Emanuele E (2006) Neurophysiology and neurobiology of the musical experience. Funct Neurol 21(4):187

Bougard D, Brandel JP, Bélondrade M, Béringue V, Segarra C, Fleury H, Laplanche JL, Mayran C, Nicot S, Green A, Welaratne A (2016) Detection of prions in the plasma of presymptomatic and symptomatic patients with variant Creutzfeldt-Jakob disease. Sci Transl Med 8(370):370ra182

Bright F, Werry EL, Dobson-Stone C, Piguet O, Ittner LM, Halliday GM, Hodges JR, Kiernan MC, Loy CT, Kassiou M, Kril JJ (2019) Neuroinflammation in frontotemporal dementia. Nat Rev Neurol 15(9):540–555

Bunt L, Stige B (2014) Music therapy: An art beyond words. Routledge, London

Cacabelos R (2008) Pharmacogenomics and therapeutic prospects in dementia. Eur Arch Psychiatry Clin Neurosci 258(1):28–47

Cantón-Habas V, Rich-Ruiz M, Romero-Saldaña M, Carrera-González MD (2020) Depression as a risk factor for dementia and Alzheimer's disease. Biomedicine 8(11):457

Chen JH, Lin KP, Chen YC (2009) Risk factors for dementia. J Formos Med Assoc 108(10):754–764

Chen JL, Penhune VB, Zatorre RJ (2008) Listening to musical rhythms recruits motor regions of the brain. Cereb Cortex 18(12):2844–2854

Chen S, Liang T, Zhou FH, Cao Y, Wang C, Wang FY, Li F, Zhou XF, Zhang JY, Li CQ (2019) Regular music exposure in juvenile rats facilitates conditioned fear extinction and reduces anxiety after foot shock in adulthood. Biomed Res Int 14:2019

Cheung AT, Li WH, Ho KY, Lam KK, Ho LL, Chiu SY, Chan GC, Chung JO (2019) Efficacy of musical training on psychological outcomes and quality of life in Chinese pediatric brain tumor survivors. Psycho-Oncology 28(1):174–180

Cho HK (2018) The effects of music therapy-singing group on quality of life and affect on persons with dementia: a randomized controlled trial. Front Med 15(5):279

Choi AN, Lee MS, Lee JS (2010) Group music intervention reduces aggression and improves self-esteem in children with highly aggressive behavior: a pilot-controlled trial. Evid Based Complement Alternat Med 1:7

Convery R, Mead S, Rohrer JD (2019) Clinical, genetic and neuroimaging features of frontotemporal dementia. Neuropathol Appl Neurobiol 45(1):6–18

Coppus AM, Evenhuis H, Verberne GJ, Visser F, Van Gool P, Eikelenboom P, Van Duijin C (2006) Dementia and mortality in persons with Down's syndrome. J Intellect Disabil Res 50(10):768–777

Coskun PE, Wyrembak J, Derbereva O, Melkonian G, Doran E, Lott IT, Head E, Cotman CW, Wallace DC (2010) Systemic mitochondrial dysfunction and the etiology of Alzheimer's disease and down syndrome dementia. J Alzheimers Dis 20(s2):S293–S310

Custodio N, Montesinos R, Lira D, Herrera-Pérez E, Bardales Y, Valeriano-Lorenzo L (2017) Mixed dementia: a review of the evidence. Dementia Neuropsychol 11(4):364–370

Davidsson P, Sjögren M, Andreasen N, Lindbjer M, Nilsson CL, Westman-Brinkmalm A, Blennow K (2002) Studies of the pathophysiological mechanisms in frontotemporal dementia by proteome analysis of CSF proteins. Mol Brain Res 109(1–2):128–133

Davies SJ, Burhan AM, Kim D, Gerretsen P, Graff-Guerrero A, Woo VL, Kumar S, Colman S, Pollock BG, Mulsant BH, Rajji TK (2018) Sequential drug treatment algorithm for agitation and aggression in Alzheimer's and mixed dementia. J Psychopharmacol 32(5):509–523

De Reuck J, Maurage CA, Deramecourt V, Pasquier F, Cordonnier C, Leys D, Bordet R (2018) Aging and cerebrovascular lesions in pure and in mixed neurodegenerative and vascular dementia brains: a neuropathological study. Folia Neuropathol 56(2):81–87

Eguchi R (2018) The Differences of Musical Self-Expression and the Interrelationships Within Group Music Improvisation Between Japan and the United States

Ewers M, Walsh C, Trojanowski JQ, Shaw LM, Petersen RC, Jack CR Jr, Feldman HH, Bokde AL, Alexander GE, Scheltens P, Vellas B (2012) Prediction of conversion from mild cognitive impairment to Alzheimer's disease dementia based upon biomarkers and neuropsychological test performance. Neurobiol Aging 33(7):1203–1214

Fang R, Ye S, Huangfu J, Calimag DP (2017) Music therapy is a potential intervention for cognition of Alzheimer's disease: a mini-review. Transl Neurodegen 6(1):2

Fink HA, Jutkowitz E, McCarten JR, Hemmy LS, Butler M, Davila H, Ratner E, Calvert C, Barclay TR, Brasure M, Nelson VA (2018) Pharmacologic interventions to prevent cognitive decline, mild cognitive impairment, and clinical Alzheimer-type dementia: a systematic review. Ann Intern Med 168(1):39–51

Fukui H, Toyoshima K (2008) Music facilitate the neurogenesis, regeneration and repair of neurons. Med Hypotheses 71(5):765–769

Fukui H, Arai A, Toyoshima K (2012) Efficacy of music therapy in treatment for the patients with Alzheimer's disease. Int J Alzheimers Dis 2012:531646. https://doi.org/10.1155/2012/531646

Fusani L, Gahr M (2006) Hormonal influence on song structure and organization: the role of estrogen. Neuroscience 138(3):939–946

Fymat AL (2019) On dementia and other cognitive disorders. Clin Res Neurol 2(1):1–4

Galla L, Redolfi N, Pozzan T, Pizzo P, Greotti E (2020) Intracellular calcium dysregulation by the Alzheimer's disease-linked protein presenilin 2. Int J Mol Sci 21(3):770

Gallego MG, García JG (2017) Music therapy and Alzheimer's disease: cognitive, psychological, and behavioural effects. Neurología (English Edition) 32(5):300–308

García JJ, Iodice R, Carro J, Sánchez JA, Palmero F, Mateos AM (2012) Improvement of autobiographic memory recovery by means of sad music in Alzheimer's disease type dementia. Aging Clin Exp Res 24(3):227–232

Gluck MA, Mercado E, Myers CE (2016) Learning and memory: from brain to behavior. Worth Publishers, New York

Gogoularadja A, Bakshi SS (2020) A randomized study on the efficacy of music therapy on pain and anxiety in nasal septal surgery. Int Arch Otorhinolaryngol 24(2):e232–e236

Grande G, Haaksma ML, Rizzuto D, Melis RJF, Marengoni A, Onder G, Welmer A-K, Fratiglioni L, Vetrano DL (2019) Co-occurrence of cognitive impairment and physical frailty, and incidence of dementia: Systematic review and meta-analysis. Neurosci Biobehav Rev 107: 96–103

Groveman BR, Foliaki ST, Orru CD, Zanusso G, Carroll JA, Race B, Haigh CL (2019) Sporadic Creutzfeldt-Jakob disease prion infection of human cerebral organoids. Acta Neuropathol Commun 7(1):1–2

Gupta S, Warner J (2008) Alcohol-related dementia: a 21st-century silent epidemic? Br J Psychiatry 193(5):351–353

Hao J, Jiang K, Wu M, Yu J, Zhang X (2020) The effects of music therapy on amino acid neurotransmitters: insights from an animal study. Physiol Behav 1(224):113024

Haslbeck FB, Bucher HU, Bassler D, Hagmann C (2017) Creative music therapy to promote brain structure, function, and neurobehavioral outcomes in preterm infants: a randomized controlled pilot trial protocol. Pilot and feasibility studies 3(1):36

Hegde S, Aucouturier JJ, Ramanujam B, Bigand E (2012) Variations in emotional experience during phases of elaboration of North Indian Raga performance. In: 12th International Conference on Music Perception and Cognition and the 8th Triennial Conference of the Cognitive Sciences of Music 2012 Jul 23, pp 412–413

Hegde S (2017) Music therapy for mental disorder and mental health: the untapped potential of Indian classical music. BJPsych int 14(2):31–33

Herman FJ, Simkovic S, Pasinetti GM (2019) Neuroimmune nexus of depression and dementia: shared mechanisms and therapeutic targets. Br J Pharmacol 176(18):3558–3584

Héroux I, Giraldo S, Ramírez R, Dubé F, Creech A, Thouin-Poppe LÉ (2020) Measuring the impacts of extra-musical elements in guitar music playing: a pilot study. Front Psychol 14(11):1964

Hulsegge G, Picavet HS, Blokstra A, Nooyens AC, Spijkerman AM, van der Schouw YT, Smit HA, Verschuren WM (2014) Today's adult generations are less healthy than their predecessors: generation shifts in metabolic risk factors: the Doetinchem cohort study. Eur J Prev Cardiol 21(9):1134–1144

Hunter S, Brayne C (2018) Understanding the roles of mutations in the amyloid precursor protein in Alzheimer disease. Mol Psychiatry 23(1):81–93

Irish M, Cunningham CJ, Walsh JB, Coakley D, Lawlor BA, Robertson IH, Coen RF (2006) Investigating the enhancing effect of music on autobiographical memory in mild Alzheimer's disease. Dement Geriatr Cogn Disord 22(1):108–120

Jackson K, Barisone GA, Diaz E, Jin LW, DeCarli C, Despa F (2013) Amylin deposition in the brain: a second amyloid in Alzheimer disease? Ann Neurol 74(4):517–526

Jellinger KA, Attems J (2007) Neuropathological evaluation of mixed dementia. J Neurol Sci 257 (1–2):80–87

Jellinger KA, Attems J (2013) Neuropathological approaches to cerebral aging and neuroplasticity. Dialogues Clin Neurosci 15(1):29

Jellinger KA, Korczyn AD (2018) Are dementia with Lewy bodies and Parkinson's disease dementia the same disease? BMC Med 16(1):1–6

Jung D, Park JW, Kim YH, You JS (2019) Neuroplastic and motor behavioral changes after intermanual transfer training of non-dominant hand: a prospective fMRI study. NeuroRehabilitation 44(1):25–35

Kalaria RN (2016) Neuropathological diagnosis of vascular cognitive impairment and vascular dementia with implications for Alzheimer's disease. Acta Neuropathol 131(5):659–685

Kerpershoek L, de Vugt M, Wolfs C, Woods B, Jelley H, Orrell M, Stephan A, Bieber A, Meyer G, Selbaek G, Handels R (2018) Needs and quality of life of people with middle-stage dementia and their family carers from the European Actifcare study. When informal care alone may not suffice. Aging Ment Health 22(7):897–902

Kessing LV, Søndergård L, Forman JL, Andersen PK (2009) Antidepressants and dementia. J Affect Disord 117(1–2):24–29

Khalfa S, Bella SD, Roy M, Peretz I, Lupien SJ (2003) Effects of relaxing music on salivary cortisol level after psychological stress. Ann NY Acad Sci 1(999):374–376

Klassen BT, Ahlskog JE (2011) Normal pressure hydrocephalus: how often does the diagnosis hold water? Neurology 77(12):1119–1125

Kloppenborg RP, van den Berg E, Kappelle LJ, Biessels GJ (2008) Diabetes and other vascular risk factors for dementia: which factor matters most? A systematic review. Eur J Pharmacol 585(1):97–108

Kok RM, Reynolds CF (2017) Management of depression in older adults: a review. JAMA 317(20):2114–2122

Kristman VL, Borg J, Godbolt AK, Salmi LR, Cancelliere C, Carroll LJ, Holm LW, Nygren-de Boussard C, Hartvigsen J, Abara U, Donovan J (2014) Methodological issues and research recommendations for prognosis after mild traumatic brain injury: results of the international collaboration on mild traumatic brain injury prognosis. Arch Phys Med Rehabil 95(3):S265–S277

Kuhn HG, Cooper-Kuhn CM, Boekhoorn K, Lucassen PJ (2007) Changes in neurogenesis in dementia and Alzheimer mouse models: are they functionally relevant? Eur Arch Psychiatry Clin Neurosci 257(5):281–289

Iadecola C (2013) The pathobiology of vascular dementia. Neuron 80(4):844–866

Large E, Snyder J (2009) Pulse and meter as neural resonance. Ann N Y Acad Sci 1169(1):46–57

Leubner D, Hinterberger T (2017) Reviewing the effectiveness of music interventions in treating depression. Front Psychol 7(8):1109

Lin JS, O'Connor E, Rossom RC, Perdue LA, Eckstrom E (2013) Screening for cognitive impairment in older adults: a systematic review for the US preventive services task force. Ann Intern Med 159(9):601–612

Lourida I, Hannon E, Littlejohns TJ, Langa KM, Hyppönen E, Kuźma E, Llewellyn DJ (2019) Association of lifestyle and genetic risk with incidence of dementia. JAMA 322(5):430–437

Lott IT, Head E (2019) Dementia in down syndrome: unique insights for Alzheimer disease research. Nat Rev Neurol 15(3):135–147

Low LF, White F, Jeon YH, Gresham M, Brodaty H (2013) Desired characteristics and outcomes of community care services for persons with dementia: what is important according to clients, service providers and policy? Australas J Ageing 32(2):91–96

Luzzi S, Snowden JS, Neary D, Coccia M, Provinciali L, Ralph MA (2007) Distinct patterns of olfactory impairment in Alzheimer's disease, semantic dementia, frontotemporal dementia, and corticobasal degeneration. Neuropsychologia 45(8):1823–1831

Martyr A, Clare L (2018) Awareness of functional ability in people with early-stage dementia. Int J Geriatr Psychiatry 33(1):31–38

Matthews S (2015) Dementia and the power of music therapy. Bioethics 29(8):573–579

McDermott O, Ridder HM, Baker FA, Wosch T, Ray K, Stige B (2018) Indirect music therapy practice and skill-sharing in dementia care. J Music Ther 55(3):255–279

McKeith I (2007) Dementia with Lewy bodies. In: Handbook of clinical neurology, vol 84. Elsevier, Amsterdam, pp 531–548

McKeith IG, Boeve BF, DIckson DW, Halliday G, Taylor JP, Weintraub D, Aarsland D, Galvin J, Attems J, Ballard CG, Bayston A (2017) Diagnosis and management of dementia with Lewy bodies. Neurology 89(1):88–100

Ménard C, Pfau ML, Hodes GE, Russo SJ (2017 Jan) Immune and neuroendocrine mechanisms of stress vulnerability and resilience. Neuropsychopharmacology 42(1):62–80

Meng N, Dong Y, Huo T, Song M, Jiang X, Xiao Y, Lv P (2020) Past exposure to cigarette smoke aggravates cognitive impairment in a rat model of vascular dementia via neuroinflammation. Cell Mol Neurobiol 6:1–4

Morell L, Shoemark H (2018) An ecological perspective on music programmes for at-risk children and families in Australia. Int J Commun Music 11(3):337–352

Mueller C, Ballard C, Corbett A, Aarsland D (2017) The prognosis of dementia with Lewy bodies. Lancet Neurol 16(5):390–398

Muresanu DF (2007) Neuroprotection and neuroplasticity–a holistic approach and future perspectives. J Neurol Sci 257(1–2):38–43

Nowrangi MA, Lyketsos CG, Rosenberg PB (2015) Principles and management of neuropsychiatric symptoms in Alzheimer's dementia. Alzheimers Res Ther 7(1):12

O'Connor K (2016) It's a long (and rocky) way to Tipperary: singing the political and one group's navigation of diverse views in song. J Irish Assoc Creat Art Ther 4(1):10–18
O'Connor R, Gray D (2014) Identifying the value of music therapy within interdisciplinary assessment: a research project. In: Facher J, Kern P, Tucek G (eds) Proceedings of the 14. World Congress of Music Therapy, vol 10, no 1. Special Issue of Music Therapy Today, pp 170–171
Olney NT, Spina S, Miller BL (2017) Frontotemporal dementia. Neurol Clin 35(2):339–374
Osler M, Christensen GT, Mortensen EL, Christensen K, Garde E, Rozing MP (2019) Hearing loss, cognitive ability, and dementia in men age 19–78 years. Eur J Epidemiol 34(2):125–130
Otera M, Saito S, Kano H, Ichie M (2020) Clinical characteristics of home-based music therapy in supporting personhood in people with dementia. Arts Psychother 7:101682
Parmar P, Puwar N (2019) Striking a chord: dementia and song. Perform Res 24(1):25–34
Payrau B, Quere N, Breton E, Payrau C (2017 Sep) Fasciatherapy and reflexology compared to hypnosis and music therapy in daily stress management. Int Therap Massge Bodywork 10(3):4
Popescu M, Otsuka A, Ioannides AA (2004) Dynamics of brain activity in motor and frontal cortical areas during music listening: a magnetoencephalographic study. NeuroImage 21(4):1622–1638
Premi E, Calhoun VD, Diano M, Gazzina S, Cosseddu M, Alberici A, Archetti S, Paternicò D, Gasparotti R, van Swieten J, Galimberti D (2019) The inner fluctuations of the brain in presymptomatic frontotemporal dementia: the chronnectome fingerprint. NeuroImage 1(189):645–654
Prince M, Bryce R, Albanese E, Wimo A, Ribeiro W, Ferri CP (2013) The global prevalence of dementia: a systematic review and metaanalysis. Alzheimers Dement 9(1):63–75
Pongan E, Tillmann B, Leveque Y, Trombert B, Getenet JC, Auguste N, Dauphinot V, El Haouari H, Navez M, Dorey JM, Krolak-Salmon P (2017) Can musical or painting interventions improve chronic pain, mood, quality of life, and cognition in patients with mild Alzheimer's disease? Evidence from a randomized controlled trial. J Alzheimers Dis 60(2):663–677
Raglio A, Bellelli G, Traficante D, Gianotti M, Ubezio MC, Villani D, Trabucchi M (2008) Efficacy of music therapy in the treatment of behavioral and psychiatric symptoms of dementia. Alzheimer Dis Assoc Disord 22(2):158–162
Ratovohery S, Baudouin A, Gachet A, Palisson J, Narme P (2018) Is music a memory booster in normal aging? The influence of emotion. Memory 26(10):1344–1354
Reybrouck M, Eerola T (2017) Music and its inductive power: a psychobiological and evolutionary approach to musical emotions. Front Psychol 4(8):494
Ridder HM, Stige B, Qvale LG, Gold C (2013) Individual music therapy for agitation in dementia: an exploratory randomized controlled trial. Aging Ment Health 17(6):667–678
Riemenschneider M, Wagenpfeil S, Vanderstichele H, Otto M, Wiltfang J, Kretzschmar H, Vanmechelen E, Förstl H, Kurz A (2003) Phospho-tau/total tau ratio in cerebrospinal fluid discriminates Creutzfeldt–Jakob disease from other dementias. Mol Psychiatry 8(3):343–347
Rocca WA, Petersen RC, Knopman DS, Hebert LE, Evans DA, Hall KS, Gao S, Unverzagt FW, Langa KM, Larson EB, White LR (2011) Trends in the incidence and prevalence of Alzheimer's disease, dementia, and cognitive impairment in the United States. Alzheimers Dement 7(1):80–93
Roy S, Chakrabarty S, De D (2017) Time-based raga recommendation and information retrieval of musical patterns in Indian classical music using neural networks. IAES Int J Artif Intell 6(1):33
Sacktor N, Robertson K (2014) Evolving clinical phenotypes in HIV-associated neurocognitive disorders. Curr Opin HIV AIDS 9(6):517
Särkämö T, Tervaniemi M, Laitinen S, Forsblom A, Soinila S, Mikkonen M, Autti T, Silvennoinen HM, Erkkilä J, Laine M, Peretz I (2008) Music listening enhances cognitive recovery and mood after middle cerebral artery stroke. Brain 131(3):866–876
Satoh M, Takeda K, Nagata K, Hatazawa J, Kuzuhara S (2003) The anterior portion of the bilateral temporal lobes participates in music perception: a positron emission tomography study. Am J Neuroradiol 24(9):1843–1848

Satoh M, Yuba T, Tabei KI, Okubo Y, Kida H, Sakuma H, Tomimoto H (2015) Music therapy using singing training improves psychomotor speed in patients with Alzheimer's disease: a neuropsychological and fMRI study. Dementia Geriatr Cognit Disord Extra 5(3):296–308

Savva GM, Wharton SB, Ince PG, Forster G, Matthews FE, Brayne C (2009) Age, neuropathology, and dementia. N Engl J Med 360(22):2302–2309

Savva GM, Stephan BC, Alzheimer's Society Vascular Dementia Systematic Review Group (2010) Epidemiological studies of the effect of stroke on incident dementia: A systematic review. Stroke 41(1):e41–e46

Schneider LS, Tariot PN, Dagerman KS, Davis SM, Hsiao JK, Ismail MS, Lebowitz BD, Lyketsos CG, Ryan JM, Stroup TS, Sultzer DL (2006) Effectiveness of atypical antipsychotic drugs in patients with Alzheimer's disease. N Engl J Med 355(15):1525–1538

Shimada H, Makizako H, Lee S, Doi T, Lee S (2018) Lifestyle activities and the risk of dementia in older Japanese adults. Geriatr Gerontol Int 18(10):1491–1496

Smith EE (2017) Clinical presentations and epidemiology of vascular dementia. Clin Sci 131(11):1059–1068

Steeman E, De Casterlé BD, Godderis J, Grypdonck M (2006) Living with early-stage dementia: a review of qualitative studies. J Adv Nurs 54(6):722–738

Stige B (2018) Partnerships for health musicking: a case for connecting music therapy and public health practices. In: Bonde LO, Theorell T (eds) Music and public health. A Nordic perspective. Springer, Berlin, pp 115–128

Stopa EG, Tanis KQ, Miller MC, Nikonova EV, Podtelezhnikov AA, Finney EM, Stone DJ, Camargo LM, Parker L, Verma A, Baird A (2018) Comparative transcriptomics of choroid plexus in Alzheimer's disease, frontotemporal dementia and Huntington's disease: implications for CSF homeostasis. Fluids and Barriers of the CNS 15(1):1

Sung WK, Zheng H, Li S, Chen R, Liu X, Li Y, Lee NP, Lee WH, Ariyaratne PN, Tennakoon C, Mulawadi FH (2012) Genome-wide survey of recurrent HBV integration in hepatocellular carcinoma. Nat Genet 44(7):765–769

Tan ZS, Spartano NL, Beiser AS, DeCarli C, Auerbach SH, Vasan RS, Seshadri S (2017) Physical activity, brain volume, and dementia risk: the Framingham study. J Gerontol Series A Biomed Sci Med Sci 72(6):789–795

Takeda S, Rakugi H, Morishita R (2020) Roles of vascular risk factors in the pathogenesis of dementia. Hypertens Res 43(3):162–167

Tang BL (2019) Amyloid precursor protein (APP) and GABAergic neurotransmission. Cell 8(6):550

Tekok-Kilic A, Benedict RH, Weinstock-Guttman B, Dwyer MG, Carone D, Srinivasaraghavan B, Yella V, Abdelrahman N, Munschauer F, Bakshi R, Zivadinov R (2007) Independent contributions of cortical gray matter atrophy and ventricle enlargement for predicting neuropsychological impairment in multiple sclerosis. NeuroImage 36(4):1294–1300

Trimble MR (2007) The soul in the brain: the cerebral basis of language, art, and belief. JHU Press, Baltimore

Tschampa HJ, Neumann M, Zerr I, Henkel K, Schröter A, Schulz-Schaeffer WJ, Steinhoff BJ, Kretzschmar HA, Poser S (2001) Patients with Alzheimer's disease and dementia with Lewy bodies mistaken for Creutzfeldt-Jakob disease. J Neurol Neurosurg Psychiatry 71(1):33–39

Tsunoda N, Hashimoto M, Ishikawa T, Fukuhara R, Yuki S, Tanaka H, Hatada Y, Miyagawa Y, Ikeda M (2018) Clinical features of auditory hallucinations in patients with dementia with Lewy bodies: a soundtrack of visual hallucinations. J Clin Psychiatr 79(3):17m11623

Umbrello M, Sorrenti T, Mistraletti G, Formenti P, Chiumello D, Terzoni S (2019) Music therapy reduces stress and anxiety in critically ill patients: a systematic review of randomized clinical trials. Minerva Anestesiol 85(8):886–898

van den Dungen P, van Marwijk HW, van der Horst HE, Moll van Charante EP, MacNeil Vroomen J, van de Ven PM, van Hout HP (2012 Apr) The accuracy of family physicians' dementia diagnoses at different stages of dementia: a systematic review. Int J Geriatr Psychiatry 27(4):342–354

Verghese PB, Castellano JM, Holtzman DM (2011) Apolipoprotein E in Alzheimer's disease and other neurological disorders. Lancet Neurol 10(3):241–252

Viswanathan A, Rocca WA, Tzourio C (2009) Vascular risk factors and dementia: how to move forward? Neurology 72(4):368–374

Vik-Mo AO (2019) Neuropsychiatric symptoms in dementia: long-term course and neuropathology. The University of Bergen, Bergen

Wall M, Duffy A (2010) The effects of music therapy for older people with dementia. Br J Nurs 19(2):108–113

Wang Z, Li Z, Xie J, Wang T, Yu C, An N (2018) Music therapy improves cognitive function and behavior in patients with moderate Alzheimer's disease. Int J Clin Exp Med 11(5):4808–4814

Warren JD, Rohrer JD, Rossor MN (2013 Aug) Frontotemporal dementia Bmj 6(347):f4827

Wood S, Ansdell G (2018) Community music and music therapy. Oxford Handbook Commun Music 1:453

Woodward MC (2005) Pharmacological treatment of challenging neuropsychiatric symptoms of dementia. J Pharm Pract Res 35(3):228–234

Yang X, Yao C, Tian T, Li X, Yan H, Wu J, Li H, Pei L, Liu D, Tian Q, Zhu LQ (2018 Feb) A novel mechanism of memory loss in Alzheimer's disease mice via the degeneration of entorhinal–CA1 synapses. Mol Psychiatry 23(2):199–210

Zarowitz BJ, O'Shea T, Nance M (2014) Clinical, demographic, and pharmacologic features of nursing home residents with Huntington's disease. J Am Med Dir Assoc 15(6):423–428

Zerr I, Parchi P (2018) Sporadic Creutzfeldt–Jakob disease. In: Handbook of clinical neurology, vol 153. Elsevier, Amsterdam, pp 155–174

Zhang W, Tarutani A, Newell KL, Murzin AG, Matsubara T, Falcon B, Vidal R, Garringer HJ, Shi Y, Ikeuchi T, Murayama S (2020) Novel tau filament fold in corticobasal degeneration. Nature 580(7802):283–287

Zhang Y, Cai J, An L, Hui F, Ren T, Ma H, Zhao Q (2017) Does music therapy enhance behavioral and cognitive function in elderly dementia patients? A systematic review and meta-analysis. Ageing Res Rev 1(35):1

Zucchella C, Sinforiani E, Tamburin S, Federico A, Mantovani E, Bernini S, Casale R, Bartolo M (2018) The multidisciplinary approach to Alzheimer's disease and dementia. A narrative review of non-pharmacological treatment. Front Neurol 13(9):1058

Beneficial Effects of Citrus Flavonoids Against Aβ Pathology in Alzheimer's Disease

19

Hayate Javed, Shreesh Kumar Ojha, and Mohammad Chand Jamali

Abstract

Amyloid β (Aβ) protein is the major component of extracellular amyloid plaques which are the main pathological feature of Alzheimer's disease (AD). Disruption in the clearance of Aβ is associated with its accumulation and aggregation that eventually leads to amyloid plaques formation. Numerous drugs have shown to possess therapeutic potential in the treatment of AD, but indeed these drugs only delay the progression of the disease and display numerous side effects. Recently, phytochemicals have drawn much attention in the treatment of neurodegenerative disorders. Among the various phytochemicals, dietary citrus flavonoids have shown potential useful effects against neurological disorders including AD. Dietary citrus flavonoids exhibit a high amount of minerals, vitamins, and antioxidant contents. The present chapter aims to explore the mechanism of Aβ peptide generation, aggregation, and accumulation as well as their prevention through dietary citrus flavonoids.

H. Javed (✉)
Department of Anatomy, College of Medicine and Health Sciences, United Arab Emirates University, Al Ain, United Arab Emirates
e-mail: h.javed@uaeu.ac.ae

S. K. Ojha
Department of Pharmacology and Therapeutics, College of Medicine and Health Sciences, United Arab Emirates University, Al Ain, United Arab Emirates

M. C. Jamali
Department of Health and Medical Sciences, Khawarizmi International College, Abu Dhabi, United Arab Emirates

G. M. Ashraf, M. S. Uddin (eds.), *Current Thoughts on Dementia*,
https://doi.org/10.1007/978-981-16-7606-2_19

Keywords

Alzheimer's disease · Dementia · Amyloid β · APP · Phytochemicals · Flavonoids · Flavanone · Flavonols

19.1 Introduction

The most common form of dementia is Alzheimer's disease (AD) and symptomatically distinguished by the loss of memory and cognitive ability along with the death of certain types of neurons in particular areas of the brain. The classical pathological feature of AD is amyloid-β (Aβ) deposits in senile plaques. Dysregulated processing of amyloid precursor protein (APP) causes abnormal production of Aβ in the form of monomers. These monomeric forms of Aβ promote the aggregation of Aβ into oligomeric forms and eventually lead to the development of Aβ plaques (Selkoe 1994). The exact function of APP remains to be elucidated but is largely known to play a key role in cell growth and maintenance (O'Brien and Wong 2011).

The basic understanding of the Aβ pathology is based on knowing the mechanisms of the Aβ monomers generation, its clearance, and further aggregation of monomeric Aβ into their oligomeric form. The nonamyloidogenic proteolysis (normal processing) of APP via α-secretase and γ-secretase generates soluble fragments (Anderson et al. 1992). However, when APP is proteolyzed by γ-secretase and erroneous β-secretase (abnormal processing), it generates insoluble fragments, i.e., Aβ peptides that further aggregate in the form of Aβ plaques in the brain (Blasko et al. 2000; Selkoe 1994). The exact role of Aβ pathology in AD remains an enigma because the accumulation of Aβ plaques takes several years before any distinctive AD diagnosis or symptoms seen in the patients. In spite of widespread and continuous research, exploring the mechanisms accountable for Aβ deposition and methods intended for the prevention of β-amyloid plaques remain unelucidated and still no effective treatment that can modify Aβ pathology in clinical aspects (Cummings et al. 2016).

There is increasing evidence showing that intake of flavonoid-rich diets could be beneficial in boosting the cognitive abilities in humans (Bakoyiannis et al. 2019; Macready et al. 2009; Socci et al. 2017). Flavonoids have been found to ameliorate the onset of AD pathology and delay the initiation and progression of the disease (Williams and Spencer 2012). Moreover, various flavonoids showed to inhibit and delay the progression of AD pathologies and these pharmacological properties of flavonoids can suppress the cognitive deficits in multiple preclinical AD models (Bakoyiannis et al. 2019; Macready et al. 2009; Spencer 2010). In the light of the present literature, this chapter focuses specifically on the mechanism of Aβ generation and its role in the development of AD and the importance of citrus flavonoids as potential therapeutics targeting Aβ for the treatment and cure of AD.

19.2 Mechanism of Aβ Production and Deposition in Alzheimer's Disease

In the past decade, the pathogenesis of AD and underlying molecular mechanisms are well explained. However, the crucial cause that directs the pathology is still argumentative. The ultimately established amyloid hypothesis indicates that neurons and parenchyma acquired the maximum amount of Aβ deposits which are the central cause for neural, synaptic, and axonal dysfunctions and consequent cognitive impairment (Calkins and Reddy 2011; Iadecola 2003; Thal et al. 2008). The formation of Aβ is mostly governed by the equilibrium between amyloidogenic and nonamyloidogenic pathways of APP processing (Canobbio et al. 2011) as shown in Fig. 19.1.

The processing of APP relies upon enzymes such as α-, β-, and γ-secretases, which produce the peptide fragments and are released extracellularly or intracellularly. Three enzymes, viz. ADAM17, ADAM10, and ADAM9, play a key role in facilitating the function of α-secretase (Manzine et al. 2019). These enzymes belong to a disintegrin and metalloprotease family (ADAM) and are membrane-bound glycoprotein (Manzine et al. 2019). Moreover, they demonstrate a significant role in matrix protein degradation, cell fusion, cell adhesion, and shedding of ectodomain (Asai et al. 2003; Black et al. 1997; Qi et al. 1999). β-Secretase, also known as beta-

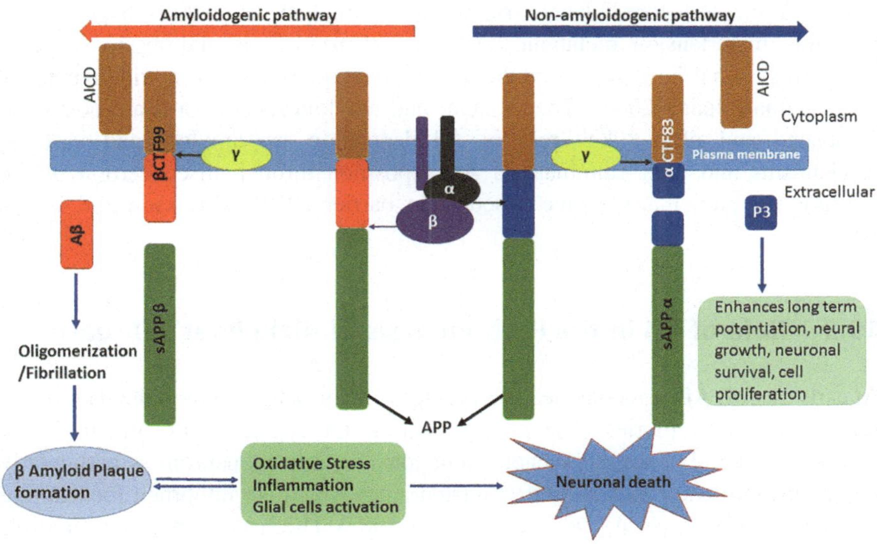

Fig. 19.1 The amyloidogenic and nonamyloidogenic pathways of amyloid precursor protein (APP) processing. The nonamyloidogenic pathway (blue arrow) causes the cleavages by α- and γ-secretases which produces a secretory form of APP (sAPPα) and C-terminal fragments of 83 amino acid residue, p3, and *APP* intracellular fragment (AICD) whereas amyloidogenic pathway (red arrow) causes the cleavages by β- and γ-secretases leading to the production of a secretory form of APP (sAPPβ), C-terminal fragments of 99 residues and Aβs and AICD. This fragment of Aβ oligomerizes and later forms the protofibrils/mature fibrils leading to Alzheimer's disease

site amyloid precursor protein cleaving enzyme 1 (BACE-1), is a rate-limiting enzyme in APP proteolytic processing. β-Secretase belongs to the pepsin family of aspartyl proteases, and moreover it is a type I transmembrane protein (Hunt and Turner 2009; Sinha et al. 1999; Yan et al. 1999). β-Secretase consists of two aspartic residues of catalytic activity, an N-terminal catalytic domain, a small cytoplamic C-terminal and harboring a transmembrane domain. γ-Secretase consists of four transmembrane proteins and a complex of heterogeneous protein. The four transmembrane proteins of γ-secretase consist of anterior pharynx-defective 1, presenilins (PS1 and PS2), nicastrin, and presenilin enhancer 2 (Bertram and Tanzi 2008; Edbauer et al. 2003; Sisodia and St George-Hyslop 2002). The APP processing by nonamyloidogenic pathway starts through α-secretase-driven cleavage of APP at residue 687 of 770 isoform of APP which releases sAPPα (soluble APPα) extracellularly. Therefore, the C terminal fragment (CTF) of APP which consists of 83 amino acid residues stays/embedded in the plasma membrane. Further, γ-secretase-mediated breakdown of CTF 83 produces p3 fragment and APP intracellular domain (AICD) which are released extracellularly and intracellularly (cytoplasm) respectively. In contrast, amyloidogenic processing of APP by β-secretase breaks APP, which produces CTF99. Further, γ-secretase cleaves the CTF 99 which causes the release of AICD in the cytoplasm and extracellular release of Aβ fragment.

Additionally, several risk factors are associated with AD that triggers the onset of disease. The risk factors linked to triggering AD are diabetes mellitus, hypercholesterolemia, hypertension, metabolic syndrome, atherosclerosis, and obesity (Orsucci et al. 2013). It has been suggested that ApoE genotype is associated with hypercholesterolemia (Huang 2006). The vascular and amyloid mechanisms are positively associated and play a key role in Aβ deposition via amyloid and vascular mechanisms and both culminate in Aβ deposition through disconcertion of Aβ transporters, particularly in the blood-brain barrier (BBB) (Erickson and Banks 2013).

19.3 Role of Aβ in the Pathogenesis of Alzheimer's Disease

At early stages, AD is recognized by cognitive impairment including the inability to recall episodic memories. The primary reason for cognitive impairment is the neuronal loss and abnormal synaptic function. The loss of neurons is particularly observed in hippocampus and other related areas which are important for learning and memory (Davies et al. 1987; Hyman et al. 1984). The synaptic plasticity controls the strength or weakness of synapses over time in reaction to their activity. Long-term depression (LTD) and long-term potentiation (LTP) are models of synaptic plasticity. The mechanism of learning and cognition is governed by synaptic plasticity. Earlier studies demonstrated that abnormal deposition of Aβ is the primary feature of AD pathology (Shankar et al. 2008; Walsh et al. 2005). The change of synaptic plasticity and abnormal proteolytic processing of APP attribute to the accumulation of Aβ simultaneously at the early onset of AD. It has also been shown that synaptic activity controls the levels of Aβ in AD, which is largely related to the regulation of APP cleavage and endocytosis.

The soluble form of Aβ is found to be necessary for memory retrieval and synaptic plasticity (Morley and Farr 2014; Puzzo and Arancio 2013). APP cleaves the soluble Aβ at synapses and releases it into the synaptic cleft along with other neurotransmitters at the time of neuronal activity (Cirrito et al. 2005; Kamenetz et al. 2003; Tampellini et al. 2009). Thereafter, Aβ acts on presynaptic neurons and enhances the possibility of additional neurotransmitter release (Fedele et al. 2015; Puzzo et al. 2015). LTP and short- and long-term memory were compromised when reduced levels of endogenous Aβ were found in rodents (Garcia-Osta and Alberini 2009). Aβ led an increased level of acetylcholine into the synaptic cleft, increases LTP, and promotes the chances of depolarization of the postsynaptic neurons.

On the other hand, soluble Aβ oligomers along with dimers which are toxic in nature are required to interrupt the proper learning and cognition (Cleary et al. 2005; Lesne et al. 2006; Shankar et al. 2008). Previous studies showed that synapse failure is initiated by dimers and trimers (Malenka 2003; Selkoe 2008; Shankar et al. 2007). Moreover, it has also been shown that LTP is extensively interrupted by trimers compared to dimers (Selkoe 2008; Townsend et al. 2006). There was the highest correlation of AD pathology associated with diffusible Aβ oligomeric assemblies which cause synaptic failure during the early onset of AD (Stephan and Phillips 2005). It is well evidenced that Aβ production is a normal physiological process in the brain; however, aggregation and senile plaque formation are caused by dysregulated production of Aβ. Prerequisite to Aβ aggregate formation, oligomerization, and fibrillation take place, and these intermediate species are more toxic to damage the neuronal cells (Sengupta et al. 2016). The soluble form of Aβ is able to interact with possible receptors and activate the downstream pathways which play key roles in a massive release of reactive oxygen species (ROS), mitochondrial impairments, and glial cells activation that further secrete various cytokines as well as inflammatory mediators and eventually leads to neurotoxicity (Fig. 19.2).

19.4 Major Risk Factors for Alzheimer's Disease

Substantial evidence showed that amyloidogenesis is caused by intrinsic properties of proteins and environmental factors. The predominant factors which are involved in AD are listed in Table 19.1. Various accessible protein conformations which are responsible for amyloidogenesis are thermodynamically affected by these factors. Though amyloid formation caused by Aβ aggregation is cytotoxic, some biological functions of amyloidogenic protein fibrils are reported in mammals, fungi, and bacteria (Kelly and Balch 2003). Multiple experimental studies have shown that oligomeric species of Aβ are more toxic instead of amyloidogenic aggregates which have been observed in AD (Lue et al. 1999; McLean et al. 1999). Thus, exploring mechanisms and factors which lead to the formation of oligomeric species and its stability as well as Aβ aggregation has drawn much attention.

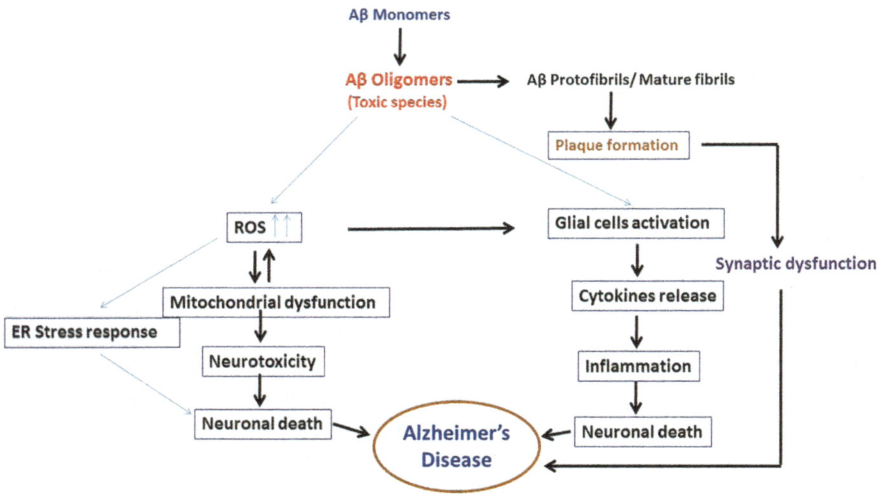

Fig. 19.2 Aβ monomers generated from the APP processing can build mid to high molecular weight oligomers/protofibrils and senile plaques whereas soluble Aβ can interact with probable receptors and activate downstream pathways to generate ROS, mitochondrial dysfunction, and glial cell activation and eventually leads to neurotoxicity and onset of AD

19.5 Flavonoids for the Treatment of Neurodegenerative Events and Alzheimer's Disease

Epidemiological-based research showed strong beneficial effects following dietary intake of phytochemicals in foods, fruits, vegetables, beverages or culinary preparations, and neurodegenerative diseases. Phytochemicals are those substances that are found in edible fruits and vegetables, which are consumed daily, show medicinal properties, and demonstrate the therapeutic potential for modifying metabolism in a way favorable to either suppress or halt the progression of human chronic and degenerative diseases. The pharmacological and physiological properties of thousands of phytochemicals have been described in a convincing number of studies. Among the phytochemicals, polyphenolic flavonoids are extensively found in citrus fruits. Flavonoids are a group of pigments found in plants and play an important role in flower and fruit coloration. The principal source of these important nutrients is citrus fruits. They contain dietary fibers, folate, vitamin C, and various plant active constituents, e.g., flavonoids and carotenoids are beneficial for the prevention or slow down the progression of neurodegenerative diseases including AD. Here, we present the antioxidant and other beneficial properties of citrus flavonoids for the prevention and treatment of AD specifically targeting Aβ aggregation, accumulation, and deposition in the in vitro and in vivo studies. Following the molecular structure, six types of flavonoids are described such as flavanones, flavones, isoflavones, flavonols, and anthocyanidins. Glycosides or aglycone forms

Table 19.1 Factors/causes for Aβ generation in Alzheimer's disease

Factors/ causes	Aβ formation	References
Protein nature	Hydrophobic residues of Aβ42 C terminals are implicated in Aβ aggregation	Kim and Hecht (2006)
Concentration	Aβ increased level leads to AD	Verdone et al. (2002)
Mutation in amino acid sequences	Mutation in APP sequences leads to AD	Buxbaum and Tagoe (2000)
Chemical modification	Oxidative modifications of Aβ	Palmblad et al. (2002)
Mutation in associated protein	Mutation in ApoE and α2-macroglobulin proteins leads to diminished clearance of Aβ	Jordan et al. (1998), Urmoneit et al. (1997)
Protein folding machinery	Impairments in molecular chaperone lead to misfolding of APP and cause excessive Aβ production	Yoo et al. (2001)
Altered proteolysis of APP	Mutation in APP and presenilins leads to Aβ generation	Cruts et al. (1998), Levy et al. (1990)
Reduced clearance	Reduced clearance of excessive generated Aβ leads to AD	Holtzman et al. (2000)
pH and ionic strength	Acidic pH and low ionic strength favor Aβ fibrils formation	Stine et al. (2003)
Other interacting factors	Metals (copper, zinc, aluminum, and iron) and proteins (lipoprotein ApoE, albumin) interact with Aβ and favor aggregation	Bush et al. (1993), Dolev and Michaelson (2004), Fasman et al. (1995), Ghribi et al. (2006)

contain citrus flavanones. The most important aglycone forms of flavanones are naringenin and hesperetin. Neohesperidosides and rutinosides are the two types of glycoside forms of flavanones. Neohesperidosides flavanones have a bitter taste such as neohesperidin, neoeriocitrin, and naringin while rutinosides flavanones with a disaccharide residue and without taste such as didymin, hesperidin, and narirutin, e.g., rutinose (rhamnosyl-a-1,6 glucose). The classic taste of citrus fruits is found in flavanones and is generally found in diglucoside form (Horowitz 1986).

Among neohesperidoside flavanones, neoeriocitrin, naringin, and neohesperidin are predominantly found in bitter orange juices, bergamot, and grapefruit whereas rutinoside flavanones such as didymin, narirutin, and hesperidin are present in mandarin, orange, lemon juices, and bergamot (Horowitz 1986). Glycosylated flavanones, naringin and neohesperidin, are found in bergamot seeds; lemon is the important source of hesperidin and eriocitrin. The composition of flavanone glycosyl of seeds and peels is relatively different than those of juices. The seed of lemon and mandarin possesses naringin but juices of lemon and mandarin are devoid of naringin (Ooghe and Detavernier 1997). In sweet oranges, usually glycosylated flavanones are not present, and its presence generally is suggestive of defilement

(Mouly et al. 1994). Citrus tissues possess a little amount of flavones and flavonols, and they have been studied to exert their antioxidant potency. Marini and Balestrieri (1995) isolated two C-glycosyl flavones from the peel of lemon fruit (Citrus Limon BURM. f.). They identified 6,8-di-C-b-glycosyldiosmin and 6-C-b-glycosyldiosmin by UV, IR, FABMS, ^{1}H NMR, and ^{13}C NMR analyses. The color of fruits and flowers is provided by anthocyanins but sometimes anthocyanin also provides color in buds, roots, and leaves (Ooghe and Detavernier 1997). Moreover, anthocyanin presents in epicarp; however it also provides a coloring compound to mesocarp of oranges. The level of fruit maturation usually determines the anthocyanin content and suggestive of the age of the fruit.

19.5.1 Inhibition of Aβ by Hesperidin, Neohesperidin, and Hesperitin

Numerous reports provided evidence that flavonoids supplementation can decrease the probability of onset of neurodegenerative diseases in humans and ameliorate toxicity caused by Aβ in experimental models of sporadic AD (SAD) (Bu and Lephart 2005; Yoon et al. 2004). There are numerous supportive data demonstrating that Aβ aggregation could be reduced by flavonoids which suggest a therapeutic basis for targeting AD. Flavanones are extensively reputed and generated interest in the past few years for their therapeutic properties to treat human pathology. Among many citrus polyphenols, hesperidin neohesperidin and hesperitin revealed to control cellular signaling pathways and regulate the cytoprotective effects. Previously, it has been found that abnormal accumulation of Aβ activates autophagy that promotes impaired metabolism of glucose/energy in the neurons. Huang et al. reported that inhibition of Aβ induced autophagy and normalization of glucose energy metabolism by hesperidin and hesperitin in neuro-2A cells (Huang et al. 2012). It is also known that mitochondrial dysfunction is caused by the cytotoxic effects of Aβ. Moreover, voltage-dependent anion channel 1 of mitochondria, which play a critical role in the release of apoptotic proteins, has possible relevance in AD neuropathology.

Hesperidin was shown to be neuroprotective in $A\beta_{25-35}$-induced neurotoxicity in PC12 cells. Hesperidin attenuated the apoptosis caused by $A\beta_{25-35}$ toxicity by ameliorating the Aβ-led mitochondrial impairments such as the opening of transition pore of mitochondria, increase in the levels of intracellular calcium and ROS release (Wang et al. 2013). Additionally, hesperidin treatment to the mouse model of AD (APP/PS1) showed significant attenuation of the Aβ load, APP-associated plaque, activation of microglia, and expression of *transforming growth factor-β* (Li et al. 2015). Reports from these studies indicate that hesperidin, neohesperidin, and hesperitin might be an important drug candidate for pharmaceutical development in the prevention and treatment of neurodegenerative diseases including AD.

19.5.2 Inhibition of Aβ by *Naringin* and *Naringenin*

Naringin and its aglycone naringenin are flavanone and considered to be biologically active and exert a positive effect on human health along with numerous therapeutic properties such as anticancer, antimutagenic, anti-inflammatory, antioxidant, antiproliferative, and antiatherogenic (Alam et al. 2014; Patel et al. 2018). Flavanone glycoside naringin is present in natural citrus fruits, particularly in grapefruit, wherein it is responsible for the bitter taste of the fruits. Naringenin is also bitter in taste, colorless flavanone, and predominant in grapefruit as well as in a variety of fruits and herbs. In experimental models of SAD, naringin and naringenin have shown beneficial effects. Naringenin was found protective against the Aβ-induced ROS using PC12 cells (Heo et al. 2004). It also protected the cell against $A\beta_{25-35}$-induced neuronal damage as a result of increasing the viability of the cell, promoting the activation of glycogen synthase kinase-3β (GSK-3β) and Akt, and preventing apoptosis as well as caspase-3 activity (Zhang et al. 2018).

Additionally, intracerebroventricular-streptozotocin (ICV-STZ) induced elevated levels of brain Aβ were found significantly lower in naringenin-treated rats by upregulation of insulin-degrading enzyme (Yang et al. 2014). ICV-STZ administration leads to severe abnormalities in brain glucose/energy metabolism and insulin signaling that attributes to the enhanced Aβ in SAD pathology. In a recent study, naringenin nanoemulsions reduced the Aβ-induced production of ROS, APP, β-secretase, total tau, and phosphorylated tau in SH-SY5Y neuroblastoma cells indicating diminished amyloidogenesis (Md et al. 2018). Microglia is the primary immune cells in the central nervous system which acts as a major inflammatory cell type in the brain. Two phenotypes of microglia are known, termed as M1 and M2. Deposition of Aβ enhances the activation of M1 phenotype activation which causes the inflammation and death of neurons. M2 phenotype of microglia is neuroprotective in nature and exhibits anti-inflammatory effects and helps in the clearance of Aβ and recovery in learning and memory in AD. Naringenin has shown to promote the activation and polarization of M2 microglia and inhibit the $A\beta_{1-42}$-induced activation of M1 type of microglia in primary culture of microglia (Yang et al. 2019).

In addition, naringin treatment to APPswe/PSΔE9 transgenic mice showed reduced senile plaques of Aβ and ameliorated impaired brain energy metabolism as well as increased GSK-3β phosphorylation (Wang et al. 2012). So, it indicates that naringin improves the learning and cognition ability in APPswe/PSΔE9 mice model of AD by attenuating the Aβ plaques and enhances the uptake of glucose via reduced activity of GSK-3β. Recently, it has been found that another derivative of naringin, known as naringin dihydrochalcone (NDC), a dietary sweetener, attenuated deposition of the Aβ in AD mouse brain (Yang et al. 2018). NDC was also found to reduce the peri-plaque-activated microglia and astrocytes that result in inhibition of neuroinflammation. The available findings demonstrate that naringenin, naringin and NDC could be a potential therapeutic agent for future drug discovery for the treatment of AD.

19.5.3 Inhibition of Aβ by *Quercetin, Myricetin,* and *Apigenin*

Flavonols such as myricetin (3,3′,4,5,5′,7-hexahydroxyflavone) and quercetin (3,3,4,5,7-penta-hydroxyflavone) are mainly present in several vegetables origin products, such as onions, red wine, tea, and cocoa as well as in plant extracts including *Ginkgo biloba* (termed EGb761 for the purified extract preparation in many studies). The anti-amyloidogenic activity of these flavonols was initially shown in vitro and thereafter evaluated for their potential therapeutic efficacy in vivo. Using a battery of biochemical and biophysical techniques including transmission electron microscopy analysis and thioflavin T (ThT) binding, these compounds were observed to inhibit the in vitro Aβ aggregates formation and to disorganize the preformed fibrils (Ono et al. 2004).

Thirty-nine different flavonoids were tested in a study, and it was found that quercetin was superior and potent than other polyphenolic compounds against Aβ fibrillation (IC_{50} 2.4 μg/mL). Further, quercetin also showed beneficial effects in opposition to cytotoxicity caused by Aβ (Kim et al. 2005). The findings of this study reveal that inhibition of Aβ fibril growth by flavonols is attributed by its 3-hydroxy, 4-keto groups in its chemical structure. Although, 3-hydroxyl group is not necessarily required for potential efficacy, however, 3′,4′-dihydroxyl group attached to the B ring is required for the inhibition of Aβ growth (Akaishi et al. 2008). The role of the chemical groups is not yet established in conferring the efficacy, but it is apparent that B ring containing more number of hydroxyl groups enhances the anti-aggregation property in case of myricetin, which possesses an extra hydroxyl group, which is considered therapeutically more potent in comparison to quercetin against the fibrillation of Aβ (Shimmyo et al. 2008). Similar to other flavonols, quercetin was found to strongly reduce the fluorescence emission of ThT (Hudson et al. 2009); therefore further in vitro experimental evidence of flavonols is needed for the anti-amyloidogenic activity.

The inhibition of aggregation of Aβ and destabilization of preformed fibrils through quercetin and myricetin were established with insulin and α-synuclein, respectively (Wang et al. 2011). Some other studies showed that quercetin prevents the Aβ fibrillization, although it failed to inhibit oligomerization (Jagota and Rajadas 2012). Thus, Aβ deposition in a model of *Caenorhabditis elegans* with increasing toxicity of Aβ; Noor et al. (2012) showed that myricetin and quercetin failed to inhibit the human islet amyloid polypeptide (hIAPP) fibrillation. However, the investigators did not show the probable effects of these two flavonols either on toxic oligomeric species or aggregate toxicity. Myricetin was further found to inhibit the hIAPP toxic aggregation dose-dependently (Zelus et al. 2012). It is still not clear how both flavonols inhibit amyloid aggregation, yet taking into consideration their activity on the identical amyloidogenic peptide.

Apart from a direct inhibitory effect on amyloid aggregation, myricetin and quercetin also showed protective effects against amyloid-induced cytotoxicity. They were observed to reduce the ROS levels (Ansari et al. 2009; Pocernich et al. 2011) and inhibit the activity of BACE-1 (Shimmyo et al. 2008). They also showed protective effects against the oxidative activity of Aβ-Cu^{2+} complex by playing their role as a competitive inhibitor. This effect of myricetin and quercetin was achieved by associating with the binding site of metal on Aβ through their α-keto enolate group (DeToma et al. 2011; Tay et al. 2013).

The flavonoid apigenin (4′,5,7-trihydroxyflavone) is abundantly found in chamomile plants, mainly in the flowers of ligulate (McKay and Blumberg 2006). The apigenin is found in lesser amounts in other plants such as grapefruit, parsley, and celery (Shukla and Gupta 2010). Apigenin is considered relatively safe due to time-tested dietary use in spices and culinary preparation and experimental studies even at high doses found to possess negligible toxicity whereas sedation and muscle relaxation have been reported with high doses of apigenin (Ross and Kasum 2002). It has been shown that toxicity caused by $A\beta_{25-35}$ in rat cerebral microvascular endothelial cells was attenuated by apigenin (Zhao et al. 2011).

Apigenin also showed neuroprotective effects in a double transgenic mouse model of AD (Zhao et al. 2013). The oral treatment with apigenin (40 mg/kg) to 4-month-old mice for 3 months resulted in the attenuation of impaired learning and cognition and a decrease in Aβ aggregation along with the reduced concentration of insoluble Aβ. In another study, apigenin (20 mg/kg) treatment resulted in improvement of spatial learning and memory in $A\beta_{25-35}$-induced amnesia in mice (Liu et al. 2011). Additionally, apigenin also showed protective effects against restricted blood supply in the brain (Liu et al. 2011). In addition to the above-mentioned flavonoids, we also explored other potential flavonoids, which have been evaluated against Aβ deposition and accumulation in AD. We have summarized the effects of additional flavonoids in Table 19.2 and their significant outcomes against the Aβ pathology.

19.6 Recent Developments and Future Perspectives

The mechanism underlying AD pathogenesis is complex and remains to be fully elucidated (Jeong 2017). The medications which are currently used for AD treatment are mostly symptomatic drugs. However, they work well to manage the symptoms of cognitive impairments and play a significant role in the cure of AD but failed in reversing the AD progress (Yiannopoulou and Papageorgiou 2013). The current research focuses on disease-modifying drugs that target the primary cause of AD and correspond to the future route of novel drug development (Yiannopoulou and Papageorgiou 2013).

In light of these views, the finding of multiple-target drugs that could have neurotrophic and neuroprotective activities is a balanced strategy in the treatment of AD. Flavonoids discussed in this chapter display several biological efficacies against AD including inhibition of Aβ fibrillation and aggregation, improved cognitive functions, decreased oxidative stress and inflammation, etc. Thus, pharmacological properties of flavonoids indicate their translational efficiency and may play a significant part in novel drug discovery and development for the treatment of AD.

Table 19.2 Flavonoids evaluated in inhibiting the aggregation of Aβ peptide in Alzheimer's disease

Flavonoids	Effects in Alzheimer's disease	Study models	References
Diosmin	• Reduced soluble and insoluble $A\beta_{1-40}$, $A\beta_{1-42}$ levels • Mitigated Aβ pathology	In vivo (Tg2576, 3 × Tg-AD mice)	Rezai-Zadeh et al. (2009), Sawmiller et al. (2016)
Luteolin	• Decreased soluble Aβ levels • Reduced GSK-3 activity • Disrupted PS1-APP association	In vivo (Tg2576 mice)	Rezai-Zadeh et al. (2009)
Didymin, poncirin, prunin	• Reduced $A\beta_{25-35}$-induced cytotoxicity • Decreased expression of BACE1, sAPPβ, and C99 • Reduced Aβ aggregation	In vitro (PC12 cells)	Ali et al. (2019)
Narirutin	• Possessed antioxidant activity • Inhibited BACE1 • Strongly inhibited Aβ aggregation	In vitro	Chakraborty and Basu (2017)
Kaempferol	• Attenuated Aβ-induced cytotoxicity • Protected oxidative stress inhibited Aβ fibrillogenesis	In vitro (PC12, SH-SY5Y)	Bhat et al. (2015), Kim et al. (2010), Sharoar et al. (2012)
Diosmetin	• Inhibited GSK-3β phosphorylation • Reduced γ-secretase activity • Reduced Aβ generation, tau hyperphosphorylation	In vitro (CHO/APP695 cells, primary neuronal cells cultured embryonic Tg2576 mice)	Sawmiller et al. (2016)

19.7 Conclusion

The studies enumerated in this chapter revealed that flavonoids mainly target AD by reducing Aβ oligomerization, aggregation, and accumulation in the brain. Besides, many flavonoids have been shown to attain the desired therapeutic concentration in the brain due to high lipophilicity and exert its therapeutic effects at the molecular level in different cellular pathways in various in vitro and in vivo models. Based on the present literature review, it can be reasonably speculated that regular dietary consumption or supplementation of flavonoids-rich foods holds promise in the prevention of neurodegenerative diseases, especially AD, and simultaneously improves cognitive and other important brain functions due to potential antioxidant and anti-inflammatory mediated pathways.

References

Akaishi T, Morimoto T, Shibao M, Watanabe S, Sakai-Kato K, Utsunomiya-Tate N, Abe K (2008) Structural requirements for the flavonoid fisetin in inhibiting fibril formation of amyloid beta protein. Neurosci Lett 444:280–285. https://doi.org/10.1016/j.neulet.2008.08.052

Alam MA, Subhan N, Rahman MM, Uddin SJ, Reza HM, Sarker SD (2014) Effect of citrus flavonoids, naringin and naringenin, on metabolic syndrome and their mechanisms of action. Adv Nutr 5:404–417. https://doi.org/10.3945/an.113.005603

Ali MY et al (2019) Flavanone glycosides inhibit beta-site amyloid precursor protein cleaving enzyme 1 and cholinesterase and reduce Abeta aggregation in the amyloidogenic pathway. Chem Biol Interact 309:108707. https://doi.org/10.1016/j.cbi.2019.06.020

Anderson JP, Chen Y, Kim KS, Robakis NK (1992) An alternative secretase cleavage produces soluble Alzheimer amyloid precursor protein containing a potentially amyloidogenic sequence. J Neurochem 59:2328–2331. https://doi.org/10.1111/j.1471-4159.1992.tb10128.x

Ansari MA, Abdul HM, Joshi G, Opii WO, Butterfield DA (2009) Protective effect of quercetin in primary neurons against Abeta(1-42): relevance to Alzheimer's disease. J Nutr Biochem 20: 269–275. https://doi.org/10.1016/j.jnutbio.2008.03.002

Asai M, Hattori C, Szabo B, Sasagawa N, Maruyama K, Tanuma S, Ishiura S (2003) Putative function of ADAM9, ADAM10, and ADAM17 as APP alpha-secretase. Biochem Biophys Res Commun 301:231–235. https://doi.org/10.1016/s0006-291x(02)02999-6

Bakoyiannis I, Daskalopoulou A, Pergialiotis V, Perrea D (2019) Phytochemicals and cognitive health: are flavonoids doing the trick? Biomed Pharmacother 109:1488–1497. https://doi.org/10.1016/j.biopha.2018.10.086

Bertram L, Tanzi RE (2008) Thirty years of Alzheimer's disease genetics: the implications of systematic meta-analyses. Nat Rev Neurosci 9:768–778. https://doi.org/10.1038/nrn2494

Bhat WF, Bhat SA, Bano B (2015) Evaluation of polyphenols as possible therapeutics for amyloidoses: comparative analysis of Kaempferol and Catechin. Int J Biol Macromol 81:60–68. https://doi.org/10.1016/j.ijbiomac.2015.07.052

Black RA et al (1997) A metalloproteinase disintegrin that releases tumour-necrosis factor-alpha from cells. Nature 385:729–733. https://doi.org/10.1038/385729a0

Blasko I, Veerhuis R, Stampfer-Kountchev M, Saurwein-Teissl M, Eikelenboom P, Grubeck-Loebenstein B (2000) Costimulatory effects of interferon-gamma and interleukin-1beta or tumor necrosis factor alpha on the synthesis of Abeta1-40 and Abeta1-42 by human astrocytes. Neurobiol Dis 7:682–689. https://doi.org/10.1006/nbdi.2000.0321

Bu L, Lephart ED (2005) Soy isoflavones modulate the expression of BAD and neuron-specific beta III tubulin in male rat brain. Neurosci Lett 385:153–157. https://doi.org/10.1016/j.neulet.2005.05.040

Bush AI et al (1993) A novel zinc(II) binding site modulates the function of the beta A4 amyloid protein precursor of Alzheimer's disease. J Biol Chem 268:16109–16112

Buxbaum JN, Tagoe CE (2000) The genetics of the amyloidoses. Annu Rev Med 51:543–569. https://doi.org/10.1146/annurev.med.51.1.543

Calkins MJ, Reddy PH (2011) Amyloid beta impairs mitochondrial anterograde transport and degenerates synapses in Alzheimer's disease neurons. Biochim Biophys Acta 1812:507–513. https://doi.org/10.1016/j.bbadis.2011.01.007

Canobbio I, Catricala S, Balduini C, Torti M (2011) Calmodulin regulates the non-amyloidogenic metabolism of amyloid precursor protein in platelets. Biochim Biophys Acta 1813(3):500–506
Chakraborty S, Basu S (2017) Multi-functional activities of citrus flavonoid narirutin in Alzheimer's disease therapeutics: an integrated screening approach and in vitro validation. Int J Biol Macromol 103:733–743. https://doi.org/10.1016/j.ijbiomac.2017.05.110
Cirrito JR et al (2005) Synaptic activity regulates interstitial fluid amyloid-beta levels in vivo. Neuron 48:913–922. https://doi.org/10.1016/j.neuron.2005.10.028
Cleary JP, Walsh DM, Hofmeister JJ, Shankar GM, Kuskowski MA, Selkoe DJ, Ashe KH (2005) Natural oligomers of the amyloid-beta protein specifically disrupt cognitive function. Nat Neurosci 8:79–84. https://doi.org/10.1038/nn1372
Cruts M et al (1998) Estimation of the genetic contribution of presenilin-1 and -2 mutations in a population-based study of presenile Alzheimer disease. Hum Mol Genet 7:43–51. https://doi.org/10.1093/hmg/7.1.43
Cummings J et al (2016) Drug development in Alzheimer's disease: the path to 2025. Alzheimers Res Ther 8:39. https://doi.org/10.1186/s13195-016-0207-9
Davies CA, Mann DM, Sumpter PQ, Yates PO (1987) A quantitative morphometric analysis of the neuronal and synaptic content of the frontal and temporal cortex in patients with Alzheimer's disease. J Neurol Sci 78:151–164. https://doi.org/10.1016/0022-510x(87)90057-8
DeToma AS, Choi JS, Braymer JJ, Lim MH (2011) Myricetin: a naturally occurring regulator of metal-induced amyloid-beta aggregation and neurotoxicity. ChemBioChem 12:1198–1201. https://doi.org/10.1002/cbic.201000790
Dolev I, Michaelson DM (2004) A nontransgenic mouse model shows inducible amyloid-beta (Abeta) peptide deposition and elucidates the role of apolipoprotein E in the amyloid cascade. Proc Natl Acad Sci U S A 101:13909–13914. https://doi.org/10.1073/pnas.0404458101
Edbauer D, Winkler E, Regula JT, Pesold B, Steiner H, Haass C (2003) Reconstitution of gamma-secretase activity. Nat Cell Biol 5:486–488. https://doi.org/10.1038/ncb960
Erickson MA, Banks WA (2013) Blood-brain barrier dysfunction as a cause and consequence of Alzheimer's disease. J Cereb Blood Flow Metab 33:1500–1513. https://doi.org/10.1038/jcbfm.2013.135
Fasman GD, Perczel A, Moore CD (1995) Solubilization of beta-amyloid-(1-42)-peptide: reversing the beta-sheet conformation induced by aluminum with silicates. Proc Natl Acad Sci U S A 92: 369–371. https://doi.org/10.1073/pnas.92.2.369
Fedele E, Rivera D, Marengo B, Pronzato MA, Ricciarelli R (2015) Amyloid beta: walking on the dark side of the moon. Mech Ageing Dev 152:1–4. https://doi.org/10.1016/j.mad.2015.09.001
Garcia-Osta A, Alberini CM (2009) Amyloid beta mediates memory formation. Learn Mem 16: 267–272. https://doi.org/10.1101/lm.1310209
Ghribi O, Golovko MY, Larsen B, Schrag M, Murphy EJ (2006) Deposition of iron and beta-amyloid plaques is associated with cortical cellular damage in rabbits fed with long-term cholesterol-enriched diets. J Neurochem 99:438–449. https://doi.org/10.1111/j.1471-4159.2006.04079.x
Heo HJ, Kim DO, Shin SC, Kim MJ, Kim BG, Shin DH (2004) Effect of antioxidant flavanone, naringenin, from Citrus junos on neuroprotection. J Agric Food Chem 52:1520–1525. https://doi.org/10.1021/jf035079g
Holtzman DM et al (2000) Apolipoprotein E isoform-dependent amyloid deposition and neuritic degeneration in a mouse model of Alzheimer's disease. Proc Natl Acad Sci U S A 97:2892–2897. https://doi.org/10.1073/pnas.050004797
Horowitz RM (1986) Taste effects of flavonoids. Prog Clin Biol Res 213:163–175
Huang Y (2006) Apolipoprotein E and Alzheimer disease. Neurology 66:S79–S85. https://doi.org/10.1212/01.wnl.0000192102.41141.9e
Huang SM, Tsai SY, Lin JA, Wu CH, Yen GC (2012) Cytoprotective effects of hesperetin and hesperidin against amyloid beta-induced impairment of glucose transport through downregulation of neuronal autophagy. Mol Nutr Food Res 56:601–609. https://doi.org/10.1002/mnfr.201100682

Hudson SA, Ecroyd H, Kee TW, Carver JA (2009) The thioflavin T fluorescence assay for amyloid fibril detection can be biased by the presence of exogenous compounds. FEBS J 276:5960–5972. https://doi.org/10.1111/j.1742-4658.2009.07307.x

Hunt CE, Turner AJ (2009) Cell biology, regulation and inhibition of beta-secretase (BACE-1). FEBS J 276:1845–1859. https://doi.org/10.1111/j.1742-4658.2009.06929.x

Hyman BT, Van Hoesen GW, Damasio AR, Barnes CL (1984) Alzheimer's disease: cell-specific pathology isolates the hippocampal formation. Science 225:1168–1170. https://doi.org/10.1126/science.6474172

Iadecola C (2003) Cerebrovascular effects of amyloid-beta peptides: mechanisms and implications for Alzheimer's dementia. Cell Mol Neurobiol 23:681–689

Jagota S, Rajadas J (2012) Effect of phenolic compounds against Abeta aggregation and Abeta-induced toxicity in transgenic C. elegans. Neurochem Res 37:40–48. https://doi.org/10.1007/s11064-011-0580-5

Jeong S (2017) Molecular and cellular basis of neurodegeneration in Alzheimer's disease. Mol Cell 40:613–620. https://doi.org/10.14348/molcells.2017.0096

Jordan J, Galindo MF, Miller RJ, Reardon CA, Getz GS, LaDu MJ (1998) Isoform-specific effect of apolipoprotein E on cell survival and beta-amyloid-induced toxicity in rat hippocampal pyramidal neuronal cultures. J Neurosci 18:195–204

Kamenetz F et al (2003) APP processing and synaptic function. Neuron 37:925–937. https://doi.org/10.1016/s0896-6273(03)00124-7

Kelly JW, Balch WE (2003) Amyloid as a natural product. J Cell Biol 161:461–462. https://doi.org/10.1083/jcb.200304074

Kim W, Hecht MH (2006) Generic hydrophobic residues are sufficient to promote aggregation of the Alzheimer's Abeta42 peptide. Proc Natl Acad Sci U S A 103:15824–15829. https://doi.org/10.1073/pnas.0605629103

Kim H, Park BS, Lee KG, Choi CY, Jang SS, Kim YH, Lee SE (2005) Effects of naturally occurring compounds on fibril formation and oxidative stress of beta-amyloid. J Agric Food Chem 53:8537–8541. https://doi.org/10.1021/jf051985c

Kim JK et al (2010) Protective effects of kaempferol (3,4′,5,7-tetrahydroxyflavone) against amyloid beta peptide (Abeta)-induced neurotoxicity in ICR mice. Biosci Biotechnol Biochem 74:397–401. https://doi.org/10.1271/bbb.90585

Lesne S et al (2006) A specific amyloid-beta protein assembly in the brain impairs memory. Nature 440:352–357. https://doi.org/10.1038/nature04533

Levy E et al (1990) Mutation of the Alzheimer's disease amyloid gene in hereditary cerebral hemorrhage, Dutch type. Science 248:1124–1126. https://doi.org/10.1126/science.2111584

Li C, Zug C, Qu H, Schluesener H, Zhang Z (2015) Hesperidin ameliorates behavioral impairments and neuropathology of transgenic APP/PS1 mice. Behav Brain Res 281:32–42. https://doi.org/10.1016/j.bbr.2014.12.012

Liu R, Zhang T, Yang H, Lan X, Ying J, Du G (2011) The flavonoid apigenin protects brain neurovascular coupling against amyloid-beta(2)(5)(-)(3)(5)-induced toxicity in mice. J Alzheimers Dis 24:85–100. https://doi.org/10.3233/JAD-2010-101593

Lue LF et al (1999) Soluble amyloid beta peptide concentration as a predictor of synaptic change in Alzheimer's disease. Am J Pathol 155:853–862. https://doi.org/10.1016/s0002-9440(10)65184-x

Macready AL, Kennedy OB, Ellis JA, Williams CM, Spencer JP, Butler LT (2009) Flavonoids and cognitive function: a review of human randomized controlled trial studies and recommendations for future studies. Genes Nutr 4:227–242. https://doi.org/10.1007/s12263-009-0135-4

Malenka RC (2003) The long-term potential of LTP Nature reviews. Neuroscience 4:923–926. https://doi.org/10.1038/nrn1258

Manzine PR et al (2019) ADAM10 in Alzheimer's disease: pharmacological modulation by natural compounds and its role as a peripheral marker. Biomed Pharmacother 113:108661. https://doi.org/10.1016/j.biopha.2019.108661

Marini D, Balestrieri F (1995) Multivariate-analysis of flavanone glycosides in citrus juices. Ital J Food Sci 7:255–264

McKay DL, Blumberg JB (2006) A review of the bioactivity and potential health benefits of chamomile tea (Matricaria recutita L.). Phytother Res 20:519–530. https://doi.org/10.1002/ptr.1900

McLean CA et al (1999) Soluble pool of Abeta amyloid as a determinant of severity of neurodegeneration in Alzheimer's disease. Ann Neurol 46:860–866

Md S, Gan SY, Haw YH, Ho CL, Wong S, Choudhury H (2018) In vitro neuroprotective effects of naringenin nanoemulsion against beta-amyloid toxicity through the regulation of amyloidogenesis and tau phosphorylation. Int J Biol Macromol 118:1211–1219. https://doi.org/10.1016/j.ijbiomac.2018.06.190

Morley JE, Farr SA (2014) The role of amyloid-beta in the regulation of memory. Biochem Pharmacol 88:479–485. https://doi.org/10.1016/j.bcp.2013.12.018

Mouly PP, Arzouyan CR, Gaydou EM, Estienne JM (1994) Differentiation of citrus juices by factorial discriminant-analysis using liquid-chromatography of flavanone glycosides. J Agric Food Chem 42:70–79. https://doi.org/10.1021/Jf00037a011

Noor H, Cao P, Raleigh DP (2012) Morin hydrate inhibits amyloid formation by islet amyloid polypeptide and disaggregates amyloid fibers. Protein Sci 21:373–382. https://doi.org/10.1002/pro.2023

O'Brien RJ, Wong PC (2011) Amyloid precursor protein processing and Alzheimer's disease. Annu Rev Neurosci 34:185–204. https://doi.org/10.1146/annurev-neuro-061010-113613

Ono K, Hasegawa K, Naiki H, Yamada M (2004) Curcumin has potent anti-amyloidogenic effects for Alzheimer's beta-amyloid fibrils in vitro. J Neurosci Res 75:742–750. https://doi.org/10.1002/jnr.20025

Ooghe WC, Detavernier CM (1997) Detection of the addition of Citrus reticulata and hybrids to Citrus sinensis by flavonoids. J Agric Food Chem 45:1633–1637. https://doi.org/10.1021/Jf9606262

Orsucci D, Mancuso M, Ienco EC, Simoncini C, Siciliano G, Bonuccelli U (2013) Vascular factors and mitochondrial dysfunction: a central role in the pathogenesis of Alzheimer's disease. Curr Neurovasc Res 10:76–80

Palmblad M, Westlind-Danielsson A, Bergquist J (2002) Oxidation of methionine 35 attenuates formation of amyloid beta -peptide 1-40 oligomers. J Biol Chem 277:19506–19510. https://doi.org/10.1074/jbc.M112218200

Patel K, Singh GK, Patel DK (2018) A review on pharmacological and analytical aspects of naringenin. Chin J Integr Med 24:551–560. https://doi.org/10.1007/s11655-014-1960-x

Pocernich CB, Lange ML, Sultana R, Butterfield DA (2011) Nutritional approaches to modulate oxidative stress in Alzheimer's disease. Curr Alzheimer Res 8:452–469

Puzzo D, Arancio O (2013) Amyloid-beta peptide: Dr. Jekyll or Mr. Hyde? J Alzheimers Dis 33 (Suppl 1):S111–S120. https://doi.org/10.3233/JAD-2012-129033

Puzzo D, Gulisano W, Arancio O, Palmeri A (2015) The keystone of Alzheimer pathogenesis might be sought in Abeta physiology. Neuroscience 307:26–36. https://doi.org/10.1016/j.neuroscience.2015.08.039

Qi H et al (1999) Processing of the notch ligand delta by the metalloprotease Kuzbanian. Science 283:91–94. https://doi.org/10.1126/science.283.5398.91

Rezai-Zadeh K et al (2009) Flavonoid-mediated presenilin-1 phosphorylation reduces Alzheimer's disease beta-amyloid production. J Cell Mol Med 13:574–588. https://doi.org/10.1111/j.1582-4934.2008.00344.x

Ross JA, Kasum CM (2002) Dietary flavonoids: bioavailability, metabolic effects, and safety. Annu Rev Nutr 22:19–34. https://doi.org/10.1146/annurev.nutr.22.111401.144957

Sawmiller D et al (2016) Diosmin reduces cerebral Abeta levels, tau hyperphosphorylation, neuroinflammation, and cognitive impairment in the 3xTg-AD mice. J Neuroimmunol 299:98–106. https://doi.org/10.1016/j.jneuroim.2016.08.018

Selkoe DJ (1994) Normal and abnormal biology of the beta-amyloid precursor protein. Annu Rev Neurosci 17:489–517. https://doi.org/10.1146/annurev.ne.17.030194.002421

Selkoe DJ (2008) Soluble oligomers of the amyloid beta-protein impair synaptic plasticity and behavior. Behav Brain Res 192:106–113. https://doi.org/10.1016/j.bbr.2008.02.016
Sengupta U, Nilson AN, Kayed R (2016) The role of amyloid-beta oligomers in toxicity, propagation, and immunotherapy. EBioMedicine 6:42–49. https://doi.org/10.1016/j.ebiom.2016.03.035
Shankar GM, Bloodgood BL, Townsend M, Walsh DM, Selkoe DJ, Sabatini BL (2007) Natural oligomers of the Alzheimer amyloid-beta protein induce reversible synapse loss by modulating an NMDA-type glutamate receptor-dependent signaling pathway. J Neurosci 27:2866–2875. https://doi.org/10.1523/JNEUROSCI.4970-06.2007
Shankar GM et al (2008) Amyloid-beta protein dimers isolated directly from Alzheimer's brains impair synaptic plasticity and memory. Nat Med 14:837–842. https://doi.org/10.1038/nm1782
Sharoar MG, Thapa A, Shahnawaz M, Ramasamy VS, Woo ER, Shin SY, Park IS (2012) Keampferol-3-O-rhamnoside abrogates amyloid beta toxicity by modulating monomers and remodeling oligomers and fibrils to non-toxic aggregates. J Biomed Sci 19:104. https://doi.org/10.1186/1423-0127-19-104
Shimmyo Y, Kihara T, Akaike A, Niidome T, Sugimoto H (2008) Multifunction of myricetin on A beta: neuroprotection via a conformational change of A beta and reduction of A beta via the interference of secretases. J Neurosci Res 86:368–377. https://doi.org/10.1002/jnr.21476
Shukla S, Gupta S (2010) Apigenin: a promising molecule for cancer prevention. Pharm Res 27: 962–978. https://doi.org/10.1007/s11095-010-0089-7
Sinha S et al (1999) Purification and cloning of amyloid precursor protein beta-secretase from human brain. Nature 402:537–540. https://doi.org/10.1038/990114
Sisodia SS, St George-Hyslop PH (2002) gamma-Secretase, Notch, Abeta and Alzheimer's disease: where do the presenilins fit in? Nat Rev Neurosci 3:281–290. https://doi.org/10.1038/nrn785
Socci V, Tempesta D, Desideri G, De Gennaro L, Ferrara M (2017) Enhancing human cognition with cocoa flavonoids. Front Nutr 4:19. https://doi.org/10.3389/fnut.2017.00019
Spencer JP (2010) The impact of fruit flavonoids on memory and cognition. Br J Nutr 104(Suppl 3): S40–S47. https://doi.org/10.1017/S0007114510003934
Stephan A, Phillips AG (2005) A case for a non-transgenic animal model of Alzheimer's disease. Genes Brain Behav 4:157–172. https://doi.org/10.1111/j.1601-183X.2004.00113.x
Stine WB Jr, Dahlgren KN, Krafft GA, LaDu MJ (2003) In vitro characterization of conditions for amyloid-beta peptide oligomerization and fibrillogenesis. J Biol Chem 278:11612–11622. https://doi.org/10.1074/jbc.M210207200
Tampellini D et al (2009) Synaptic activity reduces intraneuronal Abeta, promotes APP transport to synapses, and protects against Abeta-related synaptic alterations. J Neurosci 29:9704–9713. https://doi.org/10.1523/JNEUROSCI.2292-09.2009
Tay WM, da Silva GF, Ming LJ (2013) Metal binding of flavonoids and their distinct inhibition mechanisms toward the oxidation activity of Cu2+-beta-amyloid: not just serving as suicide antioxidants! Inorg Chem 52:679–690. https://doi.org/10.1021/ic301832p
Thal DR, Griffin WS, Braak H (2008) Parenchymal and vascular Abeta-deposition and its effects on the degeneration of neurons and cognition in Alzheimer's disease. J Cell Mol Med 12:1848–1862. https://doi.org/10.1111/j.1582-4934.2008.00411.x
Townsend M, Shankar GM, Mehta T, Walsh DM, Selkoe DJ (2006) Effects of secreted oligomers of amyloid beta-protein on hippocampal synaptic plasticity: a potent role for trimers. J Physiol 572:477–492. https://doi.org/10.1113/jphysiol.2005.103754
Urmoneit B et al (1997) Cerebrovascular smooth muscle cells internalize Alzheimer amyloid beta protein via a lipoprotein pathway: implications for cerebral amyloid angiopathy. Lab Investig 77:157–166
Verdone G et al (2002) The solution structure of human beta2-microglobulin reveals the prodromes of its amyloid transition. Protein Sci 11:487–499. https://doi.org/10.1110/ps.29002
Walsh DM et al (2005) The role of cell-derived oligomers of Abeta in Alzheimer's disease and avenues for therapeutic intervention. Biochem Soc Trans 33:1087–1090. https://doi.org/10.1042/BST20051087
Wang JB, Wang YM, Zeng CM (2011) Quercetin inhibits amyloid fibrillation of bovine insulin and destabilizes preformed fibrils. Biochem Biophys Res Commun 415:675–679. https://doi.org/10.1016/j.bbrc.2011.10.135

Wang D et al (2012) Long-term naringin consumption reverses a glucose uptake defect and improves cognitive deficits in a mouse model of Alzheimer's disease. Pharmacol Biochem Behav 102:13–20. https://doi.org/10.1016/j.pbb.2012.03.013

Wang DM, Li SQ, Zhu XY, Wang Y, Wu WL, Zhang XJ (2013) Protective effects of hesperidin against amyloid-beta (Abeta) induced neurotoxicity through the voltage dependent anion channel 1 (VDAC1)-mediated mitochondrial apoptotic pathway in PC12 cells. Neurochem Res 38:1034–1044. https://doi.org/10.1007/s11064-013-1013-4

Williams RJ, Spencer JP (2012) Flavonoids, cognition, and dementia: actions, mechanisms, and potential therapeutic utility for Alzheimer disease. Free Radic Biol Med 52:35–45. https://doi.org/10.1016/j.freeradbiomed.2011.09.010

Yan R et al (1999) Membrane-anchored aspartyl protease with Alzheimer's disease beta-secretase activity. Nature 402:533–537. https://doi.org/10.1038/990107

Yang W, Ma J, Liu Z, Lu Y, Hu B, Yu H (2014) Effect of naringenin on brain insulin signaling and cognitive functions in ICV-STZ induced dementia model of rats. Neurol Sci 35:741–751. https://doi.org/10.1007/s10072-013-1594-3

Yang W et al (2018) Naringin dihydrochalcone ameliorates cognitive deficits and neuropathology in APP/PS1 transgenic mice. Front Aging Neurosci 10:169. https://doi.org/10.3389/fnagi.2018.00169

Yang Z, Kuboyama T, Tohda C (2019) Naringenin promotes microglial M2 polarization and Abeta degradation enzyme expression. Phytother Res 33:1114–1121. https://doi.org/10.1002/ptr.6305

Yiannopoulou KG, Papageorgiou SG (2013) Current and future treatments for Alzheimer's disease. Ther Adv Neurol Disord 6:19–33. https://doi.org/10.1177/1756285612461679

Yoo BC, Kim SH, Cairns N, Fountoulakis M, Lubec G (2001) Deranged expression of molecular chaperones in brains of patients with Alzheimer's disease. Biochem Biophys Res Commun 280: 249–258. https://doi.org/10.1006/bbrc.2000.4109

Yoon I, Lee KH, Cho J (2004) Gossypin protects primary cultured rat cortical cells from oxidative stress- and beta-amyloid-induced toxicity. Arch Pharm Res 27:454–459

Zelus C, Fox A, Calciano A, Faridian BS, Nogaj LA, Moffet DA (2012) Myricetin inhibits islet amyloid polypeptide (IAPP) aggregation and rescues living mammalian cells from IAPP toxicity. Open Biochem J 6:66–70. https://doi.org/10.2174/1874091X01206010066

Zhang N et al (2018) Protective role of naringenin against Abeta25-35-caused damage via ER and PI3K/Akt-mediated pathways. Cell Mol Neurobiol 38:549–557. https://doi.org/10.1007/s10571-017-0519-8

Zhao M, Ma J, Zhu HY, Zhang XH, Du ZY, Xu YJ, Yu XD (2011) Apigenin inhibits proliferation and induces apoptosis in human multiple myeloma cells through targeting the trinity of CK2, Cdc37 and Hsp90. Mol Cancer 10:104. https://doi.org/10.1186/1476-4598-10-104

Zhao L, Wang JL, Liu R, Li XX, Li JF, Zhang L (2013) Neuroprotective, anti-amyloidogenic and neurotrophic effects of apigenin in an Alzheimer's disease mouse model. Molecules 18: 9949–9965. https://doi.org/10.3390/molecules18089949

Therapeutic Potential of Phytochemicals: Lessons Learned from Streptozotocin-Induced Sporadic Alzheimer's Disease

20

Hayate Javed and Shreesh Kumar Ojha

Abstract

Early sporadic Alzheimer's disease (SAD) caused by abnormal glucose/energy metabolism and disrupted insulin signaling is associated with a brain insulin-resistant state, which is the root cause of neurodegenerative changes. In preclinical studies, mainly in rodents, intracerebroventricular (ICV) administration of streptozotocin (STZ) at subdiabetogenic doses leads to a perturbed brain insulin-resistant state. Furthermore, ICV-STZ infused animals display memory loss, progressive cholinergic dysfunction, oxidative stress, impaired glucose metabolism, and neurodegeneration that are akin to the pathological changes in human SAD. Animal models which mimic many of these pathological features of human SAD play an important role to test the therapeutic potential of newer molecules including phytochemicals for the prevention and treatment of SAD. The present chapter aims to explore the cellular pathways involved in the ICV-STZ-induced model of SAD. Moreover, the therapeutic potential of phytochemicals which have been evaluated so far in the ICV-STZ animal model of SAD has been also included and appraised for their potential usefulness as possible molecules for pharmaceutical or dietary application in SAD.

H. Javed (✉)
Department of Anatomy, College of Medicine and Health Sciences, United Arab Emirates University, Al Ain, United Arab Emirates
e-mail: h.javed@uaeu.ac.ae

S. K. Ojha
Department of Pharmacology and Therapeutics, College of Medicine and Health Sciences, United Arab Emirates University, Al Ain, United Arab Emirates

G. M. Ashraf, M. S. Uddin (eds.), *Current Thoughts on Dementia*,
https://doi.org/10.1007/978-981-16-7606-2_20

Keywords

Dementia · Insulin · Alzheimer's disease · Amyloid-beta · Streptozotocin · Phytochemicals · Plants extracts

20.1 Introduction

Alzheimer's disease (AD) is the most common form of dementia, and most prevalent among the elderly population. AD generally affects people 65 years of age or above. Almost 35 million people are affected worldwide by AD. It is a serious health threat in society (Querfurth and LaFerla 2010), and in clinical practice, the characteristic feature of AD is decline or loss of memory and cognition as well as a deficit in cholinergic function which leads to patients' untimely death after 3–9 years of diagnosis (Querfurth and LaFerla 2010). Loss of neurons and diminished synapses in particular brain areas such as the hippocampus and cortex are the classical neuropathological signs of AD. In addition, extracellular deposition of amyloid plaques which consist of a small fragment of a peptide known as amyloid-beta (Aβ) and intracellularly neurofibrillary tangles (NFTs) made up of hyperphosphorylated tau protein are also found in the AD brain (Goedert and Spillantini 2006; Moreira et al. 2007, 2009). There are two types of AD reported based on their origin. A missense mutation in three different genes, viz., amyloid-β precursor protein (APP), presenilin-1, and presenilin-2 leads to the familial form of AD (Rocchi et al. 2003). Therefore, abnormal and continuous Aβ fragments often appear as deposits in the plaques. Though sporadic origin cases of AD are found in the majority and numerous risk factors that take part in the onset of sporadic AD (SAD) such as aging, apolipoprotein E4 (ApoE4), and type 2 diabetes as the main risk factors (Hoyer 2004a).

In SAD, early impairments in impaired metabolism of brain glucose/energy are observed in frontal and parietotemporal areas of the brain (Henneberg and Hoyer 1995; Hoyer 2002, 2004b). The occurrence of impaired brain insulin metabolism indicates a role for abnormal signaling of insulin in SAD pathogenesis (Cardoso et al. 2009). Besides, AD patients possess higher levels of plasma insulin and a lower amount of cerebrospinal fluid (CSF) (Watson et al. 2003). Additionally, the reduced density of insulin receptor (IR) and decreased activity of tyrosine kinase (Frolich et al. 1999) also reported in AD brain, suggesting impaired insulin signaling and function in the brain are the key reason which mechanistically explains the development of SAD pathology. It was shown that insulin administration to AD patients improved cognitive functions (Craft et al. 1999; Watson and Craft 2004).

In order to study the early changes in the brain of SAD, intracerebroventricular (ICV) injection of streptozotocin (STZ) in rats and mice had been employed to develop an experimental model (Hoyer 2004b). In the animal model of ICV-STZ-induced SAD, many morphological, neurochemical, and behavioral features have been observed which closely resembled with the human SAD (Grunblatt et al. 2007; Salkovic-Petrisic and Hoyer 2007). Considering the relevance of experimental model of SAD and preclinical evaluation of phytochemicals, the current chapter reviewed and critically appraised the role of abnormal signaling of insulin and the metabolism of glucose in SAD pathogenesis. Subsequently, in later part we also represented the plant extracts and other phytochemicals based therapeutic strategies evaluated in STZ induced animal model of SAD.

20.2 Function of Brain Insulin and its Signaling Pathways

Insulin receptors (IRs) and insulin are widely available in the brain (Havrankova et al. 1978) that indicate the key role of insulin in the pathophysiology of the brain disorders including SAD. Variable IRs densities are observed in the discrete areas of the brain amid maximum concentration in the cerebral cortex, hypothalamus, hippocampus, cerebellum, and olfactory bulb (Unger et al. 1989). IRs are of two types and present in the brain of adult mammals as a brain-specific IRs are maximally found on neurons and peripheral types of IRs on glial cells (Adamo et al. 1989). Though, in the brain the immunolabeled cells of insulin are well-known. However, insulin origin in the brain is controversial. In the adult brain, most of the insulin originates from the periphery and then traveled through cerebrospinal fluids (CSF) to the brain following its production in β-cells of the pancreas (Banks 2004; Burns et al. 2007; Erol 2008; Salkovic-Petrisic and Hoyer 2007).

The insulin transport across the blood–brain barrier (BBB) largely takes place through a temperature-sensitive, regulatable, carrier-mediated, and saturable active process (Banks 2004; Burns et al. 2007; Erol 2008; Salkovic-Petrisic and Hoyer 2007). Though, a lesser amount of de novo production of insulin is reported in the brain (Wozniak et al. 1993). To explore the various distribution forms of insulin and IRs, Zhao and colleagues (Zhao et al. 2004) assumed that IRs of various areas of the brain utilize the insulin from variable sources for neuronal signal transduction and cell-to-cell communication. It has been well documented that brain insulin exerts pleiotropic actions (Cardoso et al. 2009). Insulin plays a key role as a central glucose metabolism regulator in the brain and also takes part in neuroendocrine and neuromodulating signaling, thus, contribute significantly in the survival and growth of neurons (Cardoso et al. 2009; Gasparini and Xu 2003). Earlier studies showed that insulin signaling also contributes to synaptic plasticity through attenuation of the inhibitory and excitatory receptors, e.g. GABA and glutamate receptors (Zhao et al. 2004). Synaptic plasticity is also modulated by insulin signaling via activating the signal transduction pathways that lead to changes in the gene expression which affects long-term memory (Zhao et al. 2004). IR activation causes phosphorylation of many tyrosine residues, which eventually leads to intracellular substrates phosphorylation and receptor autophosphorylation, e.g. Src-homology-2-containing protein and insulin receptor substrates (IRS) (Czech and Corvera 1999; Saltiel and Pessin 2002).

Furthermore, intracellular substrates phosphorylation utilize several proteins and activate various molecular pathways and signaling events including the most notable signaling pathways, mitogen-activated protein kinase (MAPK) and phosphoinositide 3-kinase (PI3-K) (Johnston et al. 2003; Kahn and White 1988). PI3-K pathway activation, in turn, leads to protein kinase-B activation, facilitates the survival of the neurons directly by inhibiting pro-apoptotic cascades (Dudek et al. 1997; van der Heide et al. 2006). The PI3-K/Akt signaling pathway activates insulin-sensitive glucose transporter 4(GLUT-4) and translocation to membrane surface that facilitates increased uptake of glucose into cells (Bryant et al. 2002; Johnston et al. 2003). Moreover, PI3-K/Akt activation phosphorylates serine-9 residue and consequently prevents the cytosolic forms of α and β glycogen synthase kinase-3 (GSK-3) (Cross et al. 1995). It has been shown that GSK-3α controls the Aβ peptides production (Phiel et al. 2003). These altogether demonstrate that insulin controls the secretion of soluble AβPP mediating the PI3-K-dependent pathway (Solano et al. 2000). Insulin has been showed to decrease the neuronal accumulation of Aβ by enhancing APP/Aβ transport from the site of its cellular generation (trans Golgi network) to the plasma membrane (Gasparini and Xu 2003). Earlier findings demonstrated that insulin enhances Aβ levels (extracellular) by facilitating Aβ release by preventing its deprivation by the insulin-degrading enzyme (IDE) (Vekrellis et al. 2000).

In earlier studies, both IRs and insulin mRNA were showed highly abundant in the hippocampus, a brain area responsible for memory and cognition that indicated a link between cognitive function and insulin (Zhao et al. 2004). The exact mechanism (s) of insulin remains to be elucidated; although, this was reported that PI3-K and MAPK signaling pathways are equally implicated in memory and cognition. For example, the survival of neurons and synaptic plasticity is modulated by extracellular-regulated kinase 1/2 (Xia et al. 1995), both are required in learning and cognition processes (Davis and Laroche 2006). Moreover, activation of insulin-dependent MAPK signaling pathway following learning controls the expression of certain genes which is needed for long-term memory consolidation (Zhao et al. 2004). In long-term depression, the PI3-K pathway plays a key role by facilitating the internalization of glutamatergic AMPA receptors (Man et al. 2000) or by involving active GABA receptors in the postsynaptic membranes (Wan et al. 1997). It has been demonstrated that PI3-K signaling regulates the generation of nitric oxide too by the activation of endothelial nitric oxide synthase in an insulin-dependent manner (Montagnani et al. 2001), that is predominantly found in the hippocampus and is known to have significant effects in synaptic plasticity and learning (Doreulee et al. 2003).

Furthermore, it has been demonstrated that attenuation of glucose metabolism in the brain and glucose transporters (GLUTs) expression and trafficking are another processes wherein insulin affects learning and memory (McEwen and Reagan 2004). Therefore, several mechanisms demonstrated on role of insulin in regulating cerebral arteries and insulin function in the regulation of cerebrovascular function in brain insulin-resistant state. Collectively, insulin demonstrated to play an important role in the proper function of the brain. So, abnormal signaling of insulin in the brain has been shown strongly interrelated to the pathological state of AD (Fig. 20.1). The following sections present the role of abnormal metabolism of glucose/energy and disturbed insulin signaling in the brain which is implicated in AD pathophysiology.

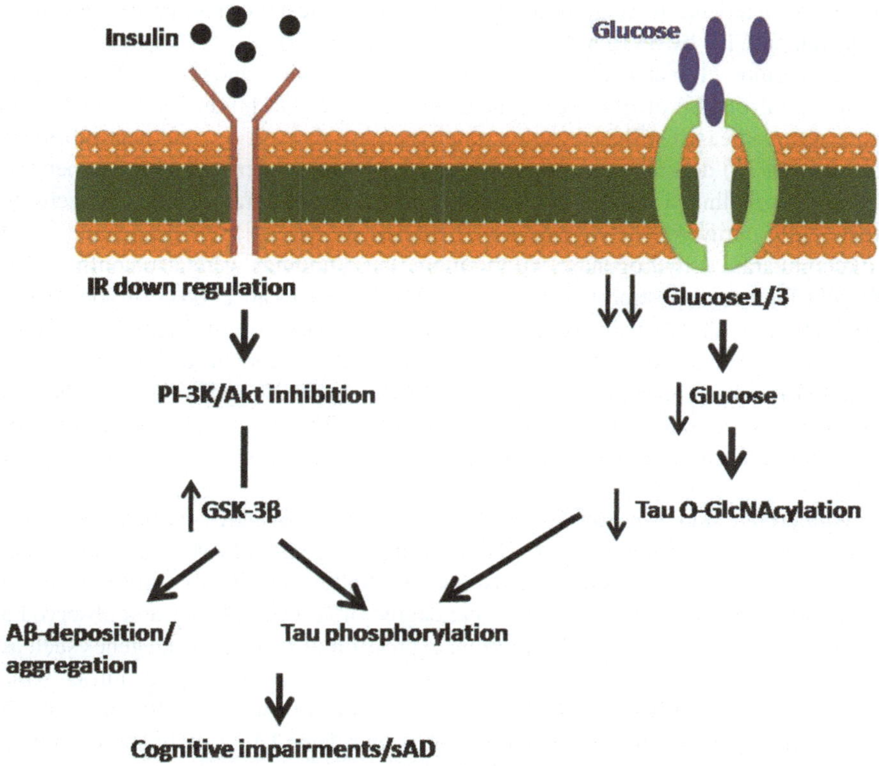

Fig. 20.1 Insulin signaling in the brain leading to the onset of cognitive impairments

20.3 Relationship between Sporadic Alzheimer's Disease and Impaired Brain Glucose/Energy Metabolism

So far, the main source of energy in the brain is glucose, thus insulin is crucial in sustaining the glucose metabolism for cerebral energy. Meanwhile, glucose is not synthesized or stored by neurons and they are solely reliant on peripheral glucose transported across BBB via GLUTs (Scheepers et al. 2004). GLUT-1 and GLUT-3 are the most prevalent isoforms of GLUTs in the brain (Vannucci et al. 1997). Endothelial cells, neurons, astrocytes, and oligodendrocytes contain mostly GLUT-1, although, neurons particularly possess GLUT-3 (Vannucci et al. 1997). Imaging studies based on positron emission tomography showed that disturbed glucose metabolism in AD patients is followed by atrophy and other neuropsychological abnormalities (Silverman et al. 2001). Furthermore, Hoyer (Hoyer 2004a) based on findings articulated that aforesaid disturbance could be a source, instead of an outcome of degenerative changes of AD. At the early AD stage, 45% reduced brain glucose consumption and 20% cerebral blood flow (CBF) has been reported.

Whereas, in advanced AD, physiological and metabolic abnormalities exaggerate to a decline in CBF by 55 to 65% (Hoyer and Nitsch 1989).

Furthermore, Liu et al. (Liu et al. 2008) showed GLUT-1 and GLUT-3 (takes part in neuronal uptake of glucose) were reduced in the AD brain. This reduction is related to decrease O-GlcNacylation, tau hyperphosphorylation, and neurofibrillary tangles (NFTs) density in the brain of humans. Impaired cerebral glucose metabolism is closely linked to the altered expression of mitochondrial key enzymes activity such as pyruvate dehydrogenase (PDH), isocitrate dehydrogenase, and α-ketoglutarate dehydrogenase. Furthermore, these findings were also confirmed in the fibroblast and postmortem brains of AD (Bubber et al. 2005). The decreased production of ATP is mainly due to the impaired metabolism of cerebral glucose at the early SAD stage that compromises the ATP-dependent cellular processes required for the proper function of the cell (Moreira et al. 2007). The metabolism of energy is centrally regulated by mitochondria. The impairment in turnover of mitochondria and its function has been believed to contribute in the pathology of AD.

Numerous studies showed that senile plaques exhibit multiple impaired mitochondria in the brain of AD (Hirai et al. 2001; Kidd 1964; Luse and Smith Jr. 1964). de la Monte and Wands (de la Monte and Wands 2006) analyzed postmortem brain tissues with a variable degree of AD complexity and observed a correlation of AD severity with abnormal expression of mitochondrial genes such as high levels of p53, complex IV, and oxidative stress markers, e.g. nitric oxide synthase and NADPH-oxidase. Therefore, impairment of mitochondria in the brain of AD elicits aggravation of oxidative stress and degradation of lipids as well as proteins and increased nucleic acid oxidation levels in the susceptible neurons (Castellani et al. 2001; Straface et al. 2005). Mitochondrial by-products formed in vulnerable neurons attribute to a higher mitochondrial turnover through autophagy and decrease turnover of proteolysis in the AD brain (Hirai et al. 2001). Altogether, the available reports demonstrated that AD begins with reduced cerebral metabolism that is evidenced by decreased usage of glucose, reduced metabolism of energy, and mitochondrial abnormalities.

In past decades, multiple hypotheses were proposed to describe cross-link between AD pathology and abnormal brain glucose and energy metabolism. Rapoport (Rapoport 2003) suggested that in the early stage of AD, altered structure and function of synapses lead to a decrease in the supply of energy in neurons which reversibly causes diminished oxidative phosphorylation. Though, as AD develops, NFTs accumulation occurs in the neuronal cytoplasm, and they prefer non-phosphorylated tau proteins for the mitochondrial transport through axons across the nucleus and synapse. At this point, energy deficiency and irreversibly down-regulated OXPHOS associated pathological processes cause cell death (Rapoport 2003). Whereas, multiple reports suggested the role of reduced insulin actions/or insulin resistance underlying abnormal metabolism of cerebral glucose in AD pathogenesis (Cardoso et al. 2009; Hoyer 2004a; Rivera et al. 2005). Frolich et al. (Frolich et al. 1998) published the first report on insulin-resistant brain states and reported reduced levels of insulin as well as tyrosine kinase activity and increased IRs density in the SAD brain. Therefore, reduced expression of insulin mRNA and protein levels (Lester-Coll et al. 2006; Rivera et al. 2005), decreased in IRS-1 and IRS-2 levels and ERK1/2 and PI3-K activities (Watson and Craft 2004) as well as IRs expression (Frolich et al. 1999) in the postmortem brain of human AD. All these biochemical, functional, and structural impairments were associated with the degree of sternness and advancement of neurodegeneration and dementia.

Rivera et al. (Rivera et al. 2005) demonstrated a gradual reduction in mRNA levels of insulin, insulin-like growth factor-1 (IGF-1), and IGF-2 as well as their receptors were related to the growing Braak stages of AD. Furthermore, a relationship has been established among Braak staging and activity of Akt or levels of protein (Akt) in the analysis of postmortem AD brains from humans, suggesting PI3-K signaling dependent changes are insulin-mediated and time-dependent (Pei et al. 2003). Moreover, AD patients reveal higher levels of fasting plasma insulin and low quantity of CSF insulin as well as a reduced ratio of CSF/plasma insulin, in addition to the enhanced levels of Aβ (Watson and Craft 2004). Altogether these indicate that decreased clearance of insulin is closely associated with an increased level of Aβ in plasma. Also, as previously described, increased IRs density observed in AD (Frolich et al. 1998) displays abnormalities in the insulin signaling pathways parallel to what is observed in type 2 diabetes. Moreover, type 2 diabetes is a high-risk factor for AD because patients of AD acquired type 2 diabetes at the greater possibility (Cole and Frautschy 2007; Moreira et al. 2009), reasonably explain that type 2 diabetes of the brain is known as SAD (Frolich et al. 1998).

20.4 Role of Oxidative Stress in the Development of Sporadic Alzheimer's Disease

Oxidative stress has been implicated in the onset and development of neurodegenerative diseases including AD (Tonnies and Trushina 2017). In addition to the accumulation of senile plaques in AD brain, oxidative stress is also a well-established characteristic of the AD brains (Tonnies and Trushina 2017). It is well evidenced that mitochondrial dysfunction occurs in AD (Beal 2005b; Dawson 1970) and impaired mitochondrial axonal transport is well described in embryonic neuronal cultures and various animal models of AD (Wang et al. 2019). Abnormal metabolism of cerebral glucose is related to impaired mitochondrial function which is observed in numerous animal models of AD and AD patients (Beal 2005a; Cardoso et al. 2004; Caspersen et al. 2005; Reiman et al. 2004; Sultana et al. 2011; Trushina et al. 2012). Mitochondrial dysfunction is elicited directly by causation of oxidative stress and synaptic dysfunction which play a key role in early disease mechanisms before the development of pathological hallmarks such as Aβ or Tau (Fig. 20.2) (Trushina et al. 2012, 2013; Valla et al. 2010).

Earlier reports showed that apart from mitochondrial reactive oxygen species (ROS) generation, imbalanced levels of metals ions such as iron (Fe), zinc (Zn), aluminum (Al), copper (Cu), manganese (Mn), and magnesium (Mg), could also be responsible in the generation of ROS and eventually contribute in the onset of AD

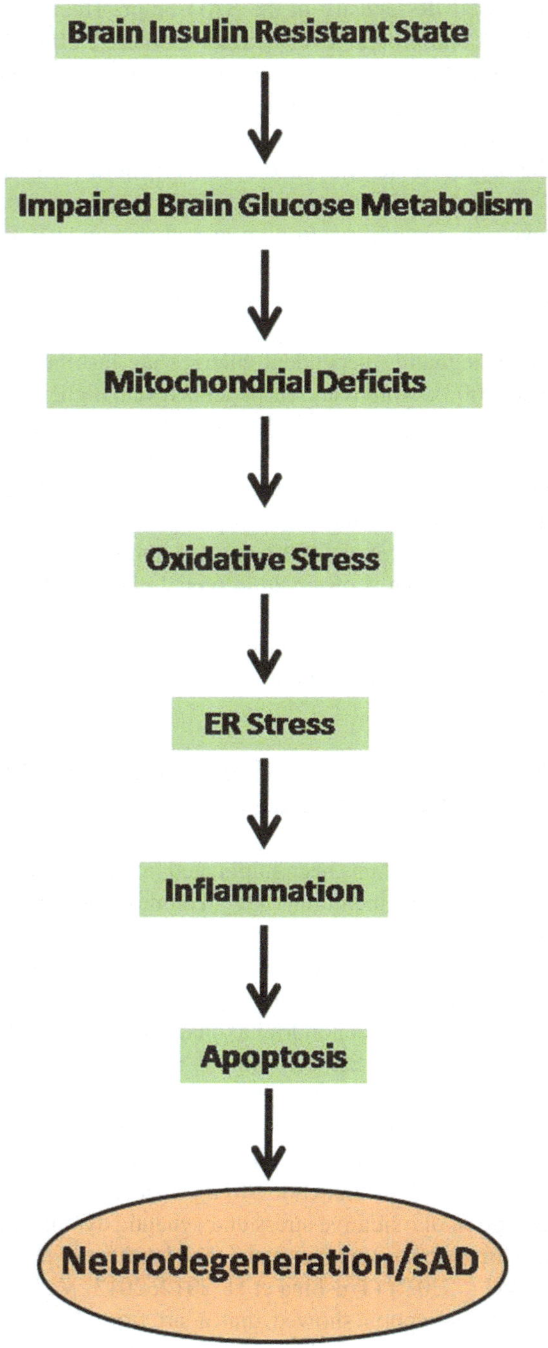

Fig. 20.2 Impaired brain glucose metabolism that causes mitochondrial dysfunction and oxidative stress which eventually leads to the development of neurodegeneration/sporadic Alzheimer's disease

(Beal 2005b; Greenough et al. 2013; Liu et al. 2006). Additional ROS generation takes place through aggregated Aβ induced activation of glial cells and enhances the inflammatory response (Nakajima and Kohsaka 2001). Moreover, accumulated Aβ also intensify ROS production by binding to the membrane of mitochondria, thus changing the mitochondrial dynamics and function, which eventually leads to impaired energy metabolism and synaptic dysfunction (Beal 2005a; Bose and Beal 2016; Gibson et al. 2008; Manczak et al. 2006; Reddy et al. 2012). Therefore, in the AD brain oxidative stress appears to be an initial event in causation, progression and development of the disease.

20.5 Streptozotocin-Induced Experimental Model of Sporadic Alzheimer's Disease for the Evaluation of Therapeutic Agents

Impairment in the brain insulin system has been shown to cause neurodegenerative processes including AD. It is well known that the brain possesses insulin and IRs and ICV-STZ infusion in rats causes dysfunction of the brain insulin system. Thus this approach has been emerged as an appropriate experimental model to study SAD in rats (Grunblatt et al. 2007; Salkovic-Petrisic and Hoyer 2007). Peripheral infusion of STZ, a diabetes-inducing drug, causes a particular loss of β-cells of the pancreas. Numerous studies previously showed that STZ causes toxicity of β-cell and this deleterious effect of STZ is associated with its chemical structure possessing glucose moiety, which facilitates the penetration of STZ into β-cells through GLUT-2 receptors (Elsner et al. 2000). Furthermore, β-cell toxicity caused by STZ occurs through the alkylation of DNA, which in turn leads to poly ADP-ribosyl activation that results in the reduction of NAD+ and ATP in cells and destruction of the primary function of β-cells including production and release of insulin (Szkudelski 2001).

Though the precise mechanism of STZ action in inducing the SAD in rodents is remained to be elucidated. However, it has been shown that a similar mechanism exists as reported in the peripheral system. GLUT-2 is abundantly present in the brain of rats, particularly in nuclei of limbic areas, highly distributed in the ventral and medial regions in the vicinity of the midline, and plays vital role in glucose recruitment as well as controlling the release of neurotransmitters (Arluison et al. 2004). A growing body of evidence suggests that multiple or a single dose of STZ infusion into lateral ventricles either unilaterally or bilaterally causes impairment in the metabolism of brain glucose, diminished long-term memory and cognition and changes in neurochemical profiles similar to that seen in the patients of AD (Grunblatt et al. 2007; Salkovic-Petrisic and Hoyer 2007). The impaired glucose/energy metabolisms in the brain, and decreased CBF have shown following ICV-STZ infusion (Tota et al. 2010). ICV-STZ infusion reported to decrease hippocampal and cortical glycolytic enzyme activities (Plaschke and Hoyer 1993) that results in altered ATP and creatine phosphate levels (Lannert and Hoyer 1998; Nitsch and Hoyer 1991). Overall, these effects culminates as a major mediator in SAD following impaired learning and memory process, lessened brain oxidative metabolism, and perturbed balance in cerebral energy (Duelli et al. 1994). Accordingly, abnormal cerebral energy metabolism, decreased synthesis of acetyl coenzyme A, and reduced cholinergic transmission have been well documented following ICV-STZ administration. Additionally, reduced hippocampal choline acetyltransferase (ChAT) activity was observed 1 and 3 weeks after ICV-STZ infusion in rats (Blokland and Jolles 1993; Ishrat et al. 2006).

Furthermore, enhanced activity of acetylcholinesterase (AChE) has been reported in ICV-STZ infused rats following elevated breakdown of acetylcholine that consequently increases the deficits of acetylcholine, an important neurotransmitter implicated in the acquisition of learning and cognition (Agrawal et al. 2009; de la Monte and Wands 2006; Ishrat et al. 2006). Moreover, ICV-STZ infusion in the hippocampus of rats causes reduced ChAT activity which is closely linked to rats' performance in the Morris water maze task for spatial learning and memory (Blokland and Jolles 1993). In the postmortem samples of AD brains, the deficits in the cholinergic system are established by decreased ACh levels and impaired AChE as well as ChAT activities (McGeer et al. 1984). Also, de la Monte and colleagues (de la Monte and Wands 2006) demonstrated that enhanced levels of ubiquitin, phospho-tau, GSK-3β, AβPP, and Aβ as well as diminished tau protein levels in ICV-STZ-infused rats. Additionally, decreased levels of GLUT-1 and GLUT-3, diminished O-GlcNAcylation levels, overactivation of GSK-3β, impaired insulin signaling pathway, and reduced tau activity for microtubule binding as well as enhanced tau phosphorylation was also observed in the brain of ICV-STZ-infused animals (Deng et al. 2009).

20.6 Phytochemicals for the Treatment of Streptozotocin-Induced Sporadic Alzheimer's Disease

Phytochemicals are plant-derived chemical compounds that are synthesized in plants for their protective or adaptive potentials and considered as secondary metabolites (Kennedy and Wightman 2011). These are devoid of nutritional properties for the plants but have been well explored and utilized for their health benefits and potential to treat numerous diseases and hence considered as a source of drug discovery apart from nutraceutical use (Kennedy and Wightman 2011). Numerous phytochemicals are found to cure chronic diseases including cardiovascular disease and cancer through ameliorating cellular abnormalities (Manach et al. 2009). The pharmacological and molecular mechanisms of a large number of phytochemicals are well studied and many of them such as curcumin, resveratrol, berberine, epigallocatechin, and cannabidiol has received enormous interest from preclinical studies to clinical development (Forni et al. 2019; Howes and Perry 2011; Upadhyay and Dixit 2015). Majority of them are reputed for their potential antioxidant and anti-inflammatory properties which also module redox immune-inflammatory signaling cascade and showed pleiotropic properties with reasonable efficacy and considerable safety.

The dietary antioxidants well explored till date are of mainly polyphenolic nature and of natural origin. Though, various experimental studies showed that these compounds exert their protective cellular effects not solely due to antioxidant property (Andrade et al. 2019). Till date many of them have been derived from common dietary plants have been shown preventive, protective, nutritive, and curative in their dietary usage and suggestive of therapeutic potential (Sofowora et al. 2013). Therefore, the majority of the natural products are perceived safe with negligible adverse effects due to dietary usage and presence in common edible plants and considered safe for human use. Like other disorders, in the management of AD and memory deficit, the plant extracts and phytochemicals play a key role, with the well-established use of pyridostigmine and galantamine. Here, we have summarized phytochemicals in Table 20.1 and plant extracts in Table 20.2 in synoptic forms which have been evaluated in the ICV-STZ model of SAD and their therapeutic outcomes.

20.7 Recent Developments and Future Perspectives

A number of pathologic features have been shown associated with risk factors for the development of AD evidenced by diabetes and aging related oxidative stress amplifies the cascade of neurodegeneration (Chen and Zhong 2014; Kamat et al. 2016; Tonnies and Trushina 2017; Tumminia et al. 2018). However, the exact mechanism remains to be elucidated for SAD development. STZ infusion in rodents' brains causes cerebral insulin resistance which further leads to developing a state of brain diabetes (de la Monte and Tong 2014; Tumminia et al. 2018). At this point, several AD-like pathologies develop which are Aβ accumulation, hyperphosphorylation of tau, oxidative stress, and loss of cognitive function (Chandra et al. 2019). Hence, STZ could bring an innovative way to study the underlying neurodegenerative mechanism of SAD.

The ongoing researches particularly focus on disease-modifying drugs which target the primary cause of AD and lead to the future way of new drug development (Cummings et al. 2016; Yiannopoulou and Papageorgiou 2013). Although searching drugs with numerous targets and antioxidant, antiinflammatory, neuroprotective activities will be a robust strategy to cure/prevent AD (Ardura-Fabregat et al. 2017; Poprac et al. 2017). The neuroprotective properties of several plant extract described herein is suggestive of potential benefits in reducing neuroinflammation, oxidative stress, Aβ aggregation, and neuronal apoptosis and improving cognitive functions, thus, these beneficial therapeutic properties of plant extracts showed their translational efficacies and may contribute to the treatment of AD.

Table 20.1 Phytochemicals evaluated in the intracerebroventricular-streptozotocin-induced animal model of sporadic Alzheimer's disease

Phytochemicals	Effects and mechanism observed	References
Epigallocatechin gallate	• Decreased oxidative stress	Biasibetti et al. (2013)
	• Improved learning and memory	
	• Reduced acetylcholinesterase activity	
Retinoic acid	• Decreased oxidative stress and inflammation	Sodhi and Singh (2013)
	• Improved learning and memory	
	• Reduced acetylcholinesterase activity	
	• Attenuated histopathological alteration and Aβ deposition	
Berberine	• Improved learning and memory	Kumar et al. (2016)
	• Decreased oxidative stress and inflammation	
	• Attenuated mitochondrial dysfunction and histological alterations	
	• Prevented calcium homeostasis alteration	
Kaempferol	• Alleviated memory impairment	Kouhestani et al. (2018)
	• Increased endogenous hippocampal antioxidants	
	• Reduced neuroinflammation	
Crocin	• Improved cognitive performance	Khalili and Hamzeh (2010); Naghizadeh et al. (2013)
	• Decreased oxidative stress	
Epicatechin	• Enhanced learning and memory	Ejaz Ahmed et al. (2013)
	• Ameliorated oxidative stress	
	• Increased expression of choline acetyltransferase	
	• Reduced levels of TNF-α, IL-1β, and iNOS	
Pterostilbene	• Improved learning and cognition	Naik et al. (2017)
	• Decreased acetylcholinesterase activity	
	• Augmented action of ATPases (Na^+, Ca^{2+}, and Mg^{2+})	
	• Augmented oxidative stress (LPO, GSH and SOD) and inflammation (nitrite, PPAR-α, TNF-α, IL-6) markers	
Gallic acid	• Prevented cognitive deficits	Mansouri et al. (2013)
	• Reduced oxidative stress	
	• Increased antioxidant enzymes in the hippocampus and cerebral cortex	
Ferulic acid	• Restored mitochondrial dynamics	Zafeer et al. (2019)
	• Mitigated bioenergetics loss	
	• Rescued memory and learning problems	

(continued)

Table 20.1 (continued)

Phytochemicals	Effects and mechanism observed	References
Rutin	• Prevented cognitive impairments	Javed et al. (2012)
	• Ameliorated oxidative stress and neuroinflammation	
	• Increased levels of endogenous antioxidants	
Myricetin	• Improved learning and memory	Ramezani et al. (2016)
	• Protected hippocampal ca3 pyramidal neurons	
Naringenin	• Prevented cognitive deficits and improved learning and memory	Baluchnejadmojarad and Roghani (2006); Khan et al. (2012); Yang et al. (2014)
	• Protected neuronal injury	
	• Reduced oxidative stress and increased antioxidant levels	
	• Decreased tau hyperphosphorylation by decreasing gsk-3β	
	• Reduced Aβ via upregulation of insulin-degrading enzyme	
Luteolin	• Ameliorated spatial learning and memory impairments	Wang et al. (2016)
	• Reduced thickness of CA1 pyramidal layer	
	• Protected the hippocampal structures	
Puerarin	• Attenuated the learning and memory impairments	Zhao et al. (2015)
	• Inhibited oxidative stress and increased endogenous antioxidants	
Caffeic acid	• Attenuated learning and memory deficits	Deshmukh et al. (2016); Kumar and Bansal (2018a); Kumar et al. (2017)
	• Inhibited oxidative stress and restored cholinergic functions	
	• Ameliorated memory loss via PI3-kinase-dependent pathway	
Ellagic acid	• Enhanced cognitive functions and ameliorated oxidative stress	Jha et al. (2018); Kumar and Bansal (2018a, b)
	• Reduced Aβ levels and neuroinflammation	
	• Prevented memory loss via PI3-kinase-enos signaling	
Naringin	• Restored cognitive deficits	Sachdeva et al. (2014)
	• Mitigated mitochondrial dysfunction mediated oxido-nitrosative stress and cytokine release	
Vanillic acid	• Improved spatial learning and memory retention	Singh et al. (2015)
	• Restored cholinergic functions and reduced oxidative stress	
	• Decreased the levels of corticosterone and TNF-α	

(continued)

Table 20.1 (continued)

Phytochemicals	Effects and mechanism observed	References
Piperine	• Attenuated the memory loss and oxidative-nitrosative stress	Khalili-Fomeshi et al. (2018); Wang et al. (2019)
	• Decreased malonaldehyde levels in CSF and hippocampus	
	• Restored neurotransmission and reduced neuroinflammation	
S-allyl cysteine	• Improved learning and memory	Javed et al. (2011)
	• Reduced oxidative stress and enhanced endogenous antioxidants	
	• Ameliorated apoptotic parameters	
Hesperidin	• Enhanced learning and memory	Javed et al. (2015)
	• Reduced oxidative stress and enhanced endogenous antioxidants	
	• Prevented neuronal death through modulation of inflammatory markers and modulated acetylcholinesterase activity	
Gingerol	• Improved in cognitive and behavioral performances	Halawany et al. (2017)
	• Decreased cerebral Aβ 1–42, β-secretase, and aph1a activity	
	• Reduced cox-2 linked neuroinflammation	
	• Increased α-secretase activity	
Taraxerol	• Attenuated memory impairment	Berte et al. (2018)
	• Inhibited hippocampal ache activity	

20.8 Conclusion

Based on the literature presented, SAD is classically recognized by abnormal metabolism of cerebral glucose and energy, and consequently, a state of the insulin-resistant brain develops which leads to massive Aβ load, NFTs formation, abnormal memory and cognition. ICV-STZ infusion into rat brain desensitizes neuronal IRs, causing abnormal metabolism of glucose in the brain and long-lasting impairment of learning and cognitive behavior similar to that found in AD patients. So far there are no curative drugs are available or the available drugs still does not provide satisfaction and seems away from perfection in the treatment of AD. Currently available drugs only delay the progression of AD and have numerous side effects. There are urgent requirements to develop a new therapeutic strategy that could slow down, and cure AD. In past few years, many natural products specifically, phytochemicals have been shown to exert pleiotropic pharmacological properties and these pharmacological properties and action could be important in achieving therapeutic benefits in AD therapy. Though, some of them have been

Table 20.2 Plants extract evaluated in intracerebroventricular-streptozotocin-induced animal model of sporadic Alzheimer's disease

Plants Extract	Observed effects/mechanisms	References
Cinnamomum zeylanicum	• Attenuated cognitive deficit	Malik et al. (2015)
	• Restored acetylcholinesterase activity and antioxidants in the brain	
Pterocarpus marsupium	• Decreased Aβ 42, total tau, phosphorylated tau	Kosaraju et al. (2014)
	• Decreased neuroinflammation	
	• Increased glucagon-like peptide-1 (GLP-1)	
	• Enhanced memory	
Eugenia jambolana	• Decreased Aβ 42, total tau, phosphorylated tau	Kosaraju et al. (2014)
	• Decreased neuroinflammation	
	• Increased glucagon-like peptide-1 (GLP-1)	
	• Cognitive enhancer	
Centella asiatica	• Prevented cognitive deficits	Veerendra Kumar and Gupta (2003)
	• Reduced oxidative stress	
	• Enhanced antioxidant enzymes	
Crocus sativus	• Improved cognitive performance	Naghizadeh et al. (2013)
	• Decreased malondialdehyde level	
	• Increased total thiol level and glutathione peroxidase activity	
Commiphora wightii	• Reduced acetylcholinesterase activity	Saxena et al. (2007)
	• Enhanced memory and cognition	
	• Decreased malondialdehyde level and increased glutathione level	
Aegle marmelos	• Improved learning and cognition	Raheja et al. (2019)
	• Reduced oxidative stress and inflammation	
	• Restored cholinergic function	
	• Increased antioxidant enzymes in the hippocampus	
Bacopa monnieir	• Enhanced learning and memory	Khan et al. (2015)
	• Restored cholinergic function	
	• Reduced oxidative stress and inflammation	
	• Increased antioxidant enzymes in the hippocampus	
	• Attenuated the morphological alterations in the hippocampus	
Bergenia ciliata	• Restored cholinergic function	Barai et al. (2018)
	• Attenuated memory loss	
	• Reduced oxidative stress	
	• Increased antioxidant enzymes in the hippocampus	
	• Attenuated the histopathological changes in the hippocampus	
Withania somnifera	• Attenuated learning and cognition	Ahmed et al. (2013)
	• Reduced oxidative stress	
	• Increased antioxidant enzymes in the hippocampus and cerebral cortex	
	• Attenuated the histopathological changes in the hippocampus	

failed and few have shown limited effectiveness in AD therapy in patients. However, still a large number of them shown efficacious in preclinical studies have not been tested in clinical studies and remain to be explored for human use supported by evidence based data. Moreover, further studies are required for these natural products to be tested in earlier phases to unravel their therapeutic potential for AD.

References

Adamo M, Raizada MK, LeRoith D (1989) Insulin and insulin-like growth factor receptors in the nervous system. Mol Neurobiol 3:71–100

Agrawal R, Tyagi E, Shukla R, Nath C (2009) A study of brain insulin receptors, AChE activity and oxidative stress in rat model of ICV STZ induced dementia. Neuropharmacology 56:779–787. https://doi.org/10.1016/j.neuropharm.2009.01.005

Ahmed ME et al (2013) Attenuation of oxidative damage-associated cognitive decline by Withania somnifera in rat model of streptozotocin-induced cognitive impairment. Protoplasma 250:1067–1078. https://doi.org/10.1007/s00709-013-0482-2

Andrade S, Ramalho MJ, Loureiro JA, Pereira MDC (2019) Natural compounds for Alzheimer's disease therapy: a systematic review of preclinical and clinical studies. Int J Molecul Sci 20. https://doi.org/10.3390/ijms20092313

Ardura-Fabregat A et al (2017) Targeting neuroinflammation to treat Alzheimer's disease. CNS Drugs 31:1057–1082. https://doi.org/10.1007/s40263-017-0483-3

Arluison M, Quignon M, Nguyen P, Thorens B, Leloup C, Penicaud L (2004) Distribution and anatomical localization of the glucose transporter 2 (GLUT2) in the adult rat brain--an immunohistochemical study. J Chem Neuroanat 28:117–136. https://doi.org/10.1016/j.jchemneu.2004.05.009

Baluchnejadmojarad T, Roghani M (2006) Effect of naringenin on intracerebroventricular streptozotocin-induced cognitive deficits in rat: a behavioral analysis. Pharmacology 78:193–197. https://doi.org/10.1159/000096585

Banks WA (2004) The source of cerebral insulin. Eur J Pharmacol 490:5–12. https://doi.org/10.1016/j.ejphar.2004.02.040

Barai P, Raval N, Acharya S, Acharya N (2018) Bergenia ciliata ameliorates streptozotocin-induced spatial memory deficits through dual cholinesterase inhibition and attenuation of oxidative stress in rats. Biomed Pharm 102:966–980. https://doi.org/10.1016/j.biopha.2018.03.115

Beal MF (2005a) Mitochondria take center stage in aging and neurodegeneration. Ann Neurol 58:495–505. https://doi.org/10.1002/ana.20624

Beal MF (2005b) Oxidative damage as an early marker of Alzheimer's disease and mild cognitive impairment. Neurobiol Aging 26:585–586. https://doi.org/10.1016/j.neurobiolaging.2004.09.022

Berte TE et al (2018) Taraxerol as a possible therapeutic agent on memory impairments and Alzheimer's disease: effects against scopolamine and streptozotocin-induced cognitive dysfunctions. Steroids 132:5–11. https://doi.org/10.1016/j.steroids.2018.01.002

Biasibetti R et al (2013) Green tea (−)epigallocatechin-3-gallate reverses oxidative stress and reduces acetylcholinesterase activity in a streptozotocin-induced model of dementia. Behav Brain Res 236:186–193. https://doi.org/10.1016/j.bbr.2012.08.039

Blokland A, Jolles J (1993) Spatial learning deficit and reduced hippocampal ChAT activity in rats after an ICV injection of streptozotocin pharmacology. Biochem Behav 44:491–494. https://doi.org/10.1016/0091-3057(93)90497-h

Bose A, Beal MF (2016) Mitochondrial dysfunction in Parkinson's disease. J Neurochem 139 (Suppl 1):216–231. https://doi.org/10.1111/jnc.13731

Bryant NJ, Govers R, James DE (2002) Regulated transport of the glucose transporter GLUT4. Nat Rev Mol Cell Biol 3:267–277. https://doi.org/10.1038/nrm782

Bubber P, Haroutunian V, Fisch G, Blass JP, Gibson GE (2005) Mitochondrial abnormalities in Alzheimer brain: mechanistic implications. Ann Neurol 57:695–703. https://doi.org/10.1002/ana.20474

Burns JM et al (2007) Peripheral insulin and brain structure in early Alzheimer disease. Neurology 69:1094–1104. https://doi.org/10.1212/01.wnl.0000276952.91704.af

Cardoso SM, Santana I, Swerdlow RH, Oliveira CR (2004) Mitochondria dysfunction of Alzheimer's disease cybrids enhances Abeta toxicity. J Neurochem 89:1417–1426. https://doi.org/10.1111/j.1471-4159.2004.02438.x

Cardoso S et al (2009) Insulin is a two-edged knife on the brain. J Alzheimer's Dis 18:483–507. https://doi.org/10.3233/JAD-2009-1155

Caspersen C et al (2005) Mitochondrial Abeta: a potential focal point for neuronal metabolic dysfunction in Alzheimer's disease. FASEB J 19:2040–2041. https://doi.org/10.1096/fj.05-3735fje

Castellani RJ et al (2001) Active glycation in neurofibrillary pathology of Alzheimer disease: N (epsilon)-(carboxymethyl) lysine and hexitol-lysine. Free Radical Biol Med 31:175–180. https://doi.org/10.1016/s0891-5849(01)00570-6

Chandra A, Valkimadi PE, Pagano G, Cousins O, Dervenoulas G, Politis M (2019) Applications of amyloid, tau, and neuroinflammation PET imaging to Alzheimer's disease and mild cognitive impairment human brain mapping 40:5424-5442. https://doi.org/10.1002/hbm.24782

Chen Z, Zhong C (2014) Oxidative stress in Alzheimer's disease. Neuroscience Bull 30:271–281. https://doi.org/10.1007/s12264-013-1423-y

Cole GM, Frautschy SA (2007) The role of insulin and neurotrophic factor signaling in brain aging and Alzheimer's disease. Exp Gerontol 42:10–21. https://doi.org/10.1016/j.exger.2006.08.009

Craft S et al (1999) Enhancement of memory in Alzheimer disease with insulin and somatostatin, but not glucose. Archiv Gen Psychiat. 56:1135–1140. https://doi.org/10.1001/archpsyc.56.12.1135

Cross DA, Alessi DR, Cohen P, Andjelkovich M, Hemmings BA (1995) Inhibition of glycogen synthase kinase-3 by insulin mediated by protein kinase B. Nature 378:785–789. https://doi.org/10.1038/378785a0

Cummings J et al (2016) Drug development in Alzheimer's disease: the path to 2025. Alzheim Res Ther 8:39. https://doi.org/10.1186/s13195-016-0207-9

Czech MP, Corvera S (1999) Signaling mechanisms that regulate glucose transport. J Biol Chem 274:1865–1868. https://doi.org/10.1074/jbc.274.4.1865

Davis S, Laroche S (2006) Mitogen-activated protein kinase/extracellular regulated kinase signalling and memory stabilization: a review. Genes Brain Behav 5(Suppl 2):61–72. https://doi.org/10.1111/j.1601-183X.2006.00230.x

Dawson RM (1970) Development of a school meal service in Newfoundland schools Canadian journal of public health = Revue canadienne de sante publique 61:257–259

Deng Y, Li B, Liu Y, Iqbal K, Grundke-Iqbal I, Gong CX (2009) Dysregulation of insulin signaling, glucose transporters, O-GlcNAcylation, and phosphorylation of tau and neurofilaments in the brain: implication for Alzheimer's disease. Am J Pathol 175:2089–2098. https://doi.org/10.2353/ajpath.2009.090157

Deshmukh R, Kaundal M, Bansal V, Samardeep A (2016) Caffeic acid attenuates oxidative stress, learning and memory deficit in intra-cerebroventricular streptozotocin induced experimental dementia in rats. Biomed Pharmacother 81:56–62. https://doi.org/10.1016/j.biopha.2016.03.017

Doreulee N et al (2003) Cortico-striatal synaptic plasticity in endothelial nitric oxide synthase deficient mice. Brain Res 964:159–163. https://doi.org/10.1016/s0006-8993(02)04121-5

Dudek H et al (1997) Regulation of neuronal survival by the serine-threonine protein kinase Akt. Science 275:661–665. https://doi.org/10.1126/science.275.5300.661

Duelli R, Schrock H, Kuschinsky W, Hoyer S (1994) Intracerebroventricular injection of streptozotocin induces discrete local changes in cerebral glucose utilization in rats. Int J Dev Neurosci 12:737–743

Ejaz Ahmed M et al (2013) Amelioration of cognitive impairment and neurodegeneration by catechin hydrate in rat model of streptozotocin-induced experimental dementia of Alzheimer's type. Neurochem Int 62:492–501. https://doi.org/10.1016/j.neuint.2013.02.006

Elsner M, Guldbakke B, Tiedge M, Munday R, Lenzen S (2000) Relative importance of transport and alkylation for pancreatic beta-cell toxicity of streptozotocin. Diabetologia 43:1528–1533. https://doi.org/10.1007/s001250051564

Erol A (2008) An integrated and unifying hypothesis for the metabolic basis of sporadic Alzheimer's disease. J Alzheimer's Dis 13:241–253

Forni C et al (2019) Beneficial role of phytochemicals on oxidative stress and age-related diseases. BioMed Res Int 2019:8748253. https://doi.org/10.1155/2019/8748253

Frolich L et al (1998) Brain insulin and insulin receptors in aging and sporadic Alzheimer's disease. J Neural Transm (Vienna) 105:423–438. https://doi.org/10.1007/s007020050068

Frolich L, Blum-Degen D, Riederer P, Hoyer S (1999) A disturbance in the neuronal insulin receptor signal transduction in sporadic Alzheimer's disease. Ann N Y Acad Sci 893:290–293. https://doi.org/10.1111/j.1749-6632.1999.tb07839.x

Gasparini L, Xu H (2003) Potential roles of insulin and IGF-1 in Alzheimer's disease trends in neurosciences 26:404-406. https://doi.org/10.1016/S0166-2236(03)00163-2

Gibson GE, Karuppagounder SS, Shi Q (2008) Oxidant-induced changes in mitochondria and calcium dynamics in the pathophysiology of Alzheimer's disease. Ann N Y Acad Sci 1147:221–232. https://doi.org/10.1196/annals.1427.038

Goedert M, Spillantini MG (2006) A century of Alzheimer's disease. Science 314:777–781. https://doi.org/10.1126/science.1132814

Greenough MA, Camakaris J, Bush AI (2013) Metal dyshomeostasis and oxidative stress in Alzheimer's disease. Neurochem Int 62:540–555. https://doi.org/10.1016/j.neuint.2012.08.014

Grunblatt E, Salkovic-Petrisic M, Osmanovic J, Riederer P, Hoyer S (2007) Brain insulin system dysfunction in streptozotocin intracerebroventricularly treated rats generates hyperphosphorylated tau protein. J Neurochem 101:757–770. https://doi.org/10.1111/j.1471-4159.2006.04368.x

Halawany AME, Sayed NSE, Abdallah HM, Dine RSE (2017) Protective effects of gingerol on streptozotocin-induced sporadic Alzheimer's disease: emphasis on inhibition of beta-amyloid, COX-2, alpha-, beta - secretases and APH1a scientific reports 7:2902. https://doi.org/10.1038/s41598-017-02961-0

Havrankova J, Roth J, Brownstein M (1978) Insulin receptors are widely distributed in the central nervous system of the rat. Nature 272:827–829. https://doi.org/10.1038/272827a0

van der Heide LP, Ramakers GM, Smidt MP (2006) Insulin signaling in the central nervous system: learning to survive. Prog Neurobiol 79:205–221. https://doi.org/10.1016/j.pneurobio.2006.06.003

Henneberg N, Hoyer S (1995) Desensitization of the neuronal insulin receptor: a new approach in the etiopathogenesis of late-onset sporadic dementia of the Alzheimer type (SDAT)? Arch Gerontol Geriatr 21:63–74

Hirai K et al (2001) Mitochondrial abnormalities in Alzheimer's disease the journal of neuroscience: the official journal of the society for. Neuroscience 21:3017–3023

Howes MJ, Perry E (2011) The role of phytochemicals in the treatment and prevention of dementia Drugs & aging 28:439–468. https://doi.org/10.2165/11591310-000000000-00000

Hoyer S (2002) The brain insulin signal transduction system and sporadic (type II) Alzheimer disease: an update. J Neural Transm (Vienna). 109:341–360. https://doi.org/10.1007/s007020200028

Hoyer S (2004a) Causes and consequences of disturbances of cerebral glucose metabolism in sporadic Alzheimer disease: therapeutic implications. Adv Exp Med Biol 541:135–152. https://doi.org/10.1007/978-1-4419-8969-7_8

Hoyer S (2004b) Glucose metabolism and insulin receptor signal transduction in Alzheimer disease. Eur J Pharmacol 490:115–125. https://doi.org/10.1016/j.ejphar.2004.02.049

Hoyer S, Nitsch R (1989) Cerebral excess release of neurotransmitter amino acids subsequent to reduced cerebral glucose metabolism in early-onset dementia of Alzheimer type. J Neural Transm 75:227–232

Ishrat T et al (2006) Coenzyme Q10 modulates cognitive impairment against intracerebroventricular injection of streptozotocin in rats. Behav Brain Res 171:9–16. https://doi.org/10.1016/j.bbr.2006.03.009

Javed H et al (2011) S-allyl cysteine attenuates oxidative stress associated cognitive impairment and neurodegeneration in mouse model of streptozotocin-induced experimental dementia of Alzheimer's type. Brain Res 1389:133–142. https://doi.org/10.1016/j.brainres.2011.02.072

Javed H et al (2012) Rutin prevents cognitive impairments by ameliorating oxidative stress and neuroinflammation in rat model of sporadic dementia of Alzheimer type. Neuroscience 210: 340–352. https://doi.org/10.1016/j.neuroscience.2012.02.046

Javed H, Vaibhav K, Ahmed ME, Khan A, Tabassum R, Islam F, Safhi MM (2015) Effect of hesperidin on neurobehavioral, neuroinflammation, oxidative stress and lipid alteration in intracerebroventricular streptozotocin induced cognitive impairment in mice. J Neurol Sci 348:51–59. https://doi.org/10.1016/j.jns.2014.10.044

Jha AB, Panchal SS, Shah A (2018) Ellagic acid: insights into its neuroprotective and cognitive enhancement effects in sporadic Alzheimer's disease pharmacology. Biochem Behav 175:33–46. https://doi.org/10.1016/j.pbb.2018.08.007

Johnston AM, Pirola L, Van Obberghen E (2003) Molecular mechanisms of insulin receptor substrate protein-mediated modulation of insulin signalling. FEBS Lett 546:32–36. https://doi.org/10.1016/s0014-5793(03)00438-1

Kahn CR, White MF (1988) The insulin receptor and the molecular mechanism of insulin action. J Clin Invest 82:1151–1156. https://doi.org/10.1172/JCI113711

Kamat PK, Kalani A, Rai S, Swarnkar S, Tota S, Nath C, Tyagi N (2016) Mechanism of oxidative stress and synapse dysfunction in the pathogenesis of Alzheimer's disease: understanding the therapeutics strategies. Mol Neurobiol 53:648–661. https://doi.org/10.1007/s12035-014-9053-6

Kennedy DO, Wightman EL (2011) Herbal extracts and phytochemicals: plant secondary metabolites and the enhancement of human brain function. Adv Nutr 2:32–50. https://doi.org/10.3945/an.110.000117

Khalili M, Hamzeh F (2010) Effects of active constituents of Crocus sativus L., crocin on streptozocin-induced model of sporadic Alzheimer's disease in male rats. Iran Biomed J 14: 59–65

Khalili-Fomeshi M et al (2018) Piperine restores streptozotocin-induced cognitive impairments: insights into oxidative balance in cerebrospinal fluid and hippocampus. Behav Brain Res 337: 131–138. https://doi.org/10.1016/j.bbr.2017.09.031

Khan MB et al (2012) Naringenin ameliorates Alzheimer's disease (AD)-type neurodegeneration with cognitive impairment (AD-TNDCI) caused by the intracerebroventricular-streptozotocin in rat model. Neurochem Int 61:1081–1093. https://doi.org/10.1016/j.neuint.2012.07.025

Khan MB, Ahmad M, Ahmad S, Ishrat T, Vaibhav K, Khuwaja G, Islam F (2015) Bacopa monniera ameliorates cognitive impairment and neurodegeneration induced by intracerebroventricular-streptozotocin in rat: behavioral, biochemical, immunohistochemical and histopathological evidences metabolic brain disease 30:115-127. doi:https://doi.org/10.1007/s11011-014-9593-5

Kidd M (1964) Alzheimer's disease--an electron Microscopical study. Brain J Neurol 87:307-320 doi:https://doi.org/10.1093/brain/87.2.307

Kosaraju J, Madhunapantula SV, Chinni S, Khatwal RB, Dubala A, Muthureddy Nataraj SK, Basavan D (2014) Dipeptidyl peptidase-4 inhibition by Pterocarpus marsupium and Eugenia jambolana ameliorates streptozotocin induced Alzheimer's disease. Behav Brain Res. 267:55–65. https://doi.org/10.1016/j.bbr.2014.03.026

Kouhestani S, Jafari A, Babaei P (2018) Kaempferol attenuates cognitive deficit via regulating oxidative stress and neuroinflammation in an ovariectomized rat model of sporadic dementia. Neural Regen Res. 13:1827–1832. https://doi.org/10.4103/1673-5374.238714

Kumar M, Bansal N (2018a) Caffeic acid phenethyl ester rescued streptozotocin-induced memory loss through PI3-kinase dependent pathway. Biomed Pharmacother 101:162-173. https://doi.org/10.1016/j.biopha.2018.02.089

Kumar M, Bansal N (2018b) Ellagic acid prevents dementia through modulation of PI3-kinase-endothelial nitric oxide synthase signalling in streptozotocin-treated rats. Naunyn-Schmiedeberg's Archiv Pharmacol 391:987-1001. https://doi.org/10.1007/s00210-018-1524-2

Kumar A, Ekavali A, Mishra J, Chopra K, Dhull DK (2016) Possible role of P-glycoprotein in the neuroprotective mechanism of berberine in intracerebroventricular streptozotocin-induced cognitive dysfunction. Psychopharmacology 233:137–152. https://doi.org/10.1007/s00213-015-4095-7

Kumar M, Kaur D, Bansal N (2017) Caffeic acid phenethyl ester (CAPE) prevents development of STZ-ICV induced dementia in rats. Pharmacognosy Magaz 13:S10-S15 doi:https://doi.org/10.4103/0973-1296.203974

Lannert H, Hoyer S (1998) Intracerebroventricular administration of streptozotocin causes long-term diminutions in learning and memory abilities and in cerebral energy metabolism in adult rats. Behav Neurosci 112:1199–1208. https://doi.org/10.1037//0735-7044.112.5.1199

Lester-Coll N, Rivera EJ, Soscia SJ, Doiron K, Wands JR, de la Monte SM (2006) Intracerebral streptozotocin model of type 3 diabetes: relevance to sporadic Alzheimer's disease. J Alzhei Dis 9:13–33

Liu G et al (2006) Metal exposure and Alzheimer's pathogenesis. J Struct Biol 155:45–51. https://doi.org/10.1016/j.jsb.2005.12.011

Liu Y, Liu F, Iqbal K, Grundke-Iqbal I, Gong CX (2008) Decreased glucose transporters correlate to abnormal hyperphosphorylation of tau in Alzheimer disease. FEBS Lett 582:359–364. https://doi.org/10.1016/j.febslet.2007.12.035

Luse SA, Smith KR Jr (1964) The ultrastructure of senile plaques. Am J Pathol 44:553–563

Malik J, Munjal K, Deshmukh R (2015) Attenuating effect of standardized lyophilized Cinnamomum zeylanicum bark extract against streptozotocin-induced experimental dementia of Alzheimer's type. J Basic Clin Physiol Pharmacol 26:275–285. https://doi.org/10.1515/jbcpp-2014-0012

Man HY et al (2000) Regulation of AMPA receptor-mediated synaptic transmission by clathrin-dependent receptor internalization. Neuron 25:649–662. https://doi.org/10.1016/s0896-6273(00)81067-3

Manach C, Hubert J, Llorach R, Scalbert A (2009) The complex links between dietary phytochemicals and human health deciphered by metabolomics Molecular nutrition & food research 53:1303–1315. https://doi.org/10.1002/mnfr.200800516

Manczak M, Anekonda TS, Henson E, Park BS, Quinn J, Reddy PH (2006) Mitochondria are a direct site of a beta accumulation in Alzheimer's disease neurons: implications for free radical generation and oxidative damage in disease progression. Hum Mol Genet 15:1437–1449. https://doi.org/10.1093/hmg/ddl066

Mansouri MT, Naghizadeh B, Ghorbanzadeh B, Farbood Y, Sarkaki A, Bavarsad K (2013) Gallic acid prevents memory deficits and oxidative stress induced by intracerebroventricular injection of streptozotocin in rats pharmacology, biochemistry, and behavior 111:90-96. https://doi.org/10.1016/j.pbb.2013.09.002

McEwen BS, Reagan LP (2004) Glucose transporter expression in the central nervous system: relationship to synaptic function. Eur J Pharmacol 490:13–24. https://doi.org/10.1016/j.ejphar.2004.02.041

McGeer PL, McGeer EG, Suzuki J, Dolman CE, Nagai T (1984) Aging, Alzheimer's disease, and the cholinergic system of the basal forebrain. Neurology 34:741–745. https://doi.org/10.1212/wnl.34.6.741

Montagnani M, Chen H, Barr VA, Quon MJ (2001) Insulin-stimulated activation of eNOS is independent of Ca2+ but requires phosphorylation by Akt at Ser(1179). J Biol Chem:276:30392-30398. https://doi.org/10.1074/jbc.M10372200

de la Monte SM, Tong M (2014) Brain metabolic dysfunction at the core of Alzheimer's disease. Biochem Pharmacol 88:548–559. https://doi.org/10.1016/j.bcp.2013.12.012

de la Monte SM, Wands JR (2006) Molecular indices of oxidative stress and mitochondrial dysfunction occur early and often progress with severity of Alzheimer's disease. J Alzheimer's Dis 9:167–181

Moreira PI, Santos MS, Oliveira CR (2007) Alzheimer's disease: a lesson from mitochondrial dysfunction. Antioxid Redox Signal 9:1621–1630. https://doi.org/10.1089/ars.2007.1703

Moreira PI, Duarte AI, Santos MS, Rego AC, Oliveira CR (2009) An integrative view of the role of oxidative stress, mitochondria and insulin in Alzheimer's disease. J Alzheimer's Dis 16:741. https://doi.org/10.3233/JAD-2009-0972

Naghizadeh B, Mansouri MT, Ghorbanzadeh B, Farbood Y, Sarkaki A (2013) Protective effects of oral crocin against intracerebroventricular streptozotocin-induced spatial memory deficit and oxidative stress in rats. Phytomedicine 20:537–542. https://doi.org/10.1016/j.phymed.2012.12.019

Naik B, Nirwane A, Majumdar A (2017) Pterostilbene ameliorates intracerebroventricular streptozotocin induced memory decline in rats Cognitive neurodynamics 11:35–49. https://doi.org/10.1007/s11571-016-9413-1

Nakajima K, Kohsaka S (2001) Microglia: activation and their significance in the central nervous system. J Biochem 130:169–175. https://doi.org/10.1093/oxfordjournals.jbchem.a002969

Nitsch R, Hoyer S (1991) Local action of the diabetogenic drug, streptozotocin, on glucose and energy metabolism in rat brain cortex. Neurosci Lett 128:199-202. https://doi.org/10.1016/0304-3940(91)90260-z

Pei JJ et al (2003) Role of protein kinase B in Alzheimer's neurofibrillary pathology. Acta Neuropathol 105:381–392. https://doi.org/10.1007/s00401-002-0657-y

Phiel CJ, Wilson CA, Lee VM, Klein PS (2003) GSK-3alpha regulates production of Alzheimer's disease amyloid-beta peptides. Nature 423:435-439 doi:https://doi.org/10.1038/nature01640

Plaschke K, Hoyer S (1993) Action of the diabetogenic drug streptozotocin on glycolytic and glycogenolytic metabolism in adult rat brain cortex and hippocampus. Int J Dev Neurosci 11:477–483

Poprac P, Jomova K, Simunkova M, Kollar V, Rhodes CJ, Valko M (2017) Targeting free radicals in oxidative stress-related human diseases. Trends Pharmacol Sci 38:592–607. https://doi.org/10.1016/j.tips.2017.04.005

Querfurth HW, LaFerla FM (2010) Alzheimer's disease. N Engl J Med 362:329–344. https://doi.org/10.1056/NEJMra0909142

Raheja S, Girdhar A, Kamboj A, Lather V, Pandita D (2019) Aegle marmelos leaf extract ameliorates the cognitive impairment and oxidative stress induced by intracerebroventricular streptozotocin in male rats. Life Sci 221:196–203. https://doi.org/10.1016/j.lfs.2019.02.032

Ramezani M, Darbandi N, Khodagholi F, Hashemi A (2016) Myricetin protects hippocampal CA3 pyramidal neurons and improves learning and memory impairments in rats with Alzheimer's disease. Neural Regener Res 11:1976-1980 doi:https://doi.org/10.4103/1673-5374.197141

Rapoport SI (2003) Coupled reductions in brain oxidative phosphorylation and synaptic function can be quantified and staged in the course of Alzheimer disease. Neurotox Res 5:385–398

Reddy PH et al (2012) Abnormal mitochondrial dynamics and synaptic degeneration as early events in Alzheimer's disease: implications to mitochondria-targeted antioxidant therapeutics. Biochim Biophys Acta 1822:639–649. https://doi.org/10.1016/j.bbadis.2011.10.011

Reiman EM et al (2004) Functional brain abnormalities in young adults at genetic risk for late-onset Alzheimer's dementia. Proc Natl Acad Sci U S A 101:284–289. https://doi.org/10.1073/pnas.2635903100

Rivera EJ, Goldin A, Fulmer N, Tavares R, Wands JR, de la Monte SM (2005) Insulin and insulin-like growth factor expression and function deteriorate with progression of Alzheimer's disease: link to brain reductions in acetylcholine. J Alzheimer's Dis 8:247–268

Rocchi A, Pellegrini S, Siciliano G, Murri L (2003) Causative and susceptibility genes for Alzheimer's disease: a review. Brain Res Bull 61:1–24. https://doi.org/10.1016/s0361-9230(03)00067-4

Sachdeva AK, Kuhad A, Chopra K (2014) Naringin ameliorates memory deficits in experimental paradigm of Alzheimer's disease by attenuating mitochondrial dysfunction. Pharmacol Biochem Behav 127:101-110. https://doi.org/10.1016/j.pbb.2014.11.002

Salkovic-Petrisic M, Hoyer S (2007) Central insulin resistance as a trigger for sporadic Alzheimer-like pathology: an experimental approach. J Neural Transm Suppl :217–233

Saltiel AR, Pessin JE (2002) Insulin signaling pathways in time and space. Trends Cell Biol 12:65–71

Saxena G, Singh SP, Pal R, Singh S, Pratap R, Nath C (2007) Gugulipid, an extract of Commiphora whighitii with lipid-lowering properties, has protective effects against streptozotocin-induced memory deficits in mice. Pharmacol Biochem Behav 86:797-805. https://doi.org/10.1016/j.pbb.2007.03.010

Scheepers A, Joost HG, Schurmann A (2004) The glucose transporter families SGLT and GLUT: molecular basis of normal and aberrant function. J Parent Enter Nutr 28:364–371 doi:https://doi.org/10.1177/0148607104028005364

Silverman DH et al (2001) Positron emission tomography in evaluation of dementia: Regional brain metabolism and long-term outcome. JAMA 286:2120–2127. https://doi.org/10.1001/jama.286.17.2120

Singh JC, Kakalij RM, Kshirsagar RP, Kumar BH, Komakula SS, Diwan PV (2015) Cognitive effects of vanillic acid against streptozotocin-induced neurodegeneration in mice. Pharmaceut Biol 53:630–636. https://doi.org/10.3109/13880209.2014.935866

Sodhi RK, Singh N (2013) All-trans retinoic acid rescues memory deficits and neuropathological changes in mouse model of streptozotocin-induced dementia of Alzheimer's type. Prog Neuro-Psychopharmacol Biol Psychiatry 40:38–46. https://doi.org/10.1016/j.pnpbp.2012.09.012

Sofowora A, Ogunbodede E, Onayade A (2013) The role and place of medicinal plants in the strategies for disease prevention. Afr J Trad Complem Altern Med 10:210-229 doi:https://doi.org/10.4314/ajtcam.v10i5.2

Solano DC, Sironi M, Bonfini C, Solerte SB, Govoni S, Racchi M (2000) Insulin regulates soluble amyloid precursor protein release via phosphatidyl inositol 3 kinase-dependent pathway. FASEB J 14:1015-1022. https://doi.org/10.1096/fasebj.14.7.1015

Straface E, Matarrese P, Gambardella L, Vona R, Sgadari A, Silveri MC, Malorni W (2005) Oxidative imbalance and cathepsin D changes as peripheral blood biomarkers of Alzheimer disease: a pilot study. FEBS Lett 579:2759–2766. https://doi.org/10.1016/j.febslet.2005.03.094

Sultana R, Mecocci P, Mangialasche F, Cecchetti R, Baglioni M, Butterfield DA (2011) Increased protein and lipid oxidative damage in mitochondria isolated from lymphocytes from patients with Alzheimer's disease: insights into the role of oxidative stress in Alzheimer's disease and initial investigations into a potential biomarker for this dementing disorder. J Alzheimer's Dis 24:77–84. https://doi.org/10.3233/JAD-2011-101425

Szkudelski T (2001) The mechanism of alloxan and streptozotocin action in B cells of the rat pancreas Physiological research 50:537–546

Tonnies E, Trushina E (2017) Oxidative stress, synaptic dysfunction, and Alzheimer's disease. J Alzheimer's Dis 57:1105–1121 doi:https://doi.org/10.3233/JAD-161088

Tota S, Awasthi H, Kamat PK, Nath C, Hanif K (2010) Protective effect of quercetin against intracerebral streptozotocin induced reduction in cerebral blood flow and impairment of memory in mice. Behav Brain Res 209:73–79. https://doi.org/10.1016/j.bbr.2010.01.017

Trushina E et al. (2012) Defects in mitochondrial dynamics and metabolomic signatures of evolving energetic stress in mouse models of familial Alzheimer's disease. PLoS One 7:e32737 doi: https://doi.org/10.1371/journal.pone.0032737

Trushina E, Dutta T, Persson XM, Mielke MM, Petersen RC (2013) Identification of altered metabolic pathways in plasma and CSF in mild cognitive impairment and Alzheimer's disease using metabolomics. PLoS One 8:e63644 doi:https://doi.org/10.1371/journal.pone.0063644

Tumminia A, Vinciguerra F, Parisi M, Frittitta L (2018) Type 2 diabetes mellitus and Alzheimer's disease: role of insulin signalling and therapeutic implications. Int J Mol Sci 19 doi:https://doi.org/10.3390/ijms19113306

Unger J, McNeill TH, Moxley RT 3rd, White M, Moss A, Livingston JN (1989) Distribution of insulin receptor-like immunoreactivity in the rat forebrain. Neuroscience 31:143–157. https://doi.org/10.1016/0306-4522(89)90036-5

Upadhyay S, Dixit M (2015) Role of polyphenols and other phytochemicals on molecular signaling oxidative medicine and cellular longevity 2015:504253 doi:https://doi.org/10.1155/2015/504253

Valla J et al. (2010) Reduced posterior cingulate mitochondrial activity in expired young adult carriers of the APOE epsilon4 allele, the major late-onset Alzheimer's susceptibility gene. J Alzheimer's Dis. 22:307-313 doi:https://doi.org/10.3233/JAD-2010-100129

Vannucci SJ, Maher F, Simpson IA (1997) Glucose transporter proteins in brain: delivery of glucose to neurons and glia. Glia 21:2–21

Veerendra Kumar MH, Gupta YK (2003) Effect of Centella asiatica on cognition and oxidative stress in an intracerebroventricular streptozotocin model of Alzheimer's disease in rats. Clin Exp Pharmacol Physiol 30:336–342 doi:https://doi.org/10.1046/j.1440-1681.2003.03842.x

Vekrellis K et al (2000) Neurons regulate extracellular levels of amyloid beta-protein via proteolysis by insulin-degrading enzyme. J Neuroscience 20:1657–1665

Wan Q et al. (1997) Recruitment of functional GABA(a) receptors to postsynaptic domains by insulin nature 388:686-690 doi:https://doi.org/10.1038/41792

Wang H, Cheng H, Che Z (2016) Ameliorating effect of luteolin on memory impairment in an Alzheimer's disease model Molecular medicine reports 13:4215–4220. https://doi.org/10.3892/mmr.2016.5052

Wang Q, Tian J, Chen H, Du H, Guo L (2019) Amyloid beta-mediated KIF5A deficiency disrupts anterograde axonal mitochondrial movement. Neurobiol Dis 127:410-418. https://doi.org/10.1016/j.nbd.2019.03.021

Watson GS, Craft S (2004) Modulation of memory by insulin and glucose: neuropsychological observations in Alzheimer's disease. Eur J Pharmacol 490:97–113. https://doi.org/10.1016/j.ejphar.2004.02.048

Watson GS et al (2003) Insulin increases CSF Abeta42 levels in normal older adults. Neurology 60: 1899-1903. https://doi.org/10.1212/01.wnl.0000065916.25128.25

Wozniak M, Rydzewski B, Baker SP, Raizada MK (1993) The cellular and physiological actions of insulin in the central nervous system. Neurochem Int 22:1–10. https://doi.org/10.1016/0197-0186(93)90062-a

Xia Z, Dickens M, Raingeaud J, Davis RJ, Greenberg ME (1995) Opposing effects of ERK and JNK-p38 MAP kinases on apoptosis. Science 270:1326-1331. https://doi.org/10.1126/science.270.5240.1326

Yang W, Ma J, Liu Z, Lu Y, Hu B, Yu H (2014) Effect of naringenin on brain insulin signaling and cognitive functions in ICV-STZ induced dementia model of rats. Neurol Sci 35:741–751. https://doi.org/10.1007/s10072-013-1594-3

Yiannopoulou KG, Papageorgiou SG (2013) Current and future treatments for Alzheimer's disease. Ther Adv Neurol Disord 6:19–33 doi:https://doi.org/10.1177/1756285612461679

Zafeer MF, Firdaus F, Anis E, Mobarak Hossain M (2019) Prolong treatment with trans-ferulic acid mitigates bioenergetics loss and restores mitochondrial dynamics in streptozotocin-induced sporadic dementia of Alzheimer's type. Neurotoxicology 73:246–257. https://doi.org/10.1016/j.neuro.2019.04.006

Zhao WQ, Chen H, Quon MJ, Alkon DL (2004) Insulin and the insulin receptor in experimental models of learning and memory. Eur J Pharmacol 490:71–81. https://doi.org/10.1016/j.ejphar.2004.02.045

Zhao SS, Yang WN, Jin H, Ma KG, Feng GF (2015) Puerarin attenuates learning and memory impairments and inhibits oxidative stress in STZ-induced SAD mice. Neurotoxicology 51:166–171. https://doi.org/10.1016/j.neuro.2015.10.010

Concept of Amnesia and Dementia in the Unani System of Medicine

21

Mohammad Zakir, Safiya Khanam, Md Tanwir Alam, Aisha Perveen, and Munawwar Husain Kazmi

Abstract

Cerebrasthenia (Du'f al-Dimāgh), loss or invalidity of intelligence (Batlan-i Aql), Humq and Nisyān, etc., are described as amnesia and dementia in the Unani system of medicine (USM). The causes of these diseases are predominance of Burūdat (coldness) and Rutūbat (moistness) in most of the cases. Inflammation and injury to the brain and accumulation of impaired humor (Khilt bad) and excess fluids in cavities or brain tissues/nerves are also described as additional causes of dementia. The preventive methods and management of Amnesia and Dementia by various modules are given in classical literature of the Unani medicine. As per principles of treatment (Usūl-i'Ilāj) the emphasis is given to remove the underlying cause of the disease. In case of impairment of temperament (Su'-i-Mizāj), the method of temperamental equilibrium (Ta'dil-o-Tabdil-i-Mizāj) is applied to correct it. The method of concoctive (Mundij) and purgative (Mushil) has been applied in case of involvement of humor (Khilt) or active substance (Mādda). External methods such as liniment (Tila), irrigation (Natūl), fumigation (Dhūnī), Turkish bath (Hammām), and sitz bath (Abzan), etc. are also advised as per need of treatment. In this chapter, the concept of amnesia and dementia is discussed in the light of USM.

M. Zakir (✉) · S. Khanam · M. H. Kazmi
National Research Institute of Unani Medicine for Skin Disorders (NRIUM-SD), Hyderabad, India

M. T. Alam
Department of Preventive & Community Medicine, Govt. Tibbi College & Hospital, Patna, India

A. Perveen
SRF, Regional Research Institute of Unani Medicine, Patna, India

G. M. Ashraf, M. S. Uddin (eds.), *Current Thoughts on Dementia*,
https://doi.org/10.1007/978-981-16-7606-2_21

Keywords

Amnesia · Dementia · Du'f al-Dimagh · Faramoshi · Humq · Nisyan · Forgetfulness · Unani

21.1 Introduction

Dementia is a syndrome with various causes and is defined as an acquired deterioration in cognitive abilities that impairs the successful performance of daily activities of humans. Alzheimer's disease (AD) a chronic neurodegenerative disease is the most common (60–70%) cause of dementia while vascular diseases have been considered as the second common cause. Other causes of dementia are Lewy bodies (DLB), mixed dementia, frontotemporal lobar degeneration (FTLD), Parkinson's disease (PD), Creutzfeldt-Jakob disease (CJD), Normal pressure hydrocephalus, etc. Various cognitive abilities can be impaired with dementia, including memory, language, reasoning, decision making, visuospatial function, attention, and orientation. Importantly, the cognitive and behavioral changes that occur with dementia interfere with work, social activities, and relationships and impair a person's ability to perform routine daily activities (e.g., driving, shopping, housekeeping, cooking, managing finances, and personal health care) (Anonymous 2019; Mc Khann et al. 2011; Kasper et al. 2019; Gilman 2010).

The major goals for the management of dementia are to treat the reversible causes and to provide comfort and support to the patient and care given. The US FDA has approved six drugs for the treatment of AD such as rivastigmine, galantamine, donepezil, memantine, memantine combined with donepezil, and tacrine (now discontinued in the USA). With the exception of memantine, these drugs temporarily improve symptoms by increasing the amount of chemicals called neurotransmitters in the brain. Memantine blocks certain receptors in the brain from excess stimulation that can damage nerve cells. Each of these compounds has only modest efficacy for AD. Cholinesterase inhibitors are relatively easy to administer, and their major side effects are gastrointestinal symptoms (nausea, diarrhea, cramps), altered sleep with unpleasant or vivid dreams, bradycardia (usually benign), and muscle cramps. There is still no effective and safe management for AD, despite hundreds of millions of dollars spent on AD research. It is the need of the hour to adopt the alternative systems of medicines for the safe and effective treatment that could be used as adjuvant treatment to conventional therapy (Anonymous 2019; Kasper et al. 2019).

The treatment modalities in Unani medicine include regimenal therapy ('Ilāj bi'l Tadbīr), dietotherapy ('Ilaj bi'l Ghizā), drug therapy ('Ilāj bi'd-Dawā), and surgical treatment ('Ilāj bi'l Yad). Regimenal therapy ('Ilāj bi'l Tadbīr) is used to correct the underlying cause of the disease by exercise, cupping (Hijama), venesection (Fasd), cauterization (Kaiyy), Turkish bath (Hammam) and other methods. Dietotherapy ('Ilaj bi'l Ghizā) is the cure of disease with the help of diet/food. It is a type of therapy which is given at first to maintain the health of a person. Modification in diet or use of specific diet having opposite qualities/Mizaj of prevailing disease condition is advised. It is advocated that initially in management of disease "Ilaj bi'l-Ghiza" should be given priority. Drug therapy ('Ilāj bi'd-Dawā) is the cure of disease with the help of drugs. The core concept of management is heteropathy ('Ilāj bi'l Didd) where drugs having opposite qualities/temperament (Mizaj) to the prevailing one in the diseases have to be used. It is used to correct the impairment in humors (Akhlat). If a disease is caused by morbidity of hot humor, drugs having cold temperament should be used. Surgical treatment ('Ilāj bi'l Yad) is the treatment by the surgery and it is placed on last place and should be used only when all methods fails (Ibn Sina 2014; Jurjani 2008; Khan 1987; Masihi 2008).

The management of diseases associated with brain and related organ is described in detail in classical literature of the Unani system of medicine (USM). The impairment of phlegmatic temperament (Balghami al-Mizaj) has been considered the main cause of diseases in brain and nervous system. The management of the impairment of phlegmatic temperament is based on Mundij (concoctive) and Mushil (purgative) therapy, a core principle of the treatment. The USM has great potential to offer safe and effective treatment for various brain and neurological disorders through its holistic approach (Ibn Sina 2014; Razi 1997; Ibn Rushd 1984). In this chapter authors have described the types, etiology as well as safe and effective management of amnesia, and dementia as per Unani concept.

21.2 Prevalence of Amnesia and Dementia

Approximately 50 million people worldwide have dementia; the total number of people with dementia is projected to reach 82 million in 2030 and 152 in 2050. Dementia is now the fifth leading cause of death. In 2015, the total global societal cost of dementia was estimated to be US $ 818 billion, equivalent to 1.1% of global gross domestic product (GDP) (World Health Organization [WHO] 2019a, 2019b). The prevalence of AD is higher among females, reflecting the longer life expectancy of women (Hebert et al. 2001). An estimated 5.8 million Americans of all ages are living with Alzheimer's dementia in 2019. This number includes an estimated 5.6 million people age 65 and older. One in 10 people (10%) age 65 and older has Alzheimer's dementia. The percentage of people with Alzheimer's dementia increases with age: 3% of people age 65–74, 17% of people age 75–84, and 32% of people age 85 and older have Alzheimer's dementia. Out of the 5.8 million people who have Alzheimer's dementia, 81% of them are of age 75 or older (Hebert et al. 2013; Barberger-Gateau et al. 2007; Tom et al. 2015; Rajan et al. 2019).

21.3 The Basic Concept of Unani System of Medicine

According to the USM temperament (Mizāj) is a quality that is produced by action and reaction of opposite qualities of elements which are broken up in small particles in order to facilitate their mixing. When these components interact among themselves, a condition is produced, which is found in equal proportion in all the particles of the compound. This new formation is known as temperament of the compound. The temperament is a resultant state of equilibrium or homeostasis emerging after the combination of more than one element (Ibn Sina 2014; Masihi 2008).

Human body contains four humors (Akhlāt), i.e. Sanguine (Dam), Phlegm (Balgham), Yellow bile (Safrā'), and Black bile (Sawdā'). Sanguine has hot and wet temperament (Mizāj Harr Ratb), Phlegm bears cold and wet temperament (Mizāj Barid Ratb), Yellow bile (Safrā') bears hot and dry temperament (Mizāj Harr Yabis), and Black bile has cold and dry temperament (Mizāj Barid & Yabis). Predominance of any one of the four humors in the human body determines the temperament/ constitution (Mizāj) of the human being (Ahmad 1983; Kabir 2014a, b; Razi 1991). The individuals with predominance of sanguine in the body have sanguineous temperament (Damwī al-Mizāj) so their temperament is hot and wet. Phlegmatic temperament (Balghami al-Mizāj) is due to predominance of phlegm in the body so their temperament is cold and wet. Predominance of yellow bile in the body results in bilious temperament (Safrāwī al-Mizāj) so their temperament is hot and dry. Melancholic temperament (Sawdāwī al-Mizāj) is due to predominance of black bile so their temperament is cold and dry (Ahmad 1983; Azmi 1995).

The temperament which is inherited by the progeny and prevails throughout the life as identity and is not likely to change is called basic temperament (Mizāj Awwal). While secondary or acquired temperament (Mizāj Thānī) is the temperament which develops in an individual under the influence of efficient causes like climate, food, and drinks, etc. (Jurjani 2008; Ahmad 1983). Moderate temperament (Mizāj Mu'tadil Tibbi) is a temperament wherein the qualities and quantities of all the participating elements in a compound are in accordance with what that compound is made for. Imbalance or alteration in temperament due to disturbed equilibrium of humors (Akhlāt) leads to imbalance in homeostasis followed by disease condition in human being (Ahmad 1983; Masihi 2008).

The secondary temperament differs in an individual in different phases of life. In growing age (Sinn-i-Numū) that extends from the birth up to thirty years body has hot and moist temperament (Mizāj Hārr Ratb). The adulthood (Sinn-i-Shabāb or Sinn-i-Wuqūf), which is from 30 to 40 years of age has moderate hot and moist temperament (Mizāj Hārr Ratb Mu'tadil). In this age body becomes fully mature and stable and has the best temperament. The age of decline (Sinn-i-Kuhūlat or Sinn-i-Inhitāt) is the period between forty to sixty years. In this age the body loses its stability and deterioration starts. Innate heat of the body gradually decreases and the body develops cold and dry temperament (Mizāj Bārid Yābis). The age of superannuation (Sinn-i-Shaykhūkhat) is the age that extends from the age of 60 years till death. In this age body heat further decreases and body develops cold and dry temperament (Mizāj Bārid Yābis) (Ahmad 1983; Kabir 2014a, b; Ibn Sina 2014) (Table 21.1; Fig. 21.1).

Every organ in the body has its own temperament and possesses specific nature according to its temperament. Some organs are hot (Hārr) in nature like heart, liver, etc.; some are cold (Bārid) in nature, i.e. bone, cartilage, etc.; some have moist (Ratb) nature, i.e. membranes, glands, nerves, etc. while some have dry (Yābis) nature, i.e. hair, teeth, tendon, etc. these nature best suits for these organs and are necessary

Table 21.1 Acquired temperament (Mizāj Thānī) according to the age or stage of human life (Jurjani 2008; Masihi 2008)

Stage in life	Period	Temperament (Mizāj)
Growing age (Sinn-i-Numū)	Birth to 30 years	Hot and moist temperament (Mizāj Hārr Ratb)
Adulthood (Sinn-i-Shabāb or Sinn-i-Wuqūf)	30–40 years	Moderate hot and moist temperament (Mizāj Hārr Ratb mu'tadil)
Age of decline (Sinn-i-Kuhūlat or Sinn-i-Inhitāt)	40–60 years	Cold and dry temperament (Mizāj Bārid Yābis)
Age of superannuation (Sinn-i-Shaykhūkhat)	60 till death	Cold and dry temperament (Mizāj Bārid Yābis)

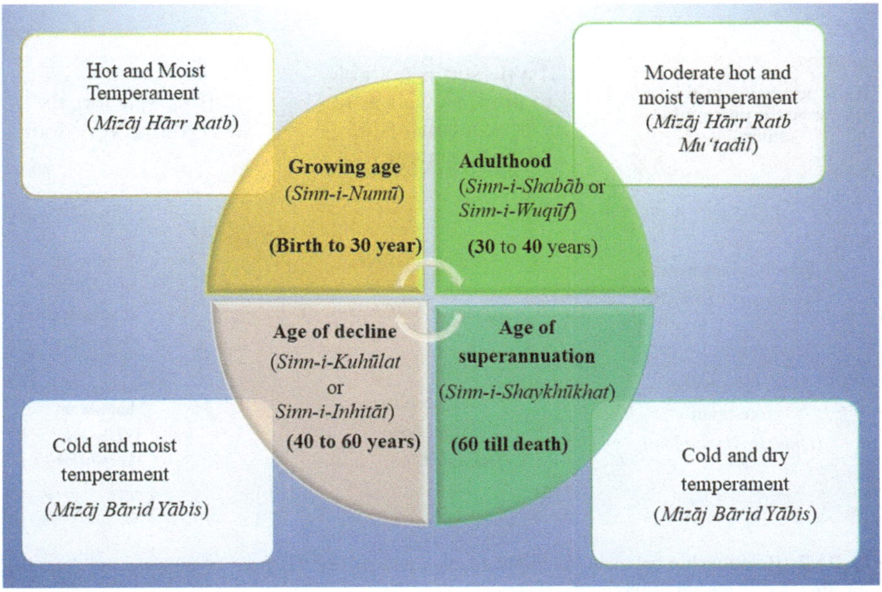

Fig. 21.1 Acquired temperament according to the stages of the human life (Jurjani 2008; Masihi 2008)

for the normal functions of these organs. If the temperament of any organ deviates from its original, then the function of that organ also disturbs. The organs which are moderate temperament (Mu'tadil Mizāj) in respect of heat, cold, moistness, and dryness called moderate organ (A'dā' Mu'tadila), skin is the best example of it. Dimāgh (brain) is one of the vital organs of the body. It is the seat of mental faculties, sensation, and movement. It has cold and moist temperament (Bārid Ratb Mizāj) (Ibn Sina 2014).

Brain (Dimāgh) has different faculties for doing its various functions effectively as mentioned here. Hiss Mushtarak (composite sense/common sense) is the power of perception; it receives all sensations, compiles them into percepts, and enables proper sensory perception. Quwwat-i-Khayāl (faculty of awareness) preserves the

Table 21.2 Association of the faculties of the brain with affected parts (Ibn Sina 2014; Jurjani 2008)

Inability/disease	Faculty associated	Part of Brain affected
If the person is not able to preserve the perceptible things	Faculty of awareness (Quwwat-i-Khayāl)	Prosencephalon/ forebrain (Dimāgh Muqaddam)
If a person is not able to think as well as to draw a conclusion from a statement	Faculty of power of justification (Quwwat Mufakkira)	Mesencephalon/ mid brain (Dimāgh Mutawassit)
If person forgets everything very soon	Faculty of orientation (Quwwat zikr)	Rhombencephalon/ hindbrain (Dimāgh Mu'akhkhar)

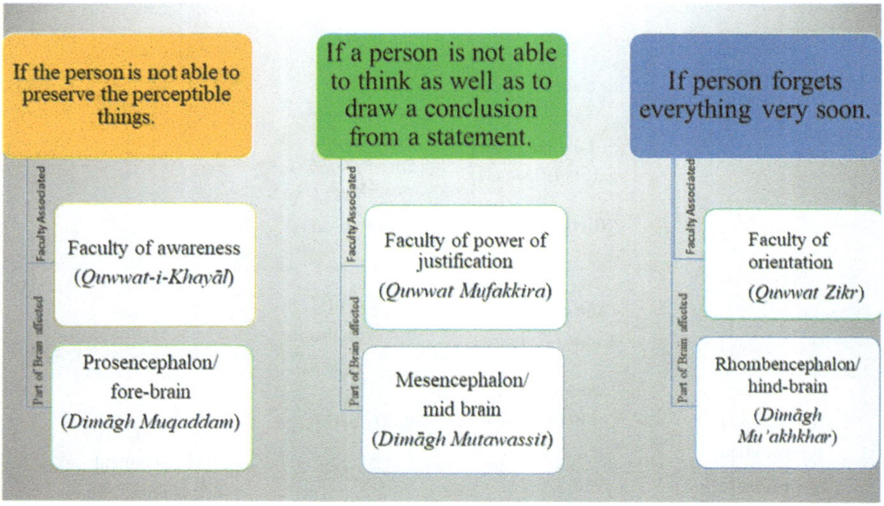

Fig. 21.2 Relationship between functions, faculty, and part of the brain in amnesia and dementia (Ibn Sina 2014; Jurjani 2008)

knowledge perceived by composite sense. New senses perceived by body have been compared with preserved knowledge by this power. Quwwat Mutasarrifa (faculty of modification) is an intellectual faculty which modifies sensory information in different ways and gives new dimensions to the preserved knowledge. Quwwat Wāhima (faculty of apprehension) decodes the meanings of those particular forms perceived by composite sense. It decides what is in favor of the individual and what is against. Quwwat Mufakkira (power of justification) is a serving faculty of apprehension and helps in justifying the perceptions of composite sense. Quwwat Nāfidha (faculty of implementation) is the faculty which helps in implementing the thoughts established by the power of justification. Quwwat Hafiza (faculty of memory) preserves the meanings derived by faculty of apprehension. This memory may be instantaneous, short term or long term (Ibn Sina 2014; Jurjani 2008) (Table 21.2; Fig. 21.2).

Table 21.3 Association of the functions of the brain with different faculties (Ibn Sina 2014; Jurjani 2008)

Functions of brain	Associated faculties of the brain
Act of imagination (fi'l-i-Takhayyul)	Faculty of awareness (Quwwat-i-Khayāl)
	Faculty of modification (Quwwat Mutasarrifa)
Act of thinking (fi'l-i-Fikr)	Perception power/ composite sense or common sense (hiss Mushtarak)
	Faculty of apprehension (Quwwat Wāhima)
	Power of justification (Quwwat Mufakkira)
Act of preservation (fi'l-i-Hifz)	Faculty of memory (Quwwat Hafiza)

The functions associated with the various parts of the brain and derived by different faculties are mentioned here. The act of imagination (Fi'l-i-Takhayyul) is performed when intellectual powers like faculty of modification (Quwwat Mutasarrifa) gives new dimensions to the perceived sensations. Act of thinking (Fi'l-i-Fikr), is the process to establish the meaning of the perceived sensations, the faculty of apprehension (Quwwat Wāhima) and power of justification (Quwwat Mufakkira) are involved in this process. The act of preservation (Fi'l-i-Hifz) is the preservation of the knowledge forwarded by the faculty of apprehension, the faculty of memory (Quwwat Hafiza) is responsible for this function. The intellect function (Fi'l-i-Tadabbur) is the function performed by the intellectual faculty and works according to intellect level and differs from person to person (Ibn Sina 2014; Jurjani 2008) (Table 21.3; Fig. 21.3).

Rutūbat (moistness) is one of the four physical properties naturally associated with matter. The body fluids (Rutūbat-i-Badan) are fluids in the body which are found in tissues, intracellular spaces, cavities, lymphatic vessels, and blood vessels (Kabir 2014a, b). The normal fluid or moisture (Rutūbat Mu'tadila) is the fluid stored in the cavities of the brain and essential for its function. The primitive fluid (Rutūbat Asli) is the fluid which is transferred by the parents to the new born that exists in the tissue elements since birth, while foreign humor (Rutūbat Ghariba) is the moistness of an organ which does not remain in equilibrium, qualitatively and quantitatively, in accordance to the temperament of that organ (Ahmad 1983).

According to the classical literature, brain and nerves have cold and moist (Bārid Ratb) and moist (Ratb) temperament and these are best suitable for the brain and nerves, respectively. In old age, the temperament gradually changes to cold and dry (Bārid Yābis) due to gradual loss of moistness (Rutūbat) from body as well as from brain. Loss of moistness mainly occurs due to increases evacuation (Istifragh) and dispersion (Tahlil). The moistness has been used for daily physical activities, movements, and other brain functions. Apart from this it is either used or become impaired in conditions such as sadness (Gham), anxiety (fikr), and fear (andesha), etc. The primitive fluid (Rutūbat Asli) always used regularly throughout life but its replacement never occurs. Diet do not provide its replacement and foreign humor (Rutūbat Ghariba) increases with time in brain as well as in whole body. When dominance of foreign humour occur then brain function especially cognitive

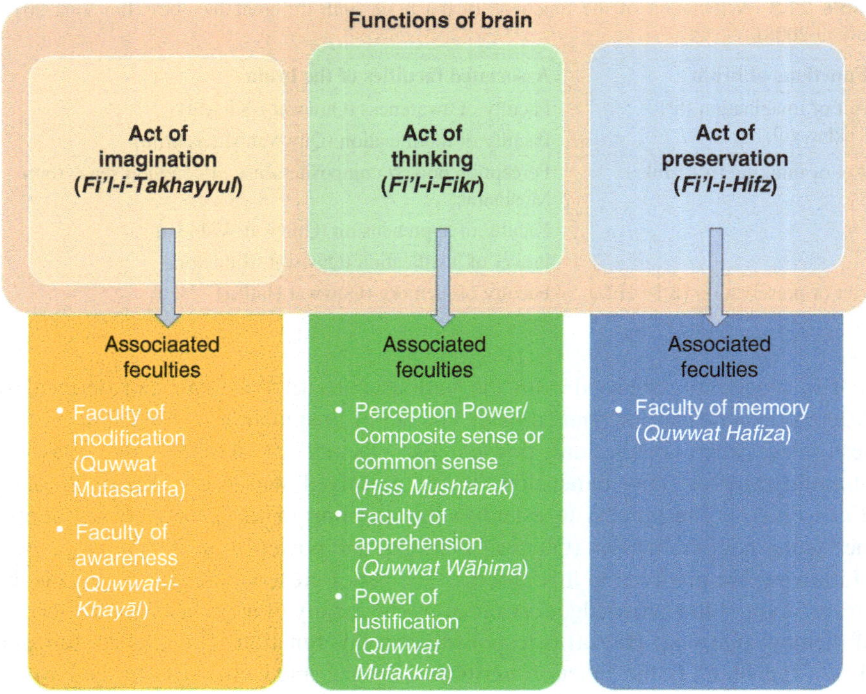

Fig. 21.3 Association of the functions and faculties of the brain (Ibn Sina 2014; Jurjani 2008)

function become weak and degree of weakness depends upon proportion of primitive fluid and foreign humour i.e. more foreign humour more disturbed function. As in the case of Alzheimer's disease which is generally occurs at old age this theory fits very well (Ibn Sina 2014; Jurjani 2008).

In case of inflammation and injury in brain the function of the brain disturbed due to alteration of the temperament and disturbance in brain tissues. Inflammation also increases the loss of primitive fluid from brain tissues. Accumulation of impaired humor (Khilt bad) or waste material (Anjaroon bad) disturbs the functions of the brain due to altered temperament caused by these waste materials. This theory resembles the modern theory of accumulation of the protein fragment beta-amyloid (plaques) outside neurons in the brain and twisted strands of the protein tau (tangles) inside neurons as in case of AD. Accumulation of excess fluids in cavities (Tajwīf) and in the brain itself, i.e. foreign humor also disturbs the cognitive function by obstructing the normal movement of the impulses, blood supply, oxygen supply, and other nutrients to the brain tissues. Accumulation of excess fluids in brain tissues/nerves (A'la hiss-o-harakat) and routes (Guzargah) in the brain is also responsible for the impaired brain function. All these factors collectively damage the brain tissues by many factors like disturbance in transmission of impulses, disturbed transport of nutrients for brain tissues, low blood supply and low oxygen supply to the affected area. The atrophy followed by damage or death of neurons occurs gradually due to above-mentioned factors (Ibn Sina 2014; Kabir 2014a, b).

21.3.1 Amnesia and Dementia

In USM the diseases associated with the forgetfulness or where it is one of the main symptoms are cerebrasthenia (Du'f al-Dimāgh), Humq, Ru'ūnat, Batlan-i Aql, Faramoshi, and Nisyān commonly. Cerebrasthenia is the weakness of brain where the blood supply to the whole brain or a part of the brain is interrupted resulting in insufficient supply of blood which produces the disturbance in the function of different faculties of the brain. This condition can also be due to impaired Mizāj (temperament) of the brain, congenital abnormality of brain, increased evacuation (Istifragh) of the body by any of evacuation method mentioned in USM (Ahmad 1983). The weakness of the thinking faculties of brain which in turn leads to distraction of mind is called Humq. The person suffering from Humq is not able to make a balanced state or equilibrium in social environment. The patient usually makes him busy in doing irrelevant chores similar to children (Ahmad 1983; Tabari 1995). Ru'ūnat is a type of weakness of the thinking faculties of brain in which the person becomes egoistic. The patient usually expresses violent behavior and feels himself superior to others (Ahmad 1983).

In case of Nuqsan-i Aql or Batlan-i Aql (loss of intelligence), the loss or invalidity of intelligence is the common symptom and the most common cause of this is predominance of coldness (Burūdat) in the temperament of the brain (Mizāj Dimāghi). Three types of coldness (Burūdat) exist in this condition, only coldness (sardi sada), combination of coldness with dryness (sardi wa khushki), and combination of coldness with moistness (sardi wa tari) (Ahmad 1983; Tabari 1995). Faramoshi which is also known as Fasād-i Zikr (impairment of orientation) is caused by coldness in temperament of hindbrain (Juzo Akhreen Dimāgh). In this case the person has full consciousness and listen perfectly but forget the things very soon. Nisyān is a simple state of forgetfulness due to impairment or weakness of the faculty of memory (Quwwat Hafiza) which is responsible for the act of preservation (Fi'l-i-Hifz). Some Unani physicians consider it an impairment or weakness of faculty of awareness (Quwwat-i-Khayāl) and power of justification (Quwwat Mufakkira) which are responsible for the act of thinking (Fi'l-i-Fikr). The dictionary meaning of Nisyān is forgetfulness; it is the chief complaint in this condition so nomenclature is according to the chief complaint (Kabir 2014a, b). Some Unani physicians described Nisyān and Faramoshi as the same disease while others have described separately with little difference. Humq is described as dementia while Nisyān as amnesia, but detail description of Nisyān showed that it encompasses dementia and have cognitive sign and symptoms.

The disturbed state of mind (Zehan ki Khrabi) is due to the impairment or weakness of the Quwwat Hafiza (faculty of memory), Quwwat-i-Khayāl (faculty of awareness), and Quwwat Mufakkira (power of justification) (Kabir 2003, 2014a, b; Khan 1869) (Table 21.4; Figure 21.4).

Table 21.4 Unani terms describing amnesia and dementia (Kabir 2003, 2014a,b; Khan 1869)

Terms	Description
Nisyān	A state of forgetfulness
Humq	The weakness of the thinking faculties of brain
Ru'ūnat	The weakness of the thinking faculties of brain in which the patient becomes egoistic
Nuqsan or Batlan-i Aql	Loss or invalidity of intelligence
Faramoshi	Patient has full consciousness and listen perfectly but forget the things very soon

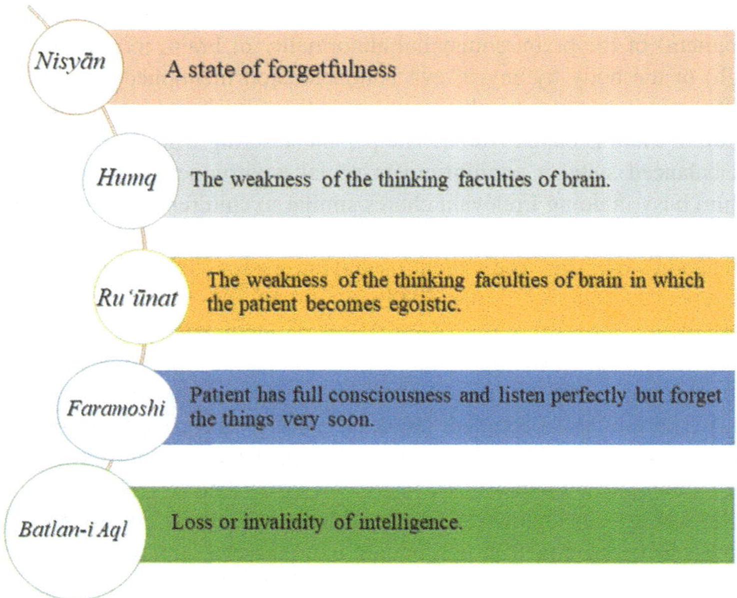

Fig. 21.4 Unani terms describing amnesia and dementia (Kabir 2003, 2014a,b; Khan 1869)

There are impairments of different functions of the brain due to various causes such as impairment of orientation function [Fasād al-Zikr (FZ)], impairment of imagination function [Fasād al-Takhayyul (FT)], and impairment of thinking function [Fasādal-Fikr (FF)]. FZ is a condition in which faculty of memory (Quwwat Hafiza) got impaired or deranged, FT is a condition where the power of justification (Quwwat Mufakkira) got impaired or affected and FF in which faculty of apprehension (Quwwat Wāhima) got affected (Kabir 2014a).

21.3.2 Etiology of Amnesia and Dementia

According to the modern system of medicine, the hallmark pathologies of Alzheimer's are the progressive accumulation of the protein fragment amyloid β (plaques) outside neurons in the brain and twisted strands of the protein tau (tangles) inside neurons. These changes are eventually accompanied by the damage and death of neurons. In typical AD, brain imaging reveals atrophy that begins in the medial temporal lobes before spreading to lateral and medial parietal and temporal lobes and lateral frontal cortex. It is a multifactorial disease, with no single cause known, and several modifiable and non-modifiable risk factors are associated with its development and progression (Igor 2014; Francis et al. 2017). There are several risk factors associated with AD; some important factors are family history of AD (Fratiglioni et al. 1993; Green et al. 2002; Lautenschlager et al. 1996), older age (Wu et al. 2008; Hebert et al. 2010), environmental factor, cardiovascular diseases (Samieri et al. 2018), smoking (Anstey et al. 2007), high blood pressure (Abell et al. 2018; Ninomiya et al. 2011; Debette et al. 2011; Gottesman et al. 2017), hypercholesterolemia (Solomon et al. 2009), obesity (Kivimaki et al. 2018; Loef and Walach 2013), and diabetes mellitus (Meng et al. 2014; Reitz et al. 2011; Gudala et al. 2013; Vagelatos and Eslick 2013) and history of traumatic brain injury (Fann et al. 2018).

According to USM brain diseases are divided into five major categories according to underlined causes i.e (Abell et al. 2018). Waram & Tafarruq Ittisāl (Inflammation and injury in brain); (Ahmad 1983) Accumulation of Khilt bad (impaired humor) or Anjaroon bad (waste material) in brain tissues; (Ali 1871) Accumulation of excess fluids in Tajwīf (cavities) and in the brain; (Ali 1860) Accumulation of excess fluids in A'la hiss-o-harakat (brain tissues/nerves) and Guzargah (routes) of the brain; and (Al-Qamri 2008) Sudā'(headache) (Jurjani 2008).

The possible causes of Fasād al-Zikr (FZ) include predominance of coldness and dryness (Ghalaba-i-Burudāt wa Yubusāt) in brain or predominance of heat and dryness (Ghalaba-i-Harārat wa Yubusāt) in brain. The accumulation of moistness without involvement of any active matter (Rutūbat bila mādda) with dryness (Yubusāt) in brain is another cause and generally associated with hindbrain (Dimāgh Mu'akhkhar) (Ibn Sina 1998). The dominance of coldness and moistness (Ghalaba-i-Burūdat wa Rutūbat) in brain tissues reduces the capacity of storage power of the brain. The impaired cold and dry temperament (Su'-i-Mizāj Bārid Yābis) of hindbrain (Dimāgh Mu'akhkhir) is also associated cause of this problem (Kabir 2014a, b).

The common causes of the Fasād al-Takhayyul (FT) are dominance of dryness (Ghalaba-i-Yubusāt), dominance of bilious humor (Ghalaba-i-Safrā), impaired hot temperament without involvement of any active matter (Su-i- Mizāj Hārr bila mādda), dominance of coldness (Ghalaba-i-Rutūbat), and an injury or disturbance in forebrain (Aafat in Dimāgh Muqaddam) (Ibn Sina 1998).

The Fasādal-Fikr (FF) is caused by an injury or disturbance in mid brain (Aafat in Dimāgh Mutawassit) (Kabir 2014a, b), dominance of coldness and moistness in brain tissues (Ghalaba-i-Burūdat wa Rutūbat), various brain diseases such as cerebrasthenia (Zu 'f al-Dimāgh), loosening of brain tissues (Talyyun al-Dimāgh), contractions in brain tissues (Tashannuj al-Dimāgh), paralysis (Fālij), obstruction or pressure by tumor leading to interruption of blood supply to brain tissues, trauma especially on left side of the brain and overwork/exertion (Tabari 1995; Kabir 2014a, b; Jurjani 2008).

The involvement of phlegmatic humor (Khilt Balgham) and coldness (Burūdat) is must for development of the Nisyān (Tabari 1995). Other causes include accumulation of viscid phlegmatic fluid in the brain, predominance of coldness (Ghalba-i Burūdat) in the brain, sometime it is also due to predominance of dryness (Yubusāt) or excess of moistness (Rutūbat) in the brain (Razi 1997). The temperament (Mizāj) of whole body becomes cold (Bārid) (Tabari 1928). Impairment of the temperament (Fasād-i Mizāj) of brain or whole body (Khan 1869), accumulation of moist phlegmatic humor (Balgham Ratb) in forebrain (Dimāgh Muqaddam) which adversely affects its ability to retain memories (Al-Qamri 2008). The reduction in brain tissues and interstitial fluids in old age due to some brain disorders may be a reason for it. Psychological disturbances like psychic power (Nafs) in erotomania ('Ishq) or sadness (Gham), feeling of dependency (Faqar), intense quest or desire for something are also precipitate the amnesia and dementia. It is also hereditary and persons having family history are more prone to develop amnesia. Some external factors are also responsible for it like accidental injury to head (Zarba-o-Saqta) (Razi 1997), excessive sleeping, and excessive awakening (Jurjani 2008).

21.3.3 Types of Amnesia and Dementia

It is classified according to the causes, sign and symptoms, and involvement of the part of the brain. As per the underlying causes it is divided into following, i.e (Abell et al. 2018). Nisyān Bārid Yābis, a state of forgetfulness caused by the predominance of coldness (Burūdat) and dryness (Yubusāt) in the brain tissues; (Ahmad 1983) Nisyān Bārid Ratab, a state of forgetfulness caused by the predominance of coldness (Burūdat) and moistness (Rutūbat)) in the brain tissues and Nisyān Hārr Yābis, a state of forgetfulness which is caused by the predominance of heat (Harārat) and dryness (Yubusāt) in the brain tissue. This state is incurable if reduction in the brain tissues occurs (Kabir 2003).

Some Unani physicians classified Nisyān in four types due to the loss of specific functions such as (Abell et al. 2018) Butlan al-Takallum (loss of speaking sense), here complete loss of speaking power occurs and person is not able to speak; (Ahmad 1983) Butlan al-Tahrir (loss of writing skill), in this condition the person is not able to write or he forget the writing skills; (Ali 1871) Ishara (loss of power of gesture), the person forget the gestures and their meaning; and (Ali 1860) Fasad-i-Fikr (impairment of the act of thinking) here person is not able to think properly to get reasoning for any act (Kabir 2014a, b).

As per the involvement of part of the brain where accumulation of viscid fluid, deranged phlegmatic matter accumulates, it is divided into three types. First type appears when phlegmatic matter (Balghami Mādda) accumulates in prosencephalon/forebrain. Second type is produced when viscid fluid (Ghalīz Rutūbat) accumulates in the cerebrum (Mukhkh) which is the seat of act of thinking. Third type appears when viscid fluid (Ghalīz Rutūbat) accumulates in rhombencephalon/rhombencephalon (Dimāgh Mu'akhkhar) (Tabari 1995).

21.3.4 Risk Factors for the Development of Amnesia and Dementia

Sleeping in day time especially in full stomach, excessive intercourse (Jima ki Ziyadti), fatigue or exertion (I'ya), long-term use of light and easily digestible food (Latif Ghiza') (Razi 1997), insomnia (Sahr) (Kabir 2003; Razi 1997), and linger intoxicated constantly (Kabir 2003).

21.3.5 Pathogenesis of amnesia and Dementia

In case of dominance of moistness (Kathrat-i-Rutūbat) in brain or whole body faculty of memory (Quwwat Hafiza) becomes weak and in turns it disturbs act of thinking (Fi'l-i-Fikr). Excessive sleep and heaviness especially on hindbrain will be present in this condition (Jurjani 2008; Kabir 2014a, b; Tabari 1928). In case of dominance of dryness (Yubusāt) with coldness (Burūdat), short-term memory becomes weak but person remembers past events. Insomnia, dryness of nose, dull sensations especially touch sensation will be present. Difficulty in speaking loudly and continuously will be present and sometimes patients feel strangulated. In case of dominance of moistness (Rutūbat) with coldness (Burūdat), person forgets the past events. Excessive sleep, dribbling of secretion from mouth, nose, and eye will be present (Ibn Sina 1998; Jurjani. 2008).

21.3.6 Diagnosis of Amnesia and Dementia

The diagnosis of the amnesia and dementia depends upon the signs and symptoms present in the patient. If person forgets everything very soon, then it is a sign of impairment of faculty of orientation (Quwwat Zikr) and site of underlying cause is rhombencephalon (Dimāgh Mu'akhkhar). If a person is not able to think as well as to draw a conclusion from a statement, then it is a sign of impairment of faculty of power of justification (Quwwat Mufakkira) and its cause lies in mesencephalon (Dimāgh Mutawassit). If the person is not able to preserve the perceptible things, not able to see dreams or forgets the dreams, then it is a sign of impairment of faculty of awareness (Quwwat-i-Khayāl) and the cause lies in prosencephalon (Dimāgh Muqaddam/) (Kabir 2003).

Excessive sleep, dribbling of secretion from mouth, nose, and eye, forgetting the past events, passage of crude white color urine (Bawl Abyad) is a sign of dominance of coldness and moistness (Burūdat wa Rutūbat) in the brain. Insomnia, dryness in mouth and nostril, forget recent events but remember past events, not able to talk for

Table 21.5 Diagnostic parameters to rule out the cause of the disease as per Unani concept (Khan 1869)

Coldness and moistness (Burūdat wa Rutūbat)	Coldness and dryness (Burūdat wa Yubusāt)
• Excessive sleep • Dribbling of secretion from mouth, nose, and eye • Forget the past events • White color crude urine (Bawl abyad)	• Insomnia • Dryness in mouth and nostril • Forget recent events but remember past events • Not able to talk for long time and loudly • White colored urine which is clear (Bawl Abyad)
Intermediate and mixed sign and symptoms will be present if only coldness (Burūdat) is there without involvement of moistness (Rutūbat) or dryness (Yubusāt).	

long time and loudly, white colored clear urine (Bawl Abyad) is a sign of dominance of coldness and dryness (Burūdat wa Yubusāt) in brain (Khan 1869) (Table 21.5).

21.4 Management of Amnesia and Dementia

21.4.1 Principles of Treatment (Usūl-i'Ilāj)

The first and foremost step toward the management is to remove the cause of the disease (Jurjani. 2008; Kabir 2014a, b). The process of moistening (Tartīb) is required to counter dryness (Yubusāt) in case of predominance of dryness in the brain (Kabir 2003). The process of calefaction (Taskhīn) is needed to counter coldness (Burūda), in case of predominance of coldness. The process of drying (Tajfīf) is essential to counter the predominance of moistness (Rutūbat) (Anonymous 2014). Hammām (Turkish bath), Abzan (sitz bath) and method to induce sleep may be used in case of dominance of coldness (Burūdat) and dryness (Yubusāt). Hot and moist paste (Dimād) and irrigation (Natūl) may be applied in case of dominance of coldness (Burūdat) and dryness (Yubusāt). Massage of moderate oil (Mu'tadil Roghan) on scalp like Roghan-i Nargis, Roghan-i Kheri, and Roghan-i Sosan may be applied in case of dominance of coldness (Burūdat) and dryness (Yubusāt) (Jurjani 2008).

The appropriate methods should be applied to induce exhilaration or mood elevation (Tafrih-ī Taba') (Anonymous 2014; Khan 1987). To improve functions of brain suitable tonics (Muqawwiyāt), protective (Muhāfizat), corrective (Muslihat), and moderators (Moaddilat) should be used (Anonymous 2014). If impairment in brain is without the involvement of active substance (Mādda), then correction of the temperament (Mizāj) is sufficient by the method of hetero therapy. In case of derangement of the temperament (Su'-i-Mizāj), temperamental equilibrium/alteration (Ta'dil-o-Tabdil-i-Mizāj) is required to correct it. If there is involvement of humor (khilt) or active substance (Mādda), then evacuation of morbid substance (Radi Mādda) is essential. The evacuation (Istifrāgh) or elimination of morbid matter (Tanqiya Dimāgh) from the brain may be done by concoctive

(Mundij) and purgative (Mushil) therapy (Ahmad 2008; Anonymous 2014; Kabir 2003; Razi 1997; Ibn Rushd 1984). In case of predominance of moistness (Rutubat) due to deranged active substance (Madda), evacuation of morbid substance should be done by appropriate concoctive (Mundij) and purgative (Mushil) therapy (Kabir 2003; Khan 1987, 2005). In case of dominance of coldness and dryness, calorific and refrigerant (Musakhkhin wa Mubarrid) drugs, and hot and moist diet (Hārr Ratb Ghizā') are beneficial. In case of dominance of hotness and dryness without involvement of any matter modification (tabrīd) by external methods like liniment (tila), irrigation (natūl), fumigation (dhūnī), etc. are sufficient (Kabir 2003).

21.4.2 Life Style Changes

Changes in lifestyle have an important role in dealing with stress and anxiety, and they can also boost cognitive ability. There are several activities which can be applied to manage amnesia and dementia such as applying of thought provoking activities, partying, and spending time with friends, indulgence in entertaining activities, listening of music, peaceful sound sleep, and smelling of flower and good aromas. The person should be well counseled if the underlying cause showed psychological factors (Anonymous 2014; Kabir 2003; Khan 1987).

21.4.3 Dietary Recommendations

Diet (Ghiza') is the substance, which after administration and digestion and metabolism (Hadm-o-Istihala), becomes the part of the body or organ and provides replacement of body constituents (Badl Ma Yatahallal). To maintain proper health, balance diet is necessary and it is also required for growth and development and fortifies immunity which is the first line defense against any derangement in the body. It is advocated that initially in management of disease Dietotherapy ('Ilaj bi'l-Ghiza') should be given priority. Several dietary items are described in classical literature for the management of every disease. In case of amnesia and dementia diet should contain one or more of the these items such as broth of chicken or quail, eggs of all animals especially fish eggs, gall bladder of all animals, kernel of coconut and almonds with dry figs, kernel of Hazelnut (Funduq), nuts (Labūb) like almond, cashew nuts, Falūda (a special dish made up of starch), and dry and less fatty meats of birds like sparrow, doves, larks, and chukar partridge (chukor) (Kabir 2003; Razi 1997; Tabari 1928; Ibn Rushd 1984; Khan 1987). Rice with milk and sugar or honey is advised in old peoples (Azmi 1995; Kabir 1935) and meat of francolin (Teetar) (Jurjani 2008; Khan 1987).

21.4.4 Dietary Restrictions

Diet plays an important role in the maintenance of homeostatis in the body and health of a person. In USM dietary restrictions are also described as per the cause of the disease and temperament (Mizaj) of the dietary items and disease. In case of amnesia and dementia the items which should be restricted from the diet includes Dhania (seeds of *Coriandrum sativum* L.), Piyaz (onion), flatulent food items like lentils, cabbage, leek, black eyed beans (Aghdhiya Mu'bakhkhira), Zafran (stigma and styles of *Crocus sativus* L.), Kahu (seeds of *Lactuca sativa* L.), Poppy seed (seeds of *Papaver somniferum* L.), and Baqla (seeds of *Vicia faba*) (Kabir 2003; Razi 1997).

In old aged persons thick or heavy (Ghalīz) and viscous or sticky (lazaj) food items should be avoided as it increases black bile (Sawdā') and phlegm (Balgham) in the body. Likewise (hot temperament diet (tez ghiza) like Ābkāma (a type of vinegar) has to be avoided. The light and easily digested food items should be used (Majusi 2010). Excessive alcohol consumption, excess of exercise or physical work, excessive anger and sadness, and other anxiety and stress should be avoided (Kabir 2003; Razi 1997).

21.4.5 Regimenal Therapy (Ilaj Bit Tadbīr)

Tadbir is an Arabic word meaning "regimen" or systemic plan, whereas Ilaj means therapy or treatment. So 'Ilaj bi'l Tadbir stands for treatment through regimen. It is a type of the therapy which is given in the form of regime to maintain the health of a person. This therapy creates changes in the obligatory causes of the health, i.e. six essentials of health on the principle of heteropathy. It deals with the rules of diet, exercise, etc. for improving health and physical or mental well-being or any intervention other than medicine that restores the health. In amnesia and dementia purgative enema (Huqna Mmushila) is very effective to evacuate morbid matters from the brain and strong enema is used for this. Gharghara (Gargle) by using appropriate drugs, 'Atus (snuff) by powder of Kundur (*Boswellia serrata* Roxb.) and Jaosheer (*Ferula galbaniflua* Boiss (Kabir 2014a, b), Natūl (irrigation) by using appropriate drugs, Saoot (Nasal drop), Riyādat (Exercise), and Hammam (Turkish Bath), etc. are used to cure the disease (Kabir 2003; Ibn Sina 1998). Qay' (vomiting) is contraindicated in this disease (Kabir 2003; Razi 1997).

21.4.6 Pharmacotherapy ('Ilāj bi'l-Dawā')

It is the management of the disease with the help of specific drugs on the basis of temperament of disease and drug. The guiding principle of the treatment is that drugs having opposite qualities/temperament (Mizaj) to the prevailing one in the diseases have to be used to correct the pathology. If a disease is caused by morbidity of hot humor, drugs having cold temperament should be used. It is used to correct the derangement in Akhlat (Humors) generally.

In the management of amnesia and dementia decoction of some specific roots (Mā'al- Usūl) is used as a first step. It is followed by oral administration of a compound formulation "Ayāraj-i Loghāziya" along with the decoction of following drugs, Halela siyāh (*Terminalia chebula* Retz.) 35 gm, Maweez Munaqqa (*Vitis vinifera* L.) 70 gm, Aftimoon (*Cuscuta epithymum* L.) 40gm, and Rewand (*Rheum emodi* Wall.) 10.5 gm (Anonymous 2014; Tabari 1928). Use of Ayāraj-i Feqra (a compound formulation) alone is also beneficial (Tabari 1928). Ayāraj is a form of compound formulation which is specific for the evacuation of morbid matters from the brain by its purgative actions. There are many formulations and named after the disease or main ingredient of particular Ayāraj, e.g. Ayāraj-i Feqra, Ayāraj-i Loghāziya, etc. their primary function is to clean the brain from waste and morbid matter and specific Ayāraj are used for specific diseases (Ali 1871; Anonymous 1967; Khan 1881).

The powder of Birhamdandi (*Echinops echinatus* Roxb.) in the dose of 12 gm with cow milk for 15 days is very effective. The oral use of kernel of Maghz Funduq (*Corylus avellana* L.) with sugar, empty stomach in the morning is useful. Consuming the water after putting some pieces of Kundur (*Boswellia serrate* Roxb.) 7.5 gm in glass of water increases the power of faculty of memory (Quwwat Hafiza) (Razi 1997).

Some oral medicines like Sharbat Banafsha, Sharbat Nilofer, Sharbat Gulab, and Arq Bedmushk, etc. with cold water (Ma' Barid) are also effective. Roughan Banafsha has moist temperament and light (latif) in nature so it helps to correct dryness of the brain and heat of the lungs. Roughan Nilofar is moister than Roughan Banafsha and it induces sleep and relieves the headache caused due to khushki (dryness) and garmi (heat) and usefull in this case (Kabir 2003).

21.4.6.1 Mufradat (Single Drug) for the Management of Amnesia and Dementia

Various single drugs mentioned below have been used to treat amnesia and dementia and produce their action directly or indirectly. These drugs can be used in accordance to the temperament of the drug and the disease. These drugs are obtained from natural sources, mostly from plants; because of this they do not have any severe adverse effects.

Abresham (silk cocoon), Amla (*Emblica officinalis* L.), Asgandh (*Withania somnifera* Dunal.), Azaraqi (*Strychnos nuxvomica* L.), Badam Shireen (*Prunus amygdalus* Batsch.), Baladur (*Semicarpus anacardium* L.f.), Balchar (*Nardostachys jatamansi* D.C), Balela (*Terminalia belerica* Gaertn.), Beesh (*Aconitum napellus* L.), Behman Surkh (*Centaurea behen* L.), Birhamdandi (*Echinops echinatus* Roxb.), Brahami booti (*Bacopa monnieri* L.), Dar-i filfil (*Piper longum* L.), Filfil siyah (*Piper nigrum* L.), Halela Hindi (*Terminalia chebula* Retz.), Jadwar (*Delphinium denudatum* Wall.), Kundur (*Boswellia serrate* Roxb.), Maghz Akhrot (Jugulans regia L.), Maghz funduq (*Corylus avellana* L.), Nakchikni (*Centipeda minima (L.) A.BR.),* Saad (*Cyperus rotundus* L.), Sankhaholi (*Evolvulus alsinoides* L.), Waj

(*Acorus calamus* L.), Zafran (*Crocus sativus* L.), Zanjabeel (*Zingiber officinale* Rosc.), and Zarnab (*Flacourtia jangomas* (Lour.) Raeusch.) (Ali 1860; Arshi 1929; Fazlullah 1877; Hasan 1865; Kabir 1955, 2003; Khare 2007; Rafiq 1985; Razi 1997;Wahid and Siddiqui 1961; Zaki 1890).

21.4.6.2 Pharmacopeial Formulations (Qarābādīni Murakkab) for the Management of Amnesia and Dementia

Different compound formulations have also been described in Unani literature that can be used as readymade medicine for the management of amnesia and dementia. These formulations have several other indications and generally used as per therapeutic actions and uses mentioned for each formulation. The formulation mentioned here are taken from the different volumes of National Formulary of Unani Medicine (NFUM) which is a government approved document for the preparation and dispensing of the multi-ingredients formulation such as Itrifal-e-Ustukhuddus, Majoon-e-Baladur, Majoon-e-Najah, Khamira Gawzaban Sada, and Khamira Ustukhuddus (Anonymous 2006a). Habb-e-Shigraf, Majoon-e-Boolis, Jawarish-e-Amla Luluvi, Jawarish-e-Shahanshahi Ambari, Itrifal-e-Kishmishi, Itrifal-e-Sana, Halwa-e-Badam, Halwa-e-Malkangani, Majoon-e-Nisyān (Anonymous 2007a). Habb-e-Jadwar Kochak, Habb-e-Kaboos and Habb-e-Qoqaya (Anonymous 2007b). Habb-e-Aafiat and Habb-e-Barmak (Anonymous 2006b). Habb-e-Khas, Qurs-e-Fizza, Qurs-e-Jawahar Mohra, Halwa-e-Salab, Itrifal-e-Muqawwi Dimagh, Jawarish-e-Amla Sada, Khamira-e-Abresham Sheera Unnabwala, Khamira Gawzaban Ambari, Khamira Sandal and Majoon-e-Brahmi (Anonymous 2008). Khamira Sadaf, Khamira Zahar Mohra, Laboob-e-Kabir Khaas, Zahbi (Anonymous 2011). Qurs-e-Marjan, Khamira Gawzaban, Qurs-e-Khabsul Hadeed, Dawa-ul-Misk Motadil, Habb-e-Ayāraj (Khan 1987).

21.4.6.3 Compound Formulations (Murakkab) Prepared as per the Need for the Management of Amnesia and Dementia

Unani physician have formulated many formulations for the treatment of amnesia and dementia, although these are not part of the pharmacopeia but still used by many practitioners and mentioned in classical literature. Few formulations have been covered here to give an idea for the effective management of amnesia and dementia. The first formulation is made up of the drugs such as Kundur (*Boswellia serrata* Roxb.), Saad (*Cyperus rotundus* L.), Filfil abyad (*Piper nigrum* L.), Zafran (*Crocus sativus* L.), Mur Makki (*Commiphora myrrha* Nees.). These drugs have been taken in equal quantity and a semi solid formulation "Majoon" with honey to be made and consumed in the dose of 5–10 gm twice daily. The other formulation is a powder form made up of Kundur (*Boswellia serrata* Roxb.) 3 gm, and Filfil (*Piper nigrum* L.) 7 gm and may be taken in the dose of 4.5 gm powder daily (Anonymous 2014; Ibn Sina 1998). *Sirka* (Vinegar) made up of *Unsul* (*Urginea indica* Roxb.) and *Sikanjabin* made up of *Unsul* (*Urginea indica* Roxb.) may also be used (Tabari 1995).

21.5 Conclusion

The detailed description of the association of different parts of the brain with various cognitive functions and diseases highlights the depth of clinical knowledge of Unani physicians. The amnesia and dementia can be compared with Nisyān and Humq respectively but the underlying causes and treatments of both are almost similar with few exceptions. The theory of primitive fluid, its role in optimal functions of the brain and dominance of foreign humor in advancing age can be compared with modern concept of the development of AD in old age. AD commonly occurs in old age and as per Unani concept foreign humors also become dominant in this age. Foreign humors adversely affect the cognitive functions of the brain and degree of weakness depends upon the proportion of primitive fluid with foreign humor, i.e. more foreign humor more disturbed function. The accumulation of impaired humor or waste material in the brain disturbs the functions of the brain due to altered temperament caused by these waste materials. This theory resembles the accumulation of the protein fragment beta-amyloid (plaques) outside neurons, in the brain and twisted strands of the protein tau (tangles) inside neurons in AD. The treatment described in this chapter may be helpful to overcome these most common causes of amnesia and dementia including AD. The use of concoctive and purgative drugs can help to remove waste materials from the brain and other regimens such as tonics, protective, corrective, and moderators may be helpful to maintain the dominance of primitive fluid for the optimal function of the brain.

References

Abell JG, Kivimäki M, Dugravot A, Tabak AG, Fayosse A, Shipley M et al (2018) Association between systolic blood pressure and dementia in the Whitehall II cohort study: role of age, duration, and threshold used to define hypertension. Eur Heart J 39(33):3119–3125

Ahmad SI (1983) Kuliyat-i Asri. New Public Aal Press, New Delhi

Ahmad HHJ (2008) Tazkirah Jaleel. Central Council for Research in Unani Medicine, New Delhi

Ali MN (1860) Nasirul Moalijeen. Matba Alwi Muhammad Ali Bakhsh, Lucknow, India

Ali HA (1871) Qarabadeen Ahsani. Kanpur, India, Matba' Nizami

Al-Qamri NM. Abu al-Mansur al-Hasan ibn Nuh [d.990 AD](2008) Ghina Muna. Central Council for Research in Unani Medicine, New Delhi

Anonymous (1967) Qarabadeen Sarkari/Unani Pharmacopoeia. Indian Medicine Department, Govt of Andhra Pradesh, Hyderabad, India

Anonymous (2006a) National Formulary of Unani medicine. Part-I. Ministry of Health & family welfare, Government of India, New Delhi

Anonymous (2006b) National Formulary of Unani Medicine. Part-IV, 1st edn. Ministry of Health & Family Welfare, Government of India, New Delhi

Anonymous (2007a) National Formulary of Unani medicine. Part-II. Ministry of Health & family welfare, Government of India, New Delhi

Anonymous (2007b) National Formulary of Unani Medicine. Part-III, 1st edn. Ministry of Health & Family Welfare, Government of India, New Delhi

Anonymous (2008) National Formulary of Unani medicine. Part-V. Ministry of Health & family welfare, Government of India, New Delhi

Anonymous (2011) National Formulary of Unani medicine. Part-VI. Ministry of Health & family welfare, Government of India, New Delhi
Anonymous (2014) Standard Unani treatment guidelines for common diseases. Central Council for Research in Unani Medicine, New Delhi
Anonymous (2019) Alzheimer's association. Alzheimer's disease facts and figures. Alzheimer's. Dementia 14(3):367–429
Anstey KJ, von Sanden C, Salim A, O'Kearney R (2007) Smoking as a risk factor for dementia and cognitive decline: a meta-analysis of prospective studies. Am J Epidemiol 166(4):367–378
Arshi MMN (1929) Mufradat-e-Arshi. Kutubkhana Arshi Dhanola, Pratap Press, Lahore, Pakistan
Azmi AA (1995) Basic concept of Unani medicine: a critical study. Hamdard Nagar, New Delhi
Barberger-Gateau P, Raffaitin C, Letenneur L, Berr C, Tzourio C, Dartigues JF et al (2007) Dietary patterns and risk of dementia: the Three-City cohort study. Neurol. 69(20):1921–1930
Debette S, Seshadri S, Beiser A, Au R, Himali JJ, Palumbo C et al (2011) Midlife vascular risk factor exposure accelerates structural brain aging and cognitive decline. Neurol. 77:461–468
Fann JR, Ribe AR, Pedersen HS, Fenger-Grøn M, Christensen J, Benros ME et al (2018) Long-term risk of dementia among people with traumatic brain injury in Denmark: a population-based observational cohort study. Lancet Psychiat 5(5):424–431
Fazlullah M (1877) Makhzanul Mufradat Maroofba Jamia ul Advia. Dar Matba Anwar Muhammad, Lucknow, India
Francis TH, Robinson M, Lee BY, Baib O, Leonenko Z, Alberta MS (2017) Recent Progress in Alzheimer's disease research, part 3: diagnosis and treatment. J Alzheim Dis. 57:645–665
Fratiglioni L, Ahlbom A, Viitanen M, Winblad B (1993) Risk factors for late-onset Alzheimer's disease: a population-based, case control study. Ann Neurol 33(3):258–266
Gilman S (2010) Oxford American handbook of neurology. Oxford University Press, Oxford
Gottesman RF, Albert MS, Alonso A, Coker LH, Coresh J, Davis SM et al (2017) Associations between midlife vascular risk factors and 25-year incident dementia in the atherosclerosis risk in communities (ARIC) cohort. JAMA Neurol 74(10):1246–1254
Green RC, Cupples LA, Go R, Benke KS, Edeki T, Griffith PA et al (2002) Risk of dementia among white and African American relatives of patients with Alzheimer disease. JAMA 287(3): 329–336
Gudala K, Bansal D, Schifano F, Bhansali A (2013) Diabetes mellitus and risk of dementia: a meta-analysis of prospective observational studies. Diab Investig 4(6):640–650
Hasan B (1865) Jamiul Mufradat. Dar Matba Hussaini, Kanpur, India
Hebert LE, Scherr PA, McCann JJ, Beckett LA, Evans DA (2001) Is the risk of developing Alzheimer's disease greater for women than for men? Am J Epidemiol 153(2):132–136
Hebert LE, Bienias JL, Aggarwal NT, Wilson RS, Bennett DA, Shah RC et al (2010) Change in risk of Alzheimer disease over time. Neurology 75:786–791
Hebert LE, Weuve J, Scherr PA, Evans DA (2013) Alzheimer's disease in the United States (2010-2015) estimated using the 2010 census. Neurolo 80:1778–1783
Ibn Rushd (1984) [1126–1198 AD]. Kitab al-Kulliyyat. Central Council for Research in Unani Medicine, New Delhi
Ibn Sina (1998) al-Sheikh al-Ra'is Abu 'Ali al-Husayn ibn 'Abd Allah [980–1037 AD]. Al-Qanun fi'l Tibb. Idara Matbuaat Sulaimani, Lahore, Pakistan
Ibn Sina (2014) al-Shaykh al-Ra'is Abu 'Ali al-Husayn ibn 'Abd Allah [980–1037 AD]. Al-Qanun fi'l Tibb. Idara Kitab al-Shifa, New Delhi
Igor K (2014) Alzheimer's disease: a clinical and basic science review. Med Stud Res J 4:24–33
Jurjani (2008) Sharaf al-Din Isma'il ibn Husayn [d.1135 AD]. Dhakhira Khawarizm Shahi. Idara Kitabul Shifa, New Delhi
Kabir D (1935) Hakim Muhammad [1894-1976AD]. Bayaz-e-Kabir, 5th edn. Shaukat Book Depot, Gujarat, India
Kabir D (1955) Hakim Muhammad [1894–1976 AD]. Makhzanul Mufradat Kitabul Adviya, 5th edn. Daftar al-Masih, Bazar Nur al-Umara, Hyderabad, India
Kabir D (2003) Hakim Muhammad [1894-1976AD]. Al-Iksir. Aijaz Publishing House, New Delhi

Kabir D (2014a) Hakim Muhammad [1894-1976AD]. Kulliyat-i Nafeesi. Idara Kitab al-Shiffa, New Delhi

Kabir D (2014b) Hakim Muhammad [1894-1976AD]. Sharah Asbab. Idara Kitab al-Shiffa, New Delhi

Kasper DL, Fauci AS, Hauser SL, Longo DL, Jameson JL, Loscalzo J (2019) Harrison's principles of internal medicine, 19th edn. McGraw-Hill, New York

Khan A (1869) Hakim Muhammad [1813–1902 AD]. Iksir-i-A'zam. Lucknow, India, Munshi Nawal Kishor

Khan HZ (1881) Qarabadeen Zakai. Matba' Munshi Naval Kishore, Lucknow, India

Khan MHA (1987) Haziq, 1st edn. Firdaus Book Depot, Hyderabad, India

Khan S (2005) Ilaj-ul Amraz. Central Council for Research in Unani Medicine, New Delhi

Khare CP (2007) Indian medicinal plants. Springer India Private Limited, New Delhi

Kivimaki M, Luukkonen R, Batty GD, Ferrie JE, Pentti J, Nyberg ST et al (2018) Body mass index and risk of dementia: analysis of individual-level data from 1.3 million individuals. Alzheimers Dement 14:601–609

Lautenschlager NT, Cupples LA, Rao VS, Auerbach SA, Becker R, Burke J et al (1996) Risk of dementia among relatives of Alzheimer's disease patients in the MIRAGE study: what is in store for the oldest old? Neurol 46(3):641–650

Loef M, Walach H (2013) Midlife obesity and dementia: meta-analysis and adjusted forecast of dementia prevalence in the United States and China. Obesity (Silver Spring) 21(1):E51–E55

Majusi A (2010) 'Ali ibn 'Abbas [d. 994 AD]. Kamil al-Sana'a al-Tibbiyya. Central Council for Research in Unani Medicine, New Delhi

Masihi AS (2008) [d. 1010 AD]. Kitab al-Mi'a. Central Council for Research in Unani Medicine, New Delhi

Mc Khann GM, Knopman DS, Chertkow H, Hyman BT, Jack CR, Kawas CH et al (2011) The diagnosis of dementia due to Alzheimer's disease: recommendations from the National Institute on Aging-Alzheimer's association workgroups on diagnostic guidelines for Alzheimer's disease. Alzheim Dement 7(3):263–269

Meng XF, Yu JT, Wang HF, Tan MS, Wang C, Tan CC, Tan L (2014) Midlife vascular risk factors and the risk of Alzheimer's disease: a systematic review and meta-analysis. J Alzheim Dis 42(4): 1295–1310

Ninomiya T, Ohara T, Hirakawa Y, Yoshida D, Doi Y, Hata J et al (2011) Midlife and late-life blood pressure and dementia in Japanese elderly: the Hisayama study. Hypertension 58(1): 22–28

Rafiq HM (1985) Kanzul Adviya Mufrada. University Publication Unit, Aligarh, India

Rajan KB, Weuve J, Barnes LL, Wilson RS, Evans DA (2019) Prevalence and incidence of clinically diagnosed Alzheimer's disease dementia from 1994 To 2012 in a population study. Alzheimers Dement 15(1):1–7

Razi M (1991) Abu Bakr Muhammad ibn Zakariyya [865–925 AD]. Kitab al-Mansuri. Central Council for Research in Unani Medicine, New Delhi

Razi M (1997) Abu Bakr Muhammad ibn Zakariyya [865–925 AD]. Kitab al-Hawi fi'l Tibb. Central Council for Research in Unani Medicine, New Delhi

Reitz C, Brayne C, Mayeux R (2011) Epidemiology of Alzheimer disease. Nat Rev Neurol 7(3): 137–152

Samieri C, Perier MC, Gaye B, Proust-Lima C, Helmer C, Dartigues JF et al (2018) Association of cardiovascular health level in older age with cognitive decline and incident dementia. JAMA 320(7):657–664

Solomon A, Kivipelto M, Wolozin B, Zhou J, Whitmer RA (2009) Midlife serum cholesterol and increased risk of Alzheimer's and vascular dementia three decades later. Dement Geriatr Cogn Disord 28(1):75–80

Tabari A (1928) Abu'l Hasan 'Ali ibn Sahl [775-890 AD]. Firdaws al-Hikmah fi'l Tibb. Idara Kitab Shifa, New Delhi

Tabari A (1995) Abu'l Hasan Ahmad Ibn Muhammad [d. 985 AD]. Mu'alajat al-Buqratia. Central Council for Research in Unani Medicine, New Delhi

Tom SE, Hubbard RA, Crane PK, Haneuse SJ, Bowen J, McCormick WC et al (2015) Characterization of dementia and Alzheimer's disease in an older population: updated incidence and life expectancy with and without dementia. Am J Public Health 105(2):408–413

Vagelatos NT, Eslick GD (2013) Type 2 diabetes as a risk factor for Alzheimer's disease: the confounders, interactions, and neuropathology associated with this relationship. Epidemiol Rev 35(1):152–160

Wahid HA, Siddiqui HH (1961) A survey of drugs: with particular reference to the Arab (Unani) medicine and Ayurveda. Institute of History of Medicine and Medical research, Hamdard Buildings, New Delhi

World Health Organization [WHO] (2019a) Dementia. https://www.who.int/news-room/fact-sheets/detail/dementia. Accessed 28 Oct 2019

World Health Organization [WHO] (2019b) The top 10 causes of death. https://www.who.int/news-room/fact-sheets/detail/the-top-10-causes-of-death. Accessed 28 Oct 2019

Wu W, Brickman AM, Luchsinger J, Ferrazzano P, Pichiule P, Yoshita M et al (2008) The brain in the age of old: the hippocampal formation is targeted differentially by diseases of late life. Ann Neurol 64:698–706

Zaki MM (1890) Makhzan-ul-Mufridat maroofba Khawasul Adviya, 2nd edn. Sheikh Muhammad Yameen Tajir Kutub, Bazar Jadeed, Saharanpur, India

Nanotechnological Applications in the Diagnosis and Treatment of Alzheimer's Dementia

22

Pradeep Singh, Garima Mishra, Faheem Hyder Pottoo, Mulugeta Molla Zeleke, and Amien Ewunetei

Abstract

Alzheimer's disease (AD) is one of the most devastating neurological disorders causing memory loss and impairment of cognitive functions. It is distinguished by the presence of extracellular amyloid beta peptides, intracellular neurofibrillary tangles, and substantial loss in the cortex and hippocampus region of the brain. AD is incurable and has significant social and economic impacts. The disease, therefore, essentially requires successful diagnostics and effective therapeutic approaches. It has been demonstrated that conventional approaches often fail to achieve excellent pharmacokinetic and pharmacodynamic properties at the target site and thus produce low therapeutic efficacy and high toxicity. Recent advances in the pharmaceutical domain have shown the development of nanosystems to overcome the limitations associated with conventional therapy. In addition, emergence of nanotechnology serves as a potential tool in understanding complex mechanisms as well as treatment strategies of AD. These nanosystems are site-specific and offer desired pharmacokinetic properties such as solubility, bioavailability, absorption, permeability across the blood-brain barrier, and better therapeutic effects. Nowadays, a plethora of nano-carriers including solid lipid carriers, liposomes, emulsions, and carbon nanotubes have been designed to attain greater therapeutic effect in AD. Furthermore, nanotechnology also contributes to the early diagnosis of AD. The current chapter

P. Singh (✉) · G. Mishra · M. M. Zeleke · A. Ewunetei
Department of Pharmacy, College of Health Sciences, Debre Tabor University, Debre Tabor, Amhara, Ethiopia

F. H. Pottoo
Department of Pharmacology, College of Clinical Pharmacy, Imam Abdulrahman Bin Faisal University, Dammam, Saudi Arabia

G. M. Ashraf, M. S. Uddin (eds.), *Current Thoughts on Dementia*,
https://doi.org/10.1007/978-981-16-7606-2_22

encompasses latest developments in nanotechnology-based diagnosis and therapeutic strategies for AD.

Keywords

Alzheimer's disease · Amyloid beta · Nanotechnology · Nano-systems · Nanocarriers · Astaxanthin · Chlorogenic acid

22.1 Introduction

Among many other neurodegenerative diseases, Alzheimer's disease (AD) is a life-devastating disorder that predominantly occurs in old age population across the globe. The risk of AD generally begins in patients over 60 years of age and continues to cause detrimental effects on quality of brain health (Hossain et al. 2019). The disease was first documented by Alois Alzheimer, a German neurologist, who witnessed the development of some pathological components in his patient Auguste D. These pathological components constitute amyloid plaque or senile plaques (extracellular) and neurofibrillary tangles (intracellular deposits) which may lead to synaptic disconnections and neuronal damage in cerebral cortex and hippocampus regions (Sajjad et al. 2018; Swerdlow 2007). Numerous hypotheses, as depicted in Fig. 22.1, have been put forth to describe the physiopathological mechanisms of AD. Of these, amyloid hypothesis is a well-accepted hypothesis (Kocahan and Doğan 2017). There are three major risk factors associated with AD development including genetic, psychosocial, and vascular parameters (Povova et al. 2012). The most common features of AD include dementia, depression, loss of intellectual ability, language impairment, and erratic behavior and personality that affect basic activities of routine life (Sharma et al. 2020).

Recent advances have been made to diagnose and cure patients of AD and dementia as early as possible. Both invasive and non-invasive diagnostic procedures are used for the detection of AD. The invasive procedure involves the biochemical estimation of amyloid beta (Aβ) peptides, hyperphosphorylated tau proteins, and total tau proteins in cerebrospinal fluid. This method is inexpensive but provides less accuracy and cause detrimental effects on lumber region. Latest approaches have implemented several neuroimaging techniques such as computed tomography (CT), positron emission tomography (PET), single photon emission computed tomography (SPECT), and magnetic resonance imaging (MRI) to improve the diagnostic accuracy and reliability (Weller and Budson 2018). Recent evidence shows the use of nanotechnology as a diagnostic tool in AD.

To date, there is no permanent therapy due to a lack of mechanistic approaches behind AD physiopathology. Although the US Food and Drug Administration has advocated the use of some cholinesterase inhibitors such as huperzine-A, rivastigmine, and galantamine as neuroprotective agents, these drugs may relieve

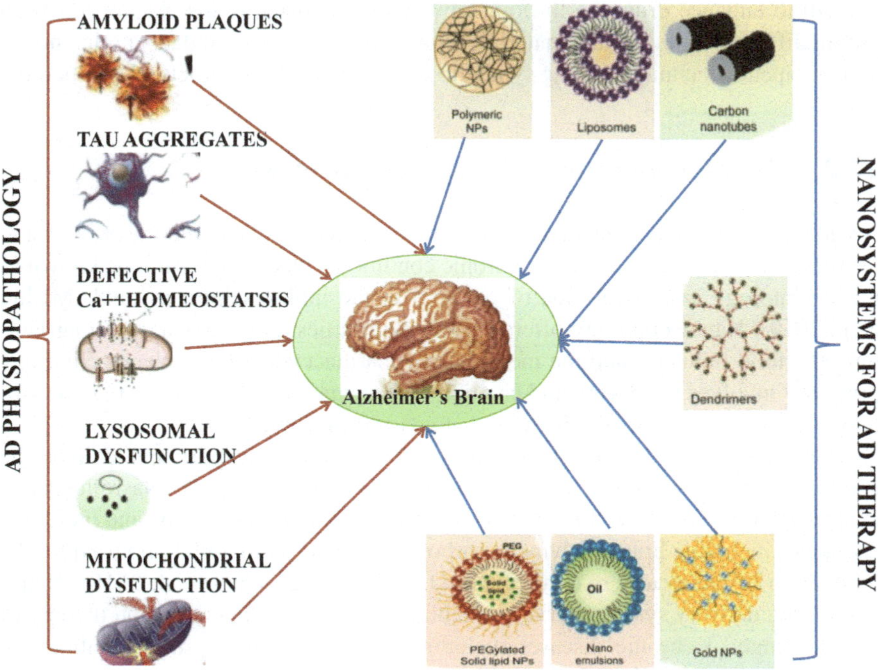

Fig. 22.1 Pathophysiology and various nanosystems used in Alzheimer's disease therapy

psychological and behavioral symptoms in AD patients. In addition, some other therapeutic strategies targeting Aβ-peptides (β-secretase and γ-secretase inhibitors/modulators), Aβ-aggregation inhibitors, M1-muscarinic agonists, tau inhibitors, and immunotherapy have also been incorporated for the treatment of AD (Hong-Qi et al. 2012).

Current research has now been centered on the development of novel therapeutic candidates of both synthetic and natural origins that demonstrate multi-target mechanisms in the treatment of AD. A plethora of natural products obtained from plants, animals, and microorganisms have been investigated for neuroprotective activity against Aβ plaques. Many of these exhibit potent antioxidant and inhibitory acetylcholinesterase (AChE) activity (Shao and Xiao 2013). More recently, in silico docking studies elaborated the use of some flavanones in the treatment of AD through inhibitory effects on acetylcholinesterase, butyrylcholinesterases (BuChE), and β-secretase 1 enzyme (BACE-1) (Lee et al. 2018). β-secretase, also called β-site amyloid precursor protein cleaving enzyme, cleaves amyloid precursor protein (APP) to generate β-amyloid peptides (Shimizu et al. 2008). It is noteworthy that the therapeutic effects of these drugs can be impeded due to poor pharmacokinetic as well as pharmacodynamic properties. Moreover, blood-brain barrier (BBB) also limits the delivery of the drug to specific site or target within the central nervous system (Pottoo et al. 2020). Due to the aforementioned factors influencing efficacy and safety of drugs, novel nanotechnology is being used in the pharmaceutical domain to enhance drug pharmacotherapy. The key focus in this chapter has been put on different types of nanomaterials along with the current updates in diagnostic and therapeutic techniques using synthetic as well as natural products in AD therapy.

22.2 Nanotechnology in Alzheimer's Disease

A wide array of therapeutic interventions are being adopted to avoid the progression as well as the treatment of various chronic conditions. One such technique known as nanotechnology is a novel area of research in the multidisciplinary field (Mishra et al. 2019). Nanotechnology offers great opportunities and applications in agriculture, the food industry, and the medical and pharmaceutical field. Nanotechnology primarily involves the design and creation of structures in the nanometers scale size range (1–100 nm) by manipulating the atoms and molecules.

These innovative nanostructures may conveniently transport and deliver the entrapped drug to the specific targeted sites safely and effectively (Nasrollahzadeh et al. 2019; Pottoo et al. 2020; Teleanu et al. 2019). Numerous organic and inorganic chemicals including lipids, polymers, carbon, or metallic compounds are used for the fabrication of nano-carriers (Harshita et al. 2020; Ramanathan et al. 2018). Nano-based drug delivery systems serve as enabling tools for the diagnosis and therapy of a wide range of chronic diseases including cancer, diabetes, asthma, pulmonary tuberculosis, atherosclerosis, cardiovascular diseases, and neurodegenerative diseases (Ansari et al. 2020a, b; Mishra et al. 2019). In particular, poorly soluble drugs with low absorption and permeability across the BBB have been encapsulated with nano-carriers for their sustained and controlled delivery and provide the desired therapeutic effect (Singh et al. 2019a, b).

The advent of nano-based systems has indeed brought about the solution of the above-mentioned problems and enhanced drug permeability across the BBB (Juillerat-Jeanneret 2008; Pottoo et al. 2020). Ideally, the nano-drug systems should have certain characteristic properties such as surface character, controlled particle size, solubility, flexibility, penetrability, and release of drug to get site-specific target and activity at a predetermined rate and time (Bennet and Kim 2014). Both natural and synthetic polymers with different physicochemical properties are being used to fabricate smart nanoparticles (Sharma et al. 2019). Natural polysaccharides such as tragacanth, scleroglucan, cellulose, chitosan & chitin, xanthan gum, guar gum, and locust bean gum have been largely investigated for the preparation of nanosystems. These natural polysaccharides offer excellent mucoadhesive as well as mechanical properties to the nanosystems. In addition, polylactides (PLA), polyglycolides (PGA), poly (lactide co-glycolides) (PLGA), polyanhydrides, poly glutamic acid, poly (N-vinylpyrrolidone), poly (methyl methacrylate), poly (vinyl alcohol), polyacrylamide, and poly ethylene glycol (PEG) are some of the examples of synthetic polymers employed in the synthesis of nanomaterials. These nanosystems offer optimum therapeutic effect with minimum side effects. Therefore, preparation techniques and route of administration of nanomaterials encapsulated with therapeutic agents play a significant role in achieving desired properties for a particular application. There are numerous methods of preparation of nanomaterial including ionic gelation, solvent evaporation, nanoprecipitation, dialysis, salting out method, emulsification and milling method which are applied for developing several types of nano-sized particulates (Brahamdutt et al. 2018).

The route of administration of nanosystems specifies the right delivery of entrapped drug at the precise target. The nanoparticles can be administered through various routes such as oral, nasal, transdermal, ocular, parenterals, pulmonary, etc., each with their merits and demerits. The oral route is one of the most common and convenient methods; but BBB, degradation, or metabolization of the drug by body fluids and hepatic clearance restricts the use of this route. Recently, intranasal administration has been proven as an excellent alternative and non-invasive method to deliver the drug at specific target especially in brain since the drugs are directly transported to CNS bypassing the BBB (Ansari et al. 2020a, b; Graff and Pollack 2005). Recent investigations have highlighted the development of numerous nano-sized particles as diagnostic and therapeutic measures for a number of chronic diseases including neurodegenerative diseases (Fonseca-Santos et al. 2015; Harilal et al. 2019; Singh et al. 2019a, b).

22.3 Nanotechnology-Based Drug Delivery Systems for Alzheimer's Disease

Many nanosystems including liposomes, polymeric and solid lipid nanoparticles (NPs), solid lipid carriers, liquid crystals, microemulsions, and hydrogels have been extensively researched and evaluated for anti-Alzheimer's activity as shown in Fig. 22.1. This section briefly describes different types of nanosystems and their applications in the field of AD therapeutics as depicted in Table 22.1 (Alyautdin et al. 2014; Amidon et al. 1995; Haque et al. 2012; Paolino et al. 2011; Stegemann et al. 2007).

22.3.1 Polymeric Nanoparticles

Polymeric nanoparticles (NPs) are composed of an aqueous or lipid core surrounded by a thin layer of polymer membrane (Guterres et al. 2007). The size of nanoparticles ranges from 1 to 1000 nm (Kreuter 1978). The polymeric chains organized on a matrix basis are called nanosphere (Couvreur et al. 2002). There are various methods of preparation of NPs including polymer polymerization (Boudad et al. 2001), spontaneous emulsification or solvent diffusion (Calvo et al. 1997), spray drying (Ali and Lamprecht 2014), ionic gelation or coacervation (Dong et al. 2013), nanoprecipitation (Mazzarino et al. 2014), emulsion solvent evaporation (Mainardes and Evangelista 2005), supercritical fluid technology (Hu et al. 2011), and particle replication in non-wetting templates (Chu et al. 2013). Various polymeric materials used are polysorbate 80-coated poly(*n*-butyl cyanoacrylate) (Wilson et al. 2010), poly ethylene gylated poly(lactic acid) polymer (Zhang et al. 2014a), PEG and

Table 22.1 Nanotechnological applications of various nanomaterials for Alzheimer's disease (Siddiqi et al. 2018)

Nanoparticles/ nanomaterials	Entrapped drugs	Applications
Polymeric nanoparticles (NPs, 1–1000 nm)	Neuroprotective peptide, rivastigmine, curcumin, estradiol, S14G-humanin, anti Aβ antibody, fibroblast growth factor, Aβ-targeting peptide, iron chelator, selegiline, clioquinol	• Drug-loaded NPs displayed specificity for Aβ plaques both in vitro and in vivo • Capable of aiding in the early diagnosis of AD
Liposomes, CPP-modified liposomes, flexible liposomes (200–500 μm)	Curcumin, phosphatidic acid, cardiolipin, XO4, glycol fused benzopyrane, anti Aβ antibody, ZnAc, epigallocatechin-3-gallate, quercetin, rivastigmine HCl, galantamine	• Stabilize therapeutic compounds, overcome obstacles to cellular and tissue uptake, and improve bio-distribution of compounds to target sites in vivo • Excellent delivery system due to their flexible physicochemical and biophysical properties, which allow easy manipulation to address different delivery considerations
Solid lipid NPs and lipid-coated microbubble/NP-derived (LCM/ND) (50–1000 nm)	Piperine galantamine, lipoylmemantine, rivastigmine HCl	• Stabilizing drugs that suffer from physicochemical or biological instability • Improve the bioavailability of drugs that cross the BBB • Increase the permeability of drugs through the BBB
Chitosan NPs (15–200 nm)	Tacrine	• Enhance the concentration of drug in the brain • Are more stable, permeable, and bioactive
Magnetite NPs (1 nm–5 μm)	Tacrine	• Serve as promising biomarkers for detecting the location and the removal of other amyloid plaques derived from different amyloidogenic proteins
Albumin NPs (40–500 nm)	Apo-E binding, tacrine	• Enhanced brain uptake of NPs by cerebral endothelium, by an endocytic mechanism, followed by transcytosis into the brain parenchyma
Gold NPs (1–150 nm)	Aβ-binding peptide	• Gold NPs dissolve toxic protein deposits of $A\beta_{1-42}$ (amyloid deposits) by the combined use of weak

(continued)

Table 22.1 (continued)

Nanoparticles/ nanomaterials	Entrapped drugs	Applications
		microwave fields and gold NPs without any bulk heating
Exosomes (30–100 nm)	BACE1-siRNA	• Penetrate the BBB and deliver drugs to the brain • Possess a suitable half-life for many diseases
Polystyrene NPs (240 nm)	Penicillamine	• Deliver D-penicillamine to the brain for the prevention of Aβ accumulation
Core-shell NPs	Thioflavin T and S	• Tools to trace and clear Aβ in the brain
Nanolipidic and microparticles (30–80 nm)	Polyphenol EGCG, donepezil	• Prevent Aβ formation. Acetylcholine esterase inhibitor with high specificity for acetylcholine esterase in the central nervous system
Trimethylated chitosan conjugated-PLGA NPs	Coenzyme Q10 (Co-Q10)	• Improved memory impairment and restoring it to a normal level
Poly(butyl) cyanoacrylate NPs	Apo-E binding	• Attachment of ApoE3 to C-PBCA NPs increased the uptake of curcumin into cells as compared to the plain solution or untargeted NPs
Nanoemulsions (10–1000 nm)	Nano-PSO, lipid-coated microbubble/NP-derived (LCM/ND)-scavenger receptor class B type I	• Good solubilization and protection of lipophilic drugs in the oil droplets and easy for large-scale production
Microemulsions (1–100 nm)	Huperzine A and ligustrazine phosphate	• Improve the cerebral cholinergic function and oxidative systems that further slow down the progression of Alzheimer's disease
Dendrimers	Amyloid β-derived diffusible ligand (ADDL), fourth (PPIG4-Mal) and fifth (PPI-G5-Mal) phosphorus-containing dendrimers	• Modulate amyloidogenesis and stop the aggregation of tau protein. Interfering with Aβ fibrilization in AD

polylactide-polyglycolide (PLGA) and *Solanum tuberosum* lectin (STL) (Zhang et al. 2014b), poly[(hexadecyl cyanoacrylate)-*co*-methoxypoly (ethylene glycol) cyanoacrylate] (Brambilla et al. 2010), apolipoprotein-E (ApoE) (Krishna et al. 2019), chitosan and *N*-carboxymethylchitosan (Amorim et al. 2010; Fazil et al. 2012).

22.3.2 Solid Lipid Carriers

The solid lipid carriers (SLNs) are spherical in shape with a dimeter of 10–1000 nm. The solid lipid core matrix of these nanostructures is made up of monoglycerides (glycerol monostearate), diglycerides (glyceryl behenate), triglycerides (tristearin), fatty acids (stearic acid), steroids (cholesterol), and waxes (cetyl palmitate) and is stabilized by surfactants which solubilize lipophilic molecules (Mehnert and Mäder 2012; Müller et al. 2002). SLNs can be prepared using high-pressure homogenization (Kovačević et al. 2014), an ultrasonication/high-shear technique (Martins et al. 2013), solvent evaporation method (Vitorino et al. 2011), solvent emulsification-diffusion method (Kumar et al. 2013), supercritical fluid method (Campardelli et al. 2013), microemulsion-based method (de Souza et al. 2012), spray-drying method (Freitas and Müllerä 1998), double emulsion method (Martins et al. 2009), and precipitation technique (Dong et al. 2012).

22.3.3 Liposomes

Liposomes are lipophilic or hydrophilic drug carriers and composed of one or more phospholipid bilayers around aqueous pockets (Gulati et al. 1998; Kumar et al. 2020; Lasic 1998). These are grouped into different classes, such as ethosomes transfersomes, niosomes, and phytosomes. Hydration of a thin lipid film followed by agitation (Chorilli et al. 2013; Ghanbarzadeh et al. 2013), sonication (Hadian et al. 2014), extrusion (Isailović et al. 2013), high-pressure homogenization (Pupo et al. 2005), or reverse-phase evaporation (Szoka Jr and Papahadjopoulos 1978) are the methods used for the preparation of liposomes.

22.3.4 Surfactant-Based Systems

Surfactant-based systems are generally prepared by self-aggregation of surfactant molecules in water. The aggregated structures depend on surfactant concentration, salts present, and temperature. More organized aggregates can be obtained by adding oil or other surfactants on the surfactant-water system. Based on the aforementioned parameters, different types of surfactant-based drug delivery systems such as microemulsions, nanoemulsions, and lyotropic liquid crystals mesophases can be produced (Ezrahi et al. 1999).

22.3.4.1 Microemulsions

Microemulsions (MEs) are isotropic liquids, produced by the simple blending of components such as oil (mixture of various hydrocarbons and olefins), water, and surfactants and are mostly thermodynamically stable (Nafisi and Maibach 2017). Specific preparatory conditions are not required in developing these systems. The size of microemulsion droplets generally occurs between 10 and 140 nm (Langevin

1988), which is essential for optically transparent and thermodynamically stable systems.

22.3.4.2 Nanoemulsions

Nanoemulsions (NEs) are conventional colloidal particulate systems of two immiscible liquids with droplet sizes ranging from 10 to 1000 nm (Jaiswal et al. 2015). These nanostructures are considered optically transparent and thermodynamically stable systems (Barkat et al. 2020; Langevin 1988; Nirale et al. 2020). In composition and nanoscale structure, nanoemulsions may be similar to some lyotropic liquid crystalline phases, also known as "micellar phases," "mesophases," and "microemulsions," but actually quite different (Chavda 2019). The most common techniques applied for the synthesis of nanoemulsions involve microfluidization, high-pressure homogenization, and the phase-inversion temperature method (Simonazzi et al. 2018).

22.3.4.3 Liquid Crystals

Liquid crystals (LCs) are elongated organic molecules that possess both structural order and mobility. Liquid crystalline structures exhibit characteristic properties of solids and liquids along with anisotropy and optical activity. Uneven distribution of electrical charges throughout their axis (dipole) leads to the special physical characteristic of liquid crystals between crystalline and liquid states. Thermotropic and lyotropic mesophases are the two classes of liquid crystals. A temperature shift results in thermotropic liquid crystalline phases, while lyotropic phases result from mixing with aqueous water. For phase transitions, thermotropic liquid crystals are dependent on temperature, whereas lyotropic liquid crystals depend on both temperature and concentration (Rajak et al. 2019).

22.3.5 Carbon Nanomaterials

22.3.5.1 Carbon Nanotube

The carbon nanotubes (CNT) were originally developed by Iijima (1991). These are mainly characterized by unique properties such as high flexibility, ultra-lightweight, low cost, and inert with thermal and electrical conductivity. Currently, these nanotubes have achieved a great position in nanotechnology field for the treatment of CNS disorders like AD, Perkinson's Disease (PD), and ischemic stroke due to their exclusive properties (Siddiqi et al. 2018). The carbon nanotubes can be divided into single-layered and multilayered nanotubes used as drug carriers. These are made up of graphite layers with higher surface area, and good thermal and electronic conductivity. The nanotubes protect the degradation of the drug during transportation as well as control the release electrically and chemically, as drug is inside the nanotube. The open ends of nanotubes are sealed by polypyrrole films (Chavda 2019).

22.3.5.2 Fullerene

Fullerenes (C60) or Buckminster fullerenes were first discovered by Kroto et al. (1985). These are spherical crystal form of carbons bound by single and double bonds in three-dimensional geodesic spheroidal shape (Zhou 2013). These unique nanoparticles can be used as carrier for neuroprotective drugs. Fullerenes exert potent antioxidant effect due to their unique basic skeleton linked with chemical groups. Some fullerene derivatives such as carboxyfullerene, hydroxyfullerene (fullerenols), and hydrated C60 fullerene (C60HyFn) have promising applicability for AD nanotherapy (Nazem and Mansoori 2011).

22.3.6 Inorganic Nanoparticles

In addition to organic materials, inorganic substances for example SiO_2 have also been investigated for diagnosis and treatment of AD. Silicon is a widely used material in the nanotechnology field due to its extremely biocompatible nature and penetrability across the BBB (Sivasankarapillai et al. 2019).

22.3.7 Magnetic Nanoparticles

In recent years, magnetic nanoparticles have been designed for both diagnostic and therapeutic purposes (Ahmad et al. 2017). When external electromagnetic fields of 28 or 79.8 mT are applied, these nanoparticles can easily cross the BBB. Furthermore, brain uptake and transport rates of magnetic nanoparticles can also be improved by applying a pulsed magnetic field (Do et al. 2016). Magnetic nanoparticles are promising carrier for active agents as these are able to provide optimal dose to the patient by active or passive targeting and can be visualized by an external magnetic field. Magnetic nanoparticles are produced from pure metals like nickel, manganese, iron, cobalt, and their alloys as well as oxides. More stable nanoparticles are obtained by modifying the chemical and physical properties of the particles. Encapsulation, adsorption, covalent binding, or electrostatic interactions are some methods employed for loading the drugs in these nanoparticles (Chavda 2019).

22.3.8 Gold Nanoparticles

Gold nanoparticles are produced by the reduction of gold salts such as AuCl (PPh_3), $HAuCl_4$ in the presence of stabilizing agents like diborane, and sodium borohydride (Alam et al. 2019). The reduction process and stabilizing agents are used to determine the particle size of gold nanoparticles (McDonald et al. 2015). Gold nanoparticles show anti-inflammatory and antioxidant properties and can be used as a promising nanotherapeutic agent for AD by reducing neuroinflammation, modulation of mitochondrial functions, and enhancing cognitive abilities (Dos Santos Tramontin et al. 2020).

22.3.9 Dendrimers

Dendrimers are spherical in shape, highly branched monodisperse polymeric macromolecules with low viscosity. The presence of large numbers of hydrophilic functional groups results in better solubility in water. Dendrimers display similar properties like proteins and are able to transport antibodies and harmones to the site of action (Aliev et al. 2019; Chavda 2019). These can be prepared from a divergent method, convergent method, or double exponential and mixed growth methods.

22.3.10 Cubosomes

Cubosomes, novel nanostructured particles, are comprised of biocompatible carriers. Cubosomes are designed in such a way that a three-dimensional lipid bilayer remains separated by two aqueous channels in which active drug and proteins come into contact with each other. These nanoparticles are thermodynamically stable and are able to load hydrophilic, hydrophobic, and amphiphilic substances. These are unique carriers for controlled release of the drug and can be delivered by several routes (Karami and Hamidi 2016).

22.3.11 Antibody-Tethered Nanoparticles

Antibodies are regarded as antigen neutralizer and potential activators of the complement system. These are nanosized biological components and are extensively researched for AD therapy. Recently, $A\beta_{1-42}$ monoclonal antibody-tethered nanoparticles have been fabricated against AD (Arruebo et al. 2009; Carradori et al. 2018). Another study reported the development of surface-modified fluorescent silica NP derivatives encapsulated with glucose (Glu) and glucose-poly(ethylene glycol) methyl ether amine (Glu-PEG) and also analyzed for their permeability across the BBB (Tamba et al. 2018). However, future studies demand designing more potent antibody nanoparticles with great efficacy and reduced side effects.

22.4 Diagnostic Approaches in Alzheimer's Disease

AD is the fifth leading cause of death in individuals over the age of 65 years. It was estimated that one new case of AD occurs every 33 s and one million in each year (Thies and Bleiler 2013). Therefore, there is a growing need for advanced and effective diagnostic methods to investigate the mild signs of AD in the early stages before dementia progresses. Currently, few diagnostic techniques for AD have been documented including neuroimaging tools and estimations of biological biomarkers in cerebrospinal fluid (CSF) as well as in the blood. Biomarkers are the substances which are measured inside or outside the body to examine the biological processes, pathological state, or pharmacologic responses to a therapeutic intervention. It has been reported that identification of biomarkers in CSF is useful for detection of dementia caused due to AD and other neurological infections such as neuroborreliosis and neurosyphilis (Fleming et al. 1995). There are three main established biomarkers measured in CSF such as Aβ, tau protein, and phosphorylated tau. CSF analysis is cost-effective, but invasive, i.e., requires lumber puncture procedure, and provides less diagnostic accuracy (85–90%) (Sharma and Singh 2016).

On the other hand, blood testing can be used to diagnose AD which estimates the amount of circulatory proteins and antibodies as biomarkers. Nevertheless, the use of specific assays and validation work is a major challenge while measuring the blood biomarkers. Furthermore, the blood brain barrier hinders the permeability of molecules between CNS and blood compartment, thereby reducing the concentrations of biomarkers in the blood (Zetterberg and Burnham 2019). Currently, panels of biomarkers are being identified to achieve satisfactory results since no single biomarker could detect AD (Eke et al. 2018). In a study, scientists identified 18 plasma proteins which were able to diagnose patients associated with AD with 90% accuracy (Ray et al. 2007). Another study reported the use of eight-protein panel namely brain-derived neurotrophic factor, angiotensinogen, insulin-like growth factor binding protein-2, osteopontin, cathepsin-D, serum amyloid P component, complement C4, and prealbumin as promising diagnostic blood biomarkers for AD with great accuracy and specificity (Cheng et al. 2018). Furthermore, other novel strategies involving the use of metabolomics, micro-RNA, and exosomes create greater interest in the field of blood biomarkers for AD (Sharma and Singh 2016). In addition, neuroimaging techniques including MRI, CT, PET, and SPECT are other beneficial diagnostic tools for AD.

22.4.1 Nanodiagnostic Approaches for Alzheimer's Disease

Currently, some studies highlighted the application of nanotechnology in combination with imaging tools in molecular detection of biomarkers using in vivo or in vitro methods for AD diagnosis. In vivo diagnostic methods include micro magnetic resonance imaging (μMRI), optical imaging, SPECT, and PET while in vitro methods include DNA-nanoparticle conjugates (Bio-Barcode Assay), nanoparticle surface plasmon resonance, scanning tunneling microscopy system, two-photon-rayleigh spectroscopy as summarized in Table 22.2 (Nazem and Mansoori 2011).

Table 22.2 Nanodiagnostic tools for AD (Fonseca-Santos et al. 2015; Nazem and Mansoori 2011)

Diagnostic mode	Method	Nanomaterial used	Biomarker detected
In vivo	Micro magnetic resonance imaging (μMRI)	Iron oxide nanoparticles	Aβ
	Optical imaging	NIAD-4 [[5′-(4-Hydroxyphenyl) [2,2′-bithiophen]-5-yl] methylene]-propanedinitrile	Aβ
	Single photon emission computed tomography (SPECT) and positron emission tomography (PET)	Polymeric nanoparticles encapsulated with radio-labeled ^{125}I-clioquinol & Polymeric NPs loaded with thioflavin-T	Aβ
In vitro	DNA-nanoparticle conjugates (bio-barcode assay)	Gold nanoparticles	Aβ-derived diffusible ligand (ADDL)
	Nanoparticle surface plasmon resonance	Silver nanoparticles	Aβ in ultra-low concentrations
	Scanning tunneling microscopy system	Gold nanoparticles	Aβ as low as 10 fg/ml concentration
	Two-photon-rayleigh spectroscopy	Gold nanoparticles	Tau protein

22.5 Nanotherapy for Alzheimer's Disease

22.5.1 Synthetic Approaches

22.5.1.1 Rivastigmine

Rivastigmine, a well known AChE and BuChE inhibitor, was approved for the treatment of mild-to-moderate AD. The drug also has antiparkisonism activity (Inglis 2002). The drug produces major side effects such as nausea, vomiting, stomach pain, and diarrhea, while overdose may cause chest pain, and irregular heartbeat and breathing. Rivastigmine has an excellent penetrability across the blood brain barrier (Corey-Bloom et al. 1998). Previously, rivastigmine has been formulated in the form of oral capsule and reformulated as transdermermal patches to reduce gastrointestinal side effects (Sadowsky et al. 2014). Currently, the drug is undertaken to develop nanosystems to enhance the bioavailability and efficacy and to reduce side effects as enlisted in Table 22.3.

Table 22.3 Rivastigmine-based nanoformulations for Alzheimer's disease (Fazil et al. 2012; Kaur et al. 2011; Ravi and Gupta 2017; Tamilselvan et al. 2014; Vijaykumar et al. 2014; Wilson et al. 2011)

Nanomaterial/polymer	Method of preparation	Effect
Chitosan & sodium triplowritten-based polymeric NPs	Ionotropic gelation	Effective carriers for the design of controlled delivery for drugs with short half-life and improved bioavailability
Poly ethylene oxide– And polyvinyl alcohol–based polymeric NPs	Solvent displacement technique	Increased bioavailability of the drug
Polymeric NPs of chitosan coated with polysorbate 80	Spontaneous emulsification	Controlled biodistribution and uptake of nanoparticles by different organs
Polymeric NPs of poly lactide-co-glycolide coated polysorbate 80	Modified nanoprecipitation method	Enhanced bioavailability for brain targeting
Solid lipid NPs of stearic acid & propylene glycol	Modified solvent emulsification technique	Faster drug release

22.5.1.2 Donepezil

Donepezil represents unique molecular structure with multiple mechanistic approaches. Donepezil has been prescribed to cure mild, moderate, and severe AD. The drug not only performs its action on neurotransmitters but also at cellular and molecular levels in the pathogenesis of AD. Donepezil inhibits glutamate-induced excitotoxity, the induction of AChE, and reduction of early expression of inflammatory cytokines and oxidative stress (Jacobson and Sabbagh 2008). In addition, it blocks ionic and peripheral anionic sites of TcAChE (Kryger et al. 1999). Donepezil possesses various side effects including muscle cramps, insomnia, loss of appetite, nausea, vomiting, and diarrhea (Rogers et al. 1998). Recent investigations reported the development of different nanoformulations such as polymeric nanoparticles, nanoemulsions, liposomes, and carbon tubes entrapped with donepezil are listed in Table 22.4.

22.5.1.3 Tacrine

Tacrine is an excellent synthetic molecule that was originally reported as a muscle relaxant antagonist and respiratory stimulant (Karis et al. 1966). In addition, the drug has been employed for the treatment of AD. In 1993, FDA first approved the drug as an anti-Alzheimer's agent and was terminated in 2013 due to serious side effects including hepatotoxicity (Tumiatti et al. 2010). Tacrine serves as a potent inhibitor of both AChE and BuChE. It binds with anionic sites of AChE predominantly with Phe330 and Trp84 (Fernández-Bachiller et al. 2010). In order to overcome the side effects and improve the bioavailability in brain as well as ability to cross blood brain barrier, various nanotechnological preparations have been researched as summarized in Table 22.5.

Table 22.4 Donepezil-based nanoformulations for Alzheimer's disease (Azalea et al. 2012; Baysal et al. 2017; Harthi et al. 2019; Kaur et al. 2020a, b; Krishna et al. 2019; Md et al. 2014; Mukhopadhyay et al. 2017; Singh et al. 2014; Tao et al. 2018)

Nanomaterial/polymer	Method of preparation	Effect
Carbon nanotubes	–	Improved bioavailability of donepezil with better-targeted delivery in brain
Polyvinyl pyrolidone (PVP), chitosan, 2-imino thiolane hydrochloride, cholesterol, dipalmitoylphosphocholine (DPPC), polyethylene glycol 600, and lactic acid glutaraldehyd–based liposomes	Reverse phase evaporation	Improved bioavailability of donepezil in brain
Abrasol-, cetyl pyridinium chloride-, and glycerol-based nanoemulsion	Homogenization and ultrasonication method	Improved bioavailability of donepezil in brain
Methoxy poly(ethylene glycol)–polycaprolactone (mPEG–PCL), polysorbate 80 and apolipoprotein E (ApoE) –based polymeric NPs	Nanoprecipitation	Improved bioavailability of donepezil in brain with dose reduction
Polymeric NPs of poly(lactic-co-glycolic acid)-block-poly (ethylene glycol) [PLGA-b-PEG]	Double emulsion method	Increased the transport of donepezil across the BBB with controlled release profile
Chitosan and sodium tripolyphos-phate-based polymeric NPs	Ionic gelation	Better brain targeting efficiency
Polymeric NPs of biomaterial from *Carica papaya* fruits	Nanoprecipitation	Increased the transport of donepezil across the BBB with controlled release profile
Polymeric NPs of cholesterol hydrophobic modified pullulan (CHP) coated with polysorbate 80	Dialysis method	Controlled release of drug-loaded CHP NPs in vitro experiments

22.5.1.4 Memantine

Memantine (1-Amino-3,5-dimethyladamantane) is an N-methyl-D-aspartate (NMDA) receptor antagonist and is used either as monotherapy or in combination with acetylcholinestarase inhibitors such as rivastigmine, donepezil, tacrine, and glutamate (Rogawski and Wenk 2003; Schneider et al. 2014). The drug was first synthesized by Eli-Lilly laboratories for diabetes but later in 1999; Forest laboratories started its use for moderate–to-severe AD (Tu et al. 2014). It acts by blocking a neurotransmitter called glutamate in brain which is responsible for neuronal excitatbility in AD (Kuns and Varghese 2019). Numerous nano-based formulations as depicted in Table 22.6 have been designed to improve the bioavailability and reduce the adverse effects associated with drug (Sánchez-López et al. 2018).

Table 22.5 Tacrine-based nanoformulations for Alzheimer's disease (Felix Joe and Sathesh Kumar 2018; Luppi et al. 2011; Sathesh Kumar and Felix Joe 2017; Wilson et al. 2008; Wilson et al. 2010)

Nanomaterial/polymer	Method of preparation	Effect
Bovine serum albumin, beta cyclodextrin, hydroxypropyl beta cyclodextrin and sulphobutylether beta cyclodextrin–based protein NPs	Coacervation method	Increased the transport of tacrine across the BBB
Polymeric NPs of methoxy poly(ethylene glycol) poly(caprolactone) (mPEG-PCL)	Emulsion polymerization	Increased the transport of tacrine across the BBB
Poly (lactide-co-glycolide) based polymeric NPs	Nano precipitation method	Improved bioavailability of tacrine in brain
Polymeric NPs of chitosan coated with polysorbate 80	Spontaneous emulsification	Improved bioavailability of tacrine in brain
Polymeric NPs of poly (n-butylcyanoacrylate) (PBCA) coated with polysorbate 80	Emulsion polymerization	Improved bioavailability of tacrine in brain

Table 22.6 Memantine-based nanoformulations for Alzheimer's disease (Kaur et al. 2020a, b; Radwan et al. 2020; Rao et al. 2018; Ruby and Pandey 2014; Sánchez-López et al. 2018)

Nanomaterial/polymer	Method of preparation	Effect
Nanoemulsion for intranasal delivery	Homogenization and ultrasonication methods	Better brain targeting efficiency
Polymeric NPs of polylactic-*co*-glycolic (PLGA) & polyethylene glycol	Double emulsion method	Enhanced the penetrability of memantine across the BBB
Chitosan- and tripolyphosphate-based polymeric NPs	Ionic gelation	Better brain targeting efficiency
Polymeric NPs of chitosan	Gamma radiation	Better brain targeting efficiency
Protein NPs of casein – sodium tripolyphosphate	Homogenization and ionically cross-linking method	Better brain targeting efficiency with controlled release of memantine-HCl

22.5.2 Herbal Approaches

22.5.2.1 Curcumin

Curcumin is a well-known dietary component obtained from turmeric root (*Curcuma longa*). It is considered as one of the potent polyphenolic antioxidants and broadly researched as a cardioprotective, anticarcinogenic, anti-inflammatory, hepatoprotective, antiarththritic, and anti-infectious molecule. Apart from that, the candidate has also been investigated as a neuroprotective agent for several neurodegenerative diseases (Del Prado-Audelo et al. 2019; Monroy et al. 2013). It possesses pleiotropic properties with an enormous range of molecular targets (Fig. 22.2). Curcumin has been highlighted as an interesting molecule in the diagnosis and treatment of AD. It is a multi-target drug that prevents the aggregation of Aβ,

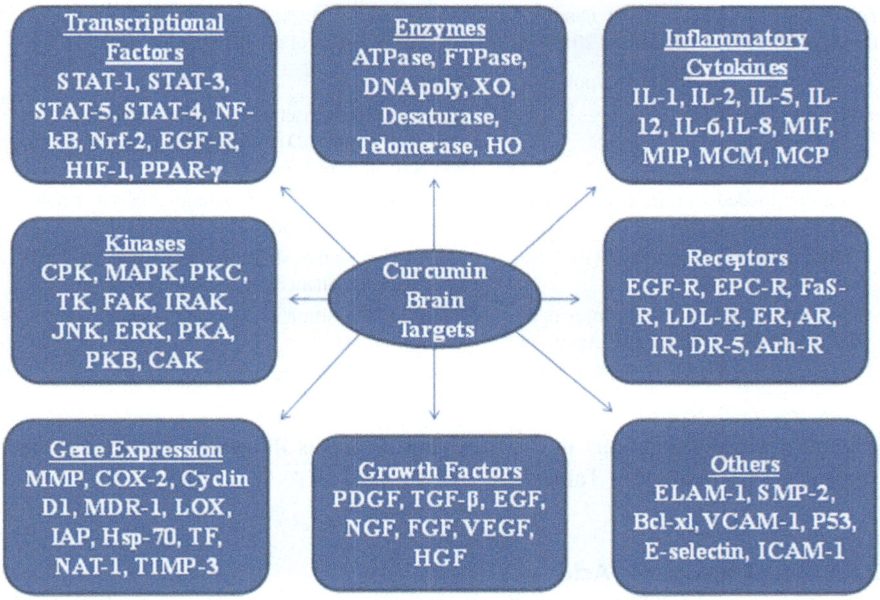

Fig. 22.2 Brain targets of curcumin in Alzheimer's disease

reduces hyperphosphorylation of tau, promotes its clearance, reduces cholesterol level and oxidative stress, inhibits acetylcholinesterase, controls insulin signaling pathway, and alters the microglial activity (Chen et al. 2018; Tang and Taghibiglou 2017).

Curcumin has poor aqueous solubility, low stability in body fluids, and bioavailability as well as rapid clearance which restrict its penetration in central nervous system. These limitations offered researchers a lot of interest in designing and improving its nano-based drug delivery systems. A recent study has shown the development of curcumin-based nanoparticle, poly (lactide-co-glycolide)-block-poly(ethylene glycol), which was further conjugated with B6 peptide. Administration of these curcumin-loaded nanoparticles in transgenic mice and HT22 cells caused significant reduction in β-amyloid aggregates and tau hyperphosphorylation (Fan et al. 2018). In another study, curcumin encapsulated with selenium and poly-lactide-co-glycolide (PLGA) nanospheres were synthesized as a novel brain target for neurodegenerative disease including AD. This nanoformulation has shown marked reduction in Aβ plaques and ameliorated the memory dysfunction (Huo et al. 2019). Neuronal damage induced by oxidative stress–related apoptosis may also trigger the development of neurodegenerative diseases. Therefore, dual drug delivery systems, namely, cubosomes and spongosomes, were prepared for neuronal regeneration. A combination of curcumin and fish oil encapsulated with lipid nanoparticles exhibited promising protective effect against H_2O_2-induced oxidative stress neurotoxicity in human SH-SY5Y cells (Rakotoarisoa et al. 2019). The use of

Table 22.7 Curcumin-based nanoformulations for Alzheimer's disease (Barbara et al. 2017; Beach et al. 1989; Cheng et al. 2013; Meng et al. 2015; Tiwari et al. 2013)

Curcumin-based nanomaterial/polymer	Effect
Curcumin-PLGA NPs	Increase gene/transcription factors including neurogenin, neuroD1, neuregulin, signal transducer and activators of transcription 3
Curcumin loaded with LDL-mimic-nano carrier with lactoferin	Slows down the level of malondialdehyde (major cause of lipid peroxidation)
Curcumin conjugated with PLGA NPs linked to g7 (cur-Nps-g7)	Decrease oxidative stress, inflammation, and Aβ aggregation; enhance the level of IkB (a protein)
Calreticulin (CRT) peptide conjugated with PLGA NPs with S1 and curcumin	Decrease deposition of Aβ, microglial activation and astrogliosis
Curcumin loaded with PEG-PLA NPs	Increase memory function

curcumin nanoformulations in the treatment of AD is illustrated by several other studies, as summarised in Table 22.7.

22.5.2.2 Chlorogenic Acid

Chlorogenic acid (CGA), one of the potent nutraceutical agents, is predominantly obtained from green coffee and tea. Chemically, it is known as 5-O-caffeoylquinic acid. It is an ester produced from cinnamic acid and quinic acid (Morimasa et al. 2018). CGA has been recognized to possess several biological properties including anti-inflammatory, antioxidant, antibacterial, antipyretic, antiviral, antihypertensive, hepatoprotective, cardioprotective, neuroprotective, etc. In addition, it is a powerful free radical scavenger and CNS stimulant (Naveed et al. 2018). Recent evidence has shown its role and applications in protecting the neurons and inhibiting the aggregation of Aβ plaques associated with AD. Besides CGA, its metabolites coumaric acid, ferulic acid, caffeic acid, and sinapic acids have also been reported to ameliorate learning and memory impairments (Morimasa et al. 2018).

Furthermore, in a randomized, double-blind, placebo-controlled clinical trial, CGA displayed beneficial effects on cognitive functions (Saitou et al. 2018). It is believed that chlorogenic acid appears to inhibit acetylcholinesterase and butyrylcholineserase enzymes along with reduction of oxidative stress responsible for neuron damage (Oboh et al. 2013). Nowadays, novel approaches are being adopted by the researchers to design nanoscale drug delivery systems for CGA. It has been demonstrated that CGA, a natural antioxidant, encapsulated with selenium nanoparticles strongly inhibited the aggregation of Aβ40 followed by providing protection to PC_{12} cells as compared to free CGA (Yang et al. 2018). Another nanoformulation of CGA has been developed with chitosan displaying promising sustain release property, bioavailability, and antioxidant activity. This formulation may be useful for the management of AD due to its antioxidant activity (Nallamuthu et al. 2015).

22.5.2.3 Asiatic Acid

Asiatic acid has achieved a great position in the pharmaceutical field due to the presence of an enormous range of therapeutic properties including antimicrobial, antidiabetic, anticancer, antioxidant, anti-inflammmatory, cardioprotective and neuroprotective, gastroprotective and hepatoprotective. It is a pentacyclic triterpenoid that abundantly occurs in various edible and medicinal plants such as *Centella asiatica*, commonly known as Gotu Kola (Lv et al. 2018; Ojha et al. 2018). Many reports have shown the beneficial effects of asiatic acid in improving learning and memory functions. In addition, it is an ideal molecule for the treatment of AD as well. Acetylcholinesterase and excitatory postsynaptic potential inhibition are the major targets of the drug (Nasir et al. 2012). The research papers have shown few preclinical studies on *Centella asiatica* extracts and clinical investigations on some formulations for brain disorders (Gray et al. 2018). However, the researchers are now fabricating some novel brain targeted drug delivery systems to improve the drug bioavailability. It has been illustrated that the combination of glutathione and asiatic acid encapsulated with bovine serum albumin nanoparticles was established which further enhanced the delivery of asiatic acid, thereby improving brain efficiency (Raval et al. 2015). On the other hand, the same nanoformulation was evaluated for its in vivo and in vitro neuroprotective potential and produced beneficial effects on brain-related impairments (Raval et al. 2018). These findings may validate the use of asiatic acid to develop nanoformulations for AD therapy.

22.5.2.4 Piperine

Piper nigrum (black pepper) popularly recognized as the king of spices that impart flavor and pungent taste of food products. In addition, the plant is known to exhibit many therapeutic actions including in central nervous system which can be attributed to one of the major components piperine. The concentration of piperine in black pepper ranges from 2% to 9% (Stojanovic-Radic et al. 2019; Takooree et al. 2019). Piperine is the alkaloid extensively investigated for many biologicalproperties such as antidiabetic, antimicrobial, anti-inflammatory, analgesic, immunomodulatory, antiallergic, antimutagenic, anticancer, gastroprotective, and neuroprotective and enzymatic activities. As a neuroprotective agent, it exhibits anticonvulsant, antianxiety, and antidepressant effects. Furthermore, the compound has been reported as a therapeutic agent for the treatment of dementia and related neurodegenerative diseases like AD and PD (Stojanovic-Radic et al. 2019). It improves memory and other cognitive deficits by inhibiting acetylcholinesterase and the lipid peroxidation process (Chonpathompikunlert et al. 2010). Interestingly, piperine is considered as a promising bioenhancer which might be credited to its binding with metabolic enzyme such as mixed function oxidases, transferases, etc. (Wadhwa et al. 2014).

In recent times, novel strategies have been applied to formulate piperine as drug delivery systems to improve its delivery to the target. Sustain release formulation of piperine was prepared using sodium alginates beads and evaluated (Madhavi et al. 2009). Piperine–based nanoformulations targeting brain have also been fabricated to cure AD. In a research, the development of piperine solid lipid nanoparticles coated with polysorbate-80 led to marked reduction in Aβ plaques and neurofibrillary tangles, the major pathological elements of AD (Yusuf et al. 2013). Another study demonstrated the preparation of mono-disperse intranasal chitosan nanoparticles which directly deliver piperine to the brain. Administration of piperine-NPs in rats ameliorated cognitive functions through AChE inhibition and antioxidant effects (Elnaggar et al. 2015a). Moreover, piperine loaded with tween-modified monoolein cubosomes was designed as an oral drug delivery system targeting the brain. It was found that in vitro release of piperine was maintained by cubosomes and the drug has shown superior neuroprotection and restored the cognitive functions of the brain (Elnaggar et al. 2015b). Recently, a combination of piperine and curcumin has shown synergistic neuroprotective effects in SH-SY$_5$Y cell model and inhibited the aggregation of Aβ_{42} proteins responsible for AD (Manap et al. 2019). This data suggest that piperine and curcumin in future might be developed as a novel drug delivery system for AD therapy.

22.5.2.5 Resveratrol

Resveratrol (RES), a potent neuroprotective agent, is naturally distributed in vegetables, grains, fruits (peanuts, mulberries, grapes, pomegranate, and bananas) tea, and wine. It is a polyphenolic compound that exists in two geometrical isomers cis-RES and trans-RES, latter one having greater biological activity. High temperature, pH and UV radiations transform trans-form to cis-form, hence protected (Gomes et al. 2018; Singh 2013). Resveratrol serves as a promising therapeutic candidate against diabetes, inflammation, obesity, cancers, pain, cardiovascular, and neurological disorders (Andrade et al. 2018; Rege et al. 2014). All the above biological effects of RES become hampered due to its poor pharmacokinetic properties and first-pass metabolism. To deal with these problems, scientists are developing numerous RES formulations. Previously formulated conventional dosage forms including injections have not revealed excellent results; therefore, attention on nanotechnology-based formulations is given to achieve superior therapeutic effects (Arora and Jaglan 2018; Summerlin et al. 2015). In this regard, different types of RES-based nanosystems including solid lipid nanoparticles, cyclodextrin complexes, RES-protein complexes, liposomes, pectinate delivery systems, and chitosan microspheres have been synthesized to increases the brain target ability of the drug (Augustin 2013).

The current study has reported the development of nano drug delivery system of resveratrol, grape skin, and seed extract using solid lipid NPs as a carrier and further assessed for anti-Alzheimer's activity. This therapeutic system could cross the BBB and prevented Aβ peptide fibrillation (Loureiro et al. 2017). More recently, resveratrol encapsulated with transferosomes and nanoemulsions were formulated and labeled with gold nanoparticles. It was observed that transferosomes, as compared with nanoemulsions, exhibited superior penetrability, and increase in learning and memory functions in experimental animals (Salem et al. 2019). These studies suggested the applicability of brain-targeted drug delivery system for the other drugs to be employed as curative therapy in the neurological disorders. Various resveratrol nanoformulations are depicted in Table 22.8.

Table 22.8 Nanoformulations of resveratrol for Alzheimer's disease (Andrade et al. 2018; Frozza et al. 2013; Loureiro et al. 2017; Lu et al. 2009; Rege et al. 2014; Salem et al. 2019; Soldatia et al. 2018)

Nanoformulations	Effect
RES-fusogenic liposomes	Antioxidant activity against cerebromicrovascular endothelial cells
RES conjugated with PEG derivatives	Increased solubility profile and bioavailability
RES-solid-lipid nanoparticles	Antioxidant effect
RES & grape extract-SLNPs	Curative effect on AD
RES-lipid core capsules of capric/caprylic acid and sorbitan monostereate	Neuroprotective effect and AD treatment
RES-polymeric micelles of poly (caprolactone) coated with PEG	Protective effect on PC12 cells
RES-SLNPs of cetyl palmitate with polysorbate 80	Inhibition of Aβ aggregation
RES-transferosomes and nanoemulsion with gold nanoparticles	AD treatment

22.5.2.6 Rosmarinic Acid

Rosmarinic acid (RA), a naturally occurring polyphenolic substance, is exclusively obtained from a wide range of medicinal plants particularly belonging to the family Lamiaceae (Habtemariam 2018). Originally, the compound was originally drawn from *Rosmarinus officinalis* L., commonly known as rosemary plant. Chemically, it is an ester form of caffeic acid and 3,4-dihydroxyphenyl lactic acid. RA contributes in several pharmacological actions including antidiabetic, anticancer, anti-inflammatory, antiviral, antioxidant, antiallergic, nephroprotective, cardioprotective, hepatoprotective, and neuroprotective (Nadeem et al. 2019). Moreover, numerous review articles demonstrated that RA plays a prominent role in curing dementia and AD and improving cognitive impairments through various mechanisms as depicted in Fig. 22.3 (Habtemariam 2018; Taram et al. 2018). Recently, RA isolated from *Isodon japonicas* (Burm.f.) and its derivatives which prevent Aβ aggregation and human islet amyloid polypeptide associated with type-2 diabetes have been reported (Sun et al. 2019). It has been documented that oxidative stress is one of the major factors that triggers neuronal damage leading to development of neurological disorders including AD.

Despite the excellent antioxidant potential of RA, its activity is hindered by poor absorption and transportation across the membrane (Hou et al. 2012). A growing attention is therefore established in developing RA brain-targeted nanosystems to improve its efficacy and therapeutic potential. It has been reported that antioxidant activity of RA was increased by developing a nanoformulation of natural extracts obtained from sage and savory encapsulated with chitosan which serve as a protective vehicle for RA (Baptista da Silva et al. 2015). Moreover, the research was extended to design a brain-targeting drug delivery system of RA loaded with polyacrylamide-chitosan-poly(lactide-co-glycolide) nanoparticles which were

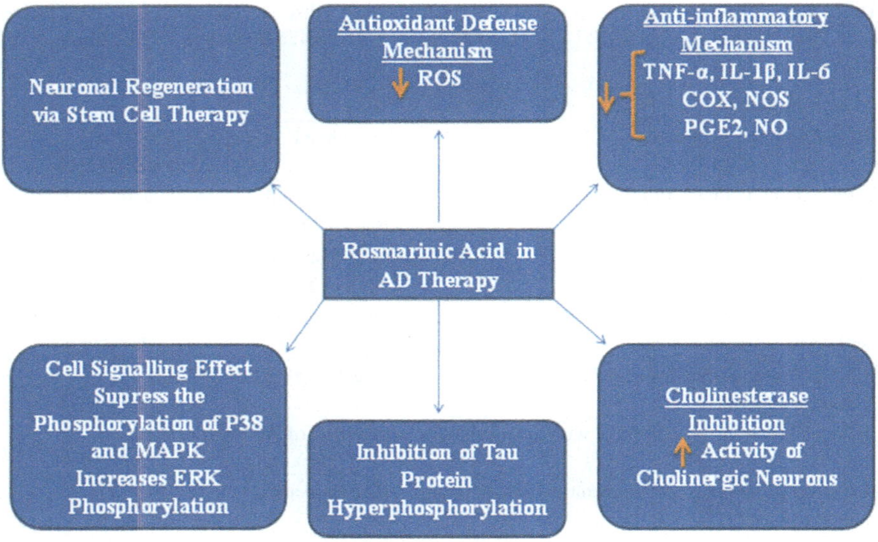

Fig. 22.3 Mechanism of action of rosmarinic acid in Alzheimer's disease

further crosslinked with CRM197 and apolipoprotein E. This nanosystem was further examined for the treatment of neurological disorders and protected SK-N-MC cells against Aβ peptides associated with AD (Kuo and Rajesh 2017). The same author also formulated a nanosystem of RA in combination with curcumin loaded with polyacrylamide-cardiolipin-poly (lactide-co-glycolide) and grafted with monoclonal antibody. This nanoformulation enhanced the permeability across the BBB along with an increase in survival rate of SK-N-MC cells affected by Aβ deposits (Kuo and Tsai 2018).

22.5.2.7 Astaxanthin

Carotenoids, a large group of natural compounds, are extensively found in plants and microorganisms. Theses natural products impart yellow, red, or orange color in fruits and leafy green vegetables (Kaulmann and Bohn 2014). In addition, they play a key role in a wide range of chronic diseases and promote human health. Recently, one such carotenoid has attracted great attention that is astaxanthin (ATX). It is naturally obtained from sea organisms, salmon, lobster, and shrimp and is responsible for several therapeutic effects, viz., anti-inflammatory, antioxidant, neuroprotective, anticancer, gastroprotective as well as cardioprotective. It is considered as an excellent nutraceutical for human brain since it has good permeability across the BBB. It produces curative effects on dementia and improves cognitive deficits (Galasso et al. 2018; Grimmig et al. 2017; Ito et al. 2018).

Recent evidence demonstrated that astaxanthin prevents insulin resistance and AD complications due to diminished levels of $A\beta_{1-42}$, TNF-α, and AChE and decreased activity of IRS-S307 and GSK-3β (Rahman et al. 2019). Besides, ATX reduces oxidative stress in mitochondria induced by Aβ oligomers and promotes

NFATc4 activation and *RyR2* gene expression downregulation (Lobos et al. 2016). To date, there is no data available on astaxanthin formulations, with few exceptions, especially for drug delivery systems. In one study, solid lipid nanoparticles loaded with ATX were fabricated to enhance the delivery in brain through nasal route. The solid lipid NPs were further evaluated for antioxidant potential using pheochromocytoma-12 cells against H_2O_2-induced toxicity (Bhatt et al. 2016). This study suggests that an ATX-based delivery system might have beneficial effects on neurodegenerative disorders caused by oxidative stress.

22.5.2.8 Epigallocatechin-3-Gallate

Green tea is known to contain various therapeutic polyphenolic components responsible for the curative effects against pathological conditions. These polyphenols include catechin, epigallocatechin, epigallocatechin-3-gallate (EGCG), epicatechin-3-gallate, etc. Of these epigallocatechin-3-gallate is the principle constituent of green tea (Chu et al. 2017). Despite a wide range of pharmacological potential, EGCG exhibits a significant role in neurodegenerative diseases, inflammatory conditions, oxidative stress, and cancer (Singh et al. 2015). These biological actions are due to the presence of antioxidant galloyl moiety in EGCG (Chu et al. 2017). This catechin is a powerful memory enhancer, and able to attenute oxidative stress in the brain. Interestingly, EGCG easily transported to different parts of the brain bypassing the BBB (Singh et al. 2008). In addition, it is free from side effects even when given at higher concentration. Many evidences have been documented in connection with its application in AD therapy. EGCG has now been promoted for clinical trials due to its immense potential for neurodegenerative diseases and cancer care (Cascella et al. 2017). Recently, there has been growing interest in EGCG to develop many formulations including nanosystems due to poor bioavailability and efficacy. Interestingly, different types of crystals of EGCG have been developed using crystalline techniques resulting in improvement in pharmacokinetic profiles such as Cmax, Tmax, AUC, half-life, and bioavailability (Smith et al. 2013).

Furthermore, EGCG loaded with selenium NPs coated with Tet-1 peptide effectively prevented the neuronal damage from Aβ plaque deposits which was mediated via antioxidant defense mechanism (Zhang et al. 2014). Another research highlighted the preparation of solid lipid NPs and nano lipid carriers of EGCG to improve its adequate release and better therapeutic effects with no risk of adverse reactions. This study can be further extended to evaluate pharmacological properties against various diseases including neurological disorders (Frias et al. 2016). It is noted that metallotoxicity contributes in AD pathogenesis as well. In this regard, EGCG nanoparticles were designed with PVA and PLA-PEG-DCM solutions and investigated for neuroprotective effects against aluminium chloride–induced neurotoxicity in rats. The findings showed that administration of EGCG nano formulation causes remarkable improvement in brain functions by inhibiting the formation of Aβ deposits and neurofibrillary tangles (Singh et al. 2018). Recently, a couple of drugs including ascorbic acid (AA) and EGCG have been used to fabricate a novel brain-targeted delivery system for the management of AD. Oral administration of dual drug delivery system brought a significant increase in synapses, suppression of

Table 22.9 Pharmacological effects of lipid nanoparticles loaded with EGCG (Granja et al. 2017)

Carrier material	Route	Effects
SLNPs (stearic acid, glycerol monostearate, lecithin, pluronic F68)	–	• Stability increses in fluids
NLCs (glyceryl tridecanoate, glyceryl tripalmitate, soy lecithin, chitosan, Kolliphor HS15)	Oral	• Improvement in stability at acidic and neutral condition • Decrease in inflammation and cholesterol level
NLCs (cetylpalmitate, /phospholipon 80, sesame oil, tween 80, pluronic acid)	Topical	• Decomposition of EGCG once exposed to UVA rays
NLCs (α-tocopherol acetate, phosphatidylcholine, Kolliphor HS15)	–	• Increase in stability of EGCG • Decrease in MCP-1 formation
Phosphatidylcholine, phosphatidic acid, phosphatidylethanolamine, phosphatydylinositol	Oral	• Increase in oral bioavailability • Promotes α-secretase activity in neurons in vitro
Cationic lipid nanocarriers (poloxamer 188, softisan100, glycerol, lipoid S75, and CTAB/DDAB)	Ocular	• Extend the duration of EGCG release

Table 22.10 Pharmacological effects of polymeric nanoparticles loaded with EGCG (Granja et al. 2017)

Carrier material	Route	Effect
Chitosan	Oral	Increase in gastrointestinal tract (GIT) permeability
Chitosan and aspartic acid	Oral	Remarkable antiatherosclerotic activity in rabbits
Chitosan casein phosphopeptides	Oral	Increase in intestinal permeability using Caco-2 manolayers
Chitosan tripolyphosphate	–	Blocks the proliferation of breast cancer cells (MCF-7)
PLGA	Topical	Prevents DNA damage
PLGA-PEG	Intravenous	Antiproliferative effect on prostate cancer

Table 22.11 Pharmacological effects of liposomes loaded with EGCG (Granja et al. 2017)

Carrier material	Route	Effects
Sorbitan monostearate and cholesterol	Oral	Increases stability at neutral pH Rise in cellular permeability in coco-2 cell
Egg lecithin and cholesterol	Topical	In vivo and in vitro increase in anti-MRSA
PhC and cholesterol	Oral	Optimum release of EGCG in the physiological fluid
Cholesterol phosphatidylcholine	–	Offer synergistic effects against breast cancer cells

neuroinflammation, and Aβ-aggregates concluding that EGCG-AA nanoparticles might have a beneficial role in AD therapy (Cano et al. 2019). Tables 22.9, 22.10, and 22.11 outline a description of EGCG nanoformulations, their carrier materials, properties, and pharmacological effects (Granja et al. 2017).

22.5.2.9 Huperzine A

Huperzine A, a potent neuroprotective candidate, is originally derived from traditional Chinese herb *Huperzia serrata* (club moss). Traditionally, this alkaloid is implicated in the treatment of pain and swelling and as an antidote against poison (Wang et al. 2006). As a neuroprotective agent, it restores cognitive deficits and improves memory impairments. In China, the compound is effectively used to cure AD and dementia. In addition, it exerts inhibitory action against AChE enzyme. Apart from that huperzine reduces oxidative stress, prevents mitochondrial damage, modulates the expression of apoptonic proteins P53, Bax, and Bcl-2, and interferes with APP metabolism (Wang and Tang 2005). Recently, anticonvulsant effect of huperzine has also been highlighted which is mediated through anti-inflammatory mechanism (slows down the level of IL-1β, TNF-α) and reducing transcriptional activation of NF-kB (Damar et al. 2016).

Huperzine-A has now received great attention in the pharmaceutical industry in order to improve this medication in the form of dosage. In a comparative clinical study, huperzine-A tablets and capsules were evaluated to determine their safety and efficacy in AD patients using double-blind, parallel, and randomized methods. The results were favorable with equal effectiveness of prepared capsules and tablets (Xu et al. 1999). However, it was noted that such formulations lack brain-targeting property and have some side effects in gastrointestinal tract (GIT) and cholinergic system. Therefore, the study was subsequently aimed at improving brain-targeted delivery systems for the direct release of the drug at a specific site. Recently, huperzine-A encapsulated with polylactide-co-glycoside nanoparticles was engineered for intranasal delivery of drug to the brain for AD therapy. The surface modification of this nanosystem was done by lactoferrin conjugated with N-trimethylated chitosan to enhance nasal adhesion and prolong retention time. The developed nanosystem has shown excellent sustain release property, and adhesion and targeting ability confirming that it could be used as a promising neuroprotective agent in AD therapy (Meng et al. 2018).

22.5.2.10 Hesperidin

Flavonoids are a large group of natural products which play an important role in various diseases as well as in preserving human health. Hesperidin (Hesp) is one such a flavanone glycoside that is largely found in citrus fruits such as sweet orange, grapefruits, and lemon. Chemically, it is composed of glycone part, rutinoside, and aglycone part hesperitin; collectively called hesperitin-7-rutinoside. An isomer of Hesp named neohesperidin imparts bitter taste to bitter orange. In addition to many pharmacological properties including anti-inflammatory antioxidant, anticancer, diuretic, antihypertensive, antihyperlipidaemic, calcium channel blocker, Hesp exerts neuroprotective potential in many neurological ailments such as AD, PD,

Huntington's disease (Hajialyani et al. 2019; Kuntiü et al. 2014). All these potent biological activities of hesperidin are due to its antioxidant nature that rely on the position and number of hydroxyl groups as well as presence of double bond at C4'–C8'. It is clear that due to the BBB, some neuroprotective agents do not reach the CNS, which in turn affects the therapeutic potential associated with neurological disorders. Hence, the emergence of novel nano drug delivery of Hesp has been designed to obtain better control release, bioavailability, and minimum side effects. It has been demonstrated that hesperidin nanoparticles exhibit excellent cerebroprotective effect against ischemic injury in rats (Praveen Kumar et al. 2020).

Furthermore, a comparative study on hesperidin and nano-hesperidin was conducted to assess the neurological activities along with the estimation of various biomarker enzymes, viz., Superoxide dismutase (SOD), catalase (CAT), glutathione reductase, glutathione peroxidase in hippocampal region of rats using the AD model. Administration of nano-Hesp led to remarkable improvement in memory impairment and increased antioxidant parameters. The study claims that nano-Hesp could be a potential candidate for the management of AD (Kheradmand et al. 2018).

22.5.2.11 Quercetin

Quercetin, a polyphenolic compound, belongs to the most diverse group of flavonoids which are not only used as a dietary supplement but also exhibit a wide range of medicinal properties. Quercetin is commonly distributed in vegetables, fruits, green tea, and red wine. An innumerable number of biological properties of quercetin have been reported including antidiabetic, anti-inflammatory, antihyperlipidemic, anticancer, antioxidant, antiviral, cardiovascular, and neuroprotective properties (Ayaz et al. 2019; Zaplatic et al. 2019). In addition, it fights against free radicals production and inhibits lipid peroxidation as well as xanthine oxidase activity (Fiorani et al. 2010). Apart from that quercetin exerts positive effects on CNS like anti-anxiety, memory-enhancing property (Williams et al. 2004) and neuroprotection against AD (Sabogal-Guáqueta et al. 2015). Recently, numerous neuroprotective mechanisms of quercetin at both cellular and molecular level have been stated involving up-and/or down regulation of cytokines through nuclear factors, protein kinase C, mitogen-activated protein kinase signaling cascades, and PI3/Akt pathways (Zaplatic et al. 2019). Many studies suggested that neuroinflammation plays an important role in AD pathogenesis due to an increase in cellular response of inflammatory mediators and activation and proliferation of glial cells. In order to increase the permeability and bioavailability, quercetin was loaded with β-cyclodextrin-dodecylcarbonate nanoparticles displaying promising anti-inflammatory action against neuroinflammation in AD brain (Testa et al. 2014). Another research demonstrated the development of PLGA, a biodegradable polymer, based quercetin NPs (QNPs) which significantly increase the survival of neuron cells by reducing the neurotoxicity of Zn^{2+}-$A\beta_{42}$system in cultured SH-SY5Y cells in-vitro. Furthermore, this drug delivery system effectively increased the therapeutic index with no side effects (Sun et al. 2016). QNPs treatment revealed a remarkable protective effect as compared to free quercetin in scopolamine-induced spatial memory impairment in rats (Palle and Neerati 2017).

In another research, quercetin nanoparticles were prepared by using sodium alginate and chitosan and studied for their neuroprotective potential against H_2O_2-induced oxidative stress in human neuroblastoma cells and 6-hydroxydopamine-induced neurotoxicity in rats. It was observed from the results that quercetin congugated with chitosan effectively improved its neuroprotective activity in both the models (Aluani et al. 2017). In addition, quercetin loaded with zein NPS was proven as a potent therapeutic agent against AD. This orally administered nanosystem ameliorated memory and other cognitive deficits in the AD model (Moreno et al. 2017). Furthermore, the neuroprotective effects of newly engineered gold-palladium nanoparticle encapsulated with quercetin have shown preventive action against neuron damage caused by Aβ via activation of autophagy (Liu et al. 2019). More recently, the bioavailability of quercetin was increased by developing a novel drug delivery system wherein quercetin was encapsulated with superparamagnetic iron oxide nanoparticles (SPION). This formulation was further evaluated to determine its effect on learning and memory functions in rats. It was found that incrassated bioavailability through QT-SPION improved learning and memory functions. In addition, quercetin exhibited better binding affinity toward various proteins such as CRK, RSK2, MSK1, FADD, Cdc42, CytC, and Apaf1 using molecular docking techniques (Amanzadeh et al. 2019). These data indicate the use of quercetin nanoparticles as a novel therapeutic strategy for AD prevention and management.

22.5.2.12 Galantamine

Galantamine, a neuroproprotective alkaloid, is predominantly found in the plants belonging to the family Amaryllidaceae including Lycoris, Hippeastrum, Amaryllis, Narcissus, Zephyranthes, etc. In addition, it has been isolated from the bulbs of *Galanthus woronowii,* commonly known as Caucasian snow-drop. The alkaloid is a potent acetylcholinesterase inhibitor and relieves from numerous neurological problems like AD and poliomyelitis (Brekov et al. 2009). The marketed product of galantamine named Razadyne® was first approved by USFDA in 2001 as anti-Alzheimer's drug (Calcul et al. 2012). Recently, researchers have been intrigued by galantamine in developing nanosystems for AD care.

The in vitro cell line study demonstrated that galantamine encapsulated with PLGA nanoparticles prepared by nanoprecipitation method offers a promising drug delivery system for AD due to its improved permeability in the brain cells (Babu 2015). Furthermore, galantamine-solid-lipid nanoparticles in combination with bone marrow–derived mesenchymal stem cells exhibited a remarkable protective effect against the intracerebroventricular-isoproterenol-induced rat model of AD (Misra et al. 2016). The development of the bovine serum albumin (BSA) nanoparticle loaded with galantamine hydrochloride as an intranasal device to release the drug into the brain has also been documented in recent research (Poddar and Sawant 2017). Moreover, the intranasal drug delivery system of galantamine encapsulated in thiolated chitosan nanoparticles has been fabricated and evaluated against scopolamine-induced amnesia in mice brain. Biochemically, acetylcholinesterase activity was also monitored. The results revealed that intranasal delivery of galantamine nanoparticles improve the bioavailability in brain when compared with oral route. Also, intranasal administration ameliorated memory and cognitive deficits suggesting its use in AD therapy (Sunena, et al. 2019).

22.5.2.13 Berberine

Berberine, an isoquinoline alkaloid, is exclusively distributed in the plant species belonging to the family Berberidaceae including *Berberis vulgaris*, *Berberis aristata*, *Berberis heterophylla*, and *Berberis aquifolium* (Mokhber-Dezfuli et al. 2014; Neag et al. 2018). Other sources include *Hydrastis Canadensis*, *Coptis chinensis*, *Phellodendron chinense*, and *Phellodendron amurense*. Berberine is a yellow color pigment and is most often used in the textile industry. Pharmaceutically, it has medicinal value due to its broad spectrum therapeutic actions like antimicrobial, antiprotozoal, antidiarrhoeal, antidiabetic, hepatoprotective, antiplatelet, anti-inflammatory, and others. In the present time, berberine has been exploited in various neurological disorders such as AD, PD, and Huntington's disease (Chander et al. 2017). It inhibits the development of amyloid plaques and neurofibrillary tangles in AD brain. Due to lipid-glucose lowering property, it is recommended as a therapeutic candidate in atherosclerosis (Singh et al. 2019a, b). There are many ayurvedic formulations of berberine including Darvyadileha, Darvyaditaila, Rasaut, Darvyadikvatha, Rasanjana, and Dasangalepa which are used to cure many pathogenic conditions (Chander et al. 2017).

However, latest research is now focused on the development of nanoformulations of berberine to achieve optimum therapeutic outcomes. Till date, a little work on berberine-based nanoformulations has been reported. A study elaborated the preparation of multiwalled carbon tubes encapsulated with berberine to relieve the symptoms associated with AD. Polysorbate and phospholipid were used as coating material in carbon nanotubes. This nanoformulation has brought significant improvement in both pharmacokinetic and pharmacodynamic parameters. In addition, memory performance was ameliorated by these carbon nanotubes which can be attributed to reduction in Aβ (Lohan et al. 2017). Further, the research on berberine might open a new door to develop other effective nano drug delivery systems for AD.

22.6 Recent Developments and Future Perspectives

Traditional therapeutic approaches appear inadequate for treating AD due to various physiological factors and complex nature of the disease. A great advancement in nanotechnology, however, contributes to increased possible therapeutic strategies to combat AD progression. The fabrication of smart and effective nanoparticles can improve the physico-chemical features of the drug in biological systems and modulates different pathways in specific sites. Despite the potential benefits, as outlines below, AD nano-therapeutics still faces certain challenges:

- Nanotechnology represents a major challenge in relation to toxicity of nanostructures; therefore, a tremendous research is required to design more biocompatible and non-toxic nanoparticles so that better therapeutic outcomes could be achieved.
- In addition, a complete understanding of molecular mechanisms associated with AD might open a new window in nanomedicine applications.
- Exploration of some newly engineered nanoparticles such as magnetic nanoparticles, antibody-tethered nanoparticles, cubosomes, etc., needs more investigations for drug transport in AD therapy.
- Future research in nanoneuromedicine may also involve advancement in diagnostic tools such as biological markers that facilitate absolute targeting without altering the normal cellular process.
- In addition to preclinical studies, the pharmacokinetic and toxicological characteristics need to be improved so that the drugs can pass clinical trials.
- It has conducted a limited range of in vivo studies of drugs as described in the chapter. Therefore, to develop a long-term efficacy in biological systems, research should concentrate on more in vivo investigation of nanoformulations.

22.7 Conclusion

AD is completely incurable due to its multifactorial physiopathologic mechanisms. While a wide array of synthetic and natural therapeutic agents have been researched, they have certain constraints such as poor solubility, low bioavailability, poor BBB penetration, non-specificity to the target site, and undefined mechanistic approaches. Many of these drawbacks have the potential to disrupt pharmacological properties and cause severe side effects. To address the above facts, nano drug delivery system serves as a beneficial tool for diagnosis and therapy of AD. Nanomaterials deliver the therapeutic agents to particular target site and improve both pharmacokinetic and pharmacodynamic profiles. However, in addition to its underlying mechanisms, further research could involve the development of more efficient and non-toxic nano drug delivery systems for AD.

References

Ahmad J, Akhter S, Rizwanullah M, Khan MA, Pigeon L, Addo RT et al (2017) Nanotechnology based theranostic approaches in Alzheimer's disease management: current status and future perspective. Curr Alzheimer Res 14(11):1164–1181

Alam MS, Javed MN, Pottoo FH, Waziri A, Almalki FA, Hasnain MS et al (2019) QbD approached comparison of reaction mechanism in microwave synthesized gold nanoparticles and their superior catalytic role against hazardous nirto-dye. Appl Organomet Chem 33(9):e5071

Ali ME, Lamprecht A (2014) Spray freeze drying for dry powder inhalation of nanoparticles. Eur J Pharm Biopharm 87(3):510–517

Aliev G, Ashraf GM, Tarasov VV, Chubarev VN, Leszek J, Gąsiorowski K et al (2019) Alzheimer's disease – future therapy based on dendrimers. Curr Neuropharmacol 17(3): 288–284

Aluani D, Tzankova V, Yordanov Y et al (2017) In vitro protective effects of encapsulated quercetin in neuronal models of oxidative stress injury. Biotechnol Biotechnol Equip 31(5): 1055–1053

Alyautdin R, Khalin I, Nafeeza MI et al (2014) Nanoscale drug delivery systems and the blood–brain barrier. Int J Nanomedicine 9:795–811

Amanzadeh E, Esmaeili A, Enteshari R, Abadi N, Kazemipour N, Pahlevanneshan Z et al (2019) Quercetin conjugated with superparamagnetic iron oxide nanoparticles improves learning and memory better than free quercetin via interacting with proteins involved in LTP. Sci Rep 9:6876

Amidon GL, Lennernäs H, Shah VP et al (1995) A theoretical basis for a biopharmaceutic drug classification: the correlation of in vitro drug product dissolution and in vivo bioavailability. Pharm Res 12(3):413–420

Amorim C d M et al (2010) Antioxidant idebenone-loaded nanoparticles based on chitosan and N-carboxymethylchitosan. Nanomedicine 6(6):745–742

Andrade S, Ramalho MJ, do Carmo Pereira M et al (2018) Resveratrol brain delivery for neurological disorders prevention and treatment. Front Pharmacol 9:1261

Ansari MA, Badrealam KF, Alam A, Tufail S, Khalique G, Equbal MJ et al (2020a) Recent nano-based therapeutic intervention of bioactive sesquiterpenes: prospects in cancer therapeutics. Curr Pharm Des 26(11):1138–1144

Ansari MA, Chung IM, Rajakumar G, Alzohairy MA, Alomary MN, Thiruvengadam M et al (2020b) Current nanoparticle approaches in nose to brain drug delivery and anticancer therapy - a review. Curr Pharm Des 26(11):1128–1137

Arora D, Jaglan S (2018) Therapeutic applications of resveratrol nanoformulations. Environ Chem Lett 16:35–31

Arruebo M et al (2009) Antibody-conjugated nanoparticles for biomedical applications. J Nanomater 2009:439389. https://doi.org/10.1155/2009/439389

Augustin MA (2013) Nano- and micro-encapsulated systems for enhancing the delivery of resveratrol. Ann N Y Acad Sci 1290(1):107–102

Ayaz M, Sadiq A, Junaid M, Ullah F, Ovais M, Ullah I et al (2019) Flavonoids as prospective neuroprotectants and their therapeutic propensity in aging associated neurological disorders. Front Aging Neurosci 11:155

Azalea DR, Mohambed M, Joji S et al (2012) Design and evaluation of chitosan nanoparticles as novel drug carriers for the delivery of donepezil. Iran J Pharm Sci 8(3):154–155

Babu GK (2015) Formulation and evaluation of galantamine loaded PLGA nanoparticles for Alzheimer's disease. J Pharma Care Health Sys 2(4):48

Baptista da Silva S, Amorim M, Fonte P, Madureira R, Ferreira D, Pintado M et al (2015) Natural extracts into chitosan nanocarriers for rosmarinic acid drug delivery. Pharm Biol 53(5):642–642

Barbara R, Belletti D, Pederzoli F et al (2017) Novel curcumin loaded nanoparticles engineered for blood-brain barrier crossing and able to disrupt Abeta aggregates. Int J Pharm 526(1–2): 413–414

Barkat MA, Harshita, Rizwanullah M, Pottoo FH, Beg S, Akhter S et al (2020) Therapeutic nanoemulsion: concept to delivery. Curr Pharm Des 26(11):1145–1166

Baysal I, Ucar G, Gultekinoglu M et al (2017) Donepezil loaded PLGA-b-PEG nanoparticles: their ability to induce destabilization of amyloid fibrils and to cross blood brain barrier in vitro. J Neural Transm (Vienna) 124(1):33–35

Beach TG et al (1989) Patterns of gliosis in Alzheimer's disease and aging cerebrum. Glia 2(6): 420–426

Bennet D, Kim S (2014) Polymer nanoparticles for smart drug delivery [cited 2020 March 11]. Available from: https://www.intechopen.com/books/application-of-nanotechnology-in-drug-delivery/polymer-nanoparticles-for-smart-drug-delivery

Bhatt PC, Srivastava P, Pandey P et al (2016) Nose to brain delivery of astaxanthin-loaded solid lipid nanoparticles: fabrication, radio labeling, optimization and biological studies. RSC Adv 6(12):10001–10010

Boudad H, Legrand P, Lebas G et al (2001) Combined hydroxypropyl-beta-cyclodextrin and poly (alkylcyanoacrylate) nanoparticles intended for oral administration of saquinavir. Int J Pharm 218(1–2):113–124

Brahamdutt, Kamboj VK, Kumar A et al (2018) Nanotechnology: various methods used for preparation of nanomaterials. Asian J Pharm Pharmacol 4(4):386–383

Brambilla D, Verpillot R, De Kimpe L et al (2010) Nanoparticles against Alzheimer's disease: PEG-PACA nanoparticles are able to link the Aβ-peptide and influence its aggregation kinetic. J Biotechnol 150(Suppl):27

Brekov S, Georgieva L, Kondakova V et al (2009) Plant sources of galathamine: phytochemical and biotechnological aspects. Biotechnol Biotechnol Equip 23:1170–1176

Calcul L, Zhang B, Jinwal UK et al (2012) Natural products as a rich source of tau-targeting drugs for Alzheimer's disease. Future Med Chem 4(13):1751–1751

Calvo P, Remuñán-López C, Vila-Jato JL et al (1997) Novel hydrophilic chitosan-polyethylene oxide nanoparticles as protein carriers. J Appl Polym Sci 63(1):125–122

Campardelli R, Cherain M, Perfetti C et al (2013) Lipid nanoparticles production by supercritical fluid assisted emulsion–diffusion. J Supercrit Fluids 82:34–40

Cano A, Ettcheto M, Chang JH, Barroso E, Espina M, Kühne BA et al (2019) Dual-drug loaded nanoparticles of Epigallocatechin-3-gallate (EGCG)/ascorbic acid enhance therapeutic efficacy of EGCG in a APPswe/PS1dE9 Alzheimer's disease mice model. J Control Release 301:62–75

Carradori D, Balducci C, Re F, Brambilla D, Le Droumaguet B, Flores O et al (2018) Antobody-functionalized polymer nanoparticles leading to memory recovery in Alzheimer's disease-like transgenic mouse model. Nanomed Nanotechnol Biol Med 14:609–618

Cascella M, Bimonte S, Muzio MR et al (2017) The efficacy of Epigallocatechin-3-gallate (green tea) in the treatment of Alzheimer's disease: an overview of pre-clinical studies and translational perspectives in clinical practice. Infect Agents Cancer 12:36

Chander V, Aswal JS, Dobhal R et al (2017) A review on pharmacological potential of Berberine; an active component of Himalayan Berberis aristata. J Phytopharmacol 6(1):53–58

Chavda VP (2019) Nanobased nano drug delivery: a comprehensive review. In: Mohapatra S, Ranjan S, Dasgupta N (eds) Applications of targeted nano drugs and delivery systems: nanoscience and nanotechnology in drug delivery. Elsevier, Amsterdam, pp 69–92

Chen M, Du ZY, Zheng X et al (2018) Use of curcumin in diagnosis, prevention, and treatment of Alzheimer's disease. Neural Regen Res 13(4):742–742

Cheng KK, Yeung CF, Ho SW et al (2013) Highly stabilized curcumin nanoparticles tested in an in vitro blood–brain barrier model and in Alzheimer's disease Tg2576 mice. AAPS J 15(2): 324–336

Cheng Z, Yin J, Yuan H, Jin C, Zhang F, Wang Z et al (2018) Blood-derived plasma protein biomarkers for Alzheimer's disease in Han Chinese. Front Aging Neurosci 10:414

Chonpathompikunlert P et al (2010) Piperine, the main alkaloid of Thai black pepper, protects against neurodegeneration and cognitive impairment in animal model of cognitive deficit like condition of Alzheimer's disease. Food Chem Toxicol 48(3):798–802

Chorilli M et al (2013) Caffeine encapsulated in small unilamellar liposomes: characerization and in vitro release profile. J Dispers Sci Technol 34(10):1465–1470

Chu KS, Hasan W, Rawal S et al (2013) Plasma, tumor and tissue pharmacokinetics of Docetaxel delivered via nanoparticles of different sizes and shapes in mice bearing SKOV-3 human ovarian carcinoma xenograft. Nanomedicine 9(5):686–683

Chu C, Deng J, Man Y et al (2017) Green tea extracts epigallocatechin-3-gallate for different treatments. Biomed Res Int 2017:5615647

Corey-Bloom J et al (1998) A randomized trial evaluating the efficacy and safety of ENA 713 (rivastigmine tartrate), a new acetylcholinesterase inhibitor, in patients with mild to moderately severe Alzheimer's disease. Int J Geriatr Psychopharmacol 1:55–65

Couvreur P et al (2002) Nanocapsule technology: a review. Crit Rev Ther Drug Carrier Syst 19(2): 99–134
Damar U, Gersner R, Johnstone JT et al (2016) Huperzine A as a neuroprotective and antiepileptic drug: a review of preclinical research. Expert Rev Neurother 16(6):671–680
de Souza ALR, Andreani T, Nunes FM et al (2012) Loading of praziquantel in the crystal lattice of solid lipid nanoparticles. J Therm Anal Calorim 108(1):353–360
Del Prado-Audelo ML, Caballero-Florán IH, Meza-Toledo JA, Mendoza-Muñoz N, González-Torres M, Florán B et al (2019) Formulations of curcumin nanoparticles for brain diseases. Biomol Ther 9(2):56
Do TD, Ul Amin F, Noh Y et al (2016) Guidance of magnetic nanocontainers for treating Alzheimer's disease using an electromagnetic. Targeted drug-delivery actuator. J Biomed Nanotechnol 12(3):569–574
Dong Y et al (2012) Solid lipid nanoparticles: continuous and potential large-scale nanoprecipitation production in static mixers. Colloids Surf B Biointerfaces 94:68–62
Dong Y, Ng WK, Shen S et al (2013) Scalable ionic gelation synthesis of chitosan nanoparticles for drug delivery in static mixers. Carbohydr Polym 94(2):940–945
Dos Santos Tramontin N, da Silva S, Arruda R, Ugioni KS, Canteiro PB, de Bem Silveira G et al (2020) Gold nanoparticles treatment reverses brain damage in Alzheimer's disease model. Mol Neurobiol 57(2):926–936
Eke CS, Jammeh E, Li X et al (2018) Identification of optimum panel of blood-based biomarkers for Alzheimer's disease diagnosis using machine learning. Conf Proc IEEE Eng Med Biol Soc 2018:3991–3994
Elnaggar YS, Etman SM, Abdelmonsif DA et al (2015a) Intranasal piperine-loaded chitosan nanoparticles as brain-targeted therapy in Alzheimer's disease: optimization, biological efficacy, and potential toxicity. J Pharm Sci 104(10):3544–3556
Elnaggar YS, Etman SM, Abdelmonsif DA et al (2015b) Novel piperine-loaded Tween-integrated monoolein cubosomes as brain-targeted oral nanomedicine in Alzheimer's disease: pharmaceutical, biological, and toxicological studies. Int J Nanomedicine 10:5459–5473
Ezrahi S et al (1999) Aggregation behavior in one-phase (Winsor IV) microemulsion systems. In: Kumar P, Mittal KL (eds) Handbook of microemulsion science and technology. Marcel Dekker, Inc, New York, pp 195–240
Fan S, Zheng Y, Liu X et al (2018) Curcumin-loaded PLGA-PEG nanoparticles conjugated with B6 peptide for potential use in Alzheimer's disease. Drug Deliv 25(1):1091–1102
Fazil M, Md S, Haque S, Kumar M, Baboota S, Sahni J et al (2012) Development and evaluation of rivastigmine loaded chitosan nanoparticles for brain targeting. Eur J Pharm Sci 47(1):6–15
Felix Joe V, Sathesh Kumar S (2018) Formulation, characterization and determination of anti-Alzheimeric activity of tacrine loaded poly (lactide-co-glycolide) nanoparticles. Int J Pharm Sci Res 9(12):5111–5120
Fernández-Bachiller MI, Pérez C, González-Muñoz GC, Conde S, López MG, Villarrova M et al (2010) Novel tacrine-8-hydroxyquinoline hybrids as multifunctional agents for the treatment of Alzheimer's disease, with neuroprotective, cholinergic, antioxidant and coppercomplexing properties. J Med Chem 53:4927–4937
Fiorani M, Guidarelli A, Blasa M, Azzolini C, Candiracci M, Piatti E et al (2010) Mitochondria accumulate large amounts of quercetin: prevention of mitochondrial damage and release upon oxidation of the extramitochondrial fraction of the flavonoid. J Nutr Biochem 21:397–404
Fleming KC et al (1995) Dementia: diagnosis and evaluation. Mayo Clin Proc 70(11):1093–1107
Fonseca-Santos B, Gremião MP, Chorilli M (2015) Nanotechnology-based drug delivery systems for the treatment of Alzheimer's disease. Int J Nanomedicine 10:4981–5003
Freitas C, Müllerä RH (1998) Spray-drying of solid lipid nanoparticles (SLN TM). Eur J Pharm Biopharm 46(2):145–151
Frias I, Neves AR, Pinheiro M et al (2016) Design, development, and characterization of lipid nanocarriers-based epigallocatechin gallate delivery system for preventive and therapeutic supplementation. Drug Des Devel Ther 10:3519–3518

Frozza RL, Bernardi A, Hoppe JB, Meneghetti AB, Matté A, Battastini AMO et al (2013) Neuroprotective effects of resveratrol against Aβ administration in rats are improved by lipid-core nanocapsules. Mol Neurobiol 47:1066–1080

Galasso C, Orefice I, Pellone P, Cirino P, Miele R, Ianora A et al (2018) On the neuroprotective role of astaxanthin: new perspectives? Mar Drugs 16(8):247

Ghanbarzadeh S et al (2013) Application of response surface methodology in development of sirolimus liposomes prepared by thin film hydration technique. Bioimpacts 3(2):75–71

Gomes BAQ, Silva JPB, Romeiro CFR et al (2018) Neuroprotective mechanisms of resveratrol in Alzheimer's disease: role of SIRT1. Oxidative Med Cell Longev 2018:8152373

Graff CL, Pollack GM (2005) Nasal drug administration: potential for targeted central nervous system delivery. J Pharm Sci 94(6):1187–1185

Granja A, Frias I, Neves AR et al (2017) Therapeutic potential of epigallocatechin gallate nanodelivery systems. Biomed Res Int 2017:5813793. https://doi.org/10.1155/2017/5813793

Gray NE, Alcazar Magana A, Lak P et al (2018) *Centella asiatica* - Phytochemistry and mechanisms of neuroprotection and cognitive enhancement. Phytochem Rev: Proceedings of the Phytochemical Society of Europe 17(1):161–194

Grimmig B, Kim SH, Nash K et al (2017) Neuroprotective mechanisms of astaxanthin: a potential therapeutic role in preserving cognitive function in age and neurodegeneration. Geroscience 39(1):19–32

Gulati M, Grover M, Singh S et al (1998) Lipophilic drug derivatives in liposomes. Int J Pharm 165(2):129–168

Guterres SS et al (2007) Polymeric nanoparticles, nanospheres and nanocapsules, for cutaneous applications. Drug Target Insights 2:147–157

Habtemariam S (2018) Molecular pharmacology of rosmarinic and salvianolic acids: potential seeds for Alzheimer's and vascular dementia drugs. Int J Mol Sci 19(2):458

Hadian Z, Sahari MA, Moghimi HR et al (2014) Formulation, characterization and optimization of liposomes containing eicosapentaenoic and docosahexaenoic acids; a methodology approach. Iran J Pharm Res 13(2):393–404

Hajialyani M, Hosein Farzaei M, Echeverría J et al (2019) Hesperidin as a neuroprotective agent: a review of animal and clinical evidence. Molecules 24(3):648

Haque S, Md S, Alam MI et al (2012) Nanostructure-based drug delivery systems for brain targeting. Drug Dev Ind Pharm 38(4):387–411

Harilal S, Jose J, Parambi DGT, Kumar R, Mathew GE, Uddin MS et al (2019) Advancements in nanotherapeutics for Alzheimer's disease: current perspectives. J Pharm Pharmacol 71(9): 1370–1383

Harshita, Barkat MA, Das SS et al (2020) Lipid-based nanosystem as intelligent carriers for versatile drug delivery applications. Curr Pharm Des 26(11):1167–1180

Harthi SA, Alavi SE, Radwan MA et al (2019) Nasal delivery of donepezil HCl-loaded hydrogels for the treatment of Alzheimer's disease. Sci Rep 9:9563

Hong-Qi Y et al (2012) Current advances in the treatment of Alzheimer's disease: focused on considerations targeting Aβ and tau. Transl Neurodegener 1:21

Hossain MF, Uddin MS, Uddin GMS, Sumsuzzman DM, Islam MS, Barreto GE et al (2019) Melatonin in Alzheimer's disease: a latent endogenous regulator of neurogenesis to mitigate Alzheimer's neuropathology. Mol Neurobiol 56(12):8255–8276

Hou Y, Wang J, Jin W et al (2012) Degradation of Laminaria japonica fucoidan by hydrogen peroxide and antioxidant activities of the degradation products of different molecular weights. Carbohydr Polym 87:153–159

Hu X, Guo Y, Wang L et al (2011) Coenzyme Q_{10} nanoparticles prepared by a supercritical fluid-based method. J Supercrit Fluids 57(1):66–62

Huo X, Zhang Y, Jin X et al (2019) A novel synthesis of selenium nanoparticles encapsulated PLGA nanospheres with curcumin molecules for the inhibition of amyloid β aggregation in Alzheimer's disease. J Photochem Photobiol B 190:98–102

Iijima S (1991) Helical microtubules of graphite carbon. Nature 354:56–58

Inglis F (2002) The tolerability and safety of cholinesterase inhibitors in the treatment of dementia. Int J Clin Pract Suppl 127:45–63

Isailović BD, Kostić IT, Zvonar A et al (2013) Resveratrol loaded liposomes produced by different techniques. Innov Food Sci Emerg Technol 19:181–189

Ito N, Saito H, Seki S et al (2018) Effects of composite supplement containing astaxanthin and sesamin on cognitive functions in people with mild cognitive impairment: a randomized, double-blind, placebo-controlled trial. J Alzheimers Dis 62(4):1767–1765

Jacobson SA, Sabbagh MN (2008) Donepezil: potential neuroprotective and disease-modifying effects. Expert Opin Drug Metab Toxicol 4:1363–1369

Jaiswal M et al (2015) Nanoemulsion: an advanced mode of drug delivery system. 3 Biotech 5(2): 123–127

Juillerat-Jeanneret L (2008) The targeted delivery of cancer drugs across the blood-brain barrier: chemical modifications of drugs or drug-nanoparticles? Drug Discov Today 13(23–24): 1099–1106

Karami Z, Hamidi M (2016) Cubosomes: remarkable drug delivery potential. Drug Discov Today 21(5):789–801

Karis JH et al (1966) The action of tacrine on neuromuscular transmission: a comparison with hexafluorenium. Brit J Anaesth 38:762–774

Kaulmann A, Bohn T (2014) Carotenoids, inflammation, and oxidative stress—implications of cellular signaling pathways and relation to chronic disease prevention. Nutr Res 34:907–929

Kaur SP, Rao R, Hussain A et al (2011) Preparation and characterization of Rivastigmine loaded chitosan nanoparticles. J Pharm Sci Res 3(5):1227–1222

Kaur A, Nigam K, Bhatnagar I, Sukhpal H, Awasthy S, Shankar S et al (2020a) Treatment of Alzheimer's diseases using donepezil nanoemulsion: an intranasal approach. Drug Deliv Transl Res. https://doi.org/10.1007/s13346-020-00754-z

Kaur A, Nigam K, Srivastava S et al (2020b) Memantine nanoemulsion: a new approach to treat Alzheimer's disease. J Microencapsul 15:1–30

Kheradmand E et al (2018) Neuroprotective effect of hesperetin and nano-hesperetin on recognition memory impairment and the elevated oxygen stress in rat model of Alzheimer's disease. Biomed Pharmacother 97:1096–1101

Kocahan S, Doğan Z (2017) Mechanisms of Alzheimer's disease pathogenesis and prevention: the brain, neural pathology, N-methyl-D-aspartate receptors, Tau protein and other risk factors. Clin Psychopharmacol Neurosci 15(1):1–8

Kovačević AB, Müller RH, Savić SD et al (2014) Solid lipid nanoparticles (SLN) stabilized with polyhydroxy surfactants: preparation, characterization and physical stability investigation. Colloids Surf A Physicochem Eng Asp 444:15–25

Kreuter J (1978) Nanoparticles and nanocapsules – new dosage forms in the nanometer size range. Pharm Acta Helv 53(2):33–39

Krishna KV, Wadhwa G, Alexander A, Kanojia N, Saha RN, Kukreti R et al (2019) Design and biological evaluation of lipoprotein-based donepezil nanocarrier for enhanced brain uptake through oral delivery. ACS Chem Neurosci 10(9):4124–4135

Kroto HW, Hearth JR, O'Brien SC, Curl RF, Smalley RE (1985) Buckminsterfullerenes. Nature 318:162–163

Kryger G et al (1999) Structure of acetylcholinesterase complexed with E2020 (Aricept): implications for the design of new anti-Alzheimer drugs. Struct 7:297–307

Kumar R, Yasir M, Saraf SA et al (2013) Glyceryl monostearate based nanoparticles of mefenamic acid: fabrication and in vitro characterization. Drug Invent Today 5(3):246–250

Kumar B, Pandey M, Pottoo FH et al (2020) Liposomes: novel drug delivery approach for targeting Parkinson's disease. Curr Pharm Des. https://doi.org/10.2174/1381612826666200128145124

Kuns B, Varghese D (2019) StatPearls: memantine. StatPearls Publishing, Florida

Kuntiü V, Brboriü J, Holclajtner-Antunoviü I et al (2014) Evaluating the bioactive effects of flavonoid hesperidin – a new literature data survey. Vojnosanit Pregl 71(1):60–65

Kuo YC, Rajesh R (2017) Targeted delivery of rosmarinic acid across the blood-brain barrier for neuronal rescue using polyacrylamide-chitosan-poly(lactide-co-glycolide) nanoparticles with surface cross-reacting material 197 and apolipoprotein E. Int J Pharm 528(1–2):228–241

Kuo YC, Tsai HC (2018) Rosmarinic acid- and curcumin-loaded polyacrylamide-cardiolipin-poly (lactide-co-glycolide) nanoparticles with conjugated 83-14 monoclonal antibody to protect β-amyloid-insulted neurons. Mater Sci Eng C Mater Biol Appl 91:445–457

Langevin D (1988) Microemulsions. Acc Chem Res 21(7):255–260

Lasic DD (1998) Novel applications of liposomes. Trends Biotechnol 16(7):307–321

Lee S, Youn K, Lim G et al (2018) In Silico docking and in vitro approaches towards BACE1 and cholinesterases inhibitory effect of citrus flavanones. Molecules 23(7):1509

Liu Y, Zhou H, Yin T, Gong Y, Yuan G, Chen L et al (2019) Quercetin-modified gold-palladium nanoparticles as a potential autophagy inducer for the treatment of Alzheimer's disease. J Colloid Interface Sci 552:388–400

Lobos P, Bruna B, Cordova A, Barattini P, Galáz JL, Adasme T et al (2016) Astaxanthin protects primary hippocampal neurons against noxious effects of Aβ-Oligomers. Neural Plast 2016: 3456783. https://doi.org/10.1155/2016/3456783

Lohan S, Raza K, Mehta SK et al (2017) Anti-Alzheimer's potential of berberine using surface decorated multi-walled carbon nanotubes: a preclinical evidence. Int J Pharm 530(1–2):263–278

Loureiro JA, Andrade S, Duarte A, Neves AR, Queiroz JF, Nunes C et al (2017) Resveratrol and grape extract-loaded solid lipid nanoparticles for the treatment of Alzheimer's disease. Molecules 22(2):277

Lu X, Ji C, Xu H, Li X, Ding H, Ye M et al (2009) Resveratrol-loaded polymeric micelles protect cells from Abeta-induced oxidative stress. Int J Pharm 375:89–86

Luppi B, Bigucci F, Corace G, Delucca A, Cerchiara T, Sorrenti M et al (2011) Albumin nanoparticles carrying cyclodextrins for nasal delivery of the anti-Alzheimer drug tacrine. Eur J Pharm Sci 44(4):559–565

Lv J, Sharma A, Zhang T et al (2018) Pharmacological review on asiatic acid and its derivatives: a potential compound. SLAS Technol 23(2):111–127

Madhavi B, Banji D, Anisetti RN et al (2009) Extraction, identification, formulation and evaluation of piperine in alginate beads. Int J Pharm Pharmaceut Sci 1(2):156–161

Mainardes RM, Evangelista RC (2005) Praziquantel-loaded PLGA nanoparticles: preparation and characterization. J Microencapsul 22(1):13–24

Manap ASA, Tan ACW, Leong WH, Chia AYY, Vijayabalan S, Arya A et al (2019) Synergistic effects of curcumin and piperine as potent acetylcholine and amyloidogenic inhibitors with significant neuroprotective activity in SH-SY5Y cells via computational molecular modeling and in vitro assay. Front Aging Neurosci 11:1–17

Martins S, Silva AC, Ferreira DC et al (2009) Improving oral absorption of Salmon calcitonin by trimyristin lipid nanoparticles. J Biomed Nanotechnol 5(1):76–73

Martins SM, Sarmento B, Nunes C et al (2013) Brain targeting effect of camptothecin-loaded solid lipid nanoparticles in rat after intravenous administration. Eur J Pharm Biopharm 85(3 Pt A):488–502

Mazzarino L, Coche-Guérente L, Labbé P et al (2014) On the mucoadhesive properties of chitosan-coated polycaprolactone nanoparticles loaded with curcumin using quartz crystal microbalance with dissipation monitoring. J Biomed Nanotechnol 10(5):787–784

McDonald TO, Siccardi M, Moss D, Liptrott N, Giardiello M, Rannard S et al (2015) The application of nanotechnology to drug delivery in medicine. In: Dolez PI (ed) Nanoengineering: global approaches to health and safety issues. Elsevier, Amsterdam, pp 173–223

Md S, Ali A, Bhatnagar A et al (2014) Design, development, optimization and characterization of donepezil loaded chitosan nanoparticles for brain targeting to treat Alzheimer's disease. Sci Adv Mater 6(4):720–735

Mehnert W, Mäder K (2012) Solid lipid nanoparticles: production, characterization and applications. Adv Drug Deliv Rev 64(Suppl):83–101

Meng F, Asghar S, Gao S et al (2015) A novel LDL-mimic nanocarrier for the targeted delivery of curcumin into the brain to treat Alzheimer's disease. Colloid Surface B 134:88–97
Meng Q, Wang A, Hua H, Jiang Y, Wang Y, Mu H et al (2018) Intranasal delivery of Huperzine A to the brain using lactoferrin-conjugated N-trimethylated chitosan surface-modified PLGA nanoparticles for treatment of Alzheimer's disease. Int J Nanomedicine 13:705–718
Mishra S, Sharma S, Javed MN, Pottoo FH, Barkat MA, Harshita et al (2019) Bioinspired nanocomposites: applications in disease diagnosis and treatment. Pharm Nanotechnol. 7(3): 206–219
Misra S, Chopra K, Saikia UN, Sinha VR, Sehgal R, Modi M et al (2016) Effect of mesenchymal stem cells and galantamine nanoparticles in rat model of Alzheimer's disease. Regen Med 11(7): 629–646
Mokhber-Dezfuli N, Saeidnia S, Gohari AR et al (2014) Phytochemistry and pharmacology of berberis species. Pharmacogn Rev 8(15):8–15
Monroy A et al (2013) Curcumin and neurodegenerative diseases. Biofactors 39(1):122–132
Moreno LCGEI, Puerta E, Suárez-Santiago JE et al (2017) Effect of the oral administration of nanoencapsulated quercetin on a mouse model of Alzheimer's disease. Int J Pharm 517(1–2): 50–57
Morimasa K, Ryuji O, Kazuya K et al (2018) Effect of chlorogenic acid intake on cognitive function in the elderly: a pilot study. Evid Based Complement Alternat Med 2018:8608497
Mukhopadhyay S et al (2017) Development and evaluation of bio-nanoparticles as novel drug carriers for the delivery of donepezil. Int J Nano Dimens 8(1):9–17
Müller RH et al (2002) Solid lipid nanoparticles (SLN) and nanostructured lipid carriers (NLC) in cosmetic and dermatological preparations. Adv Drug Deliv Rev 54(Suppl 1):S131–S155
Nadeem M, Imran M, Gondal TA, Imran A, Shahbaz M, Amir RM et al (2019) Therapeutic potential of rosmarinic acid: a comprehensive review. Appl Sci 9:3139
Nafisi S, Maibach HI (2017) Nanotechnology in cosmetics. In: Sakamoto K, Lochhead R, Maibach H, Yamashita Y (eds) Cosmetic science and technology: theoretical principles and applications. Elsevier, Amsterdam, pp 337–369
Nallamuthu I et al (2015) Chlorogenic acid loaded chitosan nanoparticles with sustained release property, retained antioxidant activity and enhanced bioavailability. Asian J Pharm Sci 10(3): 203–211
Nasir M, Abdullah J, Habsah M et al (2012) Inhibitory effect of asiatic acid on acetylcholinesterase, excitatory post synapticpotential and locomotor activity. Phytomedicine 15:311–316
Nasrollahzadeh M, Sajadi SM, Sajjadi M et al (2019) Applications of nanotechnology in daily life. In: Nasrollahzadeh M, Sajadi SM, Sajjadi M, Issaabadi Z, Atarod M (eds) Interface science and technology. Elsevier, Amsterdam, pp 113–143
Naveed M et al (2018) Chlorogenic acid (CGA): a pharmacological review and call for further research. Biomed Pharmacother 97:67–74
Nazem A, Mansoori GA (2011) Nanotechnology for Alzheimer's disease detection and treatment. Insciences J 1(4):169–193
Neag MA, Mocan A, Echeverría J, Pop RM, Bocsan CI, Crişan G et al (2018) Berberine: botanical occurrence, traditional uses, extraction methods, and relevance in cardiovascular, metabolic, hepatic, and renal disorders. Front Pharmacol 9:557
Nirale P et al (2020) Nanoemulsions for targeting the neurodegenerative diseases: Alzheimer's, Parkinson's and Prion's. Life Sci 245:117394
Oboh G, Agunloye OM, Akinyemi AJ et al (2013) Comparative study on the inhibitory effect of caffeic and chlorogenic acids on key enzymes linked to Alzheimer's disease and some pro-oxidant induced oxidative stress in rats' brain-in vitro. Neurochem Res 38(2):413–419
Ojha SK, Meeran MFN, Goyal SN et al (2018) Pharmacological properties, molecular mechanisms, and pharmaceutical development of asiatic acid: a pentacyclic triterpenoid of therapeutic promise. Front Pharmacol 9:1–35
Palle S, Neerati P (2017) Quercetin nanoparticles attenuates scopolamine induced spatial memory deficits and pathological damages in rats. Bull Fac Pharm Cairo Univ 55:101–106

Paolino D, Cosco D, Molinaro R et al (2011) Supramolecular devices to improve the treatment of brain diseases. Drug Discov Today 16(7–8):311–324

Poddar A, Sawant KK (2017) Optimization of galantamine loaded bovine serum albumin nanoparticles by quality by design and its preliminary characterizations. J Nanomed Nanotechnol 8:459

Pottoo FH, Sharma S, Javed MN, Barkat MA, Harshita, Alam MS et al (2020) Lipid-based nanoformulations in the treatment of neurological disorders. Drug Metab Rev 52(1):185–204

Povova J, Ambroz P, Bar M, Pavukova V, Sery O, Tomaskova H et al (2012) Epidemiological of and risk factors for Alzheimer's disease: a review. Biomed Pap Med Fac Univ Palacky Olomouc Czech Repub 156:108–114

Pupo E, Padrón A, Santana E, Sotolongo J, Quintana D, Dueñas S et al (2005) Preparation of plasmid DNA-containing liposomes using a high-pressure homogenization--extrusion technique. J Control Release 104(2):379–396

Radwan RR et al (2020) Gamma radiation preparation of chitosan nanoparticles for controlled delivery of memantine. J Biomater Appl 34(8):1150–1162

Rahman SO, Panda BP, Parvez S, Kaundal M, Hussain S, Akhtar M et al (2019) Neuroprotective role of astaxanthin in hippocampal insulin resistance induced by Aβ peptides in animal model of Alzheimer's disease. Biomed Pharmacother 110:47–58

Rajak P et al (2019) Liquid crystals: an approach in drug delivery. Indian J Pharm Sci 81(1):11–21

Rakotoarisoa M, Angelov B, Garamus VM et al (2019) Curcumin- and fish oil-loaded spongosome and cubosome nanoparticles with neuroprotective potential against H_2O_2-induced oxidative stress in differentiated human SH-SY5Y cells. ACS Omega 4(2):3061–3073

Ramanathan S, Archunan G, Sivakumar M, Tamil Selvan S, Fred AL, Kumar S et al (2018) Theranostic applications of nanoparticles in neurodegenerative disorders. Int J Nanomedicine 13:5561–5576

Rao SV et al (2018) Formulation, characterisation and in-vitro evaluation of novel ionically cross linked casein nanoparticles for memantine hydrochloride delivery. Int J Pharm Sci Res 9(8): 3307–3316

Raval N, Mistry T, Acharya N et al (2015) Development of glutathione-conjugated asiatic acid-loaded bovine serum albumin nanoparticles for brain-targeted drug delivery. J Pharm Pharmacol 67(11):1503–1511

Raval N, Barai P, Acharya N et al (2018) Fabrication of peptide-linked albumin nanoconstructs for receptor-mediated delivery of asiatic acid to the brain as a preventive measure in cognitive impairment: optimization, in-vitro and in-vivo evaluation. Nanomed Biotechnol 46(Suppl 3): S832–S846

Ravi G, Gupta NV (2017) Development of solid lipid nanoparticles of rivastigmine tartrate by using full factorial design for the treatment of Alzheimer's disease. J Pharm Sci Res 9(12):2447–2452

Ray S, Britschgi M, Herbert C, Takeda-Uchimura Y, Boxer A, Blennow K et al (2007) Classification and prediction of clinical Alzheimer's diagnosis based on plasma signaling proteins. Nat Med 13(11):1359–1362

Rege SD, Geetha T, Griffin GD et al (2014) Neuroprotective effects of resveratrol in Alzheimer disease pathology. Front Aging Neurosci 6:218

Rogawski MA, Wenk GL (2003) The neuropharmacological basis for the use of memantine in the treatment of Alzheimer's disease. CNS Drug Rev 9(3):275–308

Rogers SL, Farlow MR, Doody RS et al (1998) A 24 week double blind placebo controlled trial of donepezil in patients with Alzheimer's disease. Donepezil study group. Neurology 50:136–135

Ruby JJ, Pandey VP (2014) Chitosan nanoparticles as a nasal drug delivery for memantine hydrochloride. In: Material science, chemistry conference [cited on 2020 Feb 20]. Available from: https://www.semanticscholar.org/paper/CHITOSAN-NANOPARTICLES-AS-A-NASAL-DRUG-DELIVERY-FOR-Ruby-Pandey/eca1f3b82d4c87d82637c7402111fe877641164d

Sabogal-Guáqueta AM, Muñoz-Manco JI, Ramírez-Pineda JR et al (2015) The flavonoid quercetin ameliorates Alzheimer's disease pathology and protects cognitive and emotional function in aged triple transgenic Alzheimer's disease model mice. Neuropharmacology 93:134–145

Sadowsky CH, Micca JL, Grossberg GT et al (2014) Rivastigmine from capsules to patch: therapeutic advances in the management of Alzheimer's disease and Parkinson's disease dementia. Prim Care Companion CNS Disord 16(5). https://doi.org/10.4088/PCC.14r01654

Saitou K, Ochiai R, Kozuma K, Sato H, Koikeda T, Osaki N et al (2018) Effect of chlorogenic acids on cognitive function: a randomized, double-blind, placebo-controlled trial. Nutrients 10(10): 1337

Sajjad R, Arif R, Shah AA et al (2018) Pathogenesis of Alzheimer's disease: role of amyloid-β and hyperphosphorylated tau protein. Indian J Pharm Sci 80(4):581–591

Salem HF, Kharshoum RM, Abou-Taleb HA et al (2019) Brain targeting of resveratrol through intranasal lipid vesicles labelled with gold nanoparticles: in vivo evaluation and bioaccumulation investigation using computed tomography and histopathological examination. J Drug Target 27(10):1127–1124

Sánchez-López E, Ettcheto M, Egea MA, Espina M, Cano A, Calpena AC et al (2018) Memantine loaded PLGA PEGylated nanoparticles for Alzheimer's disease: in vitro and in vivo characterization. J Nanobiotechnol 16:32

Sathesh Kumar S, Felix Joe V (2017) Pharmacokinetics of tacrine loaded MPEG-PCL polymeric nanoparticles. Res J Pharm Tech 10(1):135–140

Schneider LS, Mangialasche F, Andreasen N, Feldman H, Giacobini E, Jones R et al (2014) Clinical trials and late-stage drug development for Alzheimer's disease: an appraisal from 1984 to 2014. J Intern Med 275(3):251–283

Shao R, Xiao J (2013) Natural products for treatment of Alzheimer's disease and related diseases: understanding their mechanism of action. Curr Neuropharmacol 11(4):337

Sharma N, Singh AN (2016) Exploring biomarkers for Alzheimer's disease. J Clin Diagn Res 10(7):KE01–KKE6

Sharma S, Javed MN, Pottoo FH, Rabbani SA, Barkat MA, Harshita et al (2019) Bioresponse inspired nanomaterials for targeted drug and gene delivery. Pharm Nanotechnol 7(3):220–233

Sharma P, Sharma A, Fayaz F et al (2020) Biological signatures of Alzheimer disease. Curr Top Med Chem 20(9):770–781

Shimizu H, Tosaki A, Kaneko K et al (2008) Crystal structure of an active form of BACE1, an enzyme responsible for amyloid beta protein production. Mol Cell Biol 28(11):3663–3671

Siddiqi KS, Husen A, Sohrab SS et al (2018) Recent status of nanomaterial fabrication and their potential applications in neurological disease management. Nanoscale Res Lett 13(1):231

Simonazzi A et al (2018) Nanotechnology applications in drug controlled release. In: Grumezescu AM (ed) Drug targeting and stimuli sensitive drug delivery systems. Elsevier, Amsterdam, pp 81–116

Singh N (2013) Neuroprotective properties and mechanisms of resveratrol in in vitro and in vivo experimental cerebral stroke models. ACS Chem Neurosci 4(8):1151–1162

Singh M, Arseneault M, Sanderson T et al (2008) Challenges for research on polyphenols from foods in Alzheimer's disease: bioavailability,metabolism, and cellular and molecular mechanisms. J Agric Food Chem 56:4855–4873

Singh SP, Deb CR, Kakati LN et al (2014) Carbon nanotube assisted drug delivery of donepeziland its derivatives as acetylcholinesteraseenzyme (AChE) inhibitors. Nanomed Nanobiol 2(1): 38–32

Singh NA et al (2015) Potential neuroprotective properties of epigallocatechin-3-gallate (EGCG). Nutr J 15:60

Singh NA, Bhardwaj V, Ravi C et al (2018) EGCG nanoparticles attenuate aluminum chloride induced neurobehavioral deficits, beta amyloid and Tau pathology in a rat model of Alzheimer's disease. Front Aging Neurosci 10:244

Singh AK, Singh SK, Nandi MK, Mishra G, Maurya A, Rai A et al (2019a) Berberine: a plant-derived alkaloid with therapeutic potential to combat Alzheimer's disease. Cent Nerv Syst Agents Med Chem 19:154
Singh AP, Biswas A, Shukla A et al (2019b) Targeted therapy in chronic diseases using nanomaterial-based drug delivery vehicles. Sig Transduct Target Ther 4:33
Sivasankarapillai VS, Jose J, Shanavas MS et al (2019) Silicon quantum dots: promising theranostic probes for the future. Curr Drug Targets 20(12):1255–1263
Smith AJ, Kavuru P, Arora KK, Kesani S, Tan J, Zaworotko MJ et al (2013) Crystal engineering of green tea epigallocatechin-3-gallate (EGCg) cocrystals and pharmacokinetic modulation in rats. Mol Pharm 10(8):2948–2961
Soldatia PP, Poloninia HC, Paesa CQ et al (2018) Controlled release of resveratrol from lipid nanoparticles improves antioxidant effect. IFAC-PapersOnLine 51(27):16–21
Stegemann S, Leveiller F, Franchi D et al (2007) When poor solubility becomes an issue: from early stage to proof of concept. Eur J Pharm Sci 31(5):249–261
Stojanovic-Radic Z, Pejcic M, Dimitrijevic M, Aleksic A, Anil Kumar NV, Salehi B et al (2019) Piperine-a major principle of black pepper: a review of its bioactivity and studies. Appl Sci 270: 1–29
Summerlin N, Soo E, Thakur S et al (2015) Resveratrol nanoformulations: challenges and opportunities. Int J Pharm 479(2):282–290
Sun D, Li N, Zhang W, Zhao Z, Mou Z, Huang D et al (2016) Design of PLGA-functionalized quercetin nanoparticles for potential use in Alzheimer's disease. Colloids Surf B Biointerfaces 148:116–129
Sun J, Jiang G, Shigemori H et al (2019) Inhibitory activity on amyloid aggregation of rosmarinic acid and its substructures from Isodon japonicas. Nat Prod Commun 14(5):1–5
Sunena, et al. (2019) Nose to brain delivery of galantamine loaded nanoparticles: in-vivo pharmacodynamic and biochemical study in mice. Curr Drug Deliv 16(1):51–58
Swerdlow RH (2007) Pathogenesis of Alzheimer's disease. Clin Interv Aging 2(3):347–359
Szoka F Jr, Papahadjopoulos D (1978) Procedure for preparation of liposomes with large internal aqueous space and high capture by reverse-phase evaporation. Proc Natl Acad Sci U S A 75(9): 4194–4198
Takooree H, Aumeeruddy MZ, Rengasamy KRR, Venugopala KN, Jeewon R, Zengin G et al (2019) A systematic review on black pepper (Piper nigrum L.): from folk uses to pharmacological applications. Crit Rev Food Sci 59(Supp;l 1):S210–S243
Tamba BI, Streinu V, Foltea G, Neagu AN, Dodi G, Zlei M et al (2018) Tailored surface silica nanoparticles for blood-brain barrier penetration: preparation and in-vivo investigation. Arab J Chem 11:981–990
Tamilselvan N, Raghavan CV, Balakumar K et al (2014) Brain targeted delivery of rivastigmine polymeric nanoparticles through oral administration to treat Alzheimer's disease. Int J Res Pharm Nano Sci 3(6):536–551
Tang M, Taghibiglou C (2017) The mechanisms of action of curcumin in Alzheimer's disease. J Alzheimers Dis 58(4):1003–1016
Tao X, Li Y, Hu Q, Zhu L, Huang Z, Yi J et al (2018) Preparation and drug release study of novel nanopharmaceuticals with polysorbate 80 surface adsorption. Appl Nanomater Bioeng 2018: 4718045. https://doi.org/10.1155/2018/4718045
Taram F, Ignowski E, Duval N et al (2018) Neuroprotection comparison of rosmarinic acid and carnosic acid in primary cultures of cerebellar granule neurons. Molecules 23(11):2956
Teleanu DM, Negut I, Grumezescu V et al (2019) Nanomaterials for drug delivery to the central nervous system. Nanomaterials (Basel, Switzerland) 9(3):371
Testa G, Gamba P, Badilli U, Gargiulo S, Maina M, Guina T et al (2014) Loading into nanoparticles improves quercetin's efficacy in preventing neuroinflammation induced by oxysterols. PLoS One 9(5):e96795
Thies W, Bleiler L (2013) Alzheimer's disease facts and figures. Alzheimers Dement 9(2):208–245

Tiwari SK, Agarwal S, Seth B et al (2013) Curcumin-loaded nanoparticles potently induce adult neurogenesis and reverse cognitive deficits in Alzheimer's disease model via canonical Wnt/β-catenin pathway. ACS Nano 8(1):76–103

Tu S, Okamoto S, Lipton SA et al (2014) Oligomeric Aβ-induced synaptic dysfunction in Alzheimer's disease. Mol Neurodegener 9:48

Tumiatti V, Minarini A, Bolognesi ML et al (2010) Tacrine derivatives and Alzheimer's disease. Curr Med Chem 17(17):1825–1838

Vijaykumar O et al (2014) Formulation and evaluation of rivastigmine loaded polymeric nanoparticles. J Chem Pharm Res 6(10):556–565

Vitorino C, Carvalho FA, Almeida AJ et al (2011) The size of solid lipid nanoparticles: an interpretation from experimental design. Colloids Surf B Biointerfaces 84(1):117–130

Wadhwa S, Singhal S, Rawat S et al (2014) Bioavailability enhancement by piperine: a review. Asian J Biomed Pharm Sci 4(36):1–8

Wang R, Tang XC (2005) Neuroprotective effects of huperzine A. a natural cholinesterase inhibitor for the treatment of Alzheimer's disease. Neurosignals 14(1–2):71–82

Wang R et al (2006) Progress in studies of huperzine A, a natural cholinesterase inhibitor from Chinese herbal medicine. Acta Pharmacol Sin 27:1–26

Weller J, Budson A (2018) Current understanding of Alzheimer's disease diagnosis and treatment. F1000Res 7:1161. https://doi.org/10.12688/f1000research.14506.1

Williams RJ et al (2004) Flavonoids: antioxidants or signalling molecules? Free Radic Biol Med 36: 838–849

Wilson B, Samanta MK, Santhi K et al (2008) Targeted delivery of tacrine into the brain with polysorbate 80-coated poly(n-butylcyanoacrylate) nanoparticles. Eur J Pharm Biopharm 70(1): 75–74

Wilson B, Samanta MK, Santhi K et al (2010) Chitosan nanoparticles as a new delivery system for the anti-Alzheimer drug tacrine. Nanomedicine 6(1):144–152

Wilson B et al (2011) Design and evaluation of chitosan nanoparticles as novel drug carrier for the delivery of rivastigmine to treat Alzheimer's disease. Ther Deliv 2(5):599–609

Xu SS, Cai ZY, Qu ZW, Yang RM, Cai YL, Wang GQ et al (1999) Huperzine-A in capsules and tablets for treating patients with Alzheimer disease. Acta Pharmacol Sin 20(6):486–490

Yang L, Wang N, Zheng G et al (2018) Enhanced effect of combining chlorogenic acid on selenium nanoparticles in inhibiting amyloid β aggregation and reactive oxygen species formation in vitro. Nanoscale Res Lett 13:303

Yusuf M, Khan M, Khan RA et al (2013) Preparation, characterization, in vivo and biochemical evaluation of brain targeted Piperine solid lipid nanoparticles in an experimentally induced Alzheimer's disease model. J Drug Target 21(3):300–311

Zaplatic E, Bule M, Shah SZA et al (2019) Molecular mechanisms underlying protective role of quercetin in attenuating Alzheimer's disease. Life Sci 224:109–119

Zetterberg H, Burnham SC (2019) Blood-based molecular biomarkers for Alzheimer's disease. Mol Brain 12:26

Zhang J, Zhou X, Yu Q, Yang L, Sun D, Zhou Y et al (2014) Epigallocatechin-3-Gallate (EGCG)-stabilized selenium nanoparticles coated with Tet-1 peptide to reduce amyloid-β aggregation and cytotoxicity. ACS Appl Mater Interfaces 6(11):8475–8487

Zhang C, Wan X, Zheng X et al (2014a) Dual-functional nanoparticles targeting amyloid plaques in the brains of Alzheimer's disease mice. Biomaterials 35(1):456–465

Zhang C, Chen J, Feng C et al (2014b) Intranasal nanoparticles of basic fibroblast growth factor for brain delivery to treat Alzheimer's disease. Int J Pharm 461(1–2):192–202

Zhou Z (2013) Liposome formulation of fullerene-based molecular diagnostic and therapeutic agents. Pharmaceutics 5(4):525–541

Index

G. M. Ashraf, M. S. Uddin (eds.), *Current Thoughts on Dementia*,
https://doi.org/10.1007/978-981-16-7606-2

The manufacturer's authorised representative in the EU is Springer Nature Customer Service Centre GmbH, Europaplatz 3, 69115 Heidelberg, Germany. If you have any concerns regarding our products, please contact ProductSafety@springernature.com

Printed and bound by CPI Group (UK) Ltd, Croydon, CR0 4YY

30/09/2026

02225250-0020